The Enzymes

VOLUME III

HYDROLYSIS: PEPTIDE BONDS

Third Edition

CONTRIBUTORS

D. M. BLOW
ROBERT J. DELANGE
J. DRENTH
S. D. ELLIOTT
JOSEPH FEDER
J. E. FOLK
JOSEPH S. FRUTON
A. N. GLAZER
LOWELL M. GREENBAUM
ELVIN HARPER
WILLIAM F. HARRINGTON
B. S. HARTLEY
JEAN A. HARTSUCK
GEORGE P. HESS
J. N. JANSONIUS
B. KEIL
R. KOEKOEK
J. KRAUT
MICHAEL LASKOWSKI, JR.
WILLIAM N. LIPSCOMB
TEH-YUNG LIU
STAFFAN MAGNUSSON
FRANCIS S. MARKLAND, JR.
HIROSHI MATSUBARA
WILLIAM M. MITCHELL
ROBERT W. SEALOCK
SAM SEIFTER
D. M. SHOTTON
EMIL L. SMITH
B. G. WOLTHERS

ADVISORY BOARD

THE ENZYMES

Edited by PAUL D. BOYER

Molecular Biology Institute and
Department of Chemistry
University of California
Los Angeles, California

Volume III

Hydrolysis: Peptide Bonds

THIRD EDITION

ACADEMIC PRESS New York and London 1971

ACADEMIC PRESS, INC.
111 Fifth Avenue, New York, New York 10003

United Kingdom Edition published by
ACADEMIC PRESS, INC. (LONDON) LTD.
Berkeley Square House, London W1X 6BA

LIBRARY OF CONGRESS CATALOG CARD NUMBER: 75-117107

PRINTED IN THE UNITED STATES OF AMERICA

Contents

17. Streptococcal Proteinase

Teh-Yung Liu and S. D. Elliott

18. The Collagenases

Sam Seifter and Elvin Harper

19. Clostripain

William M. Mitchell and William F. Harrington

20. Other Bacterial, Mold, and Yeast Proteases

Hiroshi Matsubara and Joseph Feder

List of Contributors

Numbers in parentheses indicate the pages on which the authors' contributions begin.

D. M. BLOW (185), Medical Research Council Laboratory of Molecular Biology, Hill Road, Cambridge, England

ROBERT J. DELANGE (81), Department of Biological Chemistry, University of California School of Medicine, Los Angeles, California

J. DRENTH (484), Laboratory of Structural Chemistry, The University, Groningen, The Netherlands

S. D. ELLIOTT (609), Department of Pathology, University of Cambridge, Cambridge, England

JOSEPH FEDER (721), New Enterprise Division, Monsanto Company, St. Louis, Missouri

J. E. FOLK (57), Section on Enzyme Chemistry, Laboratory of Biochemistry, National Institute of Dental Research, National Institutes of Health, Bethesda, Maryland

JOSEPH S. FRUTON (119), Kline Biology Tower, Yale University, New Haven, Connecticut

A. N. GLAZER (501), Department of Biological Chemistry, University of California School of Medicine, Los Angeles, California

LOWELL M. GREENBAUM (475), Department of Pharmacology, College of Physicians and Surgeons, Columbia University, New York, New York

ELVIN HARPER (649), Development Biology Laboratory, Department of Medicine, Massachusetts General Hospital, Boston, Massachusetts, and Department of Biological Chemistry, Harvard University School of Medicine, Boston, Massachusetts

WILLIAM F. HARRINGTON (699), The McCollum-Pratt Institute, Johns Hopkins University, Baltimore, Maryland

B. S. HARTLEY (323), Medical Research Council Laboratory of Molecular Biology, Cambridge, England

JEAN A. HARTSUCK (1), Department of Biochemistry, Oklahoma Medical Research Foundation, University of Oklahoma Medical Center, Oklahoma City, Oklahoma

GEORGE P. HESS (213), Section of Biochemistry and Molecular Biology, Division of Biological Sciences, Cornell University, Ithaca, New York

J. N. JANSONIUS (484), Laboratory of Structural Chemistry, The University, Groningen, The Netherlands

B. KEIL (249), Institute of Organic Chemistry and Biochemistry, Czechoslovak Academy of Sciences, Prague, Czechoslovakia

R. KOEKOEK (484), Laboratory of Structural Chemistry, The University, Groningen, The Netherlands

J. KRAUT (165, 547), Department of Chemistry, University of California at San Diego, La Jolla, California

MICHAEL LASKOWSKI, JR. (375), Department of Chemistry, Purdue University, Lafayette, Indiana

WILLIAM N. LIPSCOMB (1), Department of Chemistry, Harvard University, Cambridge, Massachusetts

TEH-YUNG LIU (609), Biology Department, Brookhaven National Laboratory, Upton, New York

STAFFAN MAGNUSSON (277), Institut for Molekylar Biologi, Aarhus Universitet, Aarhus, Denmark

FRANCIS S. MARKLAND, JR. (561), Department of Biological Chemistry, University of California School of Medicine, Los Angeles, California

HIROSHI MATSUBARA (721), Research Laboratory, Suntory Limited, Kita-ku, Osaka, Japan

WILLIAM M. MITCHELL (699), Department of Microbiology, Vanderbilt University School of Medicine, Nashville, Tennessee

ROBERT W. SEALOCK (375), Institute of General Medical Sciences, National Institutes of Health, Bethesda, Maryland

SAM SEIFTER (649), Department of Biochemistry, Albert Einstein College of Medicine, Bronx, New York

D. M. SHOTTON (323), Department of Biochemistry, University of Bristol, Bristol, England

EMIL L. SMITH (81, 501, 561), Department of Biological Chemistry, University of California School of Medicine, Los Angeles, California

B. G. WOLTHERS (484), Laboratory of Structural Chemistry, The University, Groningen, The Netherlands

Preface

Unlike an author, an editor can unabashedly extol the virtues of his volume. This volume affords such an opportunity, indeed an obligation, to the editor.

It has a readily definable unity of subject and purpose: to give, for enzymes catalyzing peptide-bond hydrolysis, the pertinent information about the enzyme protein or the reaction catalyzed at a molecular level. The coverage of this area in the second edition of this treatise, just over a decade ago, required less than half of a volume. The striking nature of the advances is well illustrated by the absence of a single presentation of X-ray or of primary sequence results in the previous edition. Such elegant structural information is given for most of the enzymes discussed in this volume.

All practitioners and students of chemistry, physics, and biology should take pride in the depth of understanding and detail of description given in the chapters. This attainment is a credit to the capability of man and to society for supporting such achievements.

The outstanding quality of the volume reflects that of its authors. The enzymes are discussed by a recognized research leader in each field. Much credit for this excellence is due to the Advisory Board for the volume, comprising C. B. Anfinsen, I. R. Lehman, Alton Meister, and Stanford Moore.

Comment is also warranted about the timeliness of the chapters. A check of reference citations will show an unusual promptness of publication. This is a continuing goal for the entire treatise. We are indebted to the authors for their cooperation. Continued appreciation is also due the staff of Academic Press and Lyda Boyer for capable editorial assistance.

PAUL D. BOYER

Contents of Other Volumes

Volume I: Structure and Control

Volume II: Kinetics and Mechanism

Volume IV: Hydrolysis (Sulfate Esters, Carboxyl Esters, Glycoside Bonds) and Phosphorolysis (Glycoside Bonds)

The Enzymes

VOLUME III

HYDROLYSIS: PEPTIDE BONDS

Third Edition

1

Carboxypeptidase A

JEAN A. HARTSUCK • WILLIAM N. LIPSCOMB

I. Introduction

Carboxypeptidase A (CPA) is a zinc-containing proteolytic enzyme which removes the C-terminal amino acid from a peptide chain if the C-terminal carboxylate is free. If the C-terminal residue of the substrate has an L configuration and a hydrophobic side chain, hydrolysis occurs readily. All studies discussed in this article were made on bovine pancreatic CPA, which was first isolated by Waldschmidt-Leitz and Purr

TABLE I
Chemical Forms of CPA

Form	Common name	N-terminus (total no. of residues)	Crystal cell constants for space group $P2_1$	Isolation	Reference for isolation
α	Cox	Ala (307)	51.41, 59.89, 47.19, 97°35′	Chromatography of juice from cannulated duct	Cox *et al.* (*6*)
β	—	Ser (305)	Not isolated	A contaminant	Sampath Kumar *et al.* (*5*)
γ	Anson[a]	Asn (300)	50.9, 57.9, 45.0, 94°40′	Selective precipitation of juice which exudes from frozen pancreas glands	Anson (*2*)
δ	Allan	Asn (300)	Same as α	Selective precipitation of dissolved acetone powders of pancreas glands	Allan *et al.* (*7*)

[a] Commercially available from Worthington Biochemical, Inc. or Calbiochem, Inc.

(*1*) and first crystallized by Anson (*2*). Porcine CPA has also been isolated by Folk and Schirmer (*3*), and the relationships of two forms of this enzyme have been established by Folk (*4*).

Four forms of active CPA arise from isolation procedures which involve enzymic cleavages of procarboxypeptidase A. These forms have been characterized by their N-terminal residues (*5*). Table I describes these forms (*2, 5–7*) and Fig. 1 shows the relation among their N-termini. Even though CPA_γ and CPA_δ are both predominantly asparagine at the N-terminus and both have some CPA_β contaminant (*5, 8*), the δ form is eight times more soluble in 1 *M* NaCl than the γ form (*7*); also, the loss of enzymic activity upon removal of zinc is reversible in the case of CPA_δ, but only partial regeneration is possible with CPA_γ (*9*). Furthermore, CPA_α and CPA_δ are alike in the properties just mentioned. The chemical sequence determination and a great many of the other chemical experiments have been performed using the commercially available γ form of the enzyme, but because of the larger and more easily formed crystals of CPA_α [prepared according to the methods described by Lipscomb *et al.* (*10*)], it has been used for the crystallographic experiments. Even though determinations of both the chemical sequence and the atomic resolution crystal structure are now complete and there are only a few apparent disagreements between these two studies (see below), the behavioral differences between CPA_α and CPA_δ are still not understood.

Also, recently, Neurath and co-workers (*11, 12*) have isolated two

Ala — Arg ⸾ Ser — Thr — Asn — Thr — Phe ⸾ Asn ———

α ⟶ β ⟶ γ, δ ⟶

FIG. 1. N-terminal sequence of CPA. The N-termini of the various forms (*5*) are indicated.

1. E. Waldschmidt-Leitz and A. Purr, *Ber. Deut. Chem. Ges.* **62**, 2217 (1929).
2. M. L. Anson, *J. Gen. Physiol.* **20**, 663 (1937).
3. J. E. Folk and E. W. Schirmer, *JBC* **238**, 3884 (1963).
4. J. E. Folk, *JBC* **238**, 3895 (1963).
5. K. S. V. Sampath Kumar, J. B. Clegg, and K. A. Walsh, *Biochemistry* **3**, 1728 (1964).
6. D. J. Cox, F. C. Bovard, J. P. Bargetzi, K. A. Walsh, and H. Neurath, *Biochemistry* **3**, 44 (1964).
7. B. J. Allan, B. J. Keller, and H. Neurath, *Biochemistry* **3**, 40 (1964).
8. P. H. Petra and H. Neurath, *Biochemistry* **8**, 2466 (1969).
9. B. L. Vallee, T. L. Coombs, and F. L. Hoch, *JBC* **235**, PC45 (1960).
10. W. N. Lipscomb, J. C. Coppola, J. A. Hartsuck, M. L. Ludwig, H. Muirhead, J. Searl, and T. A. Steitz, *JMB* **19**, 423 (1966).
11. K. A. Walsh, L. H. Ericsson, and H. Neurath, *Proc. Natl. Acad. Sci. U. S.* **56**, 1339 (1966).

allelomorphs of CPA. One form of the enzyme contains Ile, Ala, and Val at positions 179, 228, and 305; while the other has Val, Glu, and Leu at these positions. The distribution of these forms is apparently Mendelian since only animals having one of the two forms or an approximately equal mixture of the two forms have been found. Even though the valine enzyme with residue 305 Val has greater thermal stability than the leucine form 305 Leu, no enzymic difference between the two forms have been demonstrated. The crystal structure of CPA was determined from material obtained from a single cow, and the electron density map indicates Ile at 179, Ala at 228, and Val at 305. Residues 228 and 305 are on the exterior of the molecule, and neither residue makes contact with other side chains or, in fact, with any part of the molecule (*12a*). At position 179 either Ile or Val can be accommodated without disruption of the hydrophobic core in which it is located. Since none of the replacements which have been found so far justifies the observed differences in thermal stability of the two allelomorphs, perhaps a search for other replacements should be made.

Important physical properties of CPA include a molecular weight of 34,472.3 for the α (305 Val) form [computed from the sequence of Bradshaw *et al.* (*13*)] and an isoelectric point near pH 6 (*14*) at an ionic strength of 0.2. It is possible to grow crystals of CPA near pH 7.5 by gradually decreasing the ionic strength of the mother liquor. Final buffer conditions used for all crystallographic studies discussed here were 0.2 *M* LiCl, 0.02 *M* tris Cl, pH 7.5. Viscosity measurements predicted, assuming a 30% hydrated prolate ellipsoid, that the CPA molecule would have an axial ratio of 2.1:1 (*14*), whereas the crystal structure shows a more nearly spherical molecule with dimensions $50 \times 42 \times 38$ Å.

The CPA molecule is composed of a single polypeptide chain (*15*) of 307 amino acid residues (*16*) and one Zn atom (*17*). The large number

---

12. P. H. Petra, R. A. Bradshaw, K. A. Walsh, and H. Neurath, *Biochemistry* **8,** 2762 (1969).

12a. A hydrophobic interaction within the molecule by the Val 305 has been incorrectly deduced (*12*).

13. R. A. Bradshaw, L. H. Ericsson, K. A. Walsh, and H. Neurath, *Proc. Natl. Acad. Sci. U. S.* **63,** 1389 (1969).

14. F. W. Putnam and H. Neurath, *JBC* **166,** 603 (1946).

15. E. O. P. Thompson, *BBA* **10,** 633 (1953).

16. G. N. Reeke, J. A. Hartsuck, M. L. Ludwig, F. A., Quiocho, T. A. Steitz, and W. N. Lipscomb, *Proc. Natl. Acad. Sci. U. S.* **58,** 2220 (1967).

17. B. L. Vallee and H. Neurath, *JACS* **76,** 5006 (1954).

of aromatic residues, *viz.*, 8 histidines, 16 phenylalanines, 19 tyrosines, and 7 tryptophans, and the existence of two cysteine residues are notable in the amino acid analysis of CPA (*18, 19*). One cysteine (*19a*) sulfur atom was thought to be a Zn ligand (*9*), but now it is known that the two cysteine residues form a disulfide bond (*16, 21*). Recently, a complete amino acid sequence for the CPA molecule has been published (Fig. 2) (*13*), and part of the complete descriptions of this work are available (*22–24*).

ALA ARG SER THR ASN   THR PHE ASN TYR ALA   THR TYR HIS THR LEU   ASP GLU ILE TYR ASP
PHE MET ASP LEU LEU   VAL ALA GLN HIS PRO   GLU LEU VAL SER LYS   LEU GLN ILE GLY ARG
SER TYR GLU GLY ARG   PRO ILE TYR VAL LEU   LYS PHE SER THR GLY   GLY SER ASN ARG PRO
ALA ILE TRP ILE ASP   LEU GLY ILE HIS SER   ARG GLU TRP ILE THR   GLN ALA THR GLY VAL
TRP PHE ALA LYS LYS   PHE THR GLU ASN TYR   GLY GLN ASN PRO SER   PHE THR ALA ILE LEU
ASP SER MET ASP ILE   PHE LEU GLU ILE VAL   THR ASN PRO ASN GLY   PHE ALA PHE THR HIS
SER GLU ASN ARG LEU   TRP ARG LYS THR ARG   SER VAL THR SER SER   SER LEU CYS VAL GLY
VAL ASP ALA ASN ARG   ASN TRP ASP ALA GLY   PHE GLY LYS ALA GLY   ALA SER SER SER PRO
CYS SER GLU THR TYR   HIS GLY LYS TYR ALA   ASN SER GLU VAL GLU   VAL LYS SER ILE VAL
ASP PHE VAL LYS ASN   HIS GLY ASN PHE LYS   ALA PHE LEU SER ILE   HIS SER TYR SER GLN
LEU LEU LEU TYR PRO   TYR GLY TYR THR THR   GLN SER ILE PRO ASP   LYS THR GLU LEU ASN
GLN VAL ALA LYS SER   ALA VAL ALA ALA LEU   LYS SER LEU TYR GLY   THR SER TYR LYS TYR
GLY SER ILE ILE THR   THR ILE TYR GLN ALA   SER GLY GLY SER ILE   ASP TRP SER TYR ASN
GLN GLY ILE LYS TYR   SER PHE THR PHE GLU   LEU ARG ASP THR GLY   ARG TYR GLY PHE LEU
LEU PRO ALA SER GLN   ILE ILE PRO THR ALA   GLN GLU THR TRP LEU   GLY VAL LEU THR ILE
MET GLU HIS THR VAL   ASN ASN

FIG. 2. Amino acid sequence of CPA as determined by Bradshaw *et al.* (*13*).

18. E. L. Smith and A. Stockwell, *JBC* **207,** 501 (1954).

19. J. P. Bargetzi, K. S. V. Sampath Kumar, D. J. Cox, K. A. Walsh, and H. Neurath, *Biochemistry* **2,** 1468 (1963).

19a. This cysteine was believed to be contained in the "active site" cysteinyl peptide, Gly–Lys–Ala–Gly–Ala–Ser–Ser–Ser–Pro–Cys–Ser–Glu–Thr–Tyr (*20*).

20. H. Neurath, *Federation Proc.* **23,** 1 (1964).

21. W. N. Lipscomb, J. A. Hartsuck, G. N. Reeke, F. A. Quiocho, P. H. Bethge, M. L. Ludwig, T. A. Steitz, H. Muirhead, and J. C. Coppola, *Brookhaven Symp. Biol.* **21,** 24 (1968).

22. M. Nomoto, N. G. Srinivasan, R. A. Bradshaw, R. D. Wade, and H. Neurath, *Biochemistry* **8,** 2755 (1969).

23. R. A. Bradshaw, D. R. Babin, M. Nomoto, N. G. Srinivasin, L. H. Ericsson, K. A. Walsh, H. Neurath, *Biochemistry* **8,** 3859 (1969).

24. R. A. Bradshaw, *Biochemistry* **8,** 3871 (1969).

II. Results of Chemical Experiments

A. Substrate Specificity

In order that a substrate may be susceptible to lysis by CPA, certain requirements and preferences are exerted by the enzyme. These features of the substrate are listed briefly as follows: First, the peptide bond which is hydrolyzed must be adjacent to a terminal free carboxyl group (Fig. 3) (*25, 26*). Second, the rate of hydrolysis is usually enhanced if the terminal residue is aromatic or branched aliphatic (*27*). Third, dipeptides having a free amino group are hydrolyzed slowly, but if this group is blocked by *N*-acylation, the hydrolysis is rapid (*26*). Fourth, the carboxyl-terminal residue must be in the L configuration (*28–30*), and placement of a bulky D residues at R_1 (Fig. 3) greatly decreases the rate of hydrolysis (*31*). Fifth, the substitution of a methyl group (in sarcosine) or a methylene (in proline) for the H atom of the NH group of the peptide bond to be split either prohibits or greatly reduces hydrolysis (*27, 32*). Sixth, in *N*-acyl dipeptides the rate of hydrolysis is greatly decreased by substitution of sarcosine (*33*) or β-alanine (*34*) as the second amino acid. Thus, the presence and integrity of the second peptide bond are important for rapid hydrolysis. Seventh, studies of longer polypeptides indicate that the nature of the side chains of at least five residues, R_1' through R_4 (Fig. 3), influence the K_m constant and to some extent the k_{cat}. Placement of an aromatic residue at R_2 produces a pronounced decrease in

```
      R2        H   R1         H   R1'
      |         |   |        { |   |
etc.—CH—C—N—CH—C—{—N—CH—COO⁻
         ||             ||   {
         O              O
```

FIG. 3. Substrate for carboxypeptidase, to be split at the wavy line.

25. E. Waldschmidt-Leitz, *Physiol. Rev.* **11,** 358 (1931).
26. K. Hofmann and M. Bergmann, *JBC* **134,** 225 (1940).
27. M. A. Stahmann, J. S. Fruton, and J. Bergmann, *JBC* **164,** 753 (1946).
28. M. Bergmann and J. S. Fruton, *JBS* **117,** 189 (1937).
29. H. T. Hanson and E. L. Smith, *JBC* **179,** 815 (1949).
30. C. A. Dekker, S. P. Taylor, and J. S. Fruton, *JBC* **180,** 155 (1949).
31. S. Yanari and M. A. Mitz, *JACS* **79,** 1150 (1957).
32. E. L. Smith, *JBC* **175,** 39 (1948).
33. J. E. Snoke and H. Neurath, *JBC* **181,** 789 (1949).
34. H. T. Hanson and E. L. Smith, *JBC* **175,** 833 (1948).

K_m, and also substitution with an aromatic group is felt to a lesser extent at R_1, R_3, or R_4 (*35*).

B. Esterase Activity

In addition to its ability to split peptide bonds, CPA also possesses esterase activity (*36*). Not many different ester substrates have been studied, and only the L configuration at the C-terminal residue (*33*) and the presence of a metal in the enzyme (*37*) have been rigorously shown to be essential in the hydrolysis of an ester by CPA. The other requisites for good peptide substrates, *viz.*, an aromatic C-terminal side chain, a free terminal carboxyl group, and the integrity of the penultimate peptide bond, are perhaps implied by virtue of the lysis of ester substrates which contain them, but strict tests have not been made of the indispensability of these features in esters. A general statement can be made that while the values of k_{cat} for analogous derivatives of β-phenyllactic acid and phenylalanine are within a factor of five of one another, the K_m of the esters are 20- to 110-fold less than those of the peptides (*33, 38*). Recently, several kinetic anomalies and two general types of pH-rate profiles have been demonstrated for various ester substrates. These will be discussed in the appropriate sections below.

C. Kinetics

In the case of the peptidase activity of CPA, specifically toward carbo benzoxyglycyl-L-phenylalanine, the dependence of the overall rate of reaction on pH is bell-shaped with a maximum at pH 7.3 and inflections at pH 6.7 and 8.5 (*39, 40*). Values of both K_m and k_{cat} reach a minimum between pH 7.5 and 7.8, increase steadily on the acid side, and on the alkaline side increase to a maximum at pH 8.3 before decreasing (*39*). However, Lumry *et al.* (*41*) have shown that the dependence of K_m and k_{cat} on pH essentially disappears if the ionic strength of the buffer is raised to 0.5 *M*. It should also be noted that the overall reaction

35. N. Abramowitz, I. Schecter, and A. Berger, *BBRC* **29,** 862 (1967).
36. J. E. Snoke, G. W. Schwert, and H. Neurath, *JBC* **175,** 7 (1948).
37. J. E. Coleman and B. L. Vallee, *Biochemistry* **1,** 1083 (1962).
38. J. R. Whitaker, F. Menger, and M. L. Bender, *Biochemistry* **5,** 386 (1966).
39. H. Neurath and G. W. Schwert, *Chem. Rev.* **46,** 69 (1950).
40. J. F. Riordan and B. L. Vallee, *Biochemistry* **2,** 1460 (1963).
41. R. Lumry, E. L. Smith, and R. R. Glantz, *JACS* **73,** 4330 (1951).

velocity for the hydrolysis of carbobenzoxyglycyl-L-tryptophan increases rapidly with ionic strength up to a salt concentration of about 0.3 *M* where a plateau is reached (*41*). Although the data are not so complete, it seems that the rate of lysis of hippuryl-L-phenylalanine (*42*) has essentially the same relation ionic strength as that for the splitting of carbobenzoxyglycyl-L-tryptophan. A bell-shaped relation of overall reaction rate and pH has also been observed for the splitting of *N*-*trans*-3-(2-furylacryloyl)-L-phenylalanine by CPA, but in this instance the reaction rate as a function of ionic strength decreases until a plateau is reached at about 0.3 *M* salt concentration (*43*). In addition to the reaction with the synthetic peptide substrates described above, the splitting by CPA of the C-terminal isoleucine from β-lactoglobulin has been studied. The pH-rate profile is bell-shaped below pH 8.4, but above that point the rate again increases. In contrast to the hydrolysis of synthetic substrates, the reaction of CPA with β-lactoglobulin is prohibited by salt. The relative velocity drops from 50 to 0 as the ionic strength increases from 0 to 0.2 (*44*).

A diversity of relations of kinetic constants to pH have been shown for various ester substrates of CPA. For the usual assay substrate, hippuryl-L-β-phenyllactate, the overall reaction velocity increases up to a pH of about 7, is constant to pH 9.5, and then increases further with pH (*40, 45*). Later, the failure to find a decrease of activity in the regions of pH 8–9 was attributed to the restriction of these early studies to a relatively large substrate concentration (0.01 *M*), where substrate inhibition complicates the pH-rate profile (*46*). The kinetics as a function of pH of three other, poorer ester substrates of CPA, *O*-acetyl-L-mandelate (*47*), *O*-*trans*-cinnamoyl-L-β-phenyllactate (*48*), and *O*-*trans*-3-(2-furylacryloyl)-D,L-β-phenyllactate (*43*) have been studied. All three have overall reaction rate curves which are bell-shaped with a maximum near pH 7.5. This is reminiscent of the peptide substrates, but for neither the acetylmandelate nor the cinnamoylphenyllacetate are the K_m and k_{cat} curves like those of the peptide carbobenzoxyglycyl-L-phenylalanine. Apparent ionization constants (pK_{aE}) of functional groups of the enzyme, derived from analysis of k_{cat}/K_m vs. pH, are 6.9 and 6.5 for the lower leg

42. L. Gorini and J. Labouesse-Mercouroff, *BBA* **13**, 291 (1954).
43. W. O. McClure and H. Neurath, *Biochemistry* **5**, 1425 (1966).
44. E. W. Davie, C. R. Newman, and P. E. Wilcox, *JBC* **234**, 2635 (1959).
45. W. O. McClure, H. Neurath, and K. A. Walsh, *Biochemistry* **3**, 1897 (1964).
46. J. F. Riordan, M. Sokolosky, and B. L. Vallee, *Biochemistry* **6**, 3609 (1967).
47. F. W. Carson and E. T. Kaiser, *JACS* **88**, 1212 (1966).
48. P. L. Hall, B. L. Kaiser, and E. T. Kaiser, *JACS* **91**, 485 (1969).

of the bell-shaped curve and 7.5 and 9.4 for the upper leg in the *O*-acetyl-L-mandelate and *O-trans*-cinnamoyl-L-β-phenyllactate studies, respectively (*46, 48*). As will become apparent below, these p*K* values do not correlate well with those of the functional residues which have been deduced from the crystal structure and from the residue modification experiments.

Carboxypeptidase A is known to exhibit a variety of kinetic anomalies such as activation and inhibition by certain substrates. Substrate inhibition is especially apparent for aromatic *N*-acyl dipeptide substrates and their ester analogs. In addition, a variety of molecules, including both products of acyl dipeptide hydrolysis, inhibit or activate the enzyme. Substrate inhibition has been demonstrated for the peptide substrate carbobenzoxyglycyl-L-phenylalanine (*41*) and for the ester substrates hippuryl-L-phenyllactate (*38, 45, 49*) and *O*-hippuryl-L-mandelate (*50, 51*). However, neither hippuryl-L-phenylalanine (*38*) nor *O*-acetyl-L-mandelate (*47*) exhibits substrate inhibition. In addition, substrate activation by carbobenzoxyglycyl-L-phenylalanine (*38*) and by *O*-hippurylglycolate (*52*) has been demonstrated. Consistent with the existence of substrate inhibition and activation are the curve fitting experiments of Quiocho and Richards (*53*) which indicate that at least three molecules of carbobenzoxyglycyl-L-phenylalanine are bound to crystalline CPA during hydrolysis.

Many different substances have been shown to inhibit the action of carboxypeptidase A; they are grouped here as metal combining agents, substrate analogs, and products. Among the metal combining agents initially shown to be inhibitors of the CPA peptidase activity was cysteine (*54*) which was later shown to remove Zn (*55*). Also, 1,10-phenanthroline, by removing the Zn atom from CPA, inhibits both the peptidase (*56*) and the esterase (*37*) activities. In Table II a few of the inhibitors of CPA peptidase activity as well as the dissociation constants of their enzyme–inhibitor complexes (K_I) are given. Of special interest among these are β-phenylpropionate, which is the most commonly used inhibitor of CPA and which at high concentrations is not a competitive inhibitor

49. M. L. Bender, J. R. Whitaker, and F. Menger, *Proc. Natl. Acad. Sci. U. S.* **53,** 711 (1965).
50. E. T. Kaiser and F. W. Carson, *BBRC* **18,** 457 (1955).
51. S. Awazu, F. W. Carson, P. L. Hall, and E. T. Kaiser, *JACS* **89,** 3627 (1967).
52. E. T. Kaiser, S. Awazu, and F. W. Carson, *BBRC* **21,** 444 (1964).
53. F. A. Quiocho and F. M. Richards, *Biochemistry* **5,** 4062 (1966).
54. E. L. Smith and H. T. Hanson, *JBC* **179,** 803 (1949).
55. T. L. Coombs, J. P. Felber, and B. L. Vallee, *Biochemistry* **1,** 899 (1962).
56. J. P. Felber, T. L. Coombs, and B. L. Vallee, *Biochemistry* **1,** 231 (1962).

TABLE II
INHIBITORS OF CPA[a,b]

Compound	K_I[a] ($10^{-3}M$)
D-Phenylalanine	2.0
D-Histidine	20.0
β-Phenylproprionic acid	0.062
γ-Phenylbutyric acid	1.13

[a] The measurements were performed at 25°C and pH 7.5 with the substrate carbobenzoxyglycyl-L-phenylalanine.
[b] Data taken from Elkins-Kaufman and Neurath (*57*).

(*58*), and D- and L-phenylalanine whose ability to inhibit varies oppositely as a function of pH (*57*, *59*). Binding loci consistent with these results will be discussed in relation to the crystal structures of the enzyme at 2.0 Å resolution and of 6 Å studies of CPA with β-phenylpropionate and L-phenylalanine. Chloroacetate was the first product of hydrolysis shown to inhibit CPA (*26*). In contrast to chloroacetate, which inhibits in an indeterminant fashion (*57*), the α-hydroxy acids such as L-β-phenyllactate (*57*) and L-mandelate (*60*), which are split from the C-termini of esters, inhibit competitively.

The peptidase action of CPA as measured with carbobenzoxyglycyl-L-phenylalanine is activated by the product *N*-carbobenzoxyglycine (*38*). But the same compound *N*-carbobenzoxyglycine suppresses the substrate activation of *O*-hippurylglycolate (*52*) and inhibits the hydrolysis of *O*-(*trans*-cinnamoyl)-L-β-phenyllactate (*48*). Analogous to the case of *N*-carbobenzoxyglycine, several compounds have now been shown to enhance the peptidase activity, as measured with carbobenzoxyglycyl-L-phenylalanine, and to depress the esterase activity, as measured with *O*-hippuryl-L-phenyllactate. Those modifiers with the most pronounced effect are *N*-carbobenzoxyglycylglycine, cyclohexanol, cinnamic acid, and 3,3-dimethylbutanol (*61*).

Both model building experiments (see below) and chemical studies (*35*, *62*) indicate that kinetic anomalies such as substrate activation and inhibition and modifier activation are absent (or very greatly reduced) when substrates longer than acyl dipeptides are being cleaved.

57. E. Elkins-Kaufman and H. Neurath, *JBC* **178,** 645 (1949).
58. R. Lumry and E. L. Smith, *Discussions Faraday Soc.* **20,** 105 (1955).
59. H. Neurath and G. DeMaria, *JBC* **186,** 653 (1950).
60. E. T. Kaiser and F. W. Carson, *JACS* **86,** 2922 (1964).
61. R. C. Davies, D. S. Auld, and B. L. Vallee, *BBRC* **31,** 628 (1968).
62. D. S. Auld, *Federation Proc.* **27,** 781 (1968).

D. Metals

Because the activity of CPA is inhibited by cysteine and by ions such as sulfide and cyanide, it was deduced that the enzyme contains a metal ion (*62a*); however, it was incorrectly concluded that the metal is magnesium (*54, 63*). Later Vallee and Neurath found zinc in carboxypeptidase A (*17*) and it was deduced that the zinc atom of CPA is bound to cysteine sulfur because of the reaction of apocarboxypeptidase A with *p*-chloromercuribenzoate and silver ions (*9*) and because of the similarity of the spectrum of cobalt CPA and the expected cobalt–sulfur spectrum (*64*). Assuming that one of the protein zinc ligands is sulfur, it was also asserted that the second ligand to zinc is nitrogen. This assertion relied on the analogy between the affinity of CPA for various metals and the affinity of model compounds to the same metals (*65*), on the magnitudes of the dissociation constants of complexes of CPA and several metals as compared to constants for complexes with model compounds (*66*), and on the similarity of the spectra of the pyridoxal phosphate–apoCPA complex and thiazolidines (*67*). The nitrogen ligand to zinc was identified as the N-terminal α-amino group (*67*). However, the conclusions that one ligand to zinc is sulfur, that there are only two protein ligands to zinc, and that the N-terminus is a zinc ligand were shown to be incorrect when the crystal structure of the enzyme became known (*16, 21*). Now, by combining X-ray and chemical evidence, the three protein zinc ligands have been identified as His 69, Glu 72, and His 196 (*13, 68*).

The direct relation of zinc to the activity of CPA was substantiated when the percentage of total potential zinc present in samples where the metal had been partly removed and the percentage of potential enzymic activity present were shown to be equal (*56*). Moreover, when zinc is replaced by other metals, the enzymic efficiency change (Table III)

62a. The early experiments demonstrating inhibition by the anions orthophosphate, pyrophosphate, oxalate, citrate, and cyanide (*54, 63*) were later shown to be invalid (*59*). However, inhibition by cysteine was not questioned.

63. E. L. Smith and H. T. Hanson, *JBC* **176,** 997 (1948).

64. R. J. P. Williams, *Nature* **188,** 322 (1960).

65. B. L. Vallee, R. J. P. Williams, and J. E. Coleman, *Nature* **190,** 633 (1961).

66. J. E. Coleman, and B. L Vallee, *JBC* **236,** 2244 (1961)

67. T. L. Coombs, Y. Omote, and B. L. Vallee, *Biochemistry* **3,** 653 (1964).

68. W. N. Lipscomb, J. A. Hartsuck, G. N. Reeke, and F. A. Quiocho, *Proc. Natl. Acad. U. S.* **64,** 28 (1969).

TABLE III
PERCENT ACTIVITY OF METALLOCARBOXYPEPTIDASES[a–c]

Element	Peptidase	Esterase
Zn	100	100
Co	160	95
Ni	106	87
Mn	8	35
Cu	0	0
Hg	0	116
Cd	0	150
Pb	0	52

[a] Activity toward *N*-carbobenzoxyglycyl-L-phenylalanine (peptidase) and *O*-hippuryl-L-β-phenyllactate (esterase) at pH 7.5.

[b] From the data of Coleman and Vallee (*66*).

[c] Peptidase activity for Cr^{3+} CPA is reported only in the early literature (*69–71*).

(*66*, *69–71*). As might be expected, the change of activity is small if cobalt or nickel replaces zinc, but copper carboxypeptidase A is totally inactive. The exchange of the zinc ion for the radioactive ^{65}Zn ion was used by Coleman and Vallee to show that several peptide substrates of CPA prevent the binding of Zn^{2+} to apoCPA, whereas the recombination of apoCPA and Zn^{2+} is not retarded by the ester substrate hippuryl-L-β-phenyllactate (*37*, *72*). They therefore concluded that even though peptide substrates are bound to apoCPA, ester substrates are not. Similar experiments showed that amides of active peptide substrates, while not susceptible to the hydrolytic action of the enzyme, are bound to both apoCPA and ZnCPA (*37*).

More recent studies of metal ions in CPA are given in the final section of this review.

E. SIDE CHAIN MODIFICATION

Reactions of CPA with reagents which can be expected to modify tyrosine residues led to the conclusion that at least one and probably two tyrosine residues are involved in the activity of CPA (*40*, *73*). The

69. B. L. Vallee, T. L. Coombs, J. A. Rupley, and H. Neurath, *JACS* **80,** 4750 (1958).

70. H. Neurath, J. A. Rupley, and B. L. Vallee, *Abstr., 136th Am. Chem. Soc. Meeting, Atlantic City* p. 44c (1959).

71. J. E. Coleman, T. L. Coombs, and B. L. Vallee, *Abstr., 136th Am. Chem. Soc. Meeting, Atlantic City* p. 57c (1959).

72. J. E. Coleman and B. L. Vallee, *JBC* **237,** 3430 (1962).

differential iodination of the enzyme in the presence and absence of β-phenylpropionate, an experiment performed by Roholt and Pressman (*74*), showed that the tyrosine residue of the peptide Ala–Gln–Tyr–Ile is protected by this inhibitor; however, the interpretation of these results could have been ambiguous if more than one molecule of inhibitor had been bound, so that another tyrosine, unrelated to enzyme function, was also protected. The X-ray structure identified this protected residue as Tyr 248, which undergoes a large conformational change when either a substrate or an inhibitor is bound (*16*). Tyrosine residues have been acetylated, diazotized, and nitrated (Table IV) (*40, 46, 75, 76*). Unfortunately, in these experiments none of the peptides containing the modified tyrosine residues was isolated. Consequently, in Table IV, the identification of which tyrosine residues were modified has been deduced from a consideration of the crystal structure. Esterase activity increases which are tabulated in Table IV, including most of that of the acetylated enzyme, can be explained by the displacement of substrate inhibition to a new concentration range (*49, 76a*). The presence of β-phenylpropionate prohibits all of the tyrosine modifications except that of the strong diazotization. In the strong diazotization experiment (*75*), one histidine residue, which was protected by β-phenylpropionate, was modified; however, its relation to enzymic activity, if any, was not established. The concurrent effect of the modification of tyrosine residues and substitution of other metals for zinc has been studied by Coleman *et al.* (*77*).

TABLE IV
RESULTS OF TYROSINE MODIFICATION[a]

Modification	No. of Tyr modified	Esterase activity (%)	Peptidase activity (%)	Identification of Tyr residues
Acetylation	2	700	0	248,198
Diazotization	1	200	100	198
Stronger diazotization	2	200	0	248,198
Nitration	1.2	170	10	248,198

[a] Derived from the data of Riordan and Vallee (*40*), Sokolovsky and Vallee (*75*), and Riordan *et al.* (*46, 76*), except for the last column.

73. R. T. Simpson and B. L. Vallee, *Biochemistry* **5**, 1760 (1966).
74. O. A. Roholt and D. Pressman, *Proc. Natl. Acad. Sci. U. S.* **58**, 280 (1967).
75. M. Sokolovsky and B. L. Vallee, *Biochemistry* **6**, 700 (1967).
76. J. F. Riordan, M. Sokolovsky, and B. L. Vallee, *Biochemistry* **6**, 358 (1967).
76a. Observations by Riordan and Vallee consistent with this result are reported, but the conclusion was not drawn (*70*).
77. J. E. Coleman, P. Pulido, and B. L. Vallee, *Biochemistry* **5**, 2019 (1966).

Their most interesting observation was that the presence of peptide substrate does not inhibit metal exchange if CPA has been acetylated or iodinated, and they conclude that substrate binding to the modified enzyme does not occur (*77a*). This interpretation is inconsistent with the binding of peptides to acetyl CPA crystals (see below).

After the identification of Arg 145 as the binding site for the free carboxylate ion of a substrate and glutamic acid 270 as a catalytic residue (*16*, *21*), diacetyl was used to modify one to two arginine residues in CPA (*80*). This modification abolished peptidase activity and concomitantly enhanced esterase activity 3- to 4-fold. In contrast to the tyrosine modifications, the inhibitor β-phenylpropionate has only a slight effect on this reaction. Since peptides containing the modified arginine residues were not isolated, it is not possible to identify the modified residues. However, in addition to the carboxylate binding site Arg 145, arginine residues 127 and 71 are located in the active center of the enzyme, and their modification could conceivably affect the enzymic activity in the observed manner. Also, preliminary experiments modifying carboxyl groups of CPA with cyclohexylmorpholinoethyl carbodiimide show parallel decreases in peptidase and esterase activities, but again no peptides were isolated (*80*).

F. Mechanistic Suggestions before the Crystal Structure Determination (*80a*)

After deducing the importance of a metal for CPA activity, Smith and co-workers (*58*, *81*) suggested that the free carboxyl group of the substrate might be bound initially to the metal atom or that both the free carboxyl group and the carbonyl of the scissile bond could interact with the metal (Fig. 4a). They went further to speculate that later in the reaction the free carboxylate might interact with a protein side chain thereby leaving zinc to act as a strong acid on the carbonyl group of the scissile peptide bond. Also the existence of an interaction area for the C-terminal side chain was postulated. Even though there is no evidence for the initial interaction of zinc and the carboxylate ion, the rest of these

77a. Acetyl CuCPA and iodo CuCPA are inactive as esterases or peptidases (*77*), contrary to earlier reports (*78*, *79*).

78. J. E. Coleman, P. Pulido, and B. L. Vallee, *Federation Proc.* **23**, 423 (1964).

79. B. L. Vallee, *Federation Proc.* **23**, 8 (1964).

80. B. L. Vallee and J. F. Riordan, *Brookhaven Symp. Biol.* **21**, 91 (1968).

80a. A critical analysis and summary of mechanistic deductions at about this time are given by Carson and Kaiser (*47*).

81. E. L. Smith, *Proc. Natl. Acad. Sci. U. S.* **35**, 80 (1949).

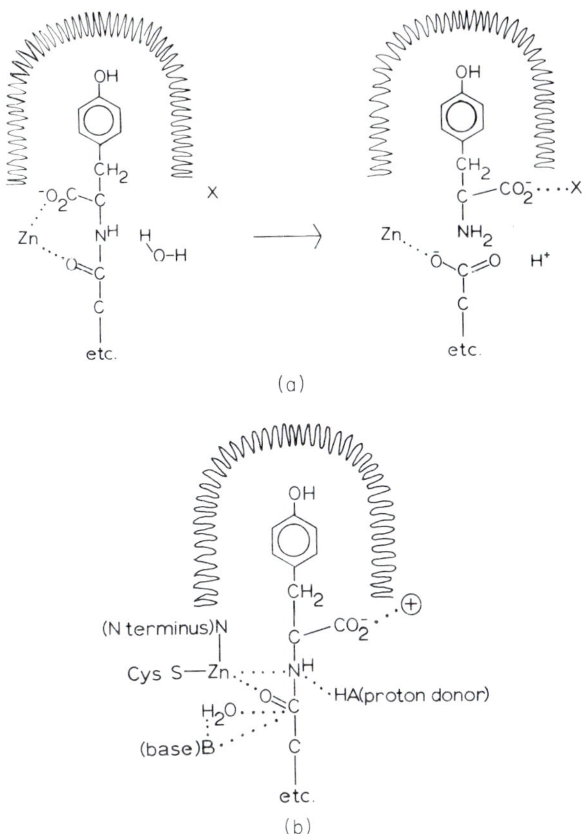

FIG. 4. (a) Lumry and Smith's tentative mechanistic scheme (*58*). The zinc was initially bonded to the CO of the susceptible peptide bond and the terminal carboxyl group of the substrate. (b) Vallee's tentative mechanism (*84*). The two proposed ligands from the protein to zinc are taken from later studies. The base B was thought to be that group which is iodinated or acetylated. The zinc was bonded to both the NH and the CO of the susceptible peptide bond. The approach of HA and the positive center for CO_2^- binding were on opposite sides of the substrate in Vallee's original proposal.

ideas are substantially in agreement with the current view of CPA catalysis. However, this early work did not recognize the importance of the catalytic protein residues known now to be Glu 270 and Tyr 248.

A suggestion that one of the ligands (then thought to be cysteine) from the protein to zinc might serve as a nucleophile was made by Williams (*82*). Now that the identities of the three zinc ligands are

82. R. J. P. Williams, *Biopolymers, Symp.* **1**, 515 (1964).

known to be His 69, Glu 72, and His 196, it is reasonable to ask if one of these could be a potential nucleophile. Very substantial rearrangement of backbone atoms of the polypeptide chain, as well as extensive motion of one or more of these side chains, would be required if any of these residues were to approach the carbon atom of the susceptible peptide bond. Even the interior β structure would be disrupted if His 196 were to participate in catalysis. There is no evidence for any of this motion in the atomic resolution structure of the complex of CPA with the substrate glycyl-L-tyrosine.

On the basis of the data then available, Vallee and co-workers (*83, 84*) presented a mechanistic proposal which is summarized in Fig. 4b. The ideas which they included and which are still thought to be correct are the existence of a protein binding area for the C-terminal side chain of the substrate, a protein positive charge which interacts with the substrate's carboxylate ion, an interaction between the carbonyl oxygen of the susceptible bond and zinc, and the existence in the enzyme of an acid and a base which participate in hydrolysis. These workers had incorrectly identified the protein zinc ligands as a cysteine sulfur and the N-terminal nitrogen, and their proposed interaction between the nitrogen atom of the scissile bond and zinc is not spatially feasible. It was deduced from the tyrosine and histidine modification reactions and their effects on peptidase and esterase activity that tyrosine (*84a*) (or histidine) is the base of the hydrolytic reaction. Now, considering the identity of the two catalytic groups Glu 270 and Tyr 248 and their steric relation to the substrate in the X-ray study, tyrosine is more likely the proton donor in the reaction than the nucleophile.

Later, because the CPA peptidase and esterase activities did not behave similarly as a function of pH, upon modification of tyrosine residues or in the presence of modifiers, Vallee and co-workers (*86*) proposed that the binding sites of ester and peptide substrates of CPA are different al-

83. B. L. Vallee, J. F. Riordan, and J. E. Coleman, *Proc. Natl. Acad. Sci. U. S.* **49,** 109 (1963).

84. B. L. Vallee, *Federation Proc.* **32,** 10 (1964).

84a. Comments on the tyrosine functions are somewhat varied. One statement indicates that two tyrosine residues are involved in the catalytic mechanism (*85*). One of these two tyrosine residues was suggested to be that tyrosine adjacent to the N-terminus of CPA_γ (*67*). Also, one tyrosine was believed to interact with Zn^{2+} (*78*). In another study one tyrosine was thought to interact with a histidine residue, which was also suggested for involvement in catalysis and/or binding of substrates (*77*). Nevertheless, the chemical evidence for participation of at least one tyrosine residue in hydrolysis of peptides by CPA is strong.

85. B. L. Vallee, J. F. Riordan, and M. Sokolovsky, *Science* **150,** 388 (1965).

86. B. L. Vallee, J. F. Riordan, J. L. Bethune, T. L. Coombs, D. S. Auld, and M. Sokolovsky, *Biochemistry* **7,** 3547 (1968).

though perhaps overlapping (*86a*). There is no firm crystallographic evidence to either support or refute this suggestion since no crystallographically suitable ester–CPA complexes have been prepared. However, in the native enzyme structure the available region near the zinc atom and extending into the pocket is so small that it cannot contain two substantially different (by more than 1 Å) binding sites. Certainly anomalous binding in the region further down the groove but overlapping the N-terminal end of a productively bound acyl dipeptide substrate is feasible, but it is difficult to see how this anomalous site could be productive.

III. Crystallography of Carboxypeptidase A

Our crystallographic studies of CPA have yielded electron density maps of the native enzyme at 6 Å (*10*), 2.8 Å (*88*), and 2.0 Å resolution (*16*, *21*, *89*). At 6 Å resolution it was possible to establish the overall shape of the enzyme molecule, the location of the N- and C-termini, the approximate helix content, the position of the zinc atom, and the general shape of the enzyme's dead-end pocket and surface groove which together compose the substrate binding site. At 2.8 Å resolution a complete trace of the polypeptide chain was possible, nearly all carbonyl oxygen atoms could be distinguished from the main chain, and if a chemical sequence had been available an atomic model could have been constructed. The 2.0-Å resolution electron density map was computed two years before the chemical sequence work was completed. During these two years, a working model, based on identifications from the X-ray map for unknown residues, was constructed. Using these atomic locations, a structure factor calculation was performed and the resulting calculated phases were used in some subsequent calculations. After the sequence was known, it was incorporated into the model and another structure factor calculation produced phases which were used to examine disagreements

86a. Earlier, because of this difference in behavior between pH 6 and 9, OH^+ ion was suggested as the base for the hydrolysis of esters (*79*), but this choice was dismissed by Bruice and Benkovic (*87*) and, at least for some esters, by Carson and Kaiser (*47*).

87. T. C. Bruice and S. J. Benkovic, "Bioorganic Mechanisms," Vol. 1, p. 5. Benjamin, New York, 1966.

88. M. L. Ludwig, J. A. Hartsuck, T. A. Steitz, H. Muirhead, J. C. Coppola, G. N. Reeke, and W. N. Lipscomb, *Proc. Natl. Acad. Sci. U. S.* **57**, 511 (1967).

89. W. N. Lipscomb, G. N. Reeke, J. A. Hartsuck, F. A. Quiocho, and P. H. Bethge, *Phil. Trans. Roy. Soc. London* **257**, 177 (1970).

between the chemical and X-ray structures. Concurrently, a study of the binding of a number of substrates and inhibitors at 6 Å resolution was under way (*90*). Subsequently, the data for the most promising of these complexes, that of glycyl-L-tyrosine with CPA, were extended to atomic resolution (*16*, *21*, *89*).

A. Determination of the Structure

Carboxypeptidase Aα crystallizes in the monoclinic space group $P2_1$, with unit cell dimensions $a = 51.41$ Å, $b = 59.89$ Å, $c = 47.19$ Å, and $\beta = 97.58$ Å. The preparation of heavy metal derivatives (Table V) (*10*) and the measurement of X-ray diffraction intensities are described by Lipscomb *et al.* (*21*). In summary, complete data (20,600 reflections) to 2.0 Å resolution were measured on the native enzyme, complete data (7200 reflections) to 2.8 Å resolution were measured on four heavy atom derivatives—Pb (2 sites), Hg (3 sites), Hg (1 site), and Pt (4 sites)—and those 6000 of the 14,000 reflections between 3.0 and 2.0 Å which gave the largest intensities in the native data set were also measured for the Pb(2) and Hg(3) derivatives. Heavy atom positional, thermal, and occupancy parameters were obtained from cycles of three-dimensional least squares refinement which alternated with centroid multiple isomorphous replacement phasing of each reflection (*21*). The refinement program used was the same as that described by Muirhead *et al.* (*91*). The heavy atom parameters from the refinement of CPA as well as the binding site on the protein of each heavy atom are listed in Table V. The 2.0-Å electron density map was computed from the "best" (*92*) multiple isomorphous replacement (MIR) phases and was displayed on illuminated Mylar sheets. Proposed locations for each of the atoms of the structure were recorded by placing colored markers on the map. Virtually all atom positions were within the electron density. As this interpretation was being made, all available chemically sequenced fragments, which constituted 214 of the 307 residues, were located in the complete sequence and assigned residue numbers. The remaining 93 residues were identified from the X-ray map alone (*89*). Subsequently, 60% of the unknown residues were shown to have been identified correctly (*68*). In retrospect, we believe that more experienced workers might have been able to identify a higher percentage of the unknown residues, but accuracy approaching that of the chemical methods will

90. T. A. Steitz, M. L. Ludwig, F. A. Quiocho, and W. N. Lipscomb, *JBC* **242**, 4662 (1967).
91. H. Muirhead, J. M. Cox, L. Mazzarella, and M. F. Perutz, *JMB* **28**, 117 (1967).
92. D. M. Blow, and F. H. C. Crick, *Acta Cryst.* **12**, 794 (1959).

TABLE V
HEAVY ATOM BINDING TO CARBOXYPEPTIDASE[a]

Atom	Z (electrons/ molecule)	B[b] (Å^2)	Heavy atom coordinates[c]			Residue
Pb_1	58	9	−0.094	0.500	−0.089	Glu 270
Pb_2	53	26	−0.089	0.540	−0.147	Citrate, not protein
Hg,g	50	16	−0.071	0.455	−0.115	His 69, Glu 72, His 196
Hg,s_1	47	13	−0.071	0.452	−0.115	His 69, Glu 72, His 169
Hg,s_2	46	31	−0.506	0.069	−0.257	His 29
Hg,s_3	48	25	−0.475	0.109	−0.136	His 29, Lys 84
Pt_1	74	71	0.341	0.430	0.034	Cys 161
Pt_2	45	71	−0.438	0.305	−0.568	Met 103
Pt_3	68	103	−0.292	0.082	0.141	N Terminus: Ala 1
Pt_4	27	69	−0.484	0.485	−0.500	His 303
Ag_1			0.238	0.498	−0.278	His 166, Ser 158
Ag_2			−0.082	0.220	0.193	His 120
Ag_3			−0.457	0.090	−0.143	His 29, (Lys 84)
Ag_4			−0.483	0.477	−0.516	His 303
Co_1			−0.500	0.500	−0.500	His 303
Co_2			−0.500	0.070	−0.130	His 29, (Lys 84)
Zn[d]			−0.087	0.443	−0.155	His 69, Glu 72, His 196

[a] All of these derivatives except Hg,g were prepared by dialyzing CPA crystals against solutions of the indicated composition.

Pb 0.003 M $PbCl_2$, 0.01 M Na citrate, 0.2 M LiCl, 0.02 M tris, pH 7.5

Hg,s Data in this paper were taken on crystals soaked against 0.0008 M $HgCl_2$, 0.2 M LiCl, 0.02 M tris, pH 7.5, but the fourth site is not occupied except under prolonged soaking against 0.003 M *p*-acetoxymercurianiline, 0.2 M Na acetate, and 0.02 M tris, pH 8. This fourth site is at $x = -0.448$, $y = 0.507$, $z = -0.575$, and is associated with His 303.

Pt 0.003 M K_2PtCl_4, 0.2 M LiCl, 0.02 M tris, pH 7.5

Ag 0.005 M $AgNO_3$, 0.2 M Na acetate, pH 8; Co 0.01 M $CoCl_2$, 0.01 M tris, pH 7.5

Hg,g 5×10^{-4} M CPA was dialyzed against 0.001 M $HgCl_2$, 1 M LiCl, 0.2 M tris, pH 8 and crystallized by dialysis against 0.18 M LiCl, 0.02 M tris, pH 8.

[b] The effective isotropic temperature factor B for the protein is 16 Å^2.

[c] The transformation to the symmetry related position is $x = -x$, $y' = \frac{1}{2} + y$, $z' = -z$.

[d] Zinc coordinates were found by interpolation of the electron density map.

probably never be attained. Of course, how good a job of sequencing a protein the crystallographer can do is related to the quality of his electron density map which in turn depends on the amount of thermal and positional disorder within the protein crystal and on the degree of isomorphism between the native and derivative crystals. Our map is certainly better than average. Extensive discussions of which residues can be most reliably identified in an X-ray map have been presented previously (*21*, *93*), and therefore we mention here only two points which may have escaped notice. We found it useful to look for hydrogen bonds formed within the molecule by amide and hydroxyl groups in order to distinguish them from spatially similar aliphatic residues and to notice the extension of glutamic acid residues into the solvent rather than the bending back toward the molecule shown by glutamine residues on the molecular surface.

The mathematical model building program of Diamond (*94*) was used to improve the atomic positions. Our use of this program has been described fully (*89*). Suffice it to say here that the positions of all atoms whose output coordinates placed them more than 0.5 Å from the input position were checked to insure that all output atoms remained within the electron density. The CPA electron density map is sufficiently reliable to conclude that there are deviations of the peptide group geometry from that of model compounds. Allowing for these deviations only as changes in the backbone angle τ (N–C_α–CO), as the Diamond program does is appropriate in most but not all instances. The average value of τ for the entire molecule is $112 \pm 5°$ rather than the expected 109.2° and τ was unrealistic (differing by more than $\pm 10°$ from tetrahedral) for 8% of the α-carbons in the molecule. The final atomic coordinates resulting from the Diamond program are available from the authors and will be published. In an additional computational study we hope to refine these by a more direct comparison of the atomic densities to the three-dimensional map of CPA.

Using the tentative total sequence, which contained 93 X-ray identifications, a structure factor (SF) calculation was performed with the atomic positions given by the Diamond program. Subsequently, the calculated structure factors were modified to include the chemical identifications of all the side chains. This updating made virtually no difference in the agreement factors. Structure factors were calculated for the most reliable 16,642 reflections. The overall standard crystallographic R factor is 0.44. The R factor and the difference between the MIR and calculated (SF) phases were examined as functions of resolution and structure

93. J. C. Kendrew, H. C. Watson, B. E. Strandberg, R. E. Dickerson, D. C. Phillips, and V. C. Shore, *Nature* **190,** 666 (1961).

94. R. Diamond, *Acta Cryst.* **21,** 253 (1966).

TABLE VI
SUMMARY OF STRUCTURE FACTOR CALCULATION[a]

Range	R factor	$<\Delta\varphi>$	$<\cos \|\Delta\varphi\|>$
Resolution (Å)			
∞–10.0	1.052	26.8	0.796
10.0–6.0	0.866	36.6	0.689
6.0–5.0	0.635	38.6	0.670
5.0–4.2	0.368	39.0	0.669
4.2–3.5	0.329	42.9	0.623
3.5–2.8	0.362	50.9	0.522
2.8–2.0	0.366	57.1	0.444
All data	0.437	50.2	0.529
Structure factor			
∞–400	0.294	29.8	0.783
400–300	0.358	34.8	0.724
300–200	0.399	44.1	0.610
200–150	0.420	56.3	0.453
150–100	0.538	64.5	0.348
100–0	1.261	73.4	0.209

[a] $R = \frac{\Sigma||F_0| - |F_c||}{\Sigma|F_0|}$ where F_0 is a scaled, observed structure factor, F_c is a calculated structure factor, and the summation is overall reflections. The phase angle differences $\Delta\varphi$ are reduced to the range $0 \leq \Delta\varphi \leq 180°$. $\Delta\varphi = |\varphi_{MIR} - \varphi_{SF}|$, where φ_{MIR} is an MIR phase and φ_{SF} is a structure factor phase.

factor magnitude (Table VI). The reflections with large interplanar spacings give poor agreement because they are most affected by the solvent structure, which has not been taken into account in the structure factor calculation. New electron density maps which showed residues whose chemical and X-ray identification had not been reconciled were computed with the observed native structure factors and calculated phases, appropriate atoms having been omitted from the phase calculation. After study of these maps, only three disagreements persist. Residues Asn 93, Phe 151, and Thr 245 appear in the X-ray maps to be Glx, Trp, and Ser, respectively (*68*). Also, because of chemical considerations, residue 256 is probably Asn rather than Asp (see *109a*).

B. Description of the Structure

1. *General Features*

The gross shape of the CPA molecule is an ellipsoid of approximate dimensions 50 × 42 × 38 Å. A view of the entire backbone of CPA is presented in Fig. 5, and photographs of the electron density map regions which contain side chains of particular interest are shown in Fig. 6. Six of the nine stretches of helix are on the left-hand surface of the molecule as viewed in Fig. 5. A twisted pleated sheet which contains both parallel and antiparallel β structure runs through the center of the molecule. Between the bank of helices and the pleated sheet there is a core of hydrophobic side chains. A similar hydrophobic trough exists to the right (Fig. 5) of the pleated sheet below the level of zinc atom. The zinc is adjacent to the pleated sheet; in fact, His 196, which is one of three zinc ligands, is a part of the β structure (Figs. 6a, d, and h). The three ligands from the protein to zinc are His 69, Glu 72, and His 196, and the fourth position of the distorted tetrahedron around zinc is occupied by a water molecule (*94a*). Associated with the zinc and lined on

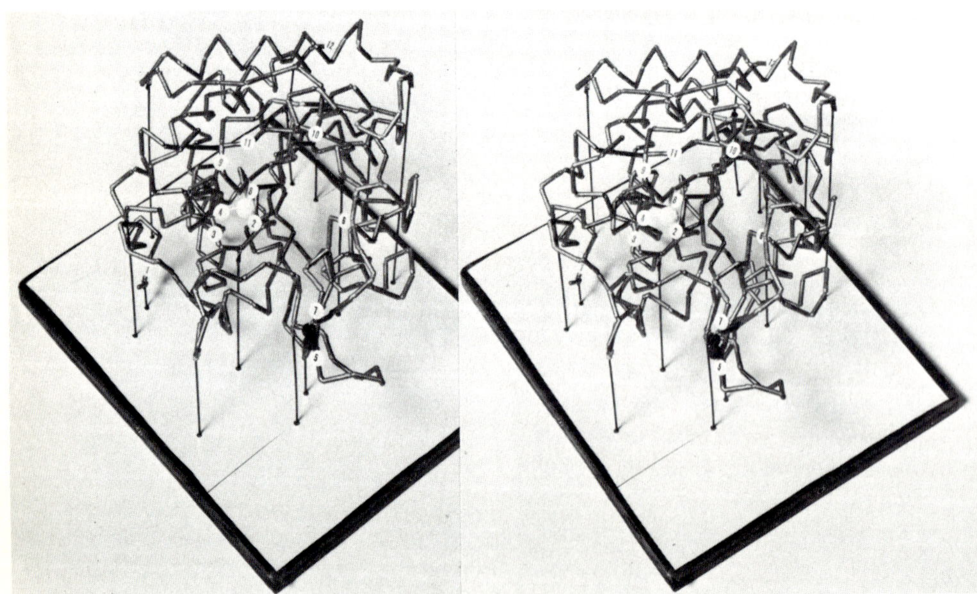

FIG. 5. Stereophotograph of the polypeptide chain of CPA. Each short rod is a peptide unit, and each intersection is one of the 307 C_α atoms. The zinc atom is the white sphere and the disulfide bridge is indicated by two black spheres. The numbers are 1,Ala 1; 2,His 69; 3,Arg 71; 4,Glu 72; 5,Cys 138; 6,Arg 145; 7,Cys 161; 8,His 196; 9,Tyr 198; 10,Tyr 248; 11,Glu 270; and 12,Asn 307; (C-terminus). (*Note:* Stereoviewers may be obtained, for example, from Ward's Natural Science Establishment, Inc., Rochester, New York, model 25 W2951. Computer drawn stereo figures were made using the program OR-TEP of Dr. Carroll Johnson.)

94a. Designating His 69 $= a$, Glu 72 $= b$, His 196 $= c$, and $H_2O = d$, bond angles about zinc are a-Zn-$b = 99°$, a-Zn-$c = 86°$, a-Zn-$d = 120°$, b-Zn-$c = 143°$, b-Zn-$d =$

one side by side chains of the pleated sheet are the groove and pocket which are essential for substrate binding (Figs. 6b and f) (*10*). With the exception of three short helical regions, the half of the molecule to the right of the pleated sheet is random coil. This coil possesses a few hydrogen bonds (approximately 10) and the disulfide bond (*16*) (Figs. 6a, e, and i), but otherwise ought to be quite flexible. Indeed, most of the conformational changes observed when a substrate molecule is bound to the enzyme are in this tortuous segment of the molecule.

2. *Helical Segments*

Table VII lists the helical segments of the molecule along with the average unit rises, unit rotations, and numbers of units per turn. The expected values for an α helix are also tabulated (*95, 96*). Included in the helix classification are those residues whose conformation is manifestly helical and which participate in at least one hydrogen bond approximately parallel to the helical axis. This criterion necessitates the breaking of the segment 72–88 into two helices because neither hydrogen bond is formed by Trp 81. The region 112–122 is an exception to this definition of a helix since it is so imperfect that only one or two of the potential hydrogen bonds exist. The averages in the table are made over

TABLE VII
PARAMETERS OF HELICES

Residues	Unit rise (Å)	Unit rotation (deg)	Units/turn
14–28	1.43	95	3.8
72–80	1.62	105	3.4
82–88	1.46	107	3.4
94–103	1.53	96	3.8
112–122	1.56	92	3.9
173–187	1.50	94	3.8
215–231	1.59	90	3.9
254–262	1.60	95	3.8
285–306	1.54	97	3.7
α Helix	1.49	100	3.6

99°, and c-Zn-d = 111°; but the estimated errors are large, especially toward the Zn–H_2O direction.

95. L. Pauling, R. B. Corey, and H. R. Branson, *Proc. Natl. Acad. Sci. U. S.* **37,** 205 (1951).

96. L. Pauling, "The Nature of the Chemical Bond," p. 499. Cornell Univ. Press, Ithaca, New York, 1960.

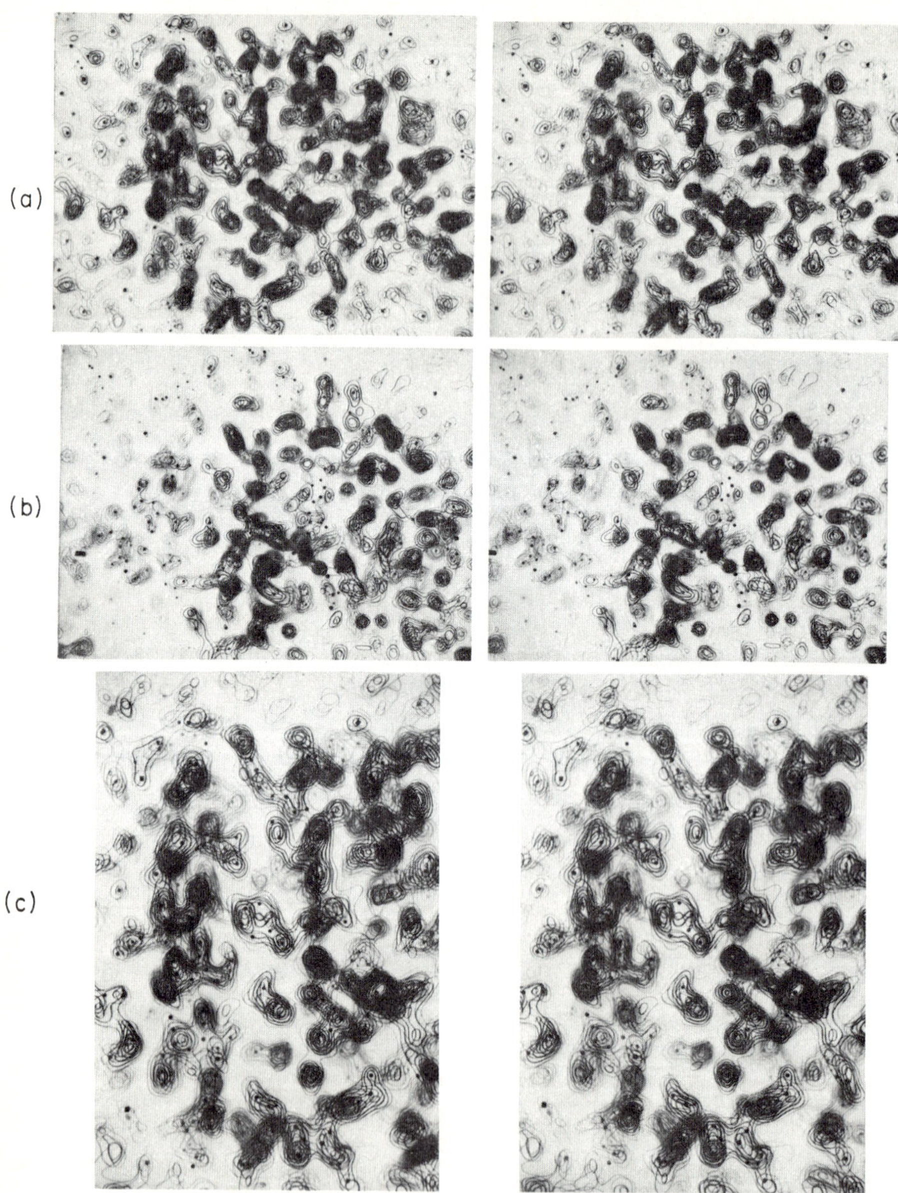

FIG. 6a–c.

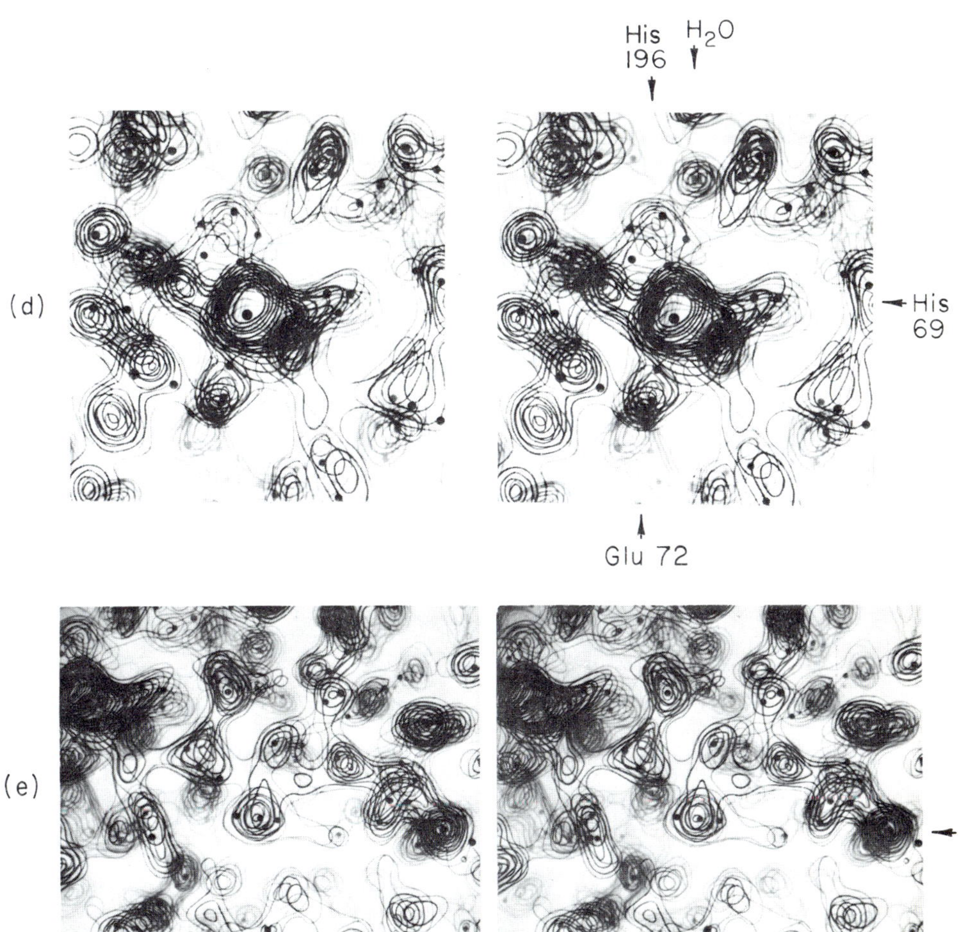

FIG. 6d, e.

the entire helical segments even though there are imperfect regions near the ends of several segments. Some of the imperfections resemble α_{II} helix (*97*), and others do not fit any of the previously described helical conformations (*98*). The C-terminal helix, residues 285–306 (Figs. 6a, c, and g), is the most nearly perfect helix in the molecule, but even here the helix axis is bent.

97. G. Nemethy, D. C. Phillips, S. J. Leach, and H. A. Scheraga, *Nature* **214,** 363 (1967).

98. L. Bragg, J. C. Kendrew, and M. F. Perutz, *Proc. Roy. Soc.* **A203,** 321 (1950).

(f)

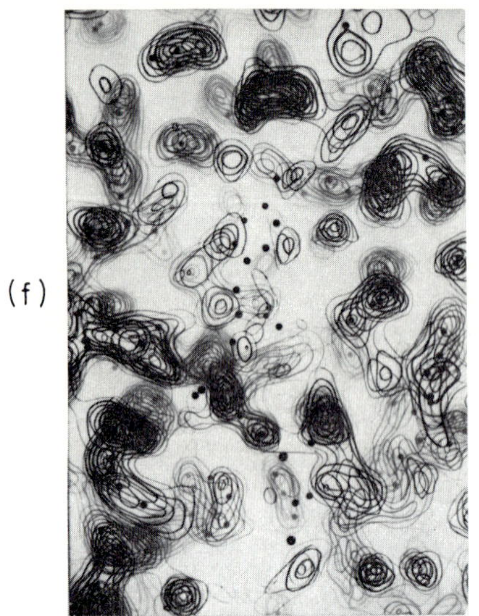

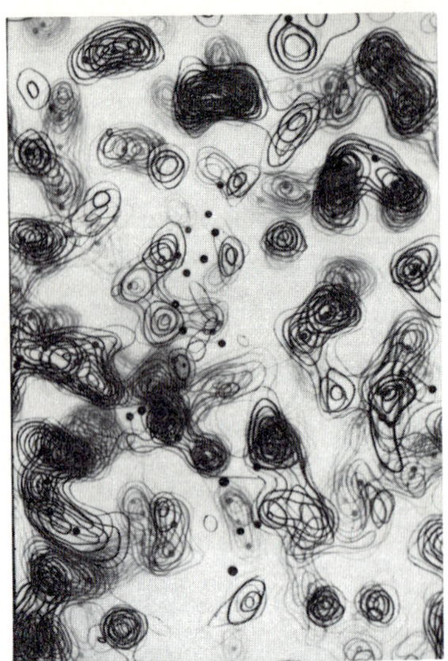

FIG. 6f.

3. β Structure

The pleated sheet, which is twisted by about 120° from the top to the bottom of the molecule (Fig. 5), contains in its four parallel and three antiparallel pairs of extended chains about 20% of the backbone atoms. However, only 45 residues, or 15% of the molecules, form at least one of the potential hydrogen bonds within the β structure (*99*). The β structure is portrayed schematically in Fig. 7. The dashed lines representing hydrogen bonds have been drawn if the oxygen–nitrogen distance is within 0.6 Å of the expected value of 2.76 Å (*96*). Under these conditions there are 33 hydrogen bonds among the 45 β-structure residues.

4. *Folding of the Polypeptide Chain*

Carboxypeptidase A is biologically synthesized as the zymogen, proCPA (*100, 101*). The folding of the CPA chain into its native conformation can be considered independently of the conformation of proCPA, however, because CPA constitutes the C-terminal portion of a proCPA chain, and because proCPA exhibits some substrate binding characteristics of active CPA (*102*), suggesting that the CPA portion of proCPA is already nearly in its active conformation in the zymogen.

99. L. Pauling and R. B. Corey, *Proc. Natl. Acad. Sci. U. S.* **37,** 729 (1951).

100. M. L. Anson, *J. Gen. Physiol.* **20,** 777 (1937).

101. J. R. Brown, R. N. Greenshields, M. Yamasaki, and H. Neurath, *Biochemistry* **2,** 867 (1963).

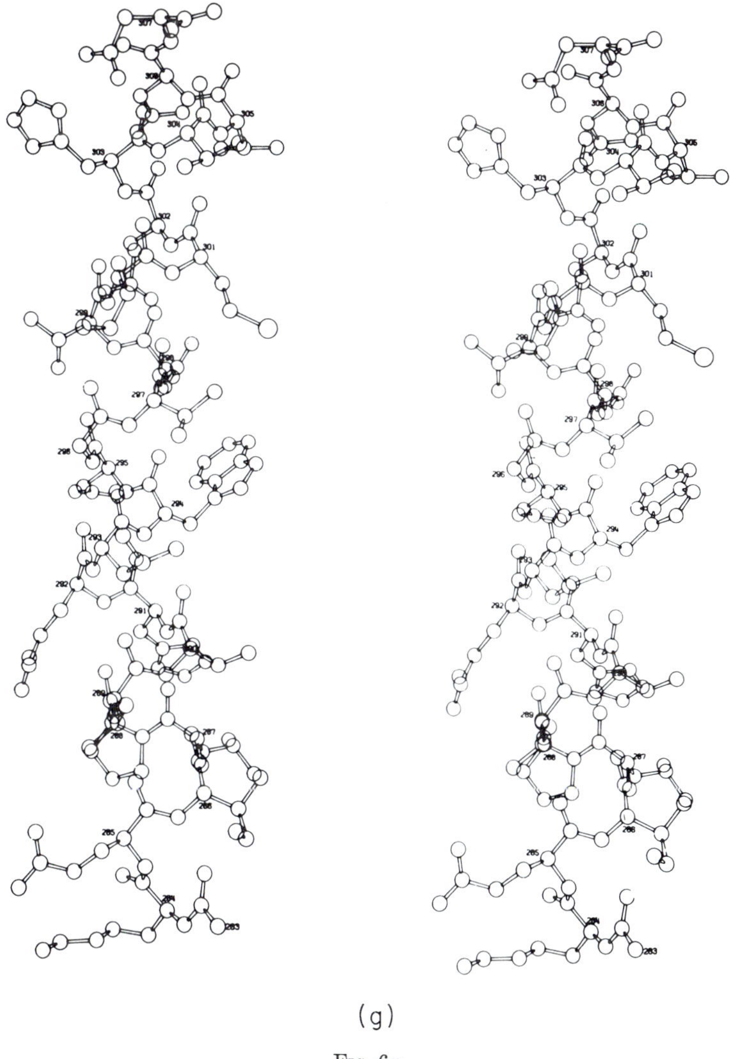

(g)

FIG. 6g.

Examining, then, the folding of CPA from its N-terminus we find that the various chains of the pleated sheet are not in sequential order (Figs. 5 and 7). Therefore, the final hydrogen bonds of extended chains 60–66 and 200–204 cannot be formed until chains 104–109 and 265–271, respectively, have been folded into place. Furthermore, to insert chain 104–109 into the pleated sheet, residues 1–103 cannot be in their final positions; for if they were, chain 104–109 would have to pass, for example, between

102. R. Piras and B. L. Vallee, *Biochemistry* **6**, 348 (1967).

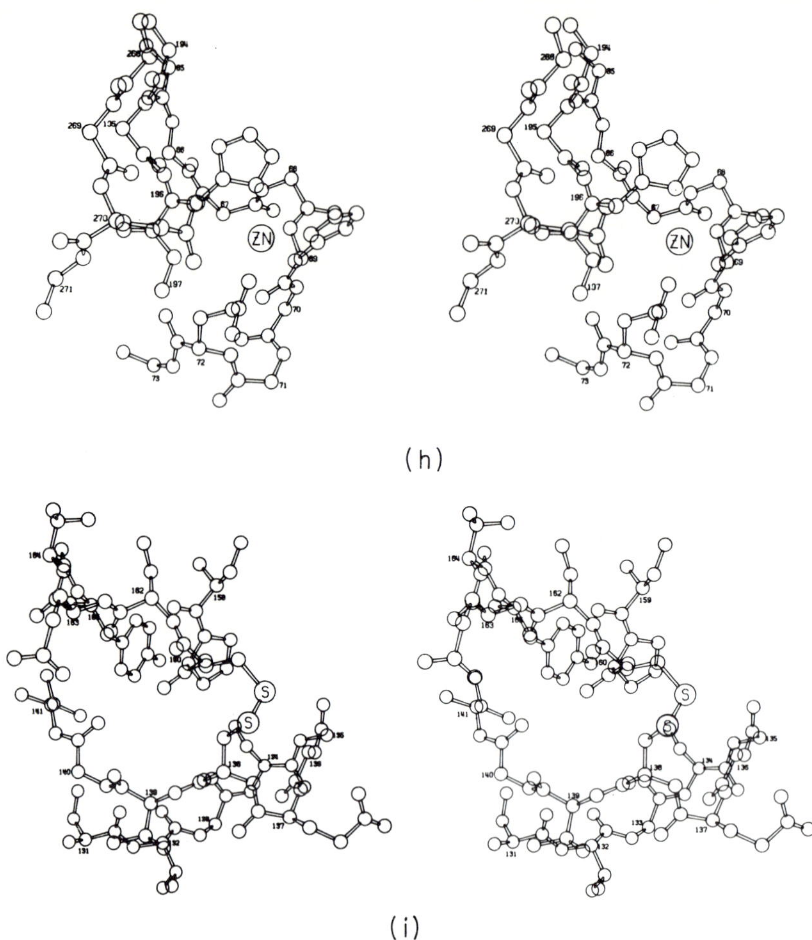

FIG. 6. (a) Stereocomposite of the electron density sections $y = 0.38$ to $y = 0.46$. From left to right the C-terminal helix, a hydrophobic core, an extended chain, the zinc atom, and the disulfide bond can be seen. (b) Stereocomposite of the electron density sections $y = 0.45$ to $y = 0.57$. To the right of the extended chain the pocket region where the substrate binds can be seen. Dots have been added to depict a bound substrate molecule. (c) Closeup view of Fig. 6a, which shows the C-terminal helix, the pleated sheet, and the zinc atom. (d) The zinc atom and its protein ligands in the electron density map. The water molecule, which completes the near-tetrahedral configuration about zinc, is in sections above those shown. (e) The disulfide bond in the electron density map at the right. The bond occurs between S atoms of Cys 138 and Cys 161. The zinc is at the left. (f) Closeup view of Fig. 6b, which shows the electron density of the protein in the region where the substrate binds. The substrate molecule is shown as dots which do not match the contours of the protein or solution. (g) OR-TEP drawing of Fig. 6a: the C-terminal helix. (h) OR-TEP drawing of Fig. 6d: zinc, its protein ligands, and Glu 270. (i) OR-TEP drawing of Fig. 6a: the region of the disulfide bond.

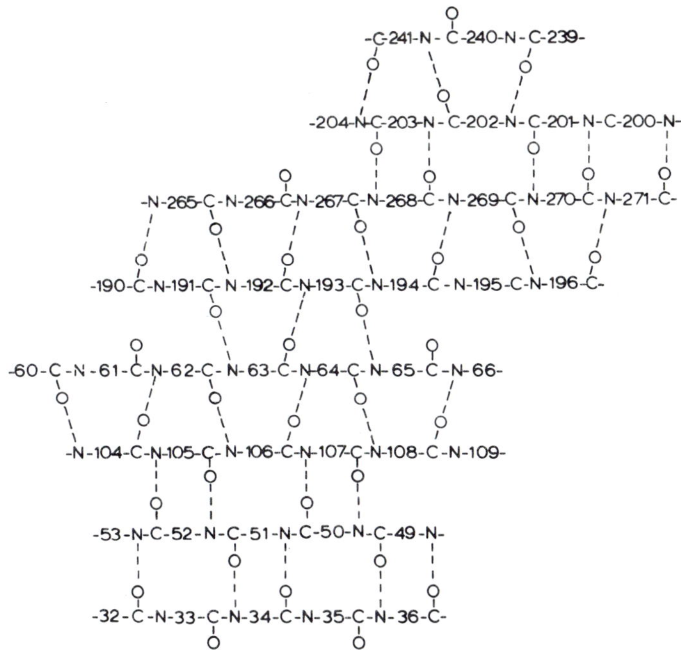

FIG. 7. Schematic drawing of the pleated sheet (β) structure of CPA. Hydrogen bonds are shown as dashed lines.

helices 14–28 and 72–88, whose axes are 12 Å apart, and between Phe 52 and Phe 86, which are 3.5 Å apart in the final structure. We have observed, in addition, three other places in the molecule where the final structure must differ from the conformation during the folding process. First, chain 249–254 must pass between two loops of random coil, which at closest approach in the final structure are only 5.5 Å apart between C_α 150 and C_α 208. Second, residues Leu 233 and Tyr 234 would interfere with placement of the C-terminal helix. Finally, we observe that the disulfide bond must be formed after residues 163–170 pass through the disulfide loop 138–161.

5. *Backbone Conformation*

A plot (*103*) of the dihedral angles of rotation ϕ (about the C_α–N bond) and ψ (about the $C_{carbonyl}$–C_α bond) (*104*) for the peptides of CPA is given in Fig. 8. We have superposed on the usual (ϕ, ψ) plot the

103. G. N. Ramachandran, C. Ramakrishnan, and V. Sasisekharan, *JMB* **7,** 95 (1963).

104. J. T. Edsall, P. J. Flory, J. C. Kendrew, A. M. Liquori, G. Nemethy, G. N. Ramachandran, and H. A. Scheraga, *JBC* **241,** 1004 (1966).

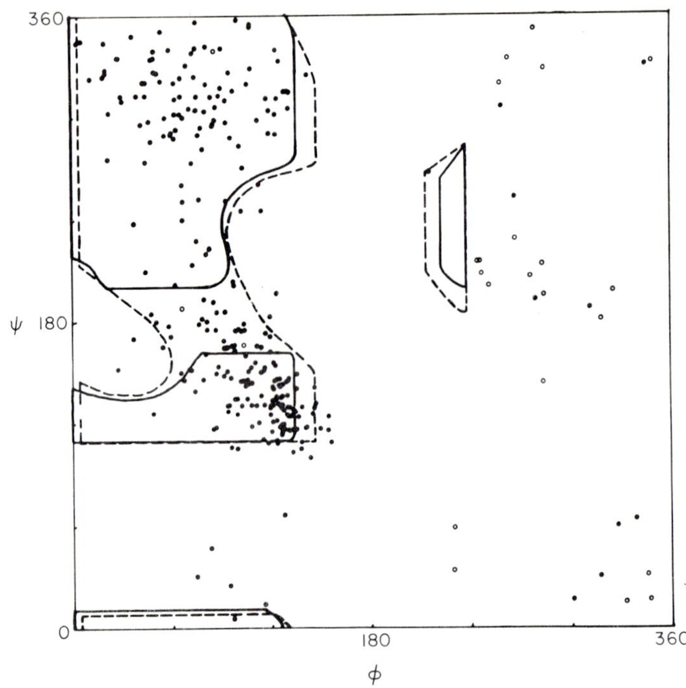

FIG. 8. Plot of dihedral angles of CPA. Solid lines demarcate allowed regions for $\tau = 110°$ and dashed lines the regions for $\tau = 115°$. Open circles are glycine residues.

boundaries of Ramachandran's "outer limit" regions allowed for $\tau = 110°$ (solid lines) and $\tau = 115°$ (dashed lines) (*105*). Of course, distortions of the peptide unit could change the shapes of the allowed regions. The distribution of points is similar to that found for lysozyme (*106*), where there are also violations of the $\tau = 110°$ limits in the regions $\psi = 180°$, $\phi = 90°$, and $\psi = 120°$, $\phi = 180°$.

At present our best interpretation of the CPA electron density map in the region of residues 196, 197, and 198 requires that the peptide bond between 197 and 198 be cis rather than trans (Fig. 9). This change in our interpretation occurred when we corrected our side chain identifications for residues 196 and 197 (*68*). Substituting His for Lys or Glx as the zinc ligand 196 required a movement of the α-carbon atom of 196 by 1.6 Å and substituting Ser for Gly at 197 required a rotation in ϕ for residue 197 of about 120° in order that the serine hydroxyl group would not interfere sterically with the zinc atom. The bond angles of the cis

105. G. N. Ramachandran and C. Ramakrishnan, *Biophys. J.* **5**, 909 (1965).
106. C. C. F. Blake, G. A. Mair, A. C. T. North, D. C. Phillips, and V. R. Sarma, *Proc. Roy. Soc.* **B167**, 365 (1967).

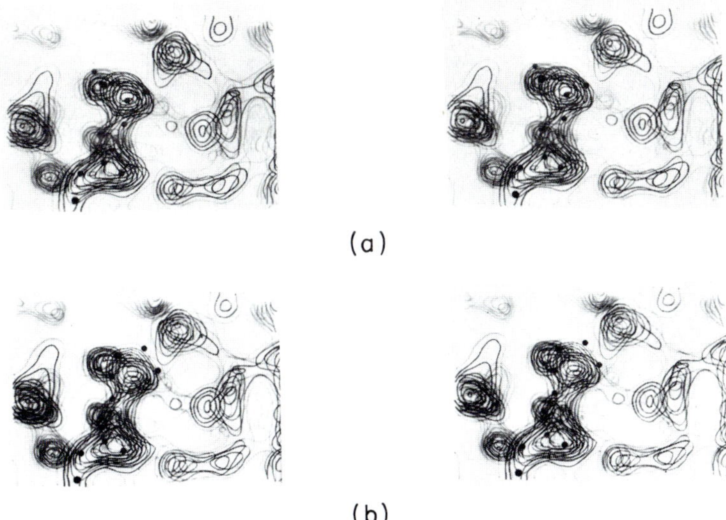

FIG. 9. Stereophotograph of native electron density map showing cis peptide between residues 197 and 198. The barrier for isomerization is estimated from the literature on small amides to be roughly 21 kcal/mole. The relation of the electron density to a cis peptide is shown in the stereopair (a), while the poorer relationship to a trans peptide is shown in (b). In (a), the OH of Ser 197 is obscured by the C_α (larger dot) of 197; hence, this OH is centrally located within this major region of electron density.

peptide are distorted so that the α-carbon atoms 197 and 198 are further apart then they would be with the bond angles of a trans peptide. This distortion was suggested by Ramachandran and Sasisekharan (*107*).

6. *Correlation of Sequence and Structure*

Table VIII which is similar to the table prepared by Cook (*108*) for hemoglobin, myoglobin, and lysozyme, is intended to reveal the existence of correlations between structural type and particular side chains. The first column lists the percentage composition of the entire CPA molecule, and the values in columns 2 through 9 are percentages of the structural type named. Certainly the results for Cys and Met, and also perhaps for His and Trp, are not significant because their percentages of the amino acid composition are small.

In regard to secondary structure, 47.9% of the molecule (147 residues) has a random conformation, 37.5% (115 residues) is contained in helix

107. G. N. Ramachandran and V. Sasisekharan, *Advan. Protein Chem.* **23**, 283 (1968).
108. D. A. Cook, *JMB* **29**, 167 (1967).

TABLE VIII
Correlation of Sequence and Structure[a]

Amino acid	1 Total molecule (%)	2 Random coil	3 Helix	4 N Helix	5 C Helix	6 β	7 Outside	8 Inside	9 Surface
Ala	6.84	6.12	11.94	—	8.33	4.44	7.31	11.53	—
Arg	3.58	7.48	—	—	—	—	5.48	—	3.07
Asn	5.53	7.48	1.49	8.33	12.50	—	7.92	2.56	3.07
Asp	3.90	2.04	4.47	12.50	4.16	4.44	6.09	1.28	1.53
Cys	0.65	1.36	—	—	—	—	1.21	—	—
Gln	3.58	2.72	4.47	4.16	8.33	2.22	5.48	1.28	1.53
Glu	4.56	2.04	5.97	12.50	8.33	4.44	5.48	2.56	4.61
Gly	7.49	11.56	4.47	—	8.33	2.22	11.58	5.12	—
His	2.60	2.72	1.49	—	8.33	2.22	1.21	—	9.23
Ile	6.84	6.12	5.97	16.66	—	8.88	2.43	16.66	6.15
Leu	7.49	3.40	8.95	4.16	4.16	22.22	4.87	11.53	9.23
Lys	4.88	2.72	7.46	4.16	4.16	8.88	7.31	—	4.61
Met	0.97	—	2.98	—	4.16	—	0.60	2.56	—
Phe	5.21	2.72	7.46	4.16	4.16	11.11	0.60	14.10	6.15
Pro	3.25	4.08	1.49	8.33	—	2.22	4.87	2.56	—
Ser	10.42	14.28	4.47	8.33	8.33	8.88	14.02	7.69	4.61
Thr	8.46	8.84	10.44	8.33	8.33	4.44	9.14	8.97	6.15
Trp	2.28	2.04	2.98	4.16	—	2.22	—	2.56	7.69
Tyr	6.18	9.52	2.98	—	—	6.66	3.04	1.28	20.00
Val	5.21	2.72	10.44	4.16	8.33	4.44	1.21	7.69	12.30

[a] The values in column 1 are percentages of the total molecule and the values in columns 2 through 9 are percentages of the structural types named. For a given residue, the entries in columns 2 through 9 would equal that of column 1 if that residue were distributed randomly among the structural types.

of varying degrees of perfection, and 14.6% (45 residues) is involved in the β structure. In the tabulation of N and C ends of helices, columns 4 and 5, three residues are included from the respective ends of each helix. The residues which appear to display a preference for a particular secondary structural type are Arg, Gly, and Tyr for random coil; Ala and Val for helix; Asp, Glu, Ile, and Pro for the N end of helix; Asn, Gln, Glu, and Val for the C end of helix; and Leu, Lys, and Phe for β structure. The presence of the two hydrophobic residues Leu and Phe in the β structure reflects more the position of the pleated sheet in the core of the CPA molecule than it does the nature of the β structure and is therefore not necessarily typical.

It is interesting to note the locations of the 10 prolines in the CPA structure. Four proline residues, *viz.*, 94, 113, 214, and 288, are at the amino ends of helices. Three other prolines, 46, 60, and 205, are at the ends of extended chains of the pleated sheet, and the remaining three prolines, 30, 160, and 282, are situated in random coil.

In regard to side chain environment 53.4% of the molecule (164 residues) is outside (i.e., all atoms of the side chain make contact with nonisolated water molecules), 25.4% (78 residues) is inside (i.e., no contact with nonisolated water molecules), and 21.2% (65 residues) is on the surface (i.e., some atoms of the side chain make contact with nonisolated water molecules and some side chain atoms do not). The preferences of Asp and Gly for the outside; Ala, Ile, Leu, and Phe for the inside; and His, Trp, Tyr, and Val for the surface appear to be significant.

7. *Side Chain Conformations and Interactions*

The conformations of the side chains of the residues are quite varied. For example, Table IX shows the various values of χ_1 (the angle of free rotation about the C_α–C_β bond) for the tyrosine residues. The staggered conformations have χ_1 equal to 60, 180, or 300 degrees; $\chi_1 = 190$ is found in the tyrosine crystal structure (*109*). Fifteen of the 19 tyrosine residues in CPA assume one of the three staggered conformations. Four other residues are closer to an eclipsed than a staggered conformation.

A consideration of that portion of the molecule which is isolated from the surrounding solvent molecules is particularly interesting. This inside region contains the side chains of 78 residues in addition to ten isolated water molecules. Of the 78 side chains, 56 (72%) are hydrophobic and 22 (28%) are capable of forming hydrogen bonds.

Within the molecule there are three charged acid residues Asp 104, Glu 108, and Glu 292. Since all three of these are effectively neutralized

109. D. W. Smits and E. H. Wiebenga, *Acta Cryst.* **6**, 531 (1953).

TABLE IX
CONFORMATION OF TYROSINE RESIDUES[a]

Residue No.	χ_1
169	47
208	84
240	61
42	179
90	181
165	185
248	202
259	185
9	314
12	295
48	276
206	273
234	300
265	290
277	289
19	267
198	102
204	27
238	255

[a] Crystallographic evidence was obtained for at least partial iodination of tyrosines 19, 42, 234, 238, and 277. Only tyrosines 19, 238, and perhaps 12 have buried hydroxyl groups, and therefore the three-dimensional structure does not support statements (*73*) that 7–8 tyrosines are at the surface and 11–12 tyrosines are located in the interior of the folded protein. These chemical results may be associated in part with hydrogen bonding. The hydrogen bond from the OH of Tyr 238 to the peptide CO of Glu 270 may play some role in the conformational change of Glu 270.

and none of them is near the active site, we do not attach catalytic significance to any of these buried charges. Aspartic acid 104 is adjacent to a water molecule, which, in turn, is next to the guanidinium group of Arg 59. Glutamic acid 108 forms a hydrogen bond with a water molecule, which is apparently a trapped H_3O^+ (see below). Glutamic acid 292 is neutralized by the positively charged guanidinium group of Arg 272.

The hydrogen bonds formed by the 22 hydrophilic residues of the molecular interior are tabulated according to side chain type (Table X). Only three of these side chains do not participate in hydrogen bonds. The Trp 63 side chain is a part of the hydrophobic core which is located to the right of the pleated sheet in Fig. 5 and below the substrate binding pocket as seen in Fig. 10. Threonine residues 289 and 304

TABLE X
INTERNAL HYDROPHILIC RESIDUES

Residue	Number	Interacts with
Asn	112	Backbone carbonyl oxygen
	146	Serine hydroxyl
Asp	104	H_2O which has Arg adjacent
Gln	76	Backbone carbonyl oxygen
Glu	108	H_3O^+
	292	Arg
Ser	41	Glutamic carboxyl
	70	Asparagine amide or carbonyl
	197	H_2O
	254,258,266	Serine hydroxyl
Thr	75	Backbone carbonyl oxygen
	78,129	Serine hydroxyl (Thr 129 also binds to a backbone amide)
	119	Tryptophan nitrogen
	289,304	Nothing
	293	H_2O
Trp	63	Nothing
	147	Backbone carbonyl oxygen
Tyr	238	H_2O

are a part of the helix at the C-terminus of the molecule and their side chains are located in the hydrophobic region between the helix and the pleated sheet (Fig. 5 and Figs. 6a and c).

Inside the protein are ten water molecules which do not make contact with the surrounding solvent. These ten do not include those water molecules located in the substrate binding pocket, some of which are displaced and some of which are trapped when a substrate is bound. Among the ten isolated water molecules, two participate in four hydrogen bonds tetrahedrally arranged. In both cases two backbone carbonyl oxygen atoms and the hydroxyl group of a serine side chain are three of the ligands to water. The fourth ligand is an amide of asparagine in one instance and water in the other. Seven of the isolated water molecules each make three hydrogen bonds. In all but one case at least one of the ligands to water is a backbone carbonyl oxygen atom. Other participants in these interactions with water are backbone imino groups, tyrosine and threonine hydroxyl groups, and side chain amides and carboxyl groups. The most interesting of the water molecules which have three ligands is that which is bound to one carboxyl oxygen of Glu 108. Because the other two ligands to this water molecule are backbone carbonyl oxygen atoms, this must be a trapped H_3O^+ ion which neutralizes the

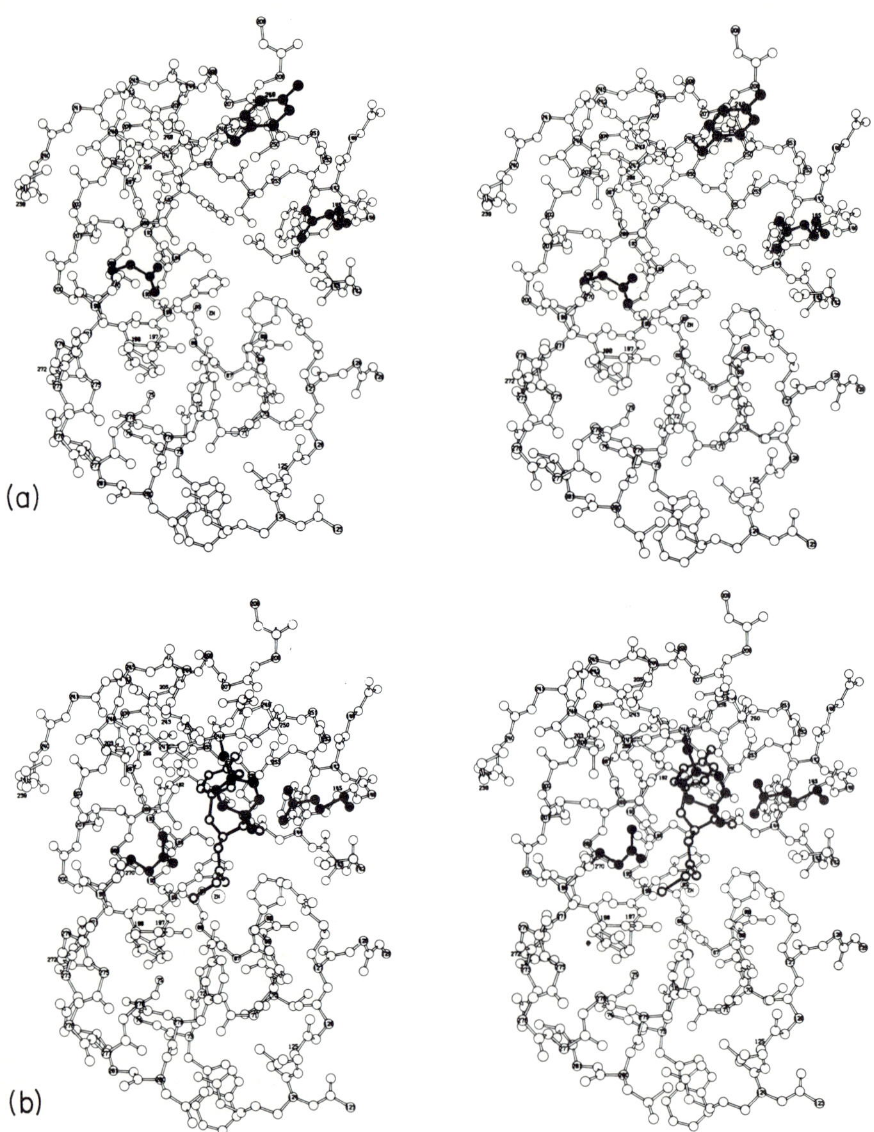

FIG. 10.

buried charge of Glu 108. We caution that the suggestion that this molecule exists as H_3O^+ rather than H_2O depends upon the chemical identification of residue 108 as Glu and not Gln. Finally, one water molecule makes only two hydrogen bonds with a backbone carbonyl oxygen and with another water.

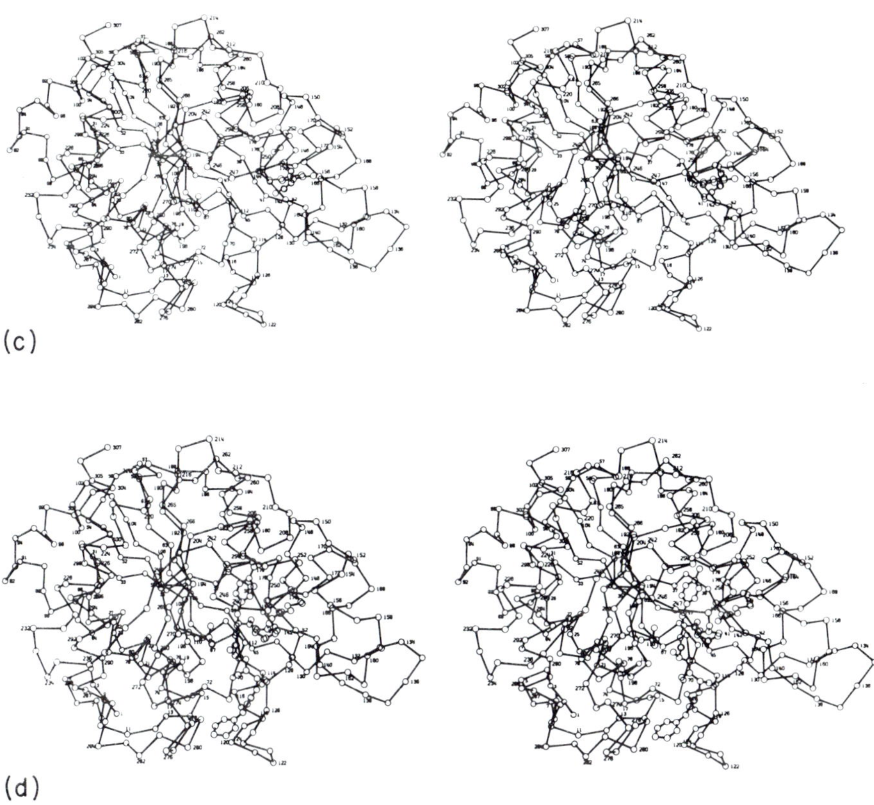

FIG. 10. (a) Stereoview (OR-TEP) of the active site region of the native enzyme before conformational changes have been made. Blackened atoms are the side chains of Arg 145, Tyr 248, and Glu 270. (b) OR-TEP view after the substrate glycyl-L-tyrosine (heavy circles) has been added and the conformational changes of Glu 270, Arg 145, and Tyr 248 have been made. (c) OR-TEP drawing of the C_a positions of CPA_a, showing side chains of Arg 145, Tyr 248, and Glu 270 before their conformational changes. (d) OR-TEP drawing of C_a positions of CPA_a, showing the substrate carbobenzoxyalanylalanyltyrosine, and the side chains of Arg 145, Tyr 248, and Glu 270 after their conformational changes.

8. *Crystalline Enzyme–Substrate and Enzyme–Inhibitor Complexes*

Several inhibitors and substrates have been diffused into CPA crystals so that the positions of the added molecules could be determined using three-dimensional difference electron density maps. With the exception of the complex of glycyl-L-tyrosine with the native enzyme, these complexes have only been studied at 6 Å resolution (*90*). The inhibitor *p*-iodo-*β*-phenylpropionate has four binding sites when diffused into crystalline CPA. Site A extends into the substrate binding pocket and

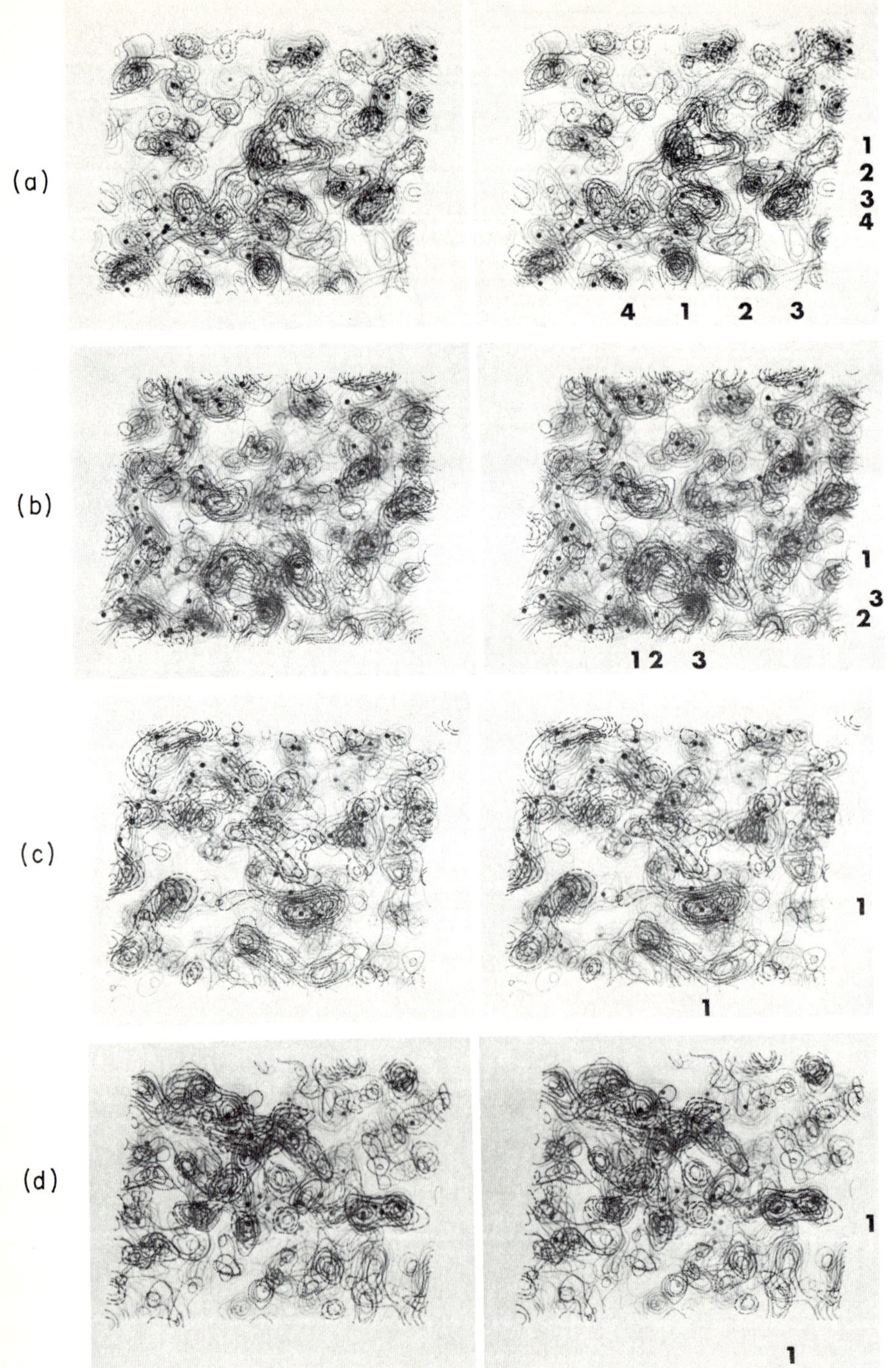
(a)
1
2
3
4
4 1 2 3
(b)
1
3
2
1 2 3
(c)
1
1
(d)
1
1

site B is in the region of the aromatic residues Tyr 198 and Phe 269 (Fig. 10). The third and fourth sites are on the molecular surface and removed from the susbstrate binding site. At 6 Å resolution the difference electron density map computed for the complex of native CPA with L-lysyl-L-tyrosinamide shows one peak which covers both sites A and B, but the occupancy appears higher in site B. The product inhibitor L-phenylalanine binds only in site A. The acyl dipeptide hippuryl-L-phenylalanine, which is a good substrate for CPA, binds to acetylated CPA crystals in both sites A and B (site B could be occupied by either the acyl end of a productively bound substrate molecule or by a second substrate molecule), whereas the dipeptide glycyl-L-tyrosine, which is a poor substrate, binds a single molecule to either apo or native CPA crystals only in site A.

The complex of glycyl-L-tyrosine with native CPA crystals was examined first at 2.8 Å resolution (*16*) and then at 2.0 Å resolution (*21*). Because the occupancy of added Gly-L-Tyr substrate was only about 30%, the difference electron density map was computed with Fourier coefficients $(|F_{SE}|-x|f_c|-(1-x)|F_N|)\exp(i\phi_{SF})$, where $|F_{ES}|$ and $|F_N|$ are scaled observed structure factors for the Gly–Tyr enzyme complex and for the native enzyme, respectively; x is the fractional occupancy of the substrate; f_c is a calculated structure factor for the native enzyme; and ϕ_{SF} is the phase angle computed from the atomic coordinates in the native enzyme (*89*). This computation was possible only after the structure factor calculation for the native enzyme had been performed (Fig. 11).

The binding of Gly–Tyr to CPA can be summarized by describing four interactions. We believe that the first three of these are characteristic of a productive enzyme–substrate complex. First, the C-terminal side

---

FIG. 11. The difference electron density function for the complex of Gly–Tyr with CPA is shown as dotted (positive) and dashed (negative) contours. Solid contours show the electron density of CPA at 2.0 Å resolution. Dots are placed at proposed atomic positions. Composite of the difference map sections $y = 0.47$ to $y = 0.52$. Near the top center the positive contours of the tyrosyl side chain of the substrate are visible [1]. To the right of the substrate are the positive [2] and negative [3] contours of the Arg 145 guanidinium group and to the left [4] are the native contours of Glu 270. (b) Composite of the difference map sections $y = 0.49$ to $y = 0.56$. Near the bottom of the picture are the positive contours of the moved Glu 270 [1], the substrate's terminal amino group [3], and the connecting water molecule (square dot, [2]). (c) Composite of the difference map sections $y = 0.56$ to $y = 0.62$ shows Tyr 248 after its conformational change in the bottom right portion of the picture [1]. (d) Composite of the difference map sections $y = 0.63$ to $y = 0.68$ shows Tyr 248 before its conformational change [1].

chain of the substrate is situated in a pocket in the enzyme; this placement necessitates the displacement of several water molecules (Fig. 11a). In agreement with the moderate but not high specificity for the C-terminal side chain of the substrate, this pocket contains no specific binding group and is large enough to accommodate a tryptophan side chain (*109a*). Second, the terminal carboxylate group of the substrate (which is essential for susceptibility to cleavage) interacts with the positively charged guanidinium group of Arg 145 (Fig. 11a). Third, the carbonyl oxygen of the scissile peptide bond replaces water as a ligand to zinc (Fig. 11a). Because the carbonyl oxygen of the substrate replaces water, the density was low in this region of early difference maps. However, it is more positive in the map computed with coefficients which take into account the low Gly–Tyr occupancy. Fourth, in an interaction possible only with dipeptide substrates, Glu 270 binds through water to the α-amino group of Gly–Tyr (*109b*) (Fig. 11b). It is probably this interaction which accounts for the unusual stability of the Gly–Tyr–CPA complex. All of these binding interactions are summarized in Fig. 12.

9. *Enzyme Conformational Changes upon Substrate Binding*

Before the computation of the high resolution difference electron density map for the Gly–Tyr complex with CPA, there were several indications that conformational changes occur upon binding of substrates and some inhibitors to CPA. Chronologically, Fujioka and Imahori (*111*) observed a Cotton effect at 295 nm in the ORD spectrum of the β-phenylpropionate complex with CPA and attributed this to a change in the enzyme, specifically stating that there occurred "orienta-

109a. Residue 256 was identified crystallographically (*89*) as Asx and in the chemical sequence (*13*) as Asp. However, because the 256 side chain is located at the back of the substrate binding pocket so that C-terminal substrate residues such as Tyr, Trp, and Phe would make contact with the charged carboxylate of Asp 256, Asn seems a more reasonable choice than Asp for 256. The 256 backbone is involved in helix so that conformational changes to expose the charge to solvent are unlikely and furthermore the Gly–Tyr difference map indicates that no motion of 256 takes place.

109b. Crystallographically, we cannot distinguish the anionic form, i.e., the species with an uncharged amino group, from the Gly–Tyr zwitterion, and the binding of either form can be rationalized. Yanari and Mitz deduced from the increased efficiency of Gly-L-Tyr and D-Leu-L-Tyr as competitive inhibitors at higher pH that only the anionic form of the dipeptide is bound to CPA (*110*).

110. S. Yanari and M. A. Mitz, *JACS* **79**, 1154 (1957).

111. H. Fujioka and K. Imahori, *JBC* **237**, 2804 (1962).

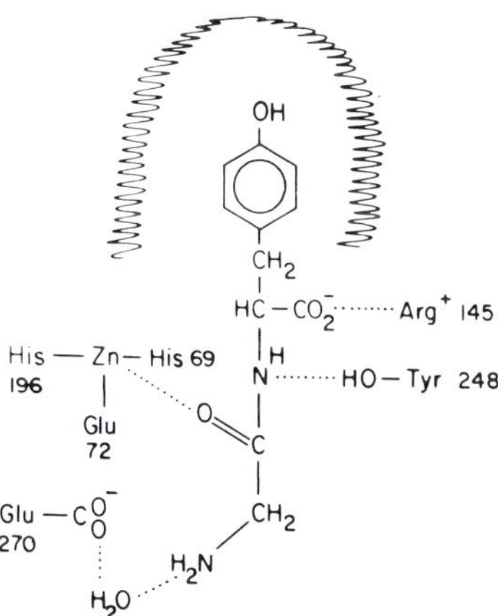

FIG. 12. Schematic drawing of the binding of Gly–Tyr to CPA. It is not certain whether the N-terminus of the Gly-Tyr is NH_2 or NH_3^+.

tion of certain aromatic side chains as a result of the strong binding with the aromatic ring of the inhibitor." Riordan *et al.* (*76*), by virtue of changes in the UV spectrum of nitrated CPA when β-phenylpropionate was added, were able to deduce an alteration in "the immediate chemical environment of the active center tyrosyl residue of CPA either by direct interaction with that group or as a result of conformational changes which move the tyrosyl group into a more hydrophobic region." The X-ray studies choose between these alternatives. Steitz *et al.* (*90*) observed consistent evidence for enzyme conformational changes in the 6 Å difference electron density maps of crystalline complexes of CPA with hippuryl-L-phenylalanine, L-phenylalanine, glycyl-L-tyrosine, and *p*-iodo-β-phenylpropionate. These changes are seen in the difference maps as negative density at the native conformation and positive density at the modified conformation.

Enumeration of the conformational changes which have been observed and interpreted at 2.0 Å resolution includes the following. First, the guanidinium group af Arg 145 moves about 2 Å by means of a rotation about the C_β–C_γ bond of the side chain (Fig. 11a). Second, the carboxylate of Glu 270 moves toward the viewer in Fig. 11b by about 2 Å. This motion results from rotations about both the C_α- - -C_β and C_β- - -C_γ bonds of Glu 270. In exception to the above statement, there is no negative density at the native Glu 270 position because this side chain was not included in the structure factor calculation. Third, the phenolic hydroxyl of Tyr 248, identified as being involved in the activity of the enzyme (*16, 74*), moves about 12 Å so that the OH group comes from the surface of the molecule to the vicinity of the peptide bond of the substrate (Figs. 11c and d). This motion involves a rotation of the side chain about its C_{α}- - -C_{β} bond by about 120° as well as a limited motion of the peptide backbone. In our interpretation, which is not accurate to more than 1 Å, the tyrosyl oxygen is located 2.7 Å from the nitrogen of the scissile peptide bond and 3.5 Å from the α-amino group of the Gly–Tyr molecule. Finally, in the tortuous portion of the native structure there is a system of four hydrogen bonds which form a link between Arg 145 and Tyr 248. This link goes from Arg 145 to backbone carbonyl 155, from backbone carbonyl 154 to Gln 249, and, finally, from Gln 249 via H_2O to the hydroxyl of Tyr 248. When Gly–Tyr is bound all these groups move, and we observe that Arg 145 then no longer interacts with backbone carbonyl 155 and that Tyr 248, having undergone an enormous conformational change, no longer interacts with Gln 249.

One of the most striking effects of the binding of the substrate to CPA is the conversion of the enzyme cavity from a water-filled to a hydrophobic region. At least four water molecules must be expelled when a substrate C-terminal side chain such as Tyr is inserted into the pocket, and one water molecule is displaced from the zinc atom when the carbonyl group of the substrate is bound. Also, the charge of Arg 145 is somewhat neutralized by its interaction with the terminal carboxylate ion of the substrate. Finally, Tyr 248, in making its conformational change, closes off the enzyme cavity so that it is not in equilibrium with the solvent. It is hard to escape the conclusion that the displacement of water upon binding the substrate and the resultant conversion of the active center of the enzyme to a hydrophobic area provide a driving force for the reaction.

The binding of glycyl-L-tyrosine, then, results in several complementary rearrangements in the enzyme structure. The most striking of these is the coordinated motion, linked together by movements of several intervening amino acids, of Arg 145 and Tyr 248, which brings a group thought to be required for catalysis into contact with the substrate. Adaptation of an enzyme to its substrate has been suggested by several workers (*112–114*); but the conformational changes in CPA are a clear example of the "induced fit" theory of Koshland (*115, 116*).

112. F. Karush, *JACS* **72,** 2705 (1950).
113. R. Lumry and H. Eyring, *J. Phys. Chem.* **58,** 110 (1954).
114. F. Vaslow, *Compt. Rend. Trav. Lab. Carlsberg* **31,** 29 (1958).
115. D. E. Koshland, *Proc. Natl. Acad. Sci. U. S.* **44,** 98 (1958).
116. D. E. Koshland, *Cold Spring Harbor Symp. Quant. Biol.* **28,** 473 (1963).

10. *Use of the CPA Model to Interpret the Enzyme Substrate Specificity*

If we assume that in the productive binding mode substrates are bound to CPA with the C-terminal side chain in the pocket, with the terminal carboxyl group salt linked to Arg 145, with the carbonyl group of the scissile peptide bond bound to zinc; and with the OH of Tyr 248 near enough to donate a proton to the NH of the susceptible peptide bond, then it is possible to build models of substrates other than Gly–Tyr interacting with CPA. These models are consistent with the chemical experiments which demonstrate the enzyme substrate specificity. First, the presence of a dead-end pocket, in addition to a groove, provides an explanation of the observation that CPA is an exopeptidase not an endopeptidase (*25*, *26*). Second the preference for uncharged side chains at the C-terminus of the substrate (*27*) is explained by the lack of charges within the pocket (see *109a*). Third, peptide substrates with a D-amino acid having a bulky side chain at the C-terminus, such as CBZ-Gly-D-Phe, are neither hydrolyzed nor bound (*117*). Here we find that if the peptide bond and carboxyl group of the substrate are appropriately positioned, the R group of a substrate with a C-terminal D residue cannot easily be directed into the pocket, and steric interactions which prevent binding may occur in the case of large side chains. Fourth, the small K_I and K_m (*118*) of Gly–Tyr as compared to the K_I of *N*-acetyl-Tyr (*110*) indicates that the former is bound more strongly than the latter. Earlier work comparing *N*-acyl dipeptides with analogous dipeptides also indicated that when a free amino group is present, the dipeptide is bound tightly but is hydrolyzed very slowly (*26*): We find from models that blocking of this amino group destroys the nonproductive binding mode, described above as an interaction of the amino group (through a water molecule) with the carboxyl group of Glu 270, thus freeing Glu 270 to serve in a catalytic role. Fifth, model building indicates that steric hindrances occur if the NH group of the sensitive peptide bond is substituted. The most severe example is CBZ-Gly-thiazolidine-4-carboxylic acid, which is neither bound nor cleaved (*32*), probably because of steric interference by Ile 247. In the case of CBZ-Gly-sarcosine and CBZ-Gly-Pro, neither of which is cleaved (*27*), model building suggests interference with the final stages of the conformational change of Tyr 248. Sixth, in *N*-acyl dipeptides the rate of hydrolysis is greatly decreased by substitution of sarcosine (*33*) or β-alanine (*34*) for the second amino acid. Thus, the presence and integrity of the second

117. H. Neurath, E. Elkins, and S. Kaufman, *JBC* **170**, 221 (1947).
118. N. Izumiya and H. Uchio, *J. Biochem.* (Tokyo) **46**, 235 (1959).

peptide bond are important for rapid hydrolysis. In the active complex, we postulate the existence of a hydrogen bond between the amide of the penultimate peptide bond of the substrate and the hydroxyl oxygen of Tyr 248 (see below). Seventh, placement of the bulky D-Leu as R_1 in several *N*-acyl-D-Leu-L-Tyr substrates is known to lower the rate of hydrolysis by a factor of about 5000 relative to that of the corresponding *N*-acyl-L-Leu-L-Tyr (*31*): If the amide of the second peptide binds to Tyr 248, the D side chain interferes sterically with the protein, but such repulsion does not occur for the L-compounds. Eighth, the experiments of Abramowitz *et al.* (*35*) have mapped carefully several effects, of which those relating primarily to the kinetic constant K_m are discussed here. Comparison of K_m for Phe–Ala–Ala with K_m for Ala–Ala–Ala yields a ratio of 72:1 in favor of this N-terminal aromatic for side chain R_2, but substitution of an aromatic group is also felt at R_1, R_3, or R_4. The effect is strongest when the relatively flexible CBZ group is at R_3, where it can most favorably be placed near Tyr 198 and Phe 279. Also, a consistently lower K_m is noted if the terminal NH_3^+ group in a tetrapeptide is blocked by acetylation or by the introduction of a urethane. Our model indicates that this substitution on NH_3^+ provides additional interactions of lone pairs of added oxygen atoms with the guanidinium group of Arg 71 and removes the possible repulsion of the NH_3^+ from Arg 71. All of these binding interactions are summarized in Fig. 13.

11. *Successes and Failures of the CPA Crystallographic Investigation*

Because the 2.0 Å resolution electron density map of CPA was available two years before the entire chemical sequence of the molecule was known, we had the opportunity to make interpretations of the density map which are not necessary if the amino acid sequence is known. A retrospective discussion of these interpretations is illuminating. Of course the electron density map allowed the total number of residues in the polypeptide chain to be determined and each residue to be assigned a number. Consequently, before the complete chemical sequence was known, knowledge of the numbers of interesting residues, such as the active center Tyr 248 which had been isolated within a tetrapeptide by Roholt and Pressman (*74*) and residues 179, 228, and 305 which are subject to genetic variation (*12*), was provided by placement of sequence fragments. We were able to correct several mistaken interpretations of chemical experiments. It had been thought that the zinc atom had two protein ligands, a cysteine sulfur and the N-terminal α-amino group (*9, 65–67*). Also, the sulfur chemistry had indicated a cysteine residue

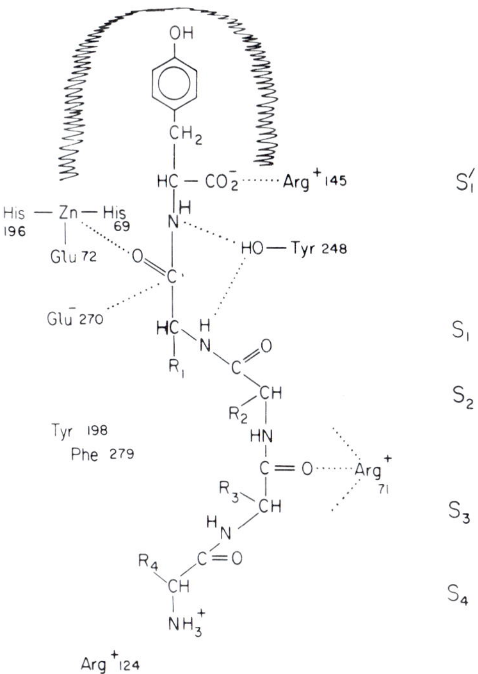

Fig. 13. Drawing of the binding of a longer substrate to CPA. Sites S_1', S_1, S_2, S_3, and S_4 correspond to the residues R_1', R_1, R_2, R_3, and R_4 of the substrate.

which reacted "atypically" (*119, 120*). The crystal structure showed three protein zinc ligands, none of which was either sulfur or the *N*-terminus. Also, the electron density map showed a disulfide bond rather than an atypical cysteine residue. Recently, the existence of the disulfide bond has been confirmed chemically (*13*). Which residues are zinc ligands (*viz.*, 69, 72, and 196) and which residues probably have functional importance (*viz.*, 145, 248, and 270—and also 71, 198, and 279) were readily derived from an examination of the CPA model. The existence of the conformational changes of Arg 145 and Tyr 248 and their interesting relation could be proved only by crystallographic experiments.

In the case of the identification from the X-ray map of important side chains for which no sequence data were available Arg 145, Glu 270,

119. K. A. Walsh, K. S. V. Sampath Kumar, J. P. Bargetzi, and H. Neurath, *Proc. Natl. Acad. Sci. U. S.* **48**, 1443 (1962).

120. K. S. V. Sampath Kumar, K. A. Walsh, J. P. Bargetzi, and H. Neurath, *in* "Aspects of Protein Structure" (G. N. Ramachandran, ed.), p. 319. Academic Press, New York, 1963.

and Tyr 198 were correctly identified. Phenylalanine 279 was left ambiguously as His or Phe. The unknown zinc ligand His 196 was not correctly identified, and because residue 265 was identified as Arg rather than Tyr it was given too much importance. As stated above, the overall success rate for residue identification was 60%. However, certain residues can be identified with far greater certainty than others. For example, we were able to examine the electron densities of the critical functional residues Arg 145, Glu 270, and Tyr 248 both before and after their conformational changes which occur when Gly–Tyr is bound to CPA. These observed conformational changes confirmed the identities of Arg 145 and Tyr 248, and eliminated two less reasonable alternatives for Glu 270. These particular alternative choices for Glu 270 could not undergo the observed conformational change.

IV. Conclusions Concerning the Mechanism of Carboxypeptidase A Action

A. Cleavage of Peptides

Now that the atomic resolution structure of the native enzyme and of the enzyme–substrate complex, as well as the identities of all 307 side chains in the enzyme molecule, are known, it is reasonable to ask which portions of the enzyme may directly affect catalytic activity and what roles these side chains may assume in the hydrolytic reaction. Actually, these matters were considered at length before the chemical sequence work was completed (*21*), and because the identities of the specific binding and catalytic residues had been deduced correctly from the X-ray maps, these early comments are still appropriate. Some of the earlier ambiguities in the roles of the catalytic residues have been removed and some extensions of the proposals have been made. Our proposed catalytic mechanism of course encompasses some, but not all, of the earlier suggestions of others, while going well beyond them. We believe that it is consistent with kinetic results under various experimental conditions, with the importance of the integrity of the amides of the first and second substrate peptide bonds, and with the effects of the modifications of various residues of the enzyme. Of course, unambiguous interpretation of all chemical experiments, some of which are conflicting, is not possible. The proposals, which are presented below, have been made with peptide substrates in mind. Their relation to the hydrolysis of artificial ester substrates will be discussed in a separate paragraph.

Because a metal ion is essential to hydrolysis by CPA (*56*) and because substrates bind over the metal in the crystal (*16*), the locus of productive substrate binding, *viz.*, at the zinc atom, has been deduced. Since the interaction of the substrate's free carboxyl group with Arg 145 is reasonable and allows placement of the scissile peptide bond over zinc, we assume that the placement of the substrate carboxylate group and peptide bond is the same in the productive complex as is seen in the Gly–Tyr complex with CPA. Insertion of the R_1' group of the substrate into the pocket of the enzyme and an interaction of Groups R_2, R_3, or R_4 with the aromatic residues Tyr 198 and Phe 279 are implied by this assumption. (These binding implications were shown above to be consistent with the enzyme specificity for substrates.) Inspection of the CPA model shows that once this binding mode is assumed, only Arg 145, Zn, Glu 270 and Tyr 248 (after its conformational change) are close enough to the active site to enter directly into catalysis.

The existence of a positive charge on the enzyme which binds the substrate's free carboxylate group was postulated by Waldschmidt-Leitz (*25*), tentatively identified by Neurath and Schwert as His or the N-terminal α-amino group (*39*), and incorporated into the mechanistic proposals of Smith (*58, 81*) and Vallee (*83, 84*). However, until the structures of both native CPA and the enzyme–substrate complex were known, it was not suspected that the interaction of the carboxylate ion with this positive charge, the guanidinium group of Arg 145, would induce a conformational change which brings a group, Tyr 248, necessary for catalysis, into contact with the substrate. (Tyr 248 may not take part in the hydrolysis of some ester substrates.) The need to induce this conformational change explains the indispensable nature of the substrate's free carboxylate group.

No hydrolytic activity has been demonstrated for apocarboxypeptidase A, and therefore the presence of a metal ion must be essential for catalysis to occur. In the complex of Gly–Tyr with CPA, the carbonyl group of the scissile peptide bond is directed toward zinc so that the C–O bond is polarized and the carbonyl carbon is rendered susceptible to nucleophilic attack. Both this mode of binding and the resultant polarization were postulated by Smith (*58, 84*). Upon consideration of the charge situation at the zinc atom (the protein ligands to zinc are His 69, Glu 72, and His 196), we suggested that induction of a dipole in the substrate by the singly positive zinc–ligand complex could be enhanced in three ways by virtue of this complex existing on the surface of the enzyme. First, when the zinc ion is covered by the apolar substrate molecule the effective charge of the ion is increased because of the lowering of the dielectric constant of the immediate surroundings of the zinc ion. Second, the lines of force from the charge tend to be localized on the carbonyl group of the substrate. Third, the presence of the negative charge of Glu 270 and of polarizable water molecules adjacent to the carbonyl carbon atom and away from zinc further enhance polarization of the carbonyl group (*68*).

Of the two remaining potential (*120a*) catalytic residues Tyr 248 and Glu 270, tyrosine, because its dissociation pK is above the pH of enzymic activity, is the probable supplier of a proton for the proteolytic reaction. After its conformational change Tyr 248 is in a position to donate a hydrogen bond to the NH of the susceptible peptide bond of the substrate and, within the limits of error of the Gly–Tyr difference map, to receive a hydrogen bond from the NH group of the penultimate peptide bond. The hydrogen bond to the N of the scissile bond tends to make the bonding around this N atom nonplanar. The system of two hydrogen bonds between Tyr 248 and the two NH groups of the substrate would also induce strain in the substrate at the susceptible C atom.

Glutamic acid 270 probably functions in promoting general base catalysis of the oxygen atom of a water molecule at the carbon atom of the susceptible carbonyl group or in nucleophilic attack on this carbon atom (see Fig. 14). Unfortunately, we cannot at present choose between the nucleophilic and the general base mechanism of attack on the carbonyl carbon. Steric aspects seem to favor the anhydride pathway. The relative orientation and separation (about 2.5 Å) of the carboxyl group of Glu 270 in its position in native CPA and the hydrolyzable bond of the substrate are ideal for nucleophilic attack. On the other hand, the placement of water for general base reaction necessitates a reorientation of the side chain of Glu 270 from its native conformation (the backbone being fixed by the β structure) in order to facilitate attack by water and in order not to violate closest acceptable distances of approach between various atoms in the enzyme, the water, and the substrate. This reorientation and positioning of water is not only possible but does occur when Gly–Tyr is bound to CPA. The failure to observe transpeptidation (*121*) or transesterification (*122*) is not a strong argument against an acyl intermediate because the anhydride intermediate would be readily susceptible to hydrolysis by water. This hydrolysis might even be facilitated by the proximity of the Zn–OH_2 complex. Moreover, the ap-

120a. Contrary to the statements in Bradshaw *et al.* (*13*), the long distances of residues 127 and 279 from the catalytic site had rendered them as unlikely or unfavorable candidates for alternative catalytic residues in the X-ray work (*21*). Also, a hydrogen bond between Tyr 265 and the enzyme's carboxylate group has been incorrectly deduced (*13*).

121. L. M. Ginodman, N. I. Mal'tsev, and V. N. Orekhovich, *Biokimiya* **31**, 1073 (1966).

122. P. L. Hall and E. T. Kaiser, *BBRC* **29**, 205 (1967).

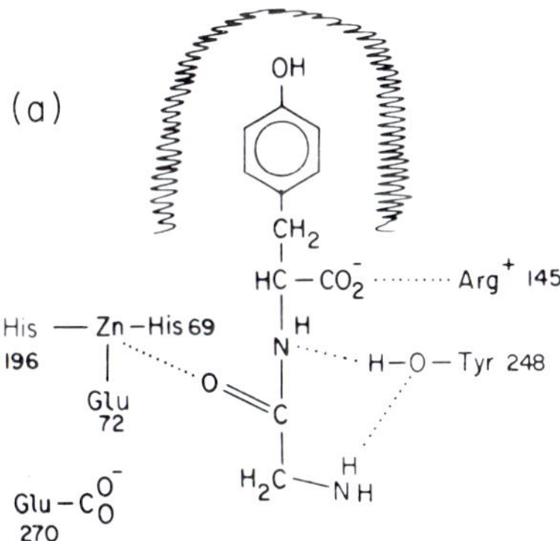

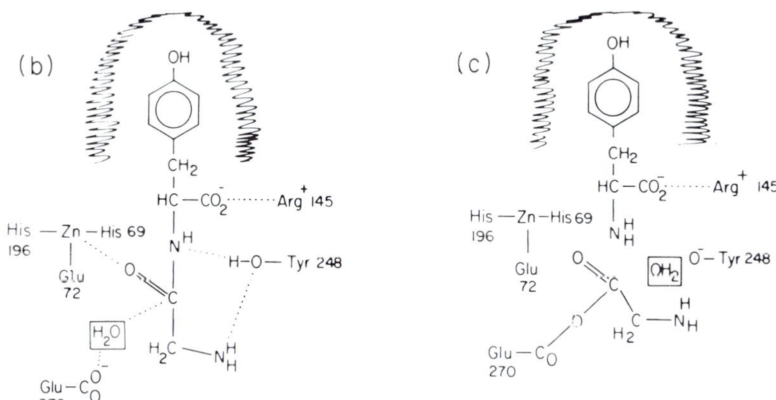

Fig. 14. Possible stages in the hydrolysis of a peptide by CPA. It is probable that the carbonyl carbon of the substrate becomes tetrahedrally bonded as the reaction proceeds, but it is uncertain at what stage of the reaction the proton is added to the NH group of the susceptible peptide bond. (a) Productive binding mode. (b) General base attack by water upon the carbonyl carbon of the substrate. (c) Nucleophilic attack by Glu 270 upon the carbonyl carbon of the substrate.

proach of another C-terminal residue for the transfer reaction would be hindered by Tyr 248. Furthermore, it is known that L-phenylalanine and L-β-phenyllactate, products of the respective peptide and ester substrates, are competitive inhibitors (*38*, *122*). Thus, the binding of the original C-terminal residue would also hinder its ready replacement by a new C-terminal residue. Although some implications can be drawn, deuterium isotope effects do not distinguish the nucleophilic attack by Glu 270 from the attack by water. Deuterium substitution effects seen in the hydrolysis by CPA of carbobenzoxyglycyl-L-phenylalanine ($k_{H_2O}/k_{D_2O} = 1.22$) (*41, 58*) or of hippuryl-L-phenylalaine (1.1) (*21, 123*) are small when compared to the effects expected in general base catalysis, which often demonstrates 2- to 5-fold higher rates in H_2O than in D_2O, as seen in the deacylation step with chymotrypsin (*124*) and in similar non-enzymic general base catalyzed reactions (*125*). However, the results observed for CPA, which are fully compatible with the nucleophilic attack, may be equally well explained in the case of any general mechanism (*125a*) where the rate determining step does not involve a proton transfer.

In summary, several conclusions concerning the hydrolysis of peptides have been drawn from a consideration of chemical experiments, the native enzyme structure and an enzyme–substrate complex structure. Substrates are bound with the R_1' side chain inserted into a pocket in the enzyme, with the terminal carboxylate ion directed toward Arg 145 and with the scissile peptide bond over the zinc ion so that the susceptible carbonyl group may be directed toward the changed zinc. Substrate binding and the resultant enzyme conformational changes bring the OH of Tyr 248 near the substrate. Then, zinc, Tyr 248, and Glu 270 are available to act on the peptide bond and to bring about cleavage. Zinc polarizes the carbonyl group, Tyr 248 donates a proton to the peptide amide, and Glu 270 either nucleophilically attacks the carbon atom of the scissile bond or promotes such an attack by water.

B. Cleavage of Esters

The mechanisms of cleavage of ester substrates by CPA are not yet clear. For the usual assay substrate hippuryl-L-β-phenyllactate, the pH-rate profile is entirely different from that of peptide substrates, the deuterium isotope effect is two (*68*), and Tyr 248 does not seem to participate in the reaction. However, substrate inhibition has a large and incompletely analyzed effect on the pH-rate profile of this "standard"

123. F. A. Quiocho, unpublished results (1968).

124. M. L. Bender, G. E. Clement, F. J. Kezdy, and H. Heck, *JACS* **86,** 3680 (1964).

125. M. L. Bender, E. J. Pollack, and M. C. Nevu, *JACS* **84,** 595 (1962).

125a. Contrary to common usage (*87*), the phrase *general base mechanism* is meant here to include all pathways where water attacks the susceptible carbonyl carbon even though proton transfer is not the rate limiting step and the mechanism is actually concerted.

ester. The substrate *O-trans*-cinnamoyl-L-β-phenyllactate has a pH-rate profile grossly like that of peptide substrates but k_{cat} depends on a pK_{a1} of 6.2 in the enzyme, k_{cat}/K_m depends on a pK_{a1E} of 6.5 and a pK_{a2E} of 9.4, and the deuterium isotope effect is two (*126*). The k_{cat} for the hydrolysis of *O*-acetyl-L-mandelate by CPA depends upon a pK_{a1} of 7.2 in the enzyme, and the k_{cat}/K_m depends on a pK_{a1E} of 6.9 and a pK_{a2E} of 7.5 (*47*); moreover, this reaction seems to require the participation of Tyr 248 since acetyl-CPA does not catalyze the hydrolysis of acetylmandelate (*21, 123*). In the light of these results, it seems safe to assume that there is more than one mechanism by which CPA splits esters. However, even the ester substrate which probably behaves most differently from peptides, *viz.*, hippuryl-L-β-phenyllactate, requires the presence of a metal, so we presume that the locus for productive binding (*126a*) is the same as that of Gly–Tyr.

The deuterium isotope effect of two for the two esters noted above indicates that proton transfer is important (*68*) and may suggest general base catalysis in a rate dominating step for these esters (*128*). However, these experiments make no distinction between the pathways shown in Figs. 14b and c. Finally, the pK_{a2E} of 9.4 is in the range expected for tyrosine, but the value of pK_{a2E} of 7.5 is abnormally low. Both values of pK_{a1E} above are abnormally high by at least 2 pK units for glutamic acid. Other pK values of interest may include that of H_2O on zinc and those associated with conformational changes. Hence, unique identifications of pK values with residues cannot be claimed at present.

C. Inhibition and Activation

In addition to their importance in the binding of longer substrates (see discussion of substrate specificity above) the region of Tyr 198, Phe 279, and Arg 71 is important in binding modes which may be responsible for some of the kinetic anomalies. So far model building experiments have elicited two abortive modes of binding for substrates which may result in substrate inhibition. In both of these cases the productive binding site which we have deduced from the Gly-L-Tyr crystallographic experiment is blocked and substrate–protein in-

126. B. L. Kaiser and E. T. Kaiser, *Proc. Natl. Acad. Sci. U. S.* **64**, 36 (1969).

126a. Some examples of metal-ion catalyzed nonenzymic hydrolyses of α-amino acid esters appear to bind the metal ion to the oxygen of the carbonyl group during the reaction (*127*).

127. R. J. Angelici and B. E. Leach, *J. Am. Chem. Soc.* **90,** 2499 (1968).

128 J. A. Thoma and D. E. Koshland, *JACS* **82,** 3329 (1960).

teraction is maximized. Furthermore, in generating these models we have assumed that Tyr 198 is associated with substrate inhibition since a consistent interpretation of the tyrosine modification experiments can be made if Tyr 248 is assumed essential for peptide hydrolysis and if Tyr 198 is associated with substrate inhibition by hippuryl-L-β-phenyllactate (Table IV). We have also been aware that only *N*-acyl dipeptides or their ester analogs with both the *N*-acyl and C-terminal R groups aromatic have so far been shown to exhibit substrate inhibition. The first abortive binding mode has the substrate (CBZ-Gly-L-Phe in Fig. 15a) shifted with respect to the productive mode, so that the C-terminal carboxyl group is bound to the zinc atom, the C-terminal side chain partly extends into the hydrophobic pocket, the aromatic acyl group has a favorable π,π interaction with Tyr 198, and the second peptide carbonyl is bound to Arg 71. The second mode of binding (Fig. 15b) has the substrate reversed so that the aromatic acyl group is in the pocket which accommodates the C-terminal side chain in productive binding. The terminal carboxyl group of the substrate can then bind to Arg 71, and the C-terminal side chain is near Tyr 198. So far we have been unsuccessful in devising a crystallographic experiment to demonstrate either of these abortive binding modes.

Of course the two binding modes discussed here do not preclude other possibilities; indeed, yet another mode must be involved in order to satisfy the kinetic data (*33, 38, 41, 45, 53*), which imply binding of more than one substrate molecule in order to inhibit one active center (*128*). One way to bind two substrate molecules is suggested in Fig. 15c, which shows a second molecule bound farther down the groove in a way that could interfere with precise placement of the first molecule in the active site for hydrolysis. In disagreement with kinetic analyses which suggest that as many as 3–5 molecules of CBZ–Gly–Pre (*41*) or hippuryl-L-phenyllactate (HPLA) (*45, 129*) are bound, we find that the binding region is probably limited to two substrate molecules of this size.

The cluster of side chains composed of Arg 71, Tyr 198, and Phe 279 may also be used to explain substrate and product activation. If *N*-carbobenzoxyglycine, which does activate the enzyme (*38*), is placed so that the free carboxyl group is bound to Arg 71, then the phenyl ring can interact favorably with Tyr 198 and with the phenyl ring of a productively bound substrate molecule (Fig. 15d). Similar interactions occur when a second molecule of hippurylglycolate or carbobenzoxy-

129. A. E. Dennard and R. J. P. Williams, *in* "Transition Metal Chemistry. A Series of Advances" (R. L. Carlin, ed.), Vol. 2, p. 139. Marcel Dekker, New York, 1960.

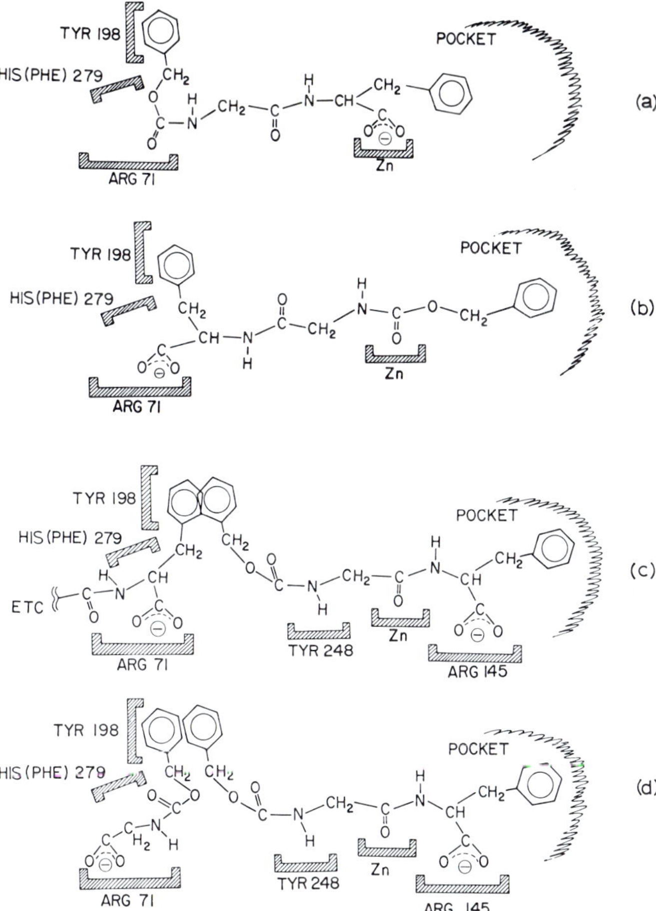

FIG. 15. Proposed modes of binding of substrate molecules showing (a) displaced binding, (b) reversed binding for CBZ-glycylphenylalanine, (c) binding of two substrate molecules giving rise to inhibition which would not be competitive, and (d) binding of a substrate and of a product molecule in an activation process.

glycyl-L-phenylalanine is bound. Substrates long enough to cover the Arg 71–Tyr 198 region are then not expected to exhibit any of the above anomalous binding modes. In fact, the data which are presently available indicate that substrates longer than acyl dipeptides are not subject to substantial substrate inhibition or to product activation (*35, 62*).

Competitive inhibition by products such as L-phenylalanine has been found in the CPA-catalyzed hydrolysis of CBZ-Gly-L-Phe at pH 9.0 (*59*). This inhibition primarily results from the anionic form of L-phenylalanine. However, it has been shown that L-Phe is a competitive inhibitor of CPA even at pH 7.5 (*38*). Moreover, L-Phe has been shown crystallographically to bind only in the pocket, and this binding is accompanied by a structural change presumably involving Tyr 248 (*90*). The detailed structural interpretation from model building of the binding of L-Phe places the α-COO^- near Arg 145. D-Phenylalanine, on the other hand, inhibits competitively and more effectively at pH 7.5 than at pH 9.0 (*57, 59*). This finding has been interpreted by assuming that the charged α-amino group of D-Phe, unlike that of L-Phe, is oriented near a negative group of the protein. Accordingly, our structural interpretation of the binding of D-Phe, based upon model building only, places the α-COO^- near zinc (*129*), and the charged α-amino group near Glu 270 (Fig. 16). The loss of proton from the α-NH_3^+ at pH 9.0 would then result in less effective binding.

The strictly competitive inhibition by β-phenylpropionate in the region of equimolecular inhibitor:CPA binding would be analogous to that depicted for L-Phe. However, mixed inhibition, competitive and noncompetitive (or uncompetitive), by β-phenylpropionate has also been found (*58*) and is probably related to the ability of this inhibitor to bind in two functional sites, as demonstrated by X-ray crystallography.

D. Recent Studies of Metals

Studies of nuclear magnetic resonance (NMR) interactions with the paramagnetic Mn^{2+} ion in MnCPA have been made by Shulman and co-workers. Relaxation times for water protons bound to oxygen in the first hydration sphere on Mn^{2+} established that one or two water molecules were complexed (*130*) to Mn^{2+} of the enzyme. The potent inhibitor β-phenylpropionate destroyed this interaction of H_2O and Mn^{2+}. Subsequent studies (*131*) of protons on inhibitors of MnCPA were consistent with the idea that the interaction of H_2O with Mn^{2+} was destroyed by displacement by H_2O in this first hydration sphere by the inhibitor. In a recent study (*132*), NMR studies of ^{19}F in solutions of MnCPA in-

130. R. G. Shulman, G. Navon, B. J. Wyluda, D. C. Douglass, and T. Yamane, *Proc. Natl. Acad. Sci. U. S.* **56,** 39 (1966).
131. G. Navon, R. G. Shulman, B. J. Wyluda, and T. Yamane, *Proc. Natl. Acad. Sci. U. S.* **60,** 86 (1968).
132. G. Navon, R. G. Shulman, B. J. Wyluda, and T. Yamane, *JMB* **51,** 15 (1970).

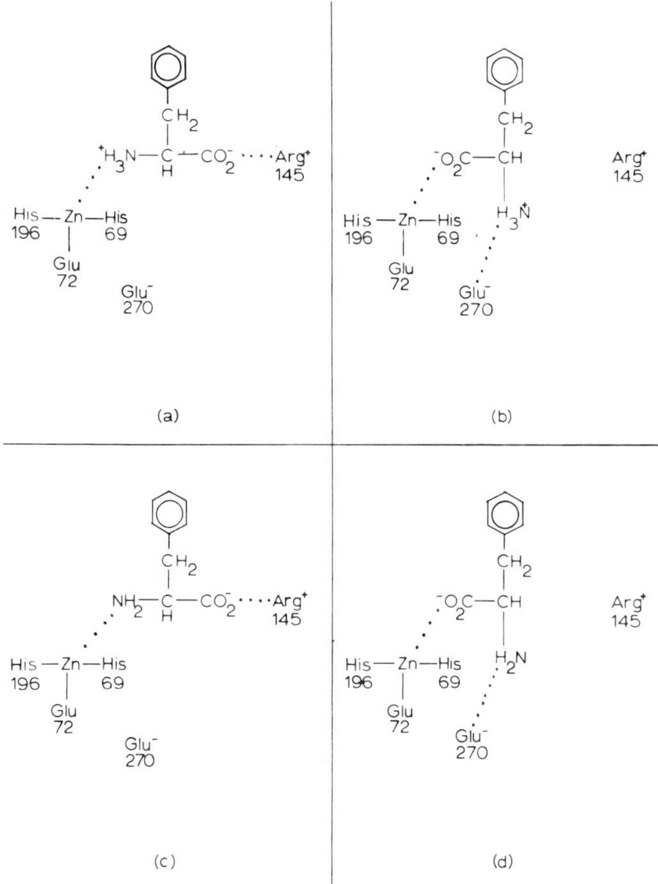

FIG. 16. Probable binding of (a) L-Phe at pH 7.5, (b) D-Phe at pH 7.5, (c) L-Phe at pH 9.0, and (d) D-Phe at pH 9.0. Inhibition (a) is one-eighth that of (b) and dependent upon phosphate buffer concentration, but at pH 9 L-Phe is about as effective as D-Phe. The binding at (a), with the CO_2^- or Arg 145 and the H_3N^+ toward $(ZnL_3)^+$, where L is a zinc ligand, is suggested by the occurrence of the Tyr 248 conformational change in an X-ray study at 6 Å resolution of the binding of L-Phe at pH 7.8.

dicate both a fluoride ion and at least one water molecule in the first coordination sphere around Mn^{2+} in MnCPA.

If we suppose that there are also three ligands (His 69, Glu 72, and His 196) bound to Mn^{2+} in MnCPA, there must be at least five ligands to Mn^{2+} here (at least when F^- is also present). On the other hand, the electron density map of CPA itself shows that a total of four ligands is present, but there is an additional H_2O molecule about 3.5 Å away from Zn^{2+} which might become the fifth neighbor when the slightly larger Mn^{2+} is substituted for the Zn^{2+} at the active site.

The Cu^{2+} of $CuPA_\gamma$ has been shown by P. R. Kirkpatrick [in A. S. Brill's laboratory (*133*)] to have two nitrogen atoms covalently bound to this metal ion. Examination of the model of CPA shows no other candidates except His 69 and His 196. An X-ray structure at atomic resolution is required to establish the detailed geometry and other ligands around Cu^{2+}, which in model compounds usually shows a preference for a near-planar configuration of four nearest neighbors. Thus a geometrical distortion may be one important factor contributing to the absence of both esterase and peptidase activities of CuCPA, as noted above.

The binding modes of L-phenylalanine (a product of peptide hydrolysis) and of L-β-phenyllactate (a product of ester hydrolysis) have been shown to be very similar in X-ray diffraction studies at 6 Å resolution of CPA and in frequency-dependent nuclear magnetic relaxation studies of Mn-CPA (*134*). It therefore seems likely that cleavage of peptides and esters occurs in a binding mode that places the similar C-terminal portions of related peptide and ester substrates in essentially the same position in the pocket of the enzyme. While the four-coordinated Zn^{2+} is not strictly comparable to the slightly larger five-coordinated Mn^{2+} in CPA, the nuclear magnetic relaxation studies (*134*) also show that the ionization of H^+ from H_2O linked to Mn in MnCPA occurs above pH 9.

Finally, we note that Bishop, Quiocho, and Richards (*135*) have shown that crystals of HgCPA have some peptidase activity. In the X-ray study (*21*) the Hg is displaced in the direction away from the β-structure by about 0.5–1.0 Å away from the position of Zn in CPA itself, but no other changes are found, except for the expected slight shifts of the ligands. In particular, the Hg is bound to the same N atoms (N_1) of both His 69 and His 196 as is the Zn in native CPA.

133. A. S. Brill, private communications (1967 and 1968).

134. F. A. Quiocho, J. F. Studebaker, R. D. Brown, and W. N. Lipscomb, unpublished investigation.

135. W. H. Bishop, F. A. Quiocho, and F. M. Richards, *Biochemistry* **5**, 4077 (1966).

2

Carboxypeptidase B

J. E. FOLK

I. Introduction

A. Historical Background

In 1931 Waldschmidt-Leitz and co-workers (*1*) showed that extracts of porcine pancreas gland contained an enzyme that catalyzed the release

1. E. Waldschmidt-Leitz, F. Ziegler, A. Schäffner, and L. Weil, *Z. Physiol. Chem.* **197**, 219 (1931).

of arginine from a variety of protamines. Their findings that an acylated protamine, the benzylidene derivative of clupein, served as a substrate, while protamine esters did not, led these workers to suggest that arginine was liberated by this enzyme from the COOH-terminal end of protamines. The name *protaminase* was chosen for this enzyme because it showed no hydrolytic action toward other proteins tested. Calvery (*2*), however, found that egg albumin, although completely resistant to digestion by a protaminase preparation, could be rendered susceptible to this enzyme by first digesting it with proteases. He suggested that perhaps lysine and histidine, as well as arginine, were liberated by protaminase. During the ensuing 20 years, protaminase was generally considered to be identical with chymotrypsin (*3, 4*), a belief based upon reports that protamine hydrolyzing activity could not be separated from proteinase activity (*5, 6*). A report in 1951 that 6× crystallized carboxypeptidase A (Chapter 1, this volume) catalyzed the hydrolysis of the protamine, salmine, raised the question of the possible identity of protaminase with carboxypeptidase A (*7*).

Two observations, namely, (a) that lysine was only slowly released from *N*-benzoylglycyl-L-lysine during incubation with high concentrations of a purified carboxypeptidase A preparation (*8*) and (b) that small amounts of an extract of commercial pancreas powder catalyzed the rapid release of lysine from this benzoyl dipeptide (*9*), prompted a series of investigations that gave subsequent proof for the existence of a second carboxypeptidase, carboxypeptidase B, of pancreas tissue and served to identify it with the protaminase of Waldschmidt-Leitz and coworkers. Early reports (*10, 11*) cited pronounced differences in the specificities of carboxypeptidases A and B and listed a series of specific competitive inhibitors for carboxypeptidase B. Carboxypeptidase B was found to act on peptides containing the basic amino acids lysine, arginine,

2. H. O. Calvery, *JBC* **102,** 73 (1933).
3. J. B. Sumner and G. F. Somers, "Chemistry and Methods of Enzymes," 2nd ed., p. 182. Academic Press, New York, 1947.
4. J. H. Northrop, M. Kunitz, and R. M. Herriott, "Crystalline Enzymes," 2nd ed., p. 118. Columbia Univ. Press, New York, 1948.
5. J. H. Northrop, cited by K. Myrbäck *in* "Die Methoden der Fermentforschung" (E. Bamann and K. Myrbäck, eds.), p. 2029, Thieme, Leipzig, 1941.
6. R. A. Portis and K. I. Altman, *JBC* **169,** 203 (1947).
7. Unpublished data cited in E. L. Smith, *in* "The Enzymes," (J. B. Sumner and K. Myrbäck, eds.) 1st ed., Vol. 1, Part 2, p. 828, Academic Press, New York, 1951.
8. K. Hofmann and M. Bergmann, *JBC* **134,** 225 (1940).
9. J. E. Folk, *ABB* **64,** 6 (1956).
10. J. E. Folk, *JACS* **78,** 3541 (1956).
11. J. E. Folk and J. A. Gladner, *JBC* **231,** 379 (1958).

ornithine, homoarginine, and S-(β-aminoethyl)cysteine as the COOH-terminal group (*11, 12*). The well-defined preferential action of carboxypeptidase A toward COOH-terminal aromatic and branched chain aliphatic amino acids is detailed by Hartsuck and Lipscomb, Chapter 1, this volume. The identity of protaminase and carboxypeptidase B became evident from a report that a protaminase fraction, prepared essentially as originally described (*1*), catalyzed the release of lysine and S-(β-aminoethyl)cysteine, as well as arginine, from the COOH-terminal position of polypeptides (*13*). The early recognition that carboxypeptidase B has a specific precursor, procarboxypeptidase B, from which it is formed by tryptic activation (*10, 11*), was confirmed by chromatographic separation of the zymogens of the two carboxypeptidases from fresh pancreatic secretions of the cow (*14*) and of the swine and dog (*15*).

The action of carboxypeptidase B on polypeptides and proteins was shown to be in complete accord with the specificity as predicted from studies on small synthetic substrates (*16*). As a result purified porcine carboxypeptidase B, because of its almost absolute specificity for COOH-terminal basic amino acids (*17*), has found wide use in end group analysis and, in conjunction with carboxypeptidase A, for sequence determination. In this regard the special value of carboxypeptidase B in releasing COOH-terminal lysine and arginine from peptides, derived by tryptic digestion of proteins, points up an early hypothesis that the particular physiological function of carboxypeptidase B is the first-step degradation of the products of tryptic digestion (*11*).

B. Distribution

Carboxypeptidase B, in the form of an inactive zymogen, probably occurs in the cellular secretions of the pancreas of most vertebrates. The proenzyme has been identified in extracts of the pancreas of the rat (*18*) as well as those of the dog (*15*), swine (*15*), and cow (*14*). The pancreatic enzyme and zymogen of the spiny Pacific dogfish have been isolated and partially characterized (*19*).

12. F. Tietze, J. A. Gladner, and J. E. Folk, *BBA* **26,** 659 (1957).
13. L. Weil, T. S. Seibles, and M. Telka, *ABB* **79,** 44 (1959).
14. P. J. Keller, E. Cohen, and H. Neurath, *JBC* **223,** 457 (1956).
15. G. Marchis-Mouren, M. Charles, A. Ben Abdeljlil, and P. Desnuelle, *BBA* **50,** 186 (1961).
16. J. A. Gladner and J. E. Folk, *JBC* **231,** 393 (1958).
17. J. E. Folk, K. A. Piez, W. R. Carroll, and J. A. Gladner, *JBC* **235,** 2272 (1960).
18. G. Marchis-Mouren, L. Paséro, and P. Desnuelle, *BBRC* **13,** 262 (1963).
19. J. W. Prahl and H. Neurath, *Biochemistry* **5,** 4137 (1966).

Activity toward carboxypeptidase B substrates has been detected in the body fluids urine (*20, 21*), blood plasma (*22*), and lymph (*20*), in kidney cortex (*23*), and in catheptic spleen preparations (*24*). Inhibitor studies indicate that the enzyme from swine blood serum is not identical to swine pancreatic carboxypeptidase B (*25*). An enzyme similar in specificity to carboxypeptidase B has been reported in the gastric juice of crayfishes (*26*).

II. Purification of the Enzyme and Zymogen; Assay

Carboxypeptidase B may be isolated from aqueous extracts of acetone powder of autolyzed porcine pancreatic glands by a process involving fractionation with ammonium sulfate, followed by chromatography on DEAE-cellulose (*17, 27*). This swine enzyme has been obtained in crystalline form (*27, 28*). Bovine carboxypeptidase B may be readily crystallized following tryptic activation of either purified or partially purified proenzyme preparations (*29, 30*).

The earliest attempts at purification of the bovine proenzyme were made by selective extraction of the euglobulin precipitate obtained from aqueous extracts of acetone powders prepared from fresh pancreas glands (*11*). The major impurity of these preparations, chymotrypsinogen B, was removed by chromatography on DEAE-cellulose (*31*). The zymogen was subsequently isolated in pure form from extracts of acetone powders by the use of several ion exchange chromatographic steps (*29*). Procarboxypeptidase B has also been purified chromatographically from

20. E. G. Erdös, E. M. Sloane, and I. M. Wohler, *Biochem. Pharmacol.* **13,** 893 (1964).
21. I. Innerfield, R. Harvey, F. Luongo, and E. Blincoe, *Proc. Soc. Exptl. Biol. Med.* **116,** 573 (1964).
22. E. G. Erdös and E. M. Sloane, *Biochem. Pharmacol.* **11,** 585 (1962).
23. I. Innerfield, F. S. Gimble, and E. Blincoe, *Life Sci.* **3,** 267 (1964).
24. L. M. Greenbaum and R. Sherman, *JBC* **237,** 1082 (1962); L. M. Greenbaum and K. Yamafuji, *in* "Hypertensive Peptides" (E. G. Erdos *et al.,* eds.), p. 252. Springer, New York, 1966.
25. E. G. Erdös, H. Y. T. Yang, L. L. Tague, and N. Manning, *Biochem. Pharmacol.* **16,** 1287 (1967).
26. R. Kleine, *Z. Vergleich. Physiol.* **56,** 142 (1967).
27. J. E. Folk, "Methods in Enzymology" (in press).
28. A. L. Baker, personal communication (1969).
29. E. Wintersberger, D. J. Cox, and H. Neurath, *Biochemistry* **1,** 1069 (1962).
30. J. H. Kycia, M. Elzinga, N. Alonzo, and C. H. W. Hirs, *ABB* **123,** 336 (1968).
31. J.-F. Pechère, G. H. Dixon, R. H. Maybury, and H. Neurath, *JBC* **233,** 1364 (1958).

fresh bovine pancreatic juice (*30*). The dogfish proenzyme has been isolated (*19*) by a procedure similar to that employed for the bovine zymogen (*29*).

The most convenient and widely used assay for carboxypeptidase B employs spectrophotometric measurement of the release of arginine from hippuryl-L-arginine (*17*, *32*). This procedure for rate measurement is based on the difference in ultraviolet absorbancy between *N*-benzoylamino acids and the corresponding carboxyl-substituted *N*-benzoylamino acids (*33*). It is applicable to a number of synthetic substrates for carboxypeptidase B and has been employed over a wide range of concentrations of substrates (*32*). In an alternate assay for carboxypeptidase B the ninhydrin procedure is used to measure arginine released from hippurylarginine (*11*).

III. Physical–Chemical Properties of Carboxypeptidase B

A. Physical Properties

Carboxypeptidase B preparations from several species have been studied in some detail. Pertinent properties of these enzymes are listed in Table I. The swine enzyme appears homogeneous by several critieria: ion exchange chromatography at pH values 6 and 8, ultracentrifugation, diffusion, and moving boundary and polyacrylamide gel electrophoresis (*17*, *34*). The mobility of the descending boundary at 0° in moving boundary electrophoresis was found to be -2.38×10^{-5} cm^2 V^{-1} sec^{-1}

TABLE I

Molecular Properties of Carboxypeptidase B from Several Species

Property	Swine	Cow	Dogfish
Molecular weight	34,300	34,000	35,000–37,000
$s_{20,w}$	3.24 S	3.10 S	3.25 S
$D_{20,w}$	8.16×10^{-7} cm^2 sec^{-1}	—	—
$\bar{V}$	0.720 cm^3 g^{-1}	—	—
Zinc	1 g-atom/mole	1 g-atom/mole	—
$E_{1\,cm}^{1\%}$	21.4 at 278 nm	21.0 at 280 nm	—
NH_2-terminal	Threonine	Threonine	Serine
COOH-terminal	Threonine	Leucine	—

32. E. C. Wolff, E. W. Schirmer, and J. E. Folk, *JBC* **237**, 3094 (1962).
33. G. W. Schwert and Y. Takenaka, *BBA* **16**, 570 (1955).
34. J. E. Folk, unpublished results (1962).

in 0.05 *M* potassium phosphate buffer, pH 7.02 (*17*). No study of the influence of pH and ionic strength on electrophoretic mobility has been conducted. The behavior of the enzyme on anion exchange cellulose under various conditions of pH, however, would suggest that the isoelectric point of the enzyme at low ionic strength (<0.1) is not far below 6.0 (*27, 34*). A molecular weight of 34,300 was calculated by the use of the sedimentation constant, $s_{20,w}$, the diffusion constant, $D_{20,w}$ and the pycnometrically determined partial specific volume, $\bar{V}$, recorded in Table I (*17*). A $\bar{V}$ of 0.73 cm^3 g^{-1} has been calculated from the amino acid composition (*17*).

The molecular weight of the bovine enzyme (*29*) and that of the dogfish (19) recorded in Table I were determined by short column sedimentation equilibrium assuming values for $\bar{V}$ of 0.730 and 0.735 cm^3 g^{-1}, respectively. The sedimentation pattern (*29*) and polyacrylamide gel electrophoretic pattern (*30*) of the crystalline bovine enzyme were found to be consistent with homogeneity. The variations in the published values for the apparent weight-average molecular weights and the *z*-average molecular weights of the dogfish enzyme (*19*) indicate some degree of heterogeneity.

Concentrated solutions of swine carboxypeptidase B in dilute tris buffer, pH 7.5, or suspensions of crystals of the enzyme in water have been stored frozen at $-10°$ for periods up to 1 year without detectable loss in enzymic activity (*27*). Crystals of the bovine enzyme reportedly may be stored indefinitely as a suspension in 0.01 *M* tris HCl, pH 8.0, at 4° (*29*).

Studies with extracts of the pancreas of swine (*35*), cow (*36*), and rat (*37*) have shown the carboxypeptidase B of these species to be antigenically distinct from other hydrolases. The enzyme has been obtained as an immunochemically pure antigen from swine pancreas (*38*). An antiserum prepared against purified swine carboxypeptidase B (*39*) was found not to react with bovine carboxypeptidase B.

B. Chemical Composition

Amino acid compositions of carboxypeptidase B preparations from the swine, cow, and dogfish are given in Table II. A redetermination of the

35. J. Uriel and S. Avrameas, *Ann. Inst. Pasteur* **106**, 396 (1964).
36. S. Avrameas and J. Uriel, *Protides Biol. Fluids, Proc. Colloq.* **12**, 225 (1964).
37. J. Pascale, S. Avrameas, and J. Uriel, *JBC* **241**, 3023 (1966).
38. S. Avrameas and J. Uriel, *Biochemistry* **4**, 1750 (1965).
39. J. T. Barrett, *Intern. Arch. Allergy Appl. Immunol.* **26**, 158 (1965).

composition of the bovine enzyme has shown one more residue each of aspartic acid and glycine, two more residues of proline, one less residue each of isoleucine and tyrosine, and three less residues of tryptophan (*30*). This composition has recently been confirmed (*41*) with the exception that 12 residues of proline were found, in agreement with the earlier analysis (*40*). Bovine carboxypeptidase B contains one —SH group which is reactive to sulfhydryl reagents only after removal of the zinc atom from the enzyme (*42*). The remaining six half-cystine residues were shown to occur as three disulfide-bonded cystines (*42*). These findings, together with the initial report that bovine carboxypeptidase B contains 1

TABLE II
AMINO ACID COMPOSITIONS OF SWINE, COW, AND DOGFISH CARBOXYPEPTIDASE B PREPARATIONS

Amino acid	Swine[a] (Residues/34,300 g)	Cow[b] (Residues/34,000 g)	Dogfish[c] (Residues/34,000 g)
Aspartic acid	32.4	25.9	31.2
Threonine	30.2	25.8	25.2
Serine	17.5	25.8	24.9
Glutamic acid	24.8	23.6	20.4
Proline	13.2	12.1	13.9
Glycine	23.0	21.4	19.8
Alanine	25.1	21.7	23.9
Half-cystine	7.6	6.8	7.2
Valine	10.8	13.7	16.8
Methionine	5.1	6.0	8.2
Isoleucine	17.2	15.7	19.6
Leucine	22.7	20.2	16.9
Tyrosine	20.4	21.8	19.6
Phenylalanine	11.9	11.8	9.2
Lysine	17.5	16.7	14.6
Histidine	5.8	7.0	4.1
Arginine	10.0	13.1	13.8
Tryptophan	9.2	9.9	9.6
Amide N	(27.8)	(22.9)	—
Total	304.4	299.0	298.9

[a] From Folk *et al.* (*17*).
[b] From Cox *et al.* (*40*).
[c] From Prahl and Neurath (*19*).

40. D. J. Cox, E. Wintersberger, and H. Neurath, *Biochemistry* **1**, 1078 (1962).
41. T. H. Plummer, Jr., *JBC* **244**, 5246 (1969).
42. E. Wintersberger, H. Neurath, T. L. Coombs, and B. L. Vallee, *Biochemistry* **4**, 1526 (1965).

g-atom of zinc (*40*), extend earlier observations that the swine enzyme, like swine (*43*) and bovine (*44*) carboxypeptidase A, contains 1 g-atom of tightly bound zinc (*17*) and that this metal serves as a structural and functional component of the enzyme (*17, 45*).

Similarities in the composition of porcine, bovine, and dogfish carboxypeptidases B, together with their general likeness in composition to bovine and porcine carboxypeptidases A (Chapter 1, this volume), except for the distribution of sulfur-containing amino acids, have prompted the suggestion of a common evolutionary origin for all of these exopeptidases (*19*). Recognition of small areas of similarity in amino acid sequences between carboxypeptidases A and B, as will be discussed below, has strengthened this suggestion and added emphasis to earlier evidence that carboxypeptidases A and B operate by a common mechanism (*11, 17, 42, 45–47*).

C. End Groups and Amino Acid Sequences

The finding of a one each NH_2- and COOH-terminal residue in porcine carboxypeptidase B (Table I) served as evidence that this enzyme is composed of a single polypeptide chain (*48*). Threonine was identified as the NH_2-terminal amino acid by the use of the fluorodinitrobenzene (FDNB) procedure. The NH_2-terminal sequence, Thr·Ser, was derived by means of the stepwise phenyl isothiocyanate method. The third residue from the NH_2-terminus was tentatively identified by this method as aspartic acid or asparagine. Treatment of the denatured enzyme protein with carboxypeptidase A showed a rapid release of threonine and asparagine followed by a slower release of valine and serine. The hydrazinolysis procedure served to identify threonine as the COOH-terminal amino acid.

The NH_2-terminal amino acid of bovine carboxypeptidase B, like that of the porcine enzyme, was identified as threonine by the use of the FDNB procedure (*40*). The suggestion from hydrazinolysis that leucine was the COOH-terminal residue of the bovine enzyme (*30*) was confirmed by sequence analysis of a 14-member COOH-terminal peptide fragment derived by cyanogen bromide cleavage of the enzyme protein (Table III) (*49*).

43. J. E. Folk and E. W. Schirmer, *JBC* **238**, 3884 (1963).
44. B. L. Vallee and H. Neurath, *JBC* **217**, 253 (1955).
45. J. E. Folk and J. A. Gladner, *BBA* **48**, 139 (1961).
46. J. E. Folk and J. A. Gladner, *BBA* **33**, 570 (1959).
47. T. H. Plummer and W. R. Lawson, *JBC* **241**, 1648 (1966).
48. J. E. Folk and J. A. Gladner, *BBA* **47**, 595 (1961).
49. M. Elzinga, C. Y. Lai, and C. H. W. Hirs, *ABB* **123**, 353 (1968).

TABLE III
SEQUENCES OF CERTAIN PEPTIDE FRAGMENTS FROM BOVINE CARBOXYPEPTIDASE B

Fragment	Sequence	Ref.
COOH-terminal	Leu·Ala·Ile·Lys·Tyr·Val·Thr·Ser·Tyr·Val·Leu·Glu·His·Leu	*(49)*
Disulfide containing loop	Asp·Cys·Gly·Phe His·Ala·Arg·Glu·Trp·Ile·Ser·Pro·Ala·Phe·Cys· Gln·Trp·Phe·Val·Arg·Glu·Ala·Val·Arg·Thr·Tyr·Gly·Arg·Glu·Ile·His·Met	*(50)*
Surrounding —SH group	Val·Leu·Pro·Glu·Ser·Gln·Ile·Gln·Pro·Thr·CySH·Glu·Glu·Thr	*(51)*
Surrounding essential tyrosine	Thr·Ile·Tyr·Pro·Ala·Ser·Gly·Gly·Ser·Asp·Asp·Trp	*(41)*

The NH_2-terminal residue of the dogfish enzyme has been identified as serine (*19*). Carboxypeptidase A was found to liberate several amino acids from the denatured enzyme protein. Examination of the rate of release of these residues suggests that leucine occupies the COOH-terminal position (*19*).

Recent studies have revealed sequences of amino acids in several important regions of bovine carboxypeptidase B. These results are given in Table III (*41, 49–51*). An early report on a systematic approach to the complete amino acid sequence of bovine carboxypeptidase B has been made (*52*). In this case 30 of the possible 36 peptides formed as a result of tryptic digestion of the reduced and aminoethylated enzyme protein were isolated. Of the 300 amino acid residues of carboxypeptidase B, 235 residues were accounted for in these peptides.

D. Early Considerations of Homology between Bovine Carboxypeptidases A and B

Early interest in comparison of peptide sequences of carboxypeptidases A and B was stimulated by the expectation that these two exopeptidases would show extensive structural homology. This concept, based upon findings of numerous common structural features among enzymes and other proteins of similar biological function (*53*), would seem valid in light of the similarities between carboxypeptidases A and B in amino acid composition, metal content, enzymic function, and competitive inhibition (*11, 17, 40*), as well as in parallel consequences of chemical modification (*47, 54*). Substrate specificity constitutes the primary difference between the enzymes (*10, 11, 40*).

First comparisons of fragmentary peptide segments of bovine carboxypeptidases A and B, however, gave no support to the concept of homology. The amino acid sequences at the COOH-terminal portion of the two enzymes were found to be distinctly different (*49*). More puzzling was the finding that the sequences surrounding the single sulfhydryl group of each of the enzyme proteins were clearly not homologous (*51*).

50. M. Elzinga and C. H. W. Hirs, *ABB* **123,** 361 (1968).
51. E. Wintersberger, *Biochemistry* **4,** 1533 (1965).
52. M. Elzinga and C. H. W. Hirs, *ABB* **123,** 343 (1968).
53. V. Bryson and H. J. Vogel, eds., "Evolving Genes and Proteins." Academic Press, New York, 1965.
54. B. L. Vallee and J. F. Riordan, *Brookhaven Symp. Biol.* **21,** 91 (1968).

More recent observations, on the other hand, give renewed interest to the possibility that indeed homology does exist, at least in important regions of these two enzyme proteins. The amino acid sequences surrounding an essential tyrosine residue in each of the enzymes

Carboxypeptidase A: [Thr · Ile · Tyr] · Gln · [Ala · Ser · Gly · Gly · Ser] · Ile · [Asp · Trp] *(55)*

Carboxypeptidase B: [Thr · Ile · Tyr] · Pro · [Ala · Ser · Gly · Gly · Ser] · Asp · [Asp · Trp] *(41)*

have been found to display a remarkable degree of similarity (*41*). Less convincing comparisons form the basis of an argument for extensive homology in the two bovine enzymes on the levels of both amino acid sequence and three-dimensional structure (*56*). This case is based upon (a) a postulated zinc-binding position within the disulfide loop of carboxypeptidase B (there are obvious similarities in the amino acid sequences of this disulfide loop and the zinc-binding loop of carboxypeptidase A), and (b) a fragment of 14 amino acid residues near the COOH-terminal end of carboxypeptidase A that contains no cysteine residue but displays a sequence similar to that surrounding the sulfhydryl group in carboxypeptidase B. These suggestions present interesting predictions based on the supposition of homology. They seem somewhat premature in support of extensive structural homology, however, when one considers that sequence data for carboxypeptidase B are fragmentary and that no information is available on its three-dimensional structure.

IV. Physical–Chemical Properties and Activation of Procarboxypeptidase B

The catalytically inactive zymogens of carboxypeptidase B from cow (*29, 40*) and dogfish (*19*) pancreas gland have been partially characterized, and the mechanism of their conversion to active enzyme has been the subject of preliminary investigations. The sedimentation coefficient, $s_{20,w}$, of the bovine proenzyme was found to be 4.0 S. A molecular weight of 57,400 $\pm$ 1,000 was estimated by short column sedimentation equilibrium with the use of an assumed partial specific volume of 0.73 cm^3 g^{-1}. The amino acid composition based on the above molecular weight is as

55. H. Neurath, cited in Plummer (*41*).
56. R. A. Bradshaw, H. Neurath, and K. A. Walsh, *Proc. Natl. Acad. Sci. U. S.* 63, 406 (1969).

follows: Asp 47, Thr 37, Ser 44, Glu 47, Pro 19, Gly 34, Ala 33, half-Cys 9, Val 34, Met 10, Ile 22, Leu 36, Tyr 27, Phe 22, Lys 27, His 17, Arg 26, Trp 14, and amide N 46; a total of 505 amino acid residues. No NH_2- or COOH-terminal residues have been identified.

Dogfish procarboxypeptidase B was found to have an $s_{20,w}$ of 3.67 S. A molecular weight in the range of 44,000–45,000 was estimated by means of sedimentation equilibrium analysis. The variations in values reported for apparent weight-average molecular weights and the *z*-average molecular weights raise some question as to the homogeneity of the isolated zymogen protein. The amino acid composition based on a molecular weight of 44,500 was determined as Asp 46, Thr 28, Ser 27, Glu 35, Pro 17, Gly 26, Ala 31, half-Cys 9, Val 27, Met 8, Ile 24, Leu 25, Tyr 19, Phe 13, Lys 21, His 5, Arg 20, and Trp 13. The NH_2-terminal residue was identified as glutamine.

Early studies on the activation of procarboxypeptidase B in extracts of fresh bovine pancreas gland and in partially purified preparations of this zymogen showed that trypsin catalyzed the rapid conversion to active enzyme (*11*). It was also evident from these studies that the activation of the zymogen of carboxypeptidase B occurs much more rapidly than that of carboxypeptidase A. The tryptic activation of purified bovine procarboxypeptidase B has been shown to involve more than a single hydrolytic event (*29, 46*). A first stage of activation that occurs over a period of 1–2 hr at low trypsin to zymogen molar ratios (1–1000) or within the first few minutes at a 1–10 trypsin to zymogen molar ratio appears to involve hydrolysis of an arginylthreonine bond in the zymogen. In this stage there is a concomitant (a) appearance of 60–70% of the maximal enzymic activity, (b) appearance of NH_2-terminal threonine, and (c) release of free arginine, supposedly originating by carboxypeptidase B–catalyzed liberation of a COOH-terminal arginine residue formed as a result of tryptic activation. No change in the sedimentation coefficient occurs during this stage of activation indicating no major fragmentation of the zymogen protein. A second stage of activation that occurs more slowly leads to formation of a protein with a lower sedimentation coefficient and with full enzymic activity. The enzyme formed appears to be identical to crystalline carboxypeptidase B. Major fragmentation of the partially activated protein probably occurs during this stage of activation, e.g., there appears to be a reduction in molecular weight from about 57,500 to 34,000.

In the case of the dogfish proenzyme activation with bovine trypsin results in reduction in molecular weight from 44,500 to 34,000 and change in the NH_2-terminal glutamine of the zymogen to NH_2-terminal serine of the active enzyme (*19*).

V. Enzymic Properties of Carboxypeptidase B

A. Specificity

Detailed studies of the substrate specificity of carboxypeptidase B have been carried out with the use of the bovine and the porcine enzymes. Porcine carboxypeptidase B shows little, if any, activity toward carboxypeptidase A substrates (*17*). The bovine and the dogfish enzymes, on the other hand, have been found to possess a large degree of intrinsic carboxypeptidase A activity (*19, 29*). Examples of these findings are given in Table IV.

TABLE IV
Activities of Several Carboxypeptidases toward Carboxypeptidase A and B Substrates[a,b]

Carboxy-peptidase	Carboxypeptidase B substrate hippuryl-L-arginine (peptide) (units/mg)	Carboxypeptidase A substrates	
		Benzyloxycarbonylglycyl-L-phenylalanine (peptide) (units/mg)	Hippuryl-D,L-phenyllactic acid (ester) (units/mg)
Porcine B[c]	18,400	Not hydrolyzed[d]	—
Bovine B[e]	10,500	0.64	0.22
Bovine A	Not hydrolyzed[f]	3.19[g]	0.21[g]
Dogfish B[h]	8,350	0.90	0.03

[a] Units—for hippurylarginine: percent hydrolysis/minute at 1 m*M* initial substrate concentration; for benzyloxycarbonylglycylphenylalanine: first order rate constant (*57*) at 20 m*M* initial substrate concentration; for hippurylphenyllactic acid: millimoles hydrolyzed/minute at 10 m*M* initial substrate concentration.

[b] Assay conditions—porcine B: 25 m*M* tris, 0.1 *M* NaCl, pH 7.65, 25°; bovine A and B, dogfish B: for hippurylarginine and hippurylphenyllactic acid, 5 m*M* sodium barbital, 45 m*M* NaCl, pH 7.5, 25°; for benzyoxycarbonylglycylphenylalanine, 20 m*M* sodium barbital, 0.1 *M* NaCl, pH 7.5, 25°.

[c] From Folk *et al.* (*17*).

[d] Reported as no hydrolysis under conditions which would have detected 0.01% of this activity in the carboxypeptidase B sample (*17*).

[e] From Wintersberger *et al.* (*29*).

[f] Reported as no hydrolysis under conditions where a specific activity of carboxypeptidase A toward this substrate, 0.3% of that of carboxypeptidase B, would have been detected (*29*).

[g] From Bargetzi *et al.* (*57*).

[h] From Prahl and Neurath (*19*).

57. J.-P. Bargetzi, K. S. V. Sampath Kumar, D. J. Cox, K. A. Walsh, and H. Neurath, *Biochemistry* **2**, 1468 (1963).

1. *Peptidase Activity*

Early studies on the peptidase activity of bovine carboxypeptidase B were conducted at a single set concentration of each of the peptide derivatives under test (Table V). These simple analyses served to identify certain COOH-terminal basic amino acids as substrates for carboxypeptidase B. More careful later analyses with the porcine enzyme, reported in Section V,B, have defined the relative sensitivity of some of these substrates to carboxypeptidase B–catalyzed hydrolysis. Comparison of the catalytic properties of carboxypeptidase B (Table V) with those of carboxypeptidase A (Chapter 1, this volume) points up the fact that, apart from their specificity toward different COOH-terminal amino acids, these two exopeptidases display striking similarities in their action. These similarities may be summarized as follows: (1) They only catalyze the hydrolysis of peptide bonds that are α to terminal free carboxyl groups; (2) they release only COOH-terminal L-amino acids; (3) they slowly hydrolyze dipeptides having a free amino group, but if this group is acylated the hydrolysis is rapid; (4) the rate at which they release COOH-terminal residues may be greatly influenced by the structure of the penultimate COOH-terminal residue; and (5) they catalyze the re-

TABLE V
ACTION OF BOVINE CARBOXYPEPTIDASE B ON SYNTHETIC PEPTIDE DERIVATIVES[a]

Peptide derivative	% Hydrolysis(min^{-1} mg^{-1})[b]
Hippuryl-L-lysine	64.3
Hippuryl-D-lysine	0
Hippuryl-L-lysine amide	0
Hippuryl-ϵ-benzyloxycarbonyl-L-lysine	0
Hippuryl-L-arginine	46.0
Hippurylnitro-L-arginine	0
Hippuryl-L-ornithine	19.0
Hippuryl-δ-benzyloxycarbonyl-L-ornithine	0
Hippuryl-L-homarginine	0.8
Benzyloxycarbonylglycyl-L-histidine	0
Glycyl-L-lysine	*c*
ϵ-*N*-(Benzyloxycarbonylglycine)-L-lysine	0
Benzoyl-β-alanyl-L-lysine	2.2
Hippuryl-L-prolyl-L-lysine	0.3
Benzoyl-L-lysylglycine	0

[a] From Folk and Gladner (*11*).

[b] Milligram enzyme as procarboxypeptidase B. Measurements with 0.025 *M* peptide derivative in 0.025 *M* tris buffer, pH 7.65 containing 2 m*M* diisopropylphosphofluoridate at 25°.

[c] Five to ten percent hydrolysis observed after 5 hr at 1 mg of enzyme per milliliter.

lease of certain amino acids that do not occur naturally but that otherwise satisfy the specificity requirements. It is noteworthy that the COOH-terminal basic amino acid, histidine, is not a substrate for carboxypeptidase B.

2. *Esterase Activity*

Carboxypeptidase B was found to catalyze the hydrolysis of hippuryl-L-argininic acid [*o*-(hippuryl)-α-L-hydroxy-δ-guanidino-*n*-valeric acid] (*46*), the ester analog of hippuryl-L-arginine. That the susceptible peptide bond of a carboxypeptidase B substrate may be replaced by an ester linkage is in accord with the well-known specific esterase activity of the other pancreatic proteolytic enzymes trypsin, chymotrypsin, and carboxypeptidase A.

B. Kinetics and Competitive Inhibition

Kinetic constants for purified porcine carboxypeptidase B with several peptide substrates and the ester substrate are listed in Table VI. No inhibition by excess substrate was observed with any of the peptide substrates. With the ester, on the other hand, inhibition by excess substrate was observed above the level of 2 m*M* (*32*, *46*). An interesting feature of the carboxypeptidase B-catalyzed reactions (Table VI) is the pronounced influence of the basic amino acid residue adjacent to the susceptible COOH-terminal residue (benzoyllsyllysine vs. hippuryllysine) and the

TABLE VI
Kinetic Constants for Procine Carboxypeptidase B[a]

Substrate	K_m ($M \times 10^3$)	k_0[b] (sec^{-1})
Peptides		
Hippuryl-L-arginine	0.21	105
Benzoyl-α-L-glutamyl-L-arginine	0.15	87
Hippuryl-L-lysine	7.70	220
Benzoyl-L-lysyl-L-lysine	0.18	86
Hippuryl-L-orthinine	125.0	255
Ester		
Hippuryl-L-argininic acid	0.04	238

[a] From Wolff *et al.* (*32*). Assays were conducted at 23° in 0.025 *M* tris chloride buffer pH 8.0.

[b] k_0(turnover number) = moles of substrate hydrolyzed per mole of enzyme per second assuming one active site per molecule of enzyme.

lack of influence of the charged carboxyl group of a residue in this adjacent position (benzoylglutamylarginine vs. hippurylarginine).

It is evident from the data of Table VI that, at saturating levels of the hippuryl substrates, ornithine and lysine are released at approximately the same rates while arginine is liberated at a significantly slower rate. It is likely that the Michaelis constants (K_m values) for these three substrates reflect their relative affinities for the enzyme in the decreasing order arginine, lysine, and ornithine (*32*).

The insensitive nature of *N*-acetyl-L-arginine as a substrate [K_m, 0.96×10^{-3} M; k_0, 0.02 sec^{-1} (*58*)] emphasizes the importance of a second peptide bond for rapid hydrolysis. The low K_m and K_I (Table VII) values for this derivative suggest that the second peptide bond plays a minor role in binding at the enzyme active site. This is further evident in light of the structural requirements for compounds that display competitive-type inhibition of carboxypeptidase B. A list of these compounds is given in Table VII together with their K_I values (inhibitor constants). The inhibition by δ-guanidinovaleric acid and ε-aminocaproic acid, together with the lack of inhibition by agmatine and cadaverine, as well as hippuric acid (*32*), shows that there are two requirements for binding at the enzyme active site, i.e., a free carboxyl group and the

TABLE VII

K_I VALUES FOR COMPETITIVE-TYPE INHIBITORS OF PORCINE CARBOXYPEPTIDASE B[a]

Inhibitor	K_I ($M \times 10^3$)
L-Arginine	0.5
D-Arginine	0.5
N-Acetyl-L-arginine	0.8
N-Acetyl-D-arginine	0.54
N-Benzoylarginine	0.04
L-Argininic acid	0.25
δ-Guanidino-*n*-valeric acid	0.39
δ-Guanidino-*n*-butyric acid	4.0
β-Guanidinopropionic acid	4.0
Guanidinoacetic acid	16.0
D,L-Homoarginine	1.6
L-Ornithine	15.0
L-Lysine	13.0
ε-Aminocaproic acid	1.1

[a] From Wolff *et al.* (*32*). Assay conditions as given in Table VI. The substrate was hippuryl-L-arginine.

58. J. E. Folk, E. C. Wolff, and E. W. Schirmer, *JBC* **237**, 3100 (1962).

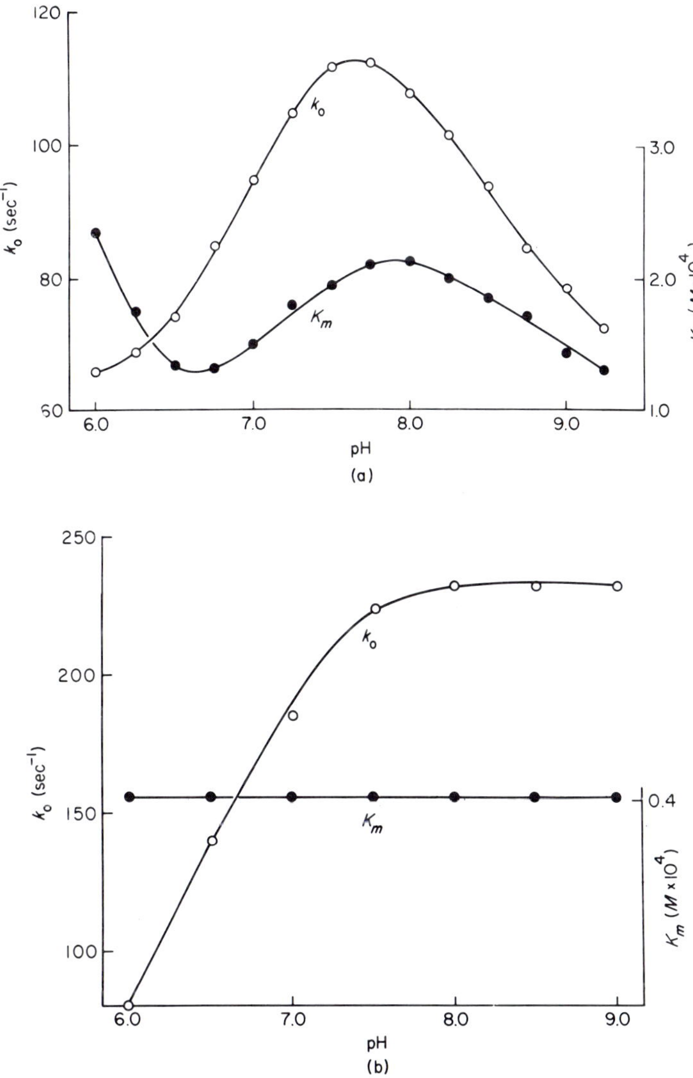

FIG. 1. Plots of K_m and k_0 against pH for the carboxypeptidase B-catalyzed hydrolysis (a) of the peptide substrate, hippuryl-L-arginine, and (b) of the ester substrate, hippuryl-L-argininic acid. Assays were conducted in 0.025 *M* tris acetate buffers at 23°. From Folk and Wolff (*59*).

59. J. E. Folk and E. C. Wolff, unpublished results (1962).

basic side chain. The length of the methylene chain separating the free carboxyl group and the basic group influences the binding properties. The effective inhibition by acetyl-D-arginine which is not a substrate (*32*) and the fact that hippuryl-D-arginine is neither a substrate nor an inhibitor (*32*) suggest that there is a spacial arrangement of groupings in the active center of the enzyme that allows ready accommodation of the improperly oriented acetyl group but not the bulky hippuryl moiety (*58*). The equally effective inhibitory properties of D- and L-arginine are in accord with this suggestion (*32*) (see also Fig. 2).

The effects of pH on the peptidase and esterase activities of carboxypeptidase B is represented in the variation of K_m and k_0 with pH (Figs. 1a and b, respectively). Variations in binding of several competitive-type inhibitors represented by K_I as a function of pH are presented in Fig. 2. Comparison of the recent X-ray solution of the structure of the carboxypeptidase A–substrate complex and model building studies of enzyme–substrate and enzyme–inhibitor interactions (*60*) (Chapter 1, this volume) with earlier speculations of carboxypeptidase A mechanism, based in part on pH effects on enzymic activity (*61, 62*), emphasizes the well-recognized pitfalls and limitations of the use of pH effects in assigning ionizable groups as participants in a catalytic mechanism. The obvious similarities between the reactions catalyzed by carboxypeptidases A and B caution against overinterpretation of the pH-activity and pH-inhibitor effects of Figs. 1 and 2. Several features of these effects, however, warrant comment.

The similarity in form of the curves of K_m for peptidase activity and of K_I for the inhibitor, argininic acid (similar curves were obtained with acetyl-D-arginine and δ-guanidinovaleric acid), suggests that K_m reflects substrate binding and that variations in substrate binding over the pH range studied are a function of changes in affinity of enzyme for the free carboxyl group and/or the basic group of substrate. The decrease in k_0 below pH 7.5 for both peptide and ester substrates suggests participation of the same group of the enzyme in both reactions; the overall forms of the k_0 curves for these two reactions, on the other hand, indicate different mechanisms of hydrolysis. The pH–K_I curves for D- and L-arginine are indicative of inhibition by the species of these amino acids in which the α-amino group is in the unionized form; a plot of $-\log K_I$

60. W. N. Lipscomb, J. A. Hartsuch, G. N. Reeke, Jr., F. A. Quiocho, P. H. Bethge, M. L. Ludwig, T. A. Steitz, H. Muirhead, and J. C. Coppola, *Brookhaven Symp. Biol.* **21,** 24 (1968).

61. B. L. Vallee, J. F. Riordan, and J. E. Coleman, *Proc. Natl. Acad. Sci. U. S.* **49,** 109 (1963).

62. F. W. Carson and E. T. Kaiser, *JACS* **88,** 1212 (1966).

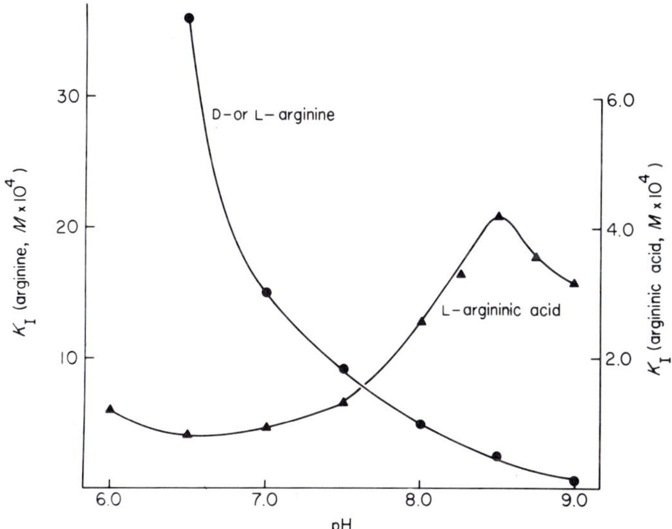

FIG. 2. Plots of K_I against pH for competitive-type inhibitors of carboxypeptidase B. The substrate was hippuryl-L-arginine. Assay conditions of Fig. 1. From Folk and Wolff (*59*).

against pH for arginine shows inflections at pH values identical to those found in this type of plot for both argininic acid and guanidinovaleric acid (*59*).

C. ACTIVATION AND INHIBITION

The inhibition of carboxypeptidase B observed with the metal-chelating agents 1,10-phenanthroline, 8-hydroxyquinoline-5-sulfonic acid, and 2,2′-dipyridyl is probably a result of binding of the essential metal ion, zinc (*17*).

The enzyme is inhibited by phosphate and citrate buffers below pH 7.5 and by pyrophosphate and borate above pH 8 (*32*). Peptidase activity is not markedly effected by limited changes in ionic strength (NaCl concentration) (*32*).

1-Butanol and several other alcohols show a pronounced influence on carboxypeptidase B activity (*63*). Approximately 0.3 *M* 1-butanol causes a 100% increase in the maximum velocity of hippuryl-L-arginine hydrol-

63. J. E. Folk, E. C. Wolff, E. W. Schirmer, and J. Cornfield, *JBC* **237**, 3105 (1962).

ysis but an 85% decrease in the maximum rate of ester hydrolysis. A kinetic analysis of this phenomenon gives evidence that these catalytic changes are a result of modification of the free enzyme rather than changes in individual rate constants for the catalytic reactions (*63*).

VI. Enzymic Properties of Metalloenzymes

Incubation of carboxypeptidase B with salts of cobalt or cadmium at pH 7.75 results in replacement of the native zinc of the enzyme with 1 g-atom per molecule of the metal used in the incubation (*45*). The cobalt and cadmium enzymes display activities toward specific peptide and ester substrates that are markedly altered from those of the native zinc enzyme (Tables VI and VIII). It is apparent from the large variation in turnover numbers for the three metalloenzymes with these substrates that the rates of the overall catalytic reactions are a function of the metal bound to enzyme at the level of 1 g-atom per molecule. In this regard hippuryl-L-arginine, which is not hydrolyzed by the cadmium enzyme, exerts an effective competitive-type inhibition against the cadmium enzyme-catalyzed esterase activity (K_I, 0.046×10^{-3} M) (*58*).

The differences in K_I values for inhibitors of the three metalloenzymes (Tables VII and IX) suggest that the affinity between enzyme and substrate is also influenced by the metal bound to enzyme. There is no reason to suggest, however, that metal ion contributes more than an extrinsic influence in this regard (*58*).

A unique feature of binding by the cadmium enzyme is the fact that hippuryl-D-arginine, which is neither a substrate nor an inhibitor for the

TABLE VIII
KINETIC CONSTANTS FOR COBALT AND CADMIUM CARBOXYPEPTIDASES B[a]

Substrate	Enzyme	K_m ($M \times 10^3$)	k_0 (sec^{-1})
Peptides			
Hippuryl-L-arginine	Cobalt	0.20	220
	Cadmium	Not hydrolyzed	
Hippuryl-L-lysine	Cobalt	5.0	269
	Cadmium	Not hydrolyzed	
Ester			
Hippuryl-L-argininic acid	Cobalt	0.04	119
	Cadmium	0.36	700

[a] From Folk *et al.* (*58*). Assay conditions as given in Table VI.

TABLE IX
K_I VALUES FOR COMPETITIVE-TYPE INHIBITORS OF COBALT AND CADMIUM CARBOXYPEPTIDASES B[a]

Inhibitor	Enzyme	K_I ($M \times 10^3$)
L-Argininic acid	Cobalt	0.91
	Cadmium	1.10
Acetyl-L-arginine	Cobalt	1.80
	Cadmium	0.78
Acetyl-D-arginine	Cobalt	1.00
	Cadmium	0.08

[a] From Folk *et al.* (*58*). Assay condition given in Table VI. The substrate was hippuryl-L-argininic acid.

zinc and cobalt enzymes and thus is not effectively bound to these enzymes, is bound by the cadmium enzyme (K_I, $0.1 \times 10^{-3}\ M$) (*58*).

VII. Comment on the Enzyme Mechanism

In consideration of the gross similarities between carboxypeptidases A and B in terms of molecular size, amino acid composition, metal content, and enzymic function, as well as the enzymic changes resulting from metal substitution and chemical modification (*47, 54*) (Chapter 1, this volume), it seems probable that many of the mechanistic features of carboxypeptidase A as determined by X-ray studies (*60*) (Chapter 1, this volume) will eventually be found common to both exopeptidases. It suffices to say at this point that the several differences in catalytic properties of these enzymes, including the obvious specificity differences, may be expected to be reflected in unique arrangements of chemical groupings in the enzymes' active sites.

VIII. Use in Structural Analysis and Modification of Proteins and Peptides

Methods for the use of carboxypeptidases A and B in structural analyses have recently been outlined in detail (*64*), as have the purposes and procedures for protein and polypeptide modification by these exo-

64. R. P. Ambler, "Methods in Enzymology," Vol. 11, p. 436, 1967.

peptidases (*65*). A few examples of the application of carboxypeptidase B for the above purposes will serve to acquaint the reader with its usefulness.

Probably the first practical application of carboxypeptidase B to end group determination was in the identification of arginine as the COOH-terminal residue of both fibrinopeptides A and B (*66*) which are released from bovine fibrinogen during its thrombin-catalyzed conversion to fibrin. This finding definitively demonstrated that the proteolytic action of thrombin on fibrinogen is limited to cleavage at four arginyl peptide bonds per molecule of fibrinogen (MW 360,000).

The identification of COOH-terminal lysine in *E. coli* β-galactosidase (*67*) presents an example of the use of carboxypeptidase B in protein end group analysis. In this case digestion by carboxypeptidase B was carried out in the presence of the denaturing agent, sodium dodecyl sulfate. The finding of one COOH-terminal lysine per monomer of molecular weight 130,000 was presented as evidence that this enzyme of molecular weight 520,000 is composed of four identical monomer subunits.

A novel procedure for identification of the COOH-terminal peptide in tryptic digests of proteins that do not have COOH-terminal basic amino acid residues has been outlined (*68*). In this method a tryptic digest, applied as a band on paper, is subjected to ionophoresis (pH 6.4). Two strips are cut from the paper parallel to the direction of ionophoresis. One of these is sprayed with a dilute buffered solution of carboxypeptidase B and digestion is allowed to proceed. After the period of digestion the two strips are sewn onto separate sheets of paper and ionophoresis (pH 6.4) is carried out in the second dimension. Because arginine or lysine is removed by carboxypeptidase B from the COOH-terminus of each tryptic peptide except the COOH-terminal one, only the COOH-terminal peptide should appear in the identical position in the digest and the control runs. As an example this method was applied successfully to the γ chain of fetal hemoglobin (*68*).

A classic example of protein modification by carboxypeptidases is the selective removal of COOH-terminal residues from the α and β chains of intact human hemoglobin (*69, 70*). The COOH-terminal sequences of human hemoglobin are

65. J. T. Potts, Jr., "Methods in Enzymology," Vol. 11, p. 648, 1967.

66. J. E. Folk, J. A. Gladner, and K. Laki, *JBC* **234,** 67 (1959); J. E. Folk, J. A. Gladner, and Y. Levin, *ibid.* p. 2317.

67. S. Koorajian and I. Zabin, *BBRC* **18,** 384 (1965).

68. M. A. Naughton and H. Hagopian, *Anal. Biochem.* **3,** 276 (1962).

α chain	·Thr·Ser·Lys·Tyr·Arg
β chain	·Ala·His·Lys·Tyr·His

Carboxypeptidase B removes only arginine from the α chain. Carboxypeptidase A liberates tyrosine and histidine from the β chain, while the two enzymes together remove four residues from the β chain and at least three residues from the α chain *(70)*. These enzymic modifications leave many of the properties of the molecule intact but cause profound and specific changes in the oxygen equilibrium *(69, 70)*. Wide variations were observed in the rates of digestion by both carboxypeptidases A and B of deoxy- and oxyhemoglobin and carboxypeptidase A- or B-modified deoxy- and oxyhemoglobin *(70)*. These findings, together with observed variations in reactivity of the sulfhydryl groups occupying position 93 of the β chains of these forms of hemoglobin, were presented in support of a thesis that irreversible specific conformational changes in the macromolecule are a result of the removal of certain amino acids near the COOH-terminus.

A recent report describes the enzymic replacement of the arginyl by a lysyl residue in the reactive site of the soybean trypsin inhibitor *(71)*. Carboxypeptidase B was first employed to remove COOH-terminal arginine from the trypsin-modified inhibtor protein (Arg 64–Ile 65 bond hydrolyzed). By a clever manipulation of reaction conditions, resulting in a reversal of the normal hydrolytic function of carboxypeptidase B, free lysine was inserted into position 64 of the des-Arg-64-modified inhibitor. This was accomplished by the use of a high concentration of carboxypeptidase B and in the presence of free trypsin. The trypsin supplied the driving force for peptide bond synthesis by reacting with the newly synthesized (Lys 64)-modified inhibitor to form trysin-inhibitor complex. Finally, the (Lys 64) inhibitor (Lys 64–Ile 65 bond intact) was obtained by dissociation of the complex with guanidine·HCl. This modified inhibitor protein was found to be comparable in trypsin inhibitory capacity with authentic soybean inhibitor.

---

69. E. Antonini, J. Wyman, R. Zito, A. Rossi-Fanelli, and A. Caputo, *JBC* **236**, PC60 (1961); M. Brunori, E. Antonini, J. Wyman, R. Zito, J. F. Taylor, and A. Rossi-Fanelli, *ibid.* **239**, 2340 (1964).
70. R. Zito, E. Antonini, and J. Wyman, *JBC* **239**, 1804 (1964).
71. R. W. Sealock and M. Laskowski, Jr., *Biochemistry* **8**, 3703 (1969).

3

Leucine Aminopeptidase and Other N-Terminal Exopeptidases

ROBERT J. DELANGE • EMIL L. SMITH

I. Introduction

It has become increasingly evident (*1–11*) that a considerable number of distinguishable enzymes, often present in the same organ or organism, may be classified as N-terminal peptidases. In some instances,

1. E. L. Smith, "The Enzymes," 2nd ed., Vol. 4, Part A, p. 1, 1960.
2. E. K. Patterson, S. Hsiao, and A. Keppel, *JBC* **238**, 3611 (1963).

such enzymes can be recognized in crude extracts or in partially purified preparations by appropriate choices of substrates; however, because there is considerable overlap in the specificity of the enzymes for various substrates, and since other types of N-terminal exopeptidases may yet be found, definitive characterization may be impossible until a homogeneous preparation is obtained. For this reason we will not attempt to review all reports of N-terminal exopeptidases but will summarize only studies made on well-characterized enzymes.

Reviews by Smith and Hill (*9*) on leucine aminopeptidase and by Smith (*1*) on peptide bond cleavage, which included a survey of several N-terminal exopeptidases, were presented in an earlier edition of this treatise. The purpose of this review is to summarize important features of these enzymes with emphasis on new developments and to include a description of studies on new enzymes.

By definition an N-terminal exopeptidase hydrolyzes peptide bonds at or near the amino-terminus but not at other bonds in the peptide chain. Generally, this type of enzyme requires a free α-amino (or α-imino) group and releases free amino acids from the N-terminus of the peptide substrate. However, N-terminal exopeptidases which release dipeptides (e.g., cathepsin C) and pyrrolidonecarboxylic acid (pyrrolidonyl peptidase) have been described (see Sections IV and VII). For the purposes of this review, an aminopeptidase is an N-terminal exopeptidase which requires a free α-amino group on the substrate; in like manner, an iminopeptidase requires a free α-imino (secondary amino) group on the substrate. In addition, other structural features in the substrate will be discussed for each enzyme.

II. Leucine Aminopeptidase

A. Introduction

The "classic" N-terminal exopeptidase, leucine aminopeptidase, was first observed in extracts of swine intestinal mucosa by Linderstrøm-Lang

3. E. K. Patterson, S. Hsiao, A. Keppel, and S. Sorof, *JBC* **240,** 710 (1965).
4. F. J. Behal, B. Asserson, F. Dawson, and J. Hardman, *ABB* **111,** 335 (1965).
5. F. J. Behal, R. A. Klein, and F. B. Dawson, *ABB* **115,** 545 (1966).
6. F. Binkley, F. Leibach, and N. King, *ABB* **128,** 397 (1968).
7. N. Marks, R. K. Datta, and A. Lajtha, *JBC* **243,** 2882 (1968).
8. C. Schwabe, *Biochemistry* **8,** 771 (1969).
9. E. L. Smith and R. L. Hill, "The Enzymes," 2nd ed. Vol. 4, Part A, p. 37, 1960.
10. K. Linderstrøm-Lang, *Z. Physiol. Chem.* **182,** 151 (1929).
11. E. L. Smith and M. Bergmann, *JBC* **153,** 627 (1944).

in 1929 (*10*). Since it catalyzed the rapid hydrolysis of L-leucylglycine and other peptides with NH_2-terminal leucine, the enzyme was called leucylpeptidase; however, the name was modified to leucine aminopeptidase when Smith and Bergmann (*11*) showed that L-leucinamide also served as a substrate. Although it was shown subsequently that the enzyme also hydrolyzes a large variety of substrates which do not contain leucyl residues (*9*), the name leucine aminopeptidase has been retained to distinguish it from other aminopeptidases and to indicate that leucyl compounds are among those most rapidly hydrolyzed by the enzyme. Enzymes with substrate specificities similar to that of the "classic" leucine aminopeptidase have been described in a variety of species and have also been called leucine aminopeptidases (*8*, *9*, *12–14*). It is obvious that these enzymes are not all identical with respect to amino acid compositions, kinetic parameters, metal requirements, etc., but the name would still appear to be descriptively useful if a thorough investigation shows specificity similar to that of the swine kidney enzyme. Until this has been done with a highly purified preparation of the enzyme, it would be preferable to refer to the enzyme simply as an aminopeptidase and to designate its origin.

In addition to the review by Smith and Hill (*9*), a review by Light (*14a*) deals primarily with the method of preparation of leucine aminopeptidase and its use in sequence studies.

B. Enzymic Assay

Although several substrates could be used to assay leucine aminopeptidase, L-leucinamide is most often employed because it is commercially available or is easily prepared (*9*). It is rapidly hydrolyzed by leucine aminopeptidase, but it is not hydrolyzed by dipeptidases or by many other proteolytic enzymes which may be present at various stages of purification of the enzyme. Other substrates, e.g., leucyl peptides or leucylnaphthyl amide, are also susceptible to hydrolysis by other enzymes.

For the assay of leucine aminopeptidase during purification procedures, the titrimetric procedure as described by Hill *et al.* (*15*) has been widely used because many samples may be conveniently assayed at one time. For the same reason this procedure has also been useful when comparing the

12. J. Uriel and S. Avrameas, *Biochemistry* **4**, 1740 (1965).
13. R. L. Joseph and W. J. Sanders, *BJ* **100**, 827 (1966).
14. H. Hanson, H. Hütter, H. Mannsfeldt, K. Kretschmer, and C. Sohr, *Z. Physiol. Chem.* **348**, 680 (1967).

14a. A. Light, "Methods in Enzymology," Vol. 11, p. 426, 1967.

15. R. L. Hill, D. H. Spackman, D. M. Brown, and E. L. Smith, *Biochem. Prep.* **6**, 35 (1958).

hydrolysis of many different substrates by the same enzyme preparation (*9*). Other similar methods of assay have also been described (*16, 17*). Although the pH-stat methods of assay (*18, 19*) are more precise, they are generally limited to a single sample at a time. Since these methods depend on differences in the pK values of substrates (dipeptides, amino acid amides, or amino acid esters) and products (amino acids, etc.) variations in pK values must be considered when comparing different substrates. Thus these methods seem most advantageous in kinetic studies or when precise comparisons of different fractions or preparations of the enzyme are to be made.

The specific activity of leucine aminopeptidase preparations is usually expressed in terms of the proteolytic coefficient C_1, which is defined as the quotient k_1/E, where k_1 is the first-order rate constant (in decimal logarithms) and E is the enzyme concentration, expressed in milligrams of protein nitrogen per milliliter of reaction mixture (*15*). Since all substrates are not hydrolyzed according to first-order kinetics, a zero-order proteolytic coefficient C_0, has been defined and is calculated when the hydrolysis of various substrates is to be compared (*9, 15*).

C. Purification

Partially purified preparations of leucine aminopeptidase were first obtained from swine intestinal mucosa by Johnson *et al.* (*20*) and later by Smith and Bergmann (*11*). Since the enzyme from intestinal mucosa is somewhat unstable, Spackman *et al.* (*21*) used swine kidney to obtain a 1700-fold purification of the enzyme, which appeared to be nearly homogeneous. Modification of this method to include zone electrophoresis on starch columns (*15*) and chromatography on DEAE-cellulose columns (*22*) gave preparations of higher specific activity.

Additional methods of purification, some of which avoid the initial step of preparing an acetone powder, have been developed (*16, 23–26*).

16. F. Binkley and C. Torres, *ABB* **86**, 201 (1960).
17. H. Fasold, P. Linhart, and F. Turba, *Biochem. Z.* **336**, 182 (1962).
18. G. F. Bryce and B. R. Rabin, *BJ* **90**, 509 (1964).
19. A. T. Matheson, S. Bjerre, and C. S. Hanes, *Can. J. Biochem. Physiol.* **41**, 1741 (1963).
20. M. J. Johnson, G. H. Johnson, and W. H. Peterson, *JBC* **116**, 515 (1936).
21. D. H. Spackman, E. L. Smith, and D. M. Brown, *JBC* **212**, 255 (1955).
22. J. E. Folk, J. A. Gladner, and T. Viswanatha, *BBA* **36**, 256 (1959).
23. H. Hanson and H. Hütter, *Z. Physiol. Chem.* **347**, 118 (1966).
24. H. Hütter, T. Barth, and I. Rychlik, *Z. Physiol. Chem.* **349**, 113 (1968).
25. S. R. Himmelhoch and E. A. Peterson, *Biochemistry* **7**, 2085 (1968).
26. M. H. Moseley and P. Melius, *Can. J. Biochem.* **45**, 1641 (1967).

and yield preparations similar to the earlier ones. Matheson *et al.* (*19*) used a method of preparation, based on one described by Binkley (*27*), which gives a nucleic acid–enzyme complex resistant to denaturation by chloroform octanol (*19*). The complex could be dissociated into nucleic acid and a more labile form of the enzyme by ion exchange chromatography or by gel filtration under appropriate conditions (*19*). The specific activity of the aminopeptidase was in the same range (C_1 = 100–120) observed for other preparations. Thus, the high specific activities (C_1 = 400–1200) reported by Binkley (*27*) for enzyme prepared by this and other methods (*28*) were not verified. Since the enzymic activity remained with the protein after dissociation, Binkley's proposal (*29*) that this form of the enzyme was nonprotein in nature could not be confirmed.

Highly purified preparations of leucine aminopeptidase were found to contain traces of contaminating endopeptidases, which could be inactivated by treating the preparations with DFP and iodoacetate (*30*). Inactivation of the contaminating enzymes is unnecessary when leucine aminopeptidase is used for complete hydrolysis of peptides (*14a*). In studies of the amino-terminal sequences of peptides this precaution must be followed to avoid errors in assignments of residues in the sequence. When native proteins were treated with preparations of leucine aminopeptidase which contained contaminating endopeptidases (unknown to the investigators) the data obtained led to the postulation of "active fragments" for enzymes like papain and enolase (*9*). It was subsequently shown that the "active fragment" of papain (*30*) consisted of undegraded native enzyme and a mixture of peptides and amino acids derived from degradation of some of the enzyme.

Himmelhoch and Peterson (*25*) have obtained preparations of the aminopeptidase apparently free of detectable endopeptidase activity. Inbred strains of hogs gave better preparations with C_1 values as high as 158. Higher specific activities have been reported by Hütter *et al.* (*24*) for a 2800-fold purified preparation of the enzyme, but differences in assay conditions could account for this as well as for differences in the C_1 values of other highly purified preparations. The lack of endopeptidase activity in preparations of the enzyme made by the method of Himmelhoch and Peterson was verified by Deftos and Potts (*31*). These investigators also demonstrated that not all of the endopeptidase activity in three commercial preparations of the enzyme could be inactivated by

27. F. Binkley, *Exptl. Cell Res.* Suppl. 2, 145 (1952).
28. F. Binkley, *JACS* **82**, 987 (1960).
29. F. Binkley, *Proc. Roy. Soc.* **B142**, 170 (1954).
30. R. Frater, A. Light, and E. L. Smith, *JBC* **240**, 253 (1965).
31. L. J. Deftos and J. T. Potts, Jr., *BBA* **171**, 121 (1969).

DFP or by iodoacetate. Evidently, there were additional contaminants in these commercial preparations that were not present in the preparations of Frater *et al.* (*30*).

Purified preparations of the enzyme have recently been obtained from swine pancreas (*12*), swine muscle (*13*), bovine dental pulp (*33*), and bovine lens (*34*). From the last tissue the enzyme was obtained in crystalline form and had C_1 values as high as 180. An enzyme with specificity similar to that of leucine aminopeptidase has also been obtained from *Aspergillus* (*35*, *36*).

Preparations of leucine aminopeptidase appear to be stable indefinitely, if stored between pH 7 and 9 in the frozen state and in the presence of $MgCl_2$ (*9*). The enzyme prepared by Himmelhoch and Peterson (*25*) was stable for at least one year. Barbital buffer has been found by Moseley and Melius (*26*) to protect the enzyme from inactivation by ethylenediaminetetraacetate (EDTA), and it is conceivable that this buffer may be useful in stabilization of the enzyme under other conditions as well. The bovine connective tissue enzyme forms a stable complex with calcium phosphate gel (*37*) which should permit long storage. The enzyme–gel complex is active in hydrolyzing small substrates and is therefore another example of an active, insoluble enzyme preparation. It will be of interest to determine whether the swine kidney and bovine lens enzymes also form such complexes.

D. Physical Properties

The enzyme prepared by Spackman *et al.* (*21*) appeared to be nearly homogeneous, but insufficient material was available for complete characterization. Preparations containing endopeptidase activity are obviously not completely homogeneous, although they appear to be by certain physical techniques. Even the preparations without endopeptidase activity (*25*) gave multiple forms of the enzyme during ion exchange chromatography, thus making an assessment of purity difficult. Multiple forms of the swine kidney and bovine connective tissue enzymes were also observed during gel electrophoresis by Schwabe (*33*). The crystalline bovine lens enzyme appears to be homogeneous (*34*).

32. P. Melius, M. H. Moseley, and D. M. Brown, *BBA* **227**, 62 (1970).
33. C. Schwabe, *Biochemistry* **8**, 783 (1969).
34. H. Hanson, D. Glässer, and H. Kirschke, *Z. Physiol. Chem.* **340**, 107 (1965).
35. K. Lehmann and H. Uhlig, *Z. Physiol. Chem.* **350**, 99 (1969).
36. J. Jollès, P. Jollès, H. Uhlig, and K. Lehmann, *Z. Physiol. Chem.* **350**, 139 (1969).
37. C. Schwabe, *Biochemistry* **8**, 795 (1969).

Leucine aminopeptidase as prepared by Spackman *et al.* (*21*) gave a single sedimenting boundary at pH 8.6 with a sedimentation constant of 12.6 S. Since a diffusion constant was not determined, the molecular weight could only be estimated as approximately 300,000, but this value has also been found for other preparations of the enzyme by gel filtration on Sephadex G-200 (*25*). Melius *et al.* (*32*) have estimated the molecular weight of the swine kidney enzyme to be 255,000 ± 5000 by sedimentation equilibrium. The sedimentation constant determined by these investigators was 13.1 S. The bovine lens enzyme has a molecular weight of 326,000, as determined by Kretschmer and Hanson (*38*), from the sedimentation coefficient (12.56 S), diffusion coefficient (3.75 F), and partial specific volume (0.751).

Melius *et al.* (*32*) have found a molecular weight of 63,500 for leucine aminopeptidase in the presence of 6 *M* guanidinium chloride and 0.5% β-mercaptoethanol by sedimentation equilibrium. From these data they proposed a four-subunit model for the swine kidney enzyme. In contrast, the bovine lens enzyme has been reported to be composed of 10 subunits with a molecular weight of about 32,000 each, and with the subunits arranged in two overlapping pentagons around a common, apparently hollow center (*39*). A molecular model of bovine lens leucine aminopeptidase, based on low angle X-ray diffraction studies by Kretschmer and Kollin (*40*), depicts the enzyme as a hollow cylinder, 66.7 Å in height and with an outer diameter of 107.2 Å. Therefore, the axial ratio is 0.62 (*40*), compared to 0.365 (oblate) or 2.65 (prolate), as calculated from hydrodynamic measurements (based on half-maximal hydration) (*38*) and compared to 0.69 as calculated from electron microscope observations (*39*). The form factor (f), as determined by X-ray studies (*40*), was 1.24 in good agreement with the frictional coefficient ($f/f_0 =$ 1.23) as determined by sedimentation diffusion (*38*). The molecule appears to be maximally hydrated, and water can apparently pass freely through the hollow center, which has an inside diameter of 32.2 Å (*40*).

The subunits of the bovine lens enzyme appear to be identical, but Kretschmer and Hanson (*41*) obtained evidence for at least two types of polypeptide chains in each subunit. One of the chains appeared to be two-thirds the size of the subunit itself. Schwabe (*33*) has indicated that the swine kidney enzyme is composed of one type of subunit and that the bovine connective tissue enzyme is composed of different types of subunits. The subunit studies by Melius *et al.* (*32*) gave no evidence for

38. K. Kretschmer and H. Hanson, *Z. Physiol. Chem.* **340,** 126 (1965).
39. K. Kretschmer, *Z. Physiol. Chem.* **349,** 715 (1968).
40. K. Kretschmer and G. Kollin, *Z. Physiol. Chem.* **350,** 431 (1969).
41. K. Kretschmer and H. Hanson, *Z. Physiol. Chem.* **349,** 831 (1968).

more than one size of subunit in the swine kidney enzyme, in support of Schwabe's observations. The presence of different types of subunits in a proteolytic enzyme, as indicated for the bovine connective tissue enzyme, would be a novel finding.

E. Chemical Properties

Since the original, single analysis by Spackman *et al.* (*21*), there have been no other studies of the amino acid composition of the enzyme from swine kidney. However, the compositions have been determined for the bovine lens enzymes (*42*) and for the various isozymes from bovine connective tissue (*33*). Kettmann *et al.* (*42*) found that the composition of the bovine lens enzyme closely resembles that of the swine kidney enzyme. Additional evidence for the similarity of the two enzymes was obtained by immunological methods (*43*). Both preparations of the enzyme formed enzyme–antibody precipitates with antibody prepared against crystalline bovine enzyme, and the enzyme–antibody complexes retained enzymic activity in both instances. However, species (or organ) differences were indicated by the spurs formed during immunoelectrophoresis. The same antibody also reacted with the enzyme prepared from lens, liver, and kidney of pig, sheep, goat and, to a lesser extent, rat, but did not react with aminopeptidase M. Since the physical characterization of the swine kidney and bovine lens enzymes indicated strikingly different structures for the two enzymes, it is clear that additional studies will be required to determine the extent of similarity that seems to exist on the basis of these studies.

It has long been assumed that leucine aminopeptidase is a metalloenzyme and that it probably contains magnesium or manganese since it is strongly activated by these metal ions (*9*). Himmelhoch (*44*) has reported that the enzyme is, indeed, a metalloenzyme, but that it contains zinc rather than magnesium or manganese. A similar finding was reported earlier by Fittkau *et al.* (*45*) on the bovine lens enzyme, but since their method of isolation involved a zinc-precipitation step, the results could not be regarded as conclusive. The swine kidney enzyme was prepared with great care to avoid contamination by zinc. The best preparations contained 4–6 g-atoms of zinc per 300,000 g of the enzyme (*44*) as com-

42. U. Kettmann, K. Kretschmer, and H. Hanson, *Z. Physiol. Chem.* **349,** 1537 (1968).
43. H. Hanson, D. Glässer, M. Ludewig, H. Mannsfeldt, M. John, and H. Nesvadba, *Z. Physiol. Chem.* **348,** 689 (1967).
44. S. R. Himmelhoch, *ABB* **134,** 597 (1969).
45. S. Fittkau, U. Kettmann, and H. Hanson, *J. Labelled Compds.* **2,** 255 (1966).

pared to 8–12 g-atoms of zinc per 326,000 g of the bovine lens enzyme (*45*). Therefore, both enzymes appear to contain about 1 g-atom of zinc per subunit of enzyme. The role of the metal will be discussed further in Section II,F,1.

The determination of end groups in leucine aminopeptidase has been unsuccessful. End groups could not be detected by the fluorodinitrobenzene procedure in the bovine lens enzyme, but after performic acid oxidation several amino acids were found as amino-terminal residues (*42*). This probably resulted from peptide bond cleavage during the oxidation procedures.

Investigations of the active site of leucine aminopeptidase have been few and have given little information concerning groups (other than metal ion) involved in the catalytic and binding sites. Frohne *et al.* (*46*) investigated the bovine lens enzyme with *N*-bromosuccinimide and $^{131}I_2$. Oxidation of trytophan with *N*-bromosuccinimide produced a 50% decrease in activity, which indicated that oxidation caused conformational changes in the protein structure or that tryptophan has some role in substrate binding. At least 15 atoms of $^{131}I_2$ could be incorporated into one molecule of the protein before any loss of activity was detected. Incorporation of higher amounts of $^{131}I_2$ produced a progressive loss of activity and all activity was lost at 60 atoms/molecule (or 6 atoms/subunit). Thus, tyrosyl residue do not appear to be directly involved at the activity site of the enzyme.

F. Enzymic Properties

Since the work subsequent to 1960 has not altered most of the basic conclusions pertaining to the enzymic properties of leucine aminopeptidase as given in the review by Smith and Hill (*9*), a brief summary of the earlier findings will be given here. Any pertinent additions or modifications will be included under the major headings, which will be the same as in the previous review.

1. *Role of Metal Ion*

Earlier studies (*9*) showed that metal ions serve to stabilize leucine aminopeptidase, with 0.005 M Mg^{2+} being most effective. Two metal ions, Mg^{2+} and Mn^{2+}, can activate the enzyme; Cd^{2+}, Cu^{+}, Hg^{2+}, and Pb^{2+} are inhibitory; and Ca^{2+}, Co^{2+}, Ni^{2+}, and Zn^{2+} have little or no effect on

46. M. Frohne, R. Michael, S. Fittkau, and H. Hanson, *Z. Physiol. Chem.* **350**, 223 (1969).

the enzyme (all metal ions tested at 0.001 M). For the swine kidney enzyme at pH 8.6 and 40° the association constant $K_a (M^{-1})$ is 1.2×10^4 for Mg^{2+}, and 3.3×10^4 for Mn^{2+}. The primary interaction of metal ions appears to be with enzyme rather than substrate, and probably only one atom of either magnesium or manganese is required per active site of the enzyme.

Himmelhoch (*44*) has found that swine kidney enzyme contains 4–6 g-atoms of zinc per mole of subunit (see Section II,E) and no other metal ions. The zinc can be removed by dialysis at pH 6.0, by metal chelators (such as *o*-phenanthroline), or by exchange with Cd^{2+}. In each case the enzyme lost activity (measured in the absence of Mg^{2+} and Mn^{2+}) in proportion to the loss of Zn^{2+}. Except with the Cd^{2+} enzyme, activity could not be restored by placing the zinc-free enzyme in a medium (pH 6–8.6) containing Zn^{2+}. With the Cd^{2+} enzyme, Zn^{2+} could displace the Cd^{2+} to restore the active enzyme, but, presumably, the zinc-free enzyme prepared by the other two procedures (dialysis at pH 6.0 and treatment with chelating agents) had undergone secondary changes in the structure of the enzyme. The zinc could also be partially removed by displacement with Mn^{2+}. This produced an increase in the activity of the enzyme in approximate proportionality to the Mn^{2+} content of the enzyme, but the experiment was complicated by the well-known instability of the enzyme in the presence of Mn^{2+}. EDTA had no effect on the activity of the Zn^{2+} enzyme. The bovine lens enzyme apparently also contains zinc (see Section II,E), and the zinc can be easily exchanged for manganese (*45*).

Bryce and Rabin (*47*) found reversible inactivation of Mg^{2+}-activated leucine aminopeptidase by EDTA. In the presence of *p*-mercuribenzoate the inactivation was irreversible, which suggested to the authors that cysteinyl residues are involved in the binding of metal ions to the protein (but see Section II,F,3). Substrates (or leucine or cysteine) also blocked complete reconstitution of the enzyme by Mg^{2+}, which suggested that metal ions are not required for substrate binding. The bound substrate evidently prevents metal ion from reaching its binding site, which must, therefore, be close to the substrate binding site. These investigators (*47*) concluded that metal ions mediate the binding of substrate to the enzyme, in agreement with earlier views (*9*), although other interactions between the enzyme and the substrate evidently permit substrate binding in the absence of metal ions.

The slow step in inactivation of the Mg^{2+}-activated enzyme by EDTA appears to be dissociation of metal ions from the enzyme (*47*). Reconstitution of the enzyme with metal ions is a much faster process. From

47. G. F. Bryce and B. R. Rabin, *BJ* **90**, 513 (1964).

the studies of Himmelhoch (*44*) the metal ion in the native enzyme is probably Zn^{2+}, but in the studies of Bryce and Rabin (*47*) the enzyme had been placed in a solution containing 0.005 *M* Mg^{2+}. Whether the Mg^{2+} had partially or wholly displaced the Zn^{2+} or had bound to the protein at entirely different sites cannot be determined, since these studies were made before the experiments which showed the presence of Zn^{2+} in the native enzyme. However, the enzyme in the presence of Mg^{2+} could be further activated by transferring it to a solution with a final concentration of 0.002 *M* Mn^{2+} (Mg^{2+} concentration is reduced to 0.0008 *M*). The activation was time dependent with the dissociation of Mg^{2+} apparently as the rate limiting step. The Arrhenius activation energy was calculated as 26 kcal/mole. Again, any involvement of Zn^{2+} in this process is unknown. Activation might simply be pictured as the displacement of Zn^{2+} in the enzyme by Mg^{2+} or Mn^{2+}, since the native enzyme has some activity (*44*), the Mg^{2+}-activated enzyme is more active, and the Mn^{2+}-activated enzyme is still more active (at least with most substrates). However, this picture may be oversimplified, and the situation is probably more complex.

Bryce and Rabin (*18*) found that for all substrates tested, V_{max} is greater for the Mn^{2+}-activated enzyme ($V_{max} = 3.38 \pm 1.65$ with leucinamide) than for the Mg^{2+}-activated enzyme ($V_{max} = 0.76 \pm 0.07$ with leucinamide). Whether Mg^{2+} or Mn^{2+} is used for activation makes little difference in the value of K_m for all substrates except L-leucinamide for which the K_m is 15.7 ± 4.1 m*M* for the Mg^{2+}-activated enzyme and 5.21 ± 0.43 m*M* for the Mn^{2+}-activated enzyme.

Bovine connective tissue leucine aminopeptidase does not appear to be activated by Mg^{2+} but is activated by Mn^{2+} (*33*). The bovine lens (*48*) and swine muscle (*13*) enzymes are similar to the swine kidney enzyme in that they are activated more strongly by Mn^{2+} than by Mg^{2+}. The enzyme from *Aspergillus* (*35*), which has a substrate specificity similar to swine kidney leucine aminopeptidase, is inhibited by EDTA and can be reactivated by Co^{2+}.

2. *Substrate Specificity and Kinetic Parameters*

Most of the information concerning specificity was obtained with the preparation of Spackman *et al.* (*9, 21, 49*) in confirmation of many earlier studies [see the review by Smith and Hill (*9*)]. Since some substrates at 0.05 *M* are hydrolyzed according to first-order kinetics and others by zero-order kinetics, the zero-order proteolytic coefficient C_0

48. H. Hanson, D. Glässer, and R. Kleine, *Z. Physiol. Chem.* **329,** 257 (1962).
49. E. L. Smith and D. H. Spackman, *JBC* **212,** 271 (1955).

was calculated from the initial rate of hydrolysis for all substrates. This permitted a comparison of the relative rates of hydrolysis of various substrates at the same concentration (*9, 49*).

The activated swine kidney enzyme has some activity from pH 6 to at least pH 10, but it is maximally active between pH 9 and 9.5 depending on the metal ion used for activation and on the substrate hydrolyzed. The status (ionized or un-ionized) of the amino group of substrates hydrolyzed by zero-order kinetics probably has little or no influence on the pH-activity profile, whereas only the un-ionized form of the substrates displaying first-order kinetics is hydrolyzed by the enzyme.

A variety of amino acid amides, dipeptides, and other substances were tested as substrates and compared with respect to relative rates of hydrolysis. All tested substances which had an NH_2-terminal L-amino acid residue (or glycine) were hydrolyzed at measurable rates. However, the relative rates varied by at least 1000-fold (e.g., glycinamide, $C_0 = 7$; L-leucinamide, $C_0 = 6600$ for the Mg^{2+}-activated enzyme). L-Alanyl-L-leucinamide was hydrolyzed most rapidly of all substrates tested ($C_0 =$ 26,000, Mn^{2+} activation; 12,300, Mg^{2+} activation), but, generally, compounds with L-leucyl residues (or the similar norleucyl or norvalyl residues) in the amino-terminal position were the best substrates. Compounds with N-terminal D-amino acid residues were not hydrolyzed at measurable rates, and compounds with penultimate D-amino acid residues were hydrolyzed at greatly reduced rates, e.g., L-leucyl-D-leucine is hydrolyzed at only 0.7% of the rate of L-leucyl-L-leucine. Leucine aminopeptidase might also be classified as an iminopeptidase since L-prolylglycine was hydrolyzed at a slow rate. However, since the preparations used for these studies may not have been completely free of other peptidases (see Section C), this observation should be regarded as tentative until verified by other studies with enzyme completely separated from other peptidases.

Bonds of the type X–Pro are not hydrolyzed by preparations of leucine aminopeptidase which are free of contaminants. Substrates with acylated amino groups, NH_2-terminal β-alanyl residues, or NH_2-terminal α-aminoisobutyryl residues are not hydrolyzed. In general, substrates with NH_2-terminal hydrophobic residues were hydrolyzed most rapidly; and, except with amino acid amides as substrates, the Mn^{2+}-activated enzyme is more active than the Mg^{2+}-activated enzyme.

It has also been reported that leucine aminopeptidase has esterase activity (*9, 50*). Compounds with large aliphatic ester groups were the best ester substrates found, but even the best of these were hydrolyzed

50. S. S. Shippey and F. Binkley, *JBC* **230**, 699 (1958).

at only 10% of the rate of the corresponding amides. No transferase activity was detected for the swine kidney enzyme *(9)* (but *vide infra*).

Although the studies reported since 1960 have generally confirmed the earlier studies briefly summarized above, only a few additional types of compounds have been investigated *(18, 23, 43, 51)* including some ester substrates. L-Leucine benzyl ester is the best ester substrate reported to date, having a C_1 value of 42.6 as compared to a value of 60–80 for leucinamide under the conditions used *(23)*. For the Mg^{2+}-activated enzyme Bryce and Rabin *(18)* found that the values for V_{max} and K_m for the ester substrate, L-leucine benzyl ester were only one-third of the values of these parameters for L-leucinamide. Thus, the ester appears to be bound more readily but hydrolyzed more slowly. The kinetic parameters varied more widely for the Mn^{2+}-activated enzyme, but they were also of a compensatory nature. The studies indicated that both substrates were hydrolyzed by the same enzyme. The bovine lens enzyme also has esterase activity as demonstrated by Hanson *et al.* *(48)*. The ratio of amidase (L-leucinamide) : esterase (L-leucine ethyl ester) was about 17:1.

The study by Schechter and Berger *(51)* on the hydrolysis of 39 peptides (di- to hexapeptides) containing only L- and D-alanine at known positions has further demonstrated the influence that residues, which are COOH-terminal to the bond to be hydrolyzed, have on the rate of hydrolysis by leucine aminopeptidase. For example, Ala_4(LDLL) was not hydrolyzed, Ala_3(LDL) was hydrolyzed very slowly, and Ala_4(LDDL) was hydrolyzed at an appreciable rate. Since all three substrates contain the same amino-terminal sequence (LD), the influence of other residues in the peptides is suggested. Generally, peptides of this type (LD) were poorly hydrolyzed, if at all. In accord with many earlier observations, peptides with N-terminal D-alanyl residues are resistant to hydrolysis. It is evident that the results of Schecter and Berger are, in general, consonant with earlier reports indicating the importance of a hydrophobic residue being linked to the NH_2-terminal residue, e.g., the high rate of hydrolysis of L-alanyl-L-leucinamide *(49)* and the much greater sensitivity of L-leucine benzyl ester as compared to the ethyl ester *(9, 18, 23)*. Further studies are desirable, particularly since no rates were measured and the commercial enzyme used was at a low level of purity *(51)*.

Table I summarizes some of the studies on the specificity of leucine aminopeptidase with a variety of substrates.

Hanson *et al.* *(43)* compared the actions of leucine aminopeptidase from bovine lens and swine kidney on amino acid amides, amino acid-β-

51. I. Schechter and A. Berger, *Biochemistry* **5**, 3371 (1966).

TABLE I

ACTION OF LEUCINE AMINOPEPTIDASE ON VARIOUS TYPES OF SUBSTRATES[a,b]

	Mn^{2+} activation		Mg^{2+} activation	
Substrate	C_0	Relative rate	C_0	Relative rate
L-Leucinamide	14,000	100	6,600	100
L-Phenylalaninamide	3,600	26	1,140	17
L-Isoleucinamide	2,800	20	1,120	17
L-Histidinamide	2,700	19	680	10
L-Lysinamide	1,000	7.1	800	12
L-Alaninamide	470	3.4	325	4.9
L-Aspartic acid diamide	410	2.9	250	3.8
L-Serinamide	106	0.8		
L-Prolinamide	100	0.7		
Glycinamide	18	0.1	7	0.1
D-Leucinamide	0	0	0	0
β-Alaninamide	0	0	0	0
L-Leucyl-L-leucine	13,900	100	3,500	53
L-Leucyl-L-isoleucine	8,900	64	3,300	50
L-Leucyl-L-alanine	9,000	64		
L-Leucyl-L-phenylalanine	3,600	26	1,160	18
L-Alanyl-L-leucine	13,000	93		
L-Alanylglycine	1,300	9.4		
Glycyl-L-leucine	1,400	10		
Glycylglycine	165	1.1		
D-Leucylglycine	0	0	0	0
L-Leucyl-β-alanine	0	0		
L-Alanyl-L-leucinamide	26,000	186	12,300	186
L-Leucyl-L-alaninamide	18,000	129	5,400	82
Glycyl-L-leucinamide	6,000	43	6,200	94
L-Leucylglycylglycine	16,800	120	4,600	70
L-Histidylglycylglycine	5,800	41		
Triglycine	330	2.4		
Acetyl-L-tyrosinamide	0	0		
Benzoyl-L-argininamide	0	0		

	Mn^{2+} activation		
Substrate	C_1[c]	Relative rate	pH
L-Leucinamide	60–80	100 (100)[d]	9.2
L-Leucine benzyl ester	43	54–72 (~25)[d]	9.15
L-Leucine ethyl ester	4.4	5.5–7.3	8.0

[a] The value for L-leucinamide is given as 100 and for the other compounds as the relative rate.

[b] Selected from Tables III, IV, and V in the review by Smith and Hill (*9*). C_0 = zero-order proteolytic coefficient.

[c] From Hanson and Hutter (*23*).

[d] Relative rates in parentheses are from Bryce and Rabin (*18*) as determined at high substrate concentration at pH 8.4.

naphthyl amides, and amino acid-*p*-nitroanilides as substrates. Although the general patterns of substrate specificity were very similar, there were some differences, e.g., the Mn^{2+}-activated kidney enzyme was more active with the β-naphthyl amide of tryptophan as substrate than was the Mn^{2+}-activated lens enzyme in comparison to the activities of these enzymes with the β-naphthyl amide of leucine as substrate. The rat kidney (*43*) and swine muscle (*13*) enzymes also appear to have substrate specificities similar to the kidney and lens enzymes, but the muscle enzyme apparently hydrolyzes L-leucylglycine and L-leucylglycylglycine at twice the rate of L-leucinamide (*13*). The Mg^{2+}-activated swine kidney enzyme hydrolyzes these peptides more slowly than it does leucinamide.

Transamidation reactions catalyzed by bovine lens leucine aminopeptidase with L-leucinamide or L-leucine-methylcellosolve ester as substrates have been reported by Hanson and Lasch (*52*); the products were identified as L-leucyl-L-leucinamide and L-leucyl-L-leucine, respectively. With leucinamide (50 μmoles/ml) as substrate at pH 7.5 the ratio of substrate hydrolyzed to substrate converted to the dipeptide amide was 1.25, and at pH 10.0 it was 3.0 under conditions in which 10 and 30%, respectively, of the substrate had reacted. The transfer accounted for an increasing proportion of the total reaction as substrate concentration was increased and as the amount of substrate acted upon was decreased. This is because the rate of the hydrolytic reaction reaches a maximum at 0.1 *M* substrate (where the rates of both reactions are approximately equal) and then decreases slightly with higher substrate concentration, whereas the rate of the transamidation reaction continues to increase with increasing substrate concentration. Thus, under conditions of high substrate concentration (0.2 *M*) and a short time of reaction (5 sec), it was possible to detect only transamidation with no apparent hydrolysis having occurred. Any differential effect on the activation of the enzyme by Mn^{2+} of Mg^{2+} was not investigated for the transamidation reaction. DFP did not affect either hydrolysis or transamidation, but both reactions were equally inhibited in proportion to the concentration of *p*-mercuribenzoate (10^{-8} *M* to 10^{-4} *M*) (see Section II,F,3). These experiments indicate that both hydrolysis and transamidation are catalyzed by the same enzyme. This property of the enzyme clearly distinguishes it from the swine kidney enzyme for which transamidation reactions could not be demonstrated under a variety of conditions and with a variety of substrates (*9*, *53*), nor is the kidney enzyme inhibited by *p*-mercuri-

52. H. Hanson and J. Lasch, *Z. Physiol. Chem.* **348**, 1525 (1967).
53. R. L. Hill and E. L. Smith, *JBC* **224**, 209 (1957).

benzoate (*9, 49*). To the best of our knowledge, this is the first report of a transfer reaction catalyzed by a metallo-exopeptidase. Together with the very different physical properties of the two enzymes (Section II,D), these findings suggest that the bovine lens and swine kidney enzymes may be different enzymes despite similar hydrolytic specificity, similar behavior with regard to metal ion activation, and some immunological cross-reaction (Section II,E).

Some kinetic parameters of the swine kidney enzyme have been determined with various substrates and are summarized in Table II. Values for K_m have also been determined for the bovine lens enzyme (*34, 52*) and are generally about 2-fold higher than for the swine kidney enzyme.

3. *Inhibitors*

As reviewed earlier (*9*), inhibitors may be classified as follows:

(1) Inhibitors which combine with the activating metal ion include EDTA and citrate, which are better inhibitors at pH 8.0 than at pH 8.5, probably because of the stronger protein–metal ion interaction at the higher pH.

(2) Many hydrophobic compounds which contain large aliphatic groups are inhibitors, evidently because they compete for the hydrophobic binding site which binds the hydrophobic side chains of the amino acid residues in substrates. For the series of alcohols, methanol to *n*-hexanol, inhibition increased almost directly proportional to the increase in aliphatic chain length (mass). Such compounds inhibit the hydrolysis of the more susceptible hydrophobic aliphatic substrates, e.g., leucinamide, to a greater degree than the more resistant, less hydrophobic ones, e.g., glycinamide.

TABLE II
KINETIC PARAMETERS OF SWINE KIDNEY LEUCINE AMINOPEPTIDASE[a,b]

	Mg^{2+} activation		Mn^{2+} activation	
Substrate	K_m (mM)	V_{max} (relative)	K_m (mM)	V_{max} (relative)
L-Leucinamide	5.21 ± 0.43	0.76 ± 0.07	15.7 ± 4.1	3.38 ± 1.65
L-Leucylglycine	1.00 ± 0.01	0.53 ± 0.01	0.81 ± 0.01	1.54 ± 0.10
L-Leucyl-L-valine	0.51 ± 0.01	0.56 ± 0.03	1.04 ± 0.02	2.34 ± 0.18
L-Leucyl-L-alanine	0.79 ± 0.04	0.50 ± 0.02	0.65 ± 0.02	1.53 ± 0.12
L-Leucine benzyl ester	1.56 ± 0.03	0.27 ± 0.01	1.10 ± 0.01	0.42 ± 0.04

[a] Determined at pH 8.4; I, 0.10; and 25°.
[b] From Bryce and Rabin (*18*).

(3) Inhibitors which compete for both the metal ion and the hydrophobic site include substances like those in (2), which also contain a functional group (such as an amino group) capable of interaction with metal ions. Thus, L-leucine, L-leucinol, isocaproic acid, α-ketoisocaproamide, and D,L-α-hydroxyisocaproamide are all inhibitors of this type. Those substances with amino groups are better inhibitors at higher pH where the concentration of un-ionized amine is highest. Since leucine is an inhibitor and a product in the hydrolysis of leucinamide, such product inhibition must be minimized in kinetic studies.

Himmelhoch (*44*) has extended the list of metal chelators that inhibit leucine aminopeptidase. Some are effective as inhibitors of the zinc enzyme as well as the Mg^{2+}- or Mn^{2+}-activated enzyme with *o*-phenanthroline being the most effective inhibitor of the native (Zn) enzyme. Other metal chelators such as EDTA do not inhibit the activity of the zinc enzyme but do inhibit the Mg^{2+}- or Mn^{2+}-activated enzyme. Bryce and Rabin (*47*) have shown that the inhibition of the Mg^{2+}-activated enzyme by EDTA can be reversed by Mg^{2+}, but substrate, leucine, or cysteine interfered with the reversal, and *p*-mercuribenzoate (0.1 m*M*) completely prevented such reversal. The authors suggested that this was because thiol groups are involved in the binding of metal ions; however, since iodoacetamide (1 m*M*) did not prevent reversal, the involvement of thiol groups is by no means certain. It should be mentioned that mercurial compounds such as *p*-mercuribenzoate have been reported to affect some enzyme systems which lack thiol groups [e.g., see Sohler *et al.* (*54*)]. Although Bryce and Rabin (*47*) found an inhibitory effect of *p*-mercuribenzoate (in the presence of Mg^{2+}), Smith and Spackman (*49*) found no inhibition by this substance (at concentrations as high as 10^{-4} *M*) or by iodoacetamide. The reason for this discrepancy is not clear but may be related to a difference in purity of the two enzyme preparations for which there is no immediate method of comparison. The study of Frohne and Hanson (*55*) on the effect of *p*-mercuribenzoate and other SH reagents on bovine lens leucine aminopeptidase is difficult to interpret. Factors such as the form of the enzyme (whether crystallized in the presence or absence of Mn^{2+}), the substrate used (leucinamide or leucine-*p*-nitroanilide), and the particular SH reagent used all strongly affected the results. However, the study did indicate that the bovine lens enzyme contains free thiol groups which could be reacted with SH reagents. The titration of these thiol groups

54. M. R. Sohler, M. A. Seibert, C. W. Kreke, and S. E. Cook, *JBC* **195,** 281 (1952).
55. M. Frohne and H. Hanson, *Z. Physiol. Chem.* **350,** 213 (1969).

on the enzyme causes changes in the activity of the enzyme: Activity decreases with leucinamide as substrate but increases with leucine-*p*-nitroanilide as substrate. From these results it is difficult to implicate thiol groups as present in the catalytic site. Indeed, all of the studies to date on "leucine aminopeptidase" from various sources seem to agree that the enzyme is not a "thiol enzyme." Indeed, as discussed above (Section II,F,2), the report that transamidase and hydrolytic activity are inhibited by a mercurial suggests that two kinds of active sites may be present in the crystalline lens enzyme, one involving Zn^{2+} and the other a thiol group.

Bryce and Rabin (*18*) also studied the inhibition of leucine aminopeptidase by 1-butanol and by glycerol. Their data indicated that inhibition was not of the simple competitive type and that factors other than hydrophobic interactions are responsible for the inhibition. No evidence for enzymic esterification of the alcohols could be detected, in agreement with earlier studies of Hill and Smith (*53*).

4. *Mechanism of Action*

Early studies led to proposals for a tentative mechanism of action by Smith and co-workers (*9, 49, 56–59*) as based on studies of substrate specificity, the action of inhibitors, and the activating role of the metal ion. These earlier views may be summarized as follows. The nature of the side chains on both the NH_2-terminal and penultimate residues of substrates is largely responsible for the rate of hydrolysis of substrates. In general, the rate of hydrolysis is increased as the side chains of these two residues are made larger and more hydrophobic, whereas smaller side chains and ionic or other hydrophilic side chains decrease the rate of hydrolysis. In dipeptides the free α-carboxyl group acts as a polar group to decrease the rate of hydrolysis as compared to the corresponding dipeptide amide. Compounds with NH_2-terminal D-amino acid residues or NH_2-terminal β-alanyl residues are not hydrolyzed; however, dipeptides with COOH-terminal D-amino acid residues or COOH-terminal β-alanyl residues are hydrolyzed. The effectiveness of inhibitors (other than those that only bind metal ions) is also dependent on their size and hydrophobic character, the larger, hydrophobic compounds being

56. E. L. Smith, N. C. Davis, E. Adams, and D. H. Spackman, *in* "Mechanism of Enzyme Action" (W. D. McElroy and B. Glass, eds.), p. 291. Johns Hopkins Press, Baltimore, Maryland, 1954.

57. E. L. Smith and R. Lumry, *Cold Spring Harbor Symp. Quant. Biol.* **14,** 168 (1950).

58. E. L. Smith, *Proc. Natl. Acad. Sci. U. S.* **35,** 80 (1949).

59. E. L. Smith, *Federation Proc.* **8,** 581 (1949).

most effective. Such inhibitors are also more effective in inhibiting the hydrolysis of the larger, more hydrophobic substrates than in inhibiting the smaller, more polar substrates. Metal ions are involved both in stabilization (Mg^{2+} particularly) and activation (Mn^{2+} better than Mg^{2+}) of the enzyme. The metal ions appear to be bound by uncharged imidazole or amino groups and might serve to form a ternary coordination complex between substrate and enzyme. From kinetic studies, apparently only one metal ion is bound per active site of the enzyme, where it may be involved not only in binding of the substrate but also as an electron sink in the catalytic reaction. In the tentative proposal of Smith and Spackman (*49*) the protein interacts with the hydrophobic side chain of the substrate, and the metal ion interacts with the α-amino group and the amide nitrogen to give a transitory complex which undergoes a nucleophilic displacement of the amide substituent by an attack of hydroxyl ion at the amide carbonyl group. The free amino acid is then displaced from the enzyme, aided by its greater affinity for hydrogen ion (as compared to the substrate). Because of the lack of additional data, several points, including the mechanism of ester hydrolysis, were left unexplained by the tentative proposal. It is perhaps worthy of mention that leucine aminopeptidase was among the first enzymes for which a possible role for metal ions in hydrolytic reactions was proposed (*58*) and for which hydrophobic interaction between side chains of substrates and proteins was implicated (*56–58*).

There have been only a few more recent studies which have contributed directly to further understanding of the mechanism of action of leucine aminopeptidase. Bryce and Rabin (*18*, *47*) made the first kinetic analysis of the hydrolysis of various substrates by the swine kidney enzyme. On the basis of these and other experiments, they proposed that the metal ion (complexed to the enzyme) interacts with the uncharged α-amino group and the carbonyl oxygen of the amide or the ester, as previously suggested by Klotz and Ming (*60*). An adduct is then formed by an attack of water (bound to a basic group on the protein, possibly an imidazole group) on the carbonyl carbon to give a tetrahedral structure, as compared to a planar configuration in the unbound peptide. The decomposition of the adduct may then be pictured as occurring with the assistance of an acidic group (possibly the same imidazole group which was protonated by water) to give the NH_2-terminal residue, bound to the protein and the metal ion, and the remainder of the substrate bound to the protein. Dissociation of the products is caused by the higher pK of the amino acid (which favors protonation of the amino group and

60. I. M. Klotz and W.-C. L. Ming, *JACS* **76**, 805 (1954).

the breakdown of the metal ion complex) and by the lack of mutual stabilization of the enzyme–substrate complex by both of the side chains in the original substrate. In addition to its greater detail, this proposal differs from the earlier view of Smith and Spackman (*49*) mainly in the nature of the second site of interaction of the metal ion and also the group which initiates attack of the bound substrate. Both proposals include an interaction of the metal ion with the uncharged α-amino group, but the scheme of Bryce and Rabin suggests an interaction of the metal ion with the carbonyl oxygen of the amide or ester groups, whereas the Smith and Spackman scheme indicates an interaction with the amide nitrogen and does not indicate what the second interaction might be with an ester substrate. Also, in the later scheme the attack on the bound substrate is initiated by enzyme-bound water, whereas in the earlier view the initial attack is by hydroxyl ion.

In the studies on the mechanism of action of carboxypeptidase A, which is also a metallo-exopeptidase containing zinc, a number of proposals on the mechanism of action have been presented (Lipscomb and Hartsuck, Chapter 1, this volume), some of which are similar to those presented above. However, one proposal (*61*) includes a very different role for the function of the metal ion, namely, as the binding site for water. Indeed, an attractive proposal is that metallo-exopeptidases, in general, have mechanisms of action which involve binding of water to the metal ion. The metal-bound water might then be pictured as the initiator of attack on the substrate or as being involved at a later step in the mechanism such as deacylation. Since the ligands to the zinc in carboxypeptidase A have been shown to be one glutamyl and two histidinyl residues (*62, 63*), this leaves one coordination position available for binding either water or substrate, but not both at one time. Therefore, these proposals are mutually exclusive for carboxypeptidase; and if there are also three protein ligands in leucine aminopeptidase, the situation would be similar. This would involve protein side chain interactions at the sensitive bond and limit the role of the metal ion. Studies on the structure and active site of the aminopeptidase are obviously needed before additional information concerning the mechanism of action is forthcoming. At the moment, the large size of the enzyme and the limited quantities that can readily be prepared have inhibited such studies.

61. B. L. Kaiser and E. T. Kaiser, *Proc. Natl. Acad. Sci. U. S.* **64,** 36 (1969).

62. W.N. Lipscomb, J. A. Hartsuck, G. N. Reeke, F. A. Quiocho, P. Bethge, M. L. Ludwig, T. A. Steitz, H. Muirhead, and J. C. Coppola, *Brookhaven Symp. Biol.* **21,** 24 (1968).

63. R. A. Bradshaw, L. H. Ericsson, K. A. Walsh, and H. Neurath, *Proc. Natl. Acad. Sci. U. S.* **63,** 1389 (1969).

G. Use in Sequence Studies

Leucine aminopeptidase is extensively used for the following types of studies in sequence analysis of proteins and peptides: (1) investigation of the sequence of amino acids at the NH_2-terminus of peptides and proteins (*9, 14a*); (2) hydrolysis of peptides in order to estimate the content of tryptophan, asparagine, glutamine (*9, 14a*) and other acid-labile derivatives, e.g., ϵ-*N*-acetyllysine (*64*); and (3) to show the distribution of amino acids in a peptide relative to the position of prolyl residues (*14a*). The enzyme has also been used to degrade a mixture of peptides leaving intact an acylated amino-terminal peptide (*65*), and, in conjunction with proteinases such as papain, to hydrolyze proteins essentially to a mixture of amino acids (*66*). These uses are discussed briefly below.

Although the Edman degradation is presently the method of choice for determining the NH_2-terminal sequence of amino acids in a peptide, the use of leucine aminopeptidase has been useful in verifying sequences and in establishing sequences where difficulties have been encountered with the Edman method. When used in this manner, it is critical that contaminating endopeptidases be inhibited or removed from the enzyme preparations. In the past this has been accomplished by use of DFP and iodoacetate, but according to Deftos and Potts (*31*) this is difficult, at least with some commercial preparations. Therefore, it seems advisable to verify that preparations are free of contaminating proteinases. Since the placement of residues in the sequence depends on the analysis of enzymic hydrolysates of the peptide at various times, hydrolytic reactions at positions in the sequence other than at the NH_2-terminus could lead to incorrect assignments of residues in the sequence.

Because the rate of release of various types of residues can vary by at least 1000-fold (see Section II,F), it is also necessary to determine the yield of each residue released at different times. To illustrate these points, for the sequence Ala–Leu–Leu– very short times of hydrolysis (and especially at high ratios of substrate to enzyme) would probably show the release of more alanine than leucine, but at longer times (and even at shorter times with low ratios of substrate to enzyme) more leucine than alanine might be released. This can lead to the incorrect assignment of sequence as Leu–Ala– or Leu(Ala, Leu). With sequences like Asp–

64. R. J. DeLange, D. M. Fambrough, E. L. Smith, and J. Bonner, *JBC* **244,** 319 (1969).
65. J. I. Harris and J. Hindley, *JMB* **3,** 117 (1961).
66. R. L. Hill and W. R. Schmidt, *JBC* **237,** 389 (1962).

Leu– it is often difficult to determine the order of residues since NH_2-terminal leucyl residues are released so much more rapidly than NH_2-terminal aspartyl residues.

Leucine aminopeptidase can be used for the complete hydrolysis of any peptide which does not contain prolyl or hydroxyprolyl residues. If combined with an enzyme, e.g., aminopeptidase P (Section IX) or preparations of prolidase which hydrolyze X–Pro– bonds (Section X), any peptide should theoretically be hydrolyzed completely. Such mild enzymic hydrolysis permits the quantitative determination of acid-labile amino acids and isolation of labile amino acid derivatives. In practice, however, complete hydrolysis has been difficult to achieve, partly because of slow hydrolysis at highly polar peptide sequences, and partly because of the occurrence of such structures as N-terminal pyrrolidonecarboxyl residues and the β or imide forms of aspartyl residues, which are not liberated by the enzyme. Nevertheless, most smaller peptides are usually readily hydrolyzed, and larger peptides or proteins can be hydrolyzed to small peptides with endopeptidases. The complete amino acid analysis of horse heart cytochrome c was determined in this manner (*67*).

Since leucine aminopeptidase does not hydrolyze X–Pro– bonds (*9, 14a*) the distribution of residues NH_2-terminal to such a dipeptide sequence may be readily determined. For example, in the peptide Leu–Ser–Ala–Phe–Gly–Pro–Val–Thr–Lys only the first four residues would be removed by the enzyme. This feature of the enzyme has been used to advantage in deciding which steps to employ in completing the sequence of peptides. With the above peptide, if the sequence Leu–Ser–Ala had been determined by the Edman degradation, the additional release of phenylalanine by leucine aminopeptidase might suggest that chymotrypsin is the enzyme of choice to cleave the peptide in order to establish the remainder of the sequence. For other examples involving the use of leucine aminopeptidase in protein sequence studies, see the review by Light (*14*a).

III. Aminopeptidase M

The usual methods of preparation of swine kidney leucine aminopeptidase include steps involving fractionation with acetone or alcohols (*9, 14a*). Pfleiderer and Celliers (*68*), using other methods of fractionation, showed that a major portion of the aminopeptidase activity in

67 E. Margoliash, J. R. Kimmel, R. L. Hill, and W. R. Schmidt, *JBC* **237,** 2148 (1962).

68. G. Pfleiderer and P. G. Celliers, *Biochem. Z.* **339,** 186 (1963).

swine kidney extracts was denatured by acetone or alcohol treatment. Whereas leucine aminopeptidase cleaved substrates with amino-terminal leucine most rapidly, the aminopeptidase which is denatured by acetone cleaved substrates with amino-terminal alanine most rapidly, and the rates for various amino-terminal residues varied by only a few fold (*69–71*) as compared to at least 1000-fold for leucine aminopeptidase (*9*). This new enzyme, aminopeptidase M, was isolated by Pfleiderer and Celliers (*68*) from the microsomal fraction from which it could be solubilized by treatment with toluene to swell the particles, and by trypsin. An improved method for preparation of the homogeneous enzyme has been described by Wachsmuth *et al.* (*70*).

A solution of 1 mg of aminopeptidase M per milliliter gives an absorbance of 1.63 at 280 nm with a 1-cm path (*70*). The enzyme is stable in 0.06 *M* phosphate buffer at pH 7.0 at temperatures up to 65° (at least for short times), and it is stable between pH 3.5 and pH 11 at room temperature (*70*). (We have found that frozen solutions of the enzyme are stable for many months.) The enzyme is not affected by sulfhydryl reagents, by divalent metal ions, by trypsin or by 6 *M* urea, but it is irreversibly denatured by alcohols or 0.5 *M* guanidinium chloride (*70*). The enzyme is unusual in that it cannot be precipitated with tricholoroacetic acid.

Aminopeptidase M has a molecular weight of about 280,000 (*72*) and is apparently composed of 10 subunits (molecular weight of 28,000 ± 3000) of two different types (*73*). The enzyme molecule contains five disulfide bridges, each of which apparently connects two subunits, since the molecular weight in 0.5 *M* guanidine at pH 2.0 is 60,000 before reduction and 30,000 after reduction (*73*). The amino acid composition has been reported (*73*). By the use of $^{131}I_2$, Wachsmuth (*74*) showed that monoiodination of 10 tyrosyl residues per enzyme molecule resulted in inactivation of the enzyme. Hydrophobic substrates reduce the rate of iodination more than hydrophilic substrates do.

At the substrate concentrations generally used in sequence work, the pH optimum is between 7.0 and 7.5, but at high substrate concentration, the pH optimum approaches 9.0 (*70*). This is because the K_m is lowest at pH 7.0–7.3 and reaches a maximum at about pH 9.0, whereas the V_{max}

69. G. Pfleiderer, P. G. Celliers, M. Stanulovic, E. D. Wachsmuth, H. Determann, and G. Braunitzer, *Biochem. Z.* **340,** 552 (1964).
70. E. D. Wachsmuth, I. Fritze, and G. Pfleiderer, *Biochemistry* **5,** 169, 175 (1966).
71. E. D. Wachsmuth, *Biochem. Z.* **344,** 361 (1966).
72. F. Auricchio and C. B. Bruni, *Biochem. Z.* **340,** 321 (1964).
73. E. D. Wachsmuth, *Biochem. Z.* **346,** 467 (1967).
74. E. D. Wachsmuth, *Biochem. Z.* **346,** 446 (1967).

is highest at pH 9.0 and considerably lower at pH 7.0–7.3 (*70*). The best preparation of the enzyme obtained had a turnover number of 10,450 moles of leucine *p*-nitroanilide per mole of aminopeptidase at 37° in 0.006 *M* phosphate buffer, pH 7.2 (*70*).

Aminopeptidase M can completely hydrolyze all peptides bearing a free α-amino or α-imino group and containing only L-amino acids (*68–71*). Therefore, it appears to be an iminopeptidase as well as an aminopeptidase, although the iminopeptidase activity could conceivably result from a contaminating enzyme. The rates of hydrolysis of substrates with free α-amino groups vary only by a factor of 3 at pH 7; however, substrates with a free α-imino group (e.g., N-terminal proline or sarcosine) are hydrolyzed much more slowly. Compounds with α-*N*-acylated residues or with N-terminal D-amino acids are not hydrolyzed, and no activity could be detected (by release of ammonia) on asparagine, glutamine, β-alaninamide, or γ-butyramide (*70*).

Our experience in using commercial preparations of the enzyme for many types of sequence studies can be summarized as follows.

(a) Generally, most smaller peptides are hydrolyzed completely when incubated for 16–24 hr in 0.1 *M* NH_4HCO_3 at pH 7.5 at 40° with 100–200 milliequivalent units of the enzyme for up to 0.1 μmole of peptide. There is no detectable release of amino acids from the enzyme preparation.

(b) Compositions of peptides containing tryptophan, amides, and other labile components are readily obtained by hydrolysis of the peptides with aminopeptidase M; however, it is also essential to analyze acid hydrolysates of the peptides in order to avoid difficulties of the type given below.

(c) The major factors preventing complete hydrolysis are: (1) The presence of aspartyl residues in the β or imide forms (*75*). In several peptides containing aspartic acid, hydrolysis of all residues amino-terminal to the aspartyl residue(s) was complete, whereas removal of aspartic acid and all residues carboxyl-terminal to it was incomplete but at the same level. (2) Insolubility of the original peptide or of a partially degraded fragment of the original peptide. (3) Cyclization of glutaminyl residues to pyrrolidonecarboxyl residues. We have encountered little difficulty in this regard except when the glutaminyl residue is the amino-terminal residue in the unhydrolyzed peptide. However, analytical values for glutamine are often low because of cyclization after hydrolysis of the peptide. (4) The presence of bulky substitutions, such as carbohydrate, on amino acid side chains. (5) The presence of prolyl residues. Although prolyl residues are removed from peptides by aminopeptidase

75. D. L. Swallow and E. P. Abraham, *BJ* **70**, 364 (1958).

M at a slow rate, it appears that a secondary reaction can occur. In several peptides containing an X–Pro sequence [Leu–Pro, Met(sulfone)–Pro, etc.] all residues, whether amino-terminal or carboxyl-terminal to the X–Pro sequence, were quantitatively released except for proline and X. This same behavior has been observed by Plummer (*76*) for a peptide (–Tyr(carboxymethyl)–Pro–) from carboxypeptidase B. The problem seems to arise primarily when a bulky hydrophobic residue (Leu, Tyr, or Trp) is adjacent to the prolyl residue, since with other peptides (e.g., –Lys–Pro–) this has not been a problem. In the case of the peptide from carboxypeptidase B, the dipeptide *O*-carboxymethyl-Tyr–Pro was isolated from the aminopeptidase M hydrolyzate (*76*). Whether this type of reaction results from aminopeptidase M itself or from contaminating endopeptidases and carboxypeptidase-like enzymes is unclear at the present time. We have obtained evidence of contaminating enzymes in commercial preparations of aminopeptidase M since the peptide Asp–Asn–Ile–Glu–Gly–Ile–Thr–Lys–Pro–Ala–Ile–Arg (*64*) yielded only isoleucine and arginine after 20 hr of hydrolysis with aminopeptidase M. This was presumably because the N-terminal aspartyl residue was in the β or imide form and because of the presence of a carboxypeptidase. After one step of the Edman degradation to remove the aspartyl residue (at least partially), complete hydrolysis of the remainder peptide was obtained. In view of these findings, commercial preparations of aminopeptidase M should be used cautiously for studies of sequential removal of residues from the N-terminus, unless it is certain that other enzymes are absent.

IV. Dipeptidyltransferase (Dipeptidyl Aminopeptidase I, Cathepsin C)

An unusual N-terminal exopeptidase, which catalyzes the cleavage of the bond involving the carbonyl group of the penultimate residue of appropriate peptide compounds, was first observed in swine kidney extracts by Gutmann and Fruton (*77*). Partially purified preparations were obtained from beef spleen by Tallan *et al.* (*78*), and in these and later studies (*79*) the enzyme was called cathespin C. Since the enzyme can liberate dipeptides and since a free α-amino or α-imino group on the substrate is required, the enzyme may be classified as an aminopeptidase (and

76. T. H. Plummer, *JBC* **244,** 5246 (1969).
77. H. R. Gutmann and J. S. Fruton, *JBC* **174,** 851 (1948).
78. H. H. Tallan, M. E. Jones, and J. S. Fruton, *JBC* **194,** 793 (1952).
79. J. S. Fruton, "The Enzymes," 2nd ed., Vol. 4, Part A, p. 233, 1960.

iminopeptidase), and the name dipeptidyl aminopeptidase I has been used (*80*). Since the enzyme can also catalyze transfer of the dipeptidyl moiety to acceptor amine compounds, it has also been called dipeptidyltransferase (*81*). From the work of McDonald *et al.* (*80, 82*) the dipeptidyltransferase (as we shall call it in this review) appears to be identical to "glucagon-degrading enzyme" (*83*) and "glucagonase" (*84*).

Highly purified preparations of dipeptidyltransferase have been obtained from bovine spleen (*81, 85, 85a*) and rat liver (*80*). The molecular weight of the spleen enzyme has been reported as 210,000 (9.55 S) (*81, 85*) and 197,000 ± 10,000 (*85a*). Metrione *et al.* obtained evidence for the presence of 8 subunits, possibly of 2 different types and probably arranged in the form of 2 tetramers in the enzyme. The possibility of different types of subunits is based on the finding of both aspartic acid (5.6 S) and leucine (2.1 S) by NH_2-terminal determinations. The bovine enzyme is an SH enzyme (*86, 87*), possibly contains carbohydrate (*81*), and has a halide ion (or nitrate or nitrite) requirement with chloride ions being most effective (*88*). The rat enzyme (*80*) has an estimated molecular weight of 200,000 and is apparently composed of 2 types of subunits, one of which is apparently three times larger than the other as judged by migration in gels containing sodium dodecyl sulfate. The possibility that the larger material represented undissociated or partially dissociated material was apparently not investigated. The rat enzyme (*80*) also requires halide ions for activity with the following order of effectiveness established at concentrations of 4 m*M* or higher: $Cl^- > Br^- > I^- > NO_3^- > NO_2^-$. Fluoride ions had little or no effect, and at lower concentrations the order of effectiveness of the various ions is changed. The rat liver enzyme (*80*) is also an SH enzyme, and activation of the enzyme requires the presence of both thiol compounds and halide ions.

80. J. K. McDonald, B. B. Zeitman, T. J. Reilly, and S. Ellis, *JBC* **244,** 2693 (1969).

81. R. M. Metrione, A. G. Neves, and J. S. Fruton, *Biochemistry* **5,** 1597 (1966).

82. J. K. McDonald, P. X. Callahan, B. B. Zeitman, and S. Ellis, *JBC* **244,** 6199 (1969).

83. S. Kakiuchi and H. H. Tomizawa, *JBC* **239,** 2160 (1964).

84. R. H. Williams, *in* "Textbook of Endocrinology" (R. H. Williams, ed.), 4th ed., pp. 630 and 670. Saunders, Philadelphia, Pennsylvania, 1968.

85. R. J. Planta and M. Gruber, *BBA* **89,** 503 (1964).

85a. R. M. Metrione, Y. Okuda, and G. F. Fairclough, Jr., *Biochemistry* **9,** 2427 (1970).

86. J. S. Fruton and M. J. Mycek, *ABB* **65,** 11 (1956).

87. I. M. Voynick and J. S. Fruton, *Biochemistry* **7,** 40 (1968).

88. J. K. McDonald, T. J. Reilly, B. B. Zeitman, and S. Ellis, *BBRC* **24,** 771 (1966).

In the early studies by Fruton and his collaborators (*79*) the specificity of dipeptidyltransferase in removing dipeptidyl units rather than amino acids from the substrates was clearly established. Ester and amide derivatives of a large variety of dipeptides were cleaved, but at widely varying rates. Of the compounds tested, those with aromatic residues in the penultimate position and with glycyl residues or other residues with small side chains in the N-terminal position were the best substrates. Thus, glycyl-L-phenylalanyl ethyl ester (or amide) was one of the best substrates found. However, none of these studies included compounds with penultimate lysyl or arginyl residues. Recently, McDonald *et al.* (*80*) have used such compounds as substrates and found that they are hydrolyzed 10–20 times faster than the corresponding compounds with aromatic residues in the penultimate position, thus making compounds like Gly-L-Arg-β-naphthyl amide and Gly-L-Lys-OEt among the best substrates found thus far. The enzyme requires either an α-amino or α-imino group for activity. Either proline or sarcosine can provide the α-imino group although the bulky α-*N*-methyl group of sarcosine greatly decreases the rate of hydrolysis (*87*). The α-*N*-dimethyl derivatives have not been tested, but compounds with α-*N*-acylated residues are not hydrolyzed (*80, 89, 90*). Compounds with N-terminal arginyl or lysyl residues or penultimate prolyl residues are not cleaved under the conditions of testing (*80, 90, 91*). Several tripeptides are hydrolyzed, although generally more slowly than similar tetrapeptides (*80*). Trialanine is one of the tripeptides not hydrolyzed. In addition to the small peptide substrates, polypeptides such as the 29-residue glucagon molecule (*82*), the 27-residue secretion molecule (*82*), the 39-residue β-corticotropin molecule, and the 30-residue B chain of oxidized insulin (*80*) were degraded by removal of 8, 10, 5, and 13 dipeptides, respectively. The continued degradation of glucagon and secretion was terminated by the appearance of an N-terminal arginyl residue, and of β-corticotropin and the B chain of oxidized insulin by the presence of a penultimate prolyl residue. Activity with denatured proteins as substrates has not been found for highly purified, well-characterized preparations of the enzyme. Tables III and IV summarize some of the studies on the kinetic parameters and specificity of hydrolysis of various substrates. Although all of the studies were not performed under identical conditions of pH, buffer, etc., the general features of the hydrolytic cleavage of such compounds appear to be clear.

Dipeptidyltransferase probably forms a dipeptidyl–enzyme intermedi-

89. D. S. Wiggans, M. Winitz, and J. S. Fruton, *Yale J. Biol. Med.* **27**, 11 (1954).
90. R. J. Planta, J. Gorter, and M. Gruber, *BBA* **89**, 511 (1964).
91. C. P. Heinrich and J. S. Fruton, *Biochemistry* **7**, 3556 (1968).

TABLE III
KINETIC PARAMETERS FOR THE HYDROLYSIS OF VARIOUS SUBSTRATES BY BOVINE SPLEEN AND RAT LIVER DIPEPTIDYLTRANSFERASE

Substrate	$K_{m\ app}$(mM)	k_{cat}(sec^{-1})	$k_{cat}/K_{m\ app}$ (mM^{-1} sec^{-1})
	Bovine Spleen Enzyme[a]		
Gly-L-Trp-OMe	1.5 ± 0.1	293 ± 13	195
Gly-L-Tyr-OEt	0.63 ± 0.05	90 ± 5	143
Gly-L-Phe-OEt	1.5 ± 0.1	98 ± 6	65
Gly-L-Phe-OMe	1.3 ± 0.1	61 ± 4	47
Gly-L-Leu-OEt	4.0 ± 0.2	126 ± 9	30
Gly-Gly-OEt	5.3 ± 0.4	74 ± 6	14
Gly-Pla[b]-OMe	19 ± 1	210 ± 12	11
Sar[b]-L-Phe-OEt	61 ± 4	157 ± 11	2.6
Gly-N-methyl-Phe-OEt[c]		Not hydrolyzed	
Diazoacetyl-Gly-OEt		Not hydrolyzed	
	Rat Liver Enzyme[d]		
Gly-L-Lys-OMe	1.6 ± 0.18	1125 ± 49	705
Gly-L-Phe-OMe	0.43 ± 0.02	282 ± 2.5	655
L-Ser-L-Met-OMe	1.1 ± 0.09	268 ± 6.2	245
L-Ala-L-Ala-OMe	1.4 ± 0.06	178 ± 2.7	127
Gly-L-Arg-βNA[e]	0.10 ± 0.006	1300 ± 30	13,000
L-Ser-L-Met-βNA	0.17 ± 0.03	510 ± 39	3,000
L-Ala-L-Ala-βNA	0.19 ± 0.01	248 ± 9	1,300
Gly-L-Phe-βNA	0.17 ± 0.01	79 ± 3	465
Gly-L-Arg-NH_2	30 ± 6.5	1160 ± 146	39
L-Ala-L-Ala-NH_2	51 ± 13	1160 ± 187	23
Gly-L-Phe-NH_2	18 ± 2.3	263 ± 14	14

[a] From Voynick and Fruton (*87*).
[b] Pla = β-phenyl-L-lactic acid; Sar = sarcosine.
[c] From Izumiya and Fruton (*91a*).
[d] From McDonald *et al.* (*80*).
[e] βNA = β-naphthyl amide.

ate which can undergo hydrolysis to the free dipeptide (plus enzyme) or can be deacylated by an appropriate acceptor molecule (*91*, *92*). The hydrolytic reaction predominates at acidic pH (e.g., pH 5.0) but usually represents only a fraction of the catalytic activity at alkaline pH (e.g., pH 7.5). The only acceptors which have been described are amines, and since the substrate itself is an amine, polymerization of the substrate is one possible reaction at alkaline pH. Other suitable amines, if present, inhibit the polymerization reaction by acting as acceptors to form the N-dipeptidyl amide derivatives.

91a. N. Izumiya and J. S. Fruton, *JBC* **218**, 59 (1956).
92. M. E. Jones, W. R. Hearn, M. Fried, and J. S. Fruton, *JBC* **195**, 645 (1952).

TABLE IV
RATES OF HYDROLYSIS OF β-NAPHTHYL AMIDE DERIVATIVES OF VARIOUS DIPEPTIDES BY RAT LIVER DIPEPTIDYLTRANSFERASE[a]

Substrate (-β-napthyl amide)	Activity	
	Specific (μmole min^{-1} mg^{-1} protein)	Relative (%)
Gly-Arg-	300	100
Ala-Arg-	215	72
Pro-Arg-	164	55
Ser-Met-	100	33
Ala-Ala-	46	15
Glu-His-	33	11
Ser-Tyr-	28.0	9.3
Phe-Arg-	23.5	7.9
His-Phe-	18.5	6.1
Gly-Phe-	17.0	5.6
His-Ser-	16.0	5.3
Leu-Ala-	10.3	3.4
Asp-Ala-	1.3	0.4
Lys-Ala-	0	0
Arg-Ala-	0	0
Gly-Pro-	0	0
Acetyl-Gly-Lys-	0	0

[a] From McDonald *et al.* (*80*).

The kinetics of the polymerization reaction are complex since hydrolysis and polymerization reactions occur simultaneously, even at alkaline pH, and since soluble products of polymerization also serve as substrates for the enzyme. An additional complication is the insolubility of some of the products, which often precipitate at the hexapeptide or octapeptide level during the reaction (*91*, *93*). Despite this, on the basis of kinetic studies over a 4-fold range of enzyme concentrations and a 10-fold range of substrate concentrations, Heinrich and Fruton (*91*) obtained evidence for the existence of a dipeptidyl–enzyme intermediate since the fraction of dipeptidyl units being converted to free dipeptide *and* tetrapeptide amide (using a dipeptidyl amide as substrate) remains relatively constant with increasing substrate concentration and at constant enzyme concentration. The proportion of dipeptidyl units appearing as free dipeptide decreases under these conditions to approximately 10% of the total reacted dipeptidyl units; this demonstrates the competition between water and the substrate for the dipeptidyl moiety on the enzyme. These investigators further propose that the polymerization reaction involves

93. K. K. Nilsson and J. S. Fruton, *Biochemistry* 3, 1220 (1964).

oligopeptidyl–enzyme intermediates, since the formation of higher oligopeptides such as hexapeptide amide or octapeptide amide increases more rapidly with increasing enzyme concentration (at constant substrate concentration) than the other reactions catalyzed by the enzyme such as hydrolysis or formation of the tetrapeptide amide. This suggests that chain elongation in polymerization reactions may involve cooperative interaction of different enzyme molecules or subunits, one of which bears the dipeptidyl moiety and the other a dipeptidyl or oligopeptidyl moiety. These observations are summarized in the following equations, where X and Y are amino acid residues, "a" represents amide, and E (or E′) is an enzyme unit (either a molecule or a subunit) (*91*):

Initiation:	$2\,XYa + E + E' \rightarrow XY\text{—}E + XY\text{—}E' + 2\,a$
Polymerization:	$XY\text{—}E + XY\text{—}E' \rightarrow (XY)_2\text{—}E + E'$
	$(XY)_n\text{—}E + XY\text{—}E' \rightarrow (XY)_{n+1}\text{—}E + E'$
Termination:	$(XY)_n\text{—}E + H_2O \rightarrow (XY)_n + E$ (hydrolysis)
	$(XY)_n\text{—}E + XYa \rightarrow (XY)a_{n+1} + E$ (transamidation)

Further investigation will be necessary to show whether this scheme applies under all circumstances. The presence of subunits in the enzyme, and possibly two types of subunits (see above), provides support for this type of mechanism.

Other enzymes similar to dipeptidyltransferase have also been described. Three enzymes from bovine anterior pituitary catalyze removal of dipeptidyl units from dipeptidyl-β-naphthyl amide substrates (*94–97*). Dipeptidyl arylamidase I (*94*) appears to be identical to dipeptidyltransferase in every way that was examined and could partially degrade adrenocorticotropic hormone. Dipeptidyl arylamidase II (dipeptidyl aminopeptidase II) (*95, 96*) has a complementary activity to dipeptidyltransferase, since it removes dipeptidyl units from compounds with N-terminal lysyl or arginyl residues such as L-Lys-L-Ala-β-naphthylamide (or methyl ester). Such compounds are not cleaved by dipeptidyltransferase (see above). The N-terminal basic residue is not a requirement, however, since a variety of other dipeptidyl substrates are hydrolyzed. In addition tripeptides (including trialanine, which is not hydrolyzed by dipeptidyltransferase) are substrates and must contain both free α-amino and free α-carboxyl groups. Thus, the enzyme displays a "carboxytripeptidase" activity. Dipeptides, tripeptide esters, and tetrapep-

94. J. K. McDonald, S. Ellis, and T. J. Reilly, *JBC* **241,** 1494 (1966).

95. J. K. McDonald, T. J. Reilly, B. B. Zeitman, and S. Ellis, *JBC* **243,** 2028 (1968).

96. J. K. McDonald, F. H. Leibach, R. E. Grindeland, and S. Ellis, *JBC* **243,** 4143 (1968).

97. S. Ellis and J. M. Nuenke, *JBC* **242,** 4623 (1967).

tides are not substrates, and it is therefore unlikely that longer peptides are substrates. The enzyme is apparently neither an SH enzyme nor a metallo-enzyme, since neither sulfhydryl reagents nor EDTA are inhibitory.

Dipeptidyl arylamidase III (*97*) removes dipeptidyl units from a variety of substrates which contain a minimum of four amino acid residues. For example, tetralysine (not hydrolyzed by dipeptidyltransferase), tetraphenylalanine, and tetra- and hexaalalanine are hydrolyzed to dipeptides. Dipeptides, tripeptides, some tetrapeptides (tetraglycine and tetraglutamic acid) and most dipeptidyl-β-naphthylamides are not substrates. The only β-naphthyl amide derivative hydrolyzed was Arg-Arg-β-naphthyl amide, and it thus serves as a rather specific substrate. The enzyme is inhibited by thiol reagents and activated by thiol compounds.

An enzyme which removes the dipeptide Gly–Pro from substrates like Gly–Pro-β-naphthyl amide, Gly–Pro–Gly–Gly, and Gly–Pro–Ala, none of which is hydrolyzed by dipeptidyltransferase, has been obtained from rat liver microsomes (*98*). Other substances such as Leu–Leu-β-naphthyl amide and Ala–Ala-β-naphthyl amide are also hydrolyzed but at slower rates; this action may result from a contaminating enzyme since the preparation is not homogeneous. The enzyme requires a free α-amino group on the substrate and is unaffected by EDTA, divalent metal ions, or sulfhydryl reagents.

The function of dipeptidyltransferase and these other related enzymes is presently unknown. Possibly, they are involved in the degradation of peptide hormones and other peptides, since at least dipeptidyltransferase has proved to be capable of such activity. It is also possible, however, that they have other functions, e.g., in synthetic reactions involving polymerization or transamidation. Until the function of these enzymes is elucidated, the primary interest in them will undoubtedly be in their potential as degradative tools for use in sequence analysis. Clearly, from the work of Callahan *et al.* (*99*), dipeptidyltransferase may prove useful in this respect.

V. Aminopeptidase A

An enzyme which hydrolyzes *N*-(α-L-glutamyl) β-naphthyl amide, *N*-(α-L-aspartyl) β-naphthyl amide, α-L-glutamyl-L-phenylalanine, and α-L-

98. V. K. Hopsu-Havu and S. R. Sarimo, *Z. Physiol. Chem.* **348**, 1540 (1967).
99. P. X. Callahan, J. K. McDonald, and S. Ellis, *Federation Proc.* **28**, 661 (1969).

aspartyl-L-arginine has been partially purified from rat kidney and called aminopeptidase A (acid α-aminopeptidase) (*100*). Hydrolysis was not observed with N-(α-D-glutamyl) β-naphthyl amide, N-(N-α-benzoyl-α-L-glutamyl) β-naphthyl amide, N-carbobenzoxy-α-L-glutamyl-L-phenylalanine, or β-naphthyl amide derivatives of other amino acids. This suggests that the enzyme is specific for amino-terminal acidic residues of the L configuration bearing a free α-amino group. The enzyme is optimally activated by 0.01 M $CaCl_2$ and in the presence of 0.01 M EDTA it has only 5% of the maximal activity obtained in the presence of Ca^{2+}.

Similar enzymic activity has been found in extracts of swine kidney and duodenum (*100*) and in human serum (*101, 102*). The serum enzyme removes the amino-terminal aspartyl residue from the octapeptide hormone, angiotensin II, apparently as the first step in the degradation of this vasopressor hormone. However, until it has been demonstrated that the enzymes from different sources are similar and that they have no endopeptidase activity, classification as N-terminal exopeptidases must be considered tentative.

VI. Aminopeptidase B

Aminopeptidase B is an N-terminal exopeptidase specific for the liberation of NH_2-terminal lysyl and arginyl residues of peptide substrates. It has been found in extracts of several organs of the rat and other species (*103–106*) and has been purified to near homogeneity from rat liver (*104*). Of the many compounds tested only the following were hydrolyzed (*104*): L-Arg-β-naphthyl amide, L-Lys-β-naphthyl amide, L-Arg-L-Val, D-Arg-D-Val, L-Arg-L-Leu, D-Arg-L-Phe, L-Lys-L-Leu, L-Lys-L-Phe, L-Lys-L-Ala, L-Lys-L-Val, L-Lys-L-Lys, Gly-L-Lys, poly-L-Lys, and poly-D,L-Lys. Ester substrates, Arg–NH_2, a variety of other di- and tripeptides, and β-naphthyl amide derivatives of other amino acids were not

100. G. G. Glenner, P. J. McMillan, and J. E. Folk, *Nature* **194,** 867 (1962).

101. I. Nagatsu, L. Gillespie, J. E. Folk, and G. G. Glenner, *Biochem. Pharmacol.* **14,** 721 (1965).

102. I. Nagatsu, L. Gillespie, J. M. George, J. E. Folk, and G. G. Glenner, *Biochem. Pharmacol.* **14,** 853 (1965).

103. V. K. Hopsu, U.-M. Kantonen, and G. G. Glenner, *Life Sci.* **3,** 1449 (1964).

104. V. K. Hopsu, K. K. Mäkinen, and G. G. Glenner, *ABB* **114,** 557 and 567 (1966).

105. V. K. Hopsu-Havu, K. K. Mäkinen, and G. G. Glenner, *Nature* **212,** 1271 (1966).

106. K. K. Mäkinen, *ABB* **126,** 803 (1968).

hydrolyzed. With one exception (Gly-L-Lys, hydrolysis unexplained) all substrates required a lysyl or arginyl residue with a free α-amino group in the amino-terminal position. Of interest is the apparent lack of stereospecificity since both D and L compounds were hydrolyzed.

Aminopeptidase B from rat liver has a molecular weight of about 95,000 (*104*), is activated by chloride ions and thiols (*104, 106*), and is inhibited by metal chelating agents (*104*), sulfhydryl reagents (*104, 106*), L-1-toluensulfonyl-L-phenylalanyl chloromethyl ketone, phenylmethane sulfonyl fluoride, and diphenylcarbamyl chloride (*106*). From these and other experiments it is postulated that both histidine and cysteine are functional residues in the active site (*106*). Further work will be required to verify this point and also to exclude the possibility of endopeptidase activity for the enzyme.

Aminopeptidase B releases the amino-terminal lysyl residue from the decapeptide kallidin-10 to form the nonapeptide bradykinin (*105*). Although bradykinin has an amino-terminal arginyl residue, it is not released, probably because it is adjacent to a prolyl residue. It has been shown for several proteolytic enzymes that bonds adjacent to proline are not hydrolyzed or are hydrolyzed at a very slow rate. Because of the action of aminopeptidase B on kallidin-10 *in vitro*, it has been postulated that aminopeptidase B may also be involved *in vivo* in the formation of bradykinin (*105*). According to this concept aminopeptidase B inside the cell (inactive because of the low chloride ion concentration) is released during tissue damage into the tissue fluid in which the higher chloride concentration activates the enzyme; there it can catalyze the formation of bradykinin from kallidin-10, which is formed in turn from serum α_2-globulin by other proteolytic enzymes. Thus, aminopeptidase B might be part of the "converting enzyme" implicated in the formation of bradykinin (*107, 108*).

VII. Pyrrolidonyl Peptidase

Pyrrolidonyl peptidase may be classified as an N-terminal exopeptidase, and, as such, is the only one known which removes residues specifically from an "N-terminus" lacking a free α-amino or α-imino group. The enzyme was purified about 100-fold by Doolittle and Armen-

107. M. E. Webster and J. V. Pierce, *Ann. N. Y. Acad. Sci.* **104,** 91 (1963).
108. E. Werle and J. Trautschold, *Ann. N. Y. Acad. Sci.* **104,** 117 (1963).

trout (*109*) from a strain of *Pseudomonas fluorescens* and later (530-fold) by Szewczuk and Mulczyk (*110*) from a strain of *Bacillus subtilis*. The enzyme from *Pseudomonas* was shown to remove pyrrolidonecarboxylic acid (Pyr) from substrates like Pyr-L-Ala, Pyr-L-Val, and fibrinopeptide B of a number of species (*109*). In the fibrinopeptides, bonds such as Pyr–His–, Pyr–Ser–, Pyr–Pro–, and Pyr–Phe– were hydrolyzed, but no bonds of other types were hydrolyzed as judged from an examination of the pyrrolidonyl peptidase hydrolyzate of the 21-residue bovine fibrinopeptide B. The *Bacillus* enzyme removed only pyrrolidonecarboxylic acid from substrates like L-pyrrolidonyl-β-naphthyl amide (the α-naphthyl amide was hydrolyzed very slowly, if at all), L-pyrrolidonylanilide, L-pyrrolidonyl-glutamic diethyl ester, and seromucoid. Other types of peptidase activity could not be detected in these preparations.

Uliana and Doolittle (*110a*) studied the effect of the penultimate residue on the rates of hydrolysis of various substrates. Pyr-L-Ala was most rapidly hydrolyzed and no hydrolysis of Pyr-L-Pro was detected.

The *Pseudomonas* enzyme was unstable in the most purified form and was inhibited by *p*-mercuriphenyl sulfonate or iodoacetamide. The *Bacillus* enzyme was stable in its purified form, was most active in the presence of mercaptoethanol and EDTA, and was inactivated by Hg^{2+}, Cu^{2+}, iodoacetate, iodoacetamide, or by shaking in air. L-Pyrrolidonecarboxylic acid and L-pyrrolidonyl-α-naphthyl amide were competitive inhibitors in the hydrolysis of L-pyrrolidonyl-β-naphthyl amide. From these studies and others it was concluded that pyrrolidonyl peptidase is a thiol enzyme specific for the removal of pyrrolidonecarboxylic acid from peptide substrates.

Jackson and Hirs (*111*) have used the enzyme from *P. fluorescens* to remove pyrrolidonecarboxylic acid from the peptide Pyr–His–Met–Asp–Pro–Asp. The peptide product with an amino-terminal histidyl residue was then sequenced by conventional methods.

Pyrrolidonyl peptidase has been detected in rat liver (*111a*) and in a number of animal, plant, and human tissues, as well as in a variety of bacteria (*111b*). Of interest is an aminopeptidase from *B. stearothermophilus* which is also able to deformylate α-formylmethioninyl peptides (*111c*).

109. R. F. Doolittle and R. W. Armentrout, *Biochemistry* **7,** 516 (1968).
110. A. Szewczuk and M. Mulczk, *European J. Biochem.* **8,** 63 (1969).
110a. J. A. Uliana and R. F. Doolittle, *ABB* **131,** 561 (1969).
111. R. L. Jackson and C. H. W. Hirs, *JBC* **245,** 624 (1970).
111a. R. W. Armentrout, *BBA* **191,** 756 (1969).
111b. A. Szewczuk and J. Kwiatkowska, *European J. Biochem.* **15,** 92 (1970).
111c. P. Moser, G. Roncari, and H. Zuber, *Intern. J. Protein Res.* **2,** 191 (1970).

VIII. Proline Iminopeptidase

An enzyme which liberates only amino-terminal prolyl residues from peptide substrates has been purified from *Escherichia coli* by Sarid *et al.* (*112, 113*). The enzyme was also detected in extracts of swine kidney and in other organisms (*113*). The enzyme has been reported to be specific for amino-terminal prolyl residues, and even amino-terminal hydroxyprolyl residues are not released. Tested substrates included Pro–Gly, Pro–Gly–Gly, polyproline, and salmine.

The iminopeptidase requires Mn^{2+} (Mg^{2+} cannot substitute) for activity and is inhibited by Ca^{2+}, Co^{2+}, Zn^{2+}, Cu^{2+}, *p*-mercuribenzoate and iodoacetamide (*113*). Methods have been described for detecting the enzyme in the presence of other enzymes acting on substrates containing proline; these methods depend on the hydrolysis of polyproline as substrate (*113*). However, since aminopeptidase P (see Section IX) can also degrade this substrate, the specificity is not unique.

IX. Aminopeptidase P

Yaron and Mlynar (*114*) have isolated an enzyme in homogeneous form from *Escherichia coli B* which releases amino-terminal residues from peptide substrates only when they are adjacent to prolyl residues. Thus, substrates include such compounds as Gly–Pro–Gly, Pro–Pro–Ala, Gly–Pro, Ala–Pro, Val–Pro, Pro–Pro, Pro–Pro–Ala–OMe, poly-L-proline, carboxymethylated papain (amino-terminal sequence Ile–Pro–) and bradykinin (amino-terminal sequence Arg–Pro–Pro–). Poly-L-hydroxyproline is not hydrolyzed which suggests that hydroxyproline cannot substitute for proline. *N*-Dinitrophenyl-polyproline is not hydrolyzed which indicates that an α-amino or α-imino group is essential for activity. The enzyme has a molecular weight of 230,000, requires Mn^{2+} for activity, and is inhibited by heavy metals and EDTA.

In some respects aminopeptidase P resembles prolidase of swine kidney (see Section X) but appears to differ mainly in that prolidase acts on compounds containing hydroxyproline residues, albeit more slowly.

In order to obtain complete hydrolysis of a peptide containing proline, aminopeptidase M can be used (see Section III) or, alternatively, leucine

112. S. Sarid, A. Berger, and E. Katchalski, *JBC* **234,** 1740 (1959).
113. S. Sarid, A. Berger, and E. Katchalski, *JBC* **237,** 2207 (1962).
114. A. Yaron and D. Mlynar, *BBRC* **32,** 658 (1968).

aminopeptidase can be used to remove all residues amino-terminal to the residue adjacent to the prolyl residue. Aminopeptidase P could then be used to remove the residue adjacent to proline, proline iminopeptidase could be used to remove the prolyl residue, and leucine aminopeptidase could complete the hydrolysis. The combination of leucine aminopeptidase with preparations of prolidase (see Section X) can also affect complete hydrolysis.

X. Dipeptidases

Glycylglycine dipeptidase, glycyl-L-leucine dipeptidase, iminodipeptidase (prolinase, hydrolyzes Pro–X), imidopeptidase (prolidase, hydrolyzes X–Pro), cysteinylglycinase, and carnosinase were all described in the previous review by Smith (*1*). Little in the way of further characterization of these enzymes has appeared. In the case of the last two enzymes mentioned it was not definitively established that these enzymes were dipeptidases.

Although prolidase was originally thought to be a dipeptidase specific for such peptides as X–Pro or X–Hypro, it was later shown by Hill and Schmidt (*66*) that preparations of the highly purified enzyme will slowly release an amino-terminal residue in larger peptides in which proline is second residue. This finding was utilized by Frater *et al.* (*30*) to release only isoleucine from papain (Ile–Pro–), by Light and Greenberg (*115*) to release glutamine from Gln–Pro–, and by Nolan and Smith (*116*) to liberate only lysine from the glycopeptide from bovine γ-globulin with the structure, Lys–Pro–Arg–Glu–Glu–Gln–Phe–Asp(carbohydrate). Whether this activity results from prolidase or a possible contaminating enzyme cannot be stated at present, but it is obvious that prolidase preparations can be used in sequence work in the same manner as aminopeptidase P (Section IX). Indeed, the alleged presence of aminopeptidase P in swine kidney extract (*114*) may be due to prolidase.

Many other dipeptidases have been described in the literature, but only one well-characterized enzyme will be mentioned here as a representative of the group. An enzyme called *renal dipeptidase* was prepared in homogeneous crystalline form from swine kidney cortex by Campbell *et al.* (*117*). The dipeptidase had a molecular weight of 47,200 and contained 1 g-atom of zinc per mole of protein. Dialysis against *o*-phenan-

115. A. Light and J. Greenberg, *JBC* **240**, 258 (1965).
116. C. Nolan and E. L. Smith, *JBC* **237**, 453 (1962).
117. B. J. Campbell, Y.-C. Lin, R. V. Davis, and E. Ballew, *BBA* **118**, 371 (1966).

throline inactivated the enzyme by removal of zinc, which could be added back to restore activity. The amino acid composition has been determined, and the kinetic parameters of the enzyme were measured as a function of pH (*118*). From these and other studies on copper-catalyzed hydrolysis of dipeptides (*119*, *120*), a mechanism of action involving the dissociation of protons from the α-amino group of the substrate dipeptide and from a water molecule coordinated to the zinc at the active center of the enzyme has been postulated (*118*). A variety of dipeptides such as L-Leu–Gly, Gly-L-His, Gly-L-Lys, L-Ala–Gly, Gly-L-Asp, L-Ser–Gly and Gly–Gly were hydrolyzed with Gly–Gly being hydrolyzed most rapidly. Tripeptides, L-leucinamide, D-Leu–Gly, carbobenzoxy-Gly-L-Phe, casein, and hemoglobin were not hydrolyzed. Thus, this dipeptidase seems clearly differentiated from other dipeptidases described in the literature [such as the glycylglycine dipeptidase (*1*, *121*)].

XI. Aminotripeptidase

Although various aminopeptidases can hydrolyze tripeptides, an enzyme previously described from many animal tissues (*1*, *121–126*) is specific for the hydrolysis of tripeptides to yield a dipeptide and a free amino acid from the amino-terminus of the tripeptide. Dipeptides and peptides larger than tripeptides are not hydrolyzed. The enzyme apparently has no metal ion requirement but is inhibited by Cd^{2+}; it does not appear to be a thiol enzyme. Highly purified preparations have been prepared from bovine thymus by Ellis and Fruton (*123*). Extensive studies of the specificity of this enzyme with respect to action on various stereoisomeric tripeptides (*123–125*), β-alanine-containing peptides (*123–125*), and even on unusual dipeptides (*126*) have been described earlier.

A recent study of aminotripeptidase obtained from rat kidney (*127*) shows that this enzyme has properties similar to those described above. Although several active fractions were obtained from cytoplasm and

118. A. M. René and B. J. Campbell, *JBC* **244,** 1445 (1969).
119. B. J. Campbell, Y.-C. Lin, and M. E. Bird, *JBC* **238,** 3632 (1963).
120. B. J. Campbell, Y.-C. Lin, and E. Ballew, *JBC* **242,** 930 (1967).
121. E. L. Smith, *Advan. Enzymol.* **12,** 191 (1951).
122. E. L. Smith, "Methods in Enzymology," Vol. 2, p. 83, 1955.
123. D. Ellis and J. S. Fruton, *JBC* **191,** 153 (1951).
124. J. S. Fruton, V. A. Smith, and P. E. Driscoll, *JBC* **173,** 457 (1948).
125. E. Adams, N. C. Davis, and E. L. Smith, *JBC* **199,** 845 (1952).
126. N. C. Davis and E. L. Smith, *JBC* **214,** 209 (1955).
127. H. Kirscke, J. Lasch, and H. Hanson, *Z. Physiol. Chem.* **350,** 1449 (1969).

microsomes, the behavior of these fractions led to the conclusion that they contained the same enzyme. The enzyme was inhibited by Cd^{2+} and was little affected by the addition of EDTA or cysteine. In accord with earlier studies, dipeptides and tetrapeptides were not hydrolyzed, but the tripeptides Gly–Gly–Pro, Gly–Pro–Gly, Pro–Gly–Gly, Ala–Gly–Gly, Leu–Gly–Gly, and Gly–Gly–Gly all served as substrates with only the N-terminal residue being released. The enzyme has also been studied recently in ox brain (*128*). A useful assay for aminotripeptidase using the difference in absorption of the Cu^{2+} substrate (Gly–Gly–Gly) complex (maximum at 550 nm) and the Cu^{2+} product (Gly–Gly and Gly) complexes (maximum at 640 nm) has recently been described (*129*).

XII. Concluding Remarks

Although most of the enzymes mentioned in this review have been only briefly described, it is evident that a great variety of N-terminal exopeptidases exists in nature. Although little is known about the precise functions of any of these enzymes, all of them are potentially useful in sequence studies of peptides and proteins, and, in fact, many of them have already been employed for such investigations. As more is learned about the functions of these enzymes, they will undoubtedly be useful in the study of the catabolism of particular types of biologically active peptides such as the polypeptide hormones.

We are certain that the list of peptidases cited in this review is far from complete and that others will be found in future investigations. Hopefully, many of these will also prove to be useful in sequence studies and will provide additional understanding of this important class of enzymes (*130*).

128. A. S. Brecher and R. E. Sobel, *BJ* **105,** 641 (1967).

129. J. F. Eccleston, *BBA* **139,** 186 (1967).

130. References to many earlier publications have been omitted for the sake of brevity. The interested reader will find many additional citations to the literature prior to 1960 in earlier reviews (*1, 9, 121*) and treatises (*131, 132*).

131. K. Oppenheimer, ed., "Die Fermente und ihre Wirkungen," various eds.; 5th ed., 4 vols., Thieme, Leipzig, 1925–1929; Supplements, 3 vols., Junk, The Hague, 1936–1939.

132. T. Bersin, *in* "Handbuch der Enzymologie" (F. F. Nord and T. Weidenhagen, eds.). Akad. Verlagsges., Leipzig, 1940.

4

Pepsin

JOSEPH S. FRUTON

I. Introduction

A. Historical Background

Although the crystallization of pepsin (*1*) followed that of urease (*2*), it was Northrop's achievement which forced biochemists to abandon the view (*3*) that enzymes are composed of small catalytic molecules adsorbed on protein carriers. In providing convincing evidence of the protein nature of pepsin, Northrop thus brought biochemical thought on the track it had followed during the latter half of the nineteenth century, when the protein nature of "unorganized ferments" such as pepsin was widely accepted (*4*). Pepsin was extensively studied at that time, both because it was recognized to be the principal catalytic agent of gastric digestion in mammals and birds and because of the great interest in the nature of enzymic action (*5, 6*). The characterization of pepsin by Schwann (*7*) had been followed by repeated efforts to purify it; the work of von Brücke (*8*) and of Pekelhäring (*9*) is especially noteworthy in relation to later studies. An important discovery made before 1900 was Langley's observation that a slightly alkaline extract of gastric mucosa contained a catalytically inactive material (pepsinogen) which was converted to pepsin upon acidification of the extract (*10*). The crystallization of pepsinogen by Herriott (*11*) made available this material in highly purified form. The work of Northrop and Herriott thus marks the beginnings of the modern study of pepsin as a protein and as a catalytic agent (*12*).

1. J. H. Northrop, *J. Gen. Physiol.* **13,** 739 (1930).
2. J. B. Sumner, *JBC* **69,** 435 (1926).
3. R. Willstätter, "Problems and Methods in Enzyme Research." Cornell Univ. Press, Ithaca, New York, 1927.
4. E. Fischer, *Z. Physiol. Chem.* **26,** 60 (1898).
5. J. R. Green, "The Soluble Ferments and Fermentation." Cambridge Univ. Press, Cambridge, 1898.
6. E. H. Starling, "Recent Advances in the Physiology of Digestion." Keener, Chicago, Illinois, 1906.
7. T. Schwann, *Arch. Anat. Physiol.* p. 90 (1836).
8. E. von Brücke, *Sitzber. Akad. Wiss. Wien* **43,** 601 (1861).
9. C. A. Pekelhäring, *Z. Physiol. Chem.* **22,** 233 (1896); **35,** 8 (1902); **75,** 282 (1911).
10. J. N. Langley, *J. Physiol.* (*London*) **3,** 246 (1882); J. N. Langley and B. Edkins, *ibid.* **7,** 371 (1886).
11. R. M. Herriott, *J. Gen. Physiol.* **21,** 501 (1938).
12. J. H. Northrop, M. Kunitz, and R. M. Herriott, "Crystalline Enzymes," 2nd ed. Columbia Univ. Press, New York, 1948.

B. Occurrence

1. *Classification and Nomenclature*

The term *pepsin* is applied to the gastric proteinases active at acid pH values (pH 1–5) and formed by partial proteolysis of their inactive zymogens, the pepsinogens. During his pioneer work on the crystallization of porcine pepsin, Northrop noted that preparations of the crystalline material differed considerably in homogeneity, as indicated by measurement of their solubility behavior. Part of the inhomogeneity was a consequence of peptide material, formed by autodigestion of pepsin. In addition, Northrop called attention to the presence of a second gastric proteinase, characterized by its exceptional ability to liquefy gelatin (*13*), and stated that it was difficult to remove this gelatinase from crystals of porcine pepsin. Subsequently, Steinhardt (*14*) performed a very careful study of the solubility properties of crystalline porcine pepsin and gave clear evidence of its inhomogeneity as a protein. After the introduction of ion exchange chromatography for the fractionation of proteins, Ryle and Porter (*15*) demonstrated the presence of two minor pepsinlike proteinases (parapepsins I and II) in extracts of porcine gastric mucosa; these components have been renamed pepsins B and C, the term pepsin A being assigned to the predominant enzyme. Ryle and his associates have studied porcine pepsinogen B and pepsin B (*16*) as well as pepsinogen C and pepsin C (*17, 18*). In addition they have shown (*19*) the existence of a fourth proteinase (pepsin D) and its zymogen (pepsinogen D); these appear to be dephosphorylated pepsin A and pepsinogen A, respectively.

A similar multiplicity of proteinases has been found for human gastric juice and for extracts of human gastric mucosa. In addition to the principal enzymic component (pepsin), a pepsinlike enzyme named *gastricsin* has been described and purified (*20–22*); about 400 mg of pepsin and 150 mg of gastricsin are present per liter of human gastric

13. J. H. Northrop, *J. Gen. Physiol.* **15,** 29 (1933).
14. J. Steinhardt, *JBC* **129,** 135 (1939).
15. A. P. Ryle and R. R. Porter, *BJ* **73,** 75 (1959).
16. A. P. Ryle, *BJ* **96,** 6 (1965).
17. A. P. Ryle, *BJ* **75,** 145 (1960).
18. A. P. Ryle and M. P. Hamilton, *BJ* **101,** 176 (1966).
19. D. Lee and A. P. Ryle, *BJ* **104,** 735 and 742 (1967).
20. J. Tang, S. Wolf, R. Caputto, and R. E. Trucco, *JBC* **234,** 1174 (1959).
21. J. Tang and K. Tang, *JBC* **238,** 606 (1963).
22. J. Tang, J. Mills, L. Chiang, and L. deChiang, *Ann. N. Y. Acad. Sci.* **140,** 688 (1967).

juice. A separate zymogen for gastricsin has not be reported. However, the chromatographic separation of three human pepsinogen fractions (denoted I, II, and III) has been described (*23*); these fractions gave rise to four pepsins, named I, IIa, IIb, and III, to indicate the zymogen from which they were derived (*24*), and pepsin I is stated to exhibit a chromatographic behavior similar to that of gastricsin. Other investigators have observed similar multiplicity of human and porcine pepsin and have used other symbols to denote the fractions separated by chromatography or starch gel electrophoresis (*25*, *26*).

The nomenclature of the pepsins is thus in a confused state, and greater clarity may emerge only after the individual components from various species have been purified and fully characterized. It may be expected that as in the case of the chymotrypsins some of the pepsins in a given species are derived from different zymogens (e.g., chymotrypsin A and B), and apparently different pepsins may arise from varying degrees of proteolysis of the same zymogen (e.g., chymotrypsin π, γ, and α). Furthermore, analogous pepsin components from different species may be expected to differ somewhat in their amino acid composition and sequence, with corresponding differences in chromatographic behavior and electrophoretic mobility. With these considerations in mind, it may be tentatively suggested that porcine pepsin C, gastricsin, and human pepsin I have much in common and are probably derived from a zymogen secreted by the pyloric mucosa (*27*). Furthermore, porcine pepsin B appears to be identical with Northrop's gelatinase. The principal gastric proteinase, derived largely from the pepsinogen of the fundic mucosa, is named pepsin A. It has been proposed that the nomenclature of the pepsins be based on the relative mobility of the fractions upon electrophoretic separation in agar gels (*28*); although this proposal, if adopted, may aid in the comparison of data from different laboratories, the method does not show that each electrophoretic band represents a different enzyme.

In this chapter, the A, B, and C nomenclature is used for the porcine pepsins; since most of the available data on the chemical structure and kinetic properties of pepsin are for porcine pepsin A and its zymogen (pepsinogen A), these forms will be termed *pepsin* and *pepsinogen* in what

23. M. J. Seijffers, H. L. Segal, and L. L. Miller, *Am. J. Physiol.* **205,** 1099 and 1106 (1963); **207,** 8 (1964).
24. M. J. Seijffers, L. L. Miller, and H. L. Segal, *Biochemistry* **3,** 1203 (1964).
25. I. Kushner, W. Rapp, and P. Burtin, *J. Clin. Invest.* **43,** 1983 (1964).
26. W. B. Hanley, S. H. Boyer, and M. A. Naughton, *Nature* **209,** 996 (1966).
27. W. H. Taylor, *Physiol. Rev.* **42,** 519 (1962).
28. D. J. Etherington and W. H. Taylor, *Nature* **216,** 279 (1967).

follows. Other pepsinlike enzymes will be designated by means of the terms used by the authors of the cited papers.

In addition to the pepsins from the pig and from man, bovine pepsin has been obtained in crystalline form (*29*), and the corresponding pepsinogen has been isolated (*30*). Highly purified or crystalline pepsin preparations have been obtained from the gastric mucosa of birds [e.g., chicken (*31*, *32*)] and of fishes [salmon (*33*), shark (*34*), and tuna (*35*)]. The multiplicity of pepsinogen observed for human and porcine gastric mucosa has also been demonstrated for the dogfish (*36*) and chicken (*36a*). Proteinase activity at acid pH values has been reported for extracts of the digestive tract of a wide variety of other vertebrates.

2. *Other Pepsinlike Enzymes*

To the pepsinlike enzymes listed above may be added other proteinases that resemble pepsin in several respects. The best known of these is rennin (chymosin), derived from the fourth stomach of the calf, and which has been crystallized and extensively characterized (*37*); its high milk-clotting activity has made rennet preparations important in the manufacture of cheese. A less well-defined proteinase (or group of proteinases) that acts optimally at acid pH values is the cathepsin D found in extracts of several animal tissues (spleen, liver, kidney, etc.); it has been considerably purified from extracts of beef spleen (*38*, *39*).

Pepsinlike enzymes have also been found in numerous fungi, and some of them have been crystallized (*39a–d*).

29. J. H. Northrop, *J. Gen. Physiol.* **16,** 615 (1933).
30. R. B. Chow and B. Kassell, *JBC* **243,** 1718 (1968).
31. T. P. Levchuk and V. N. Orekhovich, *Biokhimiya* **28,** 1004 (1963).
32. Z. Bohak, *JBC* **244,** 4638 (1969).
33. E. R. Norris and D. W. Elam, *JBC* **134,** 443 (1940).
34. G. P. Sprissler, Ph.D. Dissertation. Catholic University of America, Washington, D. C., 1942.
35. E. R. Norris and D. W. Elam, *JBC* **204,** 673 (1953).
36. T. R. Merrett, E. Bar-Eli, and H. Van Vunakis, *Biochemistry* **8,** 3696 (1969).
36a. S. T. Donta and H. Van Vunakis, *Biochemistry* **9,** 2791, 2798 (1970).
37. B. Foltmann, *Compt. Rend. Trav. Lab. Carlsberg* **35,** 143 (1966).
38. E. M. Press, R. R. Porter, and J. Cebra, *BJ* **74,** 501 (1960).
39. H. Keilova and B. Keil, *Collection Czech. Chem. Commun.* **33,** 131 (1968).
39a. J. Fukumoto, D. Tsuru, and T. Yamamoto, *Agr. Biol. Chem.* **31,** 710 (1967).
39b. J. Sodek and T. Hofmann, *JBC* **243,** 450 (1968).
39c. D. Tsuru, A. Hattori, T. Yamamoto, and J. Fukumoto, *Agr. Biol. Chem.* **33,** 1419 (1969).
39d. J. Sodek and T. Hofmann, *Can. J. Biochem.* **48,** 425 (1970).

C. Purification of Pepsinogen and Pepsin

The procedure developed by Herriott (*11*) for the crystallization of porcine pepsinogen depends largely on fractional precipitation with $(NH_4)_2SO_4$ and yields a product that is homogeneous by several criteria (*40–42*). Crystalline pepsin preparations are obtained by fractional precipitation with $MgSO_4$ (*1, 43, 44*) or by crystallization from alcohol (*45*); as indicated above, such products are heterogeneous with respect to their behavior on ion exchange columns. Some purification, especially in the removal of nonprotein contaminants, has been effected on such columns (*46, 47*). The most significant advance, however, was made by Rajagopalan *et al.* (*42*), who have used hydroxylapatite for the fractionation of commercial preparations of crystalline pepsin. They have described a valuable method for the preparation of pepsin samples that are homogeneous by chromatography on hydroxylapatite and by end group analysis; this method involves rapid activation of pepsinogen, and passage of the activation mixture through sulfoethyl Sephadex C-25 to remove the activation peptides, followed by desalting with Sephadex G-25. This procedure has been modified by the use of a long column of sulfoethyl Sephadex C-25 (*47a*).

D. Assay

The most widely used assay method for pepsin activity is that developed by Anson (*48*); acid-denatured hemoglobin is the substrate at pH 1.8 and 37°, and the release of cleavage products that are soluble in about 3% trichloroacetic acid is measured spectrophotometrically at 280 nm. One unit of pepsin activity is usually defined as the amount of enzyme that produces an increase in absorbance of 0.001 per minute under the conditions of the assay. Commercial preparations of crystalline pepsin generally contain 2500–3000 units per milligram. Pepsin prepara-

40. H. Van Vunakis and R. M. Herriott, *BBA* **23,** 600 (1957).
41. R. Arnon and G. E. Perlmann, *JBC* **238,** 653 (1963).
42. T. G. Rajagopalan, S. Moore, and W. H. Stein, *JBC* **241,** 4940 (1966).
43. J. St. L. Philpot, *BJ* **29,** 2458 (1935).
44. R. M. Herriott, V. Desreux, and J. H. Northrop, *J. Gen. Physiol.* **24,** 213 (1940).
45. J. H. Northrop, *J. Gen. Physiol.* **30,** 177 (1946).
46. K. Heirwegh and P. Edman, *BBA* **24,** 219 (1957).
47. M. A. Mitz and R. J. Schleuter, *JACS* **81,** 4024 (1959).
47a. R. Trujillo and M. Schlamowitz, *Anal. Biochem.* **31,** 149 (1969).
48. M. L. Anson, *J. Gen. Physiol.* **22,** 79 (1938).

tions obtained by the method of Rajagopalan *et al.* (*42*) assay at about 4000 units per milligram. The potential pepsin actvity of pepsinogen preparations is also assayed by the hemoglobin method; commercial preparations usually have a potential pepsin activity of 2500–3000 units per milligram of pepsinogen.

Other assay methods with protein substrates include the clotting of milk at pH 5 (*11*) and the cleavage of ^{131}I-labeled bovine serum albumin at pH 2 (*49*). A number of synthetic substrates have been used, principally acetyl-L-phenylalanyl-diiodo-L-tyrosine (Ac–Phe–TyrI_2), whose rate of cleavage is followed by measurement of the rate of release of ninhydrin-reactive material (*50, 51*). Such procedures may be automated (*52–54*). A spectrophotometric method for following the cleavage of substrates such as Ac–Phe–Tyr is based on the procedure of Schwert and Takenaka (*55*), in which the decrease in absorbance at 237 nm is measured (*56*). Another spectrophotometric method involves the measurement of the increase in absorbance at 310 nm when synthetic substrates of the type A–Phe(NO_2)–Y–B (Phe(NO_2) = *p*-nitro-L-phenylalanyl) are cleaved at the Phe(NO_2)–Y bond (*57*).

II. Molecular Properties of Pepsinogen and Pepsin

A. Molecular Weight and Shape

Several determinations of the sedimentation coefficient of crystalline porcine pepsin have given values of $s_{20,w} = 2.9–3.3$ S, and the reported values of the diffusion coefficient ($D_{20,w}$) are near 9×10^{-7} cm^2 sec^{-1}; on the assumption that the partial specific volume is 0.75 cm^3 g^{-1}, a molecular weight of approximately 35,000 has been calculated (*58–60*).

49. A. P. Klotz and M. R. Duvall, *J. Lab. Clin. Med.* **50,** 753 (1957).
50. L. E. Baker, *JBC* **193,** 809 (1951).
51. W. T. Jackson, M. Schlamowitz, and A. Shaw, *Biochemistry* **4,** 1537 (1965).
52. J. Lenard, S. L. Johnson, R. W. Hyman, and G. P. Hess, *Anal. Biochem.* **11,** 30 (1965).
53. A. J. Cornish-Bowden and J. R. Knowles, *BJ* **96,** 71P (1965).
54. T. R. Hollands and J. S. Fruton, *Biochemistry* **7,** 2045 (1968).
55. G. W. Schwert and Y. Takenaka, *BBA* **16,** 570 (1955).
56. M. S. Silver, J. L. Denburg, and J. J. Steffens, *JACS* **87,** 886 (1965).
57. K. Inouye and J. S. Fruton, *Biochemistry* **6,** 1765 (1967).
58. J. St. L. Philpot and I.-B. Erikson-Quensel, *Nature* **132,** 932 (1933).
59. H. Edelhoch, *JACS* **79,** 6100 (1957).
60. H. A. Dieu, *Bull. Soc. Chim. Belges* **65,** 603 (1956).

A recent careful determination (*61*) of the molecular weight of the homogeneous pepsin prepared by Rajagopalan *et al.* (*42*), using the sedimentation-equilibrium method, has given a value $\overline{\text{MW}} = 32{,}700 \pm 1{,}200$; in this calculation, a value of $\bar{V} = 0.726 \pm 0.008$ (based on amino acid composition) was used. This value of $\overline{\text{MW}}$ may be compared with the molecular weight of 34,163 calculated from amino acid analyses (*42*). Light scattering measurements have given values near 35,000 (*62*) and 38,000 (*63*), although a value near 32,500 has also been noted with carefully treated solutions of pepsin (*62*). Other estimates of the molecular weight of pepsin have been 35,000 [osmotic pressure (*1*)], 40,000 [X-ray diffraction (*64*)], and 34,400 [pressure of monolayers (*65*)]. From the above data, it would appear that the molecular weight of pepsin is near 33,500. It may be added that the $\overline{\text{MW}}$ value obtained by sedimentation equilibrium was independent of protein concentration over the range 0.01–0.15% (*61*), indicating that under the conditions used (pH 5.6) no significant aggregation had occurred.

The available estimates of the frictional ratio (f/f_0) and of the reduced specific viscosity of pepsin have given axial ratios (a/b) of 2.5–3.0 for an assumed prolate ellipsoid of revolution (*66*). Low-angle X-ray scattering data (*67*) are in accord with this ratio.

The molecular weight ($\overline{\text{MW}}$) of pepsinogen has been estimated (*61*) to be $40{,}400 \pm 1{,}600$, upon extrapolation of sedimentation-equilibrium data to zero concentration; a value of $\bar{V} = 0.730 \pm 0.008$ (based on amino acid composition) was used for the calculation. The molecular weight calculated from amino acid analyses is 38,944 (*42*); other determinations have given values of $\overline{\text{MW}} = 41{,}000$ for an assumed $\bar{V} = 0.75$ (*41*) and of $\bar{M}_n = 42{,}000$ (*11*). Thus, the molecular weight of porcine pepsinogen appears to be near 40,000; some of the deviations from this value may be a consequence of the concentration-dependent behavior of this protein (*61*).

The enzymes (and their zymogens) related to porcine pepsin and pepsinogen have molecular weights similar to those given above. The reported values for several of these pepsinlike enzymes are human pepsin, 34,000 (*68*); human gastricsin, 31,400 (*68*); porcine pepsin C, 36,000 (*18*); chicken pepsin, 35,000 (*32*); and rennin, 30,700 (*37*). Bovine

61. R. C. Williams, Jr. and T. G. Rajagopalan, *JBC* **241,** 4951 (1966).
62. M. J. Kronman and M. D. Stern, *J. Phys. Chem.* **59,** 969 (1955).
63. D. S. Yasnoff and H. B. Bull, *JBC* **200,** 619 (1953).
64. J. D. Bernal and D. Crowfoot, *Nature* **133,** 794 (1934).
65. H. A. Dieu and H. B. Bull, *JACS* **71,** 450 (1949).
66. F. A. Bovey and S. S. Yanari, "The Enzymes" 2nd ed., Vol. 4, Chapter 4, p. 63, 1960.
67. A. A. Vazina, V. V. Lednev, and B. K. Lemazhkin, *Biokhimiya* **31,** 720 (1966).
68. J. N. Mills and J. Tang, *JBC* **242,** 3093 (1967).

pepsinogen has a molecular weight of 38,000 (*30*); porcine pepsinogen C, 41,000 (*18*); chicken pepsinogen, 43,000 (*32*); dogfish pepsinogens A, C, and D, 41,500 (*36*); and prorennin, 36,000 (*37*).

B. Electrophoretic Mobility

As indicated above, pepsin preparations are often heterogeneous upon electrophoresis; this heterogeneity is in part the result of autolysis. The isoelectric point of porcine pepsin is below 1.0, since at this pH value highly purified pepsin still migrates as an anion (*69, 70*). As will be seen from the amino acid composition of pepsin (Section II,D), there is a predominance of carboxyl groups over cationic groups; calculation of the isoionic point based on amino acid analyses, and the assumption that the intrinsic pK_a of the side chain carboxyls is 4.0, gives a value near 3.0. This apparent discrepancy may be a consequence of adsorbed anions, of the single phosphoryl group covalently bound to pepsin (*71–73*), and of the possible presence of one or two abnormally acidic carboxyl groups (Sections II,F and V,A). Dephosphorylation of pepsin by means of potato phosphatase shifts the isoelectric point to 1.7 and does not alter the proteolytic activity (*74, 74a*). In contrast to the extremely low isoelectric point of pepsin, that of pepsinogen is about 3.7 (*11*); this difference is consistent with the cationic character of the peptides removed from the zymogen upon its conversion to pepsin (Section III,A).

Agar-gel electrophoresis has been used for the detection of pepsinlike enzymes in pepsin preparations, in gastric juice, and in extracts of gastric mucosa; the significant differences in net charge have permitted their separation by means of ion exchange chromatography (Section I,B,1).

C. Optical Properties

The specific optical rotation of pepsin at pH 4.6 and 25° ($[\alpha]_{600} = -63.5°$; $[\alpha]_{400} = -178°$) and the $\lambda_c = 216$ nm which characterizes the optical rotatory dispersion are not changed appreciably by brief exposure

69. A. Tiselius, G. E. Henschen, and H. Svensson, *BJ* **32,** 1814 (1938).
70. R. M. Herriott, V. Desreux, and J. H. Northrop, *J. Gen. Physiol.* **23,** 439 (1940).
71. G. E. Perlmann, *JACS* **74,** 6308 (1952).
72. M. Flavin, *JBC* **210,** 771 (1952).
73. G. E. Perlmann, *J. Gen. Physiol.* **41,** 441 (1958).
74. G. E. Perlmann, *Advan. Protein Chem.* **10,** 1 (1955).
74a. G. E. Clement, J. Rooney, D. Zakheim, and J. Eastman, *JACS* **92,** 186 (1970).

to 8 M urea or 3 M guanidine (*75–77*), and the enzyme remains fully active. On the other hand, the value of λ_c is decreased in the presence of LiCl and increased by raising the temperature above 60°; both changes are accompanied by the loss of enzymic activity (*78*). In contrast to the behavior of pepsin, the optical rotation of pepsinogen ($[\alpha]_{366}$) decreases from —200° to —300° upon exposure to urea, and λ_c is decreased from 236 to 216 nm (*79*); similar changes occur over the temperature range 45°–53°. The molar absorptivity at 278 nm for pepsinogen and pepsin are 56×10^3 and 51×10^3, respectively (*80, 81*). A study of the ultraviolet absorption of pepsinogen has shown that at pH 12–13, all 17 tyrosine residues are ionized normally but that at pH 8–9 two tyrosines contribute to the absorbance only after the protein is denatured with urea (*80*). Upon denaturation of pepsin with urea at pH 5.3, the absorption maximum shifts from 278 to 276 nm (*82*). The effect of denaturation on the optical rotation and ultraviolet absorption of pepsinogen and pepsin are discussed further in Section II,G.

Studies on the optical rotatory dispersion and circular dichroism of pepsin have indicated a conformational change near pH 1.1, and have also shown a small Cotton effect and dichroic bands in the region 260–290 nm (*82a*).

D. Amino Acid Composition

Porcine pepsin and pepsinogen have been subjected to amino acid analysis by several investigators; at present, the preferred data are those of Rajagopalan *et al.* (*42*), presented in Table I. It will be noted that the total number of amino acid residues per molecule in pepsinogen and pepsin are about 362 and 320, respectively; these values may be compared with the earlier values for pepsinogen of 400 (*40*) and 383 (*41*) and for pepsin of 343 (*83*). The various recent analyses agree on the presence of six half-cystine residues (corresponding to three disulfide

75. R. B. Simpson and W. Kauzmann, *JACS* **75,** 5139 (1953).
76. B. Jirgensons, *ABB* **39,** 261 (1952); **41,** 333 (1952).
77. G. E. Perlmann, *Proc. Natl. Acad. Sci. U. S.* **45,** 915 (1959).
78. G. E. Perlmann, *JMB* **6,** 452 (1963).
79. G. E. Perlmann, *Advan. Chem. Ser.* **63,** 268 (1967).
80. G. E. Perlmann, *JBC* **239,** 3762 (1964).
81. G. E. Perlmann, *JBC* **241,** 153 (1966).
82. G. E. Perlmann, *ABB* **65,** 210 (1957).
82a. G. Perlmann, *in* "Structure–Function Relationships of Proteolytic Enzymes" (P. Desnuelle, H. Neurath, and M. Otteson, eds.), p. 261. Munksgaard, Copenhagen, 1970.
83. O. O. Blumenfeld and G. E. Perlmann, *J. Gen. Physiol.* **42,** 553 (1959).

TABLE I
AMINO ACID COMPOSITIONS OF VARIOUS PEPSINOGENS AND PEPSINS

Amino acid	Porcine pepsinogen A[a]	Porcine pepsin A[a]	Bovine pepsinogen[b]	Dogfish pepsinogen A[c]	Chicken pepsinogen[d]	Chicken pepsin[d]	Porcine pepsinogen C[e]	Porcine pepsin C[e]	Human gastricsin[f]	Human pepsin[f]	Prorennin[g]	Rennin[g]
Lys	10	1	8	14	18	8	12	4	0	0	13	8
His	3	1	2	7	8	3	2	1	1	1	5	4
Arg	4	2	6	14	7	4	7	4	3	3	77	5
Asp	44	40	40	44	43	35	30	28	26	40	33	30
Thr	26	25	27	23	28	24	25	25	21	27	20	18
Ser	46	43	50	43	39	35	35	35	32	43	31	27
Glu	28	26	32	39	30	23	46	41	39	31	36	29
Pro	19	16	15	19	19	14	20	18	17	19	14	12
Gly	35	34	35	40	32	27	35	32	33	35	28	25
Ala	19	16	16	18	18	15	23	21	18	18	15	13
Half-Cys	6	6	6	7	7	7	6	6	6	6	6	6
Val	23	20	25	23	26	21	22	20	23	27	23	21
Met	4	4	4	7	9	9	5	4	5	5	7	7
Ile	25	23	32	22	23	20	16	14	13	25	19	15
Leu	33	28	25	28	30	20	40	34	25	22	26	19
Tyr	17	16	18	19	24	20	22	18	17	15	18	15
Phe	15	14	15	17	21	18	24	21	15	15	16	14
Trp	(5)	(5)	6	5	5	5	6	6	4	5	4	4
Total	362	320	362	389	387	308	376	332	298	337	351	272
Amide N	27	27	37		32	20	34	32	44	50	34	31

[a] From Rajagopalan *et al.* (*42*).
[b] From Chow and Kassell (*30*).
[c] From Merrett *et al.* (*36*).
[d] From Bohak (*32*).
[e] From Ryle and Hamilton (*18*).
[f] From Tang *et al.* (*22*).
[g] From Foltmann (*37*).

bridges) in both porcine pepsinogen and porcine pepsin, and there is reasonable agreement regarding their lysine, arginine, and histidine content. The number of tryptophan residues has been reported to be six (*41*, *83*), four (*84*), and five (*85*); no tryptophan residues appear to be removed from pepsinogen upon its activation.

Aside from the overwhelming predominance of side chain carboxyl groups in pepsin (40 Asp + 26 Glu — 27 NH_3 = 29 COOH) over the side chain cationic groups (1 Lys + 1 His + 2 Arg = 4), pepsin is characterized by the presence of a relatively large number of hydroxy amino acids (25 Thr + 43 Ser), of Pro (16), and of aromatic amino acids (16 Tyr + 14 Phe + 4–6 Trp).

It will be seen that the difference between the amino acid composition of pepsinogen and pepsin (about 41 residues) is largely in their content of basic amino acids (9 Lys + 2 His + 2 Arg), which exceed the dicarboxylic acids (4 Asp + 2 Glu), thus making the peptide removed on activation of the zymogen a predominantly cationic species (*86*, *87*).

In Table I are listed some of the available data for the amino acid composition of several pepsinogens and pepsin as well as for rennin and its zymogen (prorennin). It will be noted that chicken pepsinogen and pepsin have been found to contain seven half-cystine residues, and the presence of one sulfhydryl groups has been demonstrated (*32*). This enzyme also contains two glucosamine units per molecule (*32*). The presence of acid-labile carbohydrate (mostly glucose) has been reported for porcine pepsinogen (*88*).

E. Amino Acid Sequence

The various known pepsinogens and pepsins all appear to be constituted of a single peptide chain that is cross-linked by three disulfide bridges. In porcine pepsinogen A and pepsin A, the amino-terminal amino acid residues are leucine and isoleucine, respectively (*39*); the carboxyl-terminal residue is alanine in both proteins (*41*). This finding indicates that the partial proteolysis of the zymogen to produce pepsin involves the removal of a portion of the amino-terminal sequence of pepsinogen (Section III,A). The proposed amino acid sequence of the 42-residue amino-terminal section of pepsinogen is shown in Table II (*87*). This sequence differs considerably from the one reported earlier (*89*).

84. T. A. A. Dopheide and W. M. Jones, *JBC* **243,** 3906 (1968).
85. V. Kostka, I. Moravek, I. Kluh, and B. Keil, *BBA* **175,** 459 (1969).
86. P. V. Koehn and G. E. Perlmann, *JBC* **243,** 6099 (1968).
87. E. B. Ong and G. E. Perlmann, *JBC* **243,** 6104 (1968).
88. H. Neumann, U. Zehavi, and T. D. Tanksley, *BBRC* **36,** 151 (1969).
89. L. A. Lokshina and V. N. Orekhovich, *Dokl. Akad. Nauk SSSR* **133,** 472 (1960).

TABLE II
AMINO ACID SEQUENCES IN PORCINE PEPSINOGEN

Amino-terminal sequence
Leu-Val-Lys-Val-Pro-Leu-Val-Arg-Lys-Lys-Ser-Leu-Arg-Gln-Asn-Leu-Ile-Lys-Asp-Gly-Lys-
Leu-Lys-Asp-Phe-Leu-Lys-Thr-His-Lys-His-Asn-Pro-Ala-Ser-Lys-Tyr-Phe-Pro-Ala-Glu-Ile-

Carboxyl-terminal sequence
-Tyr-Gly-Thr-Gly-Ser-Met-Asp-Val-Pro-Thr-Ser-Ser-Gly-Glu-Leu-Trp-Ile-Leu-Gly-Asp-Val-
Phe-Ile-Arg-Gln-Tyr-Tyr-Thr-Val-Phe-Asp-Arg-Ala-Asn-Asn-Lys-Val-Gly-Leu-Ala-Pro-Val-Ala

Disulfide bridges

```
 ┌──────────────────────┐
-Cys-Ser-Ser-Leu-Ala-Cys-Ser-Asp-His-Asn-Gln-Phe-Asn-Pro-Asp-Ser-Asp-Thr-Ser-Phe-
 ┌───────────────┐
-Cys-Ser-Gly-Gly-Cys-Gln-
                -Cys-Ser-Ser-Ile-Asp-Gln-
                 |
-Glu-Asn-Asn-Ser-Cys-Thr-Ser-Asp-Ser-Asp-Ser-
```

At the present writing, the amino acid sequences of pepsinogen and pepsin have not been fully elucidated. The largest segment known is the 43 amino acid unit at the carboxyl terminus. This sequence is based on the determination of the sequence of a 37 amino acid fragment produced upon cleavage of pepsin with cyanogen bromide (*90*) and one of the fragments identified by means of the diagonal paper electrophoretic technique for methionine peptides [Tyr–Gly–Thr–Gly–Ser–Met–Asp–Val–Pro–Thr–Ser (*90a*)]. Earlier work (*91–94*) had established the sequence of shorter fragments (up to 27 residues) of the carboxyl-terminal sequence of pepsin. The most striking feature of this sequence is the presence of the two arginines and the single lysine of the entire pepsin molecule in the carboxyl-terminal 20 amino acid unit (Table II).

The application of the diagonal paper electrophoretic method for cystine peptides (*95*) to chymotryptic hydrolyzates of pepsin has shown that two of the three disulfide bridges are present as small loops; the known amino acid sequences around these bridges (*96*, *97*) are shown in Table II. These sequences differ from those published earlier (*98*). It will be noted that the single histidine of pepsin is located in one of the cystine peptides. This histidine residue is located in a 158-residue amino-terminal fragment, whose sequence is under investigation (*99*, *99a*). Sequences involving tryptophan residues have been reported (*84*) to be: Val–Phe–Asp–Asn–Leu–Trp–Asp–Gln–Gly; Leu–Trp–Val–Pro–Ser; Val–Glu–Glu–(Trp, Gln); Leu–Asn–Trp–Val–Pro.

To these sequences may be added a peptide containing the phosphorylseryl residue [Glu–Ala–Thr–Ser(P)–Glu–Glu–Leu (*72*, *100*)] and the "active-site" peptide (*101*) identified as the locus of attack of an aspartyl

90. V. Kostka, L. Moravek, and F. Sorm, *European J. Biochem.* **13,** 447 (1970).
90a. J. Tang and B. S. Hartley, personal communication (1969).
91. T. A. A. Dopheide, S. Moore, and W. H. Stein, *JBC* **242,** 1833 (1967).
92. R. N. Perham and G. M. T. Jones, *European J. Biochem.* **2,** 84 (1967).
93. Y. S. Kuznetzov, G. G. Kovaleva, and V. M. Stepanov, *BBA* **118,** 219 (1966).
94. R. A. Matveeva, V. F. Krivzov, and V. M. Stepanov, *Biokhimiya* **33,** 167 (1968).
95. J. R. Brown and B. S. Hartley, *BJ* **101,** 214 (1966).
96. B. Foltmann and B. S. Hartley, *BJ* **104,** 1064 (1967).
97. J. Tang and B. S. Hartley, *BJ* **118,** 611 (1970).
98. B. Keil, L. Moravek, and F. Sorm, *Collection Czech. Chem. Commun.* **32,** 1968 (1967).
99. V. A. Trufanov, V. Kostka, B. Keil, and F. Sorm, *European J. Biochem.* **7,** 544 (1969).
99a. E. A. Vakhitova, I. B. Pugacheva, M. M. Amirkhanyan, and V. M. Stepanov, *Biokhimiya* **34,** 1042 (1969).
100. V. M. Stepanov, E. A. Vakhitova, C. A. Egorov, and S. M. Aveeva, *BBA* **110,** 632 (1965).
101. R. S. Bayliss, J. R. Knowles, and G. R. Wybrandt, *BJ* **113,** 377 (1969).

carboxyl group by diazo compounds (Ile–Val–Asp–Thr–Gly–Thr–Ser); shorter fragments of the latter peptide have been reported (*102, 103*) (see Section II,F).

The sequences described for the carboxyl-terminal 21 amino acid fragments derived from human pepsin and from human gastricsin (*104*) are almost identical to that of porcine pepsin A. Furthermore, the 21 amino acid carboxyl-terminal peptide and the cystine-containing fragments from rennin are similar in many respects to those from porcine pepsin (*96*). The amino-terminal residues of human pepsin and human gastricsin are valine and serine, respectively (*22*), while that of rennin is alanine (*37*). The carboxyl-terminal amino acid residue of chicken pepsinogen and pepsin appears to be serine, and lysine is at the amino terminus of the zymogen; the amino-terminal residue of chicken pepsin may be serine or threonine (*32*).

F. Chemical Modification

The first studies on the chemical modification of crystalline porcine pepsin were on the effect of acetylation by means of ketene (*105–108*) and showed that increasing substitution of the phenolic hydroxyl groups of tyrosyl residues caused progressively greater inhibition of proteinase activity toward hemoglobin. The concomitant acetylation of the amino groups of pepsin does not appear to affect proteinase activity since it is restored by selective hydrolytic removal of the *O*-acetyl groups with dilute acid. The conclusion that modification of the amino groups of pepsin does not affect its proteinase activity is further supported by the finding that deamination with nitrous acid is without appreciable effect (*108*). Subsequent work (*54, 81, 109*), using the more selective acetylating agent acetyl imidazole at pH 5.5, has shown that whereas the proteinase activity of pepsin toward hemoglobin is extensively inhibited by acetylation of tyrosyl side chains, the rate of cleavage of small synthetic peptide and ester substrates is enhanced; these effects are reversed by deacetylation of the modified protein by treatment with hydroxylamine. Similar effects have been noted upon carbamylation of

102. K. T. Fry, O. K. Kim, J. Spona, and G. A. Hamilton, *BBRC* **30,** 489 (1968).
103. V. M. Stepanov and T. I. Vaganova, *BBRC* **31,** 825 (1968).
104. W. Y. Huang and J. Tang, *Federation Proc.* **28,** 2253 (1969).
105. R. M. Herriott and J. H. Northrop, *J. Gen. Physiol.* **18,** 35 (1934).
106. R. M. Herriott, *J. Gen. Physiol.* **19,** 283 (1935).
107. V. Hollander, *Proc. Soc. Exptl. Biol. Med.* **53,** 179 (1943).
108. J. St. L. Philpot and P. A. Small, *BJ* **32,** 542 (1938).
109. L. A. Lokshina and V. N. Orekhovich, *Biokhimiya* **31,** 143 (1966).

pepsin with potassium cyanate (*110*) and upon nitration of pepsin with tetranitromethane (*111*). For a further discussion of these phenomena, see Section IV,A,2.

Additional evidence for a role of tyrosyl residues in the proteinase activity of porcine pepsin was provided by the inhibition of the enzyme by iodination with I_2 at pH 5.7 and the isolation of 3-iodo-L-tyrosine (as well as 3,5-diiodo-L-tyrosine) from an alkaline hydrolysate of partially inactivated pepsin (*112*). It may be noted, however, that in contrast to the effect of acetylation with acetyl imidazole, the iodination of pepsin leads to the parallel loss of proteinase, peptidase, and esterase activity (*54*). In view of the known capacity of iodine to attack histidyl and tryptophyl side chains (*113*), this finding suggests that one or both of these types of amino acid residues may be important in the catalytic action of pepsin; the inactivation of pepsin by *N*-bromosuccinimide (*114*) is consistent with the possible involvement of tryptophyl residues.

Because of the optimal action of pepsin at acidic pH values, it has long seemed probable that enzymic carboxyl groups are directly involved in the catalytic mechanism. Chemical modification studies have given strong support to this view. In earlier work, reagents such as bis(β-chloroethyl) sulfide (mustard gas) were shown to inactivate pepsin at pH 5.5–6.0 with the parallel alkylation of enzymic carboxyl groups (*115*). Partial inactivation was effected by means of *p*-bromophenacyl bromide, and evidence was offered in favor of the conclusion that the β-carboxyl group of an aspartyl residue had been substituted; it has been reported that the bromophenacyl group could be removed by means of sulfhydryl compounds such as thiophenol (*116–118*). Furthermore, diazo alkanes such as diazomethane (*119*) and diphenyldiazomethane (*120*), reagents long known to react with carboxyl groups, were found

110. S. Rimon and G. E. Perlmann, *JBC* **243,** 3566 (1968).
111. L. V. Kozlov, G. A. Kogan, and L. L. Zavada, *Biokhimiya* **34,** 1257 (1969).
112. R. M. Herriott, *J. Gen. Physiol.* **20,** 335 (1937); **25,** 185 (1941); **31,** 19 (1947).
113. B. L. Vallee and J. F. Riordan, *Ann. Rev. Biochem.* **38,** 733 (1969).
114. L. A. Lokshina, V. N. Orekhovich, and V. N. Pandakova, *Dokl. Akad. Nauk SSSR* **142,** 471 (1962).
115. R. M. Herriott, M. Anson, and J. H. Northrop, *J. Gen. Physiol.* **30,** 185 (1946).
116. B. F. Erlanger, S. M. Vratsanos, N. Wassermann, and A. G. Cooper, *JBC* **240,** PC3447 (1965).
117. B. F. Erlanger, S. M. Vratsanos, N. Wassermann, and A. G. Cooper, *BBRC* **23,** 243 (1966).
118. E. Gross and J. L. Morell, *JBC* **241,** 2638 (1966).
119. R. M. Herriott, *Advan. Protein Chem.* **3,** 170 (1947).
120. G. R. Delpierre and J. S. Fruton, *Proc. Natl. Acad. Sci. U. S.* **54,** 1161 (1965).

to inactivate pepsin at pH 5.5 but, as with mustard gas, nearly complete inactivation required multiple reaction (about five substituent groups per pepsin molecule). In recent work, however, more selective diazo compounds have been developed which inactivate pepsin completely by the introduction of a single substituent group per molecule of pepsin. These newer reagents have been either diazo ketones of the type $RCOCHN_2$ or diazoacetamido compounds of the type $N_2CHCONHR$.

The first of the diazo ketones shown to be a specific reagent for pepsin was L-1-diazo-4-phenyl-3-tosylamidobutanone-2 (tosyl-L-phenylalanyl diazomethane), which rapidly inactivates the enzyme at pH 5.4 (*121*). As in the case of other reactions involving diazo compounds (*122*), the rate of the reaction is greatly increased by the addition of copper salts. With ^{14}C-labeled reagent (prepared from ^{14}C-L-phenylalanine) it was shown that the rate of inactivation of pepsin toward protein and peptide substrates was the same as the rate of incorporation of the tosylphenylalanylmethyl group and that complete inactivation was achieved upon the introduction of one such group per molecule of pepsin (Table III). No incorporation was observed with pepsinogen, whose potential enzymic activity was unaffected by treatment with the diazo compound; nor was the label incorporated to a significant extent into alkali-denatured pepsin. The D isomer of the reagent reacted with pepsin much more slowly than

TABLE III
INACTIVATION OF PEPSIN BY DIAZO KETONES[a]

Reaction components[b]		Fraction pepsin activity lost[c]			Equiv. Phe or Gly incorp.[d]		
		5 min	15 min	45 min	5 min	15 min	45 min
Tos-L-Phe-CHN_2	Pepsin	0.51	0.88	0.94	0.47	0.84	0.99
	Pepsinogen	0	0	0	0.01	0.01	0.02
	Denatured pepsin	—	—	—	0.03	0.10	0.10
Tos-D-Phe-CHN_2	Pepsin	0	0.20	0.61	—	—	—
Tos-Gly-CHN_2	Pepsin	0.11	0.55	0.82	0.12	0.51	0.82

[a] From Delpierre and Fruton (*121*).
[b] Protein concentration, 0.0286 m*M*; reagent, 0.143 m*M*; $CuCl_2$, 1 m*M*; pH 5.3; 15°.
[c] Substrate, hemoglobin.
[d] The diazo ketones derived from L-phenylalanine and glycine were prepared from ^{14}C-labeled samples of these amino acids.

121. G. R. Delpierre and J. S. Fruton, *Proc. Natl. Acad. Sci. U. S.* **56**, 1817 (1966).
122. H. Zollinger, "Diazo and Azo Chemistry." Wiley (Interscience), New York, 1961.

did the L compound, and the corresponding tosylglycyl diazomethane was intermediate in reactivity between the L and D forms of tosyl phenylalanyl diazomethane (*121, 123*). Subsequent reports have described the inactivation of pepsin by 1-diazo-4-phenylbutanone-2 (*124*) and by benzyloxycarbonyl-L-phenylalanyl diazomethane (*125*), indicating that the tosylamido group of tosyl-L-phenylalanyl diazomethane may be omitted or replaced by other acylamido groups. Advantage has been taken of this fact to introduce chromophoric groups (e.g., 2,4-dinitrophenyl, 1-dimethylnaphthalene-5-sulfonyl) into pepsin to serve as resonance energy acceptors from tryptophan in the protein (*126*). Furthermore, α-diazo-*p*-bromoacetophenone has been reported to inactivate pepsin completely and to react with an enzymic group different from the β-aspartyl residue that was found to be reactive toward *p*-bromophenacyl bromide (*127*).

The first of the diazo acetamido compounds to be reported as a specific inhibitor of pepsin was diazoacetyl-D,L-norleucine methyl ester, chosen to permit the determination of the norleucine content of the modified protein (*128*). The rate of reaction is accelerated by cupric ion; in its absence, the inactivation is slower and the incorporation is not stoichiometric. A study of the copper catalysis of the reaction has indicated that the reactive species is a metal-complexed carbene derived from the diazo compound (*129*). Other diazo acetamido compounds found to be pepsin inhibitors are diazoacetyl-L-phenylalanine ethyl ester (*121*), its D isomer [which reacts at the same rate as the L form (*121*)] and its D,L form (*130*), as well as diazoacetylglycine ethyl ester (*130*). Of special importance was the use of diazoacetyl-L-phenylalanine methyl ester (*131*), since this work led to the identification of the site of substitution as the β-carboxyl group of the aspartyl residue in the sequence Ile–Val–Asp–Thr–Gly–Ser (*101*). The fact that the same residue is reactive toward the diazo ketones is indicated by the finding that 1-diazo-4-phenylbutanone-2 reacts with the aspartyl residue in the sequence Ile–Val–Asp–Thr (*102*).

123. J. S. Fruton, *Advan. Enzymol.* **33**, 401 (1970).
124. G. A. Hamilton, J. Spona, and L. D. Crowell, *BBRC* **26**, 193 (1967).
125. E. B. Ong and G. E. Perlmann, *Nature* **215**, 1492 (1967).
126. R. A. Badley and F. W. J. Teale, *BJ* **116**, 341 (1970).
127. B. F. Erlanger, S. M. Vratsanos, N. Wassermann, and A. G. Cooper, *BBRC* **28**, 203 (1967).
128. T. G. Rajagopalan, W. H. Stein, and S. Moore, *JBC* **241**, 4295 (1966).
129. R. L. Lundblad and W. H. Stein, *JBC* **244**, 154 (1969).
130. L. V. Kozlov, L. M. Ginodman, and V. N. Orekhovich, *Biokhimiya* **32**, 1011 (1967).
131. R. S. Bayliss and J. R. Knowles, *Chem. Commun.* p. 196 (1968).

The pepsinlike enzyme (penicillopepsin) of *Penicillium janthinellum* is also inactivated by diazoacetyl-D,L-norleucine methyl ester (*39b*), and the site of substitution has been identified as an aspartyl residue in the sequence Ile–Ala–Asp–Thr–Gly–Thr–Thr–Leu (*39d*).

In addition to the diazo compounds mentioned above as pepsin inhibitors, the following have been reported: 1-diazo-3-(2,4-dinitrophenyl)-aminobutanone-2 (*130*), α-diazo-β-(*p*-hydroxyphenyl)propionic acid ethyl ester (*130*), and *N*-diazoacetyl-*N'*-2,4-dinitrophenylethylene diamine (*132*).

Reduction of the disulfide bridges of porcine pepsinogen leads to the loss of potential enzymic activity (*133, 134*). With chicken pepsin, which contains one sulfhydryl group, the reaction of this group with 5,5'-dithiobis(2-nitrobenzoic acid) does not lower the proteinase activity markedly (*32*).

G. Denaturation

It has long been known that porcine pepsinogen can undergo reversible denaturation after being heated to about 60° at pH 7 or after treatment with alkali (up to pH 11); the alkali denaturation has been associated with the deprotonation of at least two groups having a pK_a near 9.4 (*11*). More recent studies have shown that the thermal denaturation is associated with an increased levorotation, with a transition temperature near 50°; the change in optical rotation above 60° is much less than that observed upon denaturation with 4 *M* urea (*78*). The reversible denaturation of pepsinogen appears to be biphasic, the deprotonation step preceding that associated with marked changes in optical rotation or viscosity (*135, 136*). Examination of the optical rotatory dispersion of pepsinogen has shown the presence of a negative trough of the Cotton effect at 227 nm, which is abolished in the presence of urea, and circular dichroism studies have indicated the presence of conformation-dependent tyrosyl and tryptophyl residues (*137*). These experimental data suggest that the amino-terminal portion of pepsinogen, where the basic amino acid residues are clustered (see Section II,D), participates in the stabilization of the zymogen molecule through electrostatic interaction with carboxylate groups of the protein. This view is supported

132. V. M. Stepanov, L. S. Lobareva, and N. I. Maltsev, *BBA* **151,** 719 (1968).
133. R. M. Herriott, *J. Gen. Physiol.* **45,** Suppl. 1, 57 (1962).
134. R. F. Steiner, V. Frattali, and H. Edelhoch, *JBC* **240,** 128 (1965).
135. V. Frattali, R. F. Steiner, and H. Edelhoch, *JBC* **240,** 112 (1965).
136. H. Edelhoch, V. Frattali, and R. F. Steiner, *JBC* **240,** 122 (1965).
137. K. Grizzuti and G. E. Perlmann, *JBC* **244,** 1764 (1969).

by the effect of modifying the lysine residues of pepsinogen by treatment with the *N*-carboxyanhydride of alanine, with potassium cyanate, or with succinic anhydride (*79, 110, 138*).

In contrast to the stability of porcine pepsinogen at pH 7–8, porcine pepsin is irreversibly inactivated at pH values above 6.0. The older studies on the denaturation of pepsin by alkali have been fully summarized in the previous edition of this work (*66*). The available data suggest that at pH values above six hydrogen bonds involving carboxyl groups are broken (*139*) and that, below pH 6, pepsin is largely stabilized by hydrophobic interactions since it is unaffected by heating to 60° or treatment with 4 *M* urea or 3 *M* guanidinium chloride (*77, 140*). This stability is correlated with the finding that the negative trough of the Cotton effect at 227 nm is not abolished by urea, as in the case of pepsinogen (*79*). It may be noted that chicken pepsin exhibits considerably greater stability than does porcine pepsin at pH values up to about 8, and it has been suggested that this difference is a consequence of the smaller net negative charge of the chicken enzyme (*32*). Attention has been drawn to the increase in absorbance near 300 nm when porcine pepsin is denatured (*139*), and this effect has been attributed to changes in the electrostatic environment of tryptophyl residues of the protein (*141*). The macromolecular changes (e.g., viscosity) upon the alkaline denaturation of pepsin are consistent with the unfolding of the protein to a linear polyelectrolyte (*59*).

III. Action of Pepsin on Protein Substrates

A. Formation of Pepsin from Pepsinogen

Below pH 5, the conversion of pepsinogen is effected by pepsin in an autocatalytic process that involves the removal of the 41-residue amino-terminal portion of the zymogen (Section II,E). Several low-molecular weight products are formed, including an inhibitory peptide composed of 29 amino acid residues (*142*). This inhibitor, which was

138. A. D. Gounaris and G. E. Perlmann, *JBC* **242**, 2739 (1967).
139. H. Edelhoch, *JACS* **80**, 6640 and 6648 (1958).
140. O. O. Blumenfeld, J. Leonis, and G. E. Perlmann, *JBC* **235**, 379 (1960).
141. V. S. Ananthranarayanan and C. C. Bigelow, *Biochemistry* **8**, 3717 and 3723 (1969).
142. H. Van Vanakis and R. M. Herriott, *BBA* **22**, 537 (1956).

obtained in crystalline form, is bound to pepsin at pH 5; at more acid pH values, dissociation of the pepsin–inhibitor complex is favored, and the peptide is cleaved (*143*). The rate of activation is optimal at pH 2, is first-order with respect to both pepsinogen and pepsin, and is increased by the addition of inorganic salts (*144, 145*). Immunochemical studies on porcine pepsinogen and pepsin have shown that the two proteins have antigenic sites in common (*146–149*).

With the method of fluorescence quenching by iodide ions, it has been shown (*126*) that one or two tryptophan residues of pepsinogen that are inaccessible to iodide ions are unmasked upon activation to pepsin.

B. Autolysis of Pepsin

In acid solution, pepsin readily undergoes autolytic degradation with the formation of low-molecular weight products. For this reason, the purification of active pepsin is extremely difficult (Section I,C), and pepsin preparations usually contain variable amounts of peptide material. A careful study of the autolysis of pepsin has shown that the cleavage of the protein is more rapid than the loss of proteinase activity, indicating that partially degraded pepsin may retain its catalytic activity (*150*). If the low-molecular weight fragments are removed by dialysis, the rate of autolysis is increased, and the dialyzable peptides not only inhibit the self-digestion of pepsin but also its proteinase activity toward hemoglobin (*150*). Earlier reports (*151, 152*) on the proteolytic activity of low-molecular-weight dialyzable fragments obtained from pepsin autolysates could not be confirmed (*150*). The so-called Brücke pepsin (*8, 153*) has

143. R. M. Herriott, *J. Gen. Physiol.* **24,** 325 (1941).
144. R. M. Herriott, *J. Gen. Physiol.* **22,** 65 (1938).
145. P. T. Varandani and M. Schlamowitz, *BBA* **77,** 496 (1963).
146. H. Van Vunakis, H. I. Lehrer, W. S. Allison, and L. Levine, *J. Gen. Physiol.* **46,** 589 (1963).
147. M. Schlamowitz, P. T. Varandani, and F. C. Wissler, *Biochemistry* **2,** 238 (1963).
148. J. F. Gerstein, H. Van Vunakis, and L. Levine, *Biochemistry* **2,** 964 (1963).
149. M. Schlamowitz, A. Shaw, and W. T. Jackson, *Biochemistry* **3,** 636 (1964).
150. H. Determann, D. Jaworek, R. Kotitschke, and A. Walch, *Z. Physiol. Chem.* **350,** 379 (1969).
151. G. E. Perlmann, *Nature* **173,** 406 (1954).
152. M. Funatsu and K. Tokuyasu, *Proc. Japan Acad.* **35,** 139 (1959); *J. Biochem. (Tokyo)* **46,** 1441 (1959).
153. H. Kraut and E. Tria, *Biochem. Z.* **290,** 277 (1937).

been shown to be a mixture of relatively large partial degradation products of pepsin and is characterized by a low tyrosine content (*154*).

C. Cleavage of Proteins

In its action on some proteins (e.g., hemoglobin and serum albumin), pepsin acts optimally near pH 2, but after these proteins are denatured the pH optimum shifts to about 3.5 (*155*). This difference has been attributed to the effect of H^+ in promoting the formation of denatured substrate which, because of its unfolded state, is more susceptible to pepsin action (*156, 157*); the inhibitory effect of newly released carboxylate groups may also be a contributing factor in favoring pepsin action below pH 3, where most of these groups are protonated. It should be noted, however, that the pH dependence curves for the peptic cleavage of different protein substrates may be affected differently by changes in ionic strength (*158*). Although earlier studies (*159, 160*) suggested an all-or-none action of pepsin on protein substrates, later work has made it clear that products of intermediate size are present during the course of the hydrolysis (*161, 162*), and pepsin has been used extensively for the partial degradation of proteins in connection with the determination of their amino acid sequence (*163*). Fluorescence studies on the peptic cleavage of bovine serum albumin labeled with 1-dimethylaminonaphthalene-5-sulfonyl groups have provided evidence for a biphasic process, with a rapid initial hydrolysis of a few accessible peptide bonds to form large fragments which are slowly digested in the second phase of the process (*164*).

The use of pepsin as a reagent in the partial cleavage of proteins showed that many kinds of peptide bonds were broken and led to the conclusion that pepsin is a proteinase of broad side chain specificity (*163*). Although, in general, the sensitive bonds were present in dipeptidyl

154. R. Kotitschke, Ph.D. Dissertation, Goethe Universität, Frankfurt-am-Main, 1969.

155. M. Schlamowitz and L. U. Peterson, *JBC* **234,** 3137 (1959).

156. K. Lindestrøm-Lang, R. D. Hotchkiss, and G. Johansen, *Nature* **142,** 996 (1938).

157. L. K. Christensen, *ABB* **57,** 163 (1955).

158. M. Schlamowitz and L. U. Peterson, *BBC* **46,** 381 (1961).

159. A. Tiselius and I. B. Eriksson-Quensel, *BJ* **33,** 1752 (1939).

160. G. Haugaard and R. M. Roberts, *JACS* **64,** 2664 (1942).

161. P. Desnuelle, M. Rovery, and G. Bonjour, *BBA* **5,** 116 (1950).

162. M. Schlamowitz, L. U. Peterson, and F. C. Wissler, *ABB* **92,** 58 (1961).

163. R. L. Hill, *Advan. Protein Chem.* **20,** 37 (1965).

164. G. Weber and L. B. Young, *JBC* **239,** 1415 (1964).

units containing at least one hydrophobic amino acid residue such as Phe, Tyr, Leu, or Met (*165*), a number of exceptions to this generalization were noted. Examples are the scission of Glu–Asn bond in the A chain of bovine insulin (*166*), of a Ser–Thr bond in the α chain of human hemoglobin A (*167*), of Glu–Lys, Asp–Pro, and Ala–Gly bonds in the β chain of human hemoglobin A (*168*), and one of the Ala–Ala bonds in pancreatic ribonuclease (*169*). In considering findings of this kind, it should be recalled that in most sequence studies the primary objective has not been the delineation of the specificity of pepsin, and relatively large amounts of enzyme and prolonged incubation periods are employed, thus obscuring differences in the rates of peptic cleavage of various kinds of peptide bonds. In the case of the long-chain natural peptide most widely used for the examination of the specificity of proteinases—the 30 amino acid B chain of bovine insulin—the available data (*170, 171*) indicate that the attack of porcine pepsin A is largely at Leu–Val (11–12), Ala–Leu (14–15), Tyr–Leu (16–17), Phe–Phe (24–25), and Phe–Tyr (25–26). Clearly, where two adjacent peptide bonds are both potential sites of attack, the more rapid cleavage of one of them will render the other more resistant to peptic cleavage because of the appearance of an α-carboxyl group or an α-amino group in immediate adjacence to the less susceptible bond. As will be seen from Section IV,A,1, there apear to be no major inconsistencies between the principal sites of the peptic cleavage of the B chain of insulin and the side chain specificity of pepsin toward small synthetic peptide substrates. With regard to the apparent discrepancies between the results obtained with proteins such as human hemoglobin and with small substrates, they may be viewed as a consequence of secondary interactions of protein substrate so as to position at the catalytic site a bond that would be relatively resistant to enzymic hydrolysis if it were present in a small substrate (*123*).

With enzymes related to porcine pepsin A, subtle differences have been found in the sites of cleavage of the B chain of insulin. Thus, porcine pepsin C hydrolyzes this substrate at His–Leu (10–11), Gly–Ala (13–14), and Ala–Leu (14–15), with the major cleavages at Tyr–Leu (16–17) and Phe–Phe (24–25) (*171*). Chicken pepsin does not appear to cleave

165. J. Tang, *Nature* **199,** 1094 (1963).
166. F. Sanger and E. O. P. Thompson, *BJ* **53,** 366 (1953).
167. W. Konigsberg and R. J. Hill, *JBC* **237,** 3157 (1962).
168. W. Konigsberg, J. Goldstein, and R. J. Hill, *JBC* **238,** 2028 (1963).
169. C. H. W. Hirs, S. Moore, and W. H. Stein, *JBC* **235,** 633 (1960).
170. F. Sanger and H. Tuppy, *BJ* **49,** 481 (1951).
171. A. P. Ryle, J. Leclerc, and F. Falla, *BJ* **110,** 4P (1968).

Tyr–Leu (16–17), however (*172*). The pepsinlike enzyme rennin (*173*) cleaves the B chain with major sites of hydrolysis at Leu–Val (17–18) and Phe–Phe (24–25) and with lesser action at Glu–Ala (13–14), Leu–Tyr (15–16), and Phe–Tyr (25–26); its initial action on casein is at a Phe–Met bond (*174–176*). Cathepsin D (from beef spleen) has been found (*39*) to cleave the B chain of insulin at Ala–Leu (14–15), Leu–Tyr (15–16), Tyr–Leu (16–17), Phe–Phe (24–25), and Phe–Tyr (25–26). The principal sites of cleavage by the pepsinlike enzyme from *Rhizopus chinensis* have been reported (*39c*) to be His–Leu (10–11), Leu–Val (11–12), Val–Glu (12–13), Leu–Tyr (15–16) Gly–Phe (23–24), Phe–Phe (24–25), and Phe–Tyr (25–26).

As noted above, the proteinase activity of pepsin is inhibited by a basic 29 amino acid peptide liberated on activation of pepsinogen. Inhibition is also effected not only by the basic poly-L-lysine, which binds to the enzyme (*177, 178*), but also by the acidic amylopectin sulfate, which presumably acts largely by interaction with the protein substrate so as to render it less susceptible to proteolytic attack (*179, 180, 180a*).

IV. Action of Pepsin on Synthetic Substrates

A. Specificity of Pepsin

The first known synthetic substrates for pepsin, of which Z–Glu–Tyr and Z–Glu–Tyr–NH_2 were simple prototypes (*181–183*), are cleaved very slowly, with a pH optimum near 4.0, and are sparingly soluble at the acid end of the pH range of peptic activity. The site of enzymic action

172. T. P. Levchuk, M. I. Levyant, and V. N. Orekhovich, *Biokhimiya* **30,** 986 (1965).
173. B. Bang-Jensen, B. Foltmann, and W. Rombauts, *Compt. Rend. Trav. Lab. Carlsberg* **34,** 326 (1964).
174. R. D. Hill and R. R. Laing, *Nature* **210,** 1160 (1966).
175. J. Jolles, C. Alais, and P. Jolles, *BBA* **168,** 591 (1968).
176. R. D. Hill, *BBRC* **33,** 659 (1968).
177. E. Katchalski, A. Berger, and H. Neumann, *Nature* **173,** 998 (1954).
178. W. G. Miller, *JACS* **86,** 3918 (1964).
179. D. L. Cook, S. Eich, and P. S. Cammarata, *Arch. Intern. Pharmacodyn.* **144,** 1 (1963).
180a. P. S. Cammarata, personal communication (1969).
181. J. S. Fruton and M. Bergmann, *Science* **87,** 557 (1938).
182. J. S. Fruton and M. Bergmann, *JBC* **127,** 627 (1939).
183. M. Bergmann and J. S. Fruton, *Advan. Enzymol.* **1,** 63 (1941).

is the CO–NH bond to which a L-tyrosyl (or L-phenylalanyl) residue contributes the NH group, and similar action was later reported for the peptic cleavage of Z–Cys–Tyr (*184*) and Z–Met–Tyr (*185*). Despite their shortcomings as substrates, Z–Glu–Tyr and Z–Glu–Tyr–OEt have been used for kinetic studies (*186, 187*); at pH 4 and 31.6°, k_{cat} = about 0.001 sec^{-1} and K_m = about 2 mM for both compounds.

An important advance was made by Baker (*50, 188*) who showed that acetyl dipeptides such as Ac–Phe–Tyr are cleaved by pepsin more rapidly than is Z–Glu–Tyr, although she failed to observe significant hydrolysis of Ac–Phe–Tyr–NH_2, presumably because of its sparing solubility. Baker's work provided the first indication that the action of pepsin on small substrates is favored by the presence of the side chains of aromatic amino acids on both sides of the sensitive peptide bond. Acetyl dipeptides such as Ac–Phe–Tyr or Ac–Phe–Phe (and their esters and amides) have been widely used in recent studies on the kinetics of pepsin action (see Section V,A). Ac–Phe–Tyr is hydrolyzed at pH 2 and 37° with k_{cat} = about 0.08 sec^{-1} and K_m = about 2 mM, and at pH 4 with k_{cat} = about 0.02 sec^{-1} and K_m = about 4 mM (*189, 190*). Among the most sensitive substrates of this type is Ac–Phe–TyrI_2, for which k_{cat} = 0.2 sec^{-1} and K_m = 0.08 mM at pH 2 and 37° (*51*), and Ac–Phe–TyrBr_2, for which k_{cat} = 0.2 sec^{-1} and K_m = 0.09 mM at pH 2 and 25° (*191*).

More recently, an extensive series of cationic substrates for pepsin has been developed. If the general structure of a pepsin substrate is denoted AX–YB, where X and Y are the amino acid residues forming the sensitive bond, the A groups of the new synthetic substrates have included Z–His (*57, 192–194*), Z–Gly–His (*57, 194*), Phe–Gly–His (*194*), Z–Lys (*195*), Gly–Gly (*192, 195*), isonicotinoyl (*196, 197*), and the B groups

184. C. R. Harington and R. V. Pitt-Rivers, *BJ* **38,** 417 (1944).
185. C. A. Dekker, S. P. Taylor, and J. S. Fruton, *JBC* **180,** 155 (1949).
186. E. J. Casey and K. J. Laidler, *JACS* **72,** 2159 (1950).
187. J. Tang, *JBC* **240,** 3810 (1965).
188. L. E. Baker, *JBC* **211,** 701 (1954).
189. W. T. Jackson, M. Schlamowitz, and A. Shaw, *Biochemistry* **5,** 4105 (1966).
190. N. G. Lutsenko, L. M. Ginodman, and V. N. Orekhovich, *Biokhimiya* **32,** 223 (1967).
191. E. Zeffren and E. T. Kaiser, *JACS* **88,** 3129 (1966).
192. K. Inouye, I. M. Voynick, G. R. Delpierre, and J. S. Fruton, *Biochemistry* **5,** 2473 (1966).
193. G. E. Trout and J. S. Fruton, *Biochemistry* **8,** 4183 (1969).
194. K. Medzihradszky, I. M. Voynick, H. Medzihradszky-Schweiger, and J. S. Fruton, *Biochemistry* **9,** 1154 (1970).
195. T. R. Hollands, I. M. Voynick, and J. S. Fruton, *Biochemistry* **8,** 575 (1969).
196. D. G. Doherty and J. James, *Federation Proc.* **27,** 784 (1968).
197. G. P. Sachdev and J. S. Fruton, *Biochemistry* **8,** 4231 (1969).

have included OMe, OEt, NH_2, as well as Ala, Ala–Ala, Ala–Ala–OMe, and Ala–Phe–OMe (*194*), and pyridine alkyloxy groups (*197*). Because the imidazolium, ammonium, or pyridinium groups of such substrates are largely protonated in the pH range of pepsin action, many of the protected peptides having such hydrophilic groups are moderately soluble in aqueous buffer solutions over the pH range 1–5. The most extensive kinetic studies have been performed with substrates related to Z–His–Phe–Phe–OEt, in which the Phe–Phe bond is cleaved by pepsin at pH 4.5 (near the pH optimum) and 37° with $k_{cat} = 0.5$ sec^{-1} and $K_m = 0.2$ mM (195). It is of interest that Z–His–Phe–L–phenylalaninol is resistant to pepsin action under conditions where Z–His–Phe–Phe–OEt is hydrolyzed rapidly (*57*). The most sensitive synthetic peptide substrate found thus far is the 4-pyridine propyl ester of Z–Ala–Ala–Phe–Phe, which is cleaved at the Phe–Phe bond with $k_{cat} = 260$ sec^{-1} and $K_m = 0.04$ mM at pH 3.5 and 37° (*197*).

Other known synthetic substrates of pepsin include *N*-gluconyl-Phe–Tyr–OMe, which is about three times as soluble as Ac–Phe–Tyr–OMe but is cleaved more slowly (*198*), and poly-L-glutamic acid (*199*, *200*), which is cleaved to oligopeptides, the tripeptide being the principal product of prolonged hydrolysis.

It should be noted that the above types of synthetic substrates are those for porcine pepsin A, and significant differences have been found with other pepsinlike enzymes. Thus, porcine pepsin C fails to cleave Ac–Phe–$TyrI_2$ at a measurable rate (*15*, *171*), chicken pepsin hydrolyzes this substrate at about 3% of the rate of porcine pepsin A (*31*, *32*), and human gastricsin hydrolyzes Z–Tyr–Ala at pH 2 and 37° (k_{cat} = about 0.004 sec^{-1}; K_m = about 0.7 mM) under conditions where this substrate is resistant to the action of human pepsin (*201*). Rennin has been found to hydrolyze Z–Glu–Tyr (*202*) and Z–His–Phe(NO_2)–Phe–OMe, but in the latter case the rate is about 0.5% of that for porcine pepsin A (*54*); moderately rapid cleavage of the Phe–Met bond in Ser–Leu–Phe–Met–Ala–OMe has been reported (*176*). With cathepsin D, it has been found that Gly–Phe–Leu–Gly–Phe–Leu is hydrolyzed only at the interior Phe–Leu bond, whereas pepsin attacks the Leu–Gly and the terminal Phe–Leu bonds as well (*203*).

198. M. Schlamowitz and R. Trujillo, *BBRC* **33**, 156 (1968).
199. E. R. Simons, G. D. Fasman, and E. R. Blout, *JBC* **236**, PC64 (1961).
200. H. Neumann, N. Sharon, and E. Katchalski, *Nature* **195**, 1002 (1962).
201. W. Huang and J. Tang, *JBC* **244**, 1085 (1969).
202. J. C. Fish, *Nature* **180**, 345 (1957).
203. H. Keilova, K. Blaha, and B. Keil, *European J. Biochem.* **4**, 442 (1968).

1. *Side Chain Specificity of Pepsin*

The most extensive systematic investigation of the side chain specificity of pepsin has been performed with the series of synthetic substrates AX–YB in which A = Z–His and B = OMe; in all cases, either X or Y was a L-phenylalanyl residue, and the enzymic action was restricted to the X–Y bond (*193*). It will be noted from Table IV that among the substrates of the type Z–His–X–Phe–OMe examined thus far, the two that are cleaved most rapidly are those in which X = Phe or *p*-nitro-L-phenylalanyl [Phe(NO_2)]. The favorable effect of the aromatic and planar substituent at the *β*-carbon of the X residue is emphasized by

TABLE IV
KINETICS OF PEPSIN ACTION ON SYNTHETIC SUBSTRATES OF THE TYPE Z-His-X-Y-OMe AT pH 4.0 AND 37°[a]

Substrate	k_{cat} (sec^{-1})	K_m (mM)	k_{cat}/K_m (sec^{-1} mM^{-1})
Z-His-Gly-Phe-OMe	0.0014	1.8	0.0007
Z-His-Ala-Phe-OMe	0.0023	1.7	0.0014
Z-His-Nva-Phe-OMe	0.011	1.0	0.011
Z-His-Nle-Phe-OMe	0.016	0.3	0.053
Z-His-Leu-Phe-OMe	0.017	0.5	0.034
Z-His-Met-Phe-OMe	0.010	0.5	0.020
Z-His-Phe-Phe-OMe	0.17	0.33	0.52
Z-His-Phe(NO_2)-Phe-OMe	0.29	0.46	0.63
Z-His-Phe(NO_2)-Pla-OMe	0.77	0.40	1.93
Z-His-Cha-Phe-OMe	0.015	0.17	0.088
Z-His-Tyr-Phe-OMe	0.009	0.30	0.030
Z-His-Mpa-Phe-OMe	0.088	0.34	0.26
Z-His-Trp-Phe-OMe	0.013	0.25	0.052
Z-His-Phe-Gly-OMe	0.0021	1.6	0.0013
Z-His-Phe-Ala-OMe	0.0037	1.8	0.0021
Z-His-Phe-Nva-OMe	0.0059	0.6	0.010
Z-His-Phe-Val-OMe	0.0038	0.5	0.0076
Z-His-Phe-Nle-OMe	0.007	0.4	0.018
Z-His-Phe-Ile-OMe	0.007	0.4	0.018
Z-His-Phe-*a*Ile-OMe	0.003	0.9	0.003
Z-His-Phe-Leu-OMe	0.0052	0.5	0.010
Z-His-Phe-Met-OMe	0.014	0.9	0.016
Z-His-Phe-Cha-OMe	0.0026	0.3	0.009
Z-His-Phe-Tyr-OMe	0.17	0.29	0.59
Z-His-Phe-Trp-OMe	0.51	0.20	2.55

[a] From Inouye and Fruton (*57*) and Trout and Fruton (*193*).

the finding that where X = β-cyclohexyl-L-alanyl (Cha), the value of k_{cat} is approximately one-tenth that for X = Phe. In this respect, the specificity of pepsin appears to be more restricted than that of chymotrypsin, which cleaves Ac–Cha–NH_2 at approximately the same rate as Ac–Phe–NH_2 (*204*). Since all the substrates in which the X position is occupied by aliphatic amino acid residues such as L-norvalyl, L-norleucyl, L-leucyl, and L-methionyl are hydrolyzed more rapidly than Z–His–Ala–Phe–OMe, it would appear that the side chains of these residues can interact with a portion of the enzymic region that binds planar aromatic groups. Under the conditions where Z–His–Gly–Phe–OMe is cleaved at a measurable rate, the comparable compound with X = L-valyl or L-isoleucyl is completely resistant to pepsin action, suggesting that steric hindrance is operative; the resistance of the Ile–Y bond in synthetic substrates of pepsin appears to be reflected in the peptic cleavage of proteins (*165*).

With regard to the substrates in Table IV of the type Z–His–Phe–Y–OMe, all the compounds having an aliphatic side chain in the Y residue (including those with branching at the β-carbon) are cleaved slowly at rates that are similar. The most sensitive among this group of substrates is Z–His–Phe–Trp–OMe, and those with Y = Phe or Tyr are cleaved somewhat less rapidly. That this preference for Trp, Phe, and Tyr in the Y position may be related to their aromatic character is suggested by the slow cleavage of the compound in which Y = Cha. The high susceptibility of Z–His–Phe–Trp–OMe and the relative resistance of the isomeric Z–His–Trp–Phe–OMe may be correlated with the fact that the latter compound exhibits substrate inhibition at moderate concentrations. It appears likely, therefore, that the affinity of the enzymic region that binds the Y side chain for the indolyl group is so great that the relative resistance of Z–His–Trp–Phe–OMe may be a consequence of strong nonproductive binding that brings the side chain of the X residue into the position normally occupied by the Y residue.

It may be concluded, therefore, that with small synthetic substrates of the type AX–YB, where the X–Y bond is broken, pepsin exhibits a preference for Phe in the X position and for Trp, Phe, or Tyr in the Y position. Only limited data are available on the effect of changing the A and B groups on the side chain specificity of pepsin, but it would appear that this conclusion holds for substrates in which the A group is changed from Z–His to acetyl or to Gly–Gly (*57*, *195*). Under the conditions of the studies reported in Table IV, where initial rates were measured, the presence of either a L-phenylalanyl residue in the X posi-

204. R. Foster and C. Niemann, *JACS* **77**, 1886 (1955).

tion or of a L-tryptophyl, L-tyrosyl, or L-phenylalanyl residue in the Y position is essential for measurable cleavage; thus, Z–His–Ala–Ala–OMe was found to be completely resistant under conditions that gave measurable rates of hydrolysis of Z–His–Gly–Phe–OMe or of Z–His–Phe–Gly–OMe (*193*). Although substrates of the type Z–His–Leu–Leu–OMe were not tested, it may be expected that they would be hydrolyzed by pepsin at the Leu–Leu bond.

2. *Stereochemical Specificity of Pepsin*

In its action on the Phe–Phe bond of substrates of the type A–Phe–Phe–B, pepsin exhibits an absolute requirement for the L-enantiomer in both the X and Y positions of the substrate. This has been demonstrated with substrates in which A = Z–His and B = OEt (*57*) and those with A = Ac and B = NH_2 (*205*). Diastereoisomeric compounds such as Z–His–D–Phe–Phe–OEt or Z–His–Phe–D–Phe–OEt are competitive inhibitors of the cleavage of the L–L–L substrate, and the K_I values found (0.2–0.3 mM) are the same as the kinetically determined value of K_m for Z–His–Phe–Phe–OEt under comparable conditions (*57*). Similar data have been obtained with the diastereoisomers of Ac–Phe–Phe and of Ac–Phe–Phe–NH_2, and it has further been shown that the K_I values obtained from kinetic experiments with such inhibitors are the same as the dissociation constants estimated from equilibrium-dialysis measurements (*205, 206*). Additional data in agreement with these conclusions have been obtained with diastereoisomers of Ac–Phe–TyrI_2 (*51, 207*) and of Ac–Phe–Tyr–OMe (*208*), and with Ac-aminocinnamoyl-TyrI_2 (*209*).

These data indicate that the stereochemical specificity of pepsin with respect to the X–Y unit of a substrate AX–YB is primarily an expression of the kinetic specificity of the enzyme, rather than of binding specificity, and that the side chains of the X–Y unit play an important role in positioning the X–Y bond for catalytic attack (*123, 193, 205*). Furthermore, the equivalence of K_I for D–L or L–D inhibitors and K_m for the corresponding L–L substrates provides evidence in favor of the view that for these substrates K_m approximates the dissociation constant of the enzyme–substrate complex and that the release of the first reaction product occurs after the rate limiting step in the overall process (Section V,B).

205. J. R. Knowles, H. Sharp, and P. Greenwell, *BJ* **113**, 343 (1969).
206. R. E. Humphreys and J. S. Fruton, *Proc. Natl. Acad. Sci. U. S.* **59**, 519 (1968).
207. R. M. Herriott, *J. Gen. Physiol.* **48**, 1145 (1965).
208. G. E. Clement and S. L. Snyder, *JACS* **88**, 5338 (1966).
209. M. S. Silver, *JACS* **87**, 1627 (1965).

3. *Influence of Secondary Interactions*

Kinetic studies with an extensive series of synthetic substrates of the type A–Phe–Phe–B, in which pepsin action is restricted to the Phe–Phe bond, have shown that variation of either the A or B portion of the substrate may produce large changes in the value of k_{cat}/K_m. One of the first examples was provided by three substrates of the type A–Phe–Phe–OEt, where A = Z–His, Z–Gly–His, or Z–His–Gly, and whose k_{cat}/K_m values (in sec^{-1} mM^{-1}; pH 4, 37°) are 2.6, 7.7 and 0.2, respectively (*57*, *195*); here changes in the location of both the hydrophobic benzyloxycarbonyl group and the cationic imidazolium group in relation to the Phe–Phe unit have a significant effect on the susceptibility of the Phe–Phe bond. Since the K_m values for these three substrates are all in the range 0.2–0.3 mM, it would seem that the differences in rate are a consequence of kinetic specificity.

The most striking evidence of the effect of secondary interactions on the catalytic efficiency of pepsin has come from studies with a series of substrates of the type A–Phe–Phe–B where A = Z, Z–Gly, or Z–Gly–Gly and the B portion is a pyridinium alkoxy group (*197*), and a series of the type A–Phe(NO_2)–Phe–B where Z = Z–His, Phe–His, Phe–Gly–His, or Phe–Gly–Gly–His and B = OMe, Ala, Ala–Ala, or Ala–Ala–OMe (*194*). Some of the data for the 4-pyridinium propyl (OP4P) esters are given in Table V; it will be noted that the presence of a hydrophobic benzyl group two amino acid residues away from the amino terminus of the sensitive Phe–Phe unit has a marked effect on catalytic efficiency

TABLE V

KINETICS OF PEPSIN ACTION AT 37° ON Phe–Phe BOND OF PYRIDYLALKYL ESTER SUBSTRATES[a]

Substrate	pH	k_{cat} (sec^{-1})	K_m (mM)	k_{cat}/K_m (sec^{-1} mM^{-1})
Z-Phe-Phe-OP4P	2.0	0.49	0.7	0.70
	3.0	0.74	0.2	3.7
Z-Gly-Phe-Phe-OP4P	2.0	2.2	1.1	2.0
	3.5	3.1	0.4	7.8
Z-Gly-Gly-Phe-Phe-OP4P	2.0	56.5	0.8	70.6
	3.5	71.8	0.4	180
Z-Ala-Ala-Phe-Phe-OP4P	3.5	260	0.04	6500
Gly-Gly-Phe-Phe-OP4P	3.5	3.8	1.3	3.0

[a] From Sachdev and Fruton (*197*).

without contributing significantly to a lowering of the K_m value. A decrease in the hydrophobic character of the B group, by replacing the 4-pyridinium propyloxy group by a 4-pyridinium methoxy group lowers the catalytic efficiency (*197*). In Table VI, the data for the effect of changes in the nature and location of hydrophobic groups in the A and B portions of A–Phe(NO_2)–Phe–B give further evidence for the considerable importance of secondary interactions in the cleavage of oligopeptides by pepsin. Thus, when the B group of Z–His–Phe(NO_2)–Phe–B is converted from OMe to Ala–Ala–OMe, the value of k_{cat}/K_m is increased about 300-fold; here the value of K_m drops significantly, but most of the effect is a consequence of the increase in k_{cat}. On the other hand, if the A group of A–Phe(NO_2)–Phe–OMe is changed from Z–His to Phe–His, the rate of pepsin action is very slow, but this is markedly increased by moving the amino-terminal phenylalanyl residue away from the Phe(NO_2)–Phe unit by the insertion of a glycyl residue (*194*). It will be noted from Table VI that for the series in which B = Ala, Ala–Ala, or Ala–Ala–OMe, K_m decreases with an increase in the size and hydrophobic character of the B group. Clearly, with the oligopeptide substrates having free α-amino or α-carboxyl groups, their proximity to the sensitive Phe–Phe unit may have a significant effect both on binding and on catalysis.

It is evident that by providing A and B groups that promote the catalytic efficiency of pepsin toward the X–Y bond of AX–YB, the susceptibility of that bond can be raised to a high level, and it may be

TABLE VI
KINETICS OF PEPSIN ACTION ON SYNTHETIC PEPTIDES AT pH 4.0 AND 37°[a]

Substrate	k_{cat} (sec^{-1})	K_m (mM)	k_{cat}/K_m (sec^{-1} mM^{-1})
Z-His-Phe(NO_2)-Phe-OMe	0.29	0.46	0.63
Z-His-Phe(NO_2)-Phe-Ala	3.0	0.75	4.0
Z-His-Phe(NO_2)-Phe-Ala-OMe	3.3	0.40	8.2
Z-His-Phe(NO_2)-Phe-Ala-Ala-OMe	28.0	0.12	233
Phe-His-Phe(NO_2)-Phe-OMe	0.008	0.17	0.04
Phe-Gly-His-Phe(NO_2)-Phe-OMe	0.10	0.40	0.25
Phe-Gly-Gly-His-Phe(NO_2)-Phe-OMe	0.27	0.56	0.47
Phe-Gly-His-Phe(NO_2)-Phe-Ala	1.65	1.20	1.4
Phe-Gly-His-Phe(NO_2)-Phe-Ala-Ala	19.2	0.27	71
Phe-Gly-His-Phe(NO_2)-Phe-Ala-Ala-OMe	27.8	0.16	174

[a] From Medzihradszky *et al.* (*194*).

expected therefore that the secondary interactions of an oligopeptide with pepsin may position at the catalytic site a bond that would be relatively resistant to hydrolysis if it were present in a small substrate such as Z–His–X–Y–OMe. One example is the case of Z–His–Phe(NO_2)–Gly–Ala–Ala–OMe, which is readily cleaved at the Phe(NO_2)–Gly bond under conditions where Z–His–Phe(NO_2)–Gly–OMe is resistant to hydrolysis (*194*). Another is the cleavage of Ala–Leu–Pro–Glu–Phe at the Leu–Pro bond when high substrate and pepsin concentrations are employed (*210*). Such considerations help to explain the apparent discrepancy between the specificity of pepsin in its action on the limited number of small synthetic substrates available until recently and on long-chain protein substrates subjected to extended digestion.

The fact that there is little correlation between the apparent binding energy and catalytic efficiency is evident from Tables V and VI; a striking example is the pair of substrates in which the A group is changed from Phe–His to Phe–Gly–His and the B group is changed from OMe to Ala–Ala–OMe, with little alteration in K_m but a 3000-fold increase in k_{cat}. In considering possible explanations for such rate enhancements, the possible contribution of nonproductive binding cannot be assigned predominant importance since, when Michaelis–Menten kinetics are obeyed, such binding has the same effect on K_m and k_{cat} (*211, 212*). It has been suggested that secondary interactions lead to increased catalytic efficiency by promoting the better positioning of the catalytic groups of the enzyme in relation to the sensitive peptide bond (*123*). Such better positioning might be a consequence either of a change in the conformation of the catalytic site itself or by a redistribution of the total binding energy so that the binding of the sensitive X–Y unit is weakened, thus favoring more efficient directed proton transfer and nucleophilic attack (*213, 214*). In this connection, it is of interest that acetylation of pepsin with acetyl imidazole (Section II,F) appears to block secondary binding sites rather than the primary binding site in the immediate vicinity of the catalytic center of pepsin (*206*) and that this chemical modification markedly increases the k_{cat} value for small synthetic substrates (*54*). An additional possible factor in the rate enhancement is that the secondary interactions may promote catalysis by increasing the distortion of the sensitive peptide bond, thus promoting the utilization of the potential binding energy in the enzyme–substrate interaction to

210. H. Determann and R. Köhler, *Ann. Chem.* **690,** 197 (1965).
211. G. E. Hein and C. Niemann, *JACS* **84,** 4495 (1962).
212. M. L. Bender and J. Kezdy, *Ann. Rev. Biochem.* **34,** 49 (1965).
213. J. H. Wang, *Science* **161,** 328 (1968).
214. T. R. Hollands and J. S. Fruton, *Proc. Natl. Acad. Sci. U. S.* **62,** 1116 (1969).

lower the free energy of activation in the catalytic process (for a review, see *215*).

B. Esterase Action of Pepsin

The ability of pepsin to hydrolyze ester linkage in suitable substrates has been demonstrated unequivocally by the use of the depsipeptide derivative Z–His–Phe(NO_2)–Pla–OMe (Pla = β-phenyl-L-lactyl), which is cleaved at the Phe(NO_2)–Pla bond (*57, 216*). The K_m value for this substrate is nearly the same as that for Z–His–Phe(NO_2)–Phe–OMe, but the value of k_{cat} for the cleavage of the ester linkage is about three times that for the hydrolysis of the Phe(NO_2)–Phe bond (Table IV). This finding is relevant to the consideration of the mechanism of pepsin action (Section V,G). An earlier report (*217*) on the esterase activity of pepsin described the cleavage of Ac–Phe–Pla; the ambiguities in the preparation of the substrate (*205*) and the extremely slow rate of hydrolysis made the validity of this report uncertain.

C. Cleavage of Organic Sulfites by Pepsin

Reid and Fahrney (*218*) discovered that several organic sulfite esters [e.g., PhOS(O)OMe, $(PhO)_2SO$] are rapidly hydrolyzed by pepsin; the liberation of phenol may be followed spectrophotometrically. At pH 4 and 25°, methyl sulfite is hydrolyzed with $k_{cat} = 0.16$ sec^{-1} and $K_m = 30$ mM. As evidence for the view that the same enzymic site is involved as in the cleavage of peptide substrates, it has been reported that such substrates (Z–Phe–Tyr and Ac–Phe–$TyrBr_2$) inhibit the peptic hydrolysis of sulfites with K_I values close to the K_m values for the peptides when tested as substrates under comparable conditions (*218, 219*) and that the cleavage of sulfites is abolished by treatment of pepsin with diazoacetyl-D,L-norleucine methyl ester (*218*). The rate of cleavage of the aromatic sulfite esters depends greatly on the nature of the substituents on the phenyl group; the most sensitive of the known sulfite substrates is bis-*p*-nitrophenyl sulfite for which $k_{cat} = 143$ sec^{-1} and $K_m = 0.08$ mM

215. D. E. Koshland and K. E. Neet, *Ann. Rev. Biochem.* **37,** 359 (1968).
216. K. Inouye and J. S. Fruton, *JACS* **89,** 187 (1967).
217. L. A. Lokshina, V. N. Orekhovich, and V. A. Sklyankina, *Nature* **204,** 580 (1964).
218. T. W. Reid and D. Fahrney, *JACS* **89,** 3941 (1967).
219. E. Zeffren and E. T. Kaiser, *ABB* **126,** 955 (1968).

at pH 2 and 25° (*220*). Because of the asymmetry about the pyramidal sulfur atom, the synthetic sulfite esters represent racemic mixtures, and the first resolution of such compounds has been effected by means of pepsin in the case of phenyl tetrahydrofurfuryl sulfite (*221*); with methyl phenyl sulfite, no stereochemical discrimination was evident, presumably because of the small size of the methyl group.

V. Kinetics and Mechanism of Pepsin Action

A. pH Dependence

Studies in several laboratories on the pH dependence of the peptic cleavage of acetyl dipeptides (Ac–Phe–Phe, Ac–Phe–Tyr, Ac–Phe–$TyrI_2$, Ac–Phe–$TyrBr_2$, and Ac–Phe–Trp) have shown that the pH optimum for both k_{cat} and k_{cat}/K_m is near pH 2 (*50, 51, 190, 222–226*), and the theoretical bell-shaped curves fitted to the data have been taken to indicate the presence, in both the enzyme–substrate complex and in the free enzyme, of two catalytically important prototropic groups of pK_a about 1 and about 4. The value of K_m for such acetyl dipeptide substrates is essentially invariant over the pH range 0.5–3.0, with a marked increase at more alkaline pH values, corresponding to the deprotonation of the substrate (*222, 223*). To obviate the complicating factor of the substrate dissociation over the pH range of pepsin action, kinetic data have been obtained for neutral esters and amides such as Ac–Phe–Tyr–OEt, Ac–Phe–Phe–NH_2, and Ac–Phe–Tyr–NH_2 (*190, 223–226*), for which the pH optimum is near pH 3; theoretical bell-shaped curves that have been fitted to the data are also consistent with the presence of an enzymic group of pK_a about 1, acting as the conjugate base, and an enzymic group of pK_a about 4, acting as an acid. It should be noted, however, that the relative resistance of these substrates to peptic hydrolysis has made it difficult to obtain kinetic data of a precision adequate to establish unambiguously the validity of the theoretical curves, and the sparing solubility of the neutral substrates required the addition of organic solvents (e.g., methanol and dimethylformamide) known to in-

220. S. W. May and E. T. Kaiser, *JACS* **91**, 6491 (1969).
221. T. W. Reid, T. P. Stein, and D. Fahrney, *JACS* **89**, 7125 (1967).
222. E. Zeffren and E. T. Kaiser, *JACS* **89**, 4204 (1967).
223. J. L. Denburg, R. Nelson, and M. S. Silver, *JACS* **90**, 479 (1968).
224. G. E. Clement, S. L. Snyder, H. Price, and R. Cartmell, *JACS* **90**, 5603 (1968).
225. A. J. Cornish-Bowden and J. R. Knowles, *BJ* **113**, 353 (1969).
226. W. T. Jackson, M. Schlamowitz, and A. Shaw, *ABB* **131**, 374 (1969).

hibit the cleavage of small substrates by pepsin (*187, 222, 225*). Furthermore, the pH-dependence behavior has been observed to change with the pepsin preparation employed, with no variation in k_{cat}/K_m for acetyl dipeptide esters over the pH range 2–4 when the enzyme was freshly prepared from pepsinogen (*224*).

In the case of the organic sulfites, the pH-dependence data for methyl phenyl sulfite showed a 3-fold increase in k_{cat} over the pH range 2–4 and were fitted to a sigmoid curve with an apparent pK_a of 2.6 (*218*).

The pH-dependence curves for the cationic substrates of the type Z–His–Phe–Phe–OEt exhibited little change in k_{cat} or k_{cat}/K_m over the pH range 1–3, but these parameters then rose to a maximum near pH 4.5 (*54*). Although the data for the pH range 1–4.5 could be fitted to a curve having a midpoint at pH 3.8, suggesting an interaction of the substrate with an anion of an acid having this pK_a value, other interpretations (e.g., change in the pK_a of the imidazolium group of the substrate when it is bound to pepsin) are possible. In this connection, it is of interest that Z–Gly–His–Phe–Phe–OEt and Z–His–Gly–Phe–Phe–OEt differ considerably, not only in the sensitivity of the Phe–Phe bond but also in the shape of the pH-dependence curves; whereas the first compound resembles Z–His–Phe–Phe–OEt in that k_{cat}/K_m increases about 9-fold in going from pH 2.5 to 4.5, the change in k_{cat}/K_m for the second compound over this pH range is only 2-fold and is largely a consequence of a lowering of K_m (*195*). Furthermore, for other cationic substrates (e.g., Bz–Lys–Phe–Phe–OEt, Gly–Gly–Phe–Phe–OEt, and isonicotinoyl-Phe–Phe–OP4P), k_{cat} for the hydrolysis of the Phe–Phe bond is also relatively invariant over the pH range 2.5–4.5, and the increase in k_{cat}/K_m is again a reflection of a decrease in the value of K_m (*195, 197*).

The apparent pK_a values in the range pH 2–5 assigned to catalytically important prototropic groups in free pepsin (from the pH dependence of k_{cat}/K_m) or in the enzyme–substrate complex (from the pH dependence of k_{cat}) fall in the range usually assigned to carboxyl groups, and the apparent pK_a near pH 1 has been assigned to a carboxyl group of unusual acidity. It should be noted, however, that comparison of pH-dependence data for different substrates is appropriate only when it is known that the several substrates combine with the same groups in the enzyme (*227*). In view of the probable presence of multiple carboxyl groups near the active site of pepsin, the possibility exists that the same ones may not be involved in the interaction with substrates of widely different structure. Studies on the Cu(II)-catalyzed reaction of diazoacetyl-D,L-norleucine methyl ester with pepsin have given further in-

227. L. Peller and R. A. Alberty, *JACS* **81,** 5907 (1959).

dication of the possible involvement of multiple carboxyl groups near the active site of the enzyme (*129*).

B. Inhibition of Pepsin Action

The initial studies on the kinetics of the hydrolysis of acetyl dipeptides such as Ac–Phe–Tyr were complicated by the inhibitory action of the Ac–Phe formed as a product (*188, 228, 229*). In more recent work, values reported for the K_I for Ac–Phe at pH 2 and 37° have included 23 mM (mixed competition) (*189*) and 29 mM (noncompetitive inhibition) (*230*). As was indicated in Section IV,A,2, diastereoisomers of acetyl dipeptide substrates are competitive inhibitors, and the K_I values are near the K_m values for the corresponding L–L substrates. Equilibrium-dialysis measurements have given data for the dissociation constant of the complex of pepsin with the L–D (or D–L) compound, and it is clear that in these cases K_I represents a true dissociation constant. Furthermore, the pH dependence of binding indicates invariance of this constant over the pH range 0.2–3, with a marked rise at higher pH values, corresponding to the deprotonation of the inhibitor (*205*); with neutral inhibitors (e.g., Ac–Phe–D–Phe–NH_2) there is little change over the entire pH range studied.

In the inhibition of the hydrolysis of Z–His–Phe(NO_2)–Phe–OMe [where the initial rate of release of Z–His–Phe(NO_2) is followed spectrophotometrically], Phe–OMe is a linear competitive inhibitor ($K_I = 22$ mM at pH 4 and 37°) as are Phe–OEt (10 mM) and Phe–Phe–OMe (0.25 mM); the last value is near the K_m for Z–His–Phe–Phe–OMe under comparable conditions (*231*). The resistant analog Z–His–Phe–L–phenylalaninol also has a K_I value near 0.25 mM, and even L-phenylalanine benzyl ester is a good inhibitor ($K_I = 2$ mM) (*231*). Z–His–Gly–Gly–OMe is a very weak inhibitor ($K_I =$ about 100 mM), and Gly–Phe–Gly–OMe (about 6 mM) and Gly–Gly–Phe–OMe (about 25 mM) are intermediate in their inhibitory effect (*193*). Similarly, it has been found that at pH 2 Ac–Gly–Gly has a K_I of about 400 mM (*232*) and that Ac–Phe–Gly and Ac–Gly–Phe are weaker inhibitors than Ac–D-Phe–Phe, Ac–Phe–D–Phe, or Ac–D–Phe–D–Phe (*230*).

It is clear from the foregoing that effective binding of an inhibitor or

228. L. E. Baker, *Nature* **178,** 145 (1956).
229. N. M. Green, *Nature* **178,** 145 (1956).
230. P. Greenwell, J. R. Knowles, and H. Sharp, *BJ* **113,** 369 (1969).
231. K. Inouye and J. S. Fruton, *Biochemistry* **7,** 1611 (1968).
232. M. Schlamowitz, A. Shaw, and W. T. Jackson, *JBC* **243,** 2821 (1968).

substrate to the active site of pepsin requires the presence of two adjacent hydrophobic groups, and the previous discussion of the side chain specificity of pepsin has indicated a preference for dipeptidyl units such as Phe–Phe, Phe–Tyr, or Phe–Trp (Section IV,A,1). The hydrophobic character of the interaction of the dipeptidyl unit with the active site of pepsin is also indicated by the finding that the association constants in the binding of substrate analogs are increased by increasing the ionic strength (from 0.1 to 2.0) or by raising the temperature (from 10° to 30°), and decreased by changing the buffer species from acetate to butyrate (*232a*). The reported inhibition of the action of pepsin on small synthetic substrates by various aliphatic alcohols (*187*, *222*), acetonitrile (*220*), and dimethylformamide (*225*), and the fact that such organic solvents increase K_m, are consistent with the view that these solvents act as inhibitors of the interaction of hydrophobic groups in the substrate with complementary loci in the enzyme. An extensive series of aromatic compounds (phenols and carboxylic acids) has been tested as inhibitors of the hydrolysis of Ac–Phe–Tyr at pH 2 and 37° (*232*), and the data provide further evidence for the predominant role of hydrophobic interactions in the action of pepsin on its substrates.

With hemoglobin as a substrate, inhibition by low concentrations of organic solvents is not observed because under the usual assay conditions the enzyme is saturated with substrate (*225*). It may be added that with some protein substrates, an increase in ionic strength from 0.01 to 0.1 enhances the rate of proteolysis, presumably through an effect on the substrate (*158*); such changes in ionic strength have no significant effect on the kinetics of pepsin action on small synthetic substrates (*51*, *54*, *222*).

The near identity of the K_I values for resistant substrate analogs has been taken as evidence for the view that K_m approximates K_S, the dissociation constant of the enzyme–substrate complex, and that the rate limiting step in the overall hydrolytic process precedes the release of the first reaction product (*205*, *222*, *223*, *231*). It should be noted, however, that most of the experimental data relate to the cleavage of peptide substrates that are hydrolyzed relatively slowly by pepsin (e.g., Ac–Phe–Tyr and Z–His–Phe(NO_2)–Phe–OMe). The possibility exists that for substrates with higher k_{cat} values (above 10 sec^{-1}), the assumption that $K_m = K_S$ may be invalid, because the rate of dissociation of the enzyme–substrate complex to free enzyme and substrate may not be much greater than the rate of conversion of the complex and the

232a. E. V. Raju, R. E. Humphreys, and J. S. Fruton, unpublished experiments (1969).

release of the first product may no longer be rate limiting (*123*). In the case of the peptic hydrolysis of diphenyl sulfite at pH 2 and 25°, the value of $K_m = 0.58$ mM is much lower than its K_I value (0.16 mM) when acting as an inhibitor of the hydrolysis of bis-*p*-nitrophenyl sulfite, a finding consistent with the view that the rate constants for the release of the two products are of a similar order of magnitude (*220*).

C. Condensation Reactions

Recent work on the "plastein reaction" (*233*) has clearly demonstrated the ability of pepsin to catalyze peptide bond synthesis by the direct condensation of α-amino and α-carboxyl groups. The studies of Wieland, Determann, and their colleagues (*234–239*) have shown that at pH 4, and high substrate concentration, oligopeptides are converted to polymeric products. The monomer must be at least a tetrapeptide with L-leucine, L-phenylalanine, or L-tyrosine as the terminal residues; L-alanine may also be present as the amino-terminal residue. Thus, Tyr–Leu–Gly–Glu–Phe is converted to a product that is, on the average, a pentadecapeptide. The stereochemical specificity of pepsin with respect to both amino acid units participating in the condensation is evident from the fact that neither D–Phe–Leu–Gly–Glu–Phe nor Tyr–Leu–Gly–Glu-D–Phe is converted to polymeric products; the presence of a carboxyl-terminal L-valine or L-isoleucine also blocks condensation. These findings are in accord with the available data on the specificity of pepsin in its hydrolytic action on small synthetic substrates and emphasize the importance of secondary interactions for the positioning of the α-amino and α-carboxyl groups for condensation (see Section IV,A). The thermodynamic feasibility of such condensation reactions is evident from the estimated equilibrium constant of unity for the process Ac–Phe + Tyr–OEt = Ac–Phe–Tyr–OEt + H_2O at pH 4.0 and 25° (*240*).

233. H. Wasteneys and H. Borsook, *Physiol. Rev.* **10,** 110 (1930).

234. T. Wieland, H. Determann, and E. Albrecht, *Ann. Chem.* **633,** 185 (1960).

235. H. Determann and T. Wieland, *Makromol. Chem.* **44–46,** 312 (1961).

236. H. Determann, O. Zipp, and T. Wieland, *Ann. Chem.* **651,** 172 (1962).

237. H. Determann, K. Bonhard, R. Köhler, and T. Wieland, *Helv. Chim. Acta* **46,** 2498 (1963).

238. H. Determann, S. Eggenschwiller, and W. Michel, *Ann. Chem.* **690,** 182 (1965).

239. H. Determann, J. Heuer, and D. Jaworek, *Ann. Chem.* **690,** 189 (1965).

240. L. V. Kozlov, L. M. Ginodman, V. N. Orekhovich, and T. A. Valueva, *Biokhimiya* **31,** 315 (1966).

D. Transpeptidation Reactions

The suggestion that proteolytic enzymes may be efficient catalysis of transfer reactions, not only to water but also to other acceptor molecules, was proposed in 1950 (*241, 242*), and it has received extensive experimental support (*243, 244*). Such transfer reactions, termed *transpeptidation* or *transamidation* reactions, may be of two types: (1) acyl transfer (also termed *carboxyl transfer*), in which the RCO portion of a substrate RCO–X is transferred to an acceptor (YNH_2 or YOH) to yield RCO–NHY or RCO–OY; and (2) imino transfer (also termed *amine transfer*), in which the NHR portion of a substrate X–NHR is transferred to an acceptor (YCOOH) to yield YCO–NHR. Acyl transfer reactions are catalyzed by chymotrypsin, trypsin, papain, ficin, etc. (*243*), as well as by other esterases (*245*), the "serine" enzymes being less efficient than the "cysteine" enzymes in this regard.

Neumann *et al.* (*244*) provided the first example of an enzyme-catalyzed imino transfer in demonstrating that pepsin promotes such reactions. Thus, when a substrate such as Z–Glu–Tyr or Z–Phe–Tyr is subjected to the action of pepsin at pH 4.7, chromatographically detectable amounts of Tyr–Tyr are formed; with Z–Glu–Tyr–NH_2 as the substrate, Tyr–NH_2 was the only ninhydrin-positive component observed. Neumann *et al.* explained these results by assuming the intermediate formation of an imino enzyme in the process RCO–NHR′ + E–OH = RCOOH + E–NHR′, followed by the reaction of the postulated E–Tyr with unreacted Z–X–Tyr to form Z–X–Tyr–Tyr, which was then cleaved at the X–Tyr bond to yield Tyr–Tyr. It was also reported that activation of pepsinogen at pH 3 yields a pepsin preparation that does not catalyze transpeptidation reaction but that treatment of this preparation at pH 2 restores the transferase activity (*246*); the status of this report is obscure. Later work showed that, in its specificity toward various benzyloxycarbonyl-L-amino acids as acceptors (substrate, Ac–Phe–Tyr), pepsin exhibits a preference for Z–Phe, Z–Tyr and Z–Leu; Z–Glu is less effective, and Z-Ala, Z–Val, Z–Ser, and Z–Gly are poor acceptors (*247*). Since pepsin action on Ac–Phe–Trp is accompanied by the formation of significant amounts

241. J. S. Fruton, *Yale J. Biol. Med.* **22**, 263 (1950).
242. C. S. Hanes, F. J. R. Hird, and F. A. Isherwood, *Nature* **166**, 288 (1950).
243. J. S. Fruton, *Harvey Lectures* **51**, 64 (1957).
244. H. Neumann, Y. Levin, A. Berger, and E. Katchalski, *BJ* **73**, 33 (1959).
245. M. I. Goldberg and J. S. Fruton, *Biochemistry* **8**, 86 (1969); **9**, 3371 (1970).
246. H. Neumann and N. Sharon, *BBA* **41**, 370 (1960).
247. N. I. Maltsev, L. M. Ginodman, V. N. Orekhovich, T. A. Valueva, and L. N. Akimova, *Biokhimiya* **31**, 983 (1966).

of Trp–Trp (*226*), it may be inferred that Z–Trp might also be an efficient acceptor.

Support for the view that an imino enzyme is an intermediate in pepsin-catalyzed reactions was provided by experiments on the isotope exchange of ^{14}C-labeled products with a pepsin substrate. Thus, when Z–Tyr–Tyr was incubated with labeled Z–Tyr, extensive labeling of the residual substrate was found, whereas incubation with labeled tyrosine gave negligible isotope exchange (*248*).

The report (*249*) that incubation of pepsin with tritiated Z–Tyr–Tyr gave results consistent with the formation of an imino enzyme of the kind postulated as an intermediate in pepsin-catalyzed transpeptidation reactions has been withdrawn (*249a*).

Although the available experimental evidence for a covalently bound imino enzyme intermediate is appreciable, it must be noted that the substrates hitherto used in studies on this question are relatively resistant as compared to those now available. It will be valuable, therefore, to reexamine the conclusions drawn from earlier transpeptidation and isotope exchange experiments with more sensitive synthetic substrates.

E. ^{18}O Exchange Reactions

Sharon *et al.* (*250*) showed that pepsin catalyzes the exchange of ^{18}O from water into the carboxyl group of a "virtual substrate" such as benzyloxycarbonyl-L-phenylalanine but not of its D-isomer. Similar isotope exchange reactions had previously been found for chymotrypsin (*251*, *252*) and papain (*253*), whose mechanisms include the formation of an acyl-enzyme intermediate, thus raising the question whether pepsin catalysis also involves such an intermediate. Later studies confirmed the pepsin-catalyzed ^{18}O exchange with virtual substrates and also showed some correlation between the rate of this exchange and the effectiveness of various acylamino acids as acceptors in transpeptidation reactions

248. J. S. Fruton, S. Fujii, and M. H. Knappenberger, *Proc. Natl. Acad. Sci. U. S.* **47**, 759 (1961).
249. M. Akhtar and J. M. Al-Janabi, *Chem. Commun.* p. 859 (1969).
249a. M. Akhtar, *Chem. Commun.* p. 361 (1970).
250. N. Sharon, V. Grisaro, and H. Neumann, *ABB* **97**, 219 (1962).
251. D. B. Sprinson and D. Rittenberg, *Nature* **167**, 484 (1951).
252. D. G. Doherty and F. Vaslow, *JACS* **74**, 931 (1952).
253. V. Grisaro and N. Sharon, *BBA* **89**, 152 (1964).

(*254–257*). Thus, for a series of acetyl-L-amino acids, the rate of ^{18}O exchange at pH 4 and 19.5° was greatest with Ac–Phe, less with Ac–Nle, Ac–Leu, Ac–Trp, and Ac–Nva, and negligibly slow with Ac–Ala and Ac–Gly (*257*). It is of interest to compare these results with those given in Table IV for the side chain specificity of pepsin with respect to the X residue of Z–His–X–Phe–OMe. The possible involvement of an acyl-enzyme intermediate in the pepsin-catalyzed hydrolysis of organic sulfites has been indicated by the incorporation of ^{18}O into the bisulfite ion formed from *p*-bromophenyl methyl sulfite when this compound is cleaved by pepsin (*258*).

It may be noted here that attempts to trap an acyl enzyme by the incubation of pepsin with Ac–Phe–Phe–Gly or Ac–Phe in the presence of ^{14}C-methanol failed to show the formation of significant amounts of radioactive Ac–Phe–OMe (*259*); this kind of experiment has been successful with enzymes such as chymotrypsin and papain (*212*). The report (*260*) that, when pepsin is incubated with Z–Tyr or Z–Tyr–Tyr in the presence of CT_3OH, radioactivity is incorporated into the protein by an ester or anhydride type of linkage, has been withdrawn (*249a*, *260a*).

Of special interest in relation to the interpretation of the data on the pepsin-catalyzed transfer of ^{18}O from water to virtual substrates is the finding (*261*) that the incubation of pepsin with $H_2{}^{18}O$ at pH 4 and 20° leads to the incorporation of ^{18}O into the enzyme, and that the rate of this isotope exchange is comparable to the rate of ^{18}O exchange with Ac–Phe. The labile enzymic groups (presumably carboxyl groups) that undergo isotope exchange appear to be located in the carboxyl-terminal half of the pepsin molecule (*262*). Furthermore, evidence has been presented for the identity of the enzymic carboxyl group selectively at-

254. L. V. Kozlov, L. M. Ginodman, B. M. Zolotarev, and V. N. Orekhovich, *Dokl. Akad. Nauk SSSR* **146,** 945 (1962).

255. L. V. Kozlov, L. M. Ginodman, and V. N. Orekhovich, *Dokl. Akad. Nauk SSSR* **161,** 1455 (1965).

256. N. I. Maltsev, L. M. Ginodman, and V. N. Orekhovich, *Dokl. Akad. Nauk SSSR* **165,** 1192 (1965).

257. L. V. Kozlov, L. M. Ginodman, and V. N. Orekhovich, *Dokl. Akad. Nauk SSSR* **172,** 1207 (1967).

258. T. P. Stein and D. Fahrney, *Chem. Commun.* p. 555 (1968).

259. A. J. Cornish-Bowden, P. Greenwell, and J. R. Knowles, *BJ* **113,** 369 (1969).

260. M. Akhtar and J. M. Al-Janabi, *Chem. Commun.* p. 1002 (1969).

260a. T. M. Kitson and J. R. Knowles, *Chem. Commun.* p. 361 (1970).

261. L. S. Shkarenkova, L. M. Ginodman, L. V. Kozlov, and V. N. Orekhovich, *Biokhimiya* **33,** 154 (1968).

262. L. S. Shkarenkova and L. M. Ginodman, *Biokhimiya* **33,** 1150 (1968).

tacked by diazoacetyl-D,L-phenylalanine ethyl ester (Section II,F) and one that exchanges readily with $H_2{}^{18}O$ (*263*).

Silver *et al.* (*263a*) have shown that in the ^{18}O exchange reaction, it is the carboxylate form of the acylamino acid that is the reactive species.

F. Deuterium Isotope Effects

Kinetic data on the effect of D_2O on the rate of cleavage of Ac–Phe-Tyr–OEt by pepsin have given a ratio of $k_{cat}(H_2O)/k_{cat}(D_2O)$ near unity (*208*, *224*), and a similar absence of a deuterium isotope effect has been found for the peptic hydrolysis of methyl phenyl sulfite (*218*). On the other hand, with Gly–Gly–Gly–Phe(NO_2)–Phe–OMe as the substrate, a significant deuterium isotope effect [$k_{cat}(H_2O)/k_{cat}(D_2O)$ = about 2] was observed at pH 4.0 (pD 4.4), with no change in K_m (*214*). A satisfactory explanation of this discrepancy is not readily apparent, and further work is needed to resolve the question. It is clear that the presence of a deuterium isotope effect is consistent with the view that a proton transfer is part of the rate limiting step in the catalytic action of pepsin, and the differences in the above results may reflect differences in the nature of the rate limiting step in the cleavage of the various substrates.

G. Theories of Pepsin Action

Several hypotheses have been proposed recently regarding the mechanism of pepsin action, but none of them can be said to have won general acceptance thus far. Aside from the inherent limitations of kinetic methods to elucidate enzyme mechanisms, the present situation for pepsin is in part a consequence of the slow progress in the elucidation of the complete amino acid sequence and in the X-ray diffraction analysis of the three-dimensional structure of the active site (*264*).

As indicated in Section V,B, it is probable that for the peptic hydrolysis

263. L. M. Ginodman, T. A. Valueva, L. V. Kozlov, and L. S. Shkarenkova, *Biokhimiya* **34**, 209 (1969).

263a. M. S. Silver, M. S. Stoddard, and T. P. Stein, *JACS* **92**, 2883 (1970).

264. N. S. Andreeva, V. V. Borisov, N. N. Govorin, V. R. Malik-Adamyan, V. S. Raiz, V. A. Rostovtsev, and N. E. Shutchkever, *Dokl. Akad. Nauk SSSR* **192**, 216 (1970).

of substrates of the type Ac–Phe–Tyr or Z–His–Phe–Phe–OEt the kinetically determined value of K_m approximates the value of K_S (the dissociation constant of the enzyme–substrate complex k_{-1}/k_1); in these cases, the rate limiting step in the overall process may be associated with k_2, the rate of cleavage of the sensitive bond, and precedes the release of the first reaction product. This conclusion may not apply to the hydrolysis of more sensitive substrates, for which K_m may equal K_S $(k_3/(k_2 + k_3))$ and $k_{cat} = k_2k_3/(k_2 + k_3)$, where k_3 denotes a subsequent step, possibly the release of the second reaction product. The fact that pepsin exhibits a specificity requirement for hydrophobic L-amino acid residues on both sides of the sensitive bond raises the possibility that the cleavage of this bond may not be followed in all cases by the rapid release of the first reaction product as in the hydrolysis of the *p*-nitrophenyl esters of acetyl amino acids by chymotrypsin. It has been inferred (*230*) from the noncompetitive inhibition of Ac–Phe of the peptic hydrolysis of Ac–Phe–Phe–Gly (*230*), and the linear competitive inhibition by Phe–OMe of the hydrolysis of Z–His–Phe(NO_2)–Phe–OMe (*231*), that pepsin catalysis involves an ordered release of products (*265*), with the RCOOH product of the hydrolysis of RCO–NHR′ leaving first. However, attempts to observe a "burst" release of the RCOOH product, as would be expected if $k_2/k_3 \gg 1$, have been unsuccessful thus far with substrates such as Z–His–Phe(NO_2)–Phe–OMe (*57*) or acetyl-3,5-dinitro-L-tyrosyl-L-phenylalanine (*259*) under conditions where the substrates concentration is greater than K_m. For such substrates, therefore, the rate limiting step appears to precede the release of the first reaction product.

Although the data suggesting an ordered release of products, and the findings on transpeptidation and the exchange of labeled products (Section V,D), are consistent with the idea that a covalently bound imino-enzyme intermediate is formed, these results do make such an intermediate obligatory. The importance for specificity of the side chain of the amino acid residue contributing the NH group and the ready reversibility of peptide hydrolysis at acid pH values (*240*) raise the possibility that noncovalent interactions may be involved in holding the $R'NH_2$ product in such a way as to produce the kinetic equivalent of an imino enzyme.

A variety of experimental data (e.g., pH dependence and specific inhibition by diazo compounds) supports the view that at least two carboxyl groups of pepsin are involved in its catalytic mechanism. The simple idea that an enzymic carboxyl group reacts with the NH group of the sensitive bond to form an intermediate diacylimino compound

265. W. W. Cleland, *BBA* **67,** 104, 173, and 188 (1963).

(*266*), with the subsequent release of the RCOOH product, does not account for the ^{18}O exchange with virtual substrates. It has been suggested (*212*) that two carboxyl groups of the enzyme may reversibly form an anhydride, which can undergo a four-center exchange reaction with the peptide substrate to form an imino enzyme with the NHR′ part of the substrate and an acyl enzyme with the RCO part of the substrate; several items of experimental data are inconsistent with the idea of such an anhydride structure in the active center of the enzyme (*223*, *224*, *258*).

A more plausible hypothesis assumes the concerted reaction of an enzymic carboxylate group with the RCO portion of the substrate, and of an enzymic carboxyl group with the NH group, to form an intermediate that is both an acyl enzyme and an imino enzyme, the acyl-enzyme bond being broken first (*222*, *224*). Another hypothesis invokes the participation of three carboxyl groups and suggests that the enzymic carbonyl group that forms the imino enzyme is derived from an anhydride, a third carboxyl forming the acyl enzyme (*249*). At present, the status of the acyl-enzyme intermediate is unclear, however.

An alternative hypothesis (*120*) that does not involve an acyl-enzyme intermediate suggests that the mechanism of pepsin action on a substrate RCO–NHR′ involves the attack of an enzymic carboxylate group ($ECOO^-$) at the carbonyl carbon of the protonated amide group to form reversibly a tetrahedral intermediate which undergoes a reversible four-center exchange reaction leading to the expulsion of RCOOH and the formation of an imino enzyme. The latter intermediate, on reaction with water (or a carboxylic acid), would yield $R'NH_2$ (or lead to transamidation by imino transfer) with the regeneration of $ECOO^-$. An essential part of the proposed mechanism is the protonation of the amide group either before, or concurrently with, the formation of the tetrahedral intermediate assumed to be associated with the transition state in the catalytic process, and a deuterium isotope effect would be expected (*214*). The pH range of pepsin action suggested that a COOH group might be the proton source. This mechanism (Fig. 1) thus involves the participation of at least two enzymic carboxyl groups, one of which acts as a proton donor and the other (in its dissociated form) as a nucleophile. A mechanism of this type is formally analogous to the one proposed for the cleavage of amide substrates by chymotrypsin, with an imidazolium group as the proton donor and the alkoxide ion derived from a serine hydroxyl as a nucleophile (*213*, *267*).

266. C. Zioudrou and J. S. Fruton, *Biochemistry* **5**, 2468 (1966).
267. J. H. Wang and L. Parker, *Proc. Natl. Acad. Sci. U. S.* **58**, 2451 (1967).

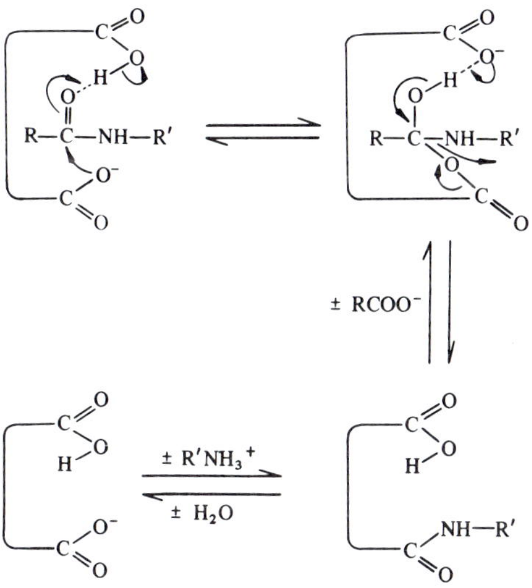

FIG. 1. A proposed mechanism of pepsin action (*120, 268*).

This hypothesis has been developed more fully by Knowles (*268*), who has adduced additional evidence in its support. He has noted that the ratio of the rates of hydrolysis of ester and amide substrates of pepsin (*57*) is consistent with a mechanism involving acid catalysis and has explained the pepsin-catalyzed ^{18}O exchange between acylamino acids and $H_2{}^{18}O$, not on the basis of an acyl-enzyme intermediate but in terms of the observation (*261*) that the enzyme itself becomes rapidly labeled with ^{18}O and that the rate of loss of this label corresponds to the rate of labeling of a suitable virtual substrate.

An additional hypothesis (*269*) of great interest assumes that pepsin has a slightly distorted peptide bond at its active center; rapid protona-

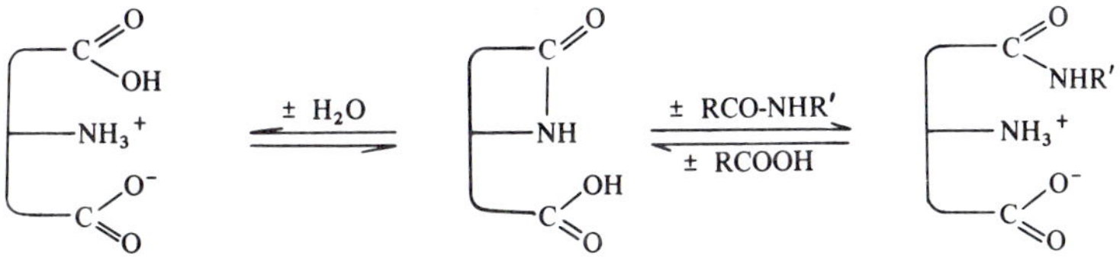

FIG. 2. A proposed mechanism of pepsin action (*269*).

268. J. R. Knowles, *Phil. Trans. Roy. Soc. London*, **B 257,** 135 (1970).

269. J. H. Wang, *in* "Structure–Function Relationships of Proteolytic Enzymes" (P. Desnuelle, H. Neurath, and M. Otteson, eds.), p. 251. Munksgaard, Copenhagen, 1970.

tion of the N of the CO–NH group by an adjacent COOH group would make the carbonyl carbon of the CO–NH group a strong electrophile, which could react either with water or with a substrate (RCO–NHR′), as shown in Fig. 2.

5

Chymotrypsinogen: X-Ray Structure

J. KRAUT

I. Introduction

The material to be covered here would logically have been incorporated into the chapter by Blow on chymotrypsin, instead of being presented as a relatively independent entity, except for the circumstance that the results described herein were still being prepared for publication while that chapter was being written. Therefore, sacrificing organizational elegance in the interest of currency, the authors of the two chapters have chosen to describe the three-dimensional structure of bovine chymotrypsinogen A and its relationship to the activation process separately. Throughout the rest of this chapter, bovine chymotrypsinogen A will be referred to simply as *chymotrypsinogen* or *the zymogen.* Almost all the material discussed here is based upon the paper by Freer *et al.* (*1*), wherein the preliminary results of the X-ray analysis of chymotrypsinogen at 2.5 Å resolution are presented. It is important to

1. S. T. Freer, J. Kraut, J. D. Robertus, H. T. Wright, and Ng. H. Xuong, *Biochemistry* 9, 1997 (1970).

emphasize the word preliminary since experience has shown that protein structures determined by X-ray analysis have a decided tendency to undergo continuous minor modification over a period of years as small errors in the original model are discovered, electron density maps are gradually improved, errors in amino acid sequences are corrected, water molecules are discovered, and, most especially, as the investigators come to see subtle but significant features that may have escaped notice in the first burst of enthusiasm over a newly determined structure. In the case of chymotrypsinogen, even as this is being written more careful comparison of the enzyme and zymogen structure is being carried out both in Cambridge and La Jolla, and improved electron density maps are being computed; thus, there is little doubt that the story presented here will soon prove to be incomplete. Nevertheless, some interesting and thought-provoking observations have already been made and will be summarized in this chapter.

The phenomenon of induction of biological activity by limited proteolysis of an inactive precursor is fairly widespread in nature. It has been found in various forms in such widely separated types of organisms as bacteria, yeast, green plants, invertebrates, and vertebrates. The subject has been reviewed recently in detail by Ottesen (*2*), and so it need not be extensively surveyed here. The point to be made is that in a sense it represents a kind of biochemical amplification and control mechanism of which the conversion of chymotrypsinogen to chymotrypsin in the vertebrate duodenum, under the influence of the enzyme trypsin, is only one relatively simple though well-known example. Another much more complex example of limited proteolysis as a biological amplification and control mechanism is seen in the clotting blood. Here, at least eight distinct factors interact in a cascading sequence that ultimately results in the conversion of soluble fibrinogen into an insoluble stabilized fibrin clot (*3*). A perhaps even more complex case familiar to the immunologist is that of the complement system. Enzymic cleavage is an essential feature of the activation mechanism of several complement components, of which 11 are now recognized (*4*). Nor is the phenomenon of activation by limited proteolysis restricted to enzymes. The hormone insulin, which plays an important role in glucose metabolism, is now known to be the product of limited proteolysis of a single-chain proinsulin molecule. The latter is synthesized in specialized cells of the pancreas, where it is activated prior to secretion by the splitting out of a polypeptide connecting

2. M. Ottesen, *Ann. Rev. Biochem.* **36,** 55 (1967).
3. E. W. Davie and O. D. Ratnoff, *in* "The Proteins" (H. Neurath, ed.), 2nd ed., Vol. 3, p. 359. Academic Press, New York, 1965.
4. H. J. Müller-Eberhard, *Ann. Rev. Biochem.* **38,** 389 (1969).

the C-terminus of the insulin B chain with the N-terminus of the A chain (*5*). Thus, with the detailed architecture of the chymotrysinogen molecule now becoming clear, and the model of chymotrypsin itself already well established, we have our first glimpse of one simple example of this important type of biological phenomenon at the level of molecular structure.

II. The Activation Process

There are several steps involved in the process of activating chymotrypsinogen to give the various recognized forms of chymotrypsin. It is worthwhile considering what is known of this process and how the various chymotrypsins are related in order to help evaluate the significance of the structural differences, to be described below, between the zymogen and the α form of the active enzyme. A review of the chemistry of the activation process as it was understood in 1960 has been given by Desnuelle (*6*) in the previous edition of "The Enzymes." Although Desnuelle's review antedated the complete sequence determination, from the point of view of classical enzyme chemistry there is not very much that is new to be said about the subject; thus it will be only briefly summarized here.

Figure 1 is reproduced from Wright *et al.* (*7*). It is a suggested modification of the scheme generally accepted until now (*6*) but will nevertheless serve as a convenient summary of the nomenclature and chemical relationships among the forms of chymotrypsin. The proposed scheme differs from the classical one in that it supposes α-chymotrypsin to be conformationally different from γ-chymotrypsin, and to be produced only by activation of the neo-chymotrypsinogen that has the dipeptide Thr 147–Asn 148 deleted, and not directly by degradation of δ-chymotrypsin. Well-established features of the activation scheme are as follows:

(1) Chymotrypsinogen is converted to the fully active π-chymotrypsin by trypsin catalyzed hydrolysis of the peptide bond between Arg 15 and Ile 16. It is this peptide-bond scission, and this one alone, that generates enzymic activity. All subsequent products result from autodegradation of the π-chymotrypsin molecule.

(2) δ-Chymotrypsin is missing the dipeptide Ser 14–Arg 15; δ-chymo-

5. D. F. Steiner, J. L. Clark, C. Nolan, A. R. Rubenstein, E. Margoliash, B. Aten, and P. E. Oyer, *Recent Progr. Hormone Res.* **25**, 207 (1969).

6. P. Desnuelle, "The Enzymes," 2nd ed., Vol. 4, p. 93, 1960.

7. H. T. Wright, J. Kraut, and P. E. Wilcox, *JMB* **37**, 363 (1968).

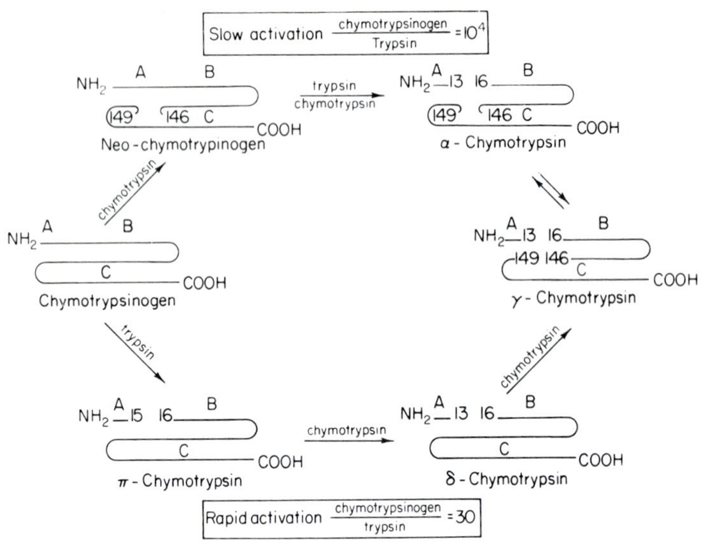

FIG. 1. Genesis of the various forms of chymotrypsin.

trypsin is the predominant form obtained when chymotrypsinogen is activated rapidly by relatively large amounts of trypsin.

(3) Both α- and γ-chymotrypsin are missing two dipeptides, Ser 14–Arg 15 and Thr 147–Asn 148; α-chymotrypsin is the predominant form obtained when chymotrypsinogen is activated slowly by relatively small amounts of trypsin, and γ-chymotrypsin is obtained from the slowly activated preparation only after standing for long times at high pH, e.g., 5 days at pH 5.6 (*8*).

(4) Conversion of γ- to α-chymotrypsin is possible but very slow, requiring several months at pH 4.0 (*8*).

(5) Neo-chymotrypsinogen designates any one of seven theoretically possible members of the class of enzymically inactive zymogen molecules that have had one or more of the following three peptide bonds broken by chymotrypsin itself: Leu 13–Ser 14, Tyr 146–Thr 147 or Asn 148–Ala 149 (*9*). Three of the seven have been characterized, and are activatable by tryptic cleavage of the critical bond Arg 15–Ile 16.

Extensive physicochemical studies have been carried out in various laboratories during the past 15 years in an effort to learn something about the structural basis of the zymogen to enzyme conversion. Neu-

8. R. B. Corey, O. Battfay, D. A. Brueckner, and F. G. Mark, *BBA* **94**, 535 (1965).
9. M. Rovery, M. Poilroux, A. Yoshida, and P. Desnuelle, *BBA* **23**, 608 (1957).

rath *et al.* (*10*) first observed that a change in optical rotation accompanies the appearance of enzymic activity during activation. Several more detailed investigations, utilizing ORD (*11, 12*) and CD (*13*) have resulted in a variety of interpretations as to whether or not additional helix is being formed. Hess and his collaborators have shown that the new free terminal amino group at Ile 16 is required to maintain the enzyme in an active conformation (*14*), and a beautiful explanation of this observation has been provided by the X-ray structure of α-chymotrypsin, as described in Chapter 6 by Blow, with the discovery that the free $-NH_3^+$ group of Ile 16 forms an internal ion pair with the side chain carboxylate group of Asp 194.

III. X-Ray Crystallographic Results

It should be emphasized at the outset that the model of chymotrypsinogen is still, at the time this is written, in a relatively unrefined state compared to the α-chymotrypsin model. Anomalous dispersion data have been measured but have not yet been included in the refinement calculations, nor has the model been matched against the electron density map with the precision made possible by development of the Richards optical comparitor (*15*). For these reasons, a table of coordinates is not included in this chapter. Nevertheless, the zymogen structure is now sufficiently detailed and reliable that many interesting observations can be made concerning the stereochemistry of the activation process, and perhaps a few meaningful questions can be asked.

An important point for the purposes of the chapter is that the three-dimensional structures of all forms of the active enzyme (that is, the α, γ, δ, and π forms) are essentially identical, at least in the crystals and to within the resolution limits of the current X-ray diffraction data. Therefore, it is assumed throughout this chapter that all significant features of the conformational changes undergone by the zymogen when it is activated will be visible upon comparison of the structures of α-chymotrypsin and the zymogen, again with the usual cautionary note about resolution. (Of course, there is no guarantee implied that, because

10. H. Neurath, J. A. Rupley, and W. J. Dreyer, *ABB* **65**, 243 (1956).
11. D. N. Raval and J. A. Schellman, *BBA* **107**, 463 (1965).
12. R. Biltonen, R. Lumry, V. Madison, and H. Parker, *Proc. Natl. Acad. Sci. U. S.* **54**, 1412 (1965).
13. G. Fasman, R. J. Foster, and S. Beychok, *JMB* **19**, 240 (1966).
14. H. L. Oppenheimer, B. Labouesse, and G. P. Hess, *JBC* **241**, 2720 (1966).
15. F. M. Richards, *JMB* **37**, 225 (1968).

they are visible, such significant conformational changes will necessarily have been noticed by the author.)

Low resolution X-ray diffraction studies have shown that PMS-π-, PMS-δ-, PMS-γ-, and active γ-chymotrypsin are crystallographically isomorphous and that all of the prominent features of difference Fourier maps among the various members of this group correspond to the expected missing dipeptides or added PMS inhibitor group (*7*, *16*). Therefore, it can safely be assumed that any conformational differences between members of the π, δ, and γ group are quite minor.

It is much more difficult to decide whether there is any conformational difference between α- and γ-chymotrypsin, or whether they are merely different crystalline forms of the same molecule, since the only well-established distinction between α- and γ-chymotrypsin is in their crystal form. The first crystallizes from 0.4 saturated $(NH_4)_2SO_4$, pH 4.0, in the monoclinic space group $P2_1$ with 2 molecules per asymmetric unit (i.e., as the dimer), and the latter crystallize from 0.4 saturated $(NH_4)_2SO_4$, pH 5.6, in the tetragonal space group $P4_22_12$ with one molecule per asymmetric unit (*8*). Figure 1 implies that there is indeed some conformational difference by showing the ends curled back at the break between residues 146 and 149 in the representation of α-chymotrypsin as well as in neo-chymotrypsinogen. The rationale for this was essentially as follows: (1) α-chymotrypsin is known to dimerize in solution at pH 4; (2) α-chymotrypsin crystallizes as dimers, in which Try 146 and Ala 149 play a role in dimer formation; (3) δ-chymotrypsin has been reported not to dimerize under the same conditions; (4) δ-chymotrypsin crystallizes in a form which does not contain a twofold axis corresponding to the local dimer axis found in α-chymotrypsin crystals; (5) as already mentioned above, crystals of PMS-δ-, PMS-γ- and active γ-chymotrypsin are isomorphous. All this suggests that there is some small conformational difference between α- and γ-chymotrypsin that expresses itself in their differing dimerization behavior. Nevertheless, X-ray crystallographic study (*7*, *17*, *18*) has thus far failed to reveal any obvious difference between the geometries of the two molecules. In order to determine that α- and γ-chymotrypsin really are different conformational states of the molecule rather than merely different crystal forms, it would be necessary to show that they were distinguishable in solution,

16. J. Kraut, H. T. Wright, M. Kellerman, and S. T. Freer, *Proc. Natl. Acad. Sci. U. S.* **58**, 304 (1967).
17. B. W. Matthews, G. H. Cohen, E. W. Silverton, H. Braxton, and D. R. Davies, *JMB* **36**, 179 (1968).
18. G. H. Cohen, E. W. Silverton, B. W. Matthews, H. Braxton, and D. R. Davies, *JMB* **44**, 129 (1969).

e.g., that α-chymotrypsin dimerizes under conditions where γ-chymotrypsin does not or, better still, that they displayed different enzyme-kinetic behavior.

To reiterate the point made at the beginning of this section, then, it appears to be well established that all forms of the active enzyme have very similar if not identical conformations. This is important because comparison between the zymogen and enzyme structures has been made using the Cambridge model of α-chymotrypsin, which is essentially an autodegraded form of the initial zymogen activation product, π-chymotrypsin. If it had happened that α-chymotrypsin assumed a different conformation from π-chymotrypsin, then we would not know which of the changes we see between the zymogen and the α molecule were due to zymogen activation, and hence presumably significant for the genesis of enzymic activity, and which were merely incidental consequences of the removal of the dipeptides Ser 14–Arg 15 and Thr 147–Asn 148 in going from π to α. In particular, one of the changes we believe may be significant, that at Arg 145, falls in a chain segment that includes the second of these dipeptides. We are fortunate that this question does not arise.

The first and most obvious result to come out of the comparison of the zymogen and enzyme structures is that the overall folding of the two molecules is very similar. No extensive rearrangement of the backbone chain has resulted from the cleavage of the critical Arg 15–Ile 16 peptide bond. In this respect, the conclusions drawn from earlier 5 Å crystallographic studies of the zymogen (*19*, *20*) and the π-, δ-, and γ-chymotrypsin group (*21*) are fully substantiated. Thus, the general description of the conformation of α-chymotrypsin given in the chapter by Blow is also applicable to the zymogen. Both molecules are composed almost entirely of a more or less fully extended polypeptide chain which often folds back on itself to form large sections of distorted antiparallel pleated sheet.

Before turning to a discussion of the details of the comparison between zymogen and enzyme structures, it would be appropriate first to dispose of one point concerning the structure of the zymogen itself. Confirming earlier conclusions based on low-resolution X-ray studies of chymotrypsinogen and the π, δ, and γ family of chymotrypsins (*21*), the two dipeptides Ser 14–Arg 15 and Thr 147–Asn 148 which are split out on con-

19. J. Kraut, L. S. Sieker, D. F. High, and S. T. Freer, *Proc. Natl. Acad. Sci. U. S.* **48,** 1417 (1962).
20. J. Kraut, D. F. High, and L. C. Sieker, *Proc. Natl. Acad. Sci. U. S.* **51,** 839 (1964).
21. J. Kraut, H. T. Wright, M. Kellerman, and S. T. Freer, *Proc. Natl. Acad. Sci. U. S.* **58,** 304 (1967).

version to δ- and α- or γ-chymotrypsin are located on the surface of the zymogen molecule and are easily accessible to enzymic attack. Indeed, the side chain of Arg 15 is pointing straight out into the surrounding medium and is an excellent candidate for attack by the activating enzyme, trypsin. Interestingly, chymotrypsinogen contains three other arginine residues, 145, 154, and 230, but the side chains of all three lie in surface crevices in our present model. However, there are in addition 14 lysine residues in chymotrypsinogen, and the side chains of most of these project outward. There is no obvious reason why some of them should not be rapidly attacked by trypsin as well.

We return now to compare the zymogen and active enzyme in detail. The high degree of similarity between the two molecules is readily apparent by comparison of Figs. 2 and 3, drawn from computer-plotted

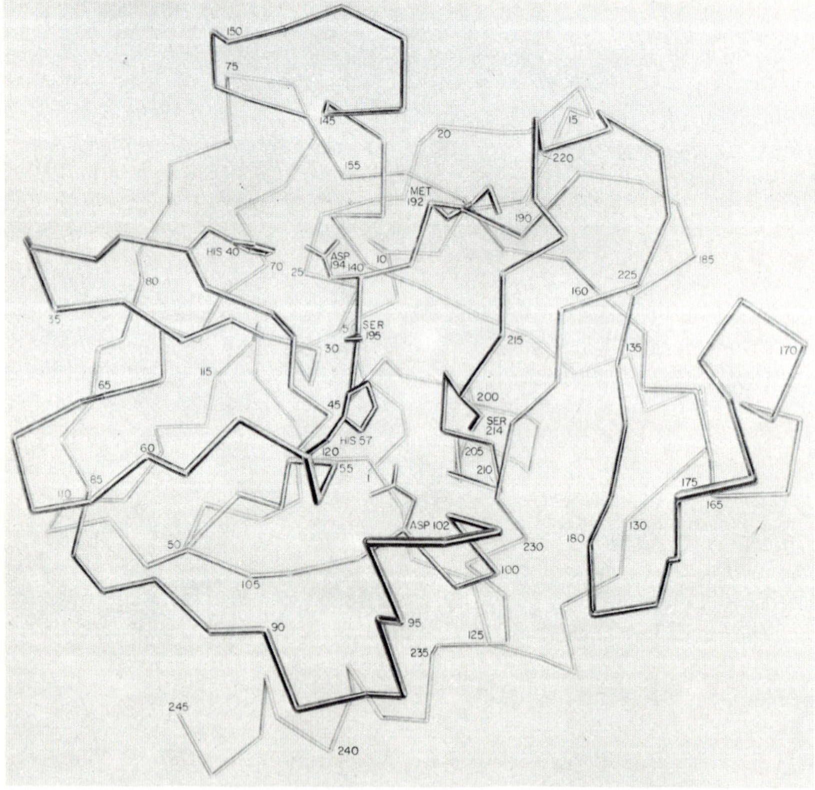

FIG. 2. Chymotrypsinogen—simplified backbone chain linking the α-carbon atoms of each amino acid residue, drawn from a vantage point looking toward the latent catalytic site region. The side chains of some important amino acid residues are also shown. The drawing was prepared from a computer-plotted perspective projection of all α-carbon atoms.

FIG. 3. α-Chymotrypsin—simplified backbone chain linking the α-carbon atoms of each amino acid residue, drawn from a vantage point equivalent to that of Fig. 2. The side chains of some important amino acid residues are also shown. This drawing and all subsequent α-chymotrypsin drawings were prepared from the coordinates of tosyl-α-chymotrypsin. Accordingly, the side chain of Ser 195 is shown in a position appropriate to the inhibited enzyme. In native α-chymotrypsin this side chain is rotated slightly about its α–β bond and assumes a position close to that shown for the zymogen in Fig. 2. The drawing was prepared from a computer-plotted perspective projection of the α-carbon atoms. A portion of the chain from residues 9–13 is not shown because of the uncertainty in position of these residues (*22*).

perspective projections of the α-carbon coordinates for the zymogen (Fig. 2) and for tosyl-α-chymotrypsin (*22*) (Fig. 3). The coordinates of the latter were first rotated to minimize the sum of squares of distances between corresponding α-carbon atoms in the two molecules. The mean displacement between all corresponding α-carbon atoms is only

22. J. J. Birktoft, B. W. Matthews, and D. M. Blow, *BBRC* **36**, 131 (1969).

1.8 Å, which may be considered as some sort of index of how slight the difference in overall conformation actually is. There are, however, a few segments of backbone chain that move considerably upon activation. Table I lists all residues for which the α-carbon atom is displaced by more than 3.6 Å, a cutoff chosen rather arbitrarily because the table can then be readily divided into groups of contiguous residues and shows which chain segments have moved the most.

Some of these movements have no apparent relationship to the activation process. Thus, segments I, III, and IV are exterior loops of backbone chain, far from the active site, in both zymogen and enzyme. Indeed, segment IV, where one of the largest movements occurs, is probably a flexible region since it is known to assume different conformations in the two molecules of the α-chymotrypsin asymmetric unit (*22*). Therefore, this particular conformational difference between the zymogen and enzyme may be simply a reflection of differing crystal packing forces and

TABLE I
RESIDUES FOR WHICH α-CARBON ATOMS DIFFER IN POSITION BY MORE THAN 3.6 Å BETWEEN THE CHYMOTRYPSINOGEN AND α-CHYMOTRYPSIN STRUCTURES

Segment	Residue	Displacement (Å)
I	Gln 7	4.8
	Pro 8	10.0[a]
II	Ile 16	11.3
	Val 17	6.6
III	Thr 37	4.0
	Gly 38	6.6
IV	Asp 72	5.6
	Gln 73	9.6
	Gly 74	9.1
	Ser 75	6.2
	Ser 76	10.1
	Ser 77	5.6
V	Thr 144	5.9
	Arg 145	8.7
	Tyr 146	4.6
	Ala 149	4.7
	Asn 150	6.7[a]
	Thr 151	4.7[a]
	Pro 152	4.6[a]
VI	Met 192	8.4
	Gly 193	6.6

[a] Tentative chymotrypsinogen coordinates.

in actuality have nothing to do with the activation process. The same is possibly true for segments I and III as well.

A conformational change not included in Table I since the greatest α-carbon movement is only 3.4 Å at residue 173 may be mentioned in passing. It involves residues 161–173. In the enzyme, this chain segment forms two turns of distorted but nevertheless unquestionable helix, which was overlooked initially but subsequently pointed out by Blow (*23*). In the zymogen this helical segment is further distorted so that it probably cannot be said to form more than one turn of helix and to include more than residues 164–168. This is the only part of the zymogen molecule where an alteration of helix content may occur upon activation.

We now turn our attention to segments II, V, and VI in Table I. These are almost certainly significant in the sense that they are required for conversion of the inactive zymogen into an active enzyme.

A. Isoleucine 16

The conformational change in segment II brings the newly formed N-terminus of Ile 16 into the interior of the molecule where it approaches the buried side chain carboxylate group of Asp 194. This movement can be seen by comparing Figs. 2 and 3 and in greater detail by comparing Figs. 4 and 5. In the latter pair of figures, note that residues 16 and 17 are above the chain segment 18–21 in the zymogen but below segment 18–21 in the enzyme. Further, in the zymogen, Asp 194 is buried to begin with, and its side chain carboxylate forms a hydrogen bond with Nε2 of His 40. In the enzyme, on the other hand, the Asp 194 side chain has swung around by about 4 Å into a position where it can hydrogen bond with the new Ile 16 N-terminus. Note also that the imidazole ring of His 40 has simultaneously turned slightly and now donates a hydrogen bond from Nε2 to the carbonyl oxygen of Gly 193. Apparently, the large movement of segment II is accomplished by rotating the main chain approximately 180° about an axis roughly defined by the α-carbon atoms of residues 19 and 21. This rotation swings the new N-terminus of the B chain out through the surrounding solvent, and then down into the interior of the molecule, burying the formerly exposed side chains of Ile 16 and Val 17 just below the surface of the enzyme molecule.

The finding that Asp 194 is buried in the zymogen as well as in the enzyme is in agreement with the chemical modification studies of Carraway *et al.* (*24*).

23. D. M. Blow, *Biochem. J.* **112**, 261 (1969).
24. K. L. Carraway, P. Spoerl, and D. E. Koshland, *JMB* **42**, 133 (1969).

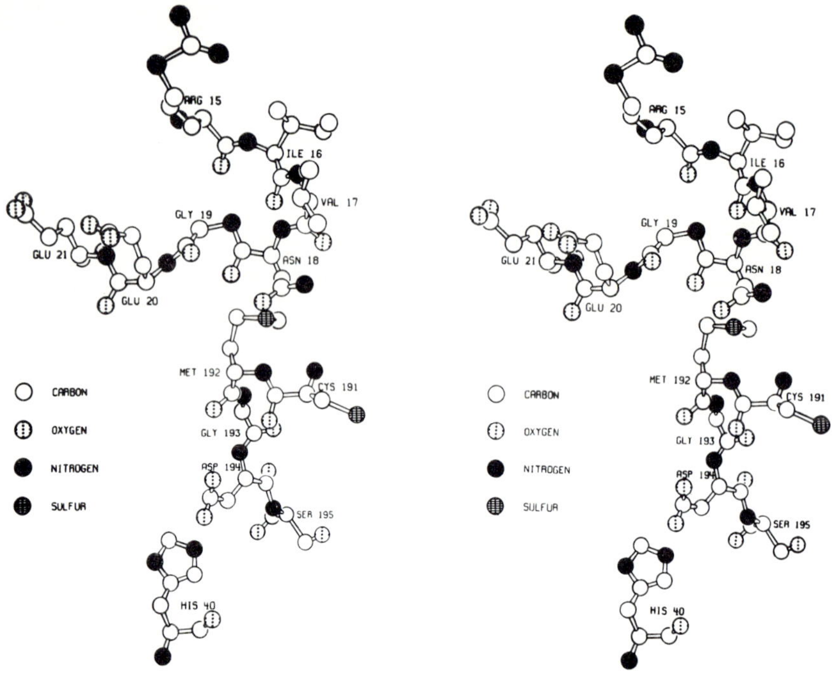

FIG. 4. Chymotrypsinogen—stereoscopic projection of amino acid residues involved in the formation of the Ile 16–Asp 194 ion pair during activation. The figure is drawn from a vantage point above the surface looking toward the interior of the molecule (from top to bottom on Fig. 2). Figures 4 through 9 were prepared from computer-plotted stereoscopic projections.

B. ARGININE 145

The most obvious result of the conformational change in segment V is to move the guanidinium side chain of Arg 145 away from the neighborhood of His 40, Asp 194, and the catalytic site Ser 195 and out into the surrounding medium. This movement is clearly evident upon comparison of Figs. 6 and 7. In the zymogen, the guanidinium group is resting on the surface of the molecule and may be close enough to the buried side chain carboxylate of Asp 194 to permit appreciable electrostatic interaction between these two oppositely charged groups. Possibly this electrostatic interaction helps to stabilize the structure. In the enzyme, on the other hand, segment V is rearranged so that the α-carbon of Arg 145 has moved by 9 Å, and its side chain has swung up to become fully extended into the solvent. That a positively charged side chain in this location may play

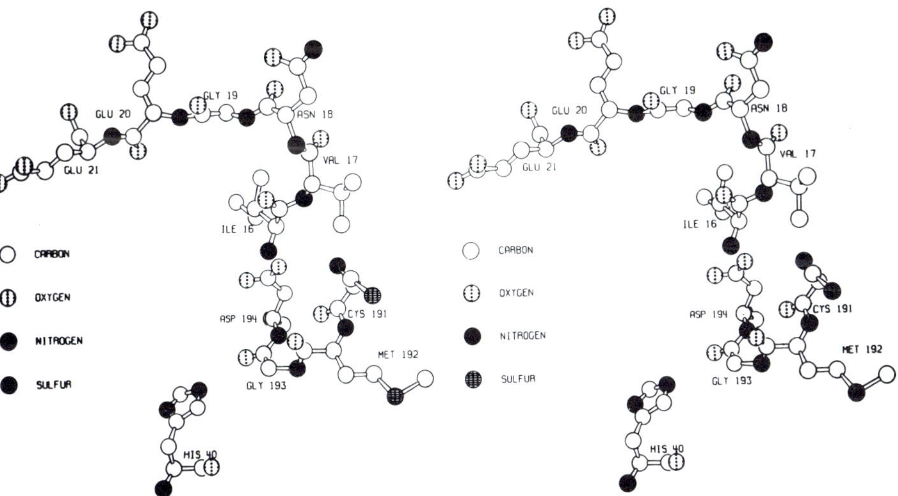

FIG. 5. α-Chymotrypsin—stereoscopic projection of amino acid residues involved in the formation of the Ile 16–Asp 194 ion pair during activation. The figure is drawn from a vantage point equivalent to that of Fig. 4.

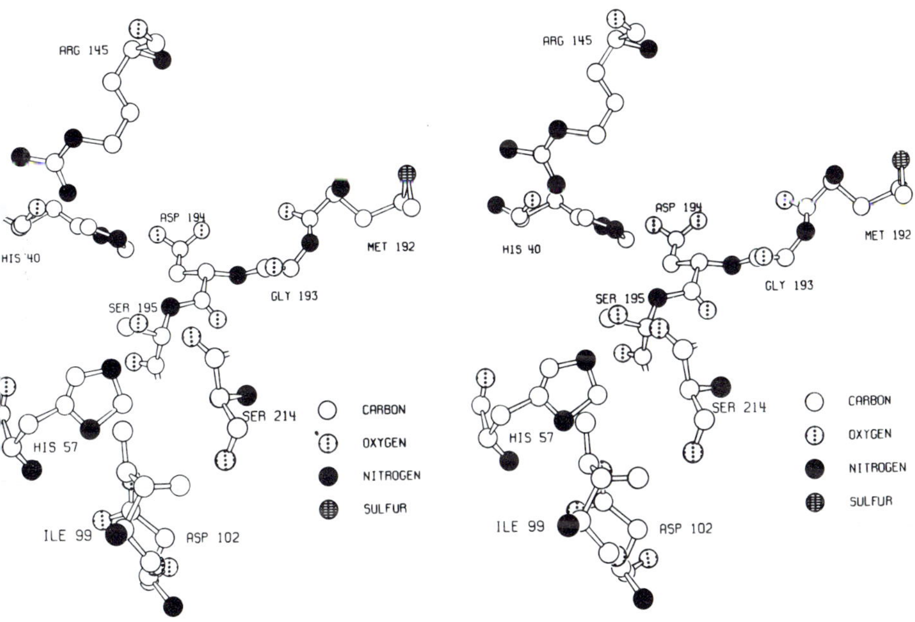

FIG. 6. Chymotrypsinogen—stereoscopic projection of the latent catalytic site region drawn from a vantage point looking into the page (see Fig. 2) at an angle inclined about 20° to the right of perpendicular.

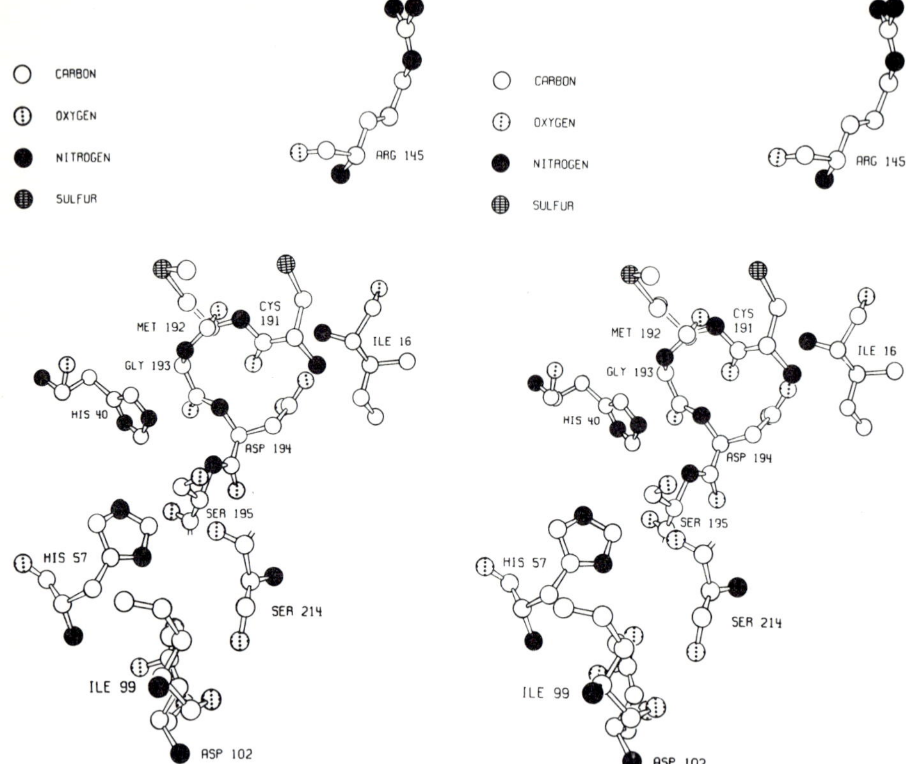

FIG. 7. α-Chymotrypsin—stereoscopic projection of the catalytic site region drawn from a vantage point equivalent to that of Fig. 6.

some role in the activation process is consistent with the fact that the corresponding residue is also an arginine in pig elastase and bovine thrombin or a lysine in bovine trypsinogen and chymotrypsinogen B (*25, 26*). It must be cautioned before drawing any final conclusions, however, that the electron density map in the region of the Arg 145 side chain is somewhat ambiguous for the zymogen and that an alternative model can be constructed with the side chain extending outward. However, the electron density in the alternative position is much weaker. Perhaps this indicates some degree of disorder with the side chain actually occupying two different positions. In any case, it is difficult to imagine

25. M. O. Dayhoff, ed., "Atlas of Protein Sequence and Structure" Vol. 4, p. D224. Natl. Biomed. Res. Found., Silver Spring, Maryland, 1969.

26. S. Magnusson, *in* "Structure–Function Relationships of Proteolytic Enzymes" (P. Desnuelle, H. Neurath, and M. Ottesen, eds.), p. 138. Munksgaard, Copenhagen, 1970.

a role for this conformational change in segment V given the alternative position for the Arg 145 side chain.

C. Methionine 192

Although segment VI in Table I includes only Met 192 and Gly 193, the conformational change represented here is of considerable interest since it is the only one to which a fairly clear-cut role in the genesis of enzymic activity can be assigned. It results in movement of the side chain of Met 192 from a deeply buried position in the zymogen out to the surface of the enzyme molecule where it forms the flexible hydrophobic lid of the specificity cavity (*27*). This movement is most clearly visualized by comparing Figs. 6 and 7 but may also be seen in the other figures as well. In the process of moving Met 192 out to the molecular surface, segment 187–193 of the main chain also becomes more extended. The consequence of these combined conformational changes is the creation of the specificity cavity, one side of which is comprised of residues 189–192 (*27*). In the zymogen this cavity does not exist, or at least it is incomplete and severely distorted. The molecular contortions involved here may be thought of in a simplified way as resulting from a rotation of the main chain about the carbonyl-carbon to α-carbon bond of Asp 194. This rotation causes the shift in position of the Asp 194 side chain from the vicinity of His 40 toward the buried N-terminus of Ile 16, as described above, and simultaneously it also moves the backbone chain carbonyl oxygen of Gly 193 into position to accept a hydrogen bond from Nϵ2 of His 40, replacing the carboxylate oxygen of Asp 194. Thus, the conformational changes involving Ile 16, Met 192, Gly 193, and Asp 194 and formation of the specificity cavity during zymogen activation can be viewed as a single unified event. An attempt has been made to depict certain aspects of the specificity cavity formation from a third vantage point in Figs. 8 and 9. In Fig. 9, the tosyl group occupying the specificity cavity in tosyl-α-chymotrypsin is shown in black.

D. Catalytic Site

As described in the chapters by Blow and by Hess on chymotrypsin, a mechanism for the bond breaking step in the catalytic pathway has been proposed which employs a "charge relay system" (*27, 28*) involving a hydrogen bond network between the side chains of Ser 195, His 57, and Asp 102. It has been very surprising therefore, to find that the spatial arrange-

27. T. A. Steitz, R. Henderson, and D. M. Blow, *JMB* **46**, 337 (1969).
28. D. M. Blow, J. J. Birktoft, and B. S. Hartley, *Nature* **221**, 337 (1969).

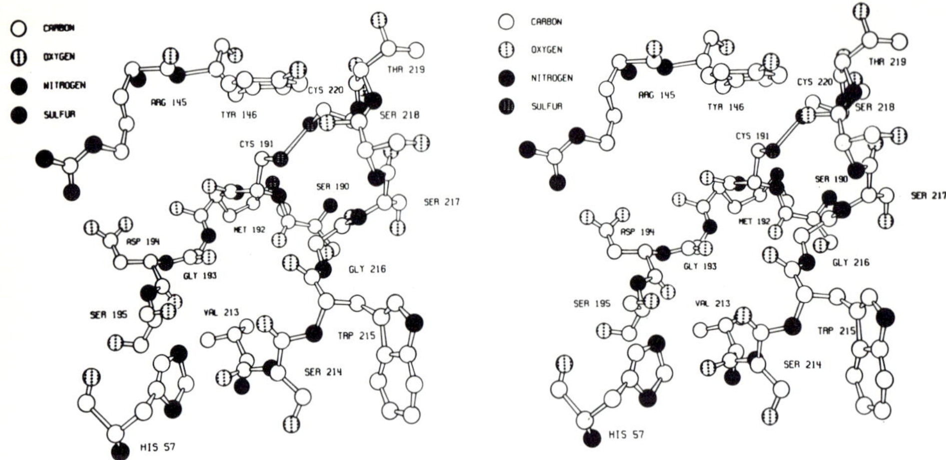

FIG. 8. Chymotrypsinogen—stereoscopic projection of the latent specificity cavity drawn from roughly the same vantage point as Fig. 2.

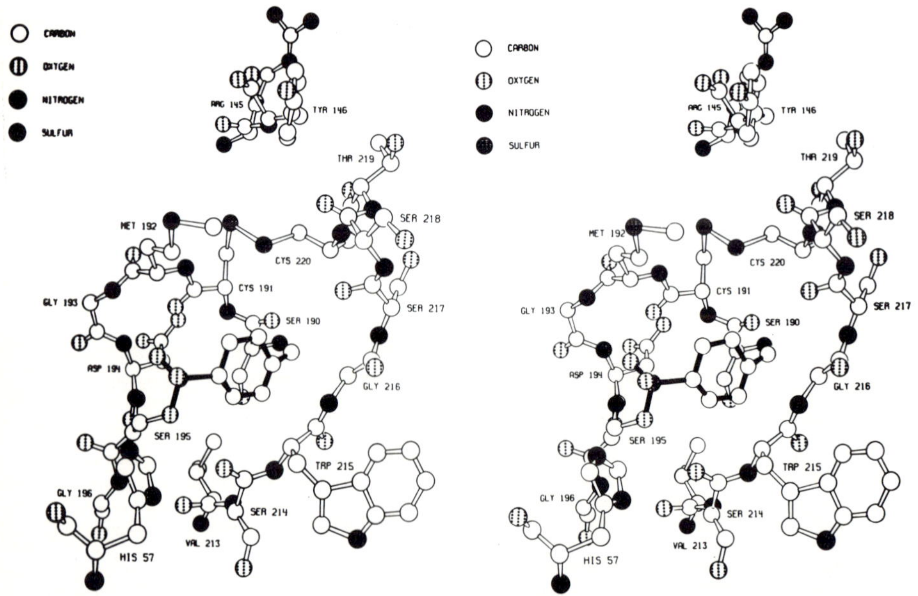

FIG. 9. Tosyl-α-chymotrypsin—stereoscopic projection of the specificity cavity drawn from a vantage point equivalent to that of Fig. 8. The tosyl group is drawn with solid bonds to contrast with the bonds of the amino acids forming the specificity cavity.

ment of these three residues in the zymogen is almost indistinguishable from that in the catalytically active enzyme. The term "almost indistinguishable" has been used here because, in fact, a very slight twist may actually occur in the His 57 side chain, and indeed such a twist may be seen by careful comparison of Figs. 2, 6, and 8 with the corresponding Figs. 3, 7, and 9. If this effect is real, it may suffice to explain the total inactivity of chymotrypsinogen since it would result in distortion of the hydrogen bond network referred to above. Wang (*29*) has argued convincingly that facilitated proton transfer along precisely aligned hydrogen bonds may play a crucial role in enzyme catalysis generally. One must be careful, however, to emphasize that the apparent twist in the His 57 side chain is so small that it could well turn out to be spurious when the zymogen electron density map is refined or when direct comparison of the two maps is carried out. Indeed, other subtle but significant structural changes in the charge relay system could also be revealed by further improvement in the data. Nevertheless, at the present stage of the X-ray study of chymotrypsinogen, one can say with reasonable certainty that the catalytic site, in contrast with the specificity cavity, is essentially preformed. The question remains, therefore, as to why the zymogen is not catalytically active, though perhaps without the specificity for hydrophobic side chains characteristic of chymotrypsin.

Further to pursue this question, it should be pointed out that two minor structural changes which may also help to explain the zymogen's inactivity are seen upon activation. These occur not in the charge relay system itself but in its immediate neighborhood: (1) the side chain of Ile 99 moves away from the imidazole ring of His 57, and (2) the hydroxyl group of Ser 214 moves into a better position to form a second hydrogen bond with Oδ2 of Asp 102. These will now be described in more detail.

The displacement of the side chain of Ile 99 upon activation is quite noticeable and readily seen on comparing Figs. 6 and 7. In the zymogen, the Ile 99 side chain is in van der Waals contact with the imidazole ring of His 57 and, together with neighboring side chains, completely blocks access of the solvent to His 57–Asp 102 portion of the catalytic site hydrogen bond network. During activation, the Ile 99 side chain rotates approximately 90° about its α-carbon to β-carbon bond and moves away from the His 57 ring. Curiously, the conformation of the main chain does not alter appreciably. Just why, or even whether or not, this change would be required for inducing enzymic activity is open to question. Such an alteration might be expected, if anything, to *decrease* the hydro-

29. J. H. Wang, *Science* **161,** 328 (1968).

phobicity of the region surrounding His 57–Asp 102, which would in fact appear to be a change in the wrong direction. It has been suggested that the hydrophobic environment of the buried Asp 102 enhances the polarization of the charge relay system and makes the reactive Ser 195 oxygen strongly nucleophilic as required by the proposed mechanism (*28*). Nevertheless, it is noteworthy that a hydrophobic residue (Ile, Leu, or Val) is conserved at the position corresponding to Ile 99 in bovine chymotrypsinogen B, bovine trypsinogen, bovine thrombin, and pig elastase (*25*, *26*).

The change at Ser 214 is more subtle and at the present resolution remains somewhat doubtful. It involves a possible movement of Oγ by perhaps 1 Å to form a better hydrogen bond with Oδ2 of Asp 102. In itself, this is probably not a sufficiently large apparent movement to be taken seriously. However, there does occur a rather obvious change in conformation of the chain segment 214–217 during activation. In the zymogen, the stretch of main chain running from the α-carbon of residue 214 to the α-carbon of residue 216 is somewhat folded, but it is fully extended in the enzyme. The effect can be observed readily by comparing Figs. 2 and 3. In light of the proposal by Steitz *et al.* (*27*) that the carbonyl oxygen of Ser 214 is involved in productive binding of substrates, one might expect to see this carbonyl group move into binding position as a result of the main chain extension, but such is not the case. The carbonyl group is in about the same orientation in both zymogen and enzyme. On the other hand, there are several observations suggesting that a second serine residue (Ser 214 in the case of chymotrypsin) forms a hydrogen bond with the same buried aspartic side chain oxygen (Oδ2 of Asp 102) that is also hydrogen bonded to the catalytic site histidine side chain (Nδ1 of His 57) in serine proteases generally. This point is discussed more fully in Chapter 15. If this view is correct, the second serine hydrogen bond may well be included in the catalytic site hydrogen bond network and play some role in the activity of these enzymes. It is all the more tempting, therefore, to suppose that distortion of this bond in chymotrypsinogen is one of the factors responsible for its lack of activity and provisionally to consider as significant the apparent slight movement of the Ser 214 side chain.

E. Activation Refolding

It will be obvious that at least as many questions about the chymotrypsinogen activation process have been raised as have been answered by the observations reported here. Indeed, the central problem of why the zymogen is inactive remains, although it seems reasonably certain that the answer will involve some subtle distortions in and around the catalytic site.

Perhaps a suitable way to conclude this chapter would be to mention briefly another, and no doubt more difficult, problem that now arises. Why does the simple act of cleaving a single exterior peptide bond between Arg 15 and Ile 16 cause the chymotrypsinogen molecule to refold in the manner observed? The answer is certainly not obvious from inspection of the two models, and we shall have to content ourselves with simply listing those structural changes which probably contribute, either positively or negatively, to the overall net free-energy decrease upon going from the zymogen to the enzyme conformation.

(1) The Arg 15–Ile 16 peptide bond is hydrolyzed to give a free $-COO^-$ and a free $-NH_3^+$ group.

(2) The free $-NH_3^+$ group tucks down inside the molecule where it interacts with the buried $-COO^-$ group of Asp 194.

(3) As a result of this process two hydrophobic side chains, at residues 16 and 17, become buried as well.

(4) The buried $-COO^-$ group of Asp 194 moves away from the His 40 side chain to which it had been hydrogen bonded and is replaced by the backbone carbonyl of Gly 193 which forms a new hydrogen bond with the His 40 side chain.

(5) The positively charged side chain of Arg 145 moves away from the surface of the molecule, where it may interact with the buried COO^- of Asp 194, and extends out into the surrounding solvent.

(6) The buried hydrophobic side chain of Met 192 moves from the molecular interior out to the surface.

Other conformational changes also occur, some of them quite large, as, for example, in segments I, III, and IV of Table I. But unlike the events enumerated above for which the situation is fairly clear, it has not yet been possible to see how they might contribute to the free energy of refolding.

6

The Structure of Chymotrypsin

D. M. BLOW

I. The Activation Products of Chymotrypsinogens A and B

Structural work on chymotrypsin has been concentrated exclusively on enzymes from the bovine pancreas. Because of the ready crystallizability of chymotrypsinogen A (*1*), a second component, chymotrypsinogen B, present in similar quantity in the pancreatic juice, has been less intensively studied (*2*).

1. J. H. Northrop, M. Kunitz, and R. Herriott, "Crystalline Enzymes." Columbia Univ. Press, New York, 1939.
2. C. K. Keith, A. Kazenko, and M. Laskowski, *JBC* **170**, 227 (1947).

The activation of the zymogen to an active chymotrypsin depends on the specific cleavage of a peptide bond by trypsin. This process is complicated by the possibility of further proteolysis by chymotrypsin or trypsin. Northrop and Kunitz (*1*) activated chymotrypsinogen A under conditions giving a good yield of α-chymotrypsin crystals. Rapid activation, using larger quantities of trypsin, gives two different species of chymotrypsin, named π- and δ-chymotrypsin by Jacobsen (*3*). Especially when activation is done slowly, chymotryptic cleavage of chymotrypsinogen is possible, leading to degraded zymogens known as neochymotrypsinogens.

The chemical events of activation were identified by end group analysis (*4–10*), by isolation of the peptides liberated (*8*, *10–12*) and by electrophoretic studies (*5*, *8*, *9*). They led to results summarized in Fig. 1. This figure also shows γ-chymotrypsin, which is identical in end group analysis to α-chymotrypsin (*13*, *14*) but has quite distinct crystallization properties (*1*). It has been shown by crystallographic studies that the transition between the α and γ forms of the enzyme is reversible (*15*). On the basis of the crystallographic similarity of δ- and γ-chymotrypsin, it has been suggested that the "slow" activation which leads to α-chymotrypsin occurs via the neochymotrypsinogens, while autolysis of δ-chymotrypsin leads directly to γ-chymotrypsin (*16*). This proposal lacks supporting evidence at present.

Chymotrypsinogen B is activated by a similar primary tryptic cleavage to chymotrypsin B_π, but the conditions of rapid activation do not cause further autolysis to an enzyme of the δ type (*17*).

For details about the structure of chymotrypsinogen and its relation to the activation process, see Chapter 5, by Kraut, this volume.

3. C. F. Jacobsen, *Compt. Rend. Trav. Lab. Carlsberg, Ser. Chim.* **25**, 325 (1947).
4. P. Desnuelle, M. Rovery, and C. Fabre, *BBA* **9**, 109 (1952).
5. J. A. Gladner and H. Neurath, *JBC* **205**, 345 (1953).
6. M. Rovery and P. Desnuelle, *BBA* **13**, 300 (1954).
7. F. R. Bettelheim, *JBC* **212**, 235 (1955).
8. F. R. Bettelheim and H. Neurath, *JBC* **212**, 241 (1955).
9. M. Rovery, M. Poilroux, A. Curnier, and P. Desnuelle, *BBA* **17**, 565 (1955).
10. M. Rovery, M. Poilroux, Y. Yoshida, and P. Desnuelle, *BBA* **23**, 608 (1957).
11. W. J. Dreyer and H. Neurath, *JBC* **217**, 527 (1955).
12. M. Rovery, M. Poilroux, A. Curnier, and P. Desnuelle, *BBA* **16**, 590 (1955).
13. M. Rovery, C. Fabre, and P. Desnuelle, *BBA* **10**, 481 (1953).
14. J. A. Gladner and H. Neurath, *JBC* **206**, 911 (1954).
15. R. B. Corey, O. Battfay, D. A. Brueckner, and F. G. Mark, *BBA* **94**, 535 (1965).
16. H. T. Wright, J. Kraut, and P. E. Wilcox, *JMB* **37**, 363 (1968).
17. O. Guy, D. Gratecos, M. Rovery, and P. Desnuelle, *BBA* **115**, 404 (1966).

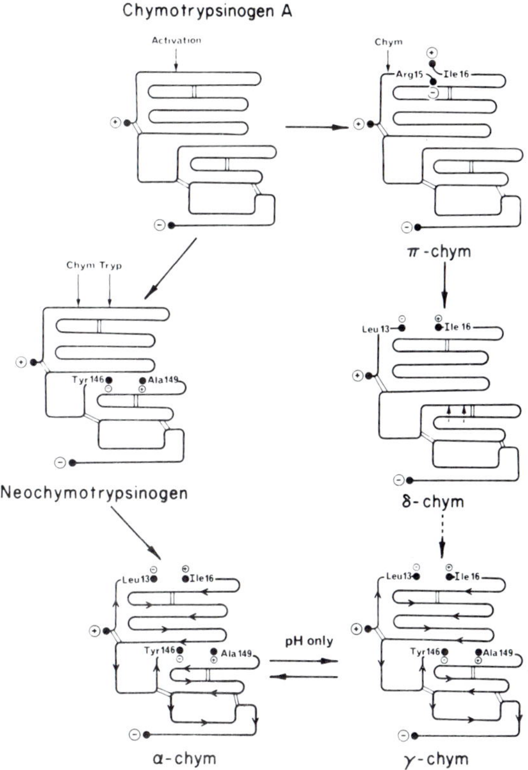

FIG. 1. Activation and autolytic reactions which give rise to the various derivatives of chymotrypsinogen A. Probably a number of different forms of neochymotrypsinogen exist. The dashed arrow represents the direct transition from δ- to γ-chymotrypsin proposed by Wright *et al.* (*16*) to take the place of the transition from δ to α previously suggested [adapted from a figure in Sigler *et al.* (*46*)].

II. Chemical Structures of Chymotrypsins

A. CHYMOTRYPSINOGEN A AND ITS DERIVATIVES

The elucidation of the terminal groups involved in the activation and autolysis steps described above (*4–14*) gave the first pieces of information about the chemical structure of chymotrypsin. Other data were obtained from the reactivity of a specific serine residue (*18, 19*). Using diisopropyl-

18. E. F. Jansen, M.-D. F. Nutting, R. Jang, and A. K. Balls, *JBC* **179,** 189 (1949).
19. N. K. Schaffer, S. C. May, and W. H. Summerson, *JBC* **202,** 67 (1953).

(^{32}P)-phosphofluoridate to label this serine, various short radioactive peptides were obtained by partial acid hydrolysis (*20, 21*). A longer radioactive peptide Gly–Asp–Ser^{32}P–Gly–Gly–Pro–Leu was obtained by hydrolysis of the labeled enzyme with a crude pancreatic extract (*22*).

Further progress beyond this point was made by the laborious procedure required for a complete amino acid sequence determination. The complete amino acid sequence of chymotrypsinogen A was announced by Hartley in 1964 (*23, 24*). Taken with determinations of the disulfide interconnections (*25–27*) this provided a complete statement of the covalent structure of the molecule. A parallel investigation of the amino acid sequence was carried out by Keil and his collaborators (*26, 28–31*). When two minor corrections had been made to Hartley's sequence (*32*), Keil's group concluded that "there are now no disagreement between our results and Hartley's" (*31*). One error, however, still remained to be detected. This involved residue 102, which the crystallographic results showed to be closely involved with the active site, and which was indicated as aspartic acid in all the homologous serine enzymes, but originally as asparagine in chymotrypsinogen. On reinvestigation it was shown to be aspartic acid in chymotrypsinogen also (*33*). The amino acid sequence of α-chymotrypsin is now believed to be as shown in Fig. 2.

This sequence shows that the serine residue labeled in the early experiments (*18, 19*) and in numerous other enzyme acylation experiments

20. F. Turba and G. Gundlach, *Biochem. Z.* **327**, 186 (1955).
21. N. K. Schaffer, L. Simet, S. Harshman, R. R. Engle, and R. W. Drisko, *JBC* **225**, 197 (1957).
22. R. A. Oosterbaan, P. Kunst, J. van Rotterdam, and J. A. Cohen, *BBA* **27**, 549 and 556 (1958).
23. B. S. Hartley, *in* "Structure and Activity of Enzymes" (T. W. Goodwin, J. I. Harris, and B. S. Hartley, eds.), p. 47. Academic Press, New York, 1964.
24. B. S. Hartley, *Nature* **201**, 1284 (1964).
25. J. R. Brown and B. S. Hartley, *BJ* **89**, 59P (1963).
26. B. Keil, Z. Prusík, and F. Šorm, *BBA* **78**, 559 (1963).
27. J. R. Brown and B. S. Hartley, *BJ* **101**, 214 (1966).
28. B. Keil, B. Meloun, J. Vaněček, V. Kostka, Z. Prusík, and F. Šorm, *BBA* **56**, 595 (1962).
29. B. Meloun, V. Kostka, B. Keil, and F. Šorm, *Collection Czech. Chem. Commun.* **28**, 2749 (1963).
30. V. Kostka, B. Meloun, and F. Šorm, *Collection Czech. Chem. Commun.* **28**, 2749 (1963).
31. B. Meloun, I. Kluh, V. Kostka, L. Moravek, Z. Prusík, J. Vaněček, B. Keil, and F. Šorm, *BBA* **130**, 543 (1966).
32. B. S. Hartley and D. L. Kauffmann, *BJ* **101**, 229 (1966).
33. D. M. Blow, J. J. Birktoft, and B. S. Hartley, *Nature* **221**, 337 (1969).

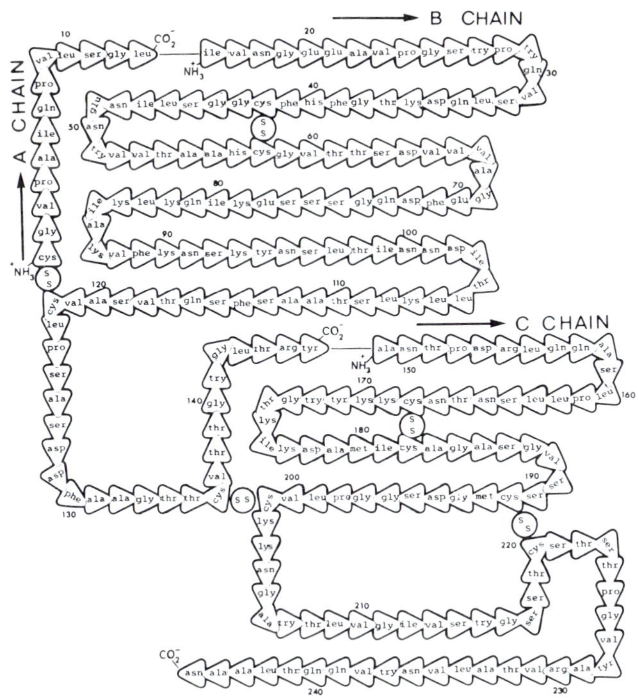

FIG. 2. The amino acid sequence of α-chymotrypsin [revised from a figure in Matthews *et al.* (*45*)].

(*34–39*) is Ser 195, and that the amino-terminal isoleucine liberated when trypsin activates the zymogen (*6*) is Ile 16.

Chemical modification experiments, reviewed in detail elsewhere in this volume (*40*), have identified several other residues which are directly involved in the active site. The activity of the enzyme is destroyed when His 57 is modified. Modification of Met 192 alters the specificity of the enzyme without decreasing its maximum rate. At least two carboxyl groups, one of which is Asp 194, are unavailable to esterifying agents and are believed to be buried.

34. A. K. Balls and H. N. Wood, *JBC* **219,** 245 (1956).
35. M. Caplow and W. P. Jencks, *Biochemistry* **1,** 883 (1962).
36. M. L. Bender, G. R. Schonbaum, and B. Zerner, *JACS* **84,** 2540 (1962).
37. B. F. Erlanger and W. Cohen, *JACS* **85,** 348 (1963).
38. D. E. Fahrney and A. M. Gold, *JACS* **85,** 997 (1963).
39. S. A. Bernhard, S. J. Lau, and M. Noller, *Biochemistry* **4,** 1108 (1965).
40. G. P. Hess, Chapter 7, this volume.

B. Chymotrypsinogen B

The amino acid sequence of the closely related enzyme, bovine chymotrypsinogen B, has been determined by Smillie and his collaborators (*41–43*). This protein, which exists in the bovine pancreas in equal quantities with chymotrypsinogen A, has almost identical activation and activity properties, and it is thus a true chymotrypsin. It appears to be an isoenzyme, the result of a gene doubling which has not yet led to any significant alteration in activity (*44*), although 20% of the individual amino acids are changed.

III. Three-Dimensional Structures Determined by X-Ray Diffraction

A. α-Chymotrypsin

An interpretable high-resolution electron density map of tosyl-α-chymotrypsin was obtained in 1967 (*45, 46*). Further calculation later produced a greatly improved electron density map, which enabled a more detailed and more certain interpretation to be made (*47*), and atomic coordinates for almost the entire molecule based on this later map have been published (*48*). Other details about the structure have been given in published lectures (*49, 50*).

Tosyl-α-chymotrypsin is a derivative of α-chymotrypsin which has been inactivated by sulfonylation of the active serine residue with a

41. B. S. Hartley, J. R. Brown, D. L. Kauffmann, L. B. Smillie, *Nature* **207,** 1157 (1965).

42. L. B. Smillie, A. G. Enenkel, and C. M. Kay, *JBC* **241,** 2097 (1966).

43. L. B. Smillie, A. Furka, N. Nagabushan, K. J. Stevenson, and C. O. Parkes, *Nature* **218,** 343 (1968).

44. B. S. Hartley, *Advan. Sci.* **23,** 47 (1966).

45. B. W. Matthews, P. B. Sigler, R. Henderson, and D. M. Blow, *Nature* **214,** 652 (1967).

46. P. B. Sigler, D. M. Blow, B. W. Matthews, and R. Henderson, *JMB* **35,** 143 (1968).

47. D. M. Blow, J. J. Birktoft, and B. S. Hartley, *Nature* **221,** 337 (1969).

48. J. J. Birktoft, B. W. Matthews, and D. M. Blow, *BBRC* **36,** 131 (1969).

49. D. M. Blow, *BJ* **112,** 261 (1969).

50. J. J. Birktoft, D. M. Blow, R. Henderson, and T. A. Steitz, *Phil. Trans. Roy. Soc. London* **B257,** 67 (1970).

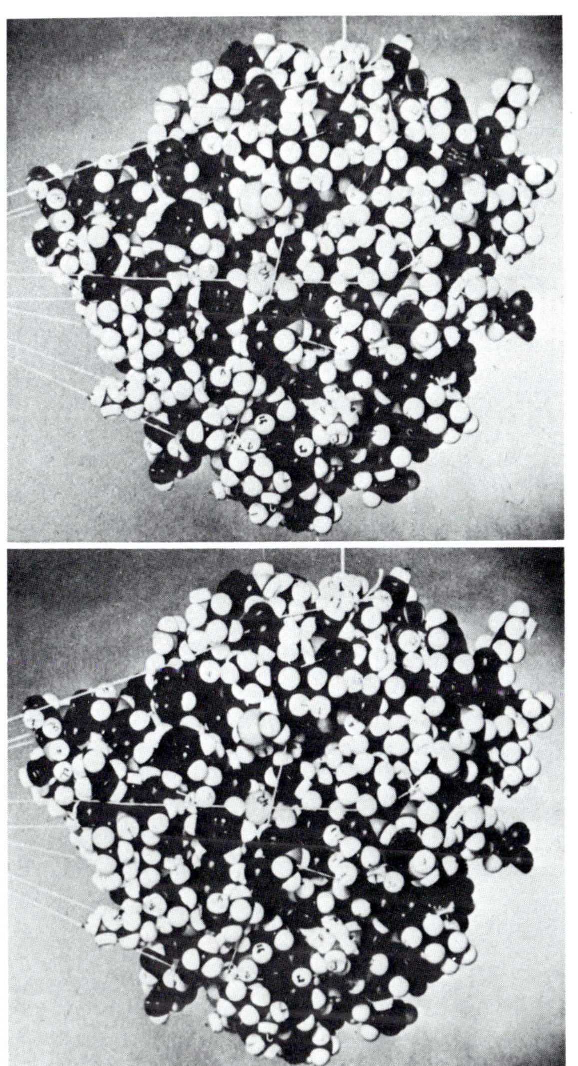

Fig. 3. Space-filling model of α-chymotrypsin. The catalytic site is just below the S atom of Cys 42. The hole referred to in the text is below and to the right of this. The S atom of Met 192 is prominently visible just above it [reproduced from Birktoft *et al.* (*50*)].

p-toluenesulfonyl group (*51, 52*). Crystalline tosyl-α-chymotrypsin is essentially isostructural with crystalline α-chymotrypsin (*53*); the small differences at the active center will be discussed in a later section. The crystals belong to space group $P2_1$ with $a = 49.3$ Å, $b = 67.3$ Å, $c = 65.9$ Å, and $\beta = 101.8°$. The crystallographic asymmetric unit contains two molecules (*54*). This means that the crystal contains two distinct sites for chymotrypsin molecules, distinguishable because they are in different environments. The two types of molecular site are related by a system of local, noncrystallographic twofold axes (*55*). The majority of the close intermolecular contacts are with molecules related by the system of local twofold axes, and these contacts are the same for all molecules in the crystal (*46, 55*). It is only a small number of relatively weak interactions which give the two molecular sites different environments. It means, however, that the X-ray diffraction results give two independent images of the chymotrypsin molecule. In only two places does it seem likely that the polypeptide backbone assumes different conformations in the two molecules—at residues 9–13 and 73–77—and in these cases the electron density is weak and hard to interpret in each molecule. Elsewhere the only differences between the molecules are in the orientations of side chains on the surface of the molecule (*48*).

The molecule is ellipsoidal, with overall dimensions 51 Å along the crystallographic a^* direction and about 40 Å along b and c (*55a*). The molecule is somewhat flattened at the active center, and there is a hole in the surface which plays an important part in the binding of specific substrates. These features can only be visualized readily in stereoscopic illustrations of a space-filling model of the molecule (Fig. 3).

There is little regular secondary structure in the molecule. Residues 235–245 at the C-terminus form a straight rod of α-helix, and there is also a short α-helixlike structure at residues 164–173. In the C-terminal helix, the last peptide bond is in an orientation which represents a tightening of the helix toward the 3.0_{10} helix structure (*56*)—an effect already

51. J. Kallos and D. Rizok, *JMB* **7,** 599 (1963).

52. D. H. Strumeyer, W. N. White, and D. E. Koshland, *Proc. Natl. Acad. Sci. U. S.* **50,** 931 (1963).

53. P. B. Sigler, B. A. Jeffery, B. W. Matthews, and D. M. Blow, *JMB* **15,** 175 (1966).

54. J. D. Bernal, I. Fankuchen, and M. F. Perutz, *Nature* **141,** 523 (1938).

55. D. M. Blow, M. G. Rossmann, and B. A. Jeffery, *JMB* **8,** 65 (1964).

55a. a^* is parallel to the local twofold axes and is shown vertical in illustrations of molecular models. It is parallel to the x-axis of the Cartesian reference system defined by Birktoft *et al.* (*48*). Contoured illustrations of the electron density map are always contoured in planes perpendicular to a^*.

56. J. Donohue, *Proc. Natl. Acad. Sci. U. S.* **39,** 470 (1953).

noted in myoglobin and lysozyme and given the name α_{II}-helix (*57*). The helix at residues 164–173 is distorted by the existence of a 3.0_{10}-like hydrogen bond in the middle of it. Careful examination of this stretch emphasizes the difficulty of deciding whether the relationship of two consecutive peptide bonds is more typical of the α-helix or the 3.0_{10} helix since the total difference in the ψ- and ϕ-dihedral angles characteristic of the two helices is only 35° (*57*). The helical regions have been emphasized by a sawtooth line in a diagram showing the probable hydrogen bonds between main chain atoms (Fig. 4).

Figure 4 also shows that structure of the antiparallel pleated-sheet type (*58*) is rather common in the chymotrypsin structure, although there are no extensive regions of regular pleated sheet. By drawing the

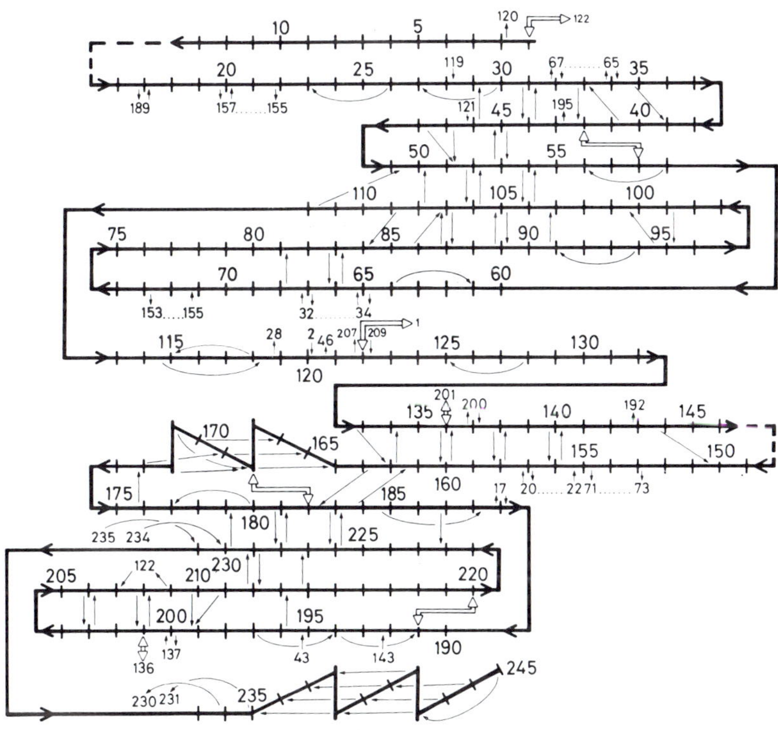

FIG. 4. Hydrogen bonds between the main peptide bonds of α-chymotrypsin. The black arrows are drawn from amido to carbonyl group. The double-headed white arrows represent disulfide linkages [reproduced from Birktoft *et al.* (*50*)].

57. G. Nemethy, D. C. Phillips, S. J. Leach, and H. A. Scheraga, *Nature* **214**, 363 (1967).

58. L. Pauling and R. B. Corey, *Proc. Natl. Acad. Sci. U. S.* **37**, 729 (1951).

main chain in a zigzag form almost all the residues which make main chain hydrogen bonds have been brought together. In every case adjacent rows of the figure represent antiparallel polypeptide chains. The pattern of zigzag lines is drawn to emphasize the existence of two folded units in the molecule from residues 27–112 and from residues 133–230. Each of these units is set out in a series of six antiparallel lines which follows the same pattern. Within each unit there are some hydrogen bonds of the antiparallel pleated-sheet type between each of these lines and its neighbors. There are further hydrogen bonds between the sixth line and the first line of the same unit, in such a way that the six lines are linked to form a hydrogen-bonded cylinder (*49*, *50*). In fact (Fig. 5) these cylinders are so distorted as to be almost unrecognizable, but they are each found to contain a core of hydrophobic residues.

Figure 4 shows that main chain hydrogen bonds from an amido group to the carbonyl of the third residue back are common. This is the type of hydrogen bond which characterizes the 3.0_{10} helix (*56*), but in chymotrypsin a single bond of this type is frequently found, especially at the ends of "loops" where the polypeptide chain turns and runs back almost parallel to its original course (*50*). This type of hydrogen bond seems to represent the most efficient way for the polypeptide chain to abruptly reverse its direction. It has been shown that there are two possible conformations at such a "corner" (*59*), one of which demands a glycine residue in the corner. Both types of corner are observed in chymotrypsin (*50*).

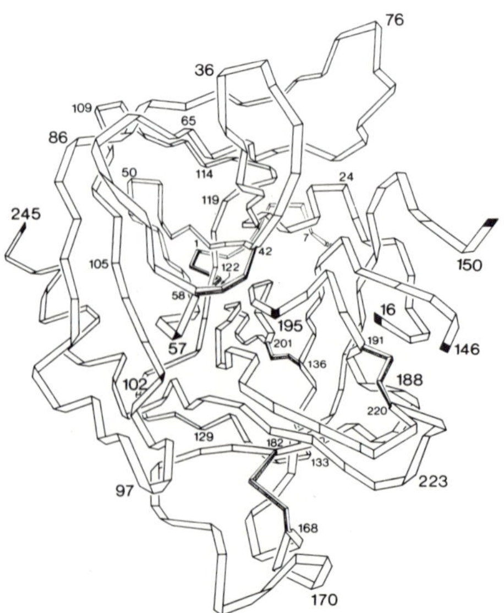

Fig. 5. α-Carbon positions in α-chymotrypsin.

TABLE I
ENVIRONMENT OF AMINO ACID SIDE CHAINS IN α-CHYMOTRYPSIN[a]

Amino acid	Side chain: External	Side chain: Surface[b]	Side chain: Internal[c]	Functional group: Group	Functional group: External	Functional group: Internal[c]	Position uncertain
Gly	11	3	8				1
Ala	13	3	6				
Val	4	8	11				
Leu	2	7	8				2
Ile	3	3	4				
Ser	20	3	3	Hydroxyl	21	5	1
Thr	15	4	3	Hydroxyl	19	3	
Asp	7	1	1	Carboxyl	7	2	
Glu	4	1	0	Carboxyl	5	0	
Asn	11	2	0	Amide	11	2	
Gln	6	3	1	Amide	8	2	
Lys	14	0	0	Amino	14	0	
Arg	3	0	0	Guanidinium	3	0	
His	0	2	0	Imidazole-N	1	3	
Phe	1	5	0				
Tyr	2	2	0	Hydroxyl	3	1	
Trp	0	6	2	Indole-N	3	5	
Pro	3	3	3				
Cys	3	4	3	Sulfur	5	5	
Met	1	0	1	Sulfur	1	1	

[a] Data from Birktoft and Blow (*60*).
[b] Surface means that only one side of the side chain (or less) is accessible to water.
[c] Internal means inaccessible to water, assuming the crystallographically observed structure to be rigid.

All charged groups are on the surface of the molecule, except for the α-amino group of Ile 16 and the carboxyl groups of Asp 102 and Asp 194, which will be discussed in relation to the active center. Although there are many hydrophobic groups at the surface of the molecule like Phe 39 and Phe 41, which stand out prominently in Fig. 3, the general tendency toward the burying of hydrophobic groups and exposure of polarizable groups noted in other proteins is observed also in chymotrypsin (Table I). There are several water molecules trapped within the structure (*60*).

The availability of tyrosine to direct iodination within the crystal has been investigated (*61*).

59. C. M. Venkatachalam, *Biopolymers* **6**, 1425 (1968).
60. J. J. Birktoft and D. M. Blow (1971) (in preparation).
61. P. B. Sigler, *Biochemistry* **9**, 3609 (1970).

B. Structure at the Active Center of α-Chymotrypsin

1. *Native Enzyme*

It was found during studies on substrate binding (*62*) that the original diffraction data, believed to represent the native form of the enzyme, were measured under conditions in which a molecule of dioxan was bound in the substrate binding site (*45, 53*). The only effect of dioxan on the molecular structure is in the repositioning of the side chain of Met 192. Revised data on the native enzyme show up the positions of ordered and partially ordered water molecules near the active site (*62*).

The conformations of some residues at the active center of α-chymotrypsin are shown in Fig. 6 (*63*). The α-amino group of Ile 16, which is produced by the tryptic cleavage that activates chymotrypsinogen (*6, 11*), and which is essential for the maintenance of enzymic activity (*64*), is turned toward the interior of the enzyme. It forms an ion pair with the carboxylate group of Asp 194, which is also internal (*45*).

The discovery of this ion pair, together with a comparison of X-ray

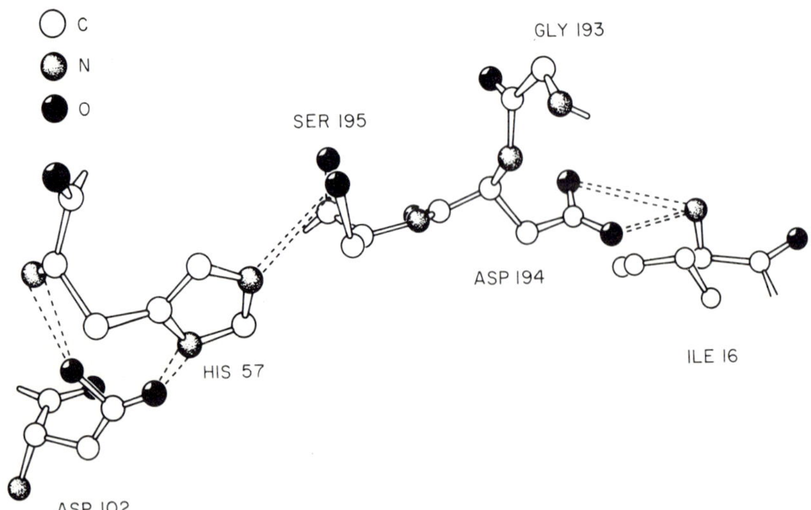

Fig. 6. The conformation of a few amino acids in the active center of α-chymotrypsin. The viewpoint is outside the surface of the molecule looking toward the interior (*63*).

62. T. A. Steitz, R. Henderson, and D. M. Blow, *JMB* **46,** 337 (1969).
63. D. M. Blow and T. A. Steitz, *Ann. Rev. Biochem.* **39,** 63 (1970).
64. H. L. Oppenheimer, B. Labouesse, and G. P. Hess, *JBC* **241,** 2720 (1966).

diffraction results for chymotrypsinogen and δ-chymotrypsin at low resolution *(65)*, led to the suggestion *(46)* that chymotrypsinogen and chymotrypsin had a similar overall tertiary structure and that the stereochemistry of the activation of the zymogen was paralleled in the enzyme by a pH-dependent structural transition *(64, 66)*. This transition depends on the titration of the α-amino group of Ile 16, with pK_{app} 8–9 *(64, 67)*, and can be recognized by its effect on the optical rotation *(68–70)* and substrate affinity *(71, 72)* of the enzyme.

At high pH, when the α-amino group is uncharged, the carboxylate of Asp 194 would seek an alternative orientation in a more polar environment. The structure of chymotrypsinogen A *(73)*, discussed in another chapter *(74)*, suggests that a probable orientation for Asp 194 in the high pH form of the enzyme is one where its carboxylate lies close to the side chain of His 40. Acylation of Ser 195 by acetyl tyrosine *(69)* or phosphorylation by diisopropylphosphofluoridate *(64)* blocks the pH-dependent transition, raising the apparent pK_a of Ile 16. These bulky acyl groups would prevent a rearrangement to a zymogen-like structure since in the zymogen the substrate binding pocket does not exist *(73, 74)*.

Serine 195, the active serine, is placed so that its O^{γ} is about 3.0 Å from $N^{\epsilon 2}$ of His 57 *(47)*. The angular relationships are consistent with the existence of a good hydrogen bond between the two, and the stereochemistry is such that the proton could be provided either by the $-N^{\epsilon 2}H$ of histidine (if this is protonated) or by the $-O^{\gamma}H$ of serine [if $N^{\epsilon 2}$(His 57) is not protonated]; $N^{\delta 1}$ (His 57) is about 2.8 Å from $O^{\delta 2}$ (Asp 102), again with proper angular relationships for a good hydrogen bond *(47)*. From the amido-N of His 57 to $O^{\delta 1}$ (Asp 102) is 2.9 Å *(48)*, implying a hydrogen bond in which the peptide –NH– group supplies the proton. Also, $O^{\delta 2}$ (Asp 102) is only 2.6 Å from O^{γ} (Ser 214), and here the orientation is convenient for the serine $-O^{\gamma}H$ to provide the proton (Fig. 6).

65. J. Kraut, L. C. Sieker, D. F. High, and S. T. Freer, *Proc. Natl. Acad. Sci. U. S.* **58,** 304 (1967).
66. G. P. Hess, J. McConn, E. Ku, and G. McConkey, *Phil. Trans. Roy. Soc. London* **B257,** 89 (1970).
67. C. Ghelis, J. Labouesse, and B. Labouesse, *BBRC* **29,** 101 (1967).
68. J. Rupley, W. J. Dreyer, and H. Neurath, *BBA* **18,** 162 (1955).
69. H. Parker and R. Lumry, *JACS* **85,** 483 (1963).
70. J. McConn, G. D. Fasman, and G. P. Hess, *JMB* **39,** 551 (1969).
71. A. Himoe, P. C. Parks, and G. P. Hess, *JBC* **242,** 919 (1967).
72. J.-R. Garel and B. Labouesse, *JMB* **47,** 41 (1970).
73. S. T. Freer, J. Kraut, J. D. Robertus, H. T. Wright, and Ng. H. Xuong, *Biochemistry* **9,** 1997 (1970).
74. J. Kraut, Chapter 5, this volume.

Aspartic acid 102 is shielded from the solvent by the side chains of Ala 55, Ala 56, His 57, Cys 58, Tyr 94, Ile 99, and Ser 214. Only one side of the imidazolium of His 57 (including $N^{\epsilon 2}$) is freely available to solvent (*47*).

Since the crystals of α-chymotrypsin were grown at pH 4.2 (*53*), the above observations were interpreted in terms of a structure in which the carboxylate of Asp 102 is ionized, while the imidazole of His 57 is protonated, forming a buried ion pair. It was pointed out that the observed arrangement of hydrogen bonds could be maintained at higher pH, after deprotonation of His 57, by using the proton of the Ser 195 hydroxyl (*47*). The implications of this arrangement for the activity of the enzyme are considered in another chapter (*40*).

2. *Enzyme–Product Complexes*

The lifetime of a complex of α-chymotrypsin with a normal substrate is too short to allow crystallographic experiments to be done under normal conditions. Complexes of crystalline α-chymotrypsin with the acylated amino acids formyl-L-tryptophan and formyl-L-phenylalanine have been studied by X-ray diffraction at 2.5 Å resolution (*62*). Such acids are termed *virtual substrates* because the carboxyl oxygen exchanges with water oxygen in the presence of the enzyme (*75*), demonstrating that they are bound in a similar mode to a true substrate.

The interpretation of electron density difference maps between the native enzyme and the complex is complicated by several factors. The displacement of firmly bound water molecules from the native enzyme introduces additional negative features into the map and almost obscures the density of the carboxyl group. The formyl group of the substrate has to push the side chain of Tyr 146 of an adjacent molecule to a new position in the crystal. There are a number of interstitial binding sites (involving two chymotrypsin molecules) with similar affinity for the virtual substrates. These are assumed to be artifacts of the crystalline state which have nothing to do with catalysis.

Despite these difficulties, a firm interpretation of the difference maps can be made (Fig. 7) (*62*). The indolyl group of formyl-L-tryptophan is embedded in the hole, close to the active site, which was pointed out in Fig. 3. It interacts with peptide bonds above and below it whose planes are almost parallel to the plane of the ring, and makes other contacts with nonpolar side chains. The amido –NH– group is pointing toward the carbonyl of Ser 214 on the enzyme surface. In crystals, the interference of Tyr 146 from an adjacent molecule probably prevents formation

75. M. L. Bender and K. C. Kemp, *JACS* **79,** 116 (1957).

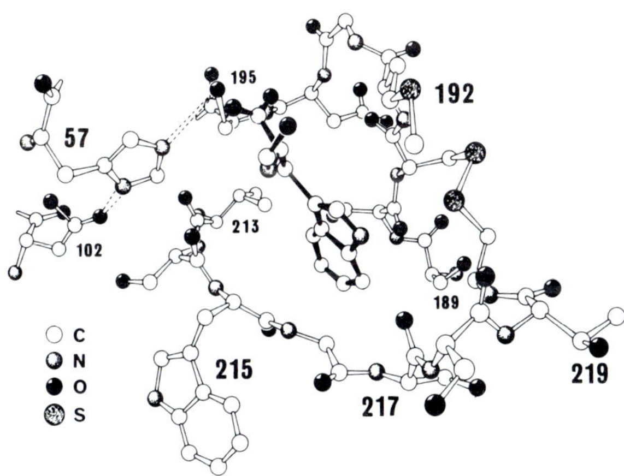

FIG. 7. The active site of α-chymotrypsin with the pseudo-substrate formyl-L-tryptophan in the position deduced from a difference electron density map (*62*). Covalent bonds of the substrate are drawn in black.

of a hydrogen bond, but in solution this is believed to be an important influence in orienting the substrate correctly (*76*). The carboxyl group lies with its carbon atom in contact with O^{γ} (Ser 195) and one C–O bond roughly parallel to the hydrogen bond joining O^{γ} (Ser 195) to $N^{\epsilon 2}$ (His 57).

Assuming that the homologous enzyme trypsin should bind substrate in a similar way, it was observed that if a lysine or arginine residue were inserted into the binding site the charged group would be adjacent to Ser 189 at the bottom of the pocket (*62*). This stimulated reinvestigations of the amino acid sequence of trypsin (*77, 78, 78a*), and it was found that the homologous residue originally designated Asn 177 (*79*) is indeed aspartic acid.

3. *Enzyme–Inhibitor Complexes*

In the same study (*62*) complexes of the crystalline enzyme with the reversibly bound inhibitors indole, β-indolepropionate, and *p*-iodo-

76. D. W. Ingles and J. R. Knowles, *BJ* **108,** 561 (1968).
77. B. S. Hartley, *Phil. Trans. Roy. Soc. London* **B257,** 77 (1970).
78. K. A. Walsh, L. L. Houston, and R. A. Kenner, *in* "Structure–Function Relationships of Proteolytic Enzymes" (P. Desnuelle, H. Neurath, and M. Ottesen, eds.), p. 56. Munksgaard, Copenhagen, 1970.
78a. A. Eyl and T. Inagami, *BBRC* **38,** 149 (1970).
79. K. A. Walsh, D. L. Kauffmann, K. S. V. S. Kuman, and H. Neurath, *Proc. Natl. Acad. Sci. U. S.* **51,** 301 (1964).

phenylacetate were prepared, and low resolution diffraction data showed that these bound in the same substrate binding pocket. As previously mentioned, the diffraction data for the complex with dioxan in this site had already been obtained (*53, 45*). High resolution diffraction studies in projection (*80*) showed that *m*-iodophenylacetate also binds in the pocket, while formyl-*p*-iodophenylalanine (L and D,L isomers) and β-(*p*-iodophenyl) propionate (*62*) bind near the active center, but not in the pocket. In this way the size of the substrate binding pocket has been carefully mapped, and it has been estimated as 10–12 Å by 5.5–6.5 Å by 3.5–4.0 Å (*62*). The exclusion of *p*-iodophenylalanine, while tyrosine is evidently accommodated, shows how precisely defined the size is.

In all these binding studies the enzyme appears to function as a rigid body. The only movement observed was a repositioning of Met 192 when the small substrate analog, dioxan, is bound (*62*). This methionine appears to function as a flexible hydrophobic lid on the substrate binding pocket.

4. *Acyl Enzymes* (*80a*)

Diisopropylphosphofluoridate (*18*) and several other compounds attack Ser 195, permanently inactivating the enzyme. Similar acyl enzyme intermediates have been shown to exist in the hydrolysis of rather poor substrates by chymotrypsin (*80b*). It is widely believed that the acyl enzyme occurs as an intermediate in the hydrolysis of "good" substrates, but its existence is so short-lived that this is hard to prove.

High resolution three-dimensional X-ray diffraction studies have been made of two acyl α-chymotrypsins. Tosyl-α-chymotrypsin (*p*-toluenesulfonyl-α-chymotrypsin) is one of a class of sulfonyl enzymes in which the active serine is permanently acylated. As already mentioned it is isomorphous with native α-chymotrypsin and was used as the parent structure in X-ray diffraction studies of α-chymotrypsin (*45, 48*).

The difference electron density map, which shows the effect of tosylation of the native enzyme (*46, 81*), indicates that the tosyl group fits into the same pocket which holds the aromatic group of substrate and reversibly bound inhibitiors; O^{γ} (Ser 195) has moved about 2.0 Å by rotation about the α–β bond in order to form a sulfonyl ester bond, and

80. R. Henderson, Ph.D. Dissertation, University of Cambridge (1969).

80a. Phosphoryl and sulfonyl esters of Ser 195 are not, of course, acyl derivatives. However, it is helpful to consider them together with the acyl enzymes. Since there seems to be no convenient generic term, the reader is asked to understand the phrase *acyl enzyme* as including them for the purposes of this section.

80b. F. J. Kezdy, G. E. Clement, and M. L. Bender, *JACS* **86,** 3690 (1964).

81. R. Henderson, *JMB* **54,** 341 (1970).

the imidazole ring of His 57 is moved by about 0.5 Å out into the solvent with a slight tilting of its plane. The movement of the serine oxygen makes a hydrogen bond to the histidine impossible. The electron density map for the tosyl enzyme shows a clear peak where a water molecule makes a hydrogen-bonded bridge from $N^{\epsilon 2}$ (His 57) to one of sulfonyl oxygens (*47*). The small movement of histidine is probably to accommodate this hydrogen bond. Other changes in the enzyme, including a movement of Met 192, are probably less than 0.3 Å (*62, 81*).

Diisopropylphosphoryl-α-chymotrypsin has been studied in projection at high resolution (*53*). The coordinates of the phosphoryl peak agree within 0.5 Å with those of the sulfonyl peak for tosyl-α-chymotrypsin.

Indoleacryloyl-α-chymotrypsin (*82*) provides an example of a short-lived acyl-chymotrypsin which, under appropriate conditions of pH, can be stabilized long enough to make crystallographic study possible. It has particular interest because its deacylation rate and spectrum are identical in solution and in crystals (*83*), giving unusually strong evidence for an identical conformation in the two phases.

An X-ray diffraction study at 2.5 Å resolution has been made of crystals of α-chymotrypsin which were approximately 60% in the indoleacryloyl form (*81*). Because of the effect of displaced water molecules, the density of the indoleacryloyl group is rather distorted but is probably in the planar trans-, S-cis-conformation (Fig. 8). The carbonyl part of the indoleacryloyl group is in a similar position to the sulfonyl group of tosyl-α-chymotrypsin, and an identical displacement of O^{γ} (Ser 195) is observed. There is a similar, but smaller, movement of His 57, and a water molecule again forms a hydrogen-bonded bridge from $N^{\epsilon 2}$ (His 57) to the carbonyl oxygen of the indoleacryloyl group. There is a significant movement of Met 192. Other movements are probably less than 0.4 Å.

The stability of these acyl enzymes is believed to result from their different geometry when compared to the acyl enzymes of "good" substrates. The stereochemistry of the active site suggests (*47*) that in the deacylation step a water molecule hydrogen bonded to $N^{\epsilon 2}$ (His 57) would be activated to attack the carbonyl carbon of the acyl enzyme. A convincing model can be built for such a system (*62, 81*), similar to the structure of indoleacryloyl-α-chymotrypsin shown in Fig. 8, but with a tetrahedral L-α-carbon replacing the trigonal 2-carbon of the acryloyl group. The water molecule attached to His 57 is in a correct orientation to attack the carbonyl carbon. In sulfonyl, phosphoryl, and acryloyl

82. S. A. Bernhard and Z. H. Tashjian, *JACS* **87**, 1806 (1965).
83. G. L. Rossi and S. A. Bernhard, *JMB* **49**, 85 (1970).

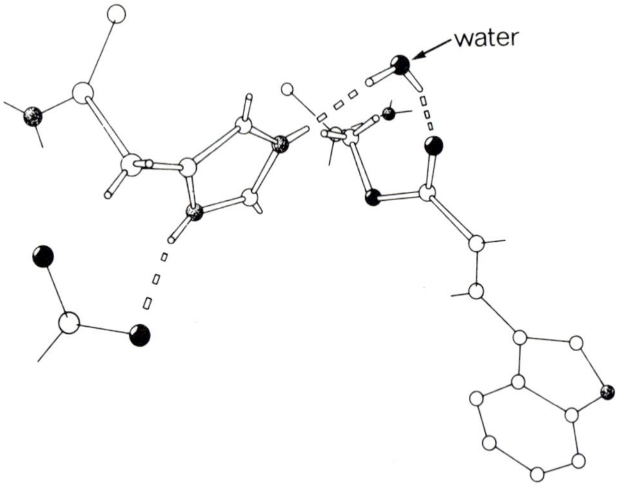

FIG. 8. Part of the active site of indoleacryloyl chymotrypsin as deduced from an electron density difference map (*81*).

chymotrypsin an oxygen atom is interposed between the activated water molecule and this carbon atom. In each case a stable hydrogen-bonded bridge can be formed, in which the lone pair of the oxygen atom is oriented away from the acylating group.

C. γ-CHYMOTRYPSIN

γ-Chymotrypsin was defined operationally by the tetragonal form of its crystals (*84*). It is prepared from the supernatant of α-chymotrypsin crystals by standing for some hours at high pH, and it was originally assumed that a further irreversible autolytic change had occurred. Later it was shown that α- and γ-chymotrypsin have the same terminal residues (*13, 14*) and that γ-chymotrypsin can be converted back to α-chymotrypsin (defined by its monoclinic crystals) over a period of months at low pH (*15*). α-Chymotrypsin and γ-chymotrypsin were thus shown to represent different structural configurations of the same protein molecule, and time seems to be an essential element in the interconversion (*15*). A similar interconversion of the diisopropylphosphoryl enzyme between crystal forms characteristic of α- and γ-chymotrypsin takes place freely and rapidly, depending only on the pH of the crystallizing medium (*14, 15, 85*).

84. M. Kunitz, *J. Gen. Physiol.* **22**, 207 (1938).
85. V. Massey and B. S. Hartley, *BBA* **21**, 361 (1956).

The three-dimensional structure of γ-chymotrypsin has been investigated at low resolution by two independent groups (*86–88*) and subsequently by Davies and his collaborators at high resolution (2.8 Å) (*89*). Both groups made use of sulfonyl fluoride inhibitors directed to the active site, using either phenylmethane sulfonyl fluoride (*38*) or tosyl fluoride (*51, 52*). γ-Chymotrypsin crystals belong to space group $P4_22_12$ with unit cell dimensions $a = 69.7$ Å and $c = 97.7$ Å (*15, 90*). The largest change of unit cell dimensions on tosylation of the crystals is only 0.3%, although difference Patterson projections against *p*-iodophenyl-sulfonyl-γ-chymotrypsin (*90*) suggest that some structural change has occurred on sulfonylation.

The low resolution maps show that the structures of γ- and α-chymotrypsin are very similar with some small change at Tyr 146, the terminal residue of the B chain (*87*). The molecular packing is quite different from that in α-chymotrypsin. The difference maps between phenylmethanesulfonyl-γ-chymotrypsin and γ-chymotrypsin (*86*) and between tosyl-γ-chymotrypsin and γ-chymotrypsin (*87*) show a different pattern of difference density than that observed for tosylation of α-chymotrypsin (*46*). This may result from the displacement of an interstitial sulfate ion on tosylation of γ-chymotrypsin (*87*). It would be located close to the activated water molecule bound to $N^{\epsilon 2}$(His 57), postulated for acyl enzymes and discussed in the previous section.

At the time of writing, no detailed information about the structure of γ-chymotrypsin determined at 2.8 Å resolution has been published. Dr. D. R. Davies has kindly provided the following information (*89*), comparing tosyl-γ- and tosyl-α-chymotrypsin. Agreement with the α-chymotrypsin coordinates is generally within 1 Å and frequently much closer. Tyrosine 146 and Met 192 are displaced from their position in α-chymotrypsin, and His 57 is moved by about 1 Å. A difference electron density map showing the changes of structure between pH 10.5 and 5.6 in γ-chymotrypsin has as its principal features a peak and hole near the position postulated for the sulfate ion in the previous paragraph.

The slow conversion of γ- to α-chymotrypsin remains something of a mystery. It seems surprising that such small and unsubtle conformational

86. J. Kraut, H. T. Wright, M. Kellerman, and S. T. Freer, *Proc. Natl. Acad. Sci. U. S.* **58,** 304 (1967).

87. B. W. Matthews, G. H. Cohen, E. W. Silverton, H. Braxton, and D. R. Davies, *JMB* **36,** 179 (1968).

88. G. H. Cohen, B. W. Matthews, and D. R. Davies, *Acta Cryst.* **B26,** 1062 (1970).

89. D. R. Davies, personal communication (1970).

90. P. B. Sigler, H. C. W. Skinner, C. L. Coulter, J. Kallos, H. Braxton, and D. R. Davies, *Proc. Natl. Acad. Sci. U. S.* **51,** 1146 (1964).

changes of Tyr 146 and Met 192 should require months to become complete, and the role of the diisopropylphosphoryl group in facilitating the transition is not obvious.

D. π- and δ-Chymotrypsin

The π and δ forms of chymotrypsin (*3*) autolyze too rapidly to be crystallized; however, inhibited derivatives can be prepared and crystallized (*91*) by using β-phenyl propionate as a competitive inhibitor during activation. The phenylmethane sulfonyl (PMS) derivatives of π-, δ-, and γ-chymotrypsin are all isomorphous with each other and have very nearly the same cell dimensions as γ-chymotrypsin (*86*).

An X-ray diffraction study of PMS-δ-chymotrypsin at 5 Å resolution, using the isomorphous replacement technique, gave an electron density distribution whose relationship to a map of chymotrypsinogen at the same resolution (*92*) was immediately obvious (*86*). This relationship is considered in detail elsewhere in this volume (*73*). A preliminary comparison with α-chymotrypsin was also made (*16*). The isomorphism of PMS-π-, PMS-δ-, and PMS-γ-chymotrypsin allowed them to be compared directly by difference maps. In this way the location of the Ser 14–Arg 15 dipeptide which differentiates π- and δ-chymotrypsin was identified, together with weaker features which were interpreted as a "slight adjustment" of the structure in the immediate neighborhood of the dipeptide (*86*). Such an adjustment appears reasonable in view of the observation that residues 10–13 are poorly ordered in α-chymotrypsin (*48*). In the same way the dipeptide Thr 147–Asn 148 which differentiates δ-and γ-chymotrypsin was identified. In this case there is relatively little readjustment of the adjacent structure (*86*). It follows that γ-chymotrypsin is probably more similar to π- and δ-chymotrypsin than it is to α-chymotrypsin, in the arrangement of the chain near Tyr 146, and it was suggested that autolysis of δ-chymotrypsin leads directly to the γ form (*16*).

Thus through comparisons of PMS-π-, PMS-δ-, and PMS-γ-chymotrypsin at 5 Å resolution, and of tosyl-γ- and tosyl-α-chymotrypsin at high resolution, a complete chain of similarities has been established, which strengthens our concept of the chymotrypsin molecule as a rather rigid molecule that only undergoes slight local conformational adjustments in different environments and in its different modifications.

91. E. Surbeck and P. E. Wilcox, cited in Kraut *et al.* (*86*).

92. J. Kraut, D. F. High, and L. C. Sieker, *Proc. Natl. Acad. Acad. Sci. U. S.* **51**, 839 (1964).

IV. Substrate Specificity of Chymotrypsin in the Light of Structural Evidence

A. Binding of the Specific Amino Acid Side Chain

Chymotrypsin primarily catalyzes the hydrolysis of amide bonds of proteins and peptides adjacent to the carbonyl group of the aromatic L-amino acid residues of tryptophan, tyrosine, and phenylalanine (*93*). This specificity was first discovered by Bergmann and Fruton (*94*), who investigated the enzymic cleavage of model peptides by proteolytic enzymes. Hydrolysis adjacent to other large hydrophobic residues occurs more slowly as, for instance, histidine (*95*), leucine (*95, 96*), and methionine (*97*). Even alanine esters are hydrolyzed 100 times faster than glycine esters (*98*). The side chain need not be aromatic since cyclohexane can be substituted for the benzene ring of phenylalanine without effect on the hydrolysis rate (*98, 99*). These effects are consistent with the observed shape and hydrophobic nature of the substrate binding pocket.

Oxygen-alkylated tyrosine (*100*) and *p*-iodo-L-phenylalanine (*80*) derivatives are hydrolyzed more slowly by the order of 10^3. This agrees exactly with the crystallographic results discussed in Section III,B,3, which suggest that these side chains are too large to fit correctly into the substrate binding pocket.

Branching at the β-carbon of the side chain (valine, isoleucine, and β,β-dimethyl phenylalanine) leads to unfavorable binding to chymotrypsin and less efficient hydrolysis than with the corresponding unbranched substrate (*101*). Methylation at the α-carbon has a stronger effect and decreases the rate of hydrolysis of typical substrates by the order of 10^5, although without strong effect on the binding affinity (*100, 102*). The conformation of formyl-L-tryptophan observed in crystallographic studies has the α-hydrogen in contact with the carbonyl oxygen

93. L. Cunningham, *Comp. Biochem.* **16,** 85 (1965).
94. M. Bergmann and J. S. Fruton, *JBC* **118,** 405 (1938).
95. N. C. Davis, *JBC* **223,** 935 (1956).
96. G. E. Hein, J. B. Jones, and C. Niemann, *BBA* **65,** 353 (1962).
97. S. Kaufman and H. Neurath, *Arch. Biochem.* **21,** 437 (1949).
98. J. B. Jones, T. Kunitake, C. Niemann, and G. E. Hein, *JACS* **87,** 1777 (1965).
99. R. R. Jennings and C. Niemann, *JACS* **75,** 4687 (1953).
100. R. L. Peterson, K. W. Hubele, and C. Niemann, *Biochemistry* **2,** 942 (1963).
101. H. I. Abrash and C. Niemann, *Biochemistry* **2,** 947 (1963).
102. H. R. Almond, Jr., D. T. Manning, and C. Niemann, *Biochemistry* **1,** 243 (1962).

of the formyl group (*62*). This observation suggests that steric hindrance prevents the acylamido group of α-methyl substrates from taking up the correct conformation about the α-carbon, resulting in incorrect orientation of the sensitive bond.

B. The Acylamido Interaction

The acylamido group attached to the specific amino acid residue has an important effect on the reactivity toward chymotrypsin, and it was recognized that the amido –NH– played an important role. Substitution of this –NH– by an ester oxygen (*103*) or a methylene group (*104*) leads to a loss of reactivity and sterospecificity, while N-methylation of the amido group can lead to even larger reductions of reactivity (*100*, *104*, *105*). In a penetrating analysis, Ingles and Knowles (*106*) isolated the various kinetic components of this specificity. They demonstrated that the improved reactivity of substrates with an amido –NH– was not primarily because of tighter binding but the result of a more favorable orientation of the specific substrate. They showed that the energy of the acylamido interaction was reasonable for a hydrogen bond but suggested that this binding energy was offset by forcing the substrate into a conformation of higher energy.

The amido group of formyl-L-tryptophan has been shown crystallographically to be oriented so it can form a hydrogen bond to the carbonyl group of Ser 214 (*62*). If the aromatic group of a specific substrate is held firmly in the substrate binding pocket, the function of the acylamido interaction would be to prevent rotation about the α–β bond of the specific amino acid residue, thus holding the susceptible bond in the proper orientation for catalysis. (Note that if the aromatic group is held firmly, and not just at one point, only one more interaction is needed to confer stereospecificity.)

As discussed in Section III,B,2, the crystallographic results on substrate binding in α-chymotrypsin are confused by the interaction with Tyr 146 from an adjacent molecule in the vicinity of the acylamido interaction. The X-ray data thus give no evidence for a substrate conformation of high energy, although they are insufficiently accurate to disprove it.

103. S. G. Cohen, J. Crossley, E. Khedouri, and R. Zand, *JACS* **84,** 4163 (1962).
104. G. E. Hein and C. Neimann, *Proc. Natl. Acad. Sci. U. S.* **47,** 1341 (1961).
105. S. G. Cohen, J. Crossley, and E. Khedouri, *Biochemistry* **2,** 820 (1963).
106. D. W. Ingles and J. R. Knowles, *BJ* **108,** 561 (1968).

C. Stereospecificity between L- and D-Amino Acids

Niemann, Cohen, and their co-workers undertook a large-scale survey of the topography of the active site of chymotrypsin, which has been summarized in some of their later papers (*104, 107, 108*). There is also an extensive review of this work by Cunningham (*93*). From this work a concept developed of a tetrahedral arrangement of four sites around the α-carbon of the specificity group. They are designated in Cohen's papers as sites which can accommodate an aryl group (*ar*), and an acylamido group (*am*), a hydrolytic site (*n*), and a restricted site only large enough to hold the α-hydrogen (*h*) (*108a*). This concept is partially vindicated by the crystallographic results. The nature of the *ar* and *am* sites has been discussed in the two preceding sections. The *n* site has been recognized as the vicinity of the His 57–Ser 195 interaction, but no crystallographic experiment has yet been done with a leaving group in this position. As already noted, no restricted *h* site is observed, and it is believed that the restriction on the size of this substituent is a consequence of steric restrictions imposed by the combination of other parts of the substrate. As we shall see, the restriction can be relaxed for some substrates. Hein and Niemann (*107*) also suggested that no specific *h* site exists.

The stereospecificity of chymotrypsin for L stereoisomers of the naturally occurring amino acids was recognized at once (*109, 110*). Hydrolysis of acetyl-D-phenylalanine ethyl ester is catalyzed too weakly by chymotrypsin to be observed, the catalytic rate constant being 10^{-6} to 10^{-7} that of the L isomer (*107*). For formylphenylalanine substrates, the corresponding factor is estimated as 10^{-4} (*111*). D Stereoisomers behave as competitive inhibitors, D esters showing slightly tighter binding to the enzyme than the corresponding L ester (*112, 113*). Niemann suggested that D substrates would bind predominantly in a mode in which the *ar* and *n* binding sites were occupied as usual, leading to exchange of the acylamido and α-hydrogen positions (*107*). The lack of hydrolysis was

107. G. E. Hein and C. Niemann, *JACS* **84,** 4495 (1962).

108. S. G. Cohen, A. Milovanović, R. M. Schultz, and S. Y. Weinstein, *JBC* **244,** 2664 (1969).

108a. The corresponding names used by Niemann's group were ρ_1 for *am*, ρ_2 for *ar*, ρ_3 for *n*, and ρ_H for *h* (*107*).

109. M. Bergmann and J. S. Fruton, *JBC* **124,** 321 (1938).

110. H. Neurath and G. W. Schwert, *Chem. Rev.* **46,** 69 (1950).

111. G. E. Hein and C. Niemann, *JACS* **84,** 4487 (1962).

112. H. T. Huang and C. Niemann, *JACS* **73,** 1541 (1951).

113. D. T. Manning and C. Niemann, *JACS* **80,** 1478 (1958).

explained on the basis that both these interactions were important for activity. Cohen suggested that the α-hydrogen must occupy its usual restricted site and that placing the amino acid side chain in the *ar* site for a D stereoisomer of a natural substrate would lead to exchange of the contents of the hydrolytic site and the aminoacyl site (*114*).

Attempts to bind D substrate analogs at the active site of α-chymotrypsin crystals have been unsuccessful (*115*), presumably because of the conditions in crystals. An alternative mode of binding now seems more probable. The *ar* and *am* sites could be occupied as usual, leaving the α-hydrogen pointing toward Ser 195 and the amide or ester bond pointing away from the molecule. The mode of binding suggested by Niemann (*107*) above, with the *ar* site occupied and with the amide or ester group correctly oriented for hydrolysis, seems possible. The proper conformation would be rare owing to the lack of orienting influence of the amido group, and binding in this mode would have to compete with the mode mentioned earlier, but if it occurs there is no obvious reason why it should not be productive. The structural results thus do not appear to make acylation by a D-amino acid substrate impossible, but the two effects mentioned might well lower the rate by a sufficient factor.

The kinetics of hydrolysis of *N*-acetyl-D,L-tryptophan *p*-nitrophenyl ester indicated the formation of a D-acyl enzyme, which is only slowly hydrolyzed (*116*). This aspect of stereospecificity has been quantitated by the study of deacylation rates of D- and L-acyl-chymotrypsin resulting from the reaction of *p*-nitrophenyl esters of some naturally occurring amino acids (*117*). The deacylation rate was shown to depend directly on the strength of binding of the amino acid side chain, increasing with binding strength for the L series and decreasing (more slowly) for the D series. The rate differences for deacylation between L-amino acids are entropic in origin (*118*), and the rate for an L-tryptophanyl derivative implies that the orientation is almost perfect for deacylation (*117*). It has been suggested (*80*) that it is possible to build a D-acyl enzyme consistent with the crystallographic results, in which the *am* and *ar* interactions are made as usual and the γ-oxygen of the acyl-Ser 195 is in the normal position for acyl enzymes. If this is done, the carbonyl oxygen of the acyl group is brought into a position where it would interfere with the attack of the activated water molecule, exactly as discussed for

114. S. G. Cohen, J. Crossley, E. Khedouri, R. Zand, and L. H. Klee, *JACS* **85**, 1686 (1963).
115. T. A. Steitz and R. Henderson, unpublished experiments (1969).
116. B. Zerner, R. P. M. Bond, and M. L. Bender, *JACS* **86**, 3714 (1964).
117. D. W. Ingles and J. R. Knowles, *BJ* **104**, 369 (1967).
118. M. L. Bender, F. J. Kézdy, and C. R. Gunter, *JACS* **86**, 3714 (1964).

indoleacryloyl chymotrypsin in Section III,B,4. The variation of deacylation rate with binding strength thus reflects the accuracy with which the acyl group is held in the proper orientation for deacylation (in the L series) or in the "wrong" position (in the D series). The results emphasize that the stereochemistry of deacylation is exactly optimized for the aromatic L-amino acids.

D. Some "Locked" Substrates

A particularly interesting series of cyclized chymotrypsin substrates has been studied by Niemann's and Cohen's groups. The methyl ester of 1-keto-3-carboxytetrahydroisoquinoline was the first to be discovered (Fig. 9a) (*119*). There is a fairly strong D specificity (K_m is 25 times less, and k_{cat} 200 times greater, for the D stereoisomer) (*111*). It was later shown that the cyclizing amide linkage at the α-carbon could be replaced by an ester linkage without impairing the very high reactivity of the

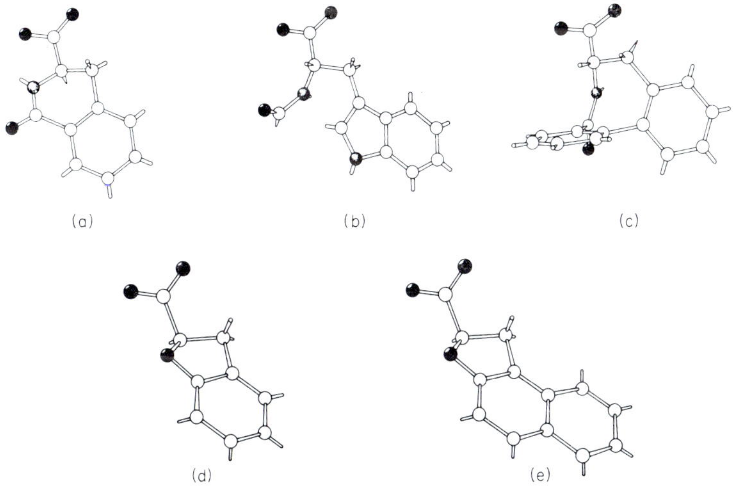

(a) (b) (c)

(d) (e)

Fig. 9. Comparison of possible conformations for several locked substrates with the conformation of formyl-L-tryptophan bound to chymotrypsin, as determined from crystallographic studies: (a) D(—)-1-keto-3-carboxytetrahydroisoquinoline, (b) formyl-L-tryptophan, (c) 2,2′-bridged biphenyl analog of benzoyl-L-phenylalanine (S–S_{eq} isomer), (d) D-hydrocoumarilic acid, (e) D-1,2-dihydronaphtho[2,1-*b*]furan-2-carboxylic acid. [Part of this figure appeared in Steitz *et al.* (*62*).]

119. G. E. Hein, R. B. McGriff, and C. Niemann, *JACS* **82**, 1830 (1960).

substrate (*120*), and it was inferred that this group was not in the normal *am* site for the acylamido group. If one follows the tetrahedral binding site theory, the most reasonable conclusion is that the aromatic part of the isoquinoline group is in the *ar* site and the methyl ester in the *n* site, which for a D substrate forces the cyclizing amide or ester linkage into the *h* site (*120*). This is much easier to understand when it is known that the *h* site is not a site of restricted volume but a site of intramolecular constraint. In Figs. 9a and b a possible conformation of the isoquinoline is compared with that deduced for formyl-L-tryptophan, and they are seen to be extremely similar (*62*). The isoquinoline ring is easily accommodated since it points away from the enzyme surface.

Figure 9c shows a bridged biphenyl analog of L-phenylalanine. The methyl ester of this compound was also discovered to be an extremely good substrate (*121*). The figure shows that the β-phenyl and carboxyl groups are again locked in a very similar conformation to that postulated crystallographically. In this case the acylamido interaction is probably also possible.

These substrates suggest that the exact conformation of the labile amide or ester bond in relation to the *overall shape* of the substrate molecule will determine the reactivity of a particular substrate. The emphasis placed by Niemann and Cohen on the tetrahedral arrangement about the α-carbon somewhat overemphasizes the significance of an absolute specificity between L and D stereochemistry. The apparent contradiction involved in D specificity for tetrahydroisoquinoline is resolved when the overall shape of the substrate molecule is considered. The rigidity of the locked substrate means that the *am* interaction is not needed because the *ar* interaction holds the whole substrate molecule in a rather rigidly defined orientation. The crucial question then is whether the hydrolytic group (represented in Fig. 9 by the carboxyl oxygen pointing to the left) is held in the correct orientation for catalysis. Since there is no specific binding site for the α- and β-carbon of the side chain, this can happen either for an L or a D configuration, depending on the geometry of the locking groups. In Fig. 9 the carbonyl carbon and the amide or ester oxygen of the locked substrate are held in the same orientation as in formyl-L-tryptophan, but the carbonyl oxygen (to the right of the carbonyl carbon) is forced into a slightly different position by the D configuration.

Another class of locked substrate is typified by methyl hydrocoumarilate [(I) Fig. 10] (*122*). In these substrates the carbon is locked in

120. S. G. Cohen and R. M. Schultz, *JBC* **243**, 2607 (1968).
121. B. Belleau and R. Chevalier, *JACS* **90**, 6864 (1968).
122. W. B. Lawson, *JBC* **242**, 3397 (1967).

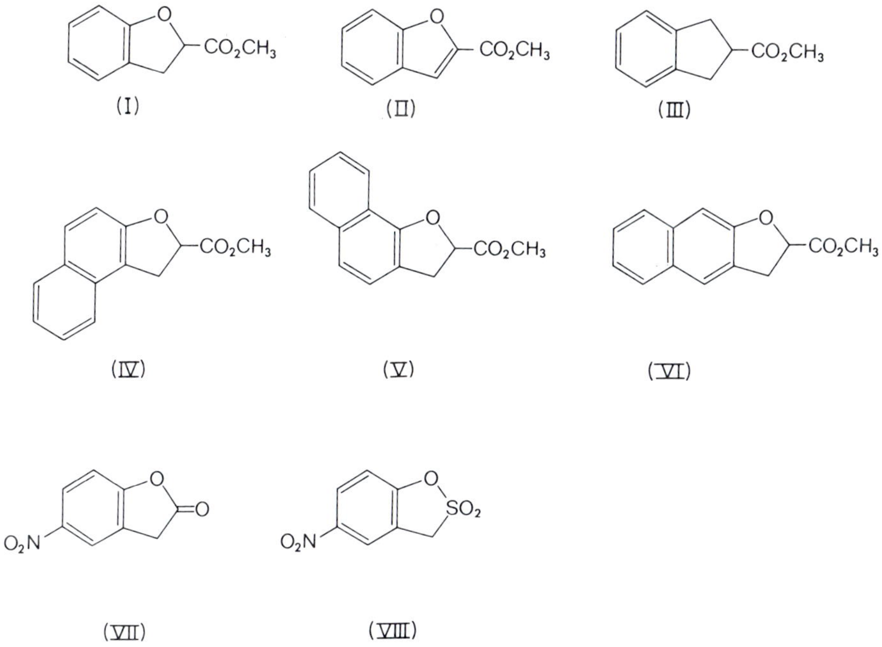

FIG. 10. Structural formulas of some locked substrates (*122, 124–126*).

place by the five-membered coumaran ring. The similar tetrahydrofuran ring can be considerably puckered (*123*). A range of conformations appears possible for hydrocoumarilate, from one in which the ester group is almost equatorial to one in which it is almost axial. Methyl hydrocoumarilate shows some D specificity as a substrate for chymotrypsin. As shown in Fig. 9d the D stereoisomer can be put into a conformation not unlike that proposed for tetrahydroisoquinoline. In this conformation, the carbonyl carbon lies almost in the plane of the six-membered ring. It would seem equally reasonable to put the L isomer into a similar conformation in which the whole molecule has been turned over (that is, the $-CH_2-$ and $-O-$ parts of the five-membered ring have been exchanged). However, this would mean placing the bridge oxygen in a buried position, while the D isomer would leave it on the surface. There could be a further contribution from a specifically favored puckering in the unsymmetrical coumaran ring, and the different bond angles at O and CH_2 will give a slightly different orientation to the ester group.

123. A. D. Mighell and R. A. Jacobson, *Acta Cryst.* **17**, 1554 (1964).

These influences must bring about the observed relatively slight D specificity. [The binding constants are similar, but $(k_{cat})_D/(k_{cat})_L = 83$ (*124*).] These results emphasize the exactness of fit which a good locked substrate must achieve. Methyl coumarilate [(II) Fig. 10] is not hydrolyzed by chymotrypsin and methyl indane-2-carboxylate [(III) Fig. 10] is a very poor substrate (*122*).

More precise information about these substrates comes from the study of a series of naphthalene analogs of hydrocoumarilate in which a second six-membered ring has been fused to the first in each of the three possible positions [(IV), (V), and (VI) Fig. 10] (*124*). The D stereoisomer of (IV) depicted in Fig. 9e is a good substrate and exhibits very high stereospecificity. The structural analogy with tryptophan, which led to the study of these naphthalene analogs (*124*), is convincing. Because of the asymmetry of the molecule, the effects of different stereochemical influences on the kinetics of this series of substrates can be separated; (V) is a poor substrate, with negligible stereospecificity, while D,L-(VI) is inert to chymotrypsin. The position of the –O– bridge seems to be an important feature in determining the reactivity.

Another type of planar locked substrate is provided by the sultones and lactones studied by Kaiser and his colleagues (*125*, *126*) [(VII) and (VIII), Fig. 10]. In these molecules, the bond to be broken is locked into a five-membered ring essentially coplanar with an aromatic ring which seems likely to fit in the binding pocket. These planar compounds are also good substrates. This must be partly because they are *p*-nitrophenol esters, and even when the nitro group is absent the intrinsic reactivity is high. Although such planar groups can be fitted into a crystallographic model of chymotrypsin in a moderately convincing way, the shape of the molecule is nothing like that of Fig. 9b, and the stereochemistry at the catalytic site is noticeably different from that assumed for a good natural substrate.

124. Y. Hayashi and W. B. Lawson, *JBC* **244**, 4158 (1969).

125. J. H. Heidema and E. T. Kaiser, *JACS* **90**, 1860 (1968).

126. P. Tobias, J. H. Heidema, K. W. Lo, E. T. Kaiser, and F. J. Kézdy, *JACS* **91**, 202 (1969).

7

Chymotrypsin—Chemical Properties and Catalysis

GEORGE P. HESS

I. Introduction

Since the publication of Desnuelle's chapter on chymotrypsin in the second edition of "The Enzymes" in 1960, the total structure of this enzyme has been determined. The amino acid sequence of chymotrypsinogen has been elucidated by Hartley (*1*, *2*) and by Meloun *et al.* (*3*). Most important for an understanding of the enzyme is the recent work

1. B. S. Hartley, *Nature* **201,** 1284 (1964).
2. B. S. Hartley and D. L. Kauffman, *BJ* **101,** 229 (1966).
3. B. Meloun, I. Kluh, V. Kostka, L. Morávek, Z. Prušik, J. Vaněcek, B. Keil, and F. Šorm, *BBA* **130,** 543 (1966).

of Blow and co-workers (*4–8*), who succeeded in determining the structure of bovine α-chymotrypsin and of α-chymotrypsin complexes from X-ray diffraction studies at 2 Å resolution. This work is discussed by Blow in Chapter 6 of this volume, and it has recently been reviewed (*8*).

This chapter concerns itself mainly with the chemical properties of the bovine enzyme and the catalytic reaction. Some of the kinetic and chemical data and the structural information which have contributed to an understanding of the enzyme have been summarized. When possible the direct contributions individual amino acid residues make to the activity of the enzyme are discussed.

The proteolytic enzyme, chymotrypsin, is synthesized in the acinous cells of the pancreas as a catalytically inert precursor, chymotrypsinogen. Chymotrypsinogen is carried by the pancreatic juice into the small intestine, where it is converted to an active enzyme, chymotrypsin, by the action of proteolytic enzymes (*9*). Tryptic cleavage of a single peptide bond between Arg 15 and Ile 16 results in a fully active enzyme (*10–12*) (Fig. 1). Subsequent chymotryptic cleavage of the bond between Leu 13 and Ser 14 liberates the dipeptide, Ser 14–Arg 15, and results in the formation of δ-chymotrypsin (*12*). Further chymotryptic cleavage liberates the dipeptide, Thr 147–Asn 148, and results in the well-known form of the enzyme, α-chymotrypsin (*12*).

α-Chymotrypsin has a molecular weight of approximately 25,000, and contains 241 amino acid residues. The molecule has three peptide chains (*13*): an A chain of 13 residues, a B chain of 131 residues, and a C chain of 97 residues. The amino acid sequence of α-chymotrypsin as determined by Hartley (*1, 2*) and by Meloun *et al.* (*3*) is given in Fig. 1. The amino acid residues which are known to participate in the catalytic reaction of chymotrypsin are indicated by circled numbers in Fig. 1; these are Asp 102, His 97, and Ser 195. Other amino acid residues

4. B. W. Matthews, P. B. Sigler, R. Henderson, and D. M. Blow. *Nature* **214,** 652 (1967).

5. P. B. Sigler, D. M. Blow, B. W. Matthews, and R. Henderson, *JMB* **35,** 143 (1968).

6. J. J. Birktoft, B. W. Matthews, and D. M. Blow, *BBRC* **36,** 131 (1969).

7. T. A. Steitz, R. Henderson, and D. M. Blow, *JMB* **46,** 337 (1969).

8. J. J. Birktoft, D. M. Blow, R. Henderson, and T. A. Steitz, *Phil. Trans. Roy. Soc. London* **B257,** 67 (1970).

9. J. H. Northrop, M. Kunitz, and R. M. Herriott, "Crystalline Enzymes" 2nd ed. Columbia Univ. Press, New York, 1948.

10. M. Rovery and P. Desnuelle, *BBA* **13,** 300 (1953).

11. W. J. Dreyer and H. Neurath, *JBC* **217,** 527 (1955).

12. For a review of this subject, see P. Desnuelle, "The Enzymes," 2nd ed., Vol. 4, p. 93, 1960.

13. B. Meedom, *BBA* **31,** 260 (1959).

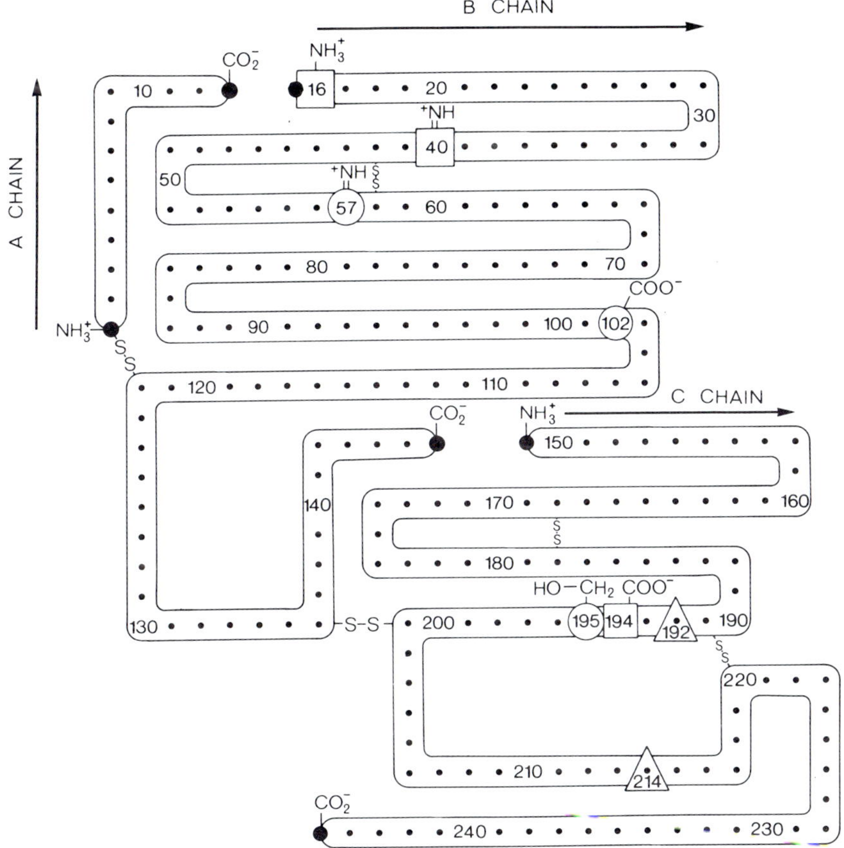

FIG. 1. The amino acid sequence of bovine α-chymotrypsin according to Hartley (*1, 2*). The numbers refer to the sequence of amino acids in bovine chymotrypsinogen A, the catalytically inactive precursor of α-chymotrypsin, and start at the amino-terminal end of that molecule. (●) Amino- and carboxyl-terminal amino acid residues; (○) amino acid residues implicated in the catalytic reaction; (□) amino acid residues implicated in the activation of chymotrypsinogen; (△) amino acid residues implicated in substrate binding. The amino acid residues which form the pocket in which specific substrates bind (*7*) are not indicated.

are indicated by dots with designations of other selected residues as noted later. The amino acid residues which are involved in the conversion of inactive chymotrypsinogen to active chymotrypsin (Ile 16, Asp 194, and His 40) have squares around them. Many other amino acid residues which are not marked are important for substrate binding and form a pocket into which the hydrophobic side chains of chymotrypsin-specific substrates fit (*7*). In addition, two other residues implicated in substrate binding lie outside this pocket; triangles are drawn around them in Fig. 1. These are Ser 214 (*7*) and Met 192 (*14–19*).

Chymotrypsin primarily catalyzes the hydrolysis of peptide bonds of proteins adjacent to the carboxyl group of the aromatic amino acids tryptophan, tyrosine, and phenylalanine (*20, 21*). Proteolysis also occurs adjacent to other large hydrophobic amino acid residues, as, for example, leucine (*22*) and methionine (*23*). The enzyme also catalyzes the hydrolysis of amides and esters of aromatic amino acids and of a large number of other hydrophobic compounds (*21, 24*). A typical substrate is *N*-acetyl-L-phenylalanine amide:

CH_3—C(=O)—NH—CH(CH_2—C_6H_5)—C(=O)—NH_2

acyl-amido part (labels CH₃—C—NH—)

chymotrypsin-specific side chain (labels the CH_2-phenyl group)

Proteolytic enzymes increase the rate of hydrolysis of proteins by a factor of about 10^5, thereby reducing the time taken for digestive processes involving these proteins from 50 years to a few hours.

The presence of a reactive serine residue important for the enzymic activity in chymotrypsin and a number of other enzymes has led to the term *serine proteases*. In many of these enzymes, the amino acid residues adjacent to this reactive serine residue are identical. The sequence Asp–Ser–Gly has been found in chymotrypsin (*25*), trypsin (*26*), throm-

14. D. E. Koshland, Jr., D. H. Strumayer, and W. J. Ray, *Brookhaven Symp. Biol.* **15,** 101 (1962).
15. H. Schachter and G. H. Dixon, *JBC* **239,** 813 (1964).
16. W. B. Lawson and H. J. Schramm, *Biochemistry* **4,** 377 (1965).
17. J. R. Knowles, *BJ* **95,** 180 (1965).
18. H. Weiner, C. W. Batt, and D. E. Koshland, Jr., *JBC* **241,** 2687 (1966).
19. F. J. Kézdy, J. Feder, and M. L. Bender, *JACS* **89,** 1009 (1967).
20. M. Bergmann and J. S. Fruton, *Advan. Enzymol.* **1,** 63 (1941).
21. For a recent review see L. Cunningham, *Comp. Biochem.* **16,** 85 (1965).
22. M. Rovery, M. Poilroux, A. Yoshida, and P. Desnuelle, *BBA* **23,** 608 (1957).
23. N. Brenner, H. R. Müller, and R. W. Pfister, *Helv. Chim. Acta* **33,** 568 (1950).
24. G. E. Hein and C. Niemann, *JACS* **84,** 4495 (1962).
25. R. A. Oosterbaan, P. Kunst, J. Van Rotterdam, and J. A. Cohen, *BBA* **27,** 556 (1958).
26. G. H. Dixon, D. L. Kauffman, and H. Neurath, *JACS* **80,** 1260 (1958).

bin (*27*), and elastase (*28*). In subtilisin the sequence is Thr–Ser–Met (*29*).

The serine proteases exhibit other structural and chemical similarities. The three-dimensional structure of three serine proteases is now known, chymotrypsin (*4–8*), elastase (*30*), and subtilisin (*31*) (see also specific chapters on these enzymes in this volume). These investigations show that enzymes which exhibit different substrate specificity are structurally quite different at the substrate binding site. However, the amino acid residues, Asp, His, and Ser, which are implicated in the catalytic reaction of chymotrypsin seem to bear a similar spatial relationship to each other in these enzymes. There is also chemical evidence which indicates that the catalytic reactions in several serine proteases proceed by a similar mechanism (*32*).

II. The Activity of Chymotrypsin and the Function of Individual Amino Acid Residues

A. Serine 195

Chymotrypsin reacts stoichiometrically with diisopropylphosphofluoridate to form diisopropylphosphoryl chymotrypsin (*33–35*). In this reaction the enzyme is inactivated and a single serine residue, identified from sequence studies (*1–3*) as Ser 195, becomes phosphorylated. This serine residue is acylated by a number of other reagents which also inhibit the enzyme (*36–41*).

27. J. A. Gladner and K. Laki, *JACS* **80,** 1263 (1958).
28. B. S. Hartley, M. A. Naughton, and F. Sanger, *BBA* **34,** 243 (1959).
29. D. C. Shaw, Ph.D. Thesis, University of Cambridge, 1962.
30. H. C. Watson, D. M. Shotton, J. M. Cox, and H. Muirhead, *Nature* **225,** 806 (1970).
31. C. S. Wright, R. A. Alden, and J. Kraut, *Nature* **221,** 235 (1969).
32. For a review on this subject, see M. L. Bender and F. J. Kézdy, *Ann. Rev. Biochem.* **34,** 49 (1965).
33. E. F. Jansen, M. D. F. Nutting, and A. K. Balls, *JBC* **179,** 201 (1949).
34. A. K. Balls and E. F. Jansen, *Advan. Enzymol.* **13,** 321 (1952).
35. N. K. Schaffer, S. C. May, Jr., and W. H. Summerson, *JBC* **206,** 201 (1954).
36. A. K. Balls and H. N. Wood, *JBC* **219,** 245 (1956).
37. M. Caplow and W. P. Jencks, *Biochemistry* **1,** 883 (1962).
38. M. L. Bender, G. R. Schonbaum, and B. Zerner, *JACS* **84,** 2540 (1962).
39. B. F. Erlanger and W. Cohen, *JACS* **85,** 348 (1963).
40. D. E. Fahrney and A. M. Gold, *JACS* **85,** 997 (1963).
41. S. A. Bernhard, S. J. Lau, and H. Noller, *Biochemistry* **4,** 1108 (1965).

1. *The Chymotrypsin-Catalyzed Hydrolysis of Esters*

The mechanism for the chymotrypsin-catalyzed hydrolysis of *p*-nitrophenyl acetate, which involves acetylation of the hydroxyl group of Ser 195 (*42*) has been proposed by Hartley and Kilby (*43*) and is shown in Fig. 2. This mechanism is based on data such as that seen in Fig. 3. The data show that in the chymotrypsin-catalyzed hydrolysis of *p*-nitrophenyl acetate, the liberation of *p*-nitrophenol is biphasic, corresponding to a rapid liberation of *p*-nitrophenol in a concentration roughly equivalent to the concentration of enzyme, and a slow, steady state liberation of *p*-nitrophenol resulting from hydrolysis of the acyl enzyme (Fig. 2).

The same reaction pathway has been observed in the catalytic hydrolysis of a series of model substrates, cinnamoyl imidazole (*38*), indoleacryloyl imidazole (*44*), and methyl cinnamate (*45*). These substrates have the specific side chain but not the acylamido part of chymotrypsin-specific amino acid derivatives and are hydrolyzed some 10^3 times more slowly (*46*).

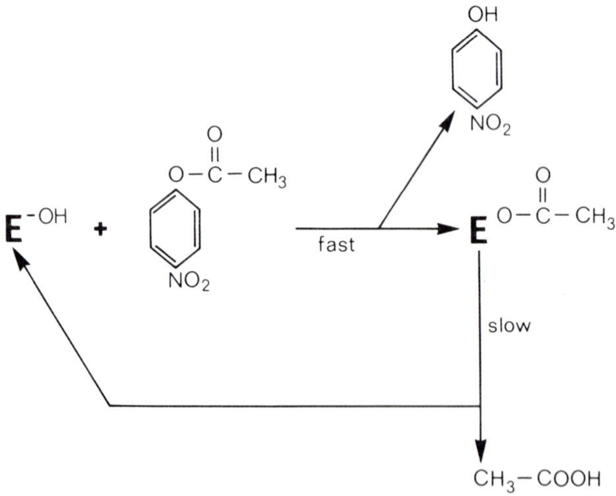

FIG. 2. The chymotrypsin-catalyzed hydrolysis of *p*-nitrophenyl acetate according to Hartley and Kilby (*43*).

42. R. A. Oosterbaan, M. van Adrichem, and J. A. Cohen, *BBA* **63,** 204 (1962).
43. B. S. Hartley and B. A. Kilby, *BJ* **56,** 288 (1954).
44. S. A. Bernhard and Z. H. Tashjian, *JACS* **87,** 1806 (1965).
45. M. L. Bender and B. Zerner, *JACS* **84,** 2550 (1962).
46. D. W. Ingles and J. R. Knowles, *BJ* **108,** 561 (1968).

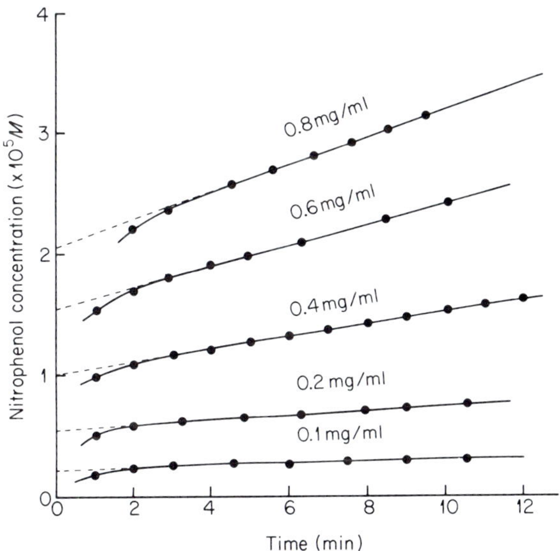

FIG. 3. The liberation of *p*-nitrophenol as a function of time and enzyme concentration (given in milligrams per milliliter) in the chymotrypsin-catalyzed hydrolysis of *p*-nitrophenyl acetate [from Hartley and Kilby (*43*)].

Gutfreund and Sturtevant (*47, 48*) have suggested that the chymotrypsin-catalyzed hydrolysis of specific substrate esters follows the same mechanism observed in the chymotrypsin-catalyzed hydrolysis of *p*-nitrophenyl acetate and other nonspecific substrates. Assuming that a reversible enzyme substrate complex is formed with specific substrates, the mechanism has been put into the general form (*47, 48*)

$$E + S \overset{K_s}{\rightleftharpoons} ES \xrightarrow[k_{23}]{\nearrow P_1} EP_2 \xrightarrow[k_{34}]{} E + P_2 \tag{1}$$

When the substrate S is *N*-acetyl-L-phenylalanine methyl ester, the first product P_1 is methanol and the second product P_2 is *N*-acetyl-L-phenylalanine; ES is the reversible enzyme–substrate complex, and K_s the enzyme–substrate dissociation constant pertaining to this complex. The acyl enzyme is EP_2, in which it is believed that the oxygen of Ser 195 is linked to the carboxyl carbon of *N*-acetyl-L-phenylalanine. In Eq. (1), k_{23} and k_{34} are the rate constants pertaining to the liberation of P_1 and P_2, respectively.

a. Steady State Kinetic Experiments. The velocity v_{obs} of the chymo-

47. H. Gutfreund and J. M. Sturtevant, *BJ* **63**, 656 (1956).
48. H. Gutfreund and J. M. Sturtevant, *Proc. Natl. Acad. Sci. U. S.* **42,** 719 (1956).

trypsin-catalyzed hydrolysis of esters has been most freqeuntly measured above pH 5.0 by continuous automatic titration of the carboxyl group which is liberated in the reaction (*49*). Several spectrophotometric methods have also been used (*50, 51*).

When the initial substrate concentration S_0 is much greater than the initial enzyme concentration E_0 appropriate plots (*52, 53*) of v_{obs} vs. S_0 allow evaluation of two steady state kinetic parameters: (1) V_{max}, the maximum velocity which can be reached in the catalytic reaction, is obtained when the enzyme is saturated with substrate. Instead of V_{max} many publications report catalytic rate constant k_{cat} [the catalytic rate constant k_{cat} is V_{max}/E_0, where E_0 refers to the molar concentration of active sites, as may be determined by the cinnamoyl imidazole method of Bender *et al.* (*38*)]. (2) $K_{m(app)}$ is the Michaelis–Menten constant. The steady state kinetic constants pertaining to Eq. (1) are

$$k_{cat} = \frac{V_{max}}{E_0} = \frac{k_{23}k_{34}}{k_{23} + k_{34}} \tag{2}$$

$$K_{m(app)} = K_s \frac{k_{34}}{k_{23} + k_{34}} \tag{3}$$

If the chymotrypsin-catalyzed hydrolysis of specific substrate esters proceeds by the same mechanism as the chymotrypsin-catalyzed hydrolysis of *p*-nitrophenyl acetate this means, in terms of the parameters of Eq. (1), that the acylation rate constant k_{23} is much greater than the deacylation rate constant k_{34}. When $k_{23} \gg k_{34}$, the steady state kinetic parameters pertaining to Eq. (1) become

$$k_{cat} = k_{34} \tag{4}$$

$$K_{m(app)} = K_s(k_{34}/k_{23}) \tag{5}$$

In the chymotrypsin-catalyzed hydrolysis of amides the acylation rate constant k_{23} was considered to be much smaller than the deacylation rate constant k_{34} (*47, 48, 51*). When $k_{34} \gg k_{23}$ the steady state kinetic parameters pertaining to Eq. (1) become

$$k_{cat} = k_{23} \tag{6}$$

$$K_{m(app)} = K_s \tag{7}$$

Bender and co-workers (*54*) presented evidence obtained from steady state kinetic experiments that the chymotrypsin-catalyzed hydrolysis

49. G. W. Schwert, H. Neurath, S. Kaufman, and J. E. Snoke, *JBC* **172,** 221 (1948).
50. G. W. Schwert and Y. Takenaka, *BBA* **16,** 570 (1955).
51. M. L. Bender, G. E. Clement, F. J. Kézdy, and H. d'A. Heck, *JACS* **86,** 3680 (1964).
52. H. Lineweaver and D. Burk, *JACS* **56,** 658 (1934).
53. G. S. Eadie, *JBC* **146,** 85 (1942).
54. B. Zerner, R. P. M. Bond, and M. L. Bender, *JACS* **86,** 3674 (1964).

of specific substrates proceeds by the mechanism shown in Eq. (1). It was observed that the k_{cat} values for the chymotrypsin-catalyzed hydrolysis of the ethyl, methyl, and *p*-nitrophenyl esters of *N*-acetylphenylalanine were similar. Assuming that chymotrypsin catalysis is a nucleophilic reaction, *p*-nitrophenyl esters are expected to be hydrolyzed much faster than ethyl or methyl esters of the corresponding acid. Identical k_{cat} values indicate, therefore, that in the catalytic hydrolyses of the three esters, the rate limiting step occurs either before or after the hydrolysis of the esters. In analogy with the chymotrypsin-catalyzed hydrolysis of *p*-nitrophenyl acetate, the rate limiting step in the chymotrypsin catalyzed hydrolysis of the three phenylalanine esters was considered (*54*) to be the decomposition of the *N*-acetyl-L-phenylalanin enzyme, intermediate EP_2 in Eq. (1). Similar results were obtained in the chymotrypsin-catalyzed hydrolysis of the ethyl and *p*-nitrophenyl esters of *N*-acetyl-tryptophan. Evidence for a common intermediate was not obtained (*54*), however, in the chymotrypsin-catalyzed hydrolysis of *N*-benzyloxycarbonyl-L-tyrosine esters. In the case of the tyrosine esters, the k_{cat} values for the ethyl and *p*-nitrophenyl esters differ by a factor of 19. Further evidence for the applicability of Eq. (1) to the chymotrypsin-catalyzed hydrolysis of specific substrates comes from the observation of spectral changes during the chymotrypsin-catalyzed hydrolysis of *N*-acetyl-L-tryptophan methyl ester at pH 2.4 (*55*).

b. Stopped Flow Experiments. The applicability of Eq. (1) to the chymotrypsin-catalyzed hydrolysis of specific ester substrates can be demonstrated directly if it can be shown that the alcohol [P_1 in Eq. (1)] is released in a fast step characterized by rate constant k_{23} [Eq. (1)] and the acid is released in a slower step.

This information cannot be obtained from steady state kinetic experiments which, for the simple mechanism shown in Eq. (1), yield only a combination of rate and equilibrium constants as shown in Eqs. (2) and (3). The individual parameters pertaining to Eq. (1) can be determined by pre-steady state kinetic investigations (*56*) of the reaction, provided intermediates in the reaction can be detected.

The problem of detecting intermediates in the chymotrypsin-catalyzed hydrolysis of specific substrate esters has been solved in several ways. It has been observed that the binding of substrates to chymotrypsin is accompanied by absorbancy changes of the enzyme near 290 nm, which can be used to measure the concentration of enzyme–substrate complexes (*57, 58*). Two important methods for detecting intermediates in chymo-

55. F. J. Kézdy, G. E. Clement, and M. L. Bender, *JACS* **86**, 3690 (1964).
56. K. G. Brandt and G. P. Hess, *BBRC* **22**, 447 (1966).
57. J. F. Wootton and G. P. Hess, *JACS* **84**, 440 (1962).
58. A. Himoe, K. G. Brandt, and G. P. Hess, *JBC* **242**, 3963 (1967).

trypsin-catalyzed reactions were developed by Bernhard and co-workers (*41*, *59*, *60*): (1) Spectral changes near 450 nm are observed (*59*) when substrates displace the chromophoric inhibitor proflavin from chymotrypsin, and (2) substrates containing the chromophoric furoylacryloyl group were prepared. Spectral changes are observed (*41*, *60*) when these substrates bind to chymotrypsin.

The spectrophotometric trace obtained in a flow experiment (*61*) during the chymotrypsin-catalyzed hydrolysis of *N*-acetyl-L-phenylalanine methyl ester is shown in Fig. 4. The absorbancy changes of the enzyme were measured near 290 nm. Four distinct phases of the reaction are seen in this reaction and in the chymotrypsin-catalyzed hydrolysis of other specific substrate esters (*62*): *N*-acetyl-L-tyrosine methyl ester, *N*-acetyl-L-tryptophan methyl ester, *N*-acetyl-L-leucine methyl ester, and *N*-furoylacryloyl-L-tyrosine ethyl ester. The four phases in the experiment shown in Fig. 4 are as follows:

(1) An initial rapid increase in absorbance seen as a jump in the oscilloscope reading. This increase is complete in less than 3 msec. The kinetics of this early phase of the reaction have been measured by temperature-jump experiments (Fig. 5). This phase of the reaction

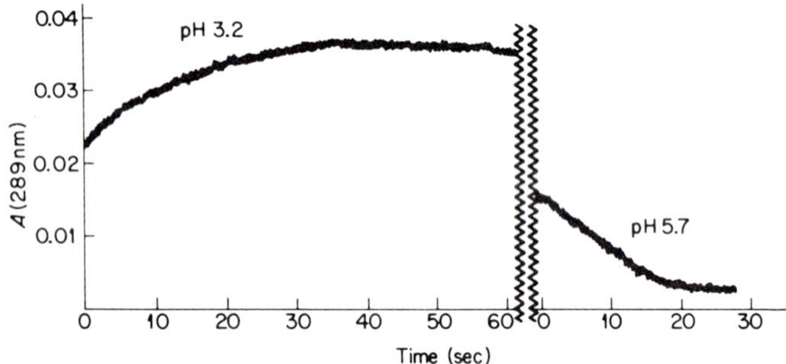

FIG. 4. Spectrophotometric trace of a stopped flow measurement, showing individual steps in the chymotrypsin-catalyzed hydrolysis of *N*-acetyl-L-phenylalanine methyl ester. Absorbance (*A*) was measured at 290 nm as a function of time on a Carey spectrophotometer at 25°C. At a specific time during the interval in which *A* is constant, the pH was changed from 3.2 to 5.7 in order to make the time-dependent spectral changes more quickly measurable. Initial solution concentrations were 37 μmoles of α-chymotrypsin and 17.2 mmoles of substrate. Ionic strength was 0.4 mole.

59. S. A. Bernhard, B. F. Lee, and Z. H. Tashjian, *JMB* **18,** 405 (1966).
60. E. Charney and S. A. Bernhard, *JACS* **89,** 2726 (1967).
61. A. Himoe, K. G. Brandt, R. J. DeSa, and G. P. Hess, *JBC* **244,** 3483 (1969).
62. G. P. Hess, J. McConn, E. Ku, and G. M. McConkey, *Phil. Trans. Roy. Soc. London* **B257,** 89 (1970).

is considered to result from the reversible formation of enzyme–substrate complexes.

(2) A slower increase in absorbance leading to the formation of a steady state intermediate considered to be EP_2 of Eq. (1). The rate of formation of this intermediate can be measured by stopped flow techniques.

(3) A period of time during which the absorbancy of the reaction does not change. This time interval corresponds to the steady state phase of the reaction.

(4) A slow decrease in absorbancy when the substrate is exhausted. This phase may be identified with the decay of the steady state intermediate to give free enzyme and product.

From progress curves such as those shown in Fig. 4, all parameters of Eq. (1) can be evaluated (*56*) when the conditions are chosen so that the initial substrate concentration is much greater than the initial enzyme concentration. Under these conditions, the observed rate constant k_{obs} for formation of the steady state intermediate is a function of initial substrate concentration and the parameters pertaining to Eq. (1) (*56*).

$$k_{obs} = \frac{k_{23}S_0}{S_0 + K_s} + k_{34} \tag{8}$$

In the chymotrypsin-catalyzed hydrolysis of all substrates listed in Table I (*58, 59, 61–66*), the rate constant for formation of the steady state intermediate k_{23} is larger than the rate constant for the decomposition of this intermediate k_{34}.

Recently, evidence was presented (*61*) that in the chymotrypsin-catalyzed hydrolysis of *N*-furoylacryloyl-L-tyrosine ethyl ester, alcohol liberation occurs concomitantly with the formation of the steady state intermediate. These conclusions are based on the experiments of Barman and Gutfreund (*67*), who succeeded in measuring the pre-steady state rate of alcohol release in the chymotrypsin-catalyzed hydrolysis of this substrate. Measurements of the rate constant for the formation of the steady state intermediate in this reaction and analysis of the data of Barman and Gutfreund indicated (*61*) that the rate constant for the two processes (formation of a steady state intermediate and pre-steady state alcohol release) are the same as required by the mechanism of Eq. (1).

63. S. A. Bernhard and H. Gutfreund, *Proc. Natl. Acad. Sci. U. S.* **53,** 1238 (1965).
64. K. G. Brandt, A. Himoe, and G. P. Hess, *JBC* **242,** 3973 (1967).
65. H. Weiner, W. N. White, D. G. Hoare, and D. E. Koshland, Jr., *JACS* **88,** 3851 (1966).
66. A. N. Glazer, *Proc. Natl. Acad. Sci. U. S.* **54,** 171 (1965).
67. T. E. Barman and H. Gutfreund, *BJ* **101,** 411 (1966).

TABLE I

RATE AND EQUILIBRIUM CONSTANTS PERTAINING TO α-CHYMOTRYPSIN-CATALYZED HYDROLYSES OF SPECIFIC SUBSTRATE ESTERS AT pH 5.0 AND 25°C[a]

Substrate	k_{23} (sec^{-1})	k_{34} (sec^{-1})	K_s (m*M*)	$K_{m(app)}$ (m*M*)	References
N-Acetyl-L-Trp ethyl ester	35 ± 9	0.84	2.1 ± 0.6	0.08	Brandt *et al.* (*64*)
N-Acetyl-L-Phe ethyl ester	13 ± 2	2.2	7.3 ± 1.5	1.3	Himoe *et al.* (*61*)
N-Acetyl-L-Tyr ethyl ester	83 ± 24	3.1	18 ± 6	0.8	Hess *et al.* (*62*)
N-Acetyl-L-Leu methyl ester	3.2 ± 0.4	0.19	93 ± 11	4.2	Hess *et al.* (*62*)
N-Furoylacryloyl-L-Tyr ethyl ester	53 ± 19	1.5	0.7	0.03	Himoe *et al.* (*61*)

[a] Stopped flow measurements yielding k_{23} and K_s [see Eq. (1)] were made by a proflavin displacement method (*58, 59, 63–66*), and steady kinetic measurements yielding k_{cat} and $K_{m(app)}$ were determined by pH-stat titration under conditions of excess initial substrate concentration.

All the specific substrate esters listed in Table I exhibit the characteristic reaction path observed in the catalytic hydrolysis of *p*-nitrophenyl acetate: the rapid formation of an intermediate with concomitant liberation of alcohol in the case measured and the rate limiting breakdown of this intermediate (*61, 62, 64*). Recent X-ray diffraction studies by Steitz *et al.* (*7*) of the complex of α-chymotrypsin and *N*-formyl-L-tryptophan at 2.5 Å resolution place the carboxyl group of the substrate in van der Waals contact with the Ser 195 hydroxyl group of the enzyme. Together with the other data available, this indicates that in the chymotrypsin-catalyzed hydrolysis of specific substrates acylation of the hydroxyl group of Ser 195 is involved as well.

2. *The Chymotrypsin-Catalyzed Hydrolysis of Amides*

Gutfreund and Sturtevant (*47, 48*) have suggested that the chymotrypsin-catalyzed hydrolysis of amides also follows the reaction pathway shown in Eq. (1) but that in contrast to ester hydrolysis the rate of formation of the acyl enzyme is rate limiting. The idea can be tested in

FIG. 5. The formation of enzyme–substrate complexes in the chymotrypsin-catalyzed hydrolysis of *N*-furoylacryloyl-L-tryptophan amide of pH 6.7 and 15°C, as measured in (a) temperature-jump and (b) stopped flow experiments [from Hess *et al.* (*62*)], and (c) decomposition of "acyl enzyme" to enzyme and products.

Temperature jump
2 mV
a

0.2 msec

Stopped flow
2 mV
b
20 msec

% [ES]
80
40
0
0
60
120
Time/s
c

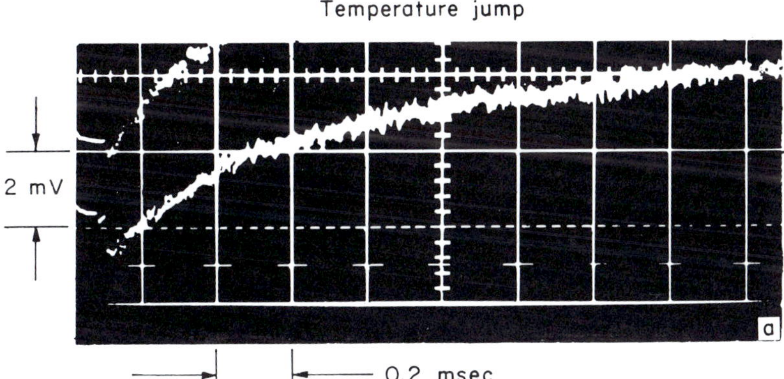

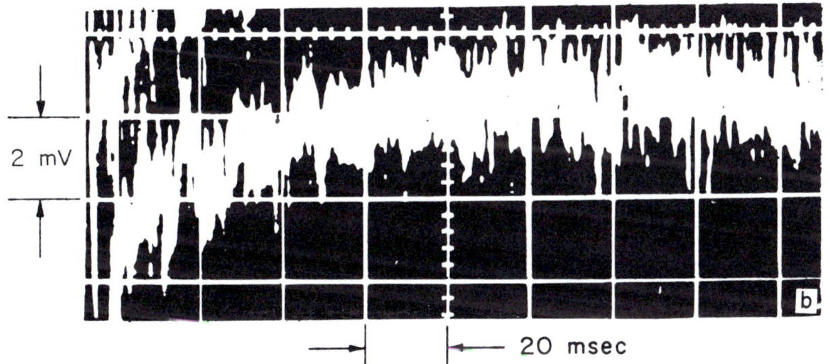

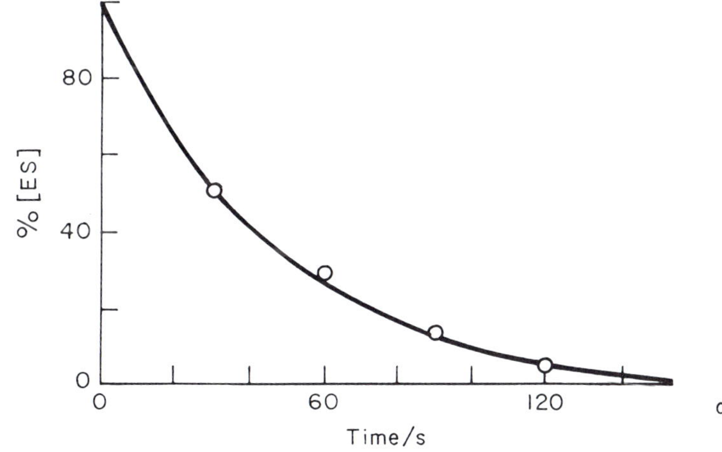

Fig. 5

part by equilibrium and kinetic experiments, since intermediates in chymotrypsin-catalyzed reactions can be detected (*41, 57–60*).

Measurements of the steady state kinetic parameters were greatly facilitated by the development of a completely automatic technique for measuring the hydrolysis of amides by Lenard *et al.* (*68*). This technique, as far as precision of measurement of enzyme concentration and time are concerned, compares favorably with the pH-stat method (*49*) used for measuring the kinetics of ester hydrolysis.

a. Equilibrium and Stopped Flow Experiments. When the acylation rate constant k_{23} is smaller than the deacylation rate constant k_{34} in the mechanism shown in Eq. (1), the steady state kinetic parameter $K_{m(\text{app})}$ is equal to K_s, the enzyme–substrate dissociation constant [see Eq. (7)]. This is quite unlike the result expected in the catalytic hydrolysis of esters for which it has been shown that $k_{23} > k_{34}$ and $K_{m(\text{app})} < K_s$ [Table I and Eq. (5)].

From the data in Table II it can be seen that in the chymotrypsin-catalyzed hydrolysis of a specific substrate amide, *N*-acetyl-L-phenylalanine amide, $K_{m(\text{app})}$ determined from steady state kinetic experiments (*69*) has the same value as K_s determined from equilibrium experiments. In the equilibrium experiments (*58*) the concentration of enzyme–substrate complexes were measured as a function of substrate concentration.

TABLE II

COMPARISON OF K_s' OBTAINED BY TWO DIFFERENT METHODS WITH STEADY STATE KINETIC PARAMETER $K_{m(\text{app})}$[a,b]

Enzyme, substrate, and method	pH	K_s' (m*M*)	$K_{m(\text{app})}$ (m*M*)
α-Chymotrypsin			
APA			
Spectral changes	8.0	28 ± 8	
Proflavin displacement	8.0	28 ± 7	
Steady state kinetics	7.9		31[c]
Spectral changes	10.0	116 ± 90	

[a] Measurements of spectral changes of the enzyme at 290 nm (*57, 58*) and of proflavin displacement (*58, 59, 63–66*) were performed at room temperature, 24°; $K_{m(\text{app})}$ values were obtained at 25 ± 0.1°.

[b] Table from Himoe *et al.* (*58*).

[c] Value is from Foster and Niemann (*69*).

68. J. Lenard, R. Hyman, S. Johnson, and G. P. Hess, *Anal. Biochem.* **11,** 30 (1965).

69. R. J. Foster and C. Niemann, *JACS* **77,** 1886 (1955).

Kinetic experiments are necessary, however, to show that in the equilibrium experiments the concentration of reversibly formed enzyme–substrate complexes and not of a steady state intermediate [such as the acyl enzyme EP_2 of Eq. (1)] was measured. In the catalyzed hydrolysis of *N*-acetyl-L-phenylalanine amide the observed rate constant for the formation of the chymotrypsin complexes was found to be greater than 200 sec^{-1} at pH 8 and 25°C (*58*), while the catalytic rate constant k_{cat} has a value of 0.07 sec^{-1} (*70*). If one assumes that the k_{obs} value of 200 sec^{-1} refers to k_{23}, the rate constant for formation of the acyl enzyme [Eq. (1)], then k_{cat} is a measure of k_{34} [Eq. (4)] and $K_{m(app)} = K_s (k_{34}/k_{23})$ [Eq. (5)]. Therefore, $K_{m(app)}$ would have to be smaller than 35×10^{-5} *M* [$K_s(0.07\ sec^{-1})/(200\ sec^{-1}) = K_s\ 35 \times 10^{-5}$]. From the data in Table II it can be seen that $K_{m(app)}$ has a value of 24×10^{-3} *M*. Therefore, the k_{obs} value of $>200\ sec^{-1}$ cannot refer to the formation of the acyl enzyme EP_2 in Eq. (1) but refers to the reversible formation of an enzyme–substrate complex.

Similar data were obtained with another chymotrypsin-specific substrate amide, *N*-acetyl-L-tryptophan amide. Analysis of these data also indicates that in terms of the mechanism of Eq. (1), the formation of an acyl enzyme is rate limiting and that the rate constant pertaining to this step, k_{23} in Eq. (1), is measured by k_{cat} [Eq. (6)].

b. Stopped Flow and Temperature-Jump Experiments. The formation of enzyme–substrate complexes involving specific substrate amides appears to be a two-step process. Figure 5 shows the results (*71*) of kinetic investigations of the chymotrypsin-catalyzed hydrolysis of *N*-furylacryloyl-L-tryptophan amide using relaxation and stopped flow techniques. The experiments shown were performed at pH 6.7 and 15°C. It is evident that two different processes are being measured. The first process, measured by the temperature-jump technique (*72*), occurs in the microsecond region of the time scale. The second process, measured by stopped flow methods, occurs in the millisecond region. The decomposition of the enzyme–substrate complex to enzyme and products occurs in a third time region which is measured in seconds.

The magnitude of the observed rate constants and their concentration dependencies help to identify the observed processes with particular steps in the reaction. The first step, measured by the temperature-jump method, is considered to be the formation of the first enzyme–substrate

70. A. Himoe, P. C. Parks, and G. P. Hess, *JBC* **242**, 919 (1967).

71. J. McConn and G. P. Hess, quoted in Hess *et al.* (*62*).

72. M. Eigen and L. DeMaeyer, *in* "Techniques of Organic Chemistry," Vol. 8, Part II, p. 895. Wiley (Interscience), New York, 1963.

complex ES_1 in Eq. (10) below. The rate constants for formation k_{12} and dissociation k_{21} of this complex were evaluated using

$$\tau^{-1} = k_{21} + k_{12}(\bar{E} + \bar{S}) \tag{9}$$

In Eq. (9), τ is the relaxation time and $\bar{E}$ and $\bar{S}$ refer to the equilibrium concentrations of E and S. The biomolecular rate constant of $6 \times 10^7\ M^{-1}\ sec^{-1}$ is typical for a diffusion-controlled process which is invariably observed in the first step in the binding of substrates to enzymes (*73, 74*). The second process, which has been measured by flow techniques, has an observed rate constant k_{obs} of 31 sec^{-1}. This k_{obs} value was found to be independent of concentration but dependent on temperature and pH. Above pH 8 the k_{obs} value is too fast to measure by stopped flow techniques.

These data were analyzed by the technique used to differentiate reversible enzyme–substrate complexes from acyl enzyme in the chymotrypsin-catalyzed hydrolysis of *N*-acetyl-L-phenylalanine amide (see Section II,A,2,a). It was concluded that the k_{obs} value of 31 sec^{-1} refers to the formation of a second reversible complex and not to the formation of an acyl enzyme with concomitant product liberation. The observed rate constants for the formation and dissociation of this second complex ES_2 are given in Eq. (10) below. They were calculated from stopped flow measurements, the total absorbancy changes which accompany the binding of substrate to enzyme, and the absorbancy changes which accompany the transformation of ES_1 to ES_2 [see Eq. (10)].

A mechanism consistent with these data and the value of the individual rate constants at pH 6.7 and 15°C is

$$E + S \underset{k_{21} = 1 \times 10^4\ sec^{-1}}{\overset{k_{12} = 6 \times 10^7\ M^{-1}\ sec^{-1}}{\rightleftharpoons}} ES_1 \underset{k_{32} = 30\ sec^{-1}}{\overset{k_{23} = 1.5\ sec^{-1}}{\rightleftharpoons}} ES_2 \rightarrow E + P \tag{10}$$

The nature of intermediates in the chymotrypsin-catalyzed hydrolysis of amide substrates is discussed next.

c. Characterization of Intermediates in the Chymotrypsin-Catalyzed Hydrolysis of Amides. Evidence for the formation of an acyl enzyme and the participation of the Ser 195 hydroxyl group in the chymotrypsin-catalyzed hydrolysis of amides is considerably weaker than in the catalytic hydrolysis of esters. The kinetic data discussed above only indicate that if an acyl enzyme is formed in the chymotrypsin-catalyzed hydrolysis of amides, the formation rate constant for this intermediate must be much smaller than the deacylation rate constant. Therefore,

73. M. Eigen and G. G. Hammes, *Advan. Enzymol.* **25,** 1 (1963).
74. G. G. Hammes, *Advan. Protein Chem.* **23,** 1 (1968).

the steady state concentration of this acyl enzyme must be very small in contrast to what is expected in ester hydrolysis.

The formation of an acyl enzyme in the chymotrypsin-catalyzed hydrolysis of a substrate could be demonstrated, however, if this intermediate could be trapped by an acyl group acceptor which could compete effectively with water.

In the chymotrypsin-catalyzed hydrolysis of esters it has been shown that hydroxylamine competes with water in the deacylation of the acyl enzyme (*75*) and increases significantly the rate of disappearance of substrate (*75*, *76*). This is to be expected for reactions in which decomposition of the acyl enzyme is rate limiting, provided the acyl enzyme reacts more rapidly with hydroxylamine than with water. In the chymotrypsin-catalyzed hydrolysis of an amide, *N*-acetyl-D,L-tyrosine *p*-nitroanilide, it was observed (*77*) that when the hydroxylamine concentration was varied from 0 to 1.6 *M*, the amount of *N*-acetyl-D,L-tyrosine hydroxamic acid which was formed in the reaction increased, while the rate of disappearance of substrate was unaffected. This is to be expected for reactions in which a rate limiting step precedes the reaction with hydroxylamine.

In the chymotrypsin-catalyzed hydrolysis of a series of different ester derivatives of hippuric acid, in the presence of the same concentration of hydroxylamine, the same yield of hydroxamic acid is obtained (*75*). This observation is easily reconciled with the formation of an identical hippuryl enzyme in the catalytic hydrolysis of different hippuryl esters. When identical concentration of glycine amide were added to chymotrypsin solutions which contained either *N*-benzoyl-L-tyrosine ethyl ester, or benzoyl-L-tyrosine-glycine amide, an unexpected result was obtained (*78*). The ratio of the rates of formation of benzoyl-L-tyrosine-glycine amide and benzoyl-L-tyrosine was two times larger in the ester-catalyzed hydrolysis than in the amide-catalyzed hydrolysis (*78*). Based on hypothesis of a common intermediate in the chymotrypsin-catalyzed hydrolysis of esters and of amides, this ratio of rates is expected to be the same. There are other reports (*79*, *80*) of experiments with trapping reagents which have alternatively given evidence for or against the acyl-enzyme hypothesis.

75. R. M. Epand and I. B. Wilson, *JBC* **238,** 1718 (1963).
76. M. L. Bender, G. E. Clement, C. R. Gunter, and F. J. Kézdy, *JACS* **86,** 3697 (1964).
77. T. Inagami and J. M. Sturtevant, *BBRC* **14,** 69 (1964).
78. R. M. Epand, *BBRC* **37,** 313 (1969).
79. S. A. Bernhard, W. C. Coles, and J. F. Nowell, *JACS* **82,** 3043 (1960).
80. M. Caplow and W. P. Jencks, *JBC* **239,** 1640 (1964).

Some information is available pertinent to the intermediate ES_2 formed [Eq. (10)] in the chymotrypsin-catalyzed hydrolysis of *N*-furoylacryloyl-L-tryptophan. Since product formation does not accompany the transformation of ES_1 to ES_2, this step presumably represents an isomerization of the protein, the substrate, or both. Protein isomerizations induced by substrate have been observed in the binding of coenzyme to triosephosphate dehydrogenase (*81*), and in reactions catalyzed by ribonuclease (*82*), lysozyme (*83*), and a number of other enzymes (*74*). Above pH 8 the rate constants for the isomerization observed in the chymotrypsin-catalyzed reaction fall within the range of rate constants observed for similar processes in other enzymes (*74*). In Eq. (10) ES_2 can also represent an isomerization of the substrate.

It has recently been suggested (*84*) that in the chymotrypsin-catalyzed hydrolysis of amides a tetrahedral intermediate accumulates rather than the acyl enzyme observed in ester hydrolysis. These speculations are shown in Fig. 6. Intermediates for which there is no evidence are printed in light type. The length of the single arrows between intermediates are in rough proportion to the magnitude of the rate constants. According

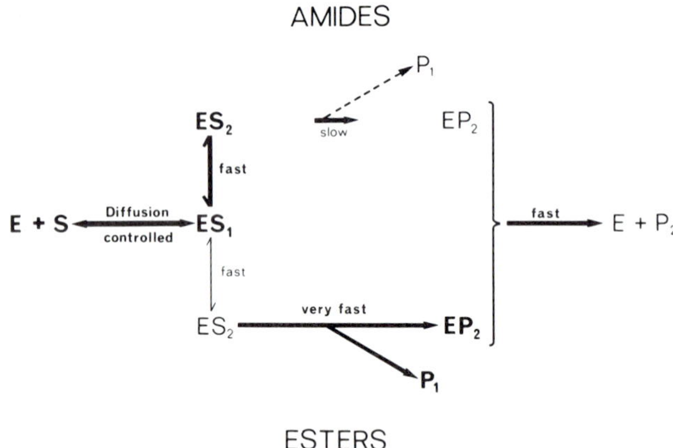

Fig. 6. Possible differences between the reaction pathways of the chymotrypsin-catalyzed hydrolysis of esters and amides. Intermediates for which there is no evidence appear in light print. It has been suggested that ES_2 is a tetrahedral intermediate (*84*).

81. K. Kirschner, M. Eigen, R. Bittman, and B. Voigt, *Proc. Natl. Acad. Sci. U. S.* **56,** 1661 (1966).
82. T. C. French and G. G. Hammes, *JACS* **87,** 4669 (1965).
83. E. Holler, J. A. Rupley, and G. P. Hess, *BBRC* **37,** 423 (1969).
84. M. Caplow, *JACS* **91,** 3639 (1969).

to these speculations (*84*) the tetrahedral intermediate accumulates in amide hydrolysis because of the rate determining breakdown of this intermediate. In esters the formation of the tetrahedral intermediate is considered to be rate limiting in the acylation reaction (Fig. 6). The suggestion for the tetrahedral intermediate in the chymotrypsin-catalyzed hydrolysis of amides is based in part on the magnitude and pH dependence of the steady state kinetic parameters of the chymotrypsin-catalyzed hydrolysis of anilide derivatives of *N*-acetyl-L-tryptophan. Apart from the difficulty of deducing mechanisms from steady state kinetic constants, which represent a combination of rate and equilibrium constants, the poor solubility of the substrates used prevented a reliable determination of these parameters. When a soluble anilide substrate was used for the investigation of trypsin-catalyzed reactions (*85*), the peculiar pH dependence reported for the chymotrypsin substrates was not observed.

The chymotrypsin-catalyzed hydrolysis of amide derivatives of chymotrypsin specific substrates is therefore considerably less well understood than the chymotrypsin-catalyzed hydrolysis of esters. The function of the Ser 195 oxygen in the reaction is consequently also not clear. Nevertheless, as a result of kinetic investigations (*58, 66*), the steady state kinetic parameters can be interpreted: $K_{m(\text{app})}$ is a measure of an overall enzyme–substrate dissociation constant and k_{cat} of the rate limiting bond breaking step which follows.

An extension of both the chemical trapping experiments and kinetic studies, including the effect of substituents of the leaving group of the substrate, may be a fruitful starting point for future investigations.

B. Histidine 57

The first indication of the importance of a histidine residue in chymotrypsin-catalyzed reactions came from kinetic studies (*86*) in which the effect of pH on the rates of these reactions was studied.

1. *The pH Dependence of Chymotrypsin-Catalyzed Reactions*

Chymotrypsin-catalyzed reactions have bell-shaped pH profiles such as the one shown in Fig. 7 for the hydrolysis of a neutral substrate, acetyl-L-tryptophan amide. Bell-shaped pH-rate profiles with neutral substrates have often been interpreted as evidence that two ionizing groups of the enzyme are important in the catalytic reaction. The mid-

85. T. Inagami, *J. Biochem. Tokyo* **66,** 277 (1969).
86. B. R. Hammond and H. Gutfreund, *BJ* **61,** 187 (1955).

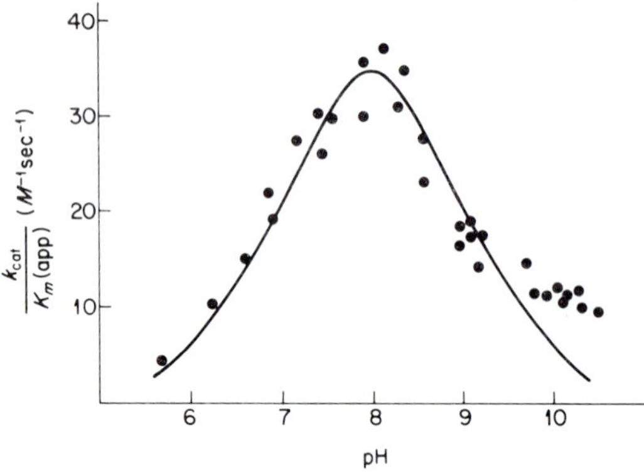

FIG. 7. A pH-rate profile typical of reactions catalyzed by chymotrypsin. This curve was drawn from steady state kinetic data on the α-chymotrypsin-catalyzed hydrolysis of *N*-acetyl-L-tryptophan amide at 25°C. Continuous measurements of released ammonia were made at each pH level by the method of Lenard *et al.* (*68*) over a range of substrate concentrations. Eadie plots (*53*) yielded values of k_{cat} and $K_{m\,(app)}$. These measurements were made at an initial enzyme concentration of from 1 to 6 μmoles and substrate concentration in the range 2–20 mmoles. Appropriately buffered solutions also contained KCl to bring the ionic strength to 0.16 mole [from Himoe *et al.* (*70*)].

point of the left-hand side of the pH-rate profile indicates that the rate of the catalytic reaction increases as an ionizing group with pK_{app} of about 7 ionizes. The midpoint of the right side of the curve in Fig. 7 indicates that the rate of the reaction decreases as an amino acid residue with pK_{app} of about 8.5 ionizes. The right-hand side of the pH-rate profile will be discussed later.

In the chymotrypsin-catalyzed hydrolysis of neutral esters it was shown that the catalytic rate constant k_{cat} also increases as an amino acid residue of the enzyme with pK_{app} of about 7 ionizes (*51, 86*). In ester hydrolysis k_{cat} is a measure of the deacylation rate constant k_{34} only [Eq. (4)]. The pH-dependence of k_{cat}, therefore, gives no information about the pH dependence of k_{23}, the acylation rate constant. The pH dependence of the acylation rate constant was obtained (*62*) from stopped flow investigations of the chymotrypsin-catalyzed hydrolysis of *N*-acetyl-L-leucine methyl ester. This substrate was chosen rather than the esters of aromatic amino acids because the rate constants involved in the catalytic hydrolysis of this substrate are somewhat smaller and can be measured with greater precision. The catalytic hydrolysis of *N*-acetyl-L-leucine methyl ester by chymotrypsin is still quite efficient (Table I), however, and the results are representative of what has been measured with less precision in the chymotrypsin-catalyzed hydrolysis of the esters of aromatic amino acids.

The acetyl-L-leucine methyl ester data shown in Fig. 8 indicate that both the measured rate constants are controlled by an amino acid residue of the enzyme with pK_{app} of about 7. As defined by the mechanism in Eq. (1), these rate constants are k_{23} (solid circles), pertaining to the formation of the steady state intermediate EP_2, and k_{34} (solid squares), pertaining to the decomposition of EP_2 to free enzyme and product. The dashed line on the bottom of the graph represents the pH dependence of k_{34} drawn on the same scale as that used for k_{23}. It is apparent that k_{34} is much smaller than k_{23}; therefore, it characterizes the rate limiting step of the reaction and is the only rate constant which is measured in steady state kinetic experiments.

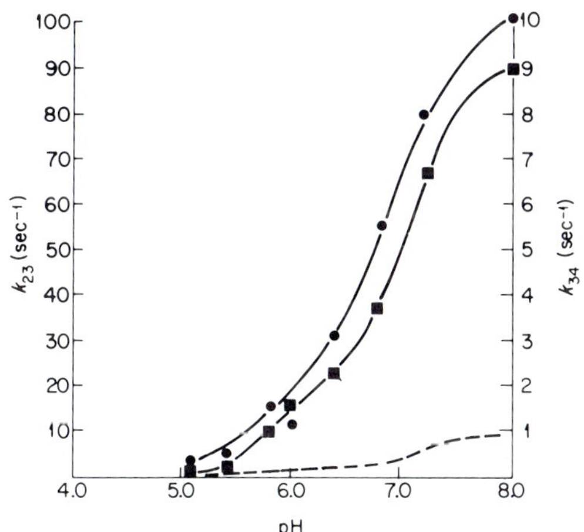

FIG. 8. The pH dependence of rate constants pertaining to bond breaking steps in the chymotrypsin-catalyzed hydrolysis of acetyl-L-leucine methyl ester. (●) k_{23}, (■) k_{24} [see Eq. (1)]. The dashed line shows k_{34} plotted on the scale drawn for k_{23}. Stopped flow measurements were performed according to a previously described proflavin displacement method (*58, 59, 63–66*) in which change in concentration of an enzyme–proflavin complex is followed at 465 nm. Solutions buffered with 0.1 mole acetate or phosphate contained KCl to give an ionic strength of 0.4 mole. Temperature was at 25°C. Values of k_{34} (and also K_S) were evaluated from plots of k_{obs} against k_{obs}/S_0 by means of a digital computer program based on the relation

$$k_{obs} = \frac{k_{23}S_0}{S_0 + K_s(1 + (F_0/K_{EF}))} + k_{34}$$

where k_{obs} is the observed rate, F_0 is initial proflavin concentration (60 μmoles), K_{EF} is the enzyme–proflavin dissociation constant, and S_0 is initial substrate concentration (2.5–80 mmoles). Data of McConn and Hess (*71*).

2. *Chemical Modification Experiments*

Ionizing groups in proteins with $pK_{(app)}$ of about 7 are usually histidine residues. Photooxidation experiments of chymotrypsin indicated the importance of such a residue in the enzymic activity (*87, 88*). Elegant proof of the importance of His 57 was obtained by Shaw and co-workers (*89, 90*) who synthesized an analog of a tosyl-L-phenylalanine substrate, L-1-tosylamino-2-phenylethylchloromethyl ketone (see Chapter 2 by Shaw in Volume I).

These investigators (*89, 90*) showed that this reagent alkylates His 57 irreversibly with concomitant loss of enzymic activity. A number of reagents are now known which react specifically with His 57 and inactivate the enzyme (*91, 92*).

3. *The Function of His 57 in the Catalytic Reaction*

There has been considerable speculation in the literature that chymotrypsin-catalyzed reactions proceed via an *N*-acyl imidazole derivative (*93, 94*). No evidence for such an intermediate was found, however, in kinetic and spectroscopic investigations of the chymotrypsin-catalyzed hydrolysis of *p*-nitrophenyl acetate (*95*). Crystallographic data indicate (*7*) that chymotrypsin substrates which bind in the specific binding site of the enzyme cannot acylate the imidazole nitrogens of His 57.

In order to explain the observation that the rate of chymotrypsin-catalyzed reactions increases as an amino acid residue with $pK_{(app)}$ of about 7 ionizes, Cunningham (*96*) suggested that the imidazoyl nitrogen of histidine acts as a hydrogen bond acceptor for the Ser 195 hydroxyl group, thereby making the Ser 195 oxygen a better nucleophile. This suggestion is consistent with present crystallographic data (*5–8*) which place the $N^{\epsilon 2}$ of His 57 and the hydroxyl group of Ser 195 into a position consistent with a hydrogen bond.

87. L. Weil, S. James, and A. R. Buchert, *ABB* **46,** 266 (1953).
88. W. J. Ray, Jr. and D. E. Koshland, Jr., *Brookhaven Symp. Biol.* **13,** 135 (1960).
89. G. Schoellman and E. Shaw, *BBRC* **7,** 36 (1962).
90. E. B. Ong, E. Shaw, and G. Schoellmann, *JACS* **86,** 1271 (1964).
91. K. J. Stevenson and L. B. Smillie, *JMB* **12,** 937 (1965).
92. Y. Nakagawa and M. L. Bender, *JACS* **91,** 1566 (1969).
93. G. H. Dixon, H. Neurath, and J. F. Perchere, *Ann. Rev. Biochem.* **27,** 489 (1958).
94. H. Neurath and B. S. Hartley, *J. Cellular Comp. Physiol.* **54,** Suppl. 1, 179 (1959).
95. J. F. Wootton and G. P. Hess, *JACS* **83,** 4234 (1962).
96. L. W. Cunningham, Jr., *Science* **125,** 1145 (1957).

C. Aspartic Acid 102

The importance of Asp 102 was deduced from X-ray diffraction measurements by Blow and co-workers (*97*). The carboxyl group of this residue is buried in a hydrophobic environment in close enough contact with $N^{\delta 1}$ of the imidazole ring of His 57 to form either a hydrogen bond or an internal ion pair, depending on the ionization state of His 57 and Asp 102 (Fig. 9). As can be seen from Fig. 9 the active center can be represented by a hydrogen-bonded network extending from Ser 195 via His 57 to Asp 102.

As indicated by X-ray diffraction studies (*30*, *31*) a similar constellation of these three residues, Ser, His, and Asp, exists in the active center of elastase, which is a highly homologous protein (*98*, *99*), and in subtilisin, which shows no common sequences with chymotrypsin and its analogs (*99*, *100*). Sequence analogies indicate that this interaction

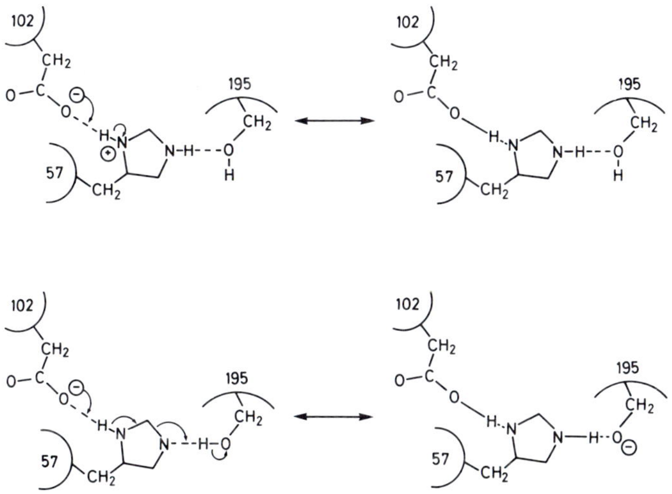

Fig. 9. Canonical forms of the active center: top line, pH 4; bottom line, pH 8. The canonical forms represented indicate purely electronic rearrangements. The protons in hydrogen bonds are represented schematically as being near the atom to which they are covalently bonded in the more "conventional" of the canonical forms, namely, the forms on the left. The forms on the right are drawn with "stretched" electronic bonds to these protons [from Blow *et al.* (*97*)].

97. D. M. Blow, J. J. Birktoft, and B. S. Hartley, *Nature* **221,** 337 (1969).
98. D. M. Shotton and B. S. Hartley, *Nature* **225,** 802 (1970).
99. B. S. Hartley, *Phil. Trans. Roy. Soc. London* **B257,** 77 (1970).
100. E. L. Smith, F. S. Markland, C. B. Kasper, R. J. De Lange, M. Landon, and W. H. Evans, *JBC* **241,** 5974 (1966).

also exists in the other serine proteases of known primary sequence, trypsin and thrombin (*97*) (see specific chapters on elastase, subtilisin, trypsin, and thrombin in this volume). It is most likely, therefore, that Asp 102 plays an important role in the action of chymotrypsin. It has been suggested (*97*) that the negatively charged carboxyl group of Asp 102 can transfer electrons through hydrogen bonds and the polarizable imidazole ring to the Ser 195 oxygen and thus make this oxygen a more powerful nucleophile.

The contribution this interaction between Asp 102 and His 57 makes to the overall efficiency of the catalytic reaction is not yet known. Blocking (*101*) of the carboxyl groups of chymotrypsin with diphenyldiazomethane has indicated the importance of two carboxyl groups in chymotrypsin. Blocking (*102, 103*) of the carboxyl groups of chymotrypsin with glycine methyl ester or glycine amide, using a modification (*104*) of the carbodiimide method for peptide synthesis of Sheehan and Hess (*105*), has indicated that the efficiency of the enzyme is decreased by 50% when all but two carboxyl groups have reacted (*106*). One of the unreactive carboxyl groups has been identified by Koshland *et al.* (*102*) as that of Asp 194. Crystallographic experiments (*4, 5*) have indicated that Asp 194 of chymotrypsin is important for maintaining the enzyme in an active conformation.

Another important function of a negative charge in the active site of chymotrypsin has been suggested by the inhibitor binding experiments of Johnson and Knowles (*107*). The authors suggested that a negative charge develops when a group with pK_{app} of about 7.3 (presumably His 57) ionizes and that the function of this negative charge is to repel the negatively charged acids which are invariably the products of chymotrypsin-catalyzed reactions.

D. Isoleucine 16

1. *The pH Dependence of Chymotrypsin-Catalyzed Reactions at Alkaline pH*

The rate of chymotrypsin-catalyzed reactions decreases at alkaline pH (*32*). The pH-rate profile shown in Fig. 6 implicates an ionizing group

101. A. A. Aboderin and J. S. Fruton, *Proc. Natl. Acad. Sci. U. S.* **56,** 1252 (1966).
102. K. L. Carraway, P. Spoerl, and D. E. Koshland, Jr., *JMB* **42,** 133 (1969).
103. J. P. Abita and M. Lazdunski, *BBRC* **35,** 707 (1969).
104. K. L. Carraway and D. E. Koshland, Jr., *BBA* **160,** 272 (1968).
105. J. C. Sheehan and G. P. Hess, *JACS* **77,** 1067 (1955).
106. Abita and Lazdunski data indicate that 15 free carboxyl groups have been

of the enzyme with pK_{app} of about 8.5. The data in Fig. 10 indicate that in the chymotrypsin-catalyzed hydrolysis of a neutral amide substrate k_{cat} is pH independent at alkaline pH while $K_{m(app)}$ is pH dependent (*70*). From equilibrium and kinetic studies (*58, 62*) it is known that $K_{m(app)}$ in the chymotrypsin-catalyzed hydrolysis of specific substrate amides is a measure of an overall enzyme–substrate dissociation constant and k_{cat} of the rate limiting bond breaking step which follows. The data in Fig. 10 can therefore be interpreted. The ionizing group of the enzyme with $pK_{(app)}$ of about 8.5 affects the enzyme–substrate dissociation constant but not the bond breaking step. Earlier kinetic investigation (*108, 109*) of the reaction of chymotrypsin with the inhibitor, diisopropylphosphofluoridate, led to the conclusion that the alkaline pH dependence of this reaction results from the effect of hydrogen ions on a step which precedes bond breaking.

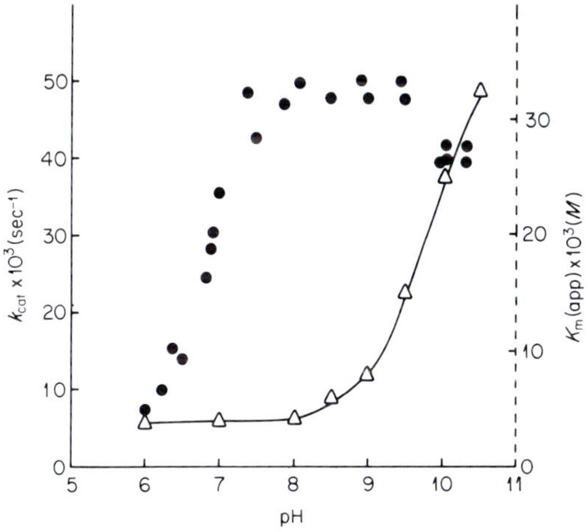

FIG. 10. The pH dependence of the steady state kinetic parameters of the α-chymotrypsin-catalyzed hydrolysis of *N*-acetyl-L-tryptophan amide at 25°C. Continuous measurements of released ammonia were made at each pH level by use of the automatic ninhydrin method of Lenard *et al.* (*68*). (●) k_{cat} and (△) $K_{m(app)}$ [from Himoe *et al.* (*70*)].

---

blocked when the efficiency of the enzyme has decreased by 50%. The authors assumed that there are only 16 free carboxyl groups in chymotrypsin, but recent experiments (*97*) have indicated that there are 17 free carboxyl groups in the enzyme.

107. C. H. Johnson and J. R. Knowles, *BJ* **101,** 56 (1966).
108. A. Y. Moon, J. Mercouroff, and G. P. Hess, *JBC* **240,** 717 (1965).
109. A. Y. Moon, J. M. Sturtevant, and G. P. Hess, *JBC* **240,** 4204 (1965).

2. *Chemical Modification of the α-Amino Group of Ile 16*

Experiments (*70, 110–112*) indicated that the ionization of the α-amino group of Ile 16 is responsible for the pH-dependent equilibrium between enzyme conformation at alkaline pH and the structure transitions which convert chymotrypsinogen into active enzyme. These experiments were confirmed by Labouesse and co-workers (*113, 114*). The newly-formed α-amino group of δ-chymotrypsin can be specifically acetylated if one first acetylates all free ε-amino groups of chymotrypsinogen (*110, 111, 113*) (see Fig. 11). Acetylated chymotrypsinogen is then activated by trypsin to yield acetylated δ-chymotrypsin, a molecule with a single free amino-terminal α-amino group. The $pK_{(app)}$ of this α-amino group is about 8.5 as indicated by the difference between the titration curves of acetylated chymotrypsinogen and of acetylated δ-chymotrypsin (*110*). As far as it has been tested (*70, 110*) the enzymic activity of acetylated δ-chymotrypsin is indistinguishable from that of the unacetylated enzyme. Acetylation (*110, 113, 115*) of acetylated δ-chymotrypsin yielded a product with a decreasing amount of amino-terminal Ile 16 α-amino groups and a corresponding

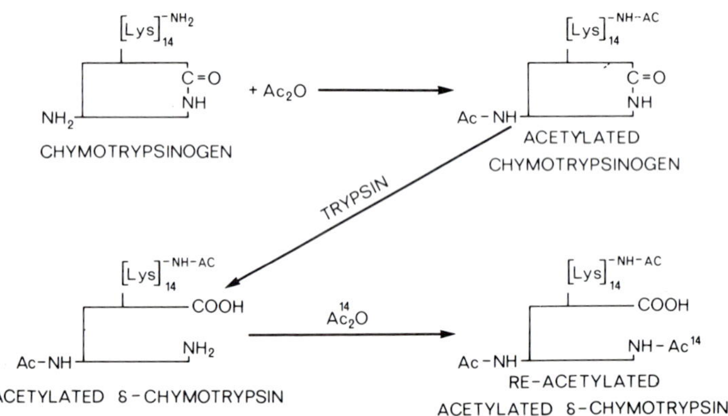

FIG. 11. Flow scheme for the preparation of acetylated δ-chymotrypsin, an active enzyme in which all ε- and α-amino groups have been acetylated, except for the amino-terminal α-amino group of IIe 16.

110. H. L. Oppenheimer, B. Labouesse, and G. P. Hess, *JBC* **241,** 2720 (1966).
111. G. P. Hess, *Brookhaven Symp. Biol.* **21,** 155 (1968).
112. J. McConn, G. D. Fasman, and G. P. Hess, *JMB* **39,** 551 (1969).
113. C. Ghelis, J. Labouesse, and B. Labouesse, *BBRC* **29,** 101 (1967).
114. D. Karibian, C. Laurent, J. Labouesse, and B. Labouesse, *European J. Biochem.* **5,** 260 (1968).
115. Data of S. Kumar, K. Dar, H. Hatano, S. Gano, and G. P. Hess, quoted in Hess (*111*).

decrease in enzymic activity (Table III). Furthermore, the amount of ^{14}C-labeled acetyl groups introduced into acetylated δ-chymotrypsin corresponded within experimental error to the amount of amino-terminal Ile which has been acetylated (Table III) (*115*). These data (*110, 111*) are essentially in agreement with the data obtained by Labouesse and co-workers (*113, 117*). These investigators reported (*113, 117*) that the concentration of the active site of the enzyme decreases less than the ability of the enzyme to catalyze the hydrolysis of *N*-acetyl-L-tyrosine ethyl ester.

TABLE III

CORRELATION OF AMINO-TERMINAL ISOLEUCINE CONTENT WITH CATALYTIC ACTIVITY IN ACETYLATED DERIVATIVES OF CHYMOTRYPSIN[a]

Material	Amino-terminal Ile content[b] (mole/mole)	^{14}C-labeled acetyl groups (mole/mole)	Activity[c] (%)
Acetylated δ-chymotrypsin	1.09	—	100
Reacetylated acetylated δ-chymotrypsin	0.25	0.80	20

[a] Data of Kumar *et al.* (*115*).

[b] Determined by the quantitative end group method of Stark and Smyth (*116*) with use of norleucine as an internal standard.

[c] A k_{cat} value of 225 sec^{-1} (measured at pH 8.0 and 25°C with acetyl-L-tyrosine ethyl ester as substrate) was taken as corresponding to 100% activity.

3. *The Function of the α-Amino Group of Ile 16 in the Conformation and Activity of the Enzyme*

The relationship between a positive charge on the α-amino group of Ile 16, the conformation of proteins, and enzymic activity are illustrated in part by the experiments (*62, 111, 112*) shown in Figs. 12 and 13:

(1) The α-amino group of Ile 16 of catalytically inactive chymotrypsinogen is in amide linkage and cannot acquire a positive charge. When this α-amino group is specifically acetylated in the active enzyme it likewise cannot acquire a positive charge. As indicated by the data in Fig. 12, the circular dichroism spectra of acetylated chymotrypsinogen and of acetylated δ-chymotrypsin in which the α-amino group of Ile 16 has been acetylated are independent of pH.

(2) Titration experiments (*110*) have indicated that at pH 10 the Ile 16 α-amino group of the active enzyme is uncharged as it is in chymotryp-

116. G. R. Stark and D. G. Smyth, *JBC* **238**, 214 (1963).
117. C. Ghélis, J. Garel, and J. Labouesse, *Biochemistry* **9**, 3902 (1970).

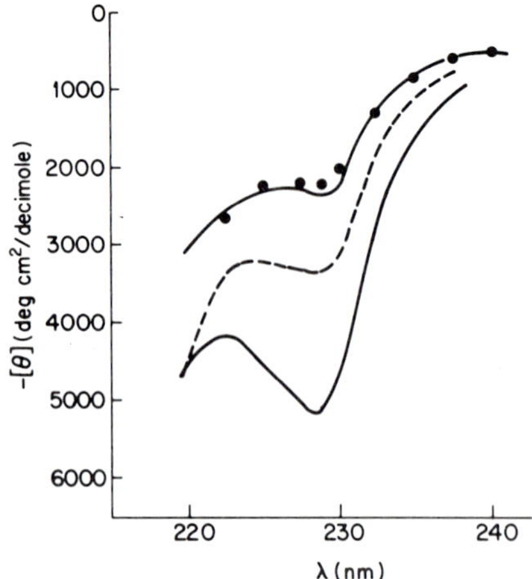

FIG. 12. Circular dichroism (CD) spectra of acetylated derivatives of chymotrypsin. Lower curve, acetylated δ-chymotrypsin at pH 6.7; middle curve, acetylated δ-chymotrypsin at pH 10; upper curve, acetylated chymotrypsinogen in the pH region 6.7–10; dots, reacetylated acetylated-δ-chymotrypsin in the pH region 6.7–10. Measurements were made at room temperature on a Cary 60 spectropolarimeter equipped with a CD attachment. Values of ellipticity θ were calculated as previously described by Fasman *et al.* (*118*). Data from McConn *et al.* (*112*).

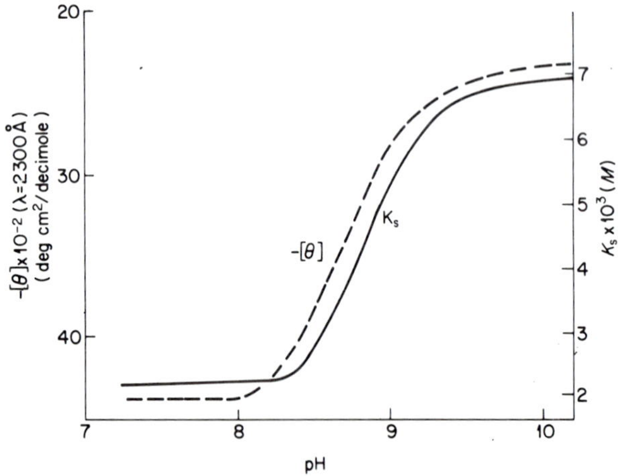

FIG. 13. The pH dependency of circular dichroism and of K_s, the enzyme–substrate dissociation constant, pertaining to δ-chymotrypsin and the neutral substrate *N*-acetyl-L-tryptophan amide. The K_s values were determined as previously described (*70*), and values of ellipticity θ at $\lambda = 230$ nm were determined on a Cary 60 recording spectropolarimeter equipped with a 6001 circular dichroism attachment [from Hess *et al.* (*62*)].

sinogen and in chymotrypsin in which this α-amino group has been specifically acetylated. The data in Fig. 12 show the similarity of the circular dichroism spectra of the high pH form of acetylated δ-chymotrypsin of acetylated chymotrypsinogen and of acetylated δ-chymotrypsin in which the α-amino group of Ile 16 has been specifically acetylated. The ellipticity band at 228 nm characteristic of the active enzyme at pH 7 (*111, 112, 118*) is missing from the circular dichroism spectra of these three proteins. There is no evidence that these proteins are able to catalyze the hydrolysis of specific substrates.

(3) In the activation of chymotrypsinogen (*10–12*), liberation of the α-amino group of Ile 16 by tryptic cleavage of the Arg 15–Ile 16 peptide bond of the enzyme produces a new ionizing group with $pK_{(app)}$ of about 8.5 which is positively charged at a neutral pH (*110*). It was observed by Neurath *et al.* (*119*) that liberation of this α-amino group is accompanied by a change in the optical rotation which parallels the appearance of enzymic activity. The low pH form of this fully active enzyme has a characteristic circular dichroism spectrum with an ellipiticity band at 228 nm (*112, 118*) (Fig. 12). As the pH is raised the circular dichroism spectra change to those characteristic of chymotrypsinogen (*111, 112, 118*) (Fig. 12).

The data (*111*) in Fig. 13 illustrate the relationship in the alkaline pH region between circular dichroism of chymotrypsin (dashed line) and the values of K_s of *N*-acetyl-L-tryptophan amide. Analysis of the data in Fig. 13 indicates that a single ionizing group of the enzyme with $pK_{(app)}$ of about 8.5 controls both the pH dependence of the conformation of chymotrypsin, as measured by circular dichroism, and of the enzyme–substrate dissociation constant. The pH-dependent changes in the circular dichroism spectra parallel earlier measurements of optical rotation changes (*110, 120, 121*).

The pH dependence of chymotrypsin-catalyzed reactions at alkaline pH appears to be a consequence of the activation mechanism of chymotrypsinogen (Fig. 14), since this process involves liberation of an ionizing group which maintains the active conformation of the enzyme when positively charged. Since the α-amino group of Ile 16 of chymotrypsin can ionize, two conformations of the enzyme are expected (*70, 110*) (Fig. 13): (1) the neutral pH conformation in which the α-amino group of Ile 16 carries a positive charge, conformation A in Fig. 14; and (2) a high pH conformation which is considered to resemble the conformation

118. G. D. Fasman, R. J. Foster, and S. Beychok, *JMB* **19**, 240 (1966).
119. H. Neurath, J. A. Rupley, and W. J. Dreyer, *ABB* **65**, 243 (1956).
120. J. A. Rupley, W. J. Dreyer, and H. Neurath, *BBA* **18**, 162 (1955).
121. H. Parker and R. Lumry, *JACS* **85**, 483 (1963).

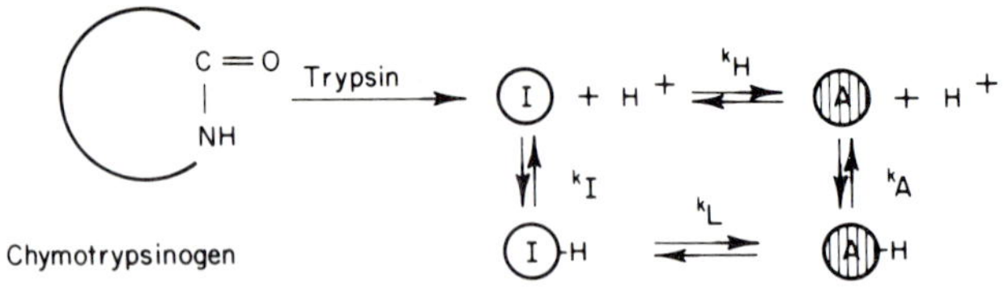

FIG. 14. Scheme for the activation of chymotrypsinogen (or acetylated chymotrypsinogen) showing equilibria between active (A) and inactive (I) conformational forms of the enzyme molecule and ionization equilibria. Conformation A corresponds to the low pH conformation which is observed in the pH region 6–7. Conformation I corresponds to the high pH conformation of the enzyme above pH 10. The high pH conformation of the enzyme is believed to be similar to the conformation of chymotrypsinogen. The constants k_H and k_L apply to the equilibria between high and low conformation of the enzyme. k_I and k_A are the acid ionization constants of α-amino group of Ile 16 in the high and low pH conformation of chymotrypsin, respectively [from Oppenheimer *et al.* (*110*)].

of chymotrypsinogen and in which the α-amino group of Ile 16 does not carry a charge. This conformation is considered (*110*) to be a catalytically inactive conformation and is designated conformation I in Fig. 14. The pH dependence of K_s is explained (*70, 110*) by the substrate's binding better to the neutral pH conformation than to the high pH conformation. The pH dependence of K_m is given by (*70*)

$$K_{m(\text{app})} = K_s(\text{H} + K_\text{E})(\text{H} + K_\text{ES})^{-1} \tag{11}$$

K_s is the enzyme substrate dissociation constant of the neutral pH conformation of the enzyme (AH in Fig. 14), and K_E and K_ES are complex constants related to the ionization constants and conformational equilibrium constants of conformations A and I in the free enzyme and the enzyme substrate complexes, respectively. The conformational equilibrium constants may depend on the structure of the substrate and may differ for various forms of the enzyme (*70*). Differences between α- and δ-chymotrypsin and between different inhibitors and the same form of the enzyme in both magnitude and pH dependence of the steady kinetic parameters have been observed (*70, 122, 123*).

Attempts have been made (*122*) to interpret the pH dependence of chymotrypsin-catalyzed reactions without considering the conformational equilibria between enzyme conformations. This treatment of data led to the conclusion (*122*) that the high pH conformations of the enzyme are catalytically active and bind substrate almost as well as the neutral

122. P. Valenzuela and M. L. Bender, *Biochemistry,* **12,** 2440 (1970).
123. J. R. Garel and B. Labouesse, *JMB* **47,** 41 (1970).

pH conformations. This conclusion, however, is not consistent with structural (*4, 5*) and chemical information (*110, 111, 117, 123*).

For the scheme in Fig. 14, the maximum increase in $K_{m(\mathrm{app})}$ at alkaline pH is given by K_E/K_{ES} (*70*) [see Eq. (11)] and the ratio of active to inactive conformations at high pH in presence of substrate is given by (K_{ES}/K_E) (S_0/K_s). When the enzyme is saturated with substrate, the active conformations predominate (*110, 111, 123*). Accordingly, at alkaline pH, substrate binding is expected to result in conversion of enzyme molecules from the high pH form (in which the α-amino group of Ile 16 is not protonated) to the neutral pH form explaining the substrate-induced conformational changes (*57, 108, 121, 123–126*) and proton uptake (*127–129*) by the enzyme observed in chymotrypsin-catalyzed reactions.

E. Aspartic Acid 194

The reason for the importance of the positively charged α-amino group of Ile 16 in maintaining the active conformation of chymotrypsin was realized when the structure of α-chymotrypsin (*4–6, 8*) and chymotrypsinogen (*130, 131*) was elucidated by X-ray diffraction experiments. Determination of the structure of α-chymotrypsin revealed the existence of an ion pair (*5, 6*) between the positively charged α-amino group of Ile 16 and the negatively charged carboxyl group of Asp 194. Sigler *et al.* (*5*) concluded that this ion pair is essential for maintaining the enzyme in an active conformation and that substrate binding prevents the disruption of the ion pair. Blocking of the carboxyl group of Asp 194 with glycine methyl ester (*102*) leads to complete loss of active sites of the enzyme. X-ray diffraction experiments have indicated the existence of an ion pair in elastase (*30*), and sequence homologies (*99*) have indicated the existence of this ion pair in trypsin.

124. B. Labouesse, B. H. Havsteen, and G. P. Hess, *Proc. Natl. Acad. Sci. U. S.* **48,** 2137 (1962).

125. B. H. Havsteen, and G. P. Hess, *JACS* **85,** 791 (1963).

126. R. Biltonen, R. Lumry, V. Madison, and H. Parker, *Proc. Natl. Acad. Sci. U. S.* **54,** 1018 (1965).

127. J. McConn, E. Ku, C. Odell, G. Czerlinski, and G. P. Hess, *Science* **161,** 274 (1968).

128. D. M. Glick, *Biochemistry* **7,** 3391 (1968).

129. M. L. Bender and F. C. Wedler, Jr., *JACS* **89,** 3052 (1967).

130. S. T. Freer, J. Kraut, J. D. Robertus, H. T. Wright, and N. H. Xuong, *Biochemistry* **9,** (9), 1997 (1970).

131. J. Kraut, "The Enzymes," 3rd ed., Vol. III, p. 547, 1971.

III. The Activation of Chymotrypsinogen

A. Unblocking the Active Site

Recent X-ray diffraction studies (*130, 131*) of chymotrypsinogen have led to the elucidation of this structure of 2.5 Å resolution (see Chapter 5 by Kraut, this volume). These studies have indicated that the substrate binding site, which is a pocket near Ser 195, is partially blocked in chymotrypsinogen. Aspartic acid 194 forms an internal ion pair with His 40 and the peptide chain segment containing Asp 194 occupies a part of the substrate binding site (*130, 131*). The key which unlocks the inactive conformation is the α-amino group of Ile 16. When this α-amino group is liberated in the conversion of chymotrypsinogen to chymotrypsin at neutral pH, it acquires a positive charge. This positively charged group induces ion pair formation with Asp 194 and the resulting movement of the peptide chains establishes a specific substrate binding site.

B. Activation of Trypsin and Elastase

Sequence analogies have indicated (*7, 99*) that trypsin also has a pocket containing at its bottom the negatively charged carboxyl group of Asp 177 which can interact with the basic side chains of trypsin substrates. The amino-terminal α-amino group of trypsin (*132*), also an Ile residue, is implicated in the maintenance of the active conformation of this enzyme (*133*). This Ile α-amino group is liberated in the activation of trypsinogen and it presumably forms an internal ion pair with a negatively charged carboxyl group in the active enzyme.

Different results are obtained with the enzyme elastase (*134*). The three-dimensional structure of this enzyme has recently been elucidated (*30*). Elastase has an ion pair analogous to that found between Asp 194 and Ile 16 in chymotrypsin except that the α-amino group of a valine residue is involved (*98*). Two important differences occur, however. In contrast to chymotrypsin and trypsin, elastase does not have a pocket into which specific substrates fit (*30*) and unlike in chymotrypsin, sub-

132. K. A. Walsh, D. L. Kauffman, K. S. V. S. Kumar, and H. Neurath, *Proc. Natl. Acad. Sci. U. S.* **51**, 301 (1964).

133. J. Chevallier, J. Yon, and J. Labouesse, *BBA* **181,** 73 (1969).

134. H. Kaplan and H. Dugas, *BBRC* **34,** 681 (1969).

strate binding to this enzyme cannot stabilize the ion pair. In elastase both the enzymic activity and the conformation of the enzyme are independent of the existence of the "ion pair" (*134*).

IV. The Activity of the Enzyme. The Interaction of the Amino Acid Residues in the Active Center of the Enzyme

The uncertainties which exist about the role of individual amino acid residues in the activity of the enzyme are increased when discussing the interaction of those amino acid residues which produce a highly efficient and specific catalyst. The most definite answers come from X-ray diffraction experiments (*4–8*) which have given convincing answers concerning the specificity of the enzyme. The stereochemical aspects of substrate specificity are discusssed in more detail by Blow, Chapter 6, this volume, and will be discussed here only insofar as it has a direct bearing on the activity of the enzyme. The side chains of specific substrates bind in the pocket of the enzyme near Ser 195 (*7*). Substrate binding experiments indicate that it is likely that the interaction of the amino group of the substrate with the enzyme is important for substrate binding (*46*). Model building suggests (*7*) a hydrogen bond between the amido –NH– of the substrate and the carbonyl oxygen of Ser 214. The carbonyl carbon of the substrate is in van der Waals contact with the Ser 195 oxygen (*7*).

The mechanism by which two groups, His 57 and Ser 195, participate in the catalytic reaction has been treated in a number of recent reviews (*21, 32, 135*). Histidine is thought to act as an acid–base catalyst, which assists in the removal of a proton from the hydroxyl group of Ser 195 or from water, thereby increasing their nucleophilicity, and as a proton donor to the leaving group of the substrate.

It has been reported that introduction of electron withdrawing groups into phenol esters increases the rate of the acylation of the enzyme (*135*). In the case of anilides the reverse was found to be true (*136–139*). An interpretation of the data for esters is that nucleophilic attack, presumably by water, on the carbonyl carbon occurs in the rate limiting step

135. T. C. Bruice and S. J. Benkovic, "Bio-organic Mechanism," Vol. I. Benjamin, New York, 1966.
136. M. L. Bender and K. Nakamura, *JACS* **84,** 2577 (1962).
137. W. F. Sager and P. C. Parks, *JACS* **85,** 2678 (1963).
138. T. Inagami, S. S. York, and A. Patchornik, *JACS* **87,** 126 (1965).
139. L. Parker and J. H. Wang, *JBC* **243,** 3729 (1968).

of the reaction. The interactions between Asp 102, His 57, and Ser 195 can increase the nucleophilicity of the serine oxygen (*97*). The conformation of the chymotrypsin-*N*-formyl-L-tryptophan complex (*7*) suggests that this nucleophilic attack may not be concerted. In order for the Ser 195 oxygen to attack the carbonyl carbon of *N*-formyl-L-tryptophan it has to rotate and break the hydrogen bond to the $N^{\epsilon 2}$ of His 57.

An interpretation of the data for anilides is that the protonation of the leaving amide group is in the rate limiting step which is thought to be the acylation of the enzyme. The structure of the *N*-formyl-L-tryptophan–α-chymotrypsin complex and model building indicates (*7*) that the –NH– of the leaving group can be near to the proton between His 57 and Ser 195. The mechanism by which such a proton can be transferred from the serine hydroxymethyl group via His 57 to the –NH– of the leaving group of the substrate has been discussed in detail by Wang *et al.* (*140, 141*) who have considered this type of proton transfer as an important aspect of enzymic mechanism.

Investigations (*142*) of the deacylation reaction of furoyl-chymotrypsin indicate that this step is general base catalyzed and involves proton transfer in the rate limiting step. Model building (*7*) shows that a water molecule can be placed between His 57 and the carbonyl carbon of the substrate in such a way that this water molecule is hydrogen bonded to the $N^{\epsilon 2}$ of His 57 with the oxygen pointing toward and near the carbonyl carbon. In ester hydrolysis, deacylation is rate limiting and activation of water by the charge relay system (*97*) consisting of Asp 102–His 57 and water can exert a powerful effect on the nucleophilicity of the water molecule.

The uncertainties which exist regarding the reaction pathway, especially in the catalytic hydrolysis of amides, make it difficult to guess which factors contribute to the high efficiency of the catalytic reaction. Various methods which an enzyme can use to increase the rate of reactions have been favored at various times (*143*): orientation effects, removal of water from the reactants, bond strain, acid–base catalysis, etc. The effect of freezing out free rotation around chemical bonds which connect two reacting groups on the rate of the reaction has been demonstrated by Bruice *et al.* (*135, 144*). The binding of the side chain of the substrate in the pocket near Ser 195 and the binding of the acylamino group of the substrate to the backbone carbonyl of Ser 214 orients

140. J. H. Wang and L. Parker, *Proc. Natl. Acad. Sci. U. S.* **58,** 2451 (1967).
141. J. H. Wang, *Science* **161,** 328 (1968).
142. P. W. Inward and W. P. Jencks, *JBC* **240,** 1986 (1965).
143. For a recent review of this subject, see W. P. Jencks, "Catalysis in Chemistry and Enzymology." McGraw-Hill, New York, 1969.

the susceptible bond of the substrate in a suitable position for hydrolysis. Inspection of steady state kinetic data makes it tempting to suggest that such orientation effects in chymotrypsin-catalyzed reactions are very important. Substrates in which the acyl amino group is not present are hydrolyzed by a factor of 10^2 to 10^4 more slowly than the corresponding substrates which have this group (*46*). High resolution X-ray diffraction studies of tosyl-α-chymotrypsin (*4, 5*) and more recent studies by Henderson (*145*), of indoleacryloyl chymotrypsin give a considerable amount of insight into this problem. Indoleacryloyl imidazole is a substrate which does not have an acylamino group (*44*). The reason the indole-acryloyl enzyme is stable appears to be mainly because of the formation of an unproductive acyl enzyme in which the carbonyl carbon of the substrate is not accessible to water (*145*). The water molecule, which is normally activated by bonding to the $N^{\epsilon 2}$ of His 57 and is in position to attack the carbonyl carbon of the acyl enzyme, is held in a stable hydrogen bonded network involving $N^{\epsilon 2}$ of His 57 and the carbonyl oxygen of the indoleacryloyl group (*145*). In order for deacylation to occur, the carbonyl oxygen of the acyl enzyme must move out of its stabilized position to allow access of water to the carbonyl carbon of the substrate. In this particular instance, the observed decrease in the deacylation rate is, therefore, not because of increased flexibility of the carbonyl carbon of the indoleacryloyl group of acyl chymotrypsin but because of a process which stabilizes the carbonyl carbon in a position in which it is not accessible to water. The interaction between the acylamino group of the substrate and the backbone carbonyl of Ser 214 prevents the substrate from binding in this unproductive mode. This interaction between the substrate's acylamino group and the enzyme is probably not possible with D-amino acids. Model building suggests (*145*) that the position of the carbonyl group of acyl enzymes of D-amino acids would have a similar nonproductive conformation, explaining the low rates with which those complexes are deacylated. The deacylation rates of indoleacryloyl chymotrypsin and acetyl-D-tryptophanyl chymotrypsin are both of the order of 3×10^{-2} sec^{-1} (*46, 146*).

At present there is no reason to believe that any single factor which can increase the rate of the catalytic reaction is of overwhelming importance. The efficiency of enzymic catalysis may be reached by a subtle balance between the advantages and disadvantages of a number of possible reaction pathways. The most obvious illustration for the subtleties of the reaction mechanism is the catalytic hydrolysis of esters,

144. T. C. Bruice and U. K. Pandit, *Proc. Natl. Acad. Sci. U. S.* **46,** 402 (1960).
145. R. Henderson, Ph.D. Thesis, University of Cambridge, 1970.
146. D. W. Ingles and J. R. Knowles, *BJ* **104,** 369 (1967).

which does not occur by direct cleavage of the ester bond by water but utilizes several bond breaking and bond making steps. In ester hydrolysis, the enzyme does not use histidine as a nucleophile but uses the hydroxyl group of serine. The nucleophilicity of this serine is further modified by a hydrogen-bonded system involving both a histidine and the buried carboxyl group of aspartic acid (*97*). This interaction makes His 57 a better base catalyst but a correspondingly poorer acid catalyst. Acid catalysis, however, may be required in the chymotrypsin-catalyzed hydrolysis of amides (*136–139*).

A binding site of the enzyme to which only specific substrates bind was proposed by Fischer over 70 years ago (*147*). All investigations of the specificity of enzymic reactions since then have been in accord with this theory. Nevertheless, discovery by X-ray diffraction studies of a slit in lysozyme (*148*) and a hole in chymotrypsin in which the substrates bind (*7*) was of great satisfaction to both protein crystallographers and enzyme chemists. The spatial orientation of *N*-formyl-L-tryptophan with regard to the other functional groups in the active site of α-chymotrypsin is the basis for the interpretation of all past and future chemical experiments. What happens after the substrate binds and which of the possibilities the enzyme uses to accelerate the reaction is still to be discovered, it is hoped to the enjoyment of all concerned.

ACKNOWLEDGMENTS

This chapter was written while the author was supported by an NIH special postdoctoral fellowship (1969–1970) MRC Laboratory of Molecular Biology, Cambridge, England. The author is grateful to Dr. Richard Henderson, MRC Laboratory of Molecular Biology, Cambridge, for lending him his Ph.D. thesis on "X-Ray Analysis of α-Chymotrypsin Substrate and Inhibitor Binding" prior to publication.

147. E. Fischer, *Chem. Ber.* **27,** 2985 (1894).

148. C. C. F. Blake, L. N. Johnson, G. A. Mair, A. C. T. North, D. C. Phillips, and V. R. Sarma, *Proc. Roy. Soc.* **B167,** 378 (1967).

8

Trypsin

B. KEIL

I. Introduction

Because there have been so many esterolytic or proteolytic enzymes described in the literature as trypsins, "trypsinlike" enzymes, "acidic" trypsins, etc., we wish to point out that this chapter will deal with a family of enzymes of molecular weights ranging from 20,000 to 25,000 which catalyze preferentially the hydrolysis of ester and peptide bonds involving the carboxyl group of basic amino acids, arginine, and lysine, and in which a serine and a histidine residue participate in the mechanism of catalysis.

Trypsin from beef pancreas was among the first proteolytic enzymes isolated in pure form in amounts sufficient for exact chemical and enzymological studies (*1*). It is therefore the best known representative of the whole trypsin family, and the term *trypsin* when used in the text refers to beef trypsin unless otherwise stated.

Thanks to modern isolation procedures, trypsins of other higher vertebrates were isolated in pure form and characterized [human (*2*), pig (*3*), sheep (*4, 5*), and turkey (*6, 7*)]. Enzymes belonging beyond doubt to the trypsin family were recently isolated from shark (*8*), crayfish (*9, 9a*), white shrimp (*10*), silk moth ("cocoonase") (*11*), and from two strains of *Streptomyces* (*12–16*).

The whole family of trypsins from different organisms is a result of

1. J. H. Northrop, M. Kunitz, and R. Herriott, "Crystalline Enzymes," 2nd ed. Columbia Univ. Press, New York, 1948.
2. J. Travis and R. C. Roberts, *Biochemistry* **8,** 2884 (1969).
3. M. Charles, M. Rovery, A. Guidoni, and P. Desnuelle, *BBA* **69,** 115 (1963).
4. J. Travis, *BBRC* **30,** 730 (1968).
5. S. Bricteux-Gregoire, R. Schyns, and M. Florkin, *BBA* **127,** 277 (1966).
6. C. A. Ryan, *ABB* **110,** 169 (1965).
7. C. A. Ryan, J. J. Clary, and Y. Tomimatsu, *ABB* **110,** 175 (1965).
8. H. Neurath, R. A. Bradshaw, and R. Arnon, *Proc. IUB/IUBS Intern. Symp., Struct.-Funct. Relationship Proteol. Enzymes, 1970* Abstr. Munksgaard, Copenhagen, 1970.
9. G. Pfleiderer, R. Zwilling, and H. Sonneborn, *Z. Physiol. Chem.* **348,** 1319 (1967).

9a. R. Zwilling and V. Tomášek, *Nature* **228,** 57 (1970).

10. B. J. Gates and J. Travis, *Biochemistry* **8,** 4483 (1969).
11. F. C. Kafatos, A. M. Tartakoff, and J. H. Law, *JBC* **242,** 1477 and 1488 (1967).
12. L. Jurasek, D. Fackre, and L. B. Smillie, *BBRC* **37,** 99 (1969).
13. K. Morihara and H. Tsuzuki, *ABB* **126,** 971 (1968).
14. M. Trop and Y. Birk, *BJ* **109,** 475 (1968).
15. S. Wählby and L. Engström, *BBA* **151,** 402 (1968).
16. S. Wählby, *BBA* **151,** 394 (1968).

the phylogenetic evolution of a much larger group of proteolytic and esterolytic enzymes whose structural and functional similarities witness their common ancestors. Discussions of these relations can also be found in chapters on α-chymotrypsin, elastase, and thrombin.

II. Chemistry of Trypsinogen and Trypsin

In pancreatic cells of vertebrates, the inactive precursor of trypsin, trypsinogen, is synthetized and subsequently converted to the active enzyme *(1, 17)*. The fact that the system of trypsinogen activation must have been established at the very early stages of the phylogenetic evolution of vertebrates is evidenced by the detection of typical trypsinogens in the pancreas of shark *(8)*. In invertebrates these precursors have not been found as yet.

A. Preparation of Trypsinogen

Bovine trypsinogen is obtained from an acid extract of pancreas by ammonium sulfate fractionation and crystallization at pH 8 *(1)*. Homogeneous trypsinogen can be prepared by recrystallization at pH 7.8 in the presence of trypsin inhibitors *(18, 19, 19a)* or by chromatography on SE Sephadex at neutral pH *(20)*. Porcine and ovine zymogens are obtained in good yields by ammonium sulfate fractionation followed by chromatography on CM-cellulose *(3, 5, 21)*.

Trypsinogen is stable between pH 2 and 4, whereas in neutral and alkaline media autocatalytic activation to trypsin takes place.

B. Activation of Trypsinogen

Active trypsin is obtained from trypsinogen by limited proteolysis at pH 8 catalyzed by small amounts of trypsin *(1, 22)*, enterokinase *(23,*

17. P. Desnuelle, "The Enzymes," 2nd ed., Vol. 4, p. 119, 1959.
18. F. Tietze, *JBC* **204,** 1 (1953).
19. L. W. Cunningham, Jr., F. Tietze, N. M. Green, and H. Neurath, *Disussions Faraday Soc.* **13,** 58 (1953).
19a. A. K. Balls, *Proc. Natl. Acad Sci. U. S.* **53,** 392 (1965).
20. D. D. Schroeder and E. Shaw, *JBC* **243,** 2943 (1968).
21. M. Charles, D. Gratecos, M. Rovery, and P. Desnuelle, *BBA* **140,** 395 (1967).
22. M. R. McDonald and M. Kunitz, *J. Gen. Physiol.* **25,** 53 (1941).
23. M. Kunitz, *J. Gen. Physiol.* **22,** 429 (1939).

24), or thrombin (*25*). Trypsinogen is also activated by a number of extracellular proteolytic enzymes from various strains of *Aspergilli* (*26*, *27*) and *Penicillia* (*28*, *29*) at low pH values where no autocatalytic activation takes place.

During autocatalytic activation in alkaline media, a significant decrease in optical rotation takes place (*30*) and a family of products is formed. The predominant product is the single chain form of the enzyme, β-trypsin (see Fig. 1), resulting from the splitting of the bond between residues 6 and 7 while a hexapeptide is released (*31*). It is noteworthy that although the splitting of this "strategic" bond is required for the formation of the first detectable active enzyme, the comparative kinetic values for trypsinogens and chymotrypsinogens show that this bond

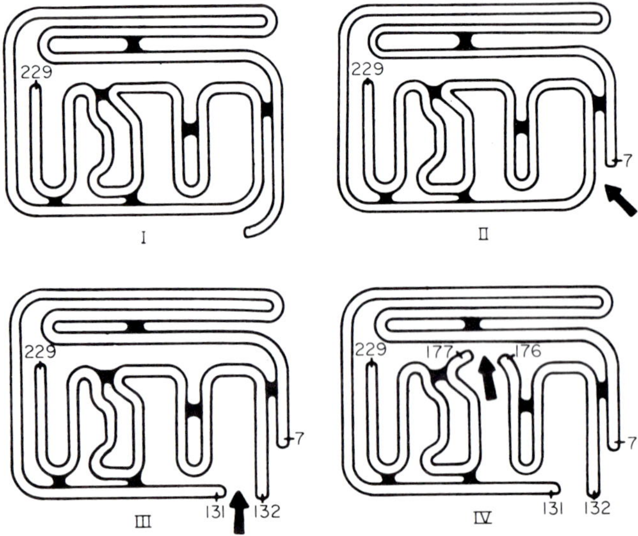

Fig. 1. Conversion of trypsinogen (I) to β-trypsin (II), α-trypsin (III), and pseudotrypsin (IV).

24. I. Yamashina, *BBA* **20**, 433 (1956).
25. A. Engel, B. Alexander, and L. Pechet, *Biochemistry* **5**, 1543 (1966).
26. C. Gabeloteau and P. Desnuelle, *BBA* **42**, 230 (1960).
27. K. Nakanishi, *J. Biochem.* (*Tokyo*) **46**, 1263 (1959).
28. T. Hofmann and R. Shaw, *BBA* **92**, 543 (1964).
29. M. Kunitz, *J. Gen. Physiol.* **21**, 601 (1938).
30. J. F. Pechère and H. Neurath, *JBC* **229**, 389 (1957).
31. E. W. Davie and H. Neurath, *JBC* **212**, 515 (1955).

is a relatively poor substrate (*32*). The adjacent four aspartyl acid residues have clearly a negative effect.

Subsequent cleavage of β-trypsin at the bond Lys 131–Ser 132 leads to α-trypsin (*20*), a two-chain structure held together by disulfide bonds. Under different conditions, the bond Arg 105–Val 106 can also be cleaved without the loss of activity (*33*). A further cleavage of α-trypsin at the bond Lys 176–Asp 177 yields another active form of trypsin, pseudotrypsin (*34*) (see Fig. 1).

All these active forms of trypsin have the same molecular weight. More degraded molecules retaining trypsin activity have been searched for (*35–38*), but direct evidence of their existence based on structural and enzymic characterization is still lacking. By contrast, other investigators have shown that the residual activity of the degraded mixtures was always that of intact high molecular weight trypsin (*39–41*).

In the process of trypsinogen activation a part of the activation mixture is converted into enzymically inactive products. Their formation is strongly influenced by divalent cations such as those of calcium. In the absence of calcium, their yields amount up to 50% of the whole protein (*22, 42*). At 0.02–0.10 M concentration, calcium ions speed up the selective splitting of the bond Lys 6–Ile 7, thus promoting β-trypsin formation; on the other hand, they prevent nonspecific cleavage (*43–45*).

Spontaneous autocatalytic activation of trypsinogen can be prevented by low molecular weight peptide inhibitors or by naturally occurring polypeptide inhibitors of trypsin. The strongest low molecular weight competitive inhibitor of trypsin, *p*-aminobenzamidine (*46*), blocks the

32. M. Lazdunski, M. Delaage, J. P. Abita, and J. P. Vincent, *Proc. IUB/IUBS Intern. Symp. Struct.-Funct. Relationship Proteol. Enzymes, 1970*, p. 42. Munksgaard, Copenhagen, 1970.
33. S. Maroux, M. Rovery, and P. Desnuelle, *BBA* **140,** 377 (1967).
34. R. L. Smith and E. Shaw, *JBC* **244,** 4704 (1969).
35. T. Viswanatha, R. C. Wong, and I. E. Liener, *BBA* **29,** 174 (1958).
36. I. E. Liener and T. Viswanatha, *BBA,* **36,** 250 (1959).
37. G. Talsky and D. Wolf, *Z. Naturforsch.* **23b,** 1389 (1968).
38. L. Weil and S. N. Timasheff, *ABB* **116,** 252 (1966).
39. J. Chevalier, Y. Jacquot-Armand, and J. Yon, *BBA* **92,** 521 (1964).
40. H. Kaufman and B. F. Erlanger, *Biochemistry* **6,** 1597 (1967).
41. S. E. Bresler, V. M. Krutjakov, and A. G. Popov, *Biokhimiya* **31,** 776 (1966).
42. M. L. Bender, J. V. Killheffer, Jr., and R. W. Roeske *BBRC* **19,** 161 (1925).
43. C. Gabeloteau and P. Desnuelle, *ABB* **69,** 475 (1957).
44. M. Delaage and M. Lazdunski, *BBRC* **28,** 390 (1967).
45. J. P. Abita, M. Delaage, M. Lazdunski, and J. Savrda, *European J. Biochem.* **8,** 314 (1969).
46. M. Mares-Guia and E. Shaw, *JBC* **240,** 1579 (1965).

activation of trypsinogen both by trypsin and enterokinase (*47*). Many other compounds containing a free basic group such as serotonin, tryptamine (*48*), or guanidinium compounds (*49, 50*) show an analogous effect. This reaction is of practical importance in human medicine for the treatment of pancreatitis.

C. Heterogeneity of Crystalline Trypsin

Crystalline bovine trypsin is commercially produced from the first trypsinogen crystals by slow activation at pH 8 and 5° in the presence of calcium chloride. Additional recrystallization are without effect on the purity. These preparations represent a mixture containing predominantly β- and α-trypsin and/or even more degraded enzyme molecules. Individual components can be obtained by chromatography on SE Sephadex columns (*20, 34*).

Activity assays of trypsin preparations are not the best criteria of its purity because of variations in the specific activities and substrate activation of β- and α-trypsin. The most convenient method of trypsin titration uses *p*-nitrophenyl-*p'*-guanidinobenzoate which reacts selectively with the active site of the enzyme (*20, 51*).

Although the single-chain β-trypsin and the double-chain α-trypsin probably have analogous tertiary structures, one is tempted to ask whether many previous data (before 1968) obtained in physicochemical and kinetic studies as well as by studies on topographical substitutions of commercial crystalline trypsin should be revised with individual enzyme forms.

D. Physicochemical Properties and Stability

The isoionic point of the bovine zymogen is 9.3 (*52*) of the porcine zymogen 7.5 (*17*). The isoionic point of both bovine and porcine trypsin is the same, 10.8 (*53*). The molecular weight calculated from the known primary structure of bovine trypsinogen (23,985) is in good agreement with the value reported earlier and obtained by physicochemical measure-

47. J. D. Geratz, *Experientia* **22,** 73 (1966).
48. J. D. Geratz, *Experientia* **21,** 699 (1966).
49. H. J. Trettin and H. Mix, *Z. Physiol. Chem.* **340,** 24 (1965).
50. J. D. Geratz, *ABB* **102,** 327 (1963).
51. T. Chase, Jr. and E. Shaw, *BBRC* **29,** 508 (1967).
52. N. M. Green and H. Neurath, *in* "The Proteins" (H. Neurath and K. Bailey, eds.), Vol. 2, Part B, p. 1057. Academic Press, New York, 1954.
53. J. Travis and I. E. Liener, *JBC* **240,** 1962 (1965).

ments, i.e., 23,560–23,700; $s_{20,w} = 2.48$, $f/f_0 = 1.154$ (*18, 54*). For routine work, the concentration of trypsinogen or trypsin can be calculated from its extinction coefficient $E_{1\%}^{1\,cm} = 15.4$ at 280 nm (*54a*).

The stability of trypsinogen and trypsin can be considered from two viewpoints, i.e., as the stability of its enzymic properties or as the stability of its molecular conformation.

With regard to the integrity of primary structure, trypsinogen is stable in acid media, whereas in the neutral and alkaline pH range the activation to trypsin takes place. Trypsin is most stable around pH 3 and retains its activity at this pH and in the cold for weeks. This does not exclude the possibility of a chemical change. Even lyophylized samples of trypsin have been observed to undergo slow autodigestion during storage at 4° (*20*). After 2 hr of treatment at 30°, trypsin (1.36×10^{-7} *M* solution in 0.01 *M* $CaCl_2$) retains 99% of its original activity at pH 5, 98% at pH 9, 90% at pH 10, and 81% at pH 10.7 (*55*).

Trypsin is denatured reversibly by high pH (above 11), by precipitation with trichloroacetic acid, or by high concentration of urea (*56, 57*). Trypsin is enzymically fully active in urea solutions up to 6.5 *M* (*58, 59*) and in 30% ethanol (*60*).

Physicochemical measurements can reveal even more subtle changes in molecular conformation of the fully active enzyme. Thus, detailed studies on the temperature and pH dependence of the conformation of trypsin have shown that both trypsinogen and trypsin pass through a stage of reversible equilibria between several conformationally distinct molecular forms stabilized by hydrophobic interactions (*32, 61*).

E. Chemical Structure

The amino acid compositions of trypsinogens and trypsins from different species are given in Table I.

The primary structure of bovine trypsinogen has been completely

54. J. Edsall, *in* "The Proteins" (H. Neurath and K. Bailey, eds.), Vol. 1, Part B, p. 549. Academic Press, New York, 1953.

54a. Worthington Manual, "Enzymes and Enzyme Reagents" (EC 3.4.4.4), 1967.

55. S.-S. Wang and F. H. Carpenter, *Biochemistry* **6**, 215 (1967).

56. M. Kunitz and J. H. Northrop, *J. Gen. Physiol.* **19**, 991 (1936).

57. J. I. Harris, *Nature* **177**, 471 (1956).

58. M. L. Anson, *J. Gen. Physiol.* **22**, 79 (1938).

59. M. Delaage and M. Lazdunski, *European J. Biochem.* **4**, 378 (1968).

60. G. W. Schwert and M. A. Eisenberg, *JBC* **179**, 665 (1949).

61. M. Lazdunski and M. Delaage, *BBA* **140**, 417 (1967).

TABLE I

AMINO ACID COMPOSITIONS OF TRYPSINOGENS AND TRYPSINS

	Trypsinogen		Trypsin						
Species	Bovine[a]	Porcine[b]	Human[c]	Ovine[d]	Turkey[e]	Shrimp[f]	Crayfish[g]	Cocoonase[h]	*Streptomyces griseus*[i]
Ala	14	15	13	17	11	16	16	16	26
Arg	2	4	6	4	2–3	3	2	6	8.4
Asp + Asn	9 + 17	28	21	20	15–16	30	30	26	18.1
Half-Cys	12	12	8	12	8	8	6	4	6.1
Glu + Gln	2 + 12	17	21	14	12	24	21	15	17.4
Gly	25	24–25	20	19	19	28	28	25	29.4
His	3	4	3	3	3	5	5	4	1
Ile	15	15	12	10	10	14	15	12	8
Leu	14	16	12	14	13–14	10	16	12	11
Lys	15	11	11	12	7	5	5	13	6.5
Met	2	2	1	2	1–2	2	2	1–2	2.7
Phe	3	5	4	5	1	6	7	5	5.7
Pro	8	11	9	9	8	11	10	9	8
Ser	34	25	24	26	24	24	17	23	14.2
Thr	10	11	10	15	11	10	15	16	16.4
Trp	4	4	3		5	3	3	3	
Tyr	10	8	7	6	10	10	11	9	8.2
Val	18	16	16	17	14–15	18	18	20	17.8

[a] From Mikeš *et al.* (*62*).
[b] From Charles *et al.* (*3*).
[c] From Travis and Roberts (*2*).
[d] From Travis (*4*).
[e] From Ryan *et al.* (*7*).
[f] From Gates and Travis (*10*).
[g] From Zwilling and Tomášek (*9a*).
[h] From Kafatos *et al.* (*11*).
[i] From Jurasek *et al.* (*12*).

established by now (*62–66*; see also, *67, 67a*). It is a single-chain structure of 229 amino acid residues, cross-linked by six disulfide bridges (*65, 68*). A schematic drawing representing bovine trypsinogen is shown in Fig. 2.

Similarities in the chemical and functional character of chymotrypsinogen and trypsinogen, two proteins produced by the same type of cells and the same organs of identical organisms, have led to the logical prediction of a similarity in the primary structures of these two zymogens

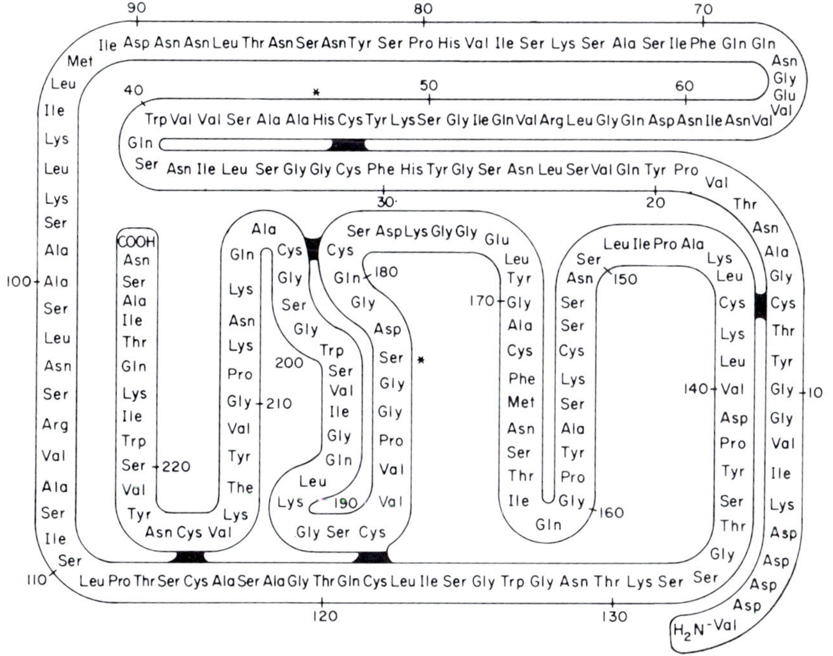

FIG. 2. Primary structure of bovine trypsinogen.

62. O. Mikeš, V. Holeyšovský, V. Tomášek, and F. Šorm, *BBRC* **24,** 346 (1966).

63. O. Mikeš, V. Tomášek, V. Holeyšovský, and F. Šorm, *BBA* **117,** 281 (1966).

64. K. A. Walsh, D. L. Kauffman, K. S. V. Sampath Kumar, and H. Neurath, *Proc. Natl. Acad. Sci. U. S.* **51,** 301 (1964).

65. V. Holeyšovský, V. Tomášek, O. Mileš, A. S. Danilova, and F. Šorm, *Collection Czech. Chem. Commun.* **30,** 3936 (1965).

66. K. A. Walsh and H. Neurath, *Proc. Natl. Acad. Sci. U. S.* **52,** 884 (1964).

67. B. Hartley, *Phil. Trans. Roy. Soc. London* **B257, 77** (1970).

67a. V. Tomášek, J. Strmeň, and F. Šorm, *FEBS Letters* **8,** 176 (1970).

68. D. L. Kauffman, *JMB* **12,** 929 (1965).

(*69, 70*). This hypothesis has been confirmed by the complete elucidation of their primary structures (*66, 71, 72*). It was found moreover that the family of structurally closely related proteinases is in fact much larger (*73*) (see also chapters on α-chymotrypsin and elastase).

Information on trypsinogens or trypsins of different species is rather meager; nevertheless, the amino acid sequences already established show far-reaching similarities in the regions critical for the function (*3–5, 12, 15, 62, 66, 74–76*) (Fig. 3).

We are lacking the data of exact X-rays diffraction studies on trypsin which would permit formulation of its three-dimensional structure. The measurement of the optical rotatory dispersion has shown that trypsin and trypsinogen are practically devoid of larger α-helical strands, in analogy to chymotrypsinogen (*77, 78*). Anti parallel pleated sheet β-conformation has been found in porcine trypsinogen by analyzing its circular dichroism spectra (*79*).

The existence of homologies in the primary structures of chymotrypsin and trypsin (*66, 69–73*), the relative position of disulfide bonds (*71, 73, 79a, 80*), and topographical considerations (*20, 81–84*) have led to an assumption of the similarity in the three-dimensional structures of the two enzymes (*73, 80, 85, 86*). On the basis of the knowledge of the com-

69. F. Šorm, B. Keil, V. Holeyšovský, B. Meloun, O. Mikeš, and J. Vaněček, *Collection Czech. Chem. Commun.* **23,** 985 (1958).
70. F. Šorm, and B. Keil, *Advan. Protein Chem.* **17,** 167 (1962).
71. F. Šorm, V. Holeyšovský, O. Mikeš, and V. Tomášek, *Collection Czech. Chem. Commun.* **30,** 2103 (1965).
72. B. Keil, *Ann. Rev. Biochem.* **34,** 175 (1965).
73. B. S. Hartley, J. R. Brown, D. L. Kauffman, and L. B. Smillie, *Nature* **207,** 1157 (1965).
74. R. A. Smith and I. E. Liener, *JBC* **242,** 4033 (1967).
75. R. A. Smith and I. E. Liener, *Federation Proc.* **26,** 601 (1967).
76. T. Kishida and I. E. Liener, *ABB* **126,** 111 (1968).
77. B. Jirgensons, *JBC* **241,** 147 (1966).
78. B. Jirgensons, *ABB* **74,** 57 (1958).
79. R. Chicheportiche and M. Lazdunski, *FEBS Letters* **3,** 195 (1969).
79a. V. Holeyšovský, B. Mesrob, V. Tomášek, O. Mikeš, and F. Šorm, *Collection Czech. Chem. Commun.* **33,** 441 (1968).
80. B. Keil and F. Šorm, *FEBS Symp.* No. 1, p. 37 (1964).
81. M. Delaage and M. Lazdunski, *BBA* **105,** 523 (1965).
82. Y. Hachimori, H. Horinishi, K. Kurihara, and K. Shibata, *BBA* **93,** 346 (1964).
83. Y. Inada, M. Kamata, A. Matsushima, and K. Shibata, *BBA* **81,** 323 (1964).
84. H. Horinishi, Y. Hachimori, K. Kurihara, and K. Shibata, *BBA* **86,** 477 (1964).
85. H. Neurath, K. A. Walsh, and W. P. Winter, *Science* **158,** 1638 (1967).
86. B. Keil, V. Dlouhá, V. Holeyšovský, and F. Šorm, *Collection Czech. Chem. Commun.* **33,** 2307 (1968).

	5
Bovine	Val·Asp·Asp·Asp·Asp·Lys·Ile·Val···
Porcine	Phe·Pro·Thr·Asp·Asp·Asp·Asp·Lys·Ile·Val···
Ovine	Phe·Pro·Val·Asp·Asp·Asp·Asp·Lys···
Turkey	Ile···

	25 30 35
Bovine	··Asn·Ser·Gly·Tyr·His·Phe·Cys·Gly·Gly·Ser·Leu···
Porcine	··Asn·Ser·Gly·Ser·His·Phe·Cys·Gly·Gly·Ser·Leu···
Turkey	··Asx·Ser·Gly·Tyr·His·Phe·Cys·Gly·Glx·Ser·Leu···
S. griseus	Ser·Met·Gly·Cys·Gly·Gly·Ala·Leu···

	45 50 55
Bovine	··Ala·Ala·His·Cys·Tyr·Lys·Ser·Gly·Ile·Gln·Val·Arg·Leu·Gly·Gln·
Porcine	··Ala·Ala·His·Cys·Tyr·Lys· ···Gly·Glx·
Turkey	··Ala·Ala·His·Cys·Tyr·Lys·Ala·Leu·Thr·His·Pro·Asx·Tyr···
S. griseus	··Ala·Ala·His·Cys·Val···

	60	80
Bovine	·—·Asp·Asn·Ile·Asn·Val··	··Ser·Ile·Val·His·Pro·Ser·Tyr···
Porcine	·His·Asx·Asx·Ile·Val·Leu··	··Ile·Ile·Thr·His·Pro·Asn·Phe···

	160	
Bovine	··Ser·Ser·Cys·Lys·Ser·Ala·Tyr·Pro·Gly·Gln··	···Met·Phe·Cys·
S. griseus	··Ala·Ala·Cys·Arg·Ser·Ala·Tyr·Gly·Asn·Glu··	···Glu·Ile·Cys·

	170 175 180
Bovine	·Ala·Gly·Tyr·—·Leu·Glu·Gly·Gly·Lys·Asp·Ser·Cys·Gln·Gly·Asp·
Porcine	Asn·Ser·Cys·Gln·Gly·Asp·
Ovine	Asn·Ser·Cys·Gln·Gly·Asp·
Turkey	
S. griseus	·Ala·Gly·Tyr·Pro·Asp·Thr·Gly·Gly·Val·Asp·Thr·Cys·Gln·Gly·Asp·

	185 190	
Bovine	·Ser·Gly·Gly·Pro·Val·Val·Cys·Ser·Gly·Lys·Leu···	··Ser·Gly·Cys·
Porcine	·Ser·Gly·Gly·Pro·Val·Val·Cys·Gly·Gln·Gln·Leu··	
Ovine	·Ser·Gly·Gly·Pro·Val·Val·Cys·Ser·Gly·Lys··	
S. griseus	·Ser·Gly·Gly·Pro·Met·Phe··	··Tyr·Gly·Cys·

	205 210	225
Bovine	·Ala·Gln·Lys·Asn·Lys·Pro·Gly·Val··	··Thr·Ile·Ala·Ser·Asn·COOH
Porcine		Thr(Ile,Gln)Ala·Asn·COOH
Turkey		Ser·Asn·COOH
S. griseus	·Ala·Arg·Pro·Gly·Tyr·Pro·Gly·Val··	

FIG. 3. Comparison of partial structures of bovine (*62, 66*), porcine (*3, 4, 74, 75*), ovine (*4, 5*), and turkey (*76*) trypsinogens, and of *Streptomyces griseus* proteinase (*12, 15*).

plete tertiary structure of α-chymotrypsin (*87*, see also Chapter 6 by Blow and Chapter 7 by Hess, this volume), the hypothetical analogous model of trypsin (*86*) can be confronted with facts emerging from further chemical and physicochemical studies (*88–92*). Although this kind of approach is stimulating, direct evidence will have to await the results of X-ray diffraction studies.

Studies on the reactivity of ionizable groups and on reversible denaturation suggest a homologous three-dimensional arrangement of the molecules of bovine and porcine trypsins (*81, 93*). With regard to more remote species, it seems more than coincidence that the three disulfide bridges of the trypsinlike protease from *Streptomyces griseus* (*12*) are homologous compared to those found in bovine trypsin. Thus, even in this case of two enzymes of so widely different phylogenetic origin, an analogous three-dimensional architecture of the molecules, especially of the regions important for the function of the enzymes, can be predicted.

III. Mechanism of Catalysis and Active Site

In the family of the pancreatic enzymes, α-chymotrypsin is the one about which we know most from the standpoint of the structure–function relationship. Although our knowledge of the three-dimensional structure of trypsin is considerably less complete, the parallelism in structure and mechanism of action with chymotrypsin as well as topographical, structural, and kinetic studies of its interaction with substrates and inhibitors have provided a certain amount of information permitting us to draw a very probable picture of the composition and the role of those parts of its molecule which are of crucial importance for its action.

Trypsin catalyzes the following reaction

$$\mathrm{E} + \mathrm{S} \underset{}{\overset{K_s}{\rightleftharpoons}} \mathrm{ES} \xrightarrow{k_2} \mathrm{ES}' + \mathrm{P}_1 \xrightarrow{k_3} \mathrm{E} + \mathrm{P}_2 \qquad (1)$$

where ES is the enzyme–substrate complex ES′ the acyl-enzyme intermediate, P_1 the leaving group of the substrate, and P_2 the carboxylic acid

87. B. W. Matthews, P. B. Sigler, R. Henderson, and D. M. Blow, *Nature* **214**, 652 (1967).
88. R. A. Kenner, K. A. Walsh, and H. Neurath, *BBRC* **33**, 353 (1968).
89. A. Light, B. C. Hardwick, L. M. Hatfield, and D. L. Sondack, *JBC* **244**, 6289 (1969).
90. B. Mesrob and V. Holeyšovský, *Collection Czech. Chem. Commun.* **33**, 1359 (1968).
91. V. Dlouhá, B. Keil, and F. Šorm, *BBRC* **31**, 66 (1968).
92. V. Holeyšovský, B. Keil, and F. Šorm, *FEBS Letters* **3**, 107 (1969).
93. M. Lazdunski and M. Delaage, *BBA* **105**, 541 (1965).

(*94*). The same equation holds true for α-chymotrypsin. The acylation step (k_2) is a nucleophilic reaction dependent on two groups, one with a pK_a of 7 and the other with a pK_a of 9 or 10.

The acyl-enzyme intermediate has been characterized as an ester of the substrate carboxyl and the hydroxyl of the serine residue in trypsin. Its decylation (k_3) is again a nucleophilic reaction dependent on one group with a pK_a of 7 (*95*). In both α-chymotrypsin and trypsin this group was found to be one of the histidine residues.

In addition to the similarities in mechanism of action (*96*), trypsin also has several structural features of its active site which are very similar to those of chymotrypsin. The loops of polypeptide chains form at the surface of the enzyme molecule a conformationally fixed grouping of amino acids acting as the catalytic site for substrate hydrolysis. Other groupings act as the specificity and binding sites for fixation of substrates or competitive inhibitors.

A. Catalytic Site

In analogy to α-chymotrypsin, Ser 183 and His 46 in trypsin apparently represent a charge relay system in which the serine hydroxyl is acylated by the substrate. Diisopropylphosphofluoridate (DFP) and related organic phosphates (*97*) as well as alkyl sulfonyl fluorides (*98, 99*) inactive trypsin by binding covalently to Ser 183 at a stoichiometric ratio 1:1 (Fig. 2). The maximum velocity of inactivation lies near the pH optimum; the reaction is, however, slower in comparison with α-chymotrypsin. Substrates and naturally occurring pancreatic trypsin inhibitors protect trypsin against this deactivation (*100*). Diisopropylphosphoryl trypsin can be reactivated by hydroxylamine and other nucleophilic reagents (*101, 102*).

Tosyl-L-lysine chloromethyl ketone (TLCK) inactivates bovine trypsin irreversibly by stoichiometric alkylation at N-3 of His 46 (*103–105*).

94. M. L. Bender and F. J. Kézdy, *JACS* **86,** 3704 (1964).
95. M. L. Bender, J. V. Killheffer, Jr., and F. J. Kézdy, *JACS* **86,** 5330 (1964).
96. M. L. Bender, J. V. Killheffer, Jr., and F. J. Kézdy, *JACS* **86,** 5331 (1964).
97. E. F. Jansen, M. D. F. Nutting, R. Jang, and A. K. Balls, *JBC* **179,** 189 (1949).
98. D. E. Fahrney and A. M. Gold, *JACS* **85,** 997 (1963).
99. P. T. Speakman and R. E. Yarwood, *Nature* **211,** 201 (1966).
100. N. M. Green, *JBC* **205,** 535 (1953).
101. W. Cohen, M. Lache, and B. F. Erlanger, *Biochemistry* **1,** 686 (1962).
102. L. A. Mounter, *BBA* **77,** 301 (1963).
103. M. Mares-Guia and E. Shaw, *Federation Proc.* **22,** 528 (1963).
104. E. Shaw and S. Springhorn, *BBRC* **27,** 391 (1967).
105. V. Tomášek, E. Severin, and F. Šorm, *BBRC* **20,** 545 (1965).

The same mechanism of inactivation was found for porcine trypsin (*106*).

The roles of serine and histidine residues in the catalytic site of trypsin were further elucidated by studies of other types of substitutions such as with bromoacetone (*107*). This small uncharged reagent has no apparent structural similarity to any of the trypsin substrates. Nevertheless, it alkylates selectively the His 46 in active trypsin. Unlike TLCK, bromoacetone reacts preferentially with this residue even in inactive DIP-trypsin. On the other hand, bromoacetone-treated trypsin is no longer reactive toward DFP. This indicates that the substitution of Ser 183 does not prevent His 46 from reacting while the modification of histidine leads to a total loss of the unusual reactivity of Ser 183 (*107*). The relay system has been apparently disconnected.

B. Specificity and Binding Sites

Although direct evidence based on X-ray diffraction measurements or on investigation of chemical substitutions of the protein is not available, a consideration of the analogy between trypsin and chymotrypsin leads to the assumption that the specificity site of trypsin, to which positively charged groups of substrates or inhibitors bind electrostatically, contains the negatively charged Asp 177 (*34, 67*). Aspartic acid at this position has been found so far only in trypsin (*67, 67a*), thrombin, and the trypsinlike proteinase from *Streptomyces griseus* (*12*). In all other serine proteinases known so far this position is occupied by neutral amino acids.

As a logical consequence of the previous considerations, the binding site can be located between the specificity site and the catalytic site. It has presumably the form of a crevice which binds by hydrophobic interactions the carbon side chains or rings of substrates and inhibitors.

Studies on the relationship between chemical composition and reactivity of low molecular weight inhibitors and substrates together with physicochemical studies (*108–117*) can serve as an experimental background for a more detailed examination of the relative locations and

106. R. A. Smith, and I. E. Liener, *Federation Proc.* **26**, 601 (1967).
107. J. G. Beeley, and H. Neurath, *Biochemistry* **7**, 1239 (1968).
108. W. B. Lawson, M. D. Leafer, A. Tewes, and G. J. S. Rao, *Z. Physiol. Chem.* **349**, 251 (1968).
109. M. Muramatu, T. Onishi, S. Makino, Y. Hayakumo, and S. Fujii, *J. Biochem.* (*Tokyo*) **58**, 214 (1965).
110. B. F. Erlanger and H. Castleman, *BBA* **85**, 507 (1964).
111. T. Inagami and T. Murachi, *JBC* **238**, PC1905 (1963).
112. T. Inagami and T. Murachi, *JBC* **239**, 1395 (1964).
113. B. M. Sanborn and G. E. Hein, *BBA* **139**, 524 (1967).

functions of the catalytic, binding, and specificity sites in trypsin. One of the remarkable outcomes of this kind of study is the assumption that the acylation step is preceded by the formation of two consecutive trans-conformational complexes between the enzyme and the substrate. The rate of formation of such a complex is approximately of the same order as that of an allosteric transition (*116, 117*). Interesting conclusions also emerge from observations that trypsin possesses two binding sites close to each other, one to which charged molecules are bound preferentially and the other for preferential binding of neutral molecules (*115, 118*). However, more experimental data, especially those provided by X-ray diffraction studies, are needed before the whole problem of the conformation of the specificity and binding sites can be properly evaluated.

IV. Substrates and Specificity

Trypsin possesses a very narrow specificity catalyzing preferentially the hydrolysis of bonds involving the carboxyl group of arginine or lysine. The maximum velocity of the reaction lies between pH 7 and 9. Studies on low molecular weight synthetic substrates and defined protein degradation products have supplied a large amount of information on this reaction. In principle, there are three kinds of substituents in the amino acid substrate which can directly influence the catalysis, e.g., the side chain, the substituent on the α-amino group, and the substituent on the carboxyl group. Additional influence can be exerted by neighboring amino acids in the primary structure of the polypeptide chain or by the tertiary structure of a macromolecular substrate.

A. Role of Side Chain

Both in synthetic (*55, 119*) and in polypeptide substrates (*55, 120, 121*), trypsin prefers the arginine side chain to the lysine side chain.

In arginine and lysine vasopressin, which are identical peptide except

114. J. D. Geratz, *ABB* **118,** 90 (1967).
115. B. M. Sanborn and W. P. Bryan, *Biochemistry* **7,** 3624 (1968).
116. G. Johannin and J. Yon, *BBRC* **25,** 320 (1966).
117. J. Chevalier and J. Yon, *BBA* **122,** 116 (1966).
118. B. M. Sanborn and G. E. Hein, *Biochemistry* **7,** 3613 (1968).
119. M. Bergmann, *Advan. Enzymol.* **2,** 49 (1942).
120. H. Keilová, V. Pliška, B. Keil, and F. Šorm, *Phys. Biochem. Physiol.* **1,** 100 (1969).
121. B. Keil and H. Keilová, *Collection Czech. Chem. Commun.* **29,** 2206 (1964).

for the side chain of the only basic amino acid, the pH optimums are different (8.0 and 7.2, respectively), but the arginine substrate is preferentially cleaved throughout the whole pH range (*120*). In the oxidized B chain of insulin, the ratio of the rate of hydrolysis at the only arginine bond to that at the only lysine bond at pH 8.0 and 30° is 25:1. This effect becomes even more pronounced at higher pH values (35:1 at pH 10.0 and 52:1 at pH 10.7) since the ϵ-amino group of lysine is considerably discharged under these conditions while the guanidinium group of arginine is not affected (*55*). The susceptibility to the cleavage decreases in the series of substrates of arginine, norarginine, homoarginine, and ornithine (*122*). *S*-2-Aminoethylcysteinyl bonds are also cleaved by trypsin (*123, 124*), the relative rates of cleavage at pH 9 and 30° being 100, 40, and 7 for amides of *N*-benzoylarginine, *N*-benzoyllysine, and *S*-2-aminoethyl cysteine, respectively (*125*). This relative rate is still higher in more alkaline media since the pK_a of the ω-amino group of the aminoethyl cysteine derivative is 9.4.

If the ϵ-amino group of the lysine side chain is monomethylated and therefore the positive charge not eliminated, tryptic hydrolysis of the substrate can occur (*126*). Substituents which eliminate the positive charge of the ϵ-amino group of lysine block at the same time trypsin catalysis.

The role of the positively charged group in the mechanism of action has been further elucidated by the study of the hydrolysis of a nonspecific substrate, acetylglycine ethyl ester. Being devoid of any positively charged side chain, this substrate is hydrolyzed at a rate three orders of magnitude lower than that of the specific reaction (*127*). This rate is increased approximately nine times by the presence of ethylammonium ions. Whereas acetylglycine ethyl ester interacts solely with the catalytic site, the ethylammonium ion is bound to the specificity-determining site. The observed increase in the rate of hydrolysis can be explained by an activation of the catalytic site induced by simultaneous binding of the positively charged molecule (*128*).

In this connection other "nonspecific" catalytic effects of trypsin which resemble the catalysis by α-chymotrypsin should be mentioned. Both enzymes catalyze the hydrolysis of esters of *m*-hydroxybenzoic acid

122. J. B. Baird, E. F. Curragh, and D. T. Elmore, *BJ* **96,** 733 (1965).
123. D. T. Elmore, D. V. Roberts, and J. J. Smyth, *BJ* **102,** 728 (1967).
124. H. Lindley, *Nature* **178,** 647 (1956).
125. S.-S. Wang and F. H. Carpenter, *JBC* **243,** 3702 (1968).
126. L. Benoiton and J. Deneault, *BBA* **113,** 613 (1966).
127. T. Inagami and H. Mitsuda, *JBC* **239,** 1388 (1964).
128. T. Inagami and H. Mitsuda, *JBC* **239,** 1395 (1964).

(*129*) and of *p*-nitrophenyl acetate (*130, 131*). The study of the interaction of trypsin with *p*-nitrophenyl acetate, which is characterized by the formation of a stable acyl-enzyme compound under simultaneous release of *p*-nitrophenol, has opened the way to direct titrations of the active site of trypsin (see Section IV,E).

Both enzymes catalyze the esterolysis of acylated phenylalanine (*132*) and tyrosine (*133*). Trypsin cleaves acetyl-L-tyrosine ethyl ester at a maximum rate, which is 13% of that observed with α-chymotrypsin (*133*). The "chymotryptic" cleavage by trypsin has stimulated experiments designed to decide whether this activity results from an impurity in the enzyme which may contain traces of α-chymotrypsin or from the far-reaching similarities in the conformation of the active sites of both enzymes. The specific chymotrypsin inhibitor, L-(1-tosylamido-2-phenyl) ethyl chloromethyl ketone, virtually eliminates the chymotryptic activity of commercial crystalline trypsin toward the B chain of insulin as substrate; however, an appreciable activity (50%) toward acetyl-L-tyrosine ethyl ester remains (*134*). The same holds true for samples of trypsin purified in different ways (*135, 136*). The chymotryptic-type cleavages by pure trypsin of high molecular weight peptides are very selective. No aromatic bond is split in the B chain of insulin; on the other hand, limited degradation was observed with glucagon and the C chain of α-chymotrypsin. These observations strongly suggest that the ability to split certain aromatic bonds is inherent to trypsin itself (*135*).

B. Substitution of α-Amino and Carboxyl Group

The blocking of the α-amino group is not a prerequisite of tryptic action since lysine ethyl ester is rapidly hydrolyzed (*137*). The same holds true for the arginine ester (*138*). The presence of one additional basic group shifts down the pH optimum of trypsin with respect to these substrates. Trypsin acts even on substrates in which the α-amino group

129. B. H. Hofstee, *BBA* **24,** 211 (1957).
130. B. S. Hartley and B. A. Kilby, *BJ* **56,** 288 (1954).
131. G. H. Dixon and H. Neurath, *JBC* **225,** 1049 (1957).
132. H. A. Ravin, P. Bernstein, and A. M. Seligman, *JBC* **208,** 1 (1954).
133. T. Inagami and J. M. Sturtevant, *JBC* **235,** 1019 (1960).
134. V. Kostka and F. H. Carpenter, *JBC* **239,** 1799 (1964).
135. S. Maroux, M. Rovery, and P. Desnuelle, *BBA* **122,** 147 (1966).
136. H. J. Schramm, *Z. Physiol. Chem.* **348,** 2321 (1967).
137. H. Werbin and A. Palm, *JACS* **73,** 1382 (1951).
138. H. Goldenberg and V. Goldenberg, *Arch. Biochem.* **29,** 154 (1950).

has been replaced by a hydroxyl group (*139*). In low molecular weight substrates used for routine assays, the α-amino group is usually substituted by a tosyl, carbobenzyloxy, or a benzoyl group.

Trypsin acts both on amides and esters, the rate of hydrolysis of a specific ester being about 300 times higher than that of an amide substrate (*52, 60, 140*). In the case of esters, the nature of the alcohol group has very little effect on the rate of hydrolysis.

C. Role of Sequence and Tertiary Structure of Polypeptide Substrate

In a polypeptide chain with a random spatial arrangement, the rate of the cleavage of a peptide bond next to a basic residue is very much reduced in the case of adjacent cystine and becomes virtually zero if the basic residue is followed by proline. Similarly, terminal basic residues with a free amino or carboxyl group are split off at a lower rate (*141–144*). In smaller peptides, the rate of the hydrolysis is affected by the number of amino acid residues (*145, 146*).

Although the majority of positively charged groups in native proteins are supposed to be located on the surface of the molecule, the three-dimensional structure imposes certain limitations on the action of trypsin. It has been observed in numerous studies that trypsin does not degrade native proteins prior to denaturation. Even though the free rotation of a polypeptide structure is a prerequisite of the trypsin action, a very specific and limited degradation at loose loops or segments of a native protein often occurs. The activations of zymogens (*1, 22*), the formation of α-trypsin and pseudotrypsin (*20, 34*), or the mechanism of formation of the complex between trypsin and polypeptide inhibitors (*147, 148*) can serve as examples.

139. J. E. Snoke and H. Neurath, *Arch. Biochem.* **21,** 351 (1949).
140. S. A. Bernhard, *BJ* **59,** 506 (1955).
141. C. H. W. Hirs, S. Moore, and W. H. Stein, *JBC* **219,** 623 (1956).
142. G. Kreil and H. Tuppy, *Nature* **192,** 1123 (1961).
143. H. Neurath and C. W. Schwert, *Chem. Rev.* **46,** 69 (1950).
144. S. G. Waley and J. Watson, *BJ* **57,** 529 (1954).
145. T. Yamamoto and N. Izumiya, *ABB* **120,** 497 (1967).
146. M. M. Nachlas, R. E. Plapinger, and A. M. Seligman, *ABB* **108,** 266 (1964).
147. K. Ozawa and M. Laskowski, Jr., *JBC* **241,** 3955 (1966).
148. M. Rigbi and L. J. Greene, *JBC* **243,** 5457 (1968).

D. Redirection of Protein Hydrolysis by Chemical Modification of Substrate

As already mentioned, the conversion of cysteine residues to *S*-2-aminoethyl cysteines gives rise to new positively charged sites susceptible to trypsin hydrolysis (*124*). The labeling of cysteine with ethylenimine is practically quantitative, and a polypeptide modified in this manner will be cleaved by trypsin at the carboxyl side of lysine, arginine, and modified cysteine residues (*149*).

The finding that a substitution which eliminates the positive charge of the ϵ-amino group of lysine blocks tryptic hydrolysis at this residue has been used extensively in studies on protein structure. After the ϵ-NH_2 groups of the lysine residues have been labeled, the polypeptide chain is cleaved by trypsin selectively only at the arginine sites and the number of the resulting fragments is considerably reduced. For this purpose, substitution by the dinitrophenyl (*150*), carbamyl (*151*), carboxymethyl (*152*), amidyl (*153*), guanidyl (*154*), trifluoroacetyl (*155*), and maleyl (*156*) groups has been employed.

Trifluoroacetylation with ethyl thioltrifluoroacetate (*155*) and maleylation with maleic anhydride (*156*) are both reversible procedures. After the cleavage at the arginyl bonds of a modified protein by trypsin and separation of the fragments thus formed, the blocking group can be removed under mild conditions and the regenerated lysyl bonds of the fragments submitted in the next step to additional tryptic hydrolysis.

E. Assays of Trypsin Activity

The amount of active trypsin in solution can be determined either from the rate of catalysis of a specific substrate or by direct titration of its active site.

Before the specific modification of the active site of trypsin became known, only the first method was used. Specific activity of a trypsin

149. M. A. Raftery and R. D. Cole, *BBRC* **10,** 467 (1963).
150. R. R. Redfield and C. B. Anfinsen, *JBC* **221,** 385 (1956).
151. G. R. Stark, W. H. Stein, and S. Moore, *JBC* **235,** 3177 (1960).
152. S. Korman and H. T. Clarke, *JBC* **221,** 133 (1956).
153. M. J. Hunter and M. L. Ludwig, *JACS* **84,** 3491 (1962).
154. G. S. Shields, R. L. Hill, and E. L. Smith, *JBC,* **234,** 1749 (1957).
155. R. F. Goldberger and C. B. Anfinsen, *Biochemistry* **1,** 401 (1962).
156. P. J. Butler, J. I. Harris, B. S. Hartley, and J. Leberman, *BJ* **112,** 679 (1969).

preparation was expressed in units of trypsin per milligram of protein. One unit of trypsin was defined as the amount of trypsin catalyzing the transformation of 1 μmole of *p*-toluensulfonyl-L-arginine methyl ester per minute at 25° and pH 8.1 in the presence of 0.01 *M* calcium ion (*157, 158*) or of some other convenient substrate. For routine assays *N*-α-benzoyl-D,L-arginine *p*-nitroanilide (BANA) [spectrophotometrical assay (*159*)], tosyl-arginine methyl ester (TAME) [spectrophotometrical (*157*) or titrimetrical (*20*)], or *N*-benzoyl-L-arginine ethyl ester (BAEE) [spectrophotometrical (*160*)] have been employed and will also serve many purposes in the future.

All of these assays based on the determination of reaction rate employ as absolute standard a "pure" trypsin, although it is well known that none of the trypsin preparations can be considered as being chemically pure because of the adsorption of impurities, water content, etc. The spectrophotometric determination of the content of trypsin in solution is not of much help since it is again based on a "pure" trypsin as absolute reference standard. Uncertainties also result from the large number of variables involved in the rate assays. Thus, for example, the hydrolysis of tosyl-arginine methyl ester does not follow the simple Michaelis–Menten kinetics, trypsin is activated by the substrate (*161*), etc.

Specific titration procedures, which employ as absolute standard low molecular weight organic compounds and which directly determine the molar concentration of the active site of trypsin in solution, clearly offer a better approach than the rate assays.

In one of the fundamental studies dealing with the determination of trypsin activity (*162*) the characteristics of an optimal titrant were defined as follows: It has to (1) be a specific substrate; (2) titrate in a few minutes in order to avoid denaturation and for the sake of convenience; (3) titrate around neutrality in order that physiological conditions can be reproduced; (4) be a stable, available, and soluble reagent whose stoichiometric reaction is easily detectable; and (5) titrate over a wide range of enzyme concentrations.

These criteria are best met by the two titrants *p*-nitrophenyl N^2-

157. B. C. W. Hummel, *Can. J. Biochem. Physiol.* **37,** 1393 (1959).

158. G. W. Schwert, H. Neurath, S. Kauffman, and J. E. Snoke, *JBC* **172,** 221 (1948).

159. B. F. Erlanger, N. Kokowsky, and W. Cohen, *ABB* **95,** 271 (1961).

160. G. W. Schwert and Y. Takenaka, *BBA* **16,** 570 (1955).

161. C. W. Trowbridge, A. Krehbiel, and M. Laskowski, Jr., *Biochemistry* **2,** 843 (1963).

162. M. L. Bender, M. L. Begué-Canton, R. L. Blakeley, L. J. Brubacher, J. Feder, C. R. Gunter, F. J. Kézdy, J. V. Killheffer, Jr., T. H. Marshall, C. G. Miller, R. W. Roeske, and J. K. Stoops, *JACS* **88,** 5890 (1966).

benzyloxycarbonyl-L-lysinate HCl (*42*) and *p*-nitrophenyl-*p*′-guanidinobenzoate HCl (*51*). In both cases *p*-nitrophenol released during the acyl-enzyme formation is determined spectrophotometrically. The titration with *p*-nitrophenyl-*p*′-guanidinobenzoate at pH 8.3 gives accurate results for trypsin concentrations from 2×10^{-6} *M* to 10^{-4} *M* and can be completed in several minutes with high precision. It cannot be used in the presence of chymotrypsin, which also cleaves the titrant. The other titrant, *p*-nitrophenyl N^2-benzyloxycarbonyl-L-lysinate, is not affected by α-chymotrypsin. It can be used also for the titration of thrombin.

Two other titrants have been proposed: α-*N*-methyl-α-*N*-toluene-*p*-sulfonyl-L-lysine β-naphthyl ester, which is affected neither by chymotrypsin nor by thrombin (*163*), and *p*-nitrophenyl-N^2-acetyl-N^1-benzylcarbazate, which does not react with thrombin either but can be used for both trypsin and chymotrypsin (*164*).

V. Chemical Modifications

The most important modifications of the amino acid residues of trypsin are those which helped to discover the key positions held by His 46 and Ser 183 in the catalytic site (see Section III,A). A large amount of information has been accumulated on the structure–function relationship of trypsin from studies on substitution reactions of various side chains of amino acid residues.

A. Amino Groups and Imidazole Rings

One can suppose that in trypsin all the ε-groups of lysine residues occupy positions on the surface and are therefore readily accessible to the modifying reagents. Trypsin, all of whose ε-amino groups of lysine residues have been acetylated by acetic anhydride, shows the same activity toward synthetic substrates as native trypsin without any observable change in the kinetics of the reactions (*165*). In this derivative both the α-amino group of the N-terminal isoleucine and the phenolic group of tyrosine residues remained free. All 14 ε-amino groups can also be acetamidated without a loss of enzymic activity (*166*). It is interesting to note that trypsin with ten lysyl residues guanidated autolyzes at neu-

163. D. T. Elmore and J. J. Smyth, *BJ* **103**, 36P (1967).
164. D. T. Elmore and J. J. Smyth, *BJ* **103**, 37P (1967).
165. J. Labouesse and M. Gervais, *European J. Biochem.* **2**, 215 (1967).
166. A. Nureddin and T. Inagami, *BBRC* **36**, 999 (1969).

tral pH at a rate comparable to that observed with the unmodified enzyme, whereas the modification of the eleventh lysine residue brings about a remarkable stabilization against autolysis in the absence of calcium ions (*166*).

When both lysine and tyrosine residues are acetylated, the specific esterolytic activity of the modified enzyme is increased (*165*, *167*). Acetylation of the α-amino group of the N-terminal isoleucine leads to a complete loss of activity. The same effect is obtained by destroying this α-amino group by nitrous acid (*168*, *169*).

From these studies it can be concluded that the ε-amino groups of trypsin are not essential for its activity; on the other hand, the presence of the α-amino group of N-terminal isoleucine is neccessary.

The reactivity of the active histidine residue has been discussed in Section III,A. Both the titration assay (*170*) and the reaction of trypsin with diazotetrazol (*84*) show that none of the three histidine residues is buried in the interior of trypsin. In trypsinogen, however, one histidine residue is not accessible to diazotetrazol (*84*).

B. Tyrosine and Tryptophan Residues

The effect of acetylation of tyrosine residues on the activity of trypsin is not quite clear. It has been shown in several studies that out of the ten residues of tyrosine in both trypsin and trypsinogen about one-half is far less accessible, presumably because they are shielded inside the tertiary structure from the outer environment (*81*, *83*, *165*, *171–175*). According to the reaction conditions, 4–7 tyrosines can be acetylated with very little effect on trypsin activity (*165*, *171*, *172*). Under different conditions, the substitution of three tyrosines induced a change in the efficiency of the catalytic site without an alteration of the binding site of trypsin (*173*).

Nitration of the tyrosine rings in trysin and trypsinogen by tetranitromethane shows clearly that the exposed and buried tyrosines 11, 28, 137,

167. H. Trenholm, W. E. Spomer, and J. F. Wootton, *JACS* **88**, 4281 (1966).
168. T. Hofmann and S. T. Scrimger, *Federation Proc.* **25**, 589 (1966).
169. T. Hofmann, H. Schechter, J. W. Dixon, C. Burrowes, A. Gerther, and S. T. Scrimger, *Federation Proc.* **26**, 828 (1967).
170. J. A. Duke, M. Bier, and F. F. Nord, *ABB* **40**, 424 (1952).
171. J. F. Riordan, W. E. C. Wacker, and B. L. Vallee, *Nature* **208**, 1209 (1965).
172. J. F. Riordan, W. E. C. Wacker, and B. L. Vallee, *Biochemistry* **4**, 1758 (1965).
173. L. L. Houston and K. A. Walsh, *BBRC* **25**, 175 (1966).
174. L. B. Smillie and C. M. Kay, *JBC* **236**, 112 (1961).
175. Y. Hachimori, A. Matsushima, M. Suzuki, and Y. Inada, *BBA* **124**, 395 (1966).

and 48 are fully nitrated (*88, 92*), Tyr 20 and 171 to approximately 60% and Tyr 82 to about 25%, whereas tyrosines 158, 212, and 218 are resistant to nitration (*92*). Nitrotrypsin is enzymically fully active.

As in the case of tyrosine residues, an unidentified portion of the four tryptophan residues of trypsin are buried in the interior of the molecule (*82, 176–179*). Only about one-third of the activity is lost after the oxidation of three out of four tryptophan residues.

C. Disulfide Bridges

One single disulfide bond connecting residues 179–203, close to the active serine residue, can be selectively reduced in trypsinogen by sodium borohydride at pH 9.1 (*89*). At a higher pH, two disulfides (one 154–168) are reduced by borohydride (*180*) while only one (154–168) is reduced by mercaptoethanol (*90*).

D. Activation of Modified Trypsinogen

Modifications of the groups surrounding the strategic bond Lys 6–Ile 7 help to better understand the process of activation. Exhaustive acetylation of the ϵ-amino groups of lysine residues leads to acetyltrypsinogen which can no longer be activated (*30, 35*). This observation can be explained by the loss of a positive charge of Lys 6. Modification of carboxylate ions of trypsinogen by glycine ethyl ester in the presence of a carbodiimide derivative resulted in a modified protein, which even in the absence of the calcium ions could be activated to about 46%, whereas with unmodified trypsinogen in the absence of calcium ions the extent of the activation did not exceed 15%. This indicates that the chief role of calcium in the process of regular activation may be that of an interaction with the carboxylate groups of Asp 2 and 5 (*181*).

Exhaustive guanidation of the ϵ-amino groups converts trypsinogen to a derivative which cannot be activated by trypsin; on the other hand, it can be activated by the proteinase of *Aspergillus oryzae,* which splits the strategic Lys–Ile bond even after the Lys 6 is modified (*182*).

176. N. M. Green and B. Witkop, *Trans. N. Y. Acad. Sci.* [2] **26,** 659 (1964).
177. T. Viswanatha, W. B. Lawson, and B. Witkop, *BBA* **40,** 216 (1960).
178. A. Previero, M. A. Coletti, and L. Galzigna, *BBRC* **16,** 195 (1964).
179. K. Hyashi, T. Imoto, and M. Funatsu, *J. Biochem.* (*Tokyo*) **55,** 516 (1964).
180. A. Light and N. K. Sinha, *JBC* **242,** 1358 (1967).
181. T. M. Radhakrishnan, K. A. Walsh, and H. Neurath, *JACS* **89,** 3059 (1967).
182. N. C. Robinson and K. A. Walsh, *Federation Proc.* **27,** 292 (1968).

Trypsinogen with three nitrated tyrosines shows only a 35% decrease in activability (*88*). After complete nitration of trypsinogen, however, its activability was considerably slower and reduced (*31b*), obviously because the strategic Lys–Ile bond is less accessible to cleavage in the nitrated product. In trypsinogen, Tyr 82 is nitrated more readily than in trypsin. This situation shows a striking analogy to the reactivity of the analogous Tyr 94 in chymotrypsinogen and chymotrypsin (*183*). The difference in accessibility of this residue seems to reflect a conformational change during the activation process in both zymogens (*31b, 92, 184*).

Carboxymethyl-trypsinogen with the disulfide bond 179–203 cleaved yields almost completely active carboxymethyl-trypsin on activation with trypsin or enterokinase (*89*).

E. Water-Insoluble Derivatives of Trypsin

Active trypsin covalently bound to an insoluble inert carrier [for a review, see Silman and Katchalski (*185*)] shows two very interesting features in comparison with the soluble enzyme: It can no longer autolyze since the individual molecules are immobile on a matrix, and its action on a substrate can be regulated by varying the flow of substrate in the liquid phase. By packing the insoluble trypsin derivative in a column, all the advantages of chromatography can be exploited.

The ε-amino groups of lysine residues are inessential for the catalytic action of trypsin. They were therefore chosen to form links to insoluble carriers such as carboxymethyl cellulose (*186*), a copolymer of maleic acid and ethylene (*187*), bromoacetyl cellulose (*188*), copolymers of amino acids (*189, 190*), or Sephadex (*191*). Active trypsin derivatives are thus obtained. The resulting catalytic capacity depends on two parameters, namely, the relative content of trypsin bound to the carrier and the loss of specific activity caused by the chemical modification of the

183. S. V. Shlyapnikov, B. Meloun, B. Keil, and F. Šorm, *Collection Czech. Chem. Commun.* **33**, 2292 (1968).
184. V. Holeyšovský, B. Keil, and F. Šorm, *FEBS Letters* **3**, 107 (1969).
185. I. H. Silman and E. Katchalski, *Ann. Rev. Biochem.* **35**, 873 (1966).
186. M. A. Mitz and L. J. Summaria, *Nature* **189**, 576 (1961).
187. Y. Levin, M. Pecht, L. Goldstein, and E. Katchalski, *Biochemistry* **3**, 1905 (1964).
188. A. Patchornik, Israel Patent 18207 (1962).
189. A. Bar-Eli and E. Katchalski, *JBC* **238**, 1690 (1963).
190. A. Bar-Eli and E. Katchalski, *Nature* **188**, 856 (1960).
191. R. Axén and J. Porath, *Nature* **210**, 367 (1966).

native enzyme or by relative orientation of the enzyme's molecules in the matrix.

Polytyrosyl trypsin coupled with the diazotized *p*-aminophenylalanine-leucine copolymer contained about 20% of the enzyme possessing a specific esterase activity corresponding up to 30% of that of crystalline trypsin (*189, 190*).

All of these trypsin derivatives are more stable than trypsin in the alkaline pH range; they can be stored in lyophilized dry form. The binding to an insoluble carrier affects less the activity toward synthetic low molecular weight substrates than that toward proteins. High molecular weight inhibitors also affect less insolubilized trypsin than the native protein (*187*). This can be explained as a result of steric hindrance caused by the carrier.

An interesting observation concerning the specificity of an insoluble trypsin derivative is that it cleaved selectively a smaller number of bonds in pepsinogen rather than in native trypsin, which means that the modification induced a more selective specificity toward the high molecular weight substrate (*192*).

VI. Inhibition

The activity of trypsin can be negatively influenced by physical parameters (temperature and pH), by conformational changes of its molecule (denaturation), by chemical modifications (substitution of amino acid residues and reduction of disulfide bridges), or by specific interactions with low molecular weight or naturally occurring polypeptide inhibitors.

Blocking of the catalytic site by DFP and TLCK may serve as an example of inhibition by chemical modification. The following section will be focused on compounds forming specific noncovalent complexes with the binding and specificity site of trypsin thus preventing trypsin from combining with the substrate.

A. Low Molecular Weight Inhibitors

The search for efficient trypsin inhibitors can be facilitated by assuming that the stereochemical arrangement of the active site of trypsin will necessarily dictate of complementary conformation of a suitable sub-

192. E. B. Ong, Y. Tsang, and G. E. Perlmann, *JBC* **241**, 5661 (1966).

strate or inhibitor. Or, vice versa, the study on substrates and inhibitors of known conformation will contribute to the understanding of the stereochemistry of the specificity and binding sites. The character of forces concerned is twofold in the case of trypsin; namely, that of an ionic interaction between the negatively charged Asp 177 and a positively charged group of the substrate (inhibitor), and that of very efficient hydrophobic bonds between the carbon side chains of the substrate (inhibitor) and the hydrophobic binding site of the enzyme.

A great number of compounds which meet these structural requirements have been assayed with positive results. *p*-Aminobenzamidine and benzamidine were found to be the most potent low molecular weight competitive inhibitors of trypsin [K_I 8.2×10^{-6} and 1.8×10^{-5} (*46*)]. They are approximately the same size as the side chain of lysine or arginine. Many other compounds with a positively charged group and a hydrophobic moiety have also been found to possess an inhibitory effect. These are alkyl and aryl guanidines (*46, 115, 193, 194*), amidine (*46, 193, 195*) and agmatine derivatives (*196, 197*), benzoyl- and tosylarginine (*198, 199*), aliphatic and aromatic amines (*200, 201*), and *p*- or ω-amino carboxylic acids (*202*) (see also Section II,B).

B. Naturally Occurring Polypeptide Inhibitors

Highly specific inhibition of proteolytic activity by naturally occurring polypeptide inhibitors is one of the regulation schemes of enzymic action which can be discussed on the molecular level.

Trypsin inhibitors from different organs and organisms have been isolated in pure form and characterized chemically [for reviews, see Werle (*203*), and Laskowski and Laskowski (*204*)]. They usually occur in groups of parallel entities with identical activity but different physicochemical

193. M. Mares-Guiy, *ABB* **127,** 317 (1968).
194. T. Inagami and S. S. York, *Biochemistry* **7,** 4045 (1968).
195. K. Tanizawa, S. Ishii, and Y. Kanaoka, *BBRC* **32,** 893 (1968).
196. N. G. Rule and L. Lorand, *BBA* **81,** 130 (1964).
197. L. Lorand and N. G. Rule, *Nature* **190,** 722 (1961).
198. J. J. Béchet, M. C. Gardiennet, and J. Yon, *BBA* **122,** 101 (1966).
199. A. d'Albis and J. J. Béchet, *BBA* **140,** 435 (1967).
200. T. Inagami, *JBC* **239,** 787 (1964).
201. H. Mix, H. J. Trattin, and M. Gülzov, *Z. Physiol. Chem.* **343,** 53 (1965).
202. M. Gülzow, H. Mix, and H. J. Trettin, *Z. Physiol. Chem.* **348,** 285 (1967).
203. E. Werle, "Natürliche Inhibitoren der Proteolytischen Enzyme." Springer, Berlin, 1966.
204. M. Laskowski and M. Laskowski, Jr., *Advan. Protein Chem.* **9,** 203 (1954).

properties or composition (*205–212*). The molecular weight of many of them is relatively low (6000–8000) and their tertiary structure very compact owing to a high content of disulfide bridges. They are therefore very stable to denaturation or to the action of proteolytic enzymes. Only rarely do they inhibit trypsin exclusively. In most cases they are polyvalent and also inhibit thrombin, chymotrypsin, kallikrein, etc.

Survey of the chemistry and inhibition specificity of naturally occurring inhibitors is beyond the scope of this chapter. The reader is referred to Chapter 11 by Laskowski and Sealock, this volume.

205. G. Jones, S. Moore, and W. H. Stein, *Biochemistry* **2**, 66 (1963).

206. D. Pospíšilová-Čechová, V. Dlouhá, and F. Šorm, *FEBS Ann. Meeting, Prague, 1968* Abstr. 971.

206a. D. Čechová, V. Jonáková-Švestková, and F. Šorm, *Collection Czech. Chem. Commun.* **35**, 3085 (1970).

207. J. J. Rockis and R. L. Anderson, *BBRC* **15**, 230 (1964).

208. H. Fritz, I. Hüller, M. Wiedeman, and E. Werle, *Z. Physiol. Chem.* **348**, 405 (1967).

209. V. Fratalli, *JBC* **244**, 274 (1969).

210. W. Ferdinand, S. Moore, and W. H. Stein, *BBA* **96**, 524 (1965).

211. V. Dlouhá, J. Neuwirthová, B. Meloun, and F. Šorm, *Collection Czech. Chem. Commun.* **30**, 1705 (1965).

212. H. Tschesche, *Z. Physiol. Chem.* **348**, 1216 (1967).

9

Thrombin and Prothrombin

STAFFAN MAGNUSSON

I. Introduction

A. General

Thrombin, like trypsin, is a serine proteinase that catalyzes the cleavage of certain arginyl and lysyl bonds. Amino acid sequence evidence clearly indicates that the B chain of thrombin and the pancreatic serine proteinases have evolved from a common ancestor. The serine proteinases concerned with digestion are omnivorous both in their choice of protein substrates and in the number of peptide bonds that they attack in a given substrate. By contrast thrombin and its substrate fibrinogen have apparently evolved together to form a highly specific enzyme–substrate system. Thus, thrombin splits only two peptide bonds per fibrinogen monomer. In comparison trypsin splits almost all the approximately 150 arginyl and lysyl bonds. Obviously the thrombin–fibrinogen system is interesting not only because of its physiological and medical implications but also as a structural problem in its own right.

B. Historical Background

In 1845, Buchanan (*1*) reported a substance in the liquid wash from the fibrin of clotted blood, which would cause clotting of hydrocele fluid. Schmidt (*2*) showed in 1872 that enzymic quantities of "fibrin ferment" or "thrombin" were sufficient to convert fibrinogen to fibrin. He could precipitate thrombin activity from serum but not from fresh blood by adding ethanol. He interpreted this to mean that thrombin exists in plasma as a zymogen that is activated to thrombin as part of the coagulation process. Arthus and Pagès (*3*) demonstrated that calcium ions are required for normal blood coagulation. Various salts that decrease the calcium ion activity either by precipitation or by complex formation prevented the normal coagulation process. When a sufficient excess of calcium chloride is added—"recalcification"—this inhibition is overcome, and the clotting time restored to normal. In 1899, Ham-

1. A. Buchanan, *London Med. Gaz.* **1,** 617 (1845); reprinted in *J. Physiol.* (*London*) **2,** 158 (1879).
2. A. Schmidt, *Arch. Ges. Physiol.* **6,** 413 (1872).
3. M. Arthus and C. Pagès, *Arch. Physiol. Norm. Pathol.* **2,** 739 (1890).

marsten (*4*) described the isolation of fibrinogen in a sufficiently pure state to show that the thrombin–fibrinogen reaction did not require calcium ions.

In Morawitz' classic theory of blood coagulation (*5*) from 1905, calcium ions and thrombokinase are required for the activation of prothrombin to thrombin. Thrombokinase was described as an enzyme released from damaged tissue cells or from blood platelets. Gradually the terms *thrombokinase* and *thromboplastin* came to be used synonymously by the majority of workers as a name for crude saline extracts of homogenized or acetone-dried tissue, usually from brain, lung, or placenta, Heparin, a naturally occurring anticoagulant from the metachromatic granules of Ehrlich's mast cells (*6*) was first discovered in 1916 by McLean (*7*). Subsequent work by Jorpes [reviewed (*8*)] led to its isolation and chemical characterization. Heparin acts with antithrombin II (heparin cofactor) as an inhibitor of the thrombin–fibrinogen reaction and as an inhibitor of the activation of prothrombin to thrombin.

Eagle and co-workers (*9*) concluded that the activation of prothrombin to thrombin and the clotting of fibrinogen to fibrin are both proteolytic reactions since they could mimic the natural reactions by trypsin catalysis of prothrombin activation and papain catalysis of fibrinogen–fibrin conversion. Dam's discovery in 1935 of vitamin K (*10*) opened up a new aspect of prothrombin. Further work showed that unless naphthoquinone compounds with vitamin K activity are supplied to vertebrates, either in the diet or by the metabolic activity of the intestinal flora, the normal plasma level of prothrombin activity is not maintained. Cattle being fed spoiled sweet clover (*Melilotus alba*) develop a bleeding disorder, and Link and co-workers (*11*) found in 1941 the cause to be Dicoumarol, presumably a breakdown product of coumarin the fragrant substance in freshly cut grass. This led to synthesis of a large number of compounds based on the 4-hydroxycoumarin structure, which is similar to that of vitamin K. These compounds act as vitamin K antagonists, decreasing the normal level of prothrombin activity in plasma. A number of them have been widely used in rat control and clinical medicine.

The availability of heparin, Dicoumarol compounds, and vitamin K

4. O. Hammarsten, *Z. Physiol. Chem.* **28,** 98 (1899).
5. P. Morawitz, *Ergeb. Physiol.* **4,** 307 (1905).
6. H. Holmgren and O. Wilander, *Z. Mikroskop.-Anat. Forsch.* **42,** 242 (1937).
7. J. McLean, *Am. J. Physiol.* **41,** 250 (1916).
8. J. E. Jorpes, "Heparin in the Treatment of Thrombosis." Oxford Univ. Press, London and New York, 1946.
9. H. Eagle and T. N. Harris, *J. Gen. Physiol.* **20,** 543 (1937).
10. H. Dam, *Vitamins Hormones* **24,** 295 (1966).
11. H. A. Campbell and K. P. Link, *JBC* **138,** 21 (1941).

for clinical use led naturally to increased interest in the clotting mechanism as a whole. Development of new assay systems for prothrombin (*12–14*) and later for other clotting factors provided the diagnostic tools for investigating bleeding disorders. One result of this has been that seven "new" clotting factors (in addition to calcium ions and thromboplastin) are now generally recognized as being involved in "normal" prothrombin activation.

The evidence that all these factors (V,X,VII,XII,XI,IX, and VIII) (*15*) actually exist (*16*) is mainly clinical and genetic. For each factor a corresponding hereditary bleeding disease has been described. In individuals that are homozygotic for the deficient gene the factor is either absent or abnormal, and their blood plasma shows no activity or virtually no activity of that factor. Heterozygotes display intermediate levels of activity. All the diseases run with definite hereditary patterns; consequently, each can be characterized as an inborn error of metabolism, affecting one particular clotting factor (*17*). These factors are generally believed to be proteins, but since they occur only in minute concentrations in blood plasma it has been exceedingly difficult to obtain them in a reasonably pure state. As a result, biochemical knowledge about them is still limited. Clear-cut evidence that the thrombin molecule is contained within the structure of prothrombin was provided by Seegers and co-workers (*18*) in 1950. They were able to achieve almost quantitative activation to thrombin by dissolving highly purified prothrombin in 25% (w/v) sodium citrate.

C. Occurrence

Prothrombin and thrombin as well as fibrinogen have been found in blood plasma in all classes of vertebrates (*19*). The essential features of the mammalian coagulation mechanism, namely, activation of prothrombin to thrombin, limited thrombic proteolysis of fibrinogen with release of fibrinopeptides (*20*), and self-assembly of the resulting fibrin

12. A. J. Quick, *JBC* **109,** P73 (1935).
13. E. D. Warner, K. M. Brinkhous, and H. P. Smith, *Am. J. Physiol.* **114,** 667 (1936).
14. P. A. Owren and K. Aas, *Scand. J. Clin. & Lab. Invest.* **3,** 201 (1951).
15. I. S. Wright, *J. Am. Med. Assoc.* **170,** 325 (1959). Nomenclature adopted by the International Committee on Nomenclature of Blood Clotting Factors.
16. M. P. Esnouf and R. G. Macfarlane, *Advan. Enzymol.* **30,** 255 (1968).
17. R. Biggs and R. G. Macfarlane, "Human Blood Coagulation and Its Disorders," 3rd ed. Blackwell, Oxford, 1962.
18. W. H. Seegers, R. I. McClaughry, and J. L. Fahey, *Blood* **5,** 421 (1950).
19. R. F. Doolittle and D. M. Surgenor, *Am. J. Physiol.* **203,** 964 (1962).
20. R. F. Doolittle, *BJ* **94,** 735 (1965).

monomer to form an insoluble polymeric fibrin gel have all been clearly described even in the lamprey eel (*Petromyzon marinus*) (*20, 21*) which is one of the most primitive vertebrates. Since the freshly formed fibrin gel can be dissolved in 5–8 *M* urea or in guanidinium solutions, polymerization of fibrin monomer presumably involves the formation of weak noncovalent bonds (*22*). In the next stage of mammalian blood coagulation the fibrin clot is reinforced by the formation of covalent interchain peptide bonds. This reaction is catalyzed by plasma *trans*-glutaminase (Factor XIII) (*23*) and the transamidation involves specific "donor" and acceptor sites. The "donor" sites are the ϵ-amino groups of two lysine residues (*24, 25*) in the α and the γ chains (*26, 27*).

The "acceptor" sites are the γ-carboxyamide groups of two (*28, 29*) glutamine (*30*) residues in the γ chain (*26, 31*). The inter-γ-cross-link involves the lysine and glutamine residues three and eight residues, respectively, removed from the C-terminus of the chain (*32*). The cross-linked fibrin clot is not soluble in 5–8 *M* urea (*22*). Although the exact nature of the cross-linking bonds in lamprey fibrin remains to be determined, it has been established that cross-linking catalyzed by plasma *trans*-glutaminase does occur in this species. Thus, it appears that both the thrombin and the fibrin–cross-linking reactions occur throughout the vertebrate phylum. Invertebrates with a circulatory system generally have a blood clotting mechanism (*33*) which in most species depends on agglutination of blood cells to form a hemostatic plug. In some arthropods, however, it also involves conversion of a soluble protein to a fibrin gel. The intracellular enzyme that catalyzes clotting in lobster resembles plasma *trans*-glutaminase (*34*) in requiring calcium and depending on SH groups. Glycinamide inhibits cross-linking in mammals (*35*) as well as clotting in the lobster (*36, 37*). It has also been found

21. R. F. Doolittle, *BJ* **94,** 742 (1965).
22. K. C. Robbins, *Am. J. Physiol.* **142,** 581 (1944).
23. K. Buluk, T. Januszko, and J. Olbromski, *Nature* **191,** 1093 (1961).
24. L. Lorand, H. H. Ong, B. Lipinski, N. G. Rule, J. Downey, and A. Jacobsen, *Biochem. Biophys. Res. Commun.* **25,** 629 (1966).
25. G. M. Fuller and R. F. Doolittle, *Biochem. Biophys. Res. Commun.* **25,** 694 (1966).
26. R. Chen and R. F. Doolittle, *Proc. Natl. Acad. Sci. U. S.* **63,** 420 (1969).
27. L. Lorand, D. Chenoweth, and R. A. Domanik, *Biochem. Biophys. Res. Commun.* **37,** 219 (1969).
28. S. Matačić and A. G. Loewy, *Biochem. Biophys. Res. Commun.* **30,** 356 (1968).
29. J. J. Pisano, J. S. Finlayson, and M. P. Peyton, *Science* **160,** 892 (1968).
30. S. Matačić and A. G. Loewy, *Biochem. Biophys. Res. Commun.* **24,** 858 (1966).
31. T. Takagi and S. Iwanaga, *Biochem. Biophys. Res. Commun.* **38,** 129 (1970).
32. R. Chen and R. F. Doolittle, *Proc. Natl. Acad. Sci. U. S.* **66,** 472 (1970).
33. C. Gregoire and H. J. Tagnon, *Com. Biochem.* 435 (1962).
34. G. Duchateau and M. Florkin, *Bull. Soc. Chim. Biol.* **36,** 295 (1954).

that lysine ε-amino groups are involved in formation of cross-links during the clotting of lobster fibrinogen (*38*). These facts indicate that clotting in this case is the result of cross-linking. There is no good evidence at present that thrombin-catalyzed clotting occurs in the invertebrate phyla.

II. Thrombin—Molecular Properties

A. Assay

Specific assay methods (*39–41*) measure the fibrinogen-clotting activity of thrombin. This clotting involves both limited proteolysis of fibrinogen and the polymerization process that follows. Thrombin catalyzes only the proteolysis. An assay that depends only on this proteolytic step has been developed (*39, 42*). It measures the increase in N-terminal glycine as the fibrinopeptidyl–fibrin monomer bonds in fibrinogen are split by thrombin. Assays based on the esterase activity of thrombin, particularly against arginyl (*43*) and lysyl (*44*) substrates with a blocked α-amino group have also been developed (*43, 45–49*).

Attempts to find a specific active site titrant for thrombin (*49–52*) have not been entirely successful. It appears that *p*-nitrophenyl-*p*′-guanidinobenzoate (*53*) is the best compound available at the moment but it reacts with trypsin (*53*) and plasmin (*52*) as well as with thrombin

---

35. L. Lorand, K. Konishi, and A. Jacobsen, *Nature* **194**, 1148 (1962).
36. R. F. Doolittle and L. Lorand, *Biol. Bull.* **123**, 481 (1962).
37. L. Lorand, R. F. Doolittle, K. Konishi, and S. K. Riggs, *ABB* **102**, 171 (1963).
38. R. F. Doolittle and G. Fuller, *Biochem. Biophys. Res. Commun.* **26**, 327 (1967).
39. S. Magnusson, *Thromb. Diath. Haemorrhag.* **4**, 169 (1960).
40. W. H. Seegers and H. P. Smith, *Am. J. Physiol.* **137**, 348 (1942).
41. S. Magnusson, "Methods in Enzymology," Vol. XIXA, p. 157, 1970.
42. E. Jorpes, T. Vrethammar, B. Öhman, and B. Blombäck, *J. Pharm. Pharmacol.* **10**, 561 (1958).
43. S. Sherry and W. Troll, *JBC* **208**, 95 (1954).
44. D. T. Elmore and E. F. Curragh, *BJ* **86**, 9P (1963).
45. S. Ehrenpreis and H. A. Scheraga, *JBC* **227**, 1043 (1957).
46. R. H. Landaburu and W. H. Seegers, *Proc. Soc. Exptl. Biol. Med.* **94**, 708 (1957).
47. C. J. Martin, J. Golubow, and A. E. Axelrod, *JBC* **234**, 1718 (1959).
48. E. F. Curragh and D. T. Elmore, *BJ* **93**, 163 (1964).
49. F. J. Kézdy, L. Lorand, and K. D. Miller, *Biochemistry* **4**, 2302 (1965).
50. D. T. Elmore and J. J. Smyth, *BJ* **107**, 97 and 103 (1968).
51. J. B. Baird and D. T. Elmore, *FEBS Letters* **1**, 343 (1968).
52. T. Chase, Jr. and E. Shaw, *Biochemistry* **8**, 2212 (1969).
53. T. Chase, Jr. and E. Shaw, *Biochem. Biophys. Res. Commun.* **29**, 508 (1967).

(*51, 52*), stressing the fact that esterase assays are less specific for thrombin than the clotting assays. The clotting activity of thrombin is generally expressed as National Institutes of Health or Iowa units. One Iowa unit equals approximately 0.6 NIH units (*54–57*). The Iowa unit is based on a clotting time of 15 sec under a given set of conditions (*13, 40, 58*). The NIH unit is based on standard thrombin (*59*). It has been difficult to prepare stable fibrinogen, and this seems to be the main reason for the present discrepancy between the two units. Waugh *et al.* (*60, 61*) have made a detailed investigation of variables in the clotting-time assay.

B. Isolation and Purity

One of the problems in the purification of thrombin in good yield is that the antithrombin system in plasma is sufficiently effective to neutralize any free thrombin in a matter of minutes, even if all available prothrombin is activated. This problem has been solved by purifying prothrombin (approximately 80–400-fold) until it is separated from antithrombin (*62*) and only then activating it and further purifying the resulting thrombin. Chromatography on the carboxylate ion exchange resin XE-64 was applied by Rasmussen (*63*) for partial purification (10–50-fold) of crude thrombin (2–3 NIH units/mg). This type of chromatography is used as the final purification step for bovine (*41, 64, 65*) and human (*57*) thrombin.

Success requires that sufficiently pure starting material (150–400 NIH units/mg) is applied to the column. Cellulose phosphate has been used as an alternative chromatographic medium (*66*). A variety of procedures

54. S. Magnusson, *Arkiv Kemi* **23,** 285 (1965).
55. K. D. Miller, personal communication (1964).
56. W. H. Seegers, "Prothrombin." Harvard Univ. Press, Cambridge, Massachusetts, 1962.
57. K. D. Miller and W. H. Copeland, *Exptl. Mol. Pathol.* **4,** 431 (1965).
58. H. P. Smith, E. D. Warner, and K. M. Brinkhous, *J. Exptl. Med.* **66,** 801 (1937).
59. "Minimum Requirements of Dried Thrombin," 2nd rev. Div. Biol. Control, Natl. Inst. Health, Bethesda, Maryland, 1946.
60. D. J. Baughman, D. F. Waugh, and C. Juvkam-Wold, *Thromb. Diath. Haemorrhag.* **20,** 477 (1968).
61. D. F. Waugh, D. J. Baughman, and C. Juvkam-Wold, *Thromb. Diath. Haemorrhag.* **20,** 497 (1968).
62. J. Mellanby, *Proc. Roy. Soc.* **B107,** 221 (1930).
63. P. S. Rasmussen, *BBA* **16,** 157 (1955).
64. W. H. Seegers, W. G. Levine, and R. S. Shepard, *Can. J. Biochem. Physiol.* **36,** 603 (1958).
65. S. Magnusson, *Arkiv Kemi* **24,** 349 (1965).
66. W. H. Seegers and R. H. Landaburu, *Can. J. Biochem. Physiol.* **38,** 1405 (1960).

has been developed recently (*67–70*) for the purification of commercially available crude (approximately 50 NIH units/mg) thrombin. In an effort to convert published figures for the specific activity of the purest thrombin preparations to NIH units per milligram of protein, using available data on tyrosine content and amino acid composition, Baughman and Waugh (*67*) found values of 2000–2300 units/mg for both bovine (*64*, *65*) and human (*57*) thrombin. Although considerably higher specific activities have been reported occasionally, it is not clear to the present author that they can be obtained reproducibly. Bovine thrombin has been observed to give a single symmetrical peak in the ultracentrifuge (*64*, *65*). It behaved as a homogeneous protein on starch gel electrophoresis (*65*) in four different buffer systems at pH 6.1–9.0, even in the presence of 7 *M* urea, as well as on disc electrophoresis on polyacrylamide gel extending the range to pH 5.1 (*71*). Immunodiffusion on Ouchterlony plates (against antibovine thrombin serum raised in rabbits) also indicated homogeneity (*72*).

Another aspect of thrombin purity is that minute contaminating quantities of other clotting factors and fibrinolytic factors may be present which would escape detection by the usual criteria of homogeneity. The caseinolytic activity before and after activation with urokinase (*73*) is so small that this type of thrombin preparation (*65*, *41*) can be used as "plasminogen-free" thrombin in fibrinolytic test systems. This indicates a very low level of either contaminating or intrinsic plasmin–plasminogen activity of the order of 0.01–0.1%. The absence of prothrombin and factors X, IX, and VII cannot be confirmed by the usual clotting-time assays since they are all short-circuited by thrombin. However, their presence as contaminants is unlikely since they behave quite differently from thrombin in the XE-64 chromatography system.

C. Physical Properties and Molecular Weight

The mobility of thrombin on starch gel and paper electrophoresis is similar to that of γ-globulins (*65*). The isoelectric point has been determined by free-boundary electrophoresis to be pH 5.3 and 5.75 at

67. D. J. Baughman and D. F. Waugh, *JBC* **242,** 5252 (1967).
68. E. T. Yin and S. Wessler, *JBC* **243,** 112 (1968).
69. T. Chase, Jr. and E. Shaw, *Biochemistry* (1970) (in press).
70. K. G. Mann and C. W. Batt, *JBC* **244,** 6555 (1969).
71. S. Magnusson, unpublished results (1966).
72. B. Magnusson, S. Magnusson, and K. D. Miller, unpublished results (1966).
73. Analysis performed by Dr. Kurt Bergström, Karolinska Institutet, Stockholm, Sweden, 1964.

ionic strengths 0.1 and 0.2, respectively, (*74*) and by paper electrophoresis to be pH 5.6 (*75*). The sedimentation and diffusion constants have been determined (*76*) as $s_{20,w} = 3.76$ S, $D_{20,w} = 8.70 \times 10^{-7}$ and 8.82×10^{-7} cm^2 sec^{-1} (calculated average for the two methods used gives 8.76×10^{-7} cm^2 sec^{-1}), respectively. The partial specific volume was measured and found to be 0.69 ml/g (*76*). These data were used to calculate a molecular weight of 33,700 for bovine thrombin. A molecular weight of 40,000 obtained by ultracentrifugation and gel filtration has also been reported (*77*). Another report of gel filtration on a Sephadex G-100 column indicates a molecular weight of 36,000 (*67*). Harmison *et al.* (*76*) have reported the following hydrodynamic data: frictional ratio 1.16, intrinsic viscosity 0.0376, and axial ratio 2.8 (length 84 Å, diameter 30 Å). A specific activity of 1160 ± 130 NIH units/absorbance unit in the ultraviolet (280 nm, 1 cm light path) at pH 7.0 has been reported (*67*).

D. Amino Acid and Carbohydrate Composition

The amino acid compositions of bovine thrombin of high specific activity (*78–81*) and of undefined specific activity (*82*, *83*) have been published. All of these analyses except one (*83*) are shown in Table I. Human thrombin of high specific activity (*57*) has been analyzed (*84*) and a comparison of the data with those from bovine thrombin showed only very small differences (*72*). Bovine thrombin (*65*) also contains 1.7% glucosamine (*85*), 1.7–1.8% sialic acid (*86*), 0.61% galactose, and 0.95% mannose (*87*). Thus, the total carbohydrate content is about 5.1%. Human thrombin (*57*) was found to contain 1.7% sialic acid (*72*).

74. W. H. Seegers, C. R. Harmison, N. Ivanovic, and D. L. Heene, *Thromb. Diath. Haemorrhag.* **15**, 343 (1966).

75. W. G. Levine and O. W. Neuhaus, *Proc. Soc. Exptl. Biol. Med.* **101**, 64 (1959).

76. C. R. Harmison, R. H. Landaburu, and W. H. Seegers, *JBC* **236**, 1693 (1961).

77. D. J. Winzor and H. A. Scheraga, *J. Phys. Chem.* **68**, 338 (1964).

78. S. Magnusson and B. S. Hartley, unpublished results (1967).

79. W. H. Seegers, L. McCoy, R. K. Kipfer, and G. Murano, *ABB* **128**, 194 (1968).

80. K. D. Miller, R. K. Brown, G. Casillas, and W. H. Seegers, *Thromb. Diath. Haemorrhag.* **3**, 362 (1959).

81. S. Magnusson and T. Hofmann, *Can. J. Biochem.* **48**, 432 (1970).

82. E. E. Schrier, C. A. Broomfield, and H. A. Scheraga, *ABB* **99**, Suppl. 1, 309 (1962).

83. K. Laki and J. A. Gladner, *Physiol. Rev.* **44**, 127 (1964).

84. K. D. Miller, personal communication (1965).

85. Minimum figure obtained on amino acid analyzer with a 20-hr hydrolyzate. S. Magnusson, J. R. Brown, and B. S. Hartley, unpublished results (1964).

TABLE I
COMPOSITION OF BOVINE THROMBIN
(g/100 g thrombin)

Residue	*a*	*b*	*c*	*d*	*e*	*f*
Alanine	2.51	2.63	3.05	2.68	3.00	3.03
Arginine	8.83	8.89	10.24	9.01	9.34	9.61
Aspartic acid + asparagine	8.10	8.07	10.12	8.17	11.34	8.72
Half-cystine	2.20	2.67 2.21[g]	1.67	2.04[g]	2.15	2.90
Glutamic acid + glutamine	10.86	11.58	12.86	11.71	13.17	12.23
Glycine	3.62	4.11	4.37	3.88	4.21	4.05
Histidine	2.18	2.38	2.74	2.36	2.41	2.60
Isoleucine	3.45	4.25	4.52	4.61	4.38	4.28
Leucine	7.70	8.80	9.38	8.11	9.55	8.57
Lysine	7.93	7.93	9.05	7.30	8.11	8.49
Methionine	1.61	1.80 1.57[h]	1.83	1.74	1.85	1.86
Phenylalanine	4.22	4.43	5.49	4.81	5.18	5.57
Proline	3.89	3.88	5.51	4.63	5.47	4.73
Serine	3.51	3.55	3.29	3.48	4.29	4.12
Threonine	3.52	3.57	3.77	3.54	4.62	3.83
Tryptophan[i]	3.24	2.54	5.95	3.67	3.93	5.29
Tyrosine	4.83	5.77	5.16	4.44	5.74	5.41
Valine	4.55	5.29	5.74	6.26	5.58	5.16
Total amino acid	87.83	98.77	104.74	92.44	104.32	100.45
Galactosamine	1.08					0.46
Glucosamine			1.68			0.69
N-Acetylneuraminic acid						1.7
Total carbohydrate		1.80				5.02

[a] K. D. Miller, R. K. Brown, G. Casillas, and W. H. Seegers, *Thromb. Diath. Haemorrhag.* **3,** 362 (1959).

[b] E. E. Schrier, C. A. Broomfield, and H. A. Scheraga, *ABB* **99,** Suppl. 1, 309 (1962).

[c] S. Magnusson and T. Hofmann, *Can. J. Biochem.* **48,** 432 (1970).

[d] S. Magnusson and B. S. Hartley, unpublished results (1967).

[e] W. H. Seegers, L. McCoy, R. K. Kipfer, and G. Murano, *ABB* **128,** 194 (1968). The "3.7 S thrombin."

[f] Same reference as given in footnote *e*. The "3.2 S thrombin."

[g] Half-cystine determined as cysteic acid.

[h] Methionine determined as methionine sulfone.

[i] Determined spectrophotometrically.

E. N-Terminal and C-Terminal Amino Acids

Miller first reported finding N-terminal threonine (*88*) by the DNP method. Quantitative analysis of PTH-amino acids showed (*65*) both N-terminal isoleucine (1 mole/30,000 g) and N-terminal threonine (1 mole/60,000 g). The presence of both isoleucine and threonine has recently been confirmed (*79*). One mole of each N-terminal was found in 25,000–30,000 g of "3.2 S thrombin" and in 34,000 g of "3.7 S thrombin," respectively. Other reports of N-terminal glutamic acid (*89, 90*) and leucine (*91*) have not been confirmed and may have resulted from misidentification of threonine and isoleucine, respectively. In human citrate thrombin (*92*) 1 mole of N-terminal isoleucine (or possibly leucine) per 26,000–32,000 g has been found (*92*).

C-terminal amino acid analysis by degradation with DFP-treated carboxypeptidases A and B gives evidence for 1 mole of arginine per 35,000 g of thrombin (*93*). A report of C-terminal tyrosine (*90*) is difficult to interpret since the ammonium thiocyanate method (*94, 95*) is not a widely recognized method for the determination of C-terminals.

F. Amino Acid Sequences of A Chain and B Chain (*96–101*)

The first sequence data were obtained from partial acid hydrolysates of DFP-inhibited pig thrombin (*102*) and showed the sequence -Asp-Ser-

---

86. B. Steele and S. Magnusson, unpublished results (1964). Expressed as *N*-acetylneuraminic acid.

87. Gas chromatographic analysis by Dr. R. J. Winzler, State University of New York, Buffalo, New York, 1967, on material supplied by the author.

88. K. D. Miller, *Vox Sanguinis* **3,** 63 (1958).

89. W. H. Seegers, G. Casillas, R. S. Shepard, W. R. Thomas, and P. Halick, *Can. J. Biochem. Physiol.* **37,** 775 (1959).

90. W. R. Thomas and W. H. Seegers, *BBA* **42,** 556 (1960).

91. R. H. Landaburu, C. R. Harmison, and W. H. Seegers, *Thromb. Diath. Haemorrhag.* **6,** 424 (1961).

92. S. Magnusson, *Arkiv Kemi* **24,** 375 (1965).

93. S. Magnusson and B. Steele, *Arkiv Kemi* **24,** 359 (1965).

94. P. Schlack and W. Kumpf, *Z. Physiol. Chem.* **154,** 125 (1926).

95. J. Tibbs, *Nature* **168,** 910 (1951).

96. S. Magnusson, *BJ* **110,** 25P (1968).

97. B. S. Hartley, *Phil. Trans. Roy. Soc. London* **B257,** 77 (1970).

98. S. Magnusson, *in* "Human Blood Coagulation" (H. C. Hemker, E. A. Loeliger, and J. J. Veltkamp, eds.), p. 18. Leiden Univ. Press, Leiden, Netherlands, 1969.

99. S. Magnusson, *Thromb. Diath. Haemorrhag.* Suppl. 38, 97 (1970).

Gly- at the active site. Edman degradation of biothrombin, citrate thrombin, biothrombin in 7 *M* urea or 2 *M* guanidinium hydrochloride, and of performic acid–oxidized biothrombin led to the conclusion (*92*) that bovine thrombin has two N-terminal sequences, one Thr- and one Ile-Val-Glu-Gly- sequence. Human citrate thrombin (*92*) had the N-terminal sequence Ile-Val-Gly-Gly-, but the amount used in the analysis was not large enough to exclude a Thr- chain. Proof that bovine thrombin contains two polypeptide chains was obtained by separation of the chains after breaking the disulfide bonds by oxidation in performic acid or reduction in mercaptoethanol (*78*). The A chain was found to consist of 49 amino acid residues and no carbohydrate. Its sequence was determined by overlapping the sequences of peptides obtained by chymotryptic and partial or complete tryptic degradation and is as follows (*96*, *103*):

Thr-Ser-Glu-Asn-His-Phe-Glu-Pro-Phe-Phe-Asn-Glu-Lys-Thr-Phe-Gly-Ala-Gly-
Glu-Ala-Asp-Cys-Gly-Leu-Arg-Pro-Leu-Phe-Glu-Lys-Lys-Gln-Val-Gln-Asp-Glu-
Thr-Gln-Lys-Glu-Leu-Phe-Glu-Ser-Tyr-Ile-Glu-Gly-Arg

The A chain thus accounts for the N-terminal threonine and the C-terminal arginine of thrombin. This chain is quite acidic with an overall negative charge of about 4.5 and a mobility of 0.35 (relative to aspartic acid) on high-voltage paper electrophoresis at pH 6.5. Its content of glutamic acid and glutamine (12 residues) and of phenylalanine (6 residues) is high, but it lacks tryptophan and methionine residues entirely.

Evidence for most of the amino acid sequence of the B chain has been obtained from tryptic digests of reduced, amino-ethylated B chain and of performic acid–oxidized thrombin, from a peptic digest of native thrombin, and from cyanogen bromide degradation of native thrombin (*78*). For the two tryptic digests 15 and 5 μmoles of thrombin, respectively, were used. Partial separation of the resulting peptides was achieved by chromatography on Dowex-50 using a pH and concentration gradient of pyridine acetate from 0.1 *M* to 2.0 *M* and from pH 3.2 to pH 5.0 followed by a step of 5.0 *M* pyridine acetate pH 6.0. The 22 cuts from the column effluent were finally purified by high-voltage paper electrophoresis and paper chromatography.

Quite a few tryptophan-containing and generally hydrophobic peptides

---

100. S. Magnusson, *Proc. IUB/IUBS Intern. Symp. Struct.-Funct. Relationships Proteol. Enzymes, 1969*, p. 138. Munksgaard, Copenhagen, 1970.

101. S. Magnusson, *BJ* **115**, 2P (1969).

102. J. A. Gladner and K. Laki, *J. Am. Chem. Soc.* **80**, 1263 (1958).

103. S. Magnusson, E. Merler, J. Wootton, and B. S. Hartley, unpublished results (1967).

were not recovered from the column effluent. To avoid this type of loss the peptides from the peptic digest of 5 μmoles of thrombin were separated in 20 m*M* acetic acid on a Sephadex G-25 column (11 cuts taken) before final purification by paper methods. Degradation with cyanogen bromide of 9 μmoles of thrombin produced six fragments from the B chain plus the intact A chain.

Good separation was obtained (*104*) in two steps on a Sephadex G-50 column in 20 m*M* acetic acid. In the first step the digest was separated with the disulfide bridges left intact. Four fractions were obtained. The first one contained the A chain and fragments B-3, B-4, and B-6 from the B chain and had N-terminal Thr-, Leu-, Phe-, and Gly-. The second fraction was fragment B-2 with 67 amino acid residues and N-terminal Leu-. The third was fragment B-1 with 17 residues and N-terminal Ile-. The fourth was fragment B-5 with 11 residues and N-terminal Lys-. Carbohydrate was found only in fragment B-2. In the second separation step on Sephadex G-50 the fraction containing fragments A, B-3, B-4, and B-6 was first reduced and carboxymethylated and then applied to the column. The following fractions were separated in order of elution: B-3 with N-terminal Leu- and 102 amino acid residues, the A chain with its N-terminal Thr- and 49 residues, B-4 with N-terminal Phe- and 26 residues, and finally B-6 with N-terminal Gly- and 37 residues. The sequence data now available are consistent with the following preliminary B-chain sequence (*78, 96–101*):

ILE-Val-Glu-Gly-Gln-Asp-Ala-Glu-Val-Gly-Leu-Ser-Pro-Trp-Gln-Val-*Met*-Leu-Phe-Arg-Lys-Ser-Pro-Gln-Glu-Leu-Leu-Cys-Gly-Ala-Ser-Leu-Ile-Ser-Asp-Arg-Trp-Val-Leu-Thr-Ala-Ala-HIS-Cys-Leu-Leu-Tyr-Pro-(Trp,Pro,Asx,Lys,Asx-CBH,Phe)-Thr-Val-Asx-Asx-Leu-Leu-(Ser,Val,Val,Glx,Trp)-Arg,Ile-Gly-Lys,His-Ser-Arg-Thr-Arg-Tyr-Glu-Arg-Lys-Val-Glu-Lys-Ile-Ser-*Met*-Leu-Asp-Lys-Ile-Tyr-Ile-His-Pro-Arg-Tyr-Asn-Trp-Lys-Glu-Asn-Leu-Asp-Arg-ASP-Ile-Ala-Leu-Leu-Lys-Leu-Lys-Arg-Pro-Ile-Glu-Leu-Ser-Asp-Tyr-Ile-His-Pro-Val-Cys-Leu-Pro-Lys,Ala-Ser-Thr-Arg,Leu-Leu-His-Ala-Gly-Phe-Lys-Gly-Arg-Val-Thr-Gly-Trp-Gly-Asn-Arg,Thr-Thr-Ser-Val-Ala-Glu-Val-Gln-Pro-Ser-Val-Leu,Gln-Thr-Ala-Ala-Lys-Leu,Gln-Val-Val-Asn-Leu-Pro-Leu,Val-Glu-Arg-Pro-Val-Cys-Lys,Arg-Ile-Arg-Ile-Thr-Asx-Asx-*Met*-Phe-Cys-Ala-Gly-Tyr-Lys-Pro-Gly-Glu-Gly-Lys-Arg-Gly-ASP-Ala-Cys-Glu-Gly-ASP-SER-Gly-Gly-Pro-Phe-Val-*Met*-Lys-Ser-Pro-Tyr-Asn-Asn-Arg-Trp-Tyr-Gln-*Met*-Gly-Ile-Val-SER-Trp-GLY-Glu-Gly-Cys-Asp-Arg-Asn-Gly-Lys-Tyr-GLY-Phe-Tyr-Thr-His-Val-Phe-Arg-Lys-Leu-Lys-Trp-Ile-Gln-Lys-Val-Ile-Asp-Arg-Leu-Gly-Ser (*105*)

The N-terminal isoleucine and the N-terminal sequence Ile-Val-Glu-Gly- as well as the active site sequence -Asp-Ser-Gly- are thus accounted

104. S. Magnusson, K. Simons, and B. S. Hartley, unpublished results (1968).

105. Residues in capital letters are homologous to ILE 16, HIS 57, ASP 102, SER (ASP in thrombin) 189, ASP 194, SER 195, SER 214, GLY 216 and GLY 226 in chymotrypsin. CBH stands for carbohydrate. *Met*—residues indicate points of cleavage by cyanogen bromide.

for by the B chain. Conventional C-terminal methods have failed so far to demonstrate the C-terminal -Ser in the B chain. The total number (309) of amino acid residues in one A chain (49 residues) plus one B chain (260 residues) fits reasonably well the amino acid composition of thrombin as given in column *d* of Table I. It also indicates a molecular weight of about 40,000.

G. Sequence Homology

The A chain shows no obvious sequence homology to the pancreatic serine proteinases. The sequence of the B chain, on the other hand, is so clearly homologous to those of chymotrypsins A (*106*) and B (*107*), trypsin (*108*), and elastase (*106, 109*) that one cannot escape the conclusion that thrombin, which is synthesized in the hepatic cells of the liver, and the pancreatic serine proteinases have all evolved from a common ancestor. Recent evidence of the complete amino acid sequence of the α-lytic proteinase from *Sorangium* (*110*), and for a sequence of 18 residues, including the active center serine, in human plasmin (*111*) clearly shows that these two serine proteinases are also homologous with thrombin. More surprisingly, it has been found (*112*) that the C-terminal chymotryptic octapeptide (-Val-Glx-Lys-Thr-Ile-Ala-Glu-Asn) from human haptoglobin β chains is also homologous with the serine proteinases. Since this β chain is the one responsible for binding hemoglobin, the interesting question arises whether the β chain is perhaps a rudimentary serine proteinase which has retained its binding specificity but lost its catalytic activity during evolution.

H. Disulfide Bridges

Analysis of bovine thrombin for -SH groups and -S-S- bridges produced evidence for 2.8 -S-S- bridges per 40,000 g (*113*). Sequence data have given evidence for 8 half-cystines, 7 in the B chain and one in the A chain. One disulfide bridge has been established (*78*) which is internal

106. B. S. Hartley, J. R. Brown, D. L. Kauffman, and L. B. Smillie, *Nature* **207,** 1157 (1965).

107. L. B. Smillie, A. Furka, N. Nagabhushan, K. J. Stevenson, and C. O. Parkes, *Nature* **218,** 343 (1968).

108. H. Neurath, K. A. Walsh, and W. P. Winter, *Science* **158,** 1638 (1967).

109. D. M. Shotton and B. S. Hartley, *Nature* **225,** 802 (1970).

110. M. O. J. Olson, N. Nagabhushan, M. Dzwiniel, L. B. Smillie, and D. R. Whitaker, *Nature* **228,** 438 (1970).

111. W. R. Groskopf, L. Summaria, and K. C. Robbins, *JBC* **244,** 3590 (1969).

112. D. R. Barnett, T.-H. Lee, and B. H. Bowman, *Nature* **225,** 938 (1970).

113. J. R. Carter and E. D. Warner, *Am. J. Physiol.* **184,** 195 (1956).

in fragment B-2. This bridge is homologous with the "histidine-loop" bridge (*97*) (Cys 42 to Cys 58) (*114*) in chymotrypsin. Since fragments A, B-3, B-4, and B-6 are joined together in the cyanogen bromide digest and B-3 and B-4 are the only ones to contain two half-cystines, one of the remaining three bridges must logically connect B-3 with B-4.

Sequence homology around the half-cystines indicates that the B-3 to B-4 bridge is the "methionine loop" bridge (Cys 168 to Cys 182 in chymotrypsin) and that thrombin also has a "serine loop" bridge (Cys 191 to Cys 220 in chymotrypsin) connecting B-4 and B-6. This leaves the half-cystine that is homologous to Cys 136 in chymotrypsin to form the A to B interchain disulfide bridge to Cys A22 in the A chain. Thus, only one (Cys 136 to Cys 201) of the four disulfide bridges common to the pancreatic serine proteinases is conspicuously absent in thrombin, since Cys 201 has been replaced by Met 201. A tryptic peptide containing the DFP-reactive serine residue has recently been isolated from reduced, amino-ethylated human DIP-thrombin and its amino acid composition determined (*115*). Since the composition is virtually identical to that of the corresponding bovine peptide it seems a fair guess that the sequence 191–202 is identical in the two species of thrombin. Unlike thrombin, plasmin conforms entirely with the "pancreatic pattern" in having both Cys 191 to Cys 201 (*111*).

I. Homology and Tertiary Structure

The tertiary structure of thrombin is not yet known, and no crystals have been reported. The tertiary structure of the B chain of thrombin however seems to be predictable to some extent from present knowledge of the thrombin sequence and of the complete sequences and tertiary structures of α-chymotrypsin (*116–119*) and elastase (*120*, *121*). Such

114. An alignment that displays the homologies in the amino acid sequences of the pancreatic serine proteinases and the B chain of bovine thrombin appears in Chapter 10 on elastase by B. S. Hartley and D. M. Shotton, this volume.

115. J. W. Fenton, personal communication (1969).

116. B. W. Matthews, P. B. Sigler, R. Henderson, and D. M. Blow, *Nature* **214**, 652 (1967).

117. P. B. Sigler, D. M. Blow, B. W. Matthews, and R. Henderson, *JMB* **35**, 143 (1968).

118. J. J. Birktoft, D. M. Blow, R. Henderson, and T. A. Steitz, *Phil. Trans. Roy. Soc. London* **B256**, 5 (1970).

119. T. A. Steitz, R. Henderson, and D. M. Blow, *JMB* **46**, 337 (1969).

120. H. C. Watson, D. M. Shotton, J. M. Cox, and H. Muirhead, *Nature* **225**, 806 (1970); D. M. Shotton and H. C. Watson, *ibid.* p. 811.

121. D. M. Shotton and H. C. Watson, *Phil. Trans. Roy. Soc. London* **B256**, 49 (1970).

predictions proved to be reasonably valid in the case of elastase. A hypothetical model of the tertiary structure of elastase had been constructed by fitting its amino acid sequence to the electron density map of chymotrypsin (*97*). When the real tertiary structure of elastase became available (*121*), the predicted structure turned out to have been essentially correct. On this basis one would predict with some confidence that thrombin has an "activation salt bridge" like the one between ILE 16 and ASP 194 in α-chymotrypsin.

Because of the pronounced sequence homology in the areas concerned, one would also predict that thrombin has the "charge relay system" involving one hydrogen bond connecting the β-hydroxyl group of SER 195 with one of the two imidazole nitrogens of HIS 57 and a second hydrogen bond connecting the β-hydroxyl group of ASP 102 with the other imidazole nitrogen of HIS 57 which was recently proposed (*122*) to be the structural basis for the catalytic mechanism of chymotrypsin. Furthermore, there is a high degree of homology involving GLY 216 and GLY 226 suggesting that a "tosyl hole," to which the side chain specificity of chymotrypsin has been attributed (*118*), is also present in thrombin. Residue 189 at the bottom of this hole is SER in the two chymotrypsins and ASP in thrombin (*98*) and trypsin (*97*). It is also obvious that at this stage no prediction can be made about the tertiary structure or functional importance of those parts of the thrombin sequence that are not homologous including the entire A chain and the two large insertions in the B chain, one of 7 residues at position 148–149 and the other of 14 residues and the carbohydrate at position 65–66.

III. Thrombin—Catalytic Properties

A. General

Available kinetic data on thrombin are compatible with the three-step catalytic reaction sequence [Eq. (1)] that was originally formulated for chymotrypsin (*123*, *124*) and is now widely recognized as applying to serine proteinases generally.

$$\mathrm{E} + \mathrm{RX} \overset{K_m}{\rightleftharpoons} \mathrm{E}\cdot\mathrm{RX} \overset{k_2}{\rightleftharpoons} \underset{\underset{\mathrm{X}}{+}}{\mathrm{ER}} \overset{k_3}{\rightleftharpoons} \mathrm{E} + \mathrm{R} \qquad (1)$$

In the first step an enzyme-substrate complex (E·RX) is formed. In the

122. D. M. Blow, J. J. Birktoft, and B. S. Hartley, *Nature* **221**, 337 (1969).
123. H. Gutfreund and J. M. Sturtevant, *Proc. Natl. Acad. Sci. U. S.* **42**, 719 (1956).
124. H. Gutfreund and J. M. Sturtevant, *BJ* **63**, 656 (1956).

second step the acyl enzyme (ER) is formed and the first product, the "leaving group," (X) is released. In the final step the enzyme (E) is regenerated, and the second product, a carboxylic acid, is released.

B. Synthetic Substrates

Synthetic substrates for serine proteinases can be generally classified as either esters or amides of carboxylic acids. Only one amide substrate for thrombin has been reported (*43*), namely, benzoyl-L-argininamide (BAA). Since its turnover rate was 3–4 orders of magnitude less than that of *p*-toluenesulfonyl-L-arginine methyl ester (TAME), and the thrombin preparation used was certainly not pure, it is doubtful whether BAA is really a substrate. The possibility that other amides are substrates for thrombin does not seem to have been well investigated. Thrombin shows the same side chain specificity as trypsin and plasmin in its preference for arginine (*43*) and lysine (*125*) ester substrates. The first such substrate, TAME, was used in a thrombin esterase assay (*43*).

In this assay and modifications of it the amount of product is titrated after incubation for 15 min or longer. With this approach K_m values for TAME in the region of $4\text{–}6 \times 10^{-3}\ M$ have been obtained (*43, 126–128*). Other substrates investigated by the same technique (*126*) are benzoyl-L-arginine methyl ester (BAME), *p*-toluenesulfonyl-L-lysine methyl ester (TLME), acetyl-L-lysine methyl ester (ALME), acetyl-L-arginine methyl ester (AAME), carbobenzoxy-L-arginine methyl ester, and carbobenzoxy-L-lysine methyl ester with K_m values of 1.3 and 1.2, 25 and 5.3, 40 and 17, 21 and 10, 5.2 and 13, 14 and $13 \times 10^{-3}\ M$, respectively, with bovine and with human thrombin, respectively. Other ester substrates that are split by thrombin at rates which, from the meager data available, seem to be about 20–800 times slower than those of TAME and TLME are L-lysine ethyl ester (*45*), L-arginine methyl ester (*129*), benzoyl-L-histidine methyl ester (*129*), L-tryptophan ethyl ester (*129*), L-histidine methyl ester (*129, 130*), L-lysine methyl ester (*129*), L-tyrosine ethyl ester (*129*), and L-phenylalanine methyl ester (*129, 131*).

Seegers *et al.* (*79*) claimed to have split off and separated a 75-residue

125. D. T. Elmore and E. F. Curragh, *BJ* **86**, 9P (1963).
126. S. Sherry, N. Alkjaersig, and A. P. Fletcher, *Am. J. Physiol.* **209**, 577 (1965).
127. E. Ronwin, *BBA* **33**, 326 (1959).
128. G. F. Lanchantin, C. A. Presant, D. W. Hart, and J. A. Friedmann, *Thromb. Diath. Haemorrhag.* **14**, 159 (1965).
129. E. R. Cole, J. L. Koppel, and J. H. Olwin, *Nature* **213**, 405 (1967).
130. E. R. Cole, *Thromb. Diath. Haemorrhag.* **19**, 321 and 334 (1968).
131. E. R. Cole, J. L. Koppel, and J. H. Olwin, *Can. J. Biochem.* **44**, 1052 (1966).

fragment, which was not recovered, however, from bovine thrombin by rechromatography of the original 3.7 S thrombin. They reported K_m values at pH 8.0 and 22° for TAME of $2.97 \times 10^{-4} M$ and $9.5 \times 10^{-5} M$ for the 3.7 S and the 3.2 S thrombins respectively, with k_{cat} values of 9.24 and 46.5×10^{-4} mole/min/mg N. Using more sophisticated kinetic methods, Curragh and Elmore (*132*) found K_m values at pH 8.4 and 25° for TAME and TLME of 1.61 and $3.21 \times 10^{-5} M$, respectively. They attributed these relatively lower K_m values to the fact that previous investigators failed to recognize substrate activation of the enzyme. Ester hydrolysis is zero order in substrate and first order in thrombin. Curragh and Elmore (*132*) also found *p*-toluenesulfonyl-L-ornithine to be a moderately good substrate and *p*-toluenesulfonyl-L-homoarginine an inhibitor. The fact that different benzoyl-L-arginine esters are hydrolyzed at the same rate was taken to indicate that deacylation is rate determining in this case (*132*). Different esters of *p*-toluenesulfonyl-L-arginine, however, are hydrolyzed at different rates, indicating that the rate of acylation contributes significantly to k_{cat} (*132*). Increasing ionic strength was found to inhibit (*132*) the hydrolysis of tosyl but not that of benzoyl esters and, therefore, presumably acylation but not deacylation. Sodium cholate and glycocholate were found to accelerate acylation and decelerate deacylation (*132*). The pH dependence of k_{cat} for tosyl-L-arginine esters indicated that dissociation of a group with pK_a about 6.6 is required (*49, 132*) for hydrolysis.

Typical "burst behavior," indicating that an acyl-thrombin intermediate is formed, has been demonstrated (*133*) with *p*-nitrophenyl esters. At pH 8.0 and 30° the K_m values found (*133*) for carbobenzoxy-L-tyrosine-*p*-nitrophenyl ester ($\Gamma/2$ 0.001–0.02) were $8.6 \times 10^{-5} M$ and $110 \times 10^{-5} M$, respectively. Relative rate constants for the *p*-nitrophenyl esters of acetic acid, carbobenzoxy-L-tyrosine, carbobenzoxy-L-phenylalanine and for TAME (at 30°, pH 8.0, $\Gamma/2$ 0.34 in aqueous 3.3% Me_2CO, 13.3% iso-PrOH) were found (*134*) to be 1, 2500, 1500, and 5000, respectively. The first attempt (*49*) to use the *p*-nitrophenol burst for determining the operational molarity of thrombin solutions led to determination of k_{cat} (0.22, 0.72, and 16.5 sec^{-1}) and K_m (7.2, 21, and $56 \times 10^{-6} M$) values at 25° for the *p*-nitrophenyl esters of CBZ-L-tyrosine and CBZ-L-lysine at pH 5.02 and for BAEE at pH 8.75, respectively. *p*-Nitrophenyl-*p'*-guanidinobenzoate (*p*-NPGB), which was developed as a burst titrant for trypsin (*53*), reacts rapidly with thrombin (*51*). The burst of *p*-nitrophenol is independent of substrate concentration

132. E. F. Curragh and D. T. Elmore, *BJ* **93**, 163 (1964).
133. C. J. Martin, J. Golubow, and A. E. Axelrod, *JBC* **234**, 1718 (1959).
134. L. Lorand, W. T. Brannen, Jr., and N. G. Rule, *ABB* **96**, 147 (1962).

TABLE II
KINETIC CONSTANTS FOR THROMBIN REACTION[a]

	EGB[b]	p-NPGB	m-NPGB
k_2	1.19×10^{-3} sec^{-1}	0.13 sec^{-1}	0.12 sec^{-1}
K_m	7.3×10^{-3} M	3.95×10^{-6} M	1.64×10^{-5} M
k_3	98×10^{-5} sec^{-1}		5.48×10^{-3} sec^{-1}

[a] T. Chase, Jr. and E. Shaw, *Biochemistry* **8,** 2212 (1969).
[b] EGB = ethyl-*p*-guanidinobenzoate.

from 16.6 to 82.0 μM and proportional to enzyme concentration between 0.8 and 7.6 μM thrombin. The first-order rate constant was 4.05 sec^{-1} at pH 8.3 and 25°. The kinetics of thrombin, trypsin, and plasmin with *p*-NPGB and ethyl *p*-guanidinobenzoate have been extensively characterized (*52*). Thrombin is acylated more slowly and deacylated at a faster rate than trypsin and plasmin (*52*). Thrombin is also an efficient esterase for *p*-nitrophenyl-*m*′-guanidinobenzoate (*m*-NPGB) which is not a good substrate for trypsin and plasmin (*52*). The kinetic constants obtained with thrombin (*52*) are shown in Table II.

C. Protein and Polypeptide Substrates

1. *General*

For most proteolytic enzymes a list of protein substrates would be virtually a catalog of all known proteins. The most fascinating property of thrombin, however, is that it is highly selective in its choice of substrate. Its natural substrate is fibrinogen and as far as other proteins and peptides are concerned strict chemical evidence for proteolysis by thrombin, showing that new N- or C-terminal residues appear as a result of thrombin action, has only been produced for the two polypeptide hormones secretin (*135*) and cholecystokinin-pancreozymin (*136*), and for thrombosthenin M (*137*).

2. *Fibrinogen* (*138*)

Evidence that fibrinogen and fibrin have different N-terminal amino acids (*139*) and that fibrinopeptides are released (*140*, *141*) proved that

135. V. Mutt, S. Magnusson, J. E. Jorpes, and E. Dahl, *Biochemistry* **4,** 2358 (1965).
136. V. Mutt and J. E. Jorpes, *European J. Biochem.* **6,** 156 (1968).
137. I. Cohen, Z. Bohak, A. deVries, and E. Katchalski, *European J. Biochem.* **10,** 388 (1969).

the conversion involves limited proteolysis of fibrinogen. Quantitative study of the N-terminal pattern during conversion (*142*) showed that the fibrinogen dimer (MW 340,000) is composed of three pairs of nonidentical chains now called α(A), β(B), and γ. The amino acid sequences of the fibrinopeptides A (13–19 residues) and B (9–21 residues) from about 30 different vertebrate species have now been determined (*143–156*). In all these cases the fibrinopeptidyl–fibrin monomer bonds split by thrombin are arginylglycyl bonds. The N-terminal sequence of the α chain, which appears as a result of thrombin action, has recently been investigated (*157*). In all of the 10 or so species so far investigated (all mammals) the sequence Gly-Pro-Arg- was found (*157*). One exception has been found, namely, the abnormal human fibrinogen$_{\text{Detroit}}$, which has Gly-Pro-Ser- (*158*). In this abnormal fibrinogen the A peptide is released normally by thrombin, but fibrin monomer polymerization is impaired.

Although the fibrinopeptide A sequences from different species are quite similar, only the C-terminal arginine is constant in all species. However, all species except one have a phenylalanine residue nine sequence steps removed from the C-terminal arginine. This phenylalanine

138. B. Blombäck, *in* "Blood Clotting Enzymology" (W. H. Seegers, ed.), p. 143. Academic Press, New York, 1967.

139. K. Bailey, F. R. Bettelheim, L. Lorand, and W. R. Middlebrook, *Nature* **167,** 233 (1951).

140. L. Lorand, *BJ* **52,** 200 (1952).

141. F. R. Bettelheim and K. Bailey, *BBA* **9,** 578 (1952).

142. B. Blombäck and I. Yamashina, *Arkiv Kemi* **12,** 299 (1958).

143. B. Blombäck, J. Sjöquist, and P. Wallén, *Acta Chem. Scand.* **13,** 819 (1959).

144. J. Sjöquist, B. Blombäck, and P. Wallén, *Arkiv Kemi* **16,** 425 (1960).

145. B. Blombäck, M. Blombäck, P. Edman, and B. Hessel, *Nature* **193,** 883 (1962).

146. J. E. Folk, J. A. Gladner, and Y. Levin, *JBC* **234,** 2317 (1959).

147. B. Blombäck and R. F. Doolittle, *Acta Chem. Scand.* **17,** 1819 (1963).

148. R. F. Doolittle and B. Blombäck, *Nature* **202,** 147 (1964).

149. B. Blombäck, M. Blombäck, N. J. Gröndahl, and E. Holmberg, *Arkiv Kemi* **25,** 411 (1966).

150. B. Blombäck, M. Blombäck, P. Edman, and B. Hessel, *BBA* **115,** 371 (1966).

151. R. F. Doolittle, D. Schubert, and S. A. Schwartz, *ABB* **118,** 456 (1967).

152. G. A. Mross and R. F. Doolittle, *ABB* **122,** 674 (1967).

153. B. Blombäck, M. Blombäck, N. J. Gröndahl, C. Guthrie, and M. Hinton, *Acta Chem. Scand.* **19,** 1788 (1965).

154. R. F. Doolittle, C. Glasgow, and G. A. Mross, *BBA* **175,** 217 (1969).

155. S. Iwanaga, B. Blombäck, N. J. Gröndahl, B. Hessel, and P. Wallén, *BBA* **160,** 280 (1968).

156. S. Iwanaga, P. Wallén, N. J. Gröndahl, A. Henschen, and B. Blombäck, *European J. Biochem.* **8,** 189 (1969).

157. B. Blombäck, personal communication (1969).

158. M. Blombäck, B. Blombäck, E. F. Mammen, and A. S. Prasad, *Nature* **218,** 314 (1968).

residue does not occur in the fibrinopeptides B. One common feature of all fibrinopeptides is the relative abundance of acidic residues, mostly aspartic and glutamic acid, but in some species tyrosine-*O*-sulfate (in the B peptide) or serine phosphate (in the A peptide) is found.

While thrombin splits only the two fibrinopeptide bonds in the fibrinogen monomer, there is a total of 100–150 trypsin-susceptible bonds (*159*) most of which are also susceptible to plasmin. The relative rates of release of the A and B peptides from fibrinogen have been followed both by chromatographic isolation of the peptides (*160*) and by direct N-terminal analysis (*161*, *162*). At both pH 6.3–6.4, which is optimal for clotting, and pH 9.0, which is optimal for the thrombin-catalyzed proteolysis, the B peptide was found to be released at a much slower rate than the A peptide. At clotting, which occurred at the lower pH but not at pH 9.0, 40% of the A peptide and only 8% of the B peptide had been released. These observations indicate that the release of fibrinopeptide A is the initial and probably rate determining event in the thrombin–fibrinogen reaction. A bovine thrombin–fibrinogen system was used, but essentially the same results have since been obtained with human (*163*) and horse (*164*) systems. The correlation between fibrin formation and release of A peptide is further stressed by the fact that lamprey thrombin and the thrombinlike enzymes from crotalid snake venoms release only the A peptide from mammalian fibrinogen (*142*, *165*, *166*) but still cause clotting. However, polymerization of the fibrin monomer takes a different course if only the A peptide has been released (*167*). It is not known at present whether this also alters the qualities of the clot as a substrate for the cross-linking enzyme, plasma *trans*-glutaminase. If this is the case, splitting off of the B peptide, although slower, may well be of as great physiological and evolutionary significance as that of the A peptide, since persons with a congenitally abnormal cross-linking enzyme present themselves as cases of severely impaired wound-healing, proving the physiological importance of the cross-linking system.

It is obvious that the thrombin–fibrinogen system is difficult to analyze

159. B. Blombäck, M. Blombäck, B. Hessel, and S. Iwanaga, *Nature* **215**, 1445 (1967).

160. B. Blombäck and A. Vestermark, *Arkiv Kemi* **12**, 173 (1958).

161. B. Blombäck, *Acta Physiol. Scand.* **43**, Suppl. 148, 1 (1958).

162. B. Blombäck, *Arkiv Kemi* **12**, 321 (1958).

163. U. Abildgaard, *Scand. J. Clin. & Lab. Invest.* **17**, 529 (1965).

164. B. Blombäck and A. C. Teger-Nilsson, *Acta Chem. Scand.* **19**, 751 (1965).

165. R. F. Doolittle, J. L. Oncley, and D. M. Surgenor, *JBC* **237**, 3123 (1962).

166. M. R. Ewart, M. Hatton, J. M. Basford, and K. S. Dodgson, *BJ* **115**, 17P (1969).

167. T. C. Laurent and B. Blombäck, *Acta Chem. Scand.* **12**, 1875 (1958).

kinetically. Waugh and associates (*168–171*), measuring fibrin yield, found that the overall reaction appears to be first order in thrombin but departs from first order in fibrinogen. Shinowara (*172*), analyzing clotting times, concluded that the reaction is second order in thrombin, first order in fibrinogen up to $2.3 \times 10^{-6}\,M$, zero order between $2.3 \times 10^{-6}\,M$ and $9.7 \times 10^{-6}\,M$ and modified zero order at higher concentration. He determined a Michaelis constant of $4.38 \times 10^{-7}\,M$. However, both Ehrenpreis and Scheraga (*173*), using techniques that involve inhibition of either polymerization or thrombin, and Blombäck (*162*), measuring the appearance of N-terminal glycine, found that the reaction is first order with respect to thrombin. It is reasonable to assume that the rate limiting event in fibrin monomer production is the cleaving of the C-terminal arginyl bond in fibrinopeptide A. A K_m value for this reaction of $1.2 \times 10^{-5}\,M$ has been calculated (*138*).

A maximum specific activity of 2400 NIH units/mg thrombin (MW 40,000) and a clotting time of 15 sec involving 40% cleavage of the A peptide at 0.2% fibrinogen (MW 150,000 for the monomer) concentration lead to the rough estimate of $k_{cat} = 24\ \text{sec}^{-1}$ for the thrombin–fibrinogen reaction. The appearance of N-terminal glycine has a pH optimum of 7–9 (*174*).

3. *Derivatives of Fibrinogen*

Peptides are released by thrombin from fibrinogen that has been about 40% acetylated (*175*). Fibrinogen subjected to sulfitolysis (*176*, *177*) or reduction followed by carboxymethylation (*178*) has been found to be a substrate for thrombin, peptides A and B being released (and no other peptides) with the same kinetics as from native fibrinogen (*179*). The separated α(A) chain (*180, 181*) from sulfitolyzed fibrinogen has a

168. D. F. Waugh and B. J. Livingstone, *Science* **113,** 121 (1951).
169. D. F. Waugh and B. J. Livingstone, *J. Phys. & Colloid Chem.* **55,** 1206 (1951).
170. D. F. Waugh and M. J. Patch, *J. Phys. Chem.* **57,** 377 (1953).
171. D. F. Waugh, *Advan. Protein Chem.* **9,** 325 (1954).
172. G. Y. Shinowara, *BBA* **113,** 359 (1966).
173. S. Ehrenpreis and H. A. Scheraga, *ABB* **79,** 43 (1959).
174. B. Blombäck and T. C. Laurent, *Arkiv Kemi* **12,** 137 (1958).
175. E. A. Caspary, *BJ* **62,** 507 (1956).
176. A. Henschen, *Acta Chem. Scand.* **16,** 1037 (1962).
177. J. B. Clegg and K. Bailey, *BBA* **63,** 525 (1962).
178. A. Henschen, *Acta Chem. Scand.* **16,** 1526 (1962).
179. A. Henschen, *Arkiv Kemi* **22,** 1 (1963).
180. A. Henschen, *Arkiv Kemi* **22,** 375 (1964).
181. P. A. McKee, L. A. Rogers, E. Marler, and R. L. Hill, *ABB* **116,** 271 (1966).

molecular weight of 47,000 (*181*) and is also as good a substrate for thrombin as native fibrinogen (*180*).

A relatively small cyanogen bromide fragment of fibrinogen has been isolated which has a molecular weight of about 26,000. Analysis of its amino acid sequence has now been nearly completed (*157*). This fragment contains the N-terminal portions of all the three different chains, α(A), β(B), and γ of fibrinogen. Again, treatment of this fragment with thrombin releases fibrinopeptides A and B with the same kinetics as from native fibrinogen (*182*). It has also been found that when fibrinogen has been digested fairly extensively (about 50 peptide bonds split) with plasmin, superaddition of thrombin to the digest will still release fibrinopeptides A and B (*183*). All these data show that the thrombin recognition site in the α(A) chain of fibrinogen does not involve more than the N-terminal 55 residue fragment of this chain that is present in the cyanogen bromide fragment, but it may well be smaller. Furthermore, thrombin recognition apparently requires neither the β(B) and γ chains nor the four disulfide bridges involved in the 55 residue sequence.

4. *Other Substrates*

Two polypeptide substrates for thrombin have been found, both gastrointestinal hormones from the duodenal mucosa. In secretin, a 27-residue polypeptide chain, the Arg 14 to Asp 15 peptide bond is cleaved, leaving intact the other three trypsin-sensitive arginyl bonds (*135*). It is interesting that secretin, like most of the fibrinopeptides A, has a phenylalanine residue (Phe 6) nine steps before the thrombin-sensitive arginine (Arg 14) in sequence. It has been suggested that rotational spectra of secretin and various polypeptide fragments from it are compatible with the existence of some helical structure between residues 6 and 13 (*184*). In cholecystokinin-pancreozymin (one chain, 33 residues) thrombin splits the Arg 6 to Val 7 bond (*136*). The C-terminal 27 residue fragment, thrombo-cholecystokinin, has full biological activity on intravenous injection (*136*).

Thrombosthenin M, the myosinlike component of the contractile protein in blood platelets, has recently been isolated. Incubation with thrombin produces one mole of C-terminal arginine per 100,000 g of substrate (*137*). All these substrates appear to be "good" substrates for thrombin

182. B. Blombäck, M. Blombäck, A. Henschen, B. Hessel, S. Iwanaga, and K. R. Woods, *Nature* **218,** 130 (1968).

183. P. Wallén and K. Bergström, *Acta Chem. Scand.* **12,** 574 (1958).

184. A. Bodanszky, M. A. Ondetti, V. Mutt, and M. Bodanszky, *J. Am. Chem. Soc.* **91,** 944 (1969).

although no detailed kinetic data are available. No thrombin-sensitive lysyl bonds have yet been demonstrated in polypeptides or proteins.

D. Purported Protein and Polypeptide Substrates

1. *Nonphysiological Substrates*

The majority of proteins and polypeptides that have been tested were found not to be substrates of thrombin. In some cases, including albumin, β-lactoglobulin, casein, gelatin, the B chain of insulin, and activation of trypsinogen, plasminogen, and chymotrypsinogen A (*185–191*), proteolysis by thrombin has been reported. Failure to produce chemical evidence for peptide bond cleavage such as new N- or C-terminals, or release of peptides, or for the purity of the thrombin used makes some of these reports rather less convincing. It is quite possible, though, that some of these proteins are indeed marginal substrates for thrombin. The fact that 12,500 units of thrombin are required to produce fibrinolysis under conditions where 0.01–0.1 unit would be sufficient to cause clotting (*192*) indicates the tremendous rate difference of 5–6 orders of magnitude between fibrinogen and fibrin as substrates for thrombin.

2. *Physiological Substrates Other Than Fibrinogen*

a. Blood Platelets (Thrombocytes). The role of platelet aggregation as an early and important event in hemostasis leading to the formation of a platelet plug has been extensively investigated in recent years. A detailed discussion of this phenomenon is outside the scope of this chapter. For recent reviews see O'Brien (*193*) and Lüscher (*194*). Thrombin (*195*), even when highly purified (*196*), as well as adenosine diphosphate (ADP) (*197*) is known to initiate platelet aggregation both *in vivo* and *in vitro*. Concentrations of 200 n*M* ADP or 1 n*M* thrombin cause aggregation in about 20 sec. Since it has been found (*198*) that

185. G. M. Thelin and R. H. Wagner, *Biochem. Biophys. Res. Commun.* **1,** 219 (1959).
186. H. E. Schultze and G. Schwick, *Z. Physiol. Chem.* **289,** 26 (1951).
187. M. Pantlitschko and E. Gründig, *Monatsh. Chem.* **88,** 253 (1957).
188. K. D. Miller, *Thromb. Diath. Haemorrhag.* **2,** 189 (1958).
189. S. Ehrenpreis, S. J. Leach, and H. A. Scheraga, *J. Am. Chem. Soc.* **79,** 6086 (1957).
190. A. Engel and B. Alexander, *Biochemistry* **5,** 3590 (1966).
191. A. Engel, B. Alexander, and L. Pechet, *Biochemistry* **5,** 1543 (1966).
192. M. M. Guest and A. G. Ware, *Science* **112,** 21 (1950).
193. J. R. O'Brien, *Ann. Rev. Med.* **17,** 275 (1966).
194. E. F. Lüscher, *Schweiz. Med. Wochschr.* **98,** 1629 (1968).

thrombin does not initiate this reaction in persons with congenital afibrinogenemia, there seems to be no compelling need at the moment to infer that any other substrate than fibrinogen is split by thrombin in this reaction.

However, the exact physiological role of thrombosthenin, which since 1958 has gradually come to be recognized as a substrate for thrombin (*199–209*) with final proof provided in 1969 (*137*), is not quite clear. It appears that in addition to its role in clot retraction thrombosthenin may well be involved in platelet aggregation. The substrate involved in thrombin-induced vasodilatation and hypotension in dogs has not yet been identified but appears to be cellbound (*210*).

b. Plasma Clotting Factors. Three of the plasmatic clotting factors are widely considered to be activated by thrombin, namely, Factors V, VIII, and XIII. None has been clearly proved to be a thrombin substrate. In the case of Factor V there are several reports indicating that activation by thrombin takes place (*211–218*), but also some to the contrary (*219–221*). With highly purified reagents, Esnouf and Jobin (*222*) found no evidence of activation by thrombin; others, however, (*215, 218, 223*) do find activation.

---

195. R. W. Shermer, R. G. Mason, R. H. Wagner, and K. M. Brinkhous, *J. Exptl. Med.* **114,** 905 (1961).

196. S. Niewiarowski and D. P. Thomas, *Nature* **212,** 1544 (1966).

197. A. Gaarder, J. Jonsen, S. Laland, A. Hellem, and P. A. Owren, *Nature* **192,** 531 (1961).

198. K. M. Brinkhous, M. S. Read, N. F. Rodman, and R. G. Mason, *J. Lab. Clin. Med.* **73,** 1000 (1969).

199. M. Bettex-Galland and E. Lüscher, *BBA* **49,** 536 (1961).

200. M. Bettex-Galland, H. Portzehl, and E. F. Lüscher, *Nature* **193,** 777 (1962).

201. M. Bettex-Galland and E. F. Lüscher, *Advan. Protein Chem.* **20,** 1 (1965).

202. J. Salmon and Y. Bounameaux, *Thromb. Diath. Haemorrhag.* **2,** 96 (1958).

203. K. Grette, *Acta Physiol. Scand.* **56,** Suppl. 195 (1962).

204. R. L. Nachman, *Blood* **25,** 703 (1965).

205. M. Bettex-Galland and E. F. Lüscher, *Haemat. Bluttransfus.* **6,** 131 (1969).

206. H. J. Weiss and S. Kochwa, *J. Lab. Clin. Med.* **71,** 153 (1968).

207. P. Ganguly, *Blood* **33,** 590 (1969).

208. P. Ganguly, *Blood* **34,** 511 (1969).

209. R. L. Nachman, A. J. Marcus, and L. B. Safier, *J. Clin. Invest.* **46,** 1380 (1967).

210. P. Olsson, J. Swedenborg, and A.-C. Teger-Nilsson, *Cardiovasc. Res.* **3,** 56 (1969).

211. A. G. Ware and W. H. Seegers, *Am. J. Physiol.* **152,** 567 (1948).

212. M. L. Lewis and A. G. Ware, *Blood* **9,** 520 (1954).

213. P. P. Hjort, S. I. Rapaport, and P. A. Owren, *Blood* **10,** 1139 (1955).

214. Q. Z. Hussain and T. F. Newcomb, *Ann. Biochem.* **23,** 569 (1963).

215. D. Papahadjapoulos, C. Hougie, and D. J. Hanahan, *Biochemistry* **3,** 264 (1964).

In the case of Factor VIII, antihemophilic factor, it seems that low concentrations of thrombin may cause activation (*216*, *217*) but higher concentrations are widely recognized to destroy Factor VIII activity.

The only convincing case for activation by thrombin is Factor XIII, the plasma *trans*-glutaminase that catalyzes cross-linking of fibrin. In this case the enzyme is completely inactive before incubation with thrombin (*23*, *224*).

E. Inhibitors

1. *Synthetic Compounds*

Inhibition of thrombin by DFP was first described by Bailey and Bettelheim (*225*). Simultaneous inhibition by DFP of both esterase and clotting activity (*226*, *227*) followed by the demonstration that DFP had combined with a single serine residue (*102*, *228*) clearly established thrombin as one of the serine proteinases with one active site for both esterase and proteolytic activities. The platelet aggregating activity of thrombin is also inhibited by DFP (*229*). The chymotrypsin inhibitor tosylphenylalanine chloromethyl ketone (TPCK) which reacts to form a covalent compound with HIS 57 (*230*) does not inhibit thrombin (*231*). The corresponding lysine derivative tosyllysine chloromethyl ketone (TLCK), which is a trypsin inhibitor (*232*), reacts with the homologous histidine residue in trypsin; TLCK also inhibits both the clotting and esterase activities of thrombin (*51*, *232–234*), and it has recently been shown (*235*) that in the thrombin reaction ^{3}H-labeled TLCK also binds

---

216. D. G. Therriault, J. L. Gray, and H. Jensen, *Proc. Soc. Exptl. Biol. Med.* **95**, 207 (1957).
217. S. I. Rapaport, S. Schiffman, M. J. Patch, and S. B. Ames, *Blood* **21**, 221 (1963).
218. R. W. Colman, *Biochemistry* **8**, 1438 (1968).
219. R. M. Hardisty, *Brit. J. Haematol.* **1**, 323 (1955).
220. D. M. Surgenor, N. A. Wilson, and A. S. Emery, *Thromb. Diath. Haemorrhag.* **5**, 1 (1960).
221. R. T. Breckenridge and O. D. Ratnoff, *J. Clin. Invest.* **44**, 302 (1965).
222. M. P. Esnouf and F. Jobin, *BJ* **102**, 660 (1967).
223. P. G. Barton and D. J. Hanahan, *BBA* **133**, 506 (1967).
224. L. Lorand and K. Konishi, *ABB* **105**, 58 (1964).
225. K. Bailey and F. R. Bettelheim, *BBA* **18**, 495 (1955).
226. K. D. Miller and H. van Vunakis, *JBC* **223**, 227 (1956).
227. J. A. Gladner and K. Laki, *ABB* **62**, 501 (1956).
228. K. Laki, J. Gladner, J. E. Folk, and D. R. Kominz, *Thromb. Diath. Haemorrhag.* **2**, 205 (1958).

to a histidine. This is consistent with the sequence evidence (*78, 97, 99–101*) which showed that thrombin has a side chain binding site identical to that of trypsin and also homologous HIS 57 and Asp 102 sequences, all located in the B chain. First-order kinetics ($k_{cat} = 4.4 \times 10^{-3}$ sec^{-1} at 25°, pH 8.3) have been reported (*51*) for the TLCK reaction with thrombin. The finding that *N*-bromosuccinimide inhibits thrombin in the absence but not in the presence of the substrate BAEE (*234*) was interpreted by Chulkova and Orekhovich as evidence that the catalytic site involves a histidine, since the loss of one histidine residue in the amino acid composition was also prevented by the presence of BAEE (*234*). Diphenylcarbamyl chloride (DPCC), which is a chymotrypsin inhibitor (*236*), and phenylmethane sulfonyl fluoride (PMSF) (*237*), which inhibits trypsin and chymotrypsin as well as acetylcholinesterase, have both (*231, 238*) been reported to inhibit thrombin. Inactivation of thrombin by reaction with nitrite leads, as expected, to loss of the α-amino groups. When the activity was protected by the presence of substrate, the N-terminal threonine was lost but not the N-terminal isoleucine (homologous to ILE 16 in chymotrypsin) (*81*).

Synthetic substrates have been shown to competitively inhibit the clotting activity of thrombin (*43*). The competitive inhibitory effects of a series of ω-guanidino and ω-amino acids on trypsin, plasmin, and kallikrein have been studied (*239, 240*). Although most of the compounds inhibited one or more of the other enzymes, none had any inhibitory effect on thrombin. In fact, stimulation of TAME hydrolysis by thrombin was observed in the presence of hexyl-ε-aminocaproate or 4 m*M* hexyl-δ-guanidinovalerate.

Derivatives of benzylamine and benzamidine have also been found to be more efficient inhibitors of trypsin and plasmin than of thrombin

---

229. A. J. Marcus and M. B. Zucker, *in* "The Physiology of Blood Platelets," p. 63. Grune & Stratton, New York, 1965.

230. G. Schoellman and E. Shaw, *Biochemistry* **2,** 252 (1963).

231. W. H. Seegers, D. Heene, E. Marciniak, N. Ivanovic, and M. J. Caldwell, *Life Sci.* **4,** 425 (1965).

232. E. Shaw, W. Mares-Guia, and W. Cohen, *Biochemistry* **4,** 2219 (1965).

233. E. Marciniak and W. H. Seegers, *Thromb. Diath. Haemorrhag.* **15,** 633 (1966).

234. T. M. Chulkova and V. N. Orekhovich, *Biokhimiya* **33,** 1222 (1968).

235. G. Glover and E. Shaw, personal communication with E. Shaw (1969).

236. B. F. Erlanger and W. Cohen, *J. Am. Chem. Soc.* **85,** 348 (1963).

237. A. M. Gold and D. E. Fahrney, *Biochemistry* **3,** 783 (1964).

238. S. Magnusson, "Studies on the Chemistry of Prothrombin and Thrombin." Norstedt, Stockholm, 1965.

239. M. Muramatu and S. Fujii. *J. Biochem.* (*Tokyo*) **64,** 807 (1968).

240. M. Muramatu and S. Fujii, *J. Biochem.* (*Tokyo*) **65,** 17 (1969).

(*241*), with one exception, namely, 4-chlorobenzylamine. The most efficient thrombin inhibitor in this series was found to be 4-amidinophenyl pyruvic acid ($K_i = 2 \times 10^{-8}\ M$).

Fibrinopeptide A is a competitive inhibitor of the thrombin–fibrinogen reaction (*242, 243*). Arginylglycine and tosylarginine are not inhibitors; tosylarginylglycine does inhibit the clotting reaction but is not a substrate (*244*). Synthetic peptides resembling the C-terminal part of fibrinopeptide A have recently been investigated (*245*). In the series of compounds Phe–Gly_{0-7}–ArgOMe the nonapeptide ester with Phe in the same position relative to Arg as in fibrinopeptide A was found to be the most efficient competitive inhibitor. It not only inhibits the thrombin–fibrinogen reaction but also prothrombin activation. The Gly_1–compound is a better inhibitor than the Gly_0– and Gly_{2-4}– compounds. In a series of tripeptide esters of the structure Phe–X–ArgOMe, the common natural amino acids were tried in position X, and valine was found to make the best inhibitor.

Treatment of thrombin with acetic anhydride has been found to give a product which has lost most of its clotting activity while retaining substantial esterase activity (*246, 247*). Although the degree of acetylation cannot be assessed from the published data, one could argue that one or more lysine ϵ-amino groups, which are not involved in binding small substrates, may be essential for fibrinogen binding; K_m and V_{max} on TAME were only slightly affected by acetylation (*79*).

2. *Polypeptide Inhibitors—Hirudin, Reduviin, and Tabanin*

Hirudin is produced in the salivary glands of the blood-sucking leech (*Hirudo medicinalis*) and prevents clotting of the ingested portion of the victim's blood (*248*). Hirudin has been purified (*249–251*) and

241. F. Markwardt, H. Landmann, and P. Walsmann, *European J. Biochem.* **6,** 502 (1968).

242. F. R. Bettelheim, *BBA* **19,** 121 (1956).

243. B. Blombäck, M. Blombäck, R. F. Doolittle, and I. Norén, *Thromb. Diath. Haemorrhag.* Suppl. **13,** 29 (1963).

244. L. Lorand and E. P. Yudkin, *BBA* **25,** 437 (1957).

245. B. Blombäck, M. Blombäck, P. Olsson, L. Svendsen, and G. Åberg, *Scand. J. Clin. & Lab. Invest.* **24,** Suppl. 107, 59 (1969).

246. R. H. Landaburu and W. H. Seegers, *Can. J. Biochem. Physiol.* **37,** 1361 (1959).

247. R. H. Landaburu and W. H. Seegers, *Can. J. Biochem. Physiol.* **38,** 613 (1960).

248. J. B. Haycraft, *Arch. Exptl. Pathol. Pharmakol.* **18,** 209 (1884).

249. F. Markwardt, *Naturwissenschaften* **42,** 537 (1955).

250. F. Markwardt, *Z. Physiol. Chem.* **308,** 147 (1957).

251. M. Justisz, A. Charbonnel-Berault, and G. Martinoli, *Bull. Soc. Chim. Biol.* **45,** 55 (1963).

found to be a 67-residue polypeptide chain (MW 10,800) with three disulfide bridges. It forms a presumably electrostatic complex with thrombin, but not with acetylated thrombin. This complex, which has no clotting or esterase activity, is electrophoretically and chromatographically stable between pH 3 and 11. A K_i of $8 \times 10^{-11}\,M$ at pH 7.4 and 20° has been reported (*252–254*). Thrombin bound to hirudin does not react with DFP (*253*), but diisopropylphosphoryl (DIP)-thrombin binds to hirudin with virtually the same K_i as native thrombin (*254*). Chemical modification of hirudin has indicated that disulfide bridges and carboxyl groups are essential for its activity (*255*). Reduviin and tabanin are two other inhibitors, also specific for thrombin, which have been isolated from the blood-sucking insects *Rhodnius prolixus*, a South American bedbug of the *Heteroptera* family Reduviidae and female *Tabanus bovinus*, cosmopolitan gadflies of the *Diptera* family Tabaniidae (*254*). The effective hirudin-binding by DIP-thrombin indicates that the catalytic site of thrombin is not part of the hirudin-binding site. Therefore, one does not expect to find that binding to thrombin leads to limited proteolysis in the inhibitor as has been observed in the case of interaction of soybean trypsin inhibitor with trypsin (*256*). In a fairly comprehensive survey (*257*) studying the activities against human trypsin, plasmin, and thrombin of a whole series of inhibitors, namely, soybean trypsin and AA inhibitors, four other bean inhibitors, colostrum, potato and pancreatic (Kazal and Kunitz) inhibitors, and 10 avian mucoid inhibitors, it was found that none of these inhibited the esterase activity of thrombin, whereas the ones that inhibit trypsin generally also inhibit plasmin. Likewise, Trasylol from bovine lung, which inhibits kallikrein, another serine proteinase in blood plasma, has been found to have no activity on thrombin (*258, 259*).

3. *Proteins—Antithrombins*

The long-recognized capacity (*5*) of blood plasma to neutralize thrombin is quite effective both in terms of the amount of thrombin and the

252. F. Markwardt, *Arch. Exptl. Pathol. Pharmakol.* **229**, 389 (1956).
253. F. Markwardt and P. Walsmann, *Z. Physiol. Chem.* **312**, 85 (1958).
254. F. Markwardt, "Blutgerinnungshemende Wirkstoffe aus blutsaugenden Tieren." Fischer, Jena, 1963.
255. C. Tertrin, P. de la Llosa, and M. Jutisz, *Bull. Soc. Chim. Biol.* **49**, 1837 (1967).
256. K. Ozawa and M. Laskowski, Jr., *JBC* **241**, 3955 (1966).
257. R. E. Feeney, G. E. Means, and J. C. Bigler, *JBC* **244**, 1957 (1969).
258. C. J. Amris, *Scand. J. Haematol.* 3, 19 (1966).
259. B. Blombäck, M. Blombäck, and P. Olsson, *in* "Neue Aspekte der Trasylol Therapie," p. 33. Schattauer, Stuttgart, 1966.

rate of its inactivation. Mainly on physiological evidence this inactivation has been attributed to several different systems (*260*), which have been called antithrombin I to VI. The chemical basis for the action of antithrombin I (adsorption of thrombin to the fibrin clot) and antithrombin VI (inhibition of thrombin by fibrinolytic degradation products of fibrin) has been recognized for some time. However, the two most important systems antithrombin II, which is also called heparin cofactor since its activity is much increased by the presence of heparin, and antithrombin III, the "progressive antithrombin" which causes a relatively slow inactivation of thrombin, have not been at all understood until fairly recently.

It is now known that at least two different plasma proteins are inhibitors of thrombin. One of these is α_2-macroglobulin (α_2M) (*261–263*), which inhibits thrombin (*261, 262*), trypsin, and plasmin (*263*) leading to slow formation of an apparently irreversible complex that has no proteolytic activity but retains esterase activity. It has been demonstrated that α_2M inhibits the release of fibrinopeptides by thrombin (*264*). The α_2M accounts for about half the "progressive antithrombin" activity in plasma *in vitro*. The second contributor is α_1-antiproteinase (MW 47,000) also active against plasmin and trypsin as well as thrombin (*265*). A third protein (MW 64,000) from the α_2 fraction of plasma (*266*) which may (*267*) or may not (*268, 269*) be identical to the heparin cofactor has also been implicated. When plasma containing labeled prothrombin is allowed to clot, the label is bound to α_2M and can be precipitated by anti-α_2M serum indicating that α_2M is the most important plasma antithrombin (*270*). It has been suggested that α_2M binds to thrombin in the manner of a substrate, but this seems unlikely in view of the residual esterase activity of the complex.

F. Mechanism

It is obvious from the above discussion of sequence data, catalytic properties, and kinetics that the catalytic site and the catalytic

260. C. Fell, N. Ivanovic, S. A. Johnson, and W. H. Seegers, *Proc. Soc. Exptl. Biol. Med.* **85**, 199 (1954).

261. G. F. Lanchantin, M. L. Plesset, J. A. Friedmann, and W. D. Hart, *Proc. Soc. Exptl. Biol. Med.* **121**, 144 (1966).

262. M. Steinbuch, C. Blatrix, and F. Josso, *Rev. Franc. Etudes Clin. Biol.* **13**, 179 (1968).

263. M. Steinbuch and C. Blatrix, *Rev. Franc. Etudes Clin. Biol.* **13**, 142 (1968).

264. U. Abildgaard, *Thromb. Diath. Haemorrhag.* **21**, 173 (1969).

265. A. Rimon, Y. Shamash, and B. Shapiro, *JBC* **241**, 5102 (1966).

mechanism of thrombin are virtually identical to those of the serine proteinases from pancreas. The activation mechanism of these enzymes is not clearly understood in structural terms at the moment, but it does involve limited proteolysis to reveal an N-terminal isoleucine (or valine) which then forms a "salt bridge" with the aspartic acid next in sequence to the active serine. To the extent that this salt bridge formation is essential for activation thrombin apparently resembles the other serine enzymes. The sequence homology between these enzymes is so striking that any comprehensive theory proposed for the activation mechanism of chymotrypsinogen that were to involve structural features not present in one of the other enzymes in this group would seem highly improbable. The side chain specificity toward small substrates is the same for thrombin as for trypsin because they presumably have identical "tosyl holes." The one characteristic feature of thrombin that we cannot explain at the moment is its highly selective binding specificity for polypeptide substrates. From the extensive data available on fibrinogen degradation products, which are still substrates for thrombin, and from the fact that secretin is a substrate, it would seem that thrombin specificity requires only a relatively short polypeptide sequence with certain characteristic features of primary structure on either or both sides of the bond to be cleaved but not very far removed from it. This implies that one or more secondary binding sites (in addition to the "tosyl hole") on the thrombin molecule are required for the binding of fibrinogen. Since we are assuming that the tertiary structure of thrombin is also homologous to those of the pancreatic proteinases, the best guess is that the two large insertions (the 7 residues at position 148–149, and the 14 residues plus the carbohydrate at position 65–66) in the B chain will appear as external loops in the tertiary structure. It then becomes conceivable that either of these insertions, or the A chain, which is also likely to be external, may contribute residues that normally mask the fibrinogen-binding site of thrombin, thus restricting its specificity to such polypeptides or proteins that have a higher binding affinity for this site than has the masking sequence. The most likely candidate for this masking function would be the C-terminal part of the A chain since it has a sequence which is to some extent analogous with that of fibrinopeptide A.

---

266. U. Abildgaard, *Scand. J. Clin. & Lab. Invest.* **24,** 23 (1969).
267. O. Egeberg, *Thromb. Diath. Haemorrhag.* **13,** 516 (1965).
268. P. Porter, M. C. Porter, and I. N. Shanberge, *Clin. Chim. Acta* **17,** 189 (1967).
269. P. O. Ganrot, *Scand. J. Clin. & Lab. Invest.* **24,** 11 (1969).
270. S. S. Shapiro and J. Cooper, *Federation Proc.* **27,** 628 (1968).

IV. Prothrombin—Molecular Properties

A. Assays

Essentially two different types of clotting assay have been used for prothrombin. In the one-stage method (*12, 14*) activation of prothrombin to thrombin, proteolysis of fibrinogen by thrombin, and polymerization of fibrin monomer to form a visible clot all occur in the same reaction mixture. In the two-stage assay (*13, 271*) the activation of prothrombin takes place in a separate mixture from which samples are assayed for thrombin activity as a function of activation time. The maximum thrombin activity found is taken to represent the prothrombin activity of the sample tested. Details of these test systems have been described elsewhere (*41*). Although the two-stage assay is not ideal, it seems to be the best specific assay available at the moment. Immunological precipitin assays are at least as sensitive (*272, 273*) as the clotting assays allowing detection of as little as 40 ng of prothrombin (*41*). However, since the immunoprecipitin assay has been claimed (*272, 273*) to detect nearly as much prothrombin in plasma from patients with vitamin K deficiency as in normal plasma, it is obviously not specific for *active* prothrombin.

B. Isolation and Purity

The concentration of prothrombin in blood plasma is about 70 mg/liter, thus accounting for less than 0.2% of the total protein and requiring 700–800 times purification to achieve complete separation from other plasma proteins. The major difficulty involved in purification is to avoid activation to thrombin. Since activation is dependent on a normal calcium concentration in plasma, the usual starting material for the preparation is blood that has been collected in oxalate or citrate solutions or through an ion exchange resin. All purification methods described so far make use of the fact that prothrombin is adsorbed to calcium phosphate (*274*) and other insoluble compounds such as $Mg(OH)_2$, $BaSO_4$, barium citrate, bentonite, $Al(OH)_3$. Although this adsorption step involves a definite

271. A. G. Ware and W. H. Seegers, *Am. J. Clin. Pathol.* **19,** 471 (1949).

272. F. Josso, J. M. Lavergne, M. Gouault, O. Prou-Wartelle, and J. P. Soulier, *Thromb. Diath. Haemorrhag.* **20,** 88 (1968).

273. P. O. Ganrot and J. E. Niléhn, *Scand. J. Clin. & Lab. Invest.* **21,** 238 (1968).

274. J. Bordet and L. Delange, *Ann. Soc. Roy. Sci. Med. Nat. Bruxelles* **72,** 87 (1914).

and unexplained loss in yield of activity as much as 100-fold purification can be obtained in this single step.

The method developed by Seegers *et al.* (*275*) based on isoelectric precipitation of euglobulins at low ionic strength (*276*), extraction of the prothrombin from the precipitate, adsorption on magnesium hydroxide (*277*), subsequent elution with carbon dioxide at elevated pressure (*277*), and ammonium sulfate fractionation and isoelectric fractionation afforded the first prothrombin preparations that were virtually pure. Equilibrium chromatography on the cation exchange resin XE-64 was introduced for final purification by Miller (*278*). Not only bovine prothrombin (*41*, *54*, *278–285*) but also human (*286–291*), horse (*292*), dog (*282*, *293*), sheep, water buffalo (*294*), and rat (*295*) prothrombin have all been obtained pure or virtually pure. Although bovine prothrombin has admittedly been studied more extensively than the other species, no striking species differences have been found except for the apparently dimeric molecular weight (130,000) and lower specific activity (700–800

275. W. H. Seegers, E. C. Loomis, and J. M. Vandenbelt, *ABB* **6**, 85 (1945).

276. J. Mellanby, *Proc. Roy. Soc.* **B107**, 221 (1930).

277. H. J. Fuchs, *Biochem. Z.* **222**, 470 (1930).

278. K. D. Miller, *JBC* **231**, 987 (1958).

279. W. H. Seegers, *Record Chem. Progr.* (*Kresge-Hooker Sci. Lib.*) **13**, 143 (1952).

280. R. Goldstein, A. LeBolloc'h, B. Alexander, and E. Zonderman, *JBC* **234**, 2857 (1959).

281. H. Kowarzyk, W. Mejbaum-Katzenellenbogen, Z. Kowarzykowa, and B. Czerwinska-Kossobudzka, *Bull. Acad. Polon. Sci., Ser. Biol.* **12**, 441 (1964).

282. H. C. Moore, S. E. Lux, O. P. Malhotra, S. Bakerman, and J. R. Carter, *BBA* **111**, 174 (1965).

283. G. H. Tishkoff, L. C. Williams, and D. M. Brown, *JBC* **243**, 4151 (1968).

284. J. S. Ingwall and H. A. Scheraga, *Biochemistry* **8**, 1860 (1969).

285. K. Andrassy, W. Krause, R. Egbring, and H. Egli, *Thromb. Diath. Haemorrhag.* **21**, 561 (1969).

286. M. L. Lewis and A. G. Ware, *Proc. Soc. Exptl. Biol. Med.* **84**, 636 (1953).

287. G. F. Lanchantin, J. A. Friedmann, J. DeGroot, and J. W. Mehl. *JBC* **238**, 238 (1963).

288. G. F. Lanchantin, D. W. Hart, J. A. Friedmann, N. V. Saavedra, and J. W. Mehl, *JBC* **243**, 5479 (1968).

289. S. Magnusson, *Arkiv Kemi* **24**, 367 (1965).

290. D. L. Aronson and D. Menaché, *Biochemistry* **5**, 2635 (1966).

291. S. S. Shapiro and D. F. Waugh, *Thromb. Diath. Haemorrhag.* **16**, 469 (1966).

292. K. D. Miller and A. W. Phelan, *Biochem. Biophys. Res. Commun.* **27**, 505 (1967).

293. D. M. Zubairov, V. N. Timerbaev, and V. M. Menshov, *Biokhimiya* **33**, 7 (1968).

294. G. J. S. Rao and N. Chandrasekhar, *Ann. Biochem. Exptl. Med.* (*Calcutta*) **21**, 233 (1961).

295. L.-F. Li and R. E. Olson, *JBC* **242**, 5611 (1967).

NIH units/mg) of horse prothrombin, and the higher specific activity (2100–2600 NIH units/mg) of human (*289*) prothrombin. Constant solubility (*275*) and homogeneity in ultracentrifugal analysis (*54*, *282–284*, *293*, *295*, *296*), immunodiffusion (*278*, *295*), starch gel electrophoresis (*297*), moving boundary electrophoresis (*282*), and disc electrophoresis (*284*, *288*, *295*) as criteria of purity indicated that the respective preparations were essentially pure.

Although physical criteria of purity have thus been reasonably well satisfied in many cases, most of these preparations are still active in test systems for Factors VII, IX, and X. These factors are all accelerators of prothrombin activation. With the exception of Factor X (*298–302*) they have not yet been satisfactorily purified. These three factors are generally considered to be three separate proteins present as trace contaminants (less than 1 or 2% of the total protein), but the possibilities that they are properties of the prothrombin molecule or derivatives of prothrombin have not yet been convincingly ruled out and certainly cannot be ruled out on the basis of available genetic evidence as is often claimed.

Attempts to separate prothrombin from these other factors by adsorption–elution chromatography on DEAE- or TEAE-cellulose (*297*, *301*, *302*) have been only partially successful leading, in spite of quantitative protein recovery, to substantial loss of prothrombin activity in terms of both yield and specific activity of the "purified" prothrombin. It could not be decided at the time whether partial denaturation of prothrombin at the very low ionic strength that was used for adsorption had led to partial loss of activity or whether the separation pattern represented products of partial activation during the purification of prothrombin. Recent investigations (*290*, *303*, *304*) on human prothrombin that had either been partially activated (*290*, *304*) or not been intentionally activated (*303*, *304*) make the latter interpretation seem the more likely. After digestion of prothrombin with thrombin, Seegers *et al.* (*305*) have ob-

296. F. Lamy and D. F. Waugh, *JBC* **203,** 489 (1953); *Thromb. Diath. Haemorrhag.* **2,** 188 (1958); *Physiol. Rev.* **34,** 722 (1954).

297. S. Magnusson, *Arkiv. Kemi* **24,** 217 (1965).

298. F. Duckert and F. Koller, *Thromb. Diath. Haemorrhag.* **7,** Suppl. 1, 226 (1962).

299. M. P. Esnouf and W. J. Williams, *BJ* **84,** 62 (1962).

300. W. H. Seegers, E. R. Cole, N. Aoki, and C. R. Harmison, *Can. J. Biochem.* **42,** 229 (1964).

301. D. Papahadjopoulos, C. Hougie, and D. J. Hanahan, *Biochemistry* **3,** 264 (1964).

302. K. Lechner and E. Deutsch, *Thromb. Diath. Haemorrhag.* **13,** 314 (1965).

303. D. L. Aronson, *Thromb. Diath. Haemorrhag.* **16,** 491 (1966).

304. G. F. Lanchantin, J. A. Friedmann, and D. W. Hart, *JBC* **243,** 476 (1968).

305. W. H. Seegers, E. Marciniak, R. K. Kipfer, and K. Yasunaga, *ABB* **121,** 372 (1967).

tained and analyzed a product that they called *prethrombin* which has very similar properties to zymogen products isolated (*290, 297, 301–303, 306–308*) and analyzed (*307*) by others. However, it is not yet clear that prethrombin is a normal intermediate in the activation of prothrombin to thrombin, and this question needs further investigation.

C. Physical Properties

The electrophoretic mobility of prothrombin is that of an α_1–α_2 protein but has recently been observed (*273*) to change on electrophoresis in a buffer containing calcium lactate to that of an α_2–β_1 protein. The isoelectric point has been found to be pH 4.2 (*18*), pH 4.25 (*309*), and pH 4.1 (*283*). The sedimentation constant has been given as $s_{25,w} = 4.89$ S (*296*), $s_{20,w} = 5.22$ S (*76*), $s_{20,w} = 4.80$ S (*284*) for bovine, $s_{20,w} = 5.0$ S (*295*) for rat, $s = 5.35$ S (*310*) for horse, and $s_{20,w} = 5.07$ S for dog (*293*) prothrombin. The diffusion constants $D_{20,w} \times 10^7$ cm^{-2} sec of 6.24 (*296*) and 5.17 (*283*) have been found for bovine, 4.93 for rat (*295*), and 3.8 for horse (*310*) prothrombin. The partial specific volume $\bar{V}$ has been determined pycnometrically (*76*) and found to be 0.70 ml g^{-1}. Calculation from physical data has given molecular weight values of 68,000 (*296*), 68,500 (*76*), 70,500 ± 2,800 (*283*), and 74,000 ± 4,100 (*284*) for bovine, 130,000 for horse (*310*), 86,000 ± 1,000 for rat (*295*), 70,900 for dog (*293*), and 70,000 for human (*288*) prothrombin. Optical rotatory dispersion data have been reported which indicate little or no change of conformation with temperature (6°, room temperature, 75°) (*284*) or as a result of dissolving prothrombin in 25% sodium citrate (*311*). The value of —35 for b_0 is consistent with a low helix content in native prothrombin (*284*).

A crystalline bovine prothrombin preparation containing 21.5% protein, 32% barium, and 8.9% citrate (*283*), as well as protein crystals (long, thin needles) of horse prothrombin (*292*) have been reported, neither of which seem suitable for an X-ray investigation of the tertiary structure.

306. T. Asada, Y. Masaki, K. Kitahara, R. Nagayama, T. Hatashita, and I. Yangagisawa, *J. Biochem.* (*Tokyo*) **49,** 721 (1961).

307. S. Magnusson, *Thromb. Diath. Haemorrhag.* **7,** Suppl. 1, 229 (1962).

308. N. R. Shulman and J. Z. Hearon, *JBC* **238,** 155 (1963).

309. W. H. Seegers, C. R. Harmison, N. Ivanovic, and D. L. Heene, *Thromb. Diath. Haemorrhag.* **15,** 343 (1966).

310. K. D. Miller and J. F. McGarrahan, *N. Y. State Dept. Health, Ann. Rept. Div. Lab. Res.* p. 69 (1958).

311. M. MacAulay, S. Bakerman, H. Moore, and J. R. Carter, *Thromb. Diath. Haemorrhag.* **11,** 289 (1964).

TABLE III
COMPOSITION OF BOVINE PROTHROMBIN
(g/100 g prothrombin)

Residue	*a*	*b*	*c*	*d*	*e*	*f*
Alanine	3.07	3.26	3.17	3.29	3.65	3.36
Arginine	7.87	7.34	8.29	7.25	8.71	7.25
Aspartic acid + asparagine	8.52	8.66	8.22	9.02	10.52	8.80
Half-cystine	2.73	2.67[g]	1.59[g]	2.40	2.77	2.57
Glutamic acid + glutamine	10.80	11.28	11.78	12.37	14.59	11.98
Glycine	3.86	3.41	3.69	3.89	4.39	3.58
Histidine	1.47	1.63	1.98	1.99	1.80	2.05
Isoleucine	2.39	2.71		2.96	3.63	2.91
Leucine	6.88	6.56	9.75[h]	6.74	6.67	6.69
Lysine	5.20	5.02	6.46	4.84	5.88	4.77
Methionine	1.28	1.71	1.02	1.15	0.80	1.36[i]
Phenylalanine	3.98	4.21	4.07	4.06	4.39	4.39
Proline	3.14	4.66	4.69	4.51	5.54	5.24
Serine	5.21	5.13	3.19	4.55	4.83	5.01
Threonine	4.47	3.95	2.82	3.96	4.33	4.20
Tryptophan[j]		3.01		2.97	4.42	
Tyrosine	3.98	3.95	4.65	4.50	3.94	3.88
Valine	4.11	4.13	5.10	4.89	5.14	4.95
Total amino acid	78.96	83.29	80.47	85.34	96.00	88.48
Hexose			3.6–5.6			
Hexosamine			2.0			
Sialic acid[k]			4.2			
Total carbohydrate			9.8–11.8			

[a] G. F. Lanchantin, D. W. Hart, J. A. Friedmann, N. V. Saavedra, and J. W. Mehl, *JBC* **243**, 5479 (1968).

[b] K. Laki, D. R. Kominz, P. Symons, L. Lorand, and W. H. Seegers, *ABB* **49**, 276 (1954).

[c] S. Magnusson, *Arkiv Kemi* **23**, 271 (1965). This analysis is based on paper chromatography of the amino acid phenylthiohydantoins, and it was performed by Dr. J. Sjöquist.

[d] W. H. Seegers, E. Marciniak, R. K. Kipfer, and K. Yasunaga, *ABB* **121**, 372 (1967).

[e] J. S. Ingwall and H. A. Scheraga, *Biochemistry* **8**, 1860 (1969).

[f] S. Magnusson and B. S. Hartley, unpublished results (1968).

[g] Half-cystine determined as cysteic acid.

[h] Isoleucine + leucine.

[i] Methionine determined as methionine sulfone.

[j] Determined spectrophotometrically.

[k] Expressed as *N*-acetylneuraminic acid.

D. Amino Acid and Carbohydrate Composition

Prothrombin contains about 11.5% carbohydrate, namely, 4.3% sialic acid, 5.4% neutral sugar (galactose and mannose in a 1:1 ratio), and 1.8% glucosamine (*54*). Degradation with neuraminidase (*312, 313*) showed that sialic acid is terminal. Partial acid hydrolysis released monosaccharide units in the order sialic acid–galactose–mannose–glucosamine. Table III gives amino acid analysis data for bovine and Table IV for human and rat prothrombin. Paper chromatographic analysis of the amino acid compositions of bovine, sheep, and water buffalo prothrombin (*294*) showed little difference between the three species, but the bovine data differ appreciably from those in Table III, probably because of the different methods used. All the analyses in Table III and IV except one were performed by the standard Moore and Stein technique.

E. Disulfide Bridges

Amperometric titration at pH 10.3 of disulfide bonds after sulfitolysis in 8 *M* urea has given evidence for 8 –S–S– bridges per mole of bovine prothrombin (*314*). Amino acid analysis of performic acid oxidized bovine prothrombin (*315*) shows 17.0 half-cystines per mole. This indicates that the "Pro" fragment(s) split off during activation account(s) for at least four bridges in addition to the four in thrombin.

F. N-Terminal and C-Terminal Analysis

One mole of N-terminal alanine per mole of prothrombin has been found both by the PTH and DNP methods in bovine (*90, 278, 316–319*) and also in human (*92, 316*) and dog (*293*) material. Horse prothrombin was found to contain two moles of N-terminal alanine per 130,000 g

312. G. Schwick and H. E. Schultze, *Clin. Chim. Acta* **4**, 26 (1960).
313. G. H. Tishkoff, L. Pechet, and B. Alexander, *Blood* **15**, 778 (1960).
314. J. R. Carter, *JBC* **234**, 1705 (1959).
315. R. H. Saundry, S. Magnusson, and B. S. Hartley, unpublished results (1969).
316. S. Magnusson, *Acta Chem. Scand.* **12**, 355 (1958).
317. V. O. Belitser, E. L. Khodorova, and A. L. Loseva, *Ukr. Biokhim. Zh.* **33**, 499 (1961).
318. D. L. Aronson, *Nature* **194**, 475 (1962).
319. S. Magnusson, *Arkiv Kemi* **23**, 271 (1965).

TABLE IV
COMPOSITION OF HUMAN AND RAT PROTHROMBIN
(g/100 g prothrombin)

Residue	*a*	*b*
Alanine	3.48	2.38
Arginine	7.94	5.34
Aspartic acid + asparagine	9.14	10.22
Half-cystine	3.35	1.36
Glutamic acid + glutamine	12.72	11.98
Glycine	3.52	3.12
Histidine	1.78	1.99
Isoleucine	3.42	3.05
Leucine	6.11	5.88
Lysine	4.81	5.31
Methionine	1.30	1.55
Phenylalanine	4.24	5.52
Proline	4.28	3.83
Serine	4.67	4.47
Threonine	5.52	6.32
Tryptophan[c]	5.34	2.82
Tyrosine	4.90	3.76
Valine	4.15	4.56
Total amino acid	90.67	83.46
Hexose	4.13	3.7
Hexosamine	2.43	0.1
Sialic acid[d]	3.40	
Total carbohydrate	9.96	

[a] Human prothrombin, G. F. Lanchantin, D. W. Hart, J. A. Hart, J. A. Friedmann, N. V. Saavedra, and J. W. Mehl, *JBC* **243,** 5479 (1968).
[b] Rat prothrombin, L.-F. Li and R. E. Olson, *JBC* **242,** 5611 (1967).
[c] Spectrophotometrically determined.
[d] Expressed as *N*-acetylneuraminic acid.

(*310*). The N-terminal sequence Ala–Asx– has been found in both human and bovine prothrombin (*92*).

C-Terminal analysis by degradation with carboxypeptidases has failed to release any significant amount of free amino acid (*93, 278*). The evidence for C-terminal serine (*90*) is of questionable reliability.

G. NUMBER OF POLYPEPTIDE CHAINS. SIZE OF "PRO" FRAGMENT(S)

The results of N- and C-terminal analysis are compatible with a one-chain structure. The specific activities found for highly purified thrombin and prothrombin allow only one mole of thrombin per mole of prothrombin. Recent evidence (*78*) obtained from gel filtration (Sephadex G-200) showed that both native prothrombin and reduced carboxymethylated prothrombin (even when maleylated) emerged from the column as one peak whether or not the buffer contained 7 *M* urea. The effluent volume of the peak was in all cases consistent with a molecular weight of 68,000. Others have found that ultracentrifugation in 6 *M* guanidinium hydrochloride in the presence of mercaptoethanol fails to produce subunits (*283*, *284*) of prothrombin. One must conclude from these data that unless the structure contains some kind of covalent crosslinking bonds other than disulfides prothrombin contains only one polypeptide chain. Accepting a one chain- model for prothrombin one can then estimate the total size and composition of the "Pro" fragment(s) that is/are split off during activation to thrombin by subtracting the composition of thrombin from that of prothrombin. On this basis the "Pro" contains a very approximate total of 22.5% carbohydrate and 208 amino acid residues. The amino acid composition thus estimated would be Ala_{17}, Arg_8, Asx_{23}, $\frac{1}{2}Cys_9$, Glx_{26}, Gly_{15}, His_3, Ile_2, Leu_{11}, Lys_3, Met_2, Phe_7, Pro_{15}, Ser_{23}, Thr_{14}, Trp_{10}, Tyr_{11}, and Val_9.

V. Prothrombin—Activation Mechanism

A. The Prothrombin Activating Enzyme

Three approaches have led to the purification of a prothrombin-activating enzyme. Milstone (*320*, *321*) was the first to achieve separation from blood plasma of such an enzyme, which he called thrombokinase. It also has TAME esterase activity (*322*) and activates chymotrypsinogen (*323*). The second line of approach followed the discovery by Marciniak and Kowarzyk (*324*) that when highly purified, and presumably pure, prothrombin preparations were activated by dissolving them in 25% sodium citrate (*18*), an enzyme could be isolated from the activation mixture which was distinct from thrombin in that it accelerates the activation of prothrombin. This enzyme was called autoprothrombin C and has

320. J. H. Milstone, *Proc. Soc. Exptl. Biol. Med.* **72,** 315 (1949).
321. J. H. Milstone, *Proc. Soc. Exptl. Biol. Med.* **103,** 361 (1960).
322. J. H. Milstone, *Proc. Soc. Exptl. Biol. Med.* **101,** 660 (1959).
323. J. H. Milstone and V. K. Milstone, *Proc. Soc. Exptl. Biol. Med.* **117,** 290 (1964).
324. E. Marciniak and H. Kowarzyk, *Polski Tygod. Lekar. Wiadomosci. Lekar.* **16,** 1941 (1961).

been purified (*325*). Not surprisingly, it was thought to be a derivative of prothrombin, a view which has lately been abandoned by Seegers in favor of considering that his prothrombin preparations contain two zymogens, one for thrombin called prethrombin and one for autoprothrombin C called autoprothrombin III, and that the two are not covalently bound to each other (*326*). The third line of approach started with the discovery by Telfer *et al.* (*327*) of a new congenital hemorrhagic disease now known as Factor X deficiency. This factor has been purified from plasma (*298, 328, 329*) after activation with a purified fraction of the venom of Russell's viper, (*Vipera russellii*). The activated factor, Factor X_a, which accelerates the activation of prothrombin, also has TAME esterase activity and is a serine proteinase as shown by inhibition with DFP (*330–332*). It is also inhibited by soybean trypsin inhibitor (*328*). Esnouf has found the activation of Factor X by Russell's viper venom to involve a change in N-terminals from a questionable Ala– to an Ile– (or possibly Leu–). Activation of Factor X can be brought about by trypsin (*333*) and also by 25% sodium citrate (*334*). Workers in this field are now in general agreement about the essential identity (*335*) of Factor X_a, thrombokinase (Milstone), and autoprothrombin C. When prothrombin is a substrate for Factor X_a the rate of thrombin formation is greatly increased (about 1000-fold) by the addition of a mixture of Factor V, phospholipids, and calcium ions. This mixture does not affect the rate of TAME cleavage. None of these three factors alone or in any mixture catalyzes prothrombin activation if Factor X_a is absent (*336*). The details of this reaction are not well understood, but evidence has been presented (*337, 338*) that a complex is formed by adsorption of

325. E. Marciniak and W. H. Seegers, *Can. J. Biochem. Physiol.* **40,** 597 (1962).
326. W. H. Seegers, *Ann. Rev. Physiol.* **31,** 269 (1969).
327. T. P. Telfer, K. W. Denson, and D. R. Wright, *Brit. J. Haemotol.* **2,** 308 (1956).
328. M. P. Esnouf and W. J. Williams, *BJ* **84,** 62 (1962).
329. D. L. Aronson and D. Menaché, *BBA* **167,** 378 (1968).
330. C. M. Jackson and D. J. Hanahan, *Biochemistry* **7,** 4506 (1968).
331. J. E. Leveson and M. P. Esnouf, *Brit. J. Haematol.* **17,** 173 (1969).
332. W. H. Seegers, E. Marciniak, and L. McCoy, *Thromb. Diath. Haemorrhag.* **22,** 32 (1969).
333. D. J. Hanahan and D. Papahadjopoulos, *Thromb. Diath. Haemorrhag.* Suppl. 17, 71 (1965).
334. T. H. Spaet and J. Cintron, *Blood* **21,** 745 (1963).
335. T. H. Spaet, *Federation Proc.* **23,** 757 (1964).
336. M. P. Esnouf, *BJ* **115,** 1P (1969).
337. D. Papahadjopoulos and D. J. Hanahan, *BBA* **90,** 346 (1964).
338. F. Jobin and M. P. Esnouf, *BJ* **102,** 666 (1967).

Factors X_a and V to phospholipid micelles in the presence of Ca^{2+}. This complex is believed at the moment to be the physiological prothrombin-activating principle. Barton and Hanahan (*339*) have recently shown that prothrombin forms a macromolecular complex with an equimolar mixture of phosphatidylserine and phosphatidylcholine in aqueous dispersion. They also found that it is attached to the prothrombin-activating complex containing Factors V and X_a. They suggest that the kinetics of prothrombin activation are consistent with a model involving catalysis of the reaction at the lipid–water interface and that thrombin, which does not adsorb under these conditions, is released into the aqueous phase. The role of the clotting factors of the "extrinsic" system, including Factor VII and tissue thromboplastin, and of the factors of the "intrinsic" system, including Factors XII, XI, IX, VIII, and platelet phospholipids, in producing the prothrombin-activating complex of Factors X_a, V, calcium, and phospholipids, is still largely obscure.

B. Structural Aspects

1. *Proteolysis of Prothrombin*

Both the activation in 25% sodium citrate, which has the kinetic characteristics of an autocatalytic reaction, and the activation in the presence of tissue thromboplastin, calcium chloride, and Factor V (the prothrombin preparation used had Factor VII and Factor X activity) were studied by N-terminal analysis (*316*, *319*). In addition to the initially present N-terminal alanine of prothrombin, two moles of isoleucine and one mole or less of some other amino acids had appeared as new N-terminals by the time thrombin activity was maximal. This shows that both types of activation involve proteolytic cleavage of the prothrombin molecule. The similarity of the patterns of new N-terminals in the two types of activation and the fact that "citrate" thrombin and "bio"-thrombin have the same N-terminal Ile- sequence indicates that the same peptide bonds are cleaved in the two types of activation. The rate of appearance of the new N-terminals paralleled the rate of thrombin formation initially, but proteolysis continued after maximal thrombin activity has been reached. Diisopropylphosphofluoridate prevented the appearance of new N-terminals as well as thrombin formation which proves the requirement for a serine proteinase in both types of activation.

The activation of prothrombin is inhibited by soybean trypsin in-

339. P. G. Barton and D. J. Hanahan, *BBA* **187**, 319 (1969).

hibitor (*340–342*) which excludes thrombin as the proteinase. The only new C-terminals found by degradation with carboxypeptidases A and B of thrombin and of the "Pro"-containing fraction were arginine and lysine (*93*) which shows the trypsinlike side chain specificity of the activating enzyme. Since Miller and van Vunakis (*226*) demonstrated that a mixture of tissue thromboplastin, calcium chloride, and dilute serum could be incubated with DFP without losing its prothrombin-activating capacity, it appears that in both types of activation the activating enzyme is formed from the prothrombin preparation itself. The activation of prothrombin with poly-L-lysine (*343*) and other polybasic compounds (*344*) has been studied kinetically in some detail and claimed to be nonenzymic (*343*). However, its pH optimum of 8.0–8.5 is still compatible with activation by a trypsinlike enzyme. No N-terminal data on this type of activation are yet available, and it appears that the DFP concentration that was used in the inhibition experiments (*343*) may not have been sufficient to inactivate Factor X_a. Until vigorous proof to the contrary is provided, there is no reason to believe that polylysine activation does not also lead to limited proteolysis of prothrombin, even if the initial reaction with polylysine that leads to activation of the activating enzyme is not proteolytic or enzymic.

2. *Models for Prothrombin*

The trypsinlike properties of the activating enzyme suggest that the B chain of thrombin with its C-terminal serine, which is unlikely to have become C-terminal through tryptic action, constitutes the C-terminal half of the prothrombin molecule. The location in the sequence of prothrombin of the A chain of thrombin must then be somewhere between the N-terminal Ala–Asx– and the B chain of thrombin. Whether in the polypeptide chain of the zymogen the C-terminal –Arg of the A chain is bound to N-terminal Ile– of the B chain is not known as yet. Three models (*345*) for the proteolytic step of prothrombin activation now appear to be consistent with the known facts.

(1) Factor X is a trace contaminant in the most active prothrombin

340. M. B. Glendening and E. W. Page, *J. Clin. Invest.* **30**, 1298 (1951).
341. N. R. Shulman and J. Z. Hearon, *JBC* **238**, 155 (1963).
342. G. F. Lanchantin, J. A. Friedmann, and D. W. Hart, *JBC* **244**, 865 (1969).
343. K. D. Miller, *JBC* **235**, PC63 (1960).
344. K. D. Miller, W. H. Copeland, and J. F. McGarrahan, *Proc. Soc. Exptl. Biol. Med.* **108**, 117 (1961).
345. S. Magnusson, *BJ* **115**, 2P (1969).

preparations. The sequence of Factor X_a is highly homologous but not identical with that of thrombin.

(2) Factor X_a is part of the structure of prothrombin and is formed from a small proportion of the prothrombin molecules which undergo activation in an alternative way to that which produces thrombin. In this hypothesis the sequence of Factor X_a is expected to contain all the structure elements of thrombin that we associate with the serine proteinase activity, that is, probably the entire B chain.

(3) Prothrombin is a double zymogen containing the sequences of two serine proteinases. The "Pro" fragment(s) will then account for the sequence of Factor X_a.

A recent attempt (*346*) to settle this question by producing tryptic fingerprints of native, and of reduced and alkylated, Factor X and prothrombin produced evidence that about a third of the tryptic peptides may be identical. Several of these peptides appear from their electrophoretic mobility to be quite large which makes the results difficult to interpret. Further knowledge of the sequence of prothrombin and of at least some vital parts of the sequence of prothrombin and of Factor X_a is clearly needed to settle this question.

Although the mechanism of activation clearly involves limited proteolysis of prothrombin, it appears that Factor X, whether it turns out to be formed from prothrombin or not, can be activated by some means other than limited proteolysis. Perhaps the initial activating event is a nonproteolytic change in the zymogen triggered by strong salt, polylysine, or at the surface of a phospholipid micelle, which makes possible the formation of an "activation salt bridge" (similar to that from ILE 16 to ASP 194 in chymotrypsin and probably trypsin and thrombin) from the N-terminal of factor X [on Model (1)] or from the N-terminal Ala– of prothrombin [on Model (3)]. This initial structural change may well be the rate limiting step at least in citrate and polylysine activation of prothrombin.

3. *Secondary Proteolysis during Activation*

It appears from recent data (*235*, *347*) that a secondary proteolytic cleavage probably occurring at Lys 79 or Lys 82 in the B chain does not lead to loss of thrombin activity. The thrombin preparation in these cases had been obtained by further purification of Parke Davis Topical thrombin. Whether this cleavage occurs during activation is not clear.

346. D. L. Aronson, A. J. Mustafa, and J. F. Mushinski, *BBA* **118,** 25 (1969).
347. K. G. Mann and C. W. Batt, *JBC* **244,** 6555 (1969).

VI. Prothrombin—Metabolism

A. Turnover Rate

The catabolic half-life of purified prothrombin labeled with ^{131}I and then injected intravenously had been found to be 2.8 days (*348*). It was also deduced that plasma prothrombin accounts for about 64% of total body prothrombin. Since no sialic acid analysis was presented, this very short half-life must be regarded with some caution. It does, however, agree with earlier data obtained by following the disappearance rate after administration of Dicoumarol.

B. Biosynthesis

Using specific fluorescent antibodies against prothrombin, Barnhart *et al.* (*349–351*) have demonstrated conclusively that prothrombin is synthesized in the hepatic cells of the liver and released from these cells into plasma. The only established function of vitamin K in vertebrates is to maintain normal plasma levels of not only prothrombin but also Factors X, VII, and IX. As mentioned previously, it has not been established whether these three activities are properties of the prothrombin molecule or of three distinct plasma proteins, separate from but very similar to prothrombin in structure and properties. It has been established that prothrombin does not contain naphthoquinone compounds (*352*). The mechanism by which vitamin K maintains the normal activity of prothrombin and the other three factors is not known, but it has been generally assumed that the rate of synthesis of prothrombin is greatly reduced in vitamin K deficiency or Dicoumarol poisoning. This view has recently been questioned (*272, 273*) since immunoprecipitin assays using antibodies against native prothrombin detect nearly normal levels of "prothrombin" in plasma even when the activity level is very low. These recent results indicate that three properties of prothrombin depend on vitamin K having been present during biosynthesis of prothrombin: namely, (1) its activation to thrombin in one- and two-

348. S. S. Shapiro and J. Martinez, *J. Clin. Invest.* **48,** 1292 (1969).
349. M. I. Barnhart, *Am. J. Physiol.* **119,** 360 (1960).
350. G. F. Anderson and M. I. Barnhart, *Am. J. Physiol.* **206,** 929 (1964).
351. M. I. Barnhart, *J. Histochem. Cytochem.* **13,** 740 (1966).
352. G. Ray, N. N. Chakravarty, and S. C. Roy, *Ann. Biochem. Exptl. Med.* **22,** 319 (1962).

stage prothrombin assays, (2) its adsorption to barium sulfate from plasma, and (3) its changing electrophoretic mobility on addition of calcium lactate to the electrophoresis buffer. From a recent review on vitamin K (*353*) it appears likely that vitamin K is involved neither in secreting the finished protein from the cell nor in attaching the carbohydrate moieties to the polypeptide chain. Probably the vitamin is required after the polypeptide chain has been synthesized in order to bring about some structural change peculiar to prothrombin and the other three factors.

353. J. W. Suttie, *Federation Proc.* **28**, 1696 (1969).

10

Pancreatic Elastase

B. S. HARTLEY • D. M. SHOTTON

I. History and Distribution

The first report of pancreatic elastase activity dates back to 1878 when Wälchli (*1*) found that ox pancreas digested ligamentum nuchae elastin, and Kühne (*2*) reported that impure trypsin preparations dissolved elastin. In the following year, Pfeiffer (*3*) observed that dried pancreas digested ligamentum nuchae elastin fibers and also ascribed

1. G. Wälchli, *J. Prakt. Chem.* **17,** 71 (1878).
2. W. Kühne, *Untersuch. Physiol. Inst. Univ. Heidelberg* **1,** 219 (1878).
3. P. Pfeiffer, *Arch. Mikroscop. Anat. Entwicklungsmech.* **16,** 17 (1879).

the effect to trypsin, while in 1896 Mall (*4*) observed that elastin fibers were rapidly dissolved by pancreatin, whereas collagen fibers were resistant to digestion. The belief that elastolytic activity was caused by trypsin persisted for many years, and it was not until 1949 that the existence of a distinct and separate pancreatic enzyme, named elastase, was established by Balo and Banga (*5*). In 1952, Banga (*6*) isolated a semipurified crystalline elastase preparation from bovine pancreas, and in 1956 Lewis *et al.* (*7*) first isolated pure elastase from porcine pancreas and established its digestive function. Although Grant and Robbins (*8*) in 1955 had demonstrated the presence of proelastase in the pancreatic juice, there was controversy as to whether proelastase was synthesized in the acinar or the islet tissue of the pancreas (*9*) until Moon and McIvor (*10*) obtained excellent immunological evidence for its acinar origin.

Elastase appears to be present in all mammals investigated, although many attempts to quantitate its presence have been complicated by a failure to appreciate fully the fact that the acinar tissue synthesizes, stores, and secretes proelastase, which only forms active elastase after tryptic activation. Elastase activity has been reported in pancreatic extracts from humans (*5, 7, 10, 11*), cows (*7, 11*), horses (*11*), pigs (*7, 10, 11*), dogs (*12*), cats (*7*), rats (*11*), guinea pigs (*10*), chickens (*7, 11*) and a teleost fish (*13*). Doubts about whether elastase activity resulted from a discrete enzyme rather than a synergistic action of other proteinases were occasioned by failure to observe the activity in ion exchange columns of bovine (*14*) or porcine pancreatic juice (*15*). This apparent discrepancy may be because the enzyme precipitates as a complex with another acidic component when extracts of pancreas are dialyzed free from salt, thus it might have been accidentally removed before ion exchange chromatography.

4. F. P. Mall, *Rept. Johns Hopkins Hosp.* **1,** 171 (1896).
5. J. Balo and I. Banga, *Schweiz. Z. Pathol. Bacteriol.* **12,** 350 (1949).
6. I. Banga, *Acta Physiol. Acad. Sci. Hung.* **3,** 317 (1952).
7. U. J. Lewis, D. E. Williams, and N. G. Brink, *JBC* **222,** 705 (1956).
8. N. H. Grant and K. C. Robbins, *Proc. Soc. Exptl. Biol. Med.* **90,** 264 (1955).
9. I. Mandl, *Advan. Enzymol.* **23,** 163 (1961).
10. H. D. Moon and B. C. McIvor, *J. Immunol.* **85,** 78 (1960).
11. P. Marrama, C. Ferrari, R. Lapiccirella, and U. Parisoli, *Ital. J. Biochem.* **8,** 280 (1959).
12. E. Kolas, I. Foldes, and I. Banga, *Acta Physiol. Acad. Sci. Hung.* **2,** 333 (1951).
13. A. I. Lansing, T. B. Rosenthal, and M. Alex, *Proc. Soc. Exptl. Biol. Med.* **84,** 689 (1953).
14. P. J. Keller, E. Cohen, and H. Neurath, *JBC* **233,** 344 (1958).
15. D. Gratecos, O. Guy, M. Rovery, and P. Desnuelle, *BBA* **175,** 82 (1969).

Uriel and Avremeas (*16*) detected small quantities of a second, quite distinct, elastolytic enzyme in porcine pancreas, in addition to elastase. Many elastolytic enzymes from plants and microorganisms have been reported [discussed, for instance, by Mandl (*9, 17*)], which will not be reviewed here. None of the vertebrate elastases except that from the porcine pancreas has been studied in detail, and nothing is known about the degree of similarity which exists between them. Consequently, the rest of this review will concentrate on a description of porcine pancreatic elastase.

II. Chemical and Enzymic Properties

A. Methods of Assay

1. *Assays using Elastin*

The ability of elastase to degrade elastin, the insoluble fibrous protein of connective tissue, to soluble peptides is not shared by the other pancreatic endopeptidases, trypsin and the chymotrypsins (*18*). This property forms the basis for many assay methods which involve the measurement of the amount of insoluble elastin solubilized by elastase digestion in a certain time. Of these, which have been well reviewed by Mandl (*9, 17*), the most convenient assays are those which involve the colorimetric determination of the amount of dye released into solution from dyed elastin substrates such as Congo red elastin (*19*) and orcein elastin (*20*). The Congo red elastin method is a logical development of the use of Congo red dyed fibrin to measure general proteolytic activity (*21*), while the orcein elastin method has the advantage that orcein is a stain specific for elastin, thus traces of collagen or other protein contaminants in the substrate do not affect the validity of the assay. They are essentially similar in principle, and procedures described for one substrate may be used with the other if desired.

Because of the peculiar cross-linked structure of elastin, the rate of release of peptides into solution is not linear with respect to the rate

16. J. Uriel and S. Avrameas, *Biochemistry* **4,** 1740 (1965).
17. I. Mandl, "Methods in Enzymology," Vol. 5, p. 665, 1962.
18. S. M. Partridge and H. F. Davis, *BJ* **61,** 21 (1955).
19. M. A. Naughton and F. Sanger, *BJ* **78,** 156 (1961).
20. L. A. Sachar, K. K. Winter, N. Sicher, and S. Frankel, *Proc. Soc. Exptl. Biol. Med.* **90,** 323 (1955).
21. N. E. Roaf, *BJ* **3,** 188 (1908).

of hydrolysis of peptide bonds in the substrate by elastase. In consequence none of these assays gives linear progress curves. The orcein elastin method (*20*) involves the measurement of the amount of digestion after an arbitrarily defined interval of 20 min. Consequent inaccuracies arise because the elastin in different assays is at various different stages of breakdown when the digestions are stopped. Calibration curves obtained by this method are far from linear. Gertler and Hofmann (*22*) have shown that the time taken to reach 50% solubilization of substrate bears a close inverse ratio to the elastase activity, and Shotton (*23*) has proposed a standard assay.

The problems involved in these assays make it difficult to compare the results of different workers. Moreover, the presence of proelastase, which binds strongly to elastase (*24*), could affect the assay in crude extracts; and Gertler and Birk (*24*) have shown that trypsin and chymotrypsin, although unable to initiate elastin digestion, can help to solubilize the partially split products. Assays using specific synthetic substrates may avoid some of these problems.

2. *Assays using Synthetic Substrates*

The action of elastase on the B chain of oxidized insulin (*25, 26*) reveals no marked residue specificity such as that shown by trypsin and chymotrypsin, but a preference for uncharged nonaromatic side chains is indicated. Kaplan and Dugas (*27*) have shown that *N*-benzoyl-L-alanine methyl ester is hydrolyzed rapidly by elastase, and this substrate has considerable promise for a specific elastase assay since it is hydrolyzed by trypsin and chymotrypsin at only 1–2% of the rate for elastase (*23, 28*). A standard assay based on the pH-stat method of Kaplan and Dugas (*27*) has been proposed (*23*), and Geneste and Bender (*29*) have used *N*-furylacroyl L-alanine methyl ester in a spectrophotometric assay. An even more promising substrate is *N*-acetyl-L-alanyl-L-alanyl-L-alanyl methyl ester (*30*) which gives pseudo-first-order rate curves since its solubility greatly exceeds its K_m of 0.43 mM. With purified enzyme, active site titrations can be carried out by measuring the

22. A. Gertler and T. Hofmann, *JBC* **242,** 2522 (1967).
23. D. M. Shotton, "Methods in Enzymology," Vol. 19, p. 113, 1970.
24. A. Gertler and Y. Birk, *European J. Biochem.* **12,** 170 (1970).
25. M. A. Naughton and F. Sanger, *BJ* **70,** 4P (1958).
26. A. Sampath Narayanan and R. A. Anwar, *BJ* **114,** 11 (1969).
27. H. Kaplan and H. Dugas, *BBRC* **34,** 681 (1969).
28. G. E. Hein and C. Niemann, *JACS* **84,** 4487 (1962).
29. P. Geneste and M. L. Bender, *Proc. Natl. Acad. Sci. U. S.* **64,** 683 (1970).
30. A. Gertler and T. Hofmann, *Can. J. Biochem.* **48,** 384 (1970).

nitrophenol released after reaction with diethyl-*p*-nitrophenyl phosphate (*31*) or the incorporation of radioactivity by reaction with [^{35}S] pipsyl or tosyl fluoride (*32*).

B. Purification and Purity

1. *Purification Methods*

Elastase is prepared either from an acetone powder of porcine pancreas or from fresh pancreas itself. Early methods of purification used various ion exchange resins and even elastin itself to absorb elastase from the initial pancreatic extracts, but the procedure which has been most widely employed is that of Lewis *et al.* (*7*). This involves fractional ammonium sulfate precipitation of the initial sodium acetate extract of the pancreatic powder, followed by extensive dialysis of the redissolved ammonium sulfate precipitate against water to yield a euglobulin precipitate. This euglobulin contains elastase plus another major and several minor components which are not elastolytic (*7, 19, 33, 34*). It can be crystallized to yield material containing mainly elastase (50–80%) plus a second more acidic nonelastolytic component. By free boundary electrophoresis of this "crystalline elastase," Lewis *et al.* (*7*) isolated fairly pure elastase, which was freely soluble in water and dilute salt solutions (*34*).

Only small amounts of enzyme can be obtained in this way, thus procedures involving chromatography on CM-cellulose (*19*) or DEAE-cellulose (*33, 35*) were developed. Batchwise adsorption of the acidic components of the euglobulin precipitate on DEAE-Sephadex followed by chromatography on CM-cellulose (*36, 37*) were found to give high yields of pure material. This technique was further improved, with some loss of yield, to give a product containing less than 0.5% of trypsin or chymotrypsin (*22*).

31. M. L. Bender, M. L. Begué-Cantón, R. L. Blakeley, L. J. Brubacher, J. Feder, C. R. Gunter, F. J. Kézdy, J. V. Killheffer, Jr., T. H. Marshall, C. G. Miller, R. W. Roeske, and J. K. Stoops, *JACS* **88,** 5890 (1966).
32. D. M. Shotton, Ph.D. Dissertation, University of Cambridge, 1969.
33. U. J. Lewis and E. H. Thiele, *JACS* **79,** 755 (1957).
34. U. J. Lewis, D. E. Williams, and N. G. Brink, *JBC* **234,** 2304 (1959).
35. J. S. Baumstark, W. A. Bardawil, A. J. Sbarra, and N. Hayes, *BBA* **77,** 676 (1963).
36. L. B. Smillie and B. S. Hartley, *BJ* **101,** 232 (1966).
37. V. Ling and R. A. Anwar, *BBRC* **24,** 593 (1966).

The recent discovery (*38*) that elastase can be easily crystallized from dilute salt solutions in which the contaminating proteins are fully soluble has led to a simple and highly selective procedure for obtaining very pure recrystallized elastase in a yield of approximately 2.8 g from 500 g of a commercial acetone powder of pig pancreas (Trypsin 1-300, Nutritional Biochemicals Corp., Cleveland, Ohio, United States). A more detailed discussion of the development of the preparative methods, the early stages of which have also been reviewed by Mandl (*9*, *17*), and full details of the method for preparing recrystallized elastase are given elsewhere (*23*).

2. *Criteria of Purity*

Recrystallized elastase sediments as a single sharp band in the analytical ultracentrifuge, with no sign of contamination or aggregation, having a sedimentation coefficient $s^{\circ}_{20,\mathrm{w}}$ of 2.58 ± 0.02 S, which agrees well with the value of 2.6 S reported for electrophoretically purified elastase by Lewis *et al.* (*7*).

Polyacrylamide gel electrophoresis of recrystallized elastase by the pH 4.5 method of Reisfield *et al.* (*39*) in 8 *M* urea at high loading (100 μg) gives only a single band, while electrophoresis of chromatographically purified elastase (*36*) reveals the presence of at least two minor components.

The most persistent contaminants of the elastase preparation are the other pancreatic endopeptidases, trypsin and the chymotrypsins. Although these enzymes will slowly hydrolyze synthetic substrates of elastase, elastase is unable to hydrolyze *N*-benzoyl-L-arginine ethyl ester (BAEE) or *N*-acetyl-L-tyrosine ethyl ester (ATEE), specific ester substrates for trypsin and the chymotrypsins, respectively, to any significant extent. Assaying for BAEEase and ATEEase activities is therefore an extremely sensitive method of measuring contaminating traces of these enzymes, especially if these activities are measured in a pH stat, where massive quantities of purified enzyme can be used to obtain accurate values of very low level contaminants. Such assays reveal the presence of traces of trypsin and chymotrypsin in chromatographically purified elastase. Recrystallized elastase, however, shows less than 0.001% contamination by carboxypeptidase, leucine aminopeptidase, ribonuclease, and deoxyribonuclease (*39a*), and only 0.03% (w/w) chymotryptic and 0.006% (w/w) tryptic contamination (*32*) it is there-

38. D. M. Shotton, B. S. Hartley, N. Camerman, T. Hofmann, S. C. Nyburg, and L. Rao, *JMB* **32**, 155 (1968).
39. R. A. Reisfield, U. J. Lewis, and D. E. Williams, *Nature* **195**, 281 (1962).
39a. A. C. Reimer, personal communication (1969).

TABLE I
AMINO ACID ANALYSIS AND COMPOSITION OF ELASTASE

Amino acid	Amino acid analysis [Ref. (40)][a]	[Ref. (22)][b]	[Ref. (32)][c]	Composition from sequence and X-ray studies
Aspartic acid	23.4	24.5	23.4	6
Asparagine	23.4	24.5	23.4	18
Threonine[d]	18.9[a]	18.9	18.8	19
Serine[d]	22.8	22.2	23.2	22
Glutamic acid	19.6	20.0	19.0	4
Glutamine	19.6	20.0	19.0	15
Proline	7.2	7.5	7.2	7
Glycine	26.8	25.2	25.6	25
Alanine	17.6	17.0	16.8	17
½-Cystine	7.9[e]	6.5	6.0	8
Valine[d]	27.0	27.7	26.0	27
Methionine	1.8	1.9	1.7	2
Isoleucine[d]	9.9	9.9	9.9	10
Leucine	17.8	18.1	17.9	18
Tyrosine	10.6	10.6	11.3	11
Phenylalanine	3.4	3.1	3.3	3
Tryptophan	7.7[f]	7.3[g]	—	7
Lysine	2.9	3.1	3.0	3
Histidine	5.0	6.2	6.0	6
Arginine	10.8	12.1	11.7	12
Total				240

[a] Diisopropylphosphoryl (DIP)-elastase (mean of 5 analyses). Recalculated to give the best fit with the amino acid composition (published values × 1.039).

[b] Chromatographically purified elastase (mean of 4 analyses). Recalculated to give the best fit with the amino acid composition (published values × 1.062).

[c] Recrystallized elastase (mean of 5 analyses).

[d] Corrected for incomplete hydrolysis or hydrolysis losses.

[e] Determined as cysteic acid after oxidation.

[f] Determined spectrophotometrically on a peptic digest (*41*).

[g] Determined colorimetrically on an enzymic digest.

fore a preparation of extremely high purity, probably better than 99.9%. This is borne out by the excellent agreement obtained between the amino acid analysis of elastase and its amino acid composition determined from sequence studies, as described in Table I.

C. Physicochemical Properties

The amino acid composition of bovine pancreatic elastase, determined from the sequence and crystallographic studies described below, is shown in Table I. The results of amino acid analysis of three different preparations of elastase are also indicated. The preparation of diisopropylphosphoryl (DIP)-elastase analyzed by Brown *et al.* (*40*) contained traces of other proteinases, but the preparations of Gertler and Hofmann (*22*) and of Shotton (*32*) appeared to be pure by gel electrophoresis. The good agreement between the analytical and sequence results shows that elastase contains 240 amino acid residues. Reaction of DIP-elastase with phenyl isothiocyanate showed only *N*-terminal Val–Val– in 90% yield (*40*), and reaction of elastase with cyanate gave 92% yield of *N*-terminal Val– plus 3–15% yield of other *N*-terminal groups which may have arisen by autolysis during the carbamylation (*22*). A single *N*-terminal peptide sequence Val–Val–Gly–Gly–Thr–Glu– was found after reaction with fluorodinitrobenzene and digestion with pepsin (*40*).

The half-cystine analysis (*22, 31*) is not reliable, but the analysis of cystine after conversion to cysteic acid showed eight residues per molecule, which were all disulfide bridged. Hence elastase is a single polypeptide chain of 240 residues cross-linked by four disulfide bridges.

The molecular weight of elastase, calculated from the amino acid composition, is 25,900; Lewis *et al.* (*7*) estimated it to be 25,000 from sedimentation diffusion data, Naughton and Sanger (*19*) calculated 28,500 from specific activity studies of ^{32}P-DIP-elastase, and Gertler and Hofmann (*22*) suggested 24,860 from amino acid analyses. Lewis *et al.* (*7*) reported the sedimentation constant to be 2.6 S, the isoelectric point to be pH 9.5 ± 0.5, and the partial specific volume to be 0.73 $cm^3\ g^{-1}$, a value which agrees well with the theoretical value of 0.726 $cm^3\ g^{-1}$ calculated from the amino acid composition by the method of Cohn and Edsall (*42*).

The molar and specific extinction coefficients for a pure recrystallized elastase solution, whose molarity was determined by hydrolysis and amino acid analysis, have been calculated in a variety of solvents. In 0.05 *M* sodium acetate, pH 5.0, $\epsilon_{1\,cm,\,280\,nm} = 5.23 \times 10^4$ and $E^{1\%}_{1\,cm,\,280\,nm} = 20.2$; in 0.1 *M* sodium hydroxide, $\epsilon_{1\,cm,\,280\,nm} = 5.74 \times 10^4$ and $E^{1\%}_{1\,cm,\,280\,nm} = 22.2$; and in 0.1 *M* sodium hydroxide, following thorough digestion with pepsin, $\epsilon_{1\,cm,\,280\,nm} = 6.11 \times 10^4$ and $E^{1\%}_{1\,cm,\,280\,nm} = 23.6$. Previous reports in undefined solvents (*19, 27, 43*) agree reasonably with these values except that of Lewis *et al.* (*7*) which is rather low.

The crystallographic studies described below have revealed elastase to be a compact globular molecule having dimensions of 5.5 by 4.0 by

40. J. R. Brown, D. L. Kauffman, and B. S. Hartley, *BJ* **103,** 497 (1967).
41. G. H. Beaven and E. R. Holiday, *Advan. Protein Chem.* **7,** 319 (1952).
42. E. J. Cohn and J. T. Edsall, *in* "Proteins, Amino Acids and Peptides," p. 370. Reinhold, New York, 1943.
43. S. Wasi and T. Hofmann, *BJ* **106,** 926 (1968).

3.8 nm, which agree reasonably well with the frictional ratio f/f_0 of 1.2 estimated by Lewis *et al.* (*7*).

D. PROELASTASE

The presence of proelastase, the inactive precursor of elastase, in porcine pancreas was first demonstrated by Grant and Robbins (*8, 44*), and partial purification was reported by Lamy and Tauber (*45*) and Gertler and Hofmann (*46*). Gertler and Birk (*24*) have recently isolated pure proelastase from the euglobulin precipitate of Lewis *et al.* (*7*) by preparing it from fresh frozen pancreas in the presence of added soybean trypsin inhibitor. Proelastase was removed from the redissolved euglobulin precipitate by absorption onto elastin at pH 8.5, eluted at pH 3.6, and then chromatographed on CM-cellulose at pH 4.5. This proelastase was homogeneous by acrylamide gel electrophoresis and ultracentrifugation, and contained only traces of chymotrypsinogen and no trypsinogen, as measured by ATEE and BAEE assays after activation. The sedimentation coefficient was 2.66 S, slightly greater than that of the active enzyme.

Activation of proelastase to elastase is induced by tryptic cleavage of a single peptide bond in the inactive zymogen in an activation process closely resembling those of trypsinogen and chymotrypsinogen, which results in the removal from the *N*-terminal end of the molecule of a small activation peptide, enabling the enzyme to adopt its active conformation. A comparison of the amino acid analyses of elastase and proelastase suggests that the activation peptide contains lysine, arginine, aspartic acid, glutamic acid, proline, alanine, phenylalanine, and probably serine (*24*). It has not yet been isolated and sequenced.

Uram and Lamy (*47*) have recently reported the isolation of two "proelastases" differing in their isoelectric points; that with the isoelectric point of 10.7, on tryptic activation, shows high elastolytic activity but very low ATEEase activity, and it is thus probably identical with the zymogen of elastase isolated by Gertler and Birk (*24*). The other, whose isoelectric point is 9.6, shows both elastolytic and ATEEase activity after tryptic activation and may therefore be the zymogen of the uncharacterized minor elastolytic pancreatic enzyme isolated by Uriel and Avrameas (*16*).

The strong specific binding of proelastase to elastin shows that the

44. N. H. Grant and K. C. Robbins, *ABB* **66,** 396 (1957).
45. F. Lamy and S. Tauber, *JBC* **238,** 939 (1963).
46. A. Gertler and T. Hofmann, *Israel J. Chem.* **5,** 132 (1967).
47. M. Uram and F. Lamy, *BBA* **194,** 102 (1969).

substrate binding site is already formed and accessible in the zymogen and suggests that the structures of proelastase and elastase are very similar.

E. Stability

Elastase is readily soluble in water and dilute salt solutions at concentrations up to 50 mg/ml between pH 4 and 10.5, and within this pH range at 2°C elastase solutions are stable for prolonged periods below pH 6.0 and reasonably stable at higher pH values. It is, however, a powerful proteolytic enzyme and will rapidly autolyze to a mixture of peptides if incubated at room temperature at or near its optimum pH of pH 8.8. A certain degree of care is therefore required in the design of experiments which involve the digestion of elastase with other enzymes possessing similar pH optima (for instance, trypsin), or the investigation of the properties of elastase at alkaline pH values, in order to avoid autolytic cleavages which would invalidate the results obtained. At pH 5.0 the proteolytic activity is very slight, and solutions at this pH can be used at room temperature with little autolysis.

Brief titration of elastase down to pH 2.6 is fully reversible (*43*), but prolonged incubation at acid pH values leads to irreversible denaturation with loss of enzymic activity (*7*). Lewis *et al.* (*7*) found that elastase could be incubated at pH values up to 12.0 for 24 hr at 5°C and still show full activity when assayed for elastolytic activity at pH 8.8, indicating that high pH-induced conformational changes (see below) up to this pH are fully reversible.

Crystals of elastase are stable indefinitely in 1.2 *M* sodium sulfate, buffered at pH 5.0 with 0.01 *M* sodium acetate, at room temperature, and show full activity when redissolved and assayed. Freeze-dried elastase prepared after dialysis against 1 m*M* acetic acid is stable indefinitely at —10°C with no loss of proteolytic activity (*32*).

F. Enzymic Activity and Substrate Specificity

1. *Action on Proteins*

It seems odd, in retrospect, that it has taken so long to recognize that elastase is a typical pancreatic endopeptidase. Perhaps its tendency to precipitate on dialysis of pancreatic extracts is responsible for its absence from the classic book "Crystalline Enzymes," (*48*) and for its subsequent

48. J. H. Northrop, M. Kunitz, and R. M. Herriott, "Crystalline Enzymes." Columbia Univ. Press, New York, 1948.

humble history compared to trypsin and chymotrypsin. Some early studies seem bedeviled by the assumption that elastase should hydrolyze only elastin and not other proteins, and reports that the enzyme has mucolytic (*49*) or lipolytic (*50*) activity are understandable in view of the difficulty of preparing elastin free from contaminants.

Partridge and Davis (*18*) were the first to show that elastase split peptide bonds in elastin, and Lewis *et al.* (*7*) showed that the pure enzyme could digest many other proteins. Naughton and Sanger (*19, 25*) showed that elastase split the A and B chains of oxidized insulin at bonds adjacent to uncharged nonaromatic residues, but recent studies with more highly purified elastase suggest that some of the cleavages reported by them may have resulted from impurities and that the elastase specificity is directed mainly to the peptide bonds on the C-terminal side of Ala 14, Val 18, and Gly 23 in the B chain and Ala 8 and Ser 12 in the A chain (*26*). Klee (*51*) has also shown that elastase splits a Ser–Ala bond, among others, in ribonuclease. This fairly broad specificity with a preference for Ala bonds is borne out by studies with synthetic peptides (see below). The physiological activity of elastase should be looked on as that of a broad specificity proteinase complementing the action of trypsin and chymotrypsin, but its action on elastin may have significance beyond simply being the result of the high content of aliphatic side chains in this cross-linked substrate. The peculiar pattern of digestion in which the enzyme appears to bind strongly to elastin and split several bonds before releasing soluble peptides is also reflected in the ability of the zymogen to bind strongly to elastin (*8*). Moreover, Kaplan *et al.* (*52*) have shown that the enzyme causes rapid lysis of cells of *Arthrobacter globiformis*—a gram-positive soil bacterium with L-alanine residues in its cell wall. It seems quite possible that the specificity of elastase will prove more complex than a simple preference for small hydrophobic side chains.

2. *Action on Synthetic Substrates*

Useful synthetic substrates for elastase have only recently been discovered, and kinetic parameters for these are summarized in Table II.

49. J. W. Czerkawski, *Abstr. 4th Intern. Congr. Biochem., Vienna, 1958* Vol. 15, p. 43. Pergamon Press, Oxford, 1960.

50. A. I. Lansing, T. B. Rosenthal, M. Alex, and E. W. Dempsey, *Anat. Record* **114,** 555 (1952).

51. W. A. Klee, *JBC* **240,** 2900 (1965).

52. H. Kaplan, V. B. Symonds, H. Dugas, and D. R. Whitaker, *Can. J. Biochem.* **48,** 649 (1970).

53. T. Hofmann, personal communication (1969).

TABLE II
ACTIVITY OF ELASTASE AGAINST SYNTHETIC SUBSTRATES

Substrate[a]	k_{cat}/K_m (M^{-1} sec^{-1})	k_{cat} (sec^{-1})	K_m (mM)	E_0 (μM)	S_0 (mM)	pH	Ref.
N-Benzoyl-L-alanine methyl ester	638	12.3	19.3	0.051	1.26–5.06	8.0–10.0	*b,c*
N-Benzoyl-L-alanine methyl ester	600	—	—	8.0–10.0	0.3–0.5	8.0–10.0	*d*
N-Benzoyl-L-valine methyl ester	19	—	—	—	—	8.0	*c*
N-Benzoyl-L-leucine methyl ester	21	—	—	—	—	8.0	*c*
N-Benzoyl-L-isoleucine methyl ester	16	—	—	—	—	8.0	*c*
N-Benzoyl-D-alanine methyl ester	2.8	—	—	0.38	4.83	8.0	*c*
N-Benzoylglycine methyl ester	2.9	—	—	1.04	1.95–5.05	8.0	*c*
N-Acetyl-L-Ala-L-Ala-L-Ala methyl ester[h]	170,000	73	0.43	0.2–0.4	0.2–7.3	8.0–10.5	*e*
N-Acetyl-L-alanine methyl ester	43.7	6.7	153	0.2–0.4	0.2–7.3	8.0	*e*
N-Acetyl-L-alanine methyl ester	63.6	—	—	0.423	1.84–4.86	8.0	*c*
N-Furylacryloyl-L-alanine methyl ester	132	—	—	4.0–5.0	0.08–0.12	8.0	*d*
N-Methyloxycarbonyl-L-alanine methyl ester	33	0.44	13	—	—	8.0	*c*
N-Ethyloxycarbonyl-L-alanine methyl ester	46	0.99	22	—	—	8.0	*c*
N-*n*-Butyloxycarbonyl-L-alanine methyl ester	260	5.5	21	—	—	8.0	*c*
N-*i*-Butyloxycarbonyl-L-alanine methyl ester	220	7.8	36	—	—	8.0	*c*
N-*t*-Butyloxycarbonyl-L-alanine methyl ester	120	4.7	41	—	—	8.0	*c*
N-Benzyloxycarbonyl-L-alanine methyl ester	300	3.9	13	—	—	8.0	*c*
N-*n*-Propyloxycarbonyl-L-alanine methyl ester	160	3.5	21	—	—	8.0	*c*
N-*n*-Propyloxycarbonyl-L-alanine ethyl ester	190	3.6	19	—	—	8.0	*c*

N-*n*-Propyloxycarbonyl-L-alanine isobutyl ester	340	2.1	6.3	—	—	8.0	*c*
N-*n*-Propyloxycarbonyl-L-alanine isopentyl ester	570	~5	~9	—	—	8.0	*c*
N-*n*-Propyloxycarbonyl-L-alanine phenylethyl ester	580	—	—	—	—	8.0	*c*
N-Benzyloxycarbonyl-L-alanine *p*-nitrophenyl ester	185,000	110	0.6	—	—	7.90	*d*
N-Benzyloxycarbonyl-L-leucine *p*-nitrophenyl ester	30,400	—	—	0.1–0.5	0.005	7.79	*f*
N-Benzyloxycarbonylglycine *p*-nitrophenyl ester	16,300	—	—	—	—	7.82	*d*
N-Benzyloxycarbonyl-L-norleucine *p*-nitrophenyl ester	5,800	—	—	—	—	7.82	*d*
N-Benzyloxycarbonyl-L-valine *p*-nitrophenyl ester	2,200	—	—	—	—	7.82	*d*
N-Benzyloxycarbonyl-L-isoleucine *p*-nitrophenyl ester	280	—	—	—	—	7.86	*d*
N-*t*-Butyloxycarbonyl-L-alanine *p*-nitrophenyl ester	—	—	0.16	—	—	8.0	*g*
p-Nitrophenyl formate[h]	1,690	0.0224	0.0133[i]	0.5–2.6	0.001–0.02	7.69	*f*
p-Nitrophenyl isobutyrate[h]	1,620	0.0340	0.0162[i]	0.7–2.6	0.007–0.15	7.68	*f*
p-Nitrophenyl acetate[h]	410	0.0210	0.0512[i]	7.0	0.16	7.43	*f*
p-Nitrophenyl trimethylacetate[h]	123	0.0018	0.0143[i]	5.0	0.055	7.33	*f*

[a] First-order kinetics were observed for all substrates except where indicated.
[b] Kaplan and Dugas (*27*). Assays in 0.1 *M* KCl at 25°.
[c] Kaplan *et al.* (*52*). Assays in 0.1 *M* KCl at 25°.
[d] Geneste and Bender (*29*). Assays in phosphate buffer, $I = 0.1$ at 25°.
[e] Gertler and Hofmann (*30*). Assays in 1.0 m*M* tris-HCl at 25°.
[f] Bender *et al.* (*31*). Assays in phosphate buffer, $I = 0.05$ at 25°.
[g] Visser and Blout (*57*).
[h] Michaelis–Menten kinetics.
[i] Here, K_m substitute represents the ratio of the deacylation and the acylation rate constants, rather than a binding constant.

Kaplan and Dugas (*27*) showed that *N*-benzoyl-L-alanine methyl ester was hydrolyzed at a sufficient rate to allow a steady state kinetic analysis, and Geneste and Bender (*29*) have explored side chain specificity using *p*-nitrophenyl esters of a series of *N*-benzyloxycarbonyl amino acids. Alanyl esters were hydrolyzed much more rapidly than glycyl esters or derivatives with longer side chains, but the insolubility and high spontaneous hydrolysis of these *p*-nitrophenyl esters made it difficult to obtain reliable values of K_m and k_{cat}.

Kaplan *et al.* (*52*) have recently carried out a systematic study of the specificity of elastase toward esters of *N*-acylamino acids with aliphatic side chains. A strong preference for an alanyl residue is again apparent, and it seems likely that the methyl side chain is important in orienting the substrate at the catalytic site, since glycyl esters, D-alanyl esters, and valyl, leucyl, or isoleucyl esters are hydrolyzed much less rapidly. There is, however, considerable specificity for the *N*-acyl substituent as is apparent from the differences in rate between *N*-acetyl, *N*-benzoyl, and a series of *N*-alkyl-oxycarbonyl derivatives of L-alanine methyl ester (*52*). This suggests that residues N-terminal to the bond split in a peptide might contribute to the specificity, and the fact that *N*-acetyl-L-alanyl-L-alanyl-L-alanyl methyl ester (*30*) is the best synthetic substrate so far for elastase makes this assumption even more probable.

As expected, the substrates with more powerful leaving groups are hydrolyzed faster, and the difference in k_{cat} between *N*-benzyloxycarbonyl-L-alanine methyl ester and *p*-nitrophenyl ester makes it unlikely that hydrolysis of an acyl-enzyme intermediate could be rate limiting for the methyl ester as it is assumed to be for specific ester substrates of chymotrypsin (*54*). However, hydrolysis of an acyl enzyme is probably rate limiting for the *p*-nitrophenyl esters of fatty acids since a "burst" of *p*-nitrophenol is observed with *p*-nitrophenyl trimethylacetate (*55*), as was found for chymotrypsin and *p*-nitrophenyl acetate (*56*), and the turnover rates for *o*-, *m*-, and *p*-nitrophenyl trimethylacetate are identical.

The pH dependence for hydrolysis of *N*-benzoyl-L-alanine methyl ester (*27*) or *N*-3(2-furylacryloyl)-L-alanine methyl ester (*29*) shows that k_{cat} is controlled by an ionizing group in the enzyme with p*K* 6.5–6.85, as with other serine enzymes, but remains constant from pH 8 to at least pH 10.0 (*52*). From pH 5 to pH 10, K_m is constant. Above this pH a drop in activity occurs, but interpretation of this is difficult because

54. G. P. Hess, Chapter 7, this volume.
55. M. L. Bender and T. H. Marshall, *JACS* **90**, 201 (1968).
56. B. S. Hartley and B. A. Kilby, *BJ* **50**, 672 (1952).
57. L. Visser and E. R. Blout, *Federation Proc.* **28**, 407 (1969).

of nonenzymic hydrolysis of the substrate and enzyme denaturation (*29*). The pH dependence for the elastase hydrolysis of *p*-nitrophenyl trimethylacetate showed that k_{cat} was controlled by an ionizing group in the enzyme with pK_a 6.7 (*55*). Since in this case deacylation must be rate limiting, whereas with benzoyl-L-alanine methyl ester acylation is probably rate limiting (see above), the ionizing group must be essential for both acylation and deacylation, but have no effect on substrate binding. Thus far, elastase behaves exactly as chymotrypsin, but chymotrypsin has a group with pK_a 8.5 which controls the conformation of the enzyme and hence K_m and this ionization is absent in elastase (*27*). This point will be discussed below.

3. *Inhibitors and Activators*

The elastolytic activity of elastase is appreciably affected by salts (*17*). Sodium chloride at 50–70 m*M* causes 50% inhibition, and similar effects are observed with potassium chloride, ammonium sulfate or sodium cyanide. Copper sulfate (0.01 m*M*) gave 50% inhibition but millimolar concentrations of zinc, manganese, cobalt, magnesium, or calcium had no effect (*7*). An observation that salt inhibition at high ionic strengths was reduced by cysteine (*58*) might result from its chelating action on heavy metals, but 1 m*M* ethylenediaminetetraacetic acid had no effect (*59*).

In contrast, the activity of elastase toward benzoyl-L-alanine methyl ester, while being inhibited 14% by 100 m*M* sodium chloride, is greatly stimulated by sodium sulfate or by Tris (*32*). When assayed at a substrate concentration of 5 m*M* in the pH stat in 0.1 *M* potassium chloride, pH 8.0, at 25°, 1.2 *M* sodium sulfate gave a threefold stimulation in rate which resulted from a decrease in $K_{m(app)}$ while k_{cat} was unaffected, and 10 m*M* Tris gave a 25% increase in rate. A similar activation by Tris is reported for the elastase-catalyzed hydrolysis of *p*-nitrophenyl trimethylacetate (*55*). One wonders whether the Tris might itself be acylated in this reaction.

Various serum and intestinal nondialyzable inhibitors of elastase have been reported, but none of these has been well characterized and their biological function is not understood. Attempts to correlate the amounts of elastase and elastase inhibitors with the onset of various diseases, especially arteriosclerosis, have unfortunately yielded ambiguous but essentially negative results. The effects of these nondialyzable inhibitors and various salts on the activity of elastase, and the relationships be-

58. J. Thomas and S. M. Partridge, *BJ* **74**, 600 (1960).
59. F. Lamy, C. P. Craig, and S. Tauber, *JBC* **236**, 86 (1961).

tween elastase and arteriosclerosis are reviewed more fully by Mandl (*9*). Soybean trypsin inhibitor does not inhibit elastase (*24*).

Hofmann (*53*) has studied competitive inhibition of the elastase-catalyzed hydrolysis of *N*-benzyloxycarbonyl-glycine *p*-nitrophenyl ester by some synthetic peptides and derivatives of aliphate L-amino acids. The tripeptide Ala–Ala–Ala, and *N*-benzyloxycarbonyl or *N*-acetyl derivatives of Ala–Ala–Ala, Ala–Ala, Val–Val, Leu–Val, Val–Leu, Leu–Leu, Val–Ile, and Ile–Val showed inhibition with K_i between 1 and 10 m*M*, whereas the dipeptides or *N*-formyl dipeptides were ineffective. These studies once again focus attention on possible binding sites in elastase substrates distant from the bond which is hydrolyzed.

4. *Irreversible Inhibitors*

Naughton and Sanger (*19*) showed that elastase was inhibited by stoichiometric reaction with diisopropylphosphorofluoridate (DFP). Hartley *et al.* (*60*) showed that the patterns of radioactive peptides produced in partial acid hydrolysates of ^{32}P-DIP-elastase, ^{32}P-DIP-chymotrypsin, and ^{32}P-DIP-trypsin were identical and concluded that these arose from a common –Gly–Asp–Ser(DIP)–Gly– sequence in each, but Naughton *et al.* (*61*) showed that this evidence allowed one to postulate only a common –Asp–Ser(DIP)–Gly– sequence.

As in other serine proteinases, the active center serine residue of elastase appears to behave as a uniquely powerful nucleophile. Like chymotrypsin (*62*) it reacts with *p*-nitrophenyl diethyl phosphate to liberate one mole of nitrophenol per mole of enzyme, and this reaction has been proposed as an assay for the number of catalytic sites (*31*). This serine residue also reacts stoichiometrically with sulfonyl fluorides such as tosyl fluoride (*63, 63a*) just like Ser 195 in chymotrypsin (*64*).

Chymotrypsin or trypsin can be specifically inhibited by reacting His 57 with substrate analog reagents such as tosyl-L-phenylalanine chloromethyl ketone (*65*) or tosyl-L-lysyl chloromethyl ketone (*66*), but no comparable reaction has been found for elastase. No inhibition was observed after reaction with tosyl-L-alanine chloro- or bromomethyl ketone

60. B. S. Hartley, M. A. Naughton, and F. Sanger, *BBA* **34,** 243 (1959).
61. M. A. Naughton, F. Sanger, B. S. Hartley, and D. C. Shaw, *BJ* **77,** 149 (1960).
62. B. S. Hartley and B. A. Kilby, *Nature* **166,** 784 (1950).
63. D. M. Shotton and B. S. Hartley, *Nature* **225,** 802 (1970).
63a. H. C. Watson, D. M. Shotton, J. M. Cox, and H. Muirhead, *Nature* **225,** 806 (1970).
64. A. M. Gold and D. E. Fahrney, *BBRC* **10,** 55 (1963).
65. G. Schoellmann and E. Shaw, *Biochemistry* **2,** 252 (1963).
66. E. Shaw, M. Mares-Guia, and W. Cohen, *Biochemistry* **4,** 2219 (1965).

or tosyl-L-valine chloromethyl ketone (*52*). Inhibition by 1-bromo-4-(2,4-dinitrophenyl)-2-butanone has been reported, but this was ascribed to reaction with a glutamic acid residue (*57*).

III. Primary Structure

A. Determination of the Amino Acid Sequence

1. *Disulfide Bridged Peptides*

The diagonal electrophoretic technique of Brown and Hartley (*67*) allows selective purification of cysteic acid peptides from an enzymic digest and indicates the manner in which these peptides were originally disulfide bridged. Smillie and Hartley (*68*) were thus rapidly able to determine the sequence around two disulfide bridges of elastase and to show that one of these, bridge A in Fig. 1, is very homologous with a similar disulfide bridge in chymotrypsin A, chymotrypsin B, and trypsin. This bridge, which joins Cys 42 and Cys 58, brings the only two histidine residues of chymotrypsin close together in primary structure; thus hypotheses that both His 40 and His 57 might participate in the catalytic were thereby encouraged (*69, 70*).

The complete pattern of disulfide bridges in elastase was subsequently determined by similar methods by Brown *et al.* (*40*), as shown in Fig. 1, and these sequences together with the N-terminal sequence Val–Val–Gly–Gly–Thr–Glu– (*40*) accounted for 82 of the 240 residues in the single polypeptide chain. Fifty-two percent of these residues were identical to corresponding residues in bovine chymotrypsin A, chymotrypsin B, or trypsin; and the four bridges of elastase were homologous with the four bridges which were common to chymotrypsin A, chymotrypsin B, and trypsin. Since the N-terminal sequence was homologous, it seemed very likely that proelastase would be activated in a similar fashion to the other enzymes. Moreover, the striking similarity in sequence around His 57 and Ser 195 and each of the disulfide bridges allowed Hartley *et al.* to conclude in 1965 (*71*) that not only did these enzymes show

67. J. R. Brown and B. S. Hartley, *BJ* **101,** 214 (1966).
68. L. B. Smillie and B. S. Hartley, *JMB* **10,** 183 (1964).
69. B. S. Hartley, *in* "Structure and Activity of Enzymes" (T. W. Goodwin, J. I. Harris, and B. S. Hartley, eds), p. 47. Academic Press, New York, 1964.
70. M. L. Bender and F. J. Kézdy, *JACS* **86,** 3704 (1964).
71. B. S. Hartley, J. R. Brown, D. L. Kauffman, and L. B. Smillie, *Nature* **207,** 1157 (1965).

```
                                                                          170 170
            38  39  40  41  42  43  44  45  46              166 167 168 169 170  A   B  171
            Trp-Ala-His-Thr-CyS-Gly-Gly-Thr-Leu             Ala-Ile-CyS-Ser-Ser-Ser-Ser-Tyr
Bridge A                    |                                       |                          Bridge B
            54  55  56  57  58  59  60  61  62              180 181 182 183 184 185 186
            Thr-Ala-Ala-His-CyS-Val-Asp-Arg-Glu             Met-Val-CyS-Ala-Gly-Gly-Asn

                                            131 132 133 134 135 136 137
                                            Ala-Asn-Asn-Ser-Pro-CyS-Tyr
                188                                             |                              Bridge D
            187 188  A  189 190 191 192 193 194 195 196  197 198 199 200 201 202 203 204 205 206 207
            Gly-Val-Arg-Ser-Gly-CyS-Gln-Gly-Asp-Ser-Gly(Gly,Pro)Leu-His-CyS-Leu-Val-Asn-Gly-Gln-Tyr
Bridge C        217             |       221
            216 217  A  218 219 220 221  A  222 223 224 225 226 227 228
            Val-Ser-Arg-Leu-Gly-CyS-Asn-Val-Thr-Arg-Lys-Pro-Thr-Val-Phe
```

Fig. 1. Disulfide bridges of elastase (*40*). Residue numbers have been added, using the chymotrypsinogen A numbering scheme. "Insertions" in the elastase sequence relative to that of chymotrypsinogen A are numbered **170A**, **170B**, etc. A serine residue has been deleted from the original sequence of Brown *et al.* (*40*) at position **170C**, a glycine residue has been added at position **205**, and residue **186** has been changed from aspartic acid to asparagine on the basis of subsequent evidence (*63*).

unmistakable signs of descent from a common ancestor but also they must retain large elements of tertiary structure in common.

2. *Complete Amino Acid Sequence*

Since these pancreatic proteinases form a group with common catalytic activity but differences in specificity, it was important to determine the complete pattern of identities and differences between them. By 1965 the sequences of bovine chymotrypsinogen A (*72*) and trypsinogen (*73*) were almost complete, and application of the disulfide bridge diagonal procedure had yielded about one-third of the sequence of bovine chymotrypsinogen B (*68*). In 1966, Shotton and Hartley (*63*) therefore began to determine the complete sequence of porcine elastase.

Tryptic peptides were first isolated from DIP-elastase in which the disulfide bridges had been reduced and reacted with ethylenimine in order to introduce additional trypsin-sensitive bonds. Fractionation on Zeo-Karb 225 ion exchange resin followed by high voltage paper electrophoresis yielded peptides accounting for about 75% of the sequence, but the remainder could not be eluted from the ion exchange column. These missing fragments were isolated by gel filtration of tryptic digests of DIP-elastase which had been reduced, reacted with ^{14}C-iodoacetic acid, and then cleaved at the two methionine residues by reaction with cyanogen bromide. One tryptic peptide, from the C-terminus, was insoluble in most solvents including 8 *M* urea and was sequenced from peptic fragments of the insoluble "core." The tryptic peptides were overlapped by isolating fragments from peptic digests of elastase or from a combined tryptic plus chymotryptic digest of reduced amino-ethylated DIP-elastase in which the lysine residues had been acetylated with ^{14}C-acetic anhydride. The amide groups of asparagine and glutamine residues were unambiguously located by measurements of the electrophoretic mobility of peptides containing them.

These studies completely determined the sequence shown in Fig. 2. Confirmation is given by four independent lines of evidence. First, tryptic peptides from each of the purified cyanogen bromide fragments of elastase were identified and agreed with those predicted from local overlaps. Second, the sequence agrees well with the amino acid analysis of elastase (Table I). Third, the extensive homologies with other serine proteinases would be greatly lessened by any reordering of the elastase peptides. Fourth, a model of the elastase molecule, built from this

72. B. S. Hartley, *Nature* **201,** 1284 (1964).

73. K. A. Walsh, D. L. Kauffman, K. S. V. Sampath Kumar, and H. Neurath, *Proc. Natl. Acad. Sci. U. S.* **51,** 301 (1964).

16	*17*	*18*	*19*	*20*	*21*	*22*	*23*	*24*	*25*	*26*	*27*	*28*	*29*	*30*	*31*
1	2	3	4	5	6	7	8	9	10	11	12	13	14	15	16
Val-	Val-	Gly	Gly-	Thr	Glu-	Ala-	Gln-	Arg-	Asn-	Ser-	Trp-	Pro-	Ser-	Gln-	Ile-

32	*33*	*34*	*35*	*36*	*36A*	*36B*	*36C*	*37*	*38*	*39*	*40*	*41*	*42*	*43*	*44*
17	18	19	20	21	22	23	24	25	26	27	28	29	30	31	32
Ser-	Leu-	Gln-	Tyr-	Arg-	Ser-	Gly-	Ser-	Ser-	Trp-	Ala-	His-	Thr-	Cys-	Gly-	Gly-

45	*46*	*47*	*48*	*49*	*50*	*51*	*52*	*53*	*54*	*55*	*56*	*57*	*58*	*59*	*60*
33	34	35	36	37	38	39	40	41	42	43	44	45	46	47	48
Thr-	Leu-	Ile-	Arg-	Gln-	Asn-	Trp-	Val-	Met-	Thr-	Ala-	Ala-	His-	Cys-	Val-	Asp-

61	*62*	*63*	*64*	*65*	*65A*	*66*	*67*	*68*	*69*	*70*	*71*	*72*	*73*	*74*	*75*
49	50	51	52	53	54	55	56	57	58	59	60	61	62	63	64
Arg-	Glu-	Leu-	Thr-	Phe-	Arg-	Val-	Val-	Val-	Gly-	Glu-	His-	Asn-	Leu-	Asn-	Gln-

76	*77*	*78*	*79*	*80*	*81*	*82*	*83*	*84*	*85*	*86*	*87*	*88*	*89*	*90*	*91*
65	66	67	68	69	70	71	72	73	74	75	76	77	78	79	80
Asn-	Asn-	Gly-	Thr-	Glu-	Gln-	Tyr-	Val-	Gly-	Val-	Gln-	Lys-	Ile-	Val-	Val-	His-

92	*93*	*94*	*95*	*96*	*97*	*98*	*99*	*99A*	*99B*	*100*	*101*	*102*	*103*	*104*	*105*
81	82	83	84	85	86	87	88	89	90	91	92	93	94	95	96
Pro-	Tyr-	Trp-	Asn-	Thr-	Asp-	Asp-	Val-	Ala-	Ala-	Gly-	Tyr-	Asp-	Ile-	Ala-	Leu-

106	*107*	*108*	*109*	*110*	*111*	*112*	*113*	*114*	*115*	*116*	*117*	*118*	*119*	*120*	*121*
97	98	99	100	101	102	103	104	105	106	107	108	109	110	111	112
Leu-	Arg-	Leu-	Ala-	Gln-	Ser-	Val-	Thr-	Leu-	Asn-	Ser-	Tyr-	Val-	Gln-	Leu-	Gly-

122	*123*	*124*	*125*	*126*	*127*	*128*	*129*	*130*	*131*	*132*	*133*	*134*	*135*	*136*	*137*
113	114	115	116	117	118	119	120	121	122	123	124	125	126	127	128
Val-	Leu-	Pro-	Arg-	Ala-	Gly-	Thr-	Ile-	Leu-	Ala-	Asn-	Asn-	Ser-	Pro-	Cys-	Tyr-

138	*139*	*140*	*141*	*142*	*143*	*144*	*145*	*147*	*148*	*149*	*150*	*151*	*152*	*153*	*154*
129	130	131	132	133	134	135	136	137	138	139	140	141	142	143	144
Ile-	Thr-	Gly-	Trp-	Gly-	Leu-	Thr-	Arg-	Thr-	Asn-	Gly-	Gln-	Leu-	Ala-	Gln-	Thr-

155	*156*	*157*	*158*	*159*	*160*	*161*	*162*	163	*164*	*165*	*166*	*167*	*168*	*169*	*170*
145	146	147	148	149	150	151	152	153	154	155	156	157	158	159	160
Leu-	Gln-	Gln-	Ala-	Tyr-	Leu-	Pro-	Thr-	Val-	Asp-	Tyr-	Ala-	Ile-	Cys-	Ser-	Ser-

170 A	*170 B*	*171*	*172*	*173*	*174*	*175*	*176*	*177*	*178*	*179*	*180*	*181*	*182*	*183*	*184*
161	162	163	164	165	166	167	168	169	170	171	172	173	174	175	176
Ser-	Ser-	Tyr-	Trp-	Gly-	Ser-	Thr-	Val-	Lys-	Asn-	Ser-	Met-	Val-	Cys-	Ala-	Gly-

185	*186*	*187*	*188*	*188 A*	*189*	*190*	*191*	*192*	*193*	*194*	*195*	*196*	*197*	*198*	*199*
177	178	179	180	181	182	183	184	185	186	187	188	189	190	191	192
Gly-	Asn-	Gly-	Val-	Arg-	Ser-	Gly-	Cys-	Gln-	Gly-	Asp-	Ser-	Gly-	Gly-	Pro-	Leu-

200	*201*	*202*	*203*	*204*	*205*	*206*	*207*	*208*	*209*	*210*	*211*	*212*	*213*	*214*	*215*
193	194	195	196	197	198	199	200	201	202	203	204	205	206	207	208
His-	Cys-	Leu-	Val-	Asn-	Gly-	Gln-	Tyr-	Ala-	Val-	His-	Gly-	Val-	Thr-	Ser-	Phe-

216	*217*	*217 A*	*218*	*219*	*220*	*221*	*221 A*	*222*	*223*	*224*	*225*	*226*	*227*	*228*	*229*
209	210	211	212	213	214	215	216	217	218	219	220	221	222	223	224
Val-	Ser-	Arg-	Leu-	Gly-	Cys-	Asn-	Val-	Thr-	Arg-	Lys-	Pro-	Thr-	Val-	Phe-	Thr-

230	*231*	*232*	*233*	*234*	*235*	*236*	*237*	*238*	*239*	*240*	*241*	*242*	*243*	*244*	*245*
225	226	227	228	229	230	231	232	233	234	235	236	237	238	239	240
Arg-	Val-	Ser-	Ala-	Tyr-	Ile-	Ser-	Trp-	Ile-	Asn-	Asn-	Val-	Ile-	Ala-	Ser-	Asn

FIG. 2. The amino acid sequence of porcine pancreatic elastase. The numbering in italics is that of bovine chymotrypsinogen A.

sequence, satisfactorily accounts for all the electron density in an electron density map of tosyl elastase, there being no point at which crystallographic and sequence evidence are in conflict (*63, 63a, 100*).

B. SEQUENCE HOMOLOGIES IN MAMMALIAN SERINE PROTEINASES

1. *The A Chains or Activation Peptides*

Pancreatic juice contains four zymogens of serine proteinases in addition to proelastase. Chymotrypsinogen A, chymotrypsinogen B, and chymotrypsinogen C are precursors of enzymes with specificity for aromatic residues (*74*), while trypsinogen gives an enzyme which attacks lysyl or arginyl bonds. In cows, chymotrypsinogen C appears to form a complex with the zymogen of carboxypeptidase A (*75*). Figure 3 shows that the N-terminal sequences of all the chymotrypsinogens are homologous. The bovine enzymes have Gln 7 compared with Pro 7 in the porcine enzymes, but the chymotrypsinogens A and B are otherwise almost identical. The chymotrypsinogens C differ about 40% from chymotrypsinogens A and B, but there are only three differences between the porcine and bovine sequences. The N-terminal sequences of trypsinogens from several species are quite different from those of the chymotrypsinogens and show species differences (*15, 72, 73, 76–81*). Sheep have two trypsinogens which differ in their activation peptides. The composi-

74. J. E. Folk and W. E. Schirmer, *JBC* **240,** 181 (1965).

75. J. R. Brown, R. N. Greenshields, M. Yamasaki, and H. Neurath, *Biochemistry* **2,** 867 (1963).

76. M. Charles, M. Rovery, A. Guidoni, and P. Desnuelle, *BBA* **69,** 115 (1963).

77. S. Bricteux-Gregoire, R. Schyns, and M. Florkin, *BBA* **127,** 277 (1966).

78. M. Charles, D. Gratecos, M. Rovery, and P. Desnuelle, *BBA* **140,** 395 (1967).

79. L. B. Smillie, A. Furka, N. Nagabhushan, K. J. Stevenson, and C. O. Parkes, *Nature* **218,** 343 (1968).

80. C. T. Tobita and J. E. Folk, *Abstr. 7th Intern. Congr. Biochem., Tokyo, 1967* Vol. 4, p. 599. Sci. Council Japan, Tokyo, 1968.

81. R. J. Peaneasky, D. Gratecos, J. Baratti, and M. Rovery, *BBA* **181,** 82 (1969).

	1	2	3	4	5	6	7	8	9	10	11	12	13	14	15
PT:								Phe-	Pro-	Thr-	Asp-	Asp-	Asp-	Asp-	Lys-
OTA:								"	"	Val	"	"	"	"	"
OTB:										Val	"	"	"	"	"
BT:										Val	"	"	"	"	"
PCA:	Cys-	Gly-	Val-	Pro-	Ala-	Ile-	Pro-	Pro-	Val-	Leu-	Ser-	Gly-	Leu-	Ser-	Arg-
BCA:	"	"	"	"	"	"	Gln	"	"	"	"	"	"	"	"
PCB:	"	"	"	"	"	"	Pro	"	"	"	"	"	"	"	"
BCB:	"	"	"	"	"	"	Gln	"	"	"	"	"	"	Ala	"
PCC:	"	"	"	"	Ser-	Phe-	Pro	"	Asn	"	"	——	——	Ala	"
BCC:	"	"	Ala	"	Ile-	Phe-	Gln	"	Asn	"	"	——	——	Ala	"

FIG. 3. N-Terminal sequences of trypsinogens and chymotrypsinogens. Only residues which differ from the sequence in porcine trypsinogen or porcine chymotrypsinogen A are indicated: PT, porcine trypsinogen (*76*); OTA and OTB, two ovine trypsinogens (*77*); BT, bovine trypsinogen (*73*); PCA, porcine chymotrypsinogen A (*78*); BCA, bovine chymotrypsinogen A (*72*); PCB, porcine chymotrypsinogen B (*15*); BCB, bovine chymotrypsinogen B (*79*); PCC, porcine chymotrypsinogen C (*80*); and BCC, "bovine chymotrypsinogen C" (procarboxypeptidase endopeptidase fraction II′) (*81*).

tion of proelastase (*24*) indicates that a peptide of approximate composition (Lys, Arg, Asp, Glu, Ser, Pro_3, Ala, and Phe) may be split off during the activation.

2. *The B Chains*

Homologies between bovine chymotrypsin A, chymotrypsin B, trypsin, and porcine elastase are more apparent when we study the sequence from residue 16 to the C-terminus. Figure 4 illustrates this and also shows the partial sequence of bovine thrombin, a representative of a class of serine proteinases synthesized in the liver and occurring in the plasma which is concerned with the complex processes of blood clotting and fibrinolysis (*63, 72, 73, 79, 82–89*). The structure and properties of thrombin will be discussed elsewhere in this volume (*90*), but the sequence is included

82. B. S. Hartley, *in* "British Biochemistry—Past and Present" (T. W. Goodwin, ed.), p. 29. Academic Press, New York, 1970.

83. B. S. Hartley and D. L. Kauffman, *BJ* **101,** 229 (1966).

84. B. Meloun, I. Kluh, V. Kostka, L. Morávek, Z. Prusík, J. Vaněček, B. Keil, and F. Šorm, *BBA* **130,** 543 (1966).

85. D. M. Blow, J. J. Birktoft, and B. S. Hartley, *Nature* **221,** 337 (1969).

86. O. Mikeš, V. Tomášek, V. Holeyšovský, and F. Šorm, *BBA* **117,** 281 (1966).

87. B. S. Hartley, *Phil. Trans. Roy. Soc. London* **B257,** 77 (1970).

88. S. M. Magnusson and B. S. Hartley, unpublished evidence (1970).

89. J. J. Birktoft, D. M. Blow, R. Henderson, and T. A. Steitz, *Phil. Trans. Roy. Soc. London* **B257,** 67 (1970).

90. S. M. Magnusson, Chapter 9, this volume.

16 17 18 19 20 21 22 23 24 25 26 27 28 29 30 31
CA: ILE-VAL-Asn -GLY-GLU-GLU-ALA-Val -Pro-GLY-SER -TRP-PRO-TRP-GLN-VAL
CB: " " " " " ASP " " " " " " " " " "
T: ILE-VAL-GLY-GLY-Tyr -Thr -Cys -Gly -Ala-ASN-THR-Val -PRO-TYR-GLN-VAL
E: VAL-VAL-GLY-GLY-Thr -GLU-ALA-GLN-Arg-ASN-SER -TRP-PRO-Ser -GLN-ILE
Th: ILE-VAL-Glu -GLY-GLN-ASP -ALA-GLU- Val-GLY-Leu -Ser -PRO-TRP-GLN-VAL

32 33 34 35 36 36A 36B 36C 37 38 39 40 41 42 43 44
CA: SER-LEU-GLN-Asp-LYS — — — THR-GLY-PHE-HIS-PHE-CYS-GLY-GLY
CB: " " " " Ser — — — " " " " " " " "
T: SER-LEU-ASN — — — — — SER-GLY-TYR-HIS-PHE-CYS-GLY-GLY
E: SER-LEU-GLN-Tyr-ARG-Ser-Gly-SER-SER -Trp -Ala -HIS-Thr -CYS-GLY-GLY
Th: Met-LEU-Phe -Arg-LYS — — SER-Pro -Gln -Glu -Leu -Leu -CYS-GLY-Ala

45 46 47 48 49 50 51 52 53 54 55 56 **57** 58 59 60
CA: SER-LEU-ILE-ASN-GLU- ASN-TRP-VAL-VAL-THR-ALA-ALA-HIS-CYS-Gly-Val
CB: " " " SER- " ASP " " " " " " " " " "
T: SER-LEU-ILE-ASN-Ser -GLN-TRP-VAL-VAL-SER -ALA-ALA-HIS-CYS-Tyr-Lys
E: THR-LEU-ILE-Arg -GLN-ASN -TRP-VAL-Met -THR-ALA-ALA-HIS-CYS-Val -Asp
Th: SER-LEU-ILE-SER-ASP -Arg -TRP-VAL-Leu -THR-ALA-ALA-HIS-CYS-Leu-Leu

61 62 63 64 65 65A 65B 65C 65D 65E 65F 65G 65H 65I 65J 65K
CA: THR-Thr-Ser -Asp-Val — — — — — — — — — — —
CA: " " " " " — — — — — — — — — — —
T: SER-Gly-ILE -Gln — — — — — — — — — — — —
E: Arg-Glu-LEU-Thr-Phe-ARG — — — — — — — — — —
Th: Tyr-Pro(Trp, Pro, Asx, LYS,Asx,Phe)Thr Val-Val-Asx-Asx-Leu-Leu(Ser,
CBH

65L 65M 65N 66 67 68 69 70 71 72 73 74 75 76 77 78
CA: — — — VAL-VAL -Ala -GLY-GLU-Phe -ASP -Gln -Gly -Ser -Ser -SER -Glu
CB: — — — " " " " " " " " " Leu-Glu-THR "
T: — — — VAL-ARG-LEU-GLY-GLN-Asp -ASN-ILE -ASN-Val -Val -GLU-GLY
E: — — — VAL-VAL -VAL-GLY-GLU-HIS-ASN-LEU-ASN-Gln-Asn-Asn -GLY
Th: Val,Val,Glx,Trp)ARG,ILE -GLY-Lys ,HIS-Ser -Arg -Thr -Arg-Tyr-GLU-Arg

79 80 81 82 83 84 84A 85 86 87 88 89 90 91 92 93
CA: LYS-ILE -GLN-LYS -LEU-Lys — ILE-Ala -LYS-VAL-PHE-Lys -Asn -Ser -LYS
CB: ASP-Thr " Val " " — " Gly " " " " " PRO "
T: ASN-GLN-GLN-PHE-ILE -SER — Ala -Ser -LYS-Ser -ILE -VAL-HIS-PRO-Ser
E: Thr-GLU-GLN-TYR-VAL-Gly — VAL-Gln-LYS-ILE - VAL-VAL-HIS-PRO-Tyr
Th: LYS-VAL -GLU-LYS -ILE -SER-Met-LEU-Asp-LYS-ILE -TYR-ILE -HIS-PRO-ARG

	94	95	96	97	98	99	99 A	99 B	100	101	**102**	103	104	105	106	107
CA:	TYR	ASN	SER	Leu	THR	ILE	—	—	ASN	ASN	ASP	ILE	Thr	LEU	LEU	LYS
CB:	PHE	Ser	Ile	“	“	VAL	—	—	Arg	“	“	“	“	“	“	“
T:	TYR	ASN	SER	ASN	THR	LEU	—	—	ASN	ASN	ASP	ILE	Met	LEU	ILE	LYS
E:	TRP	ASN	THR	ASP	ASP	VAL	Ala	Ala	Gly	Tyr	ASP	ILE	ALA	LEU	LEU	ARG
Th:	TYR	ASN	Trp	Lys	GLU	Asn	Leu	—	ASP	Arg	ASP	ILE	ALA	LEU	LEU	LYS

	108	109	110	111	112	113	114	115	116	117	118	119	120	121	122	123
CA:	LEU	Ser	THR	ALA	ALA	SER	Phe	SER	GLN	Thr	VAL	Ser	Ala	VAL	CYS	LEU
CB:	“	ALA	“	PRO	“	GLN	“	“	GLU	“	“	“	“	“	“	“
T:	LEU	LYS	SER	ALA	ALA	SER	LEU	ASN	SER	Arg	VAL	Ala	Ser	ILE	Ser	LEU
E:	LEU	ALA	Gln	Ser	VAL	THR	LEU	ASN	SER	TYR	VAL	Gln	Leu	Gly	Val	LEU
Th:	LEU	LYS	Arg	PRO	ILE	GLU	LEU	SER	ASP	TYR	ILE	His	Pro	VAL	CYS	LEU

	124	125	126	127	128	128 A	128 B	129	130	131	132	133	134	135	136	137
CA:	PRO	SER	ALA	SER	ASP	—	—	Asp	Phe	ALA	ALA	GLY	THR	Thr	CYS	Val
CB:	“	“	“	Asp	GLU	—	—	“	“	Pro	“	“	Met	Leu	“	Ala
T:	PRO	THR	—	SER	Cys	—	—	Ala	Ser	—	ALA	GLY	THR	Gln	CYS	Leu
E:	PRO	ARG	ALA	Gly	Thr	—	—	ILE	LEU	ALA	Asn	Asn	SER	Pro	CYS	Tyr
Th:	PRO	LYS	,Gln	THR	Ala	Ala	Lys	LEU	LEU	His	ALA	GLY	Phe	Lys	Gly	Arg

	138	139	140	141	142	143	144	145	146	147	148	148 A	149	150	151	152
CA:	Thr	THR	GLY	TRP	GLY	LEU	THR	ARG	Tyr	THR	Asn	—	Ala	ASN	Thr	PRO
CB:	“	“	“	“	“	Lys	“	LYS	“	ASN	Ala	—	Leu	Lys	“	“
T:	ILE	SER	GLY	TRP	GLY	ASN	THR	LYS	SER	SER	GLY	—	Thr	Ser	Tyr	PRO
E:	ILE	THR	GLY	TRP	GLY	LEU	THR	ARG	THR	ASN	GLY	—	—	GLN	Leu	Ala
Th:	VAL	THR	GLY	TRP	GLY	ASN	Arg	,Thr	THR	SER	Val	Ala	Glu	Val	Gln	PRO

	153	154	155	156	157	158	159	160	161	162	163	164	165	166	167	168
CA:	ASP	ARG	LEU	GLN	GLN	ALA	SER	LEU	PRO	LEU	LEU	SER	ASN	THR	ASN	CYS
CB:	“	LYS	“	“	“	“	THR	“	“	ILE	VAL	“	“	“	ASP	“
T:	ASP	VAL	LEU	Lys	Cys	Leu	Lys	Ala	PRO	ILE	LEU	SER	ASN	SER	Ser	CYS
E:	Gln	Thr	LEU	GLN	GLN	ALA	Tyr	LEU	PRO	Thr	VAL	ASP	Tyr	Ala	ILE	CYS
Th:	Ser	VAL	LEU	,GLN	Val	Val	Asn	LEU	PRO	LEU	VAL	GLU	Arg	Pro	VAL	CYS

	169	170	170 A	170 B	171	172	173	174	175	176	177	178	179	180	181	182
CA:	LYS	Lys	—	—	TYR	TRP	GLY	THR	LYS	ILE	LYS	ASP	Ala	MET	ILE	CYS
CB:	ARG	“	—	—	“	“	“	SER	ARG	VAL	THR	“	Val	“	“	“
T:	LYS	SER	—	—	Ala	TYR	Pro	Gly	Gln	ILE	THR	Ser	ASN	MET	PHE	CYS
E:	Ser	SER	Ser	Ser	TYR	TRP	GLY	SER	Thr	VAL	LYS	ASN	Ser	MET	VAL	CYS
Th:	LYS ,	Ala	—	—	Ser	Thr	Arg,	Ile	ARG	ILE	THR	ASX	ASX	MET	PHE	CYS

	183	184	184A	185	186	187	188	188A	188B	188C	188D	**189**	**190**	**191**	192	193
CA:	ALA	GLY	—	Ala	Ser	GLY	VAL	—	—	—	—	SER	SER	CYS	Met	GLY
CB:	"	"	—	"	"	"	"	—	—	—	—	"	"	"	"	"
T:	ALA	GLY	TYR	Leu	Glu	GLY	Gly	LYS	—	—	—	ASP	SER	CYS	GLN	GLY
E:	ALA	GLY	—	Gly	Asn	GLY	VAL	ARG	—	—	—	SER	Gly	CYS	GLN	GLY
Th:	ALA	GLY	TYR	Lys	Pro	GLY	Glu	Gly	Lys	Arg	Gly	ASP	Ala	CYS	GLU	GLY

	194	**195**	196	197	198	199	200	201	202	203	204	204A	204B	205	206	207
CA:	ASP	SER	GLY	GLY	PRO	LEU	VAL	CYS	LYS	Lys	ASN	—	—	GLY	Ala	TRP
CB:	"	"	"	"	"	"	"	"	Gln	"	"	—	—	"	"	"
T:	ASP	SER	GLY	GLY	PRO	Val	VAL	CYS	Ser	Gly	Lys	—	—	—	—	—
E:	ASP	SER	GLY	GLY	PRO	LEU	His	CYS	Leu	Val	ASN	—	—	GLY	Gln	TYR
Th:	ASP	SER	GLY	GLY	PRO	Phe	VAL	Met	LYS	Ser	Pro	Tyr	Asn	Asn	Arg	TRP

	208	209	210	211	212	**213**	214	215	**216**	217	217A	218	219	**220**	221	221A
CA:	Thr	LEU	Val	GLY	ILE	VAL	SER	TRP	GLY	SER	—	Ser	Thr	CYS	Ser	—
CB:	"	"	Ala	"	"	"	"	"	"	"	—	"	"	"	"	"
T:	—	LEU	Gln	GLY	ILE	VAL	SER	TRP	GLY	SER	—	—	GLY	GYS	Ala	Gln
E:	Ala	Val	His	GLY	VAL	Thr	SER	PHE	Val	SER	Arg	Leu	GLY	CYS	ASN	Val
Th:	Tyr	Gln	Met	GLY	ILE	VAL	SER	TRP	GLY	Glu	—	—	GLY	CYS	ASP	Arg

	222	223	224	225	**226**	227	**228**	229	230	231	232	233	234	235	236	237
CA:	THR	Ser	Thr	PRO	GLY	VAL	TYR	Ala	ARG	VAL	THR	ALA	Leu	VAL	Asn	TRP
CB:	"	"	"	"	Ala	"	"	"	"	"	"	"	"	Met	Pro	"
T:	Lys	Asn	LYS	PRO	GLY	VAL	TYR	THR	LYS	VAL	Cys	Asn	TYR	VAL	SER	TRP
E:	THR	Arg	LYS	PRO	Thr	VAL	PHE	THR	ARG	VAL	SER	ALA	TYR	ILE	SER	TRP
Th:	Asn	Gly	LYS	Tyr	GLY	Phe	TYR	THR	His	VAL	Phe	Arg	(Lys,LEU)		Lys	TRP

	238	239	240	241	242	243	244	245	245A	245B
CA:	VAL	GLN	GLN	THR	LEU	ALA	Ala	ASN	—	—
CB:	"	"	GLU	"	"	"	"	"	—	—
T:	ILE	Lys	GLN	THR	ILE	ALA	SER	ASN	—	—
E:	ILE	ASN	ASN	VAL	ILE	ALA	SER	ASN	—	—
Th:	ILE	GLN	Lys	VAL	ILE	Asp	Arg	Leu	Gly	Ser

FIG. 4. Sequence homologies in the "B chains" of mammalian serine proteinases (*82*): CA, bovine chymotrypsinogen A (residues 16–245) (*72, 83–85*); CB, bovine chymotrypsinogen B (residues 16–245) (*79*); T, bovine trypsinogen (*73, 86, 87*); E, porcine elastase (*63*); and Th, bovine thrombine (*88*). Numbering is that of CA. Insertions are numbered 36A, 36B, etc. Numbers underlined show residues internal in the tertiary structure of α-chymotrypsin and boldface numbers show residues important in the catalytic site and forming the substrate binding pocket of chymotrypsin (*85, 89*). Only differences between CA and CB are shown. Residues which are chemically similar (defined as in Table III) in any pair of CA or CB and T, E, or Th are in capital letters and those identical are underlined.

TABLE III
OVERALL HOMOLOGIES BETWEEN MAMMALIAN SERINE PROTEINASES

Enzyme	Percent identity (or chemical similarity[a]) with: Chymotrypsin B[b] (%)	Trypsin[b] (%)	Elastase[b] (%)	Thrombin[b] (%)	Any other of these five[c] (%)	Chain length (No. of residues)
Chymotrypsin A[d]	78 (85)	43 (53)	39 (51)	32 (39)	63 (76)[e]	230
Chymotrypsin B[d]	—	38 (49)	38 (47)	31 (38)	58 (71)[f]	230
Trypsin		—	35 (48)	32 (38)	61 (74)	223
Elastase			—	27 (39)	55 (72)	240
Thrombin[d]				—	46 (56)	265

[a] Chemical similarities are defined as Arg = Lys, Asp = Asn, Glu = Gln, Asp = Glu, Asn = Gln, Ser = Thr, Val = Ile, Ile = Leu, Tyr = Phe = Trp, in addition to identities.

[b] Calculated as percent of the minimum length required to accommodate the sequences being compared when aligned as in Fig. 4, discounting common deletions.

[c] Calculated as percent of the individual chain length.

[d] Residues 16–245 only (chymotrypsin A numbering).

[e] Excluding chymotrypsin B.

[f] Excluding chymotrypsin A.

here because of the light which it throws on the sequences of the pancreatic serine proteinases.

The overall pattern is clearly that of clusters of small sections of homologous sequence which occur at intervals along the whole peptide chain. These are separated by regions in which the side chains show little resemblance in any pair of enzymes, and in such regions "deletions" or "insertions" of several residues are necessary in one enzyme or another in order to bring homologous sequences into register. The relative extent of similarity between each of the enzymes is summarized in Table III. The isoenzymes, chymotrypsin A and chymotrypsin B, are 78% identical and may be assumed to have arisen by a relatively recent gene doubling (compare the α and β chains of hemoglobin with 44% identity). The pancreatic enzymes show 40–45% identity or 50–55% chemical similarity, and no pair is significantly closer than any other. The B chain of thrombin is only slightly less similar to the pancreatic enzymes than they are to each other, except for the large insertion which occurs at position 65–66 (which carries the carbohydrate moiety of thrombin).

3. *Disulfide Bridges*

The homology is further emphasized by comparing the positions of the disulfide bridges (Fig. 5) (*91*). Three bridges are common to each en-

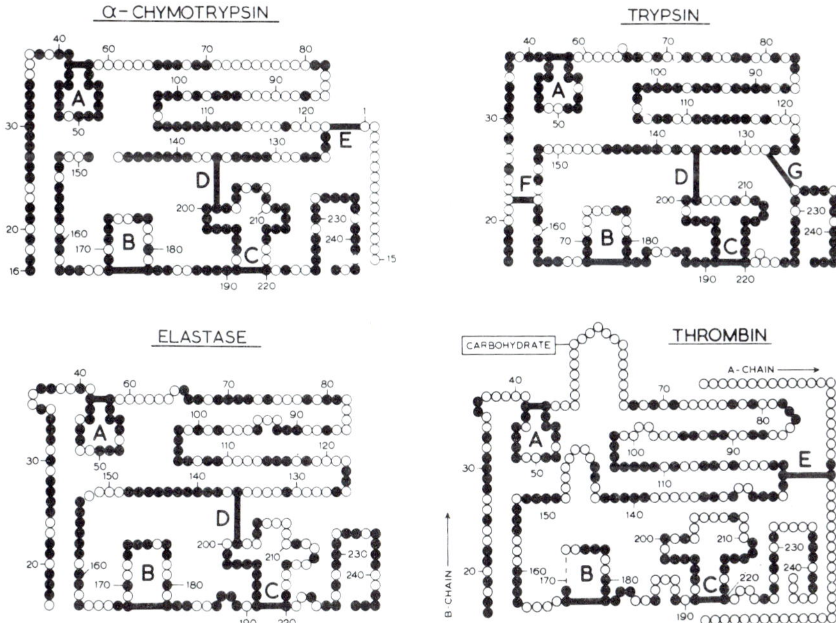

FIG. 5. Homologies of disulfide bridges in mammalian serine proteinases (*91*). Residues identical in any pair are shown as solid circles. Numbering is that of chymotrypsinogen. Insertions in one sequence relative to another are shown as kinks or small loops. The thrombin pattern is still tentative being based on homology.

zyme; A, the "histidine loop" (42–58); B, the "methionine loop" (168–182); and, C, the "serine loop" (191–220). Bridge D (136–201) is absent in thrombin, but bridge E (1–122) which joins the A chain of chymotrypsin to the B chain is paralleled exactly by a bridge at position 122 in thrombin which joins its A chain to its B chain. Two bridges, F (22–157) and G (128–232), are unique to trypsin.

4. *Hypothetical Models of Trypsin and Elastase*

The homologies of sequence and disulfide bridges clearly suggested that much of the tertiary structures of these enzymes could be identical. Determination of the tertiary structure of bovine α-chymotrypsin (*92*)

91. B. S. Hartley, *in* "Homologies in Enzymes and Metabolic Pathways." Springer, Berlin, 1970 (in press).

92. B. W. Mathews, P. B. Sigler, R. Henderson, and D. M. Blow, *Nature* **214**, 652 (1967).

made it possible to compare the homologies even more significantly. In Fig. 4 boldface numbers indicate side chains which appear inaccessible to solvent in the α-chymotrypsin A model of Matthews *et al.* (*92*). Figure 6 shows that the homologies are greatest in such regions, and there appeared to be no reason why all these enzymes should not have similar hydrophobic cores. Table IV, for example, shows that 45% of the "internal" residues are invariant in each of these enzymes, compared with only 6% of the "external" residues, and inspection of Fig. 4 will show that where replacements occur in internal positions, these are almost invariably of chemically similar amino acids.

It therefore appeared worthwhile to build hypothetical models for trypsin and elastase on the assumption that the conformation of the peptide chains and of internal hydrophobic residues was the same as that in chymotrypsin. The process of model building confirmed this hypothesis, since it was never necessary to distort the peptide chain in order to accommodate different side chains in trypsin or elastase (*87*). Even where considerable changes occur, such as residues 29, 45, 53, 200, 209,

TABLE IV
INTERNAL[a] AND EXTERNAL HOMOLOGIES BETWEEN CHYMOTRYPSINOGEN A AND TRYPSIN, ELASTASE, OR THROMBIN

Degree of homology	Internal positions	External[b] positions	Total
Identical in all four enzymes	36 (45%)	12 (6%)	48 (18%)
Identical or similar[c] in all four enzymes	48 (60%)	20 (10%)	68 (25%)
Identical in at least three enzymes	50 (63%)	31 (16%)	81 (30%)
Identical or similar in at least three enzymes	67 (84%)	65 (34%)	132 (49%)
Identical in at least two enzymes	62 (78%)	77 (40%)	139 (52%)
Identical or similar in at least two enzymes	79 (99%)	132 (69%)	211 (78%)
Positions showing no homology[d]	1 (1%)	58 (31%)	59 (22%)
Totals	80 (100%)	190 (100%)	270 (100%)

[a] Judged to be inaccessible to solvent in the α-chymotrypsin model (*92*).
[b] Insertions relative to chymotrypsinogen are treated as external.
[c] Defined as in Table III.
[d] Deletions are scored as differences. Figure 6 gives a diagrammatic representation of these results.

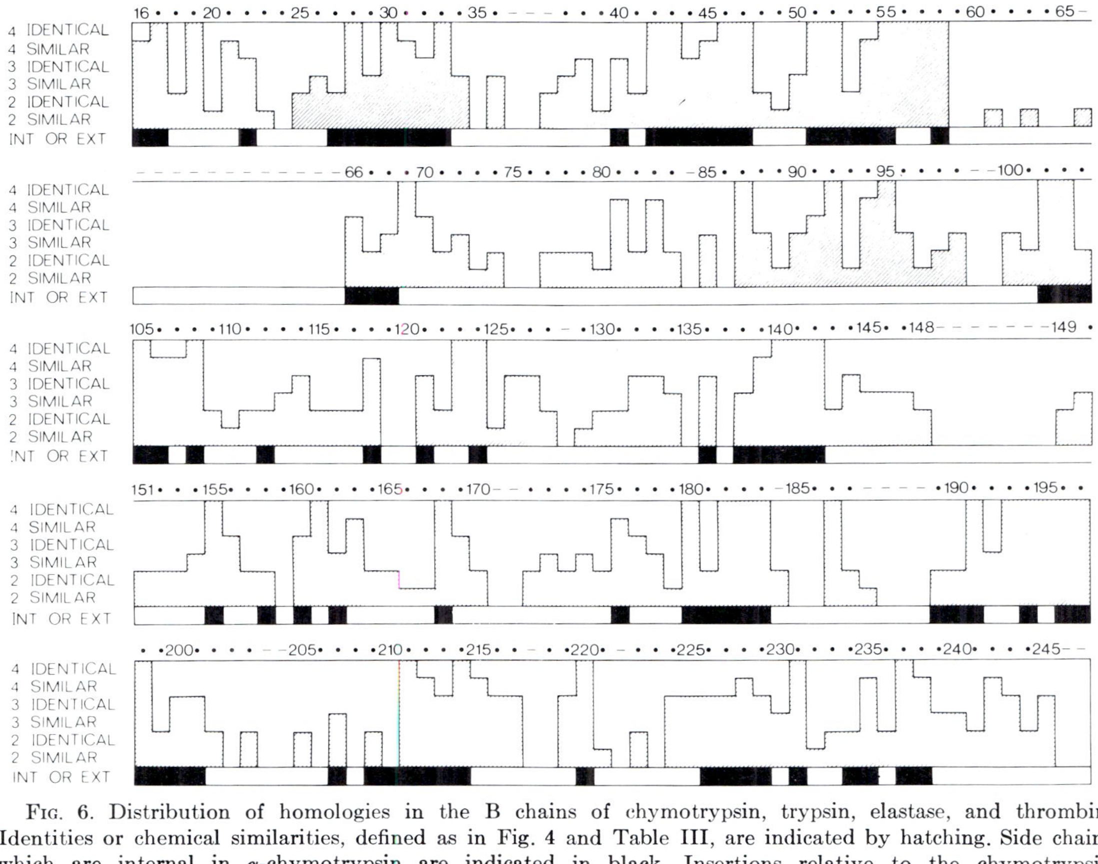

FIG. 6. Distribution of homologies in the B chains of chymotrypsin, trypsin, elastase, and thrombin. Identities or chemical similarities, defined as in Fig. 4 and Table III, are indicated by hatching. Side chains which are internal in α-chymotrypsin are indicated in black. Insertions relative to the chymotrypsin sequence are indicated by hyphens. The numbering is that of bovine chymotrypsinogen A (see Table IV).

210, and 212, which formed an internal cluster in the back of the molecule, the models were found to provide alternative arrangements of close-packed hydrophobic groups. The deletions in trypsin relative to chymotrypsin, such as residues 205–208, were found merely to remove an external loop, and no distortion of the main chain was necessary in any case. The two additional disulfide bridges of trypsin (22–157 and 128–132) were easily made in the model, as had been previously pointed out by Sigler *et al.* (*93*).

The residues of the catalytic site, Asp 102, His 57, and Ser 195, could obviously be built in the same conformation as in chymotrypsin, but the most significant conclusions from the models concerned the substrate binding site. Steitz *et al.* (*94*) had shown that formyl-L-tryptophan binds to the crystal of α-chymotrypsin in a pocket lined by the side chains of Ser 189, Ser 190, Cys 191, Val 213, Gly 216, Cys 220, Gly 226, and Tyr 228. In the trypsin model, each of these side chains was identical except that Asn 189 (*86, 94a*) replaced Ser 189. This hardly seemed to explain the different specificity of trypsin; thus, an error in the trypsin sequence was suspected. Reinvestigation of the trypsin sequence showed that residue 189 was aspartic acid (*87*). The model then allowed formyl-L-arginine to bind in the same conformation as formyl-L-tryptophan binds to chymotrypsin (*87, 94*) and an internal ion pair could readily be envisaged within the hydrophobic pocket between the carboxyl group of Asp 189 and the guanidino group of the substrate. Thus a single replacement of Ser 189 for Asp 189 seems to be all that is necessary to turn chymotrypsin into trypsin.

No such obvious argument could be made about elastase specificity from the model at that time. A replacement of Val 216 for Gly 216 and Thr 226 for Gly 226 appeared to block the entrance to the substrate binding site so that only a small side chain such as alanine or glycine could be accommodated (*87*). The subsequent discovery (*27*) that the best synthetic substrates contained alanyl residues is readily explained on this model, but at the time the model was puzzling since it did not appear to allow tosyl or diisopropylphosphoryl groups to bind in the same way as they did to chymotrypsin, yet both DFP and tosyl halides were known to react well with elastase. It was necessary to presume that both these substituents must protrude into the solution.

93. P. B. Sigler, D. M. Blow, B. W. Matthews, and R. Henderson, *JMB* **35,** 143 (1968).

94. T. A. Steitz, R. Henderson, and D. M. Blow, *JMB* **46,** 337 (1969).

94a. K. A. Walsh and H. Neurath, *Proc. Natl. Acad. Sci. U. S.* **52,** 884 (1964).

IV. The Tertiary Structure of Elastase

A. X-Ray Crystallographic Investigations

1. *Elastase Crystals*

Although the hypothetical models of trypsin and elastase seemed very plausible, it was clearly desirable to know the true structure if only to expose the limitations of this type of model building. No effort has been made up till this time to crystallize the enzyme, but it was discovered almost simultaneously in two independent laboratories that large well-ordered crystals could be readily obtained from solutions of the enzyme at pH 5 by cooling or addition of low concentrations of sodium sulfate (*38*). These crystals (Fig. 7) were orthorhombic, space group $P2_12_12_1$ with cell dimensions $a = 5.15$ nm, $b = 5.80$ nm, and $c = 7.55$ nm corresponding to a unit cell of four molecules, with one molecule per asymmetric unit. Figure 8 is a 22½% precession photograph of the $0kl$ projection of native elastase, which shows that the intensity of reflections remains strong out to a resolution of better than 0.2 nm. Moreover, the

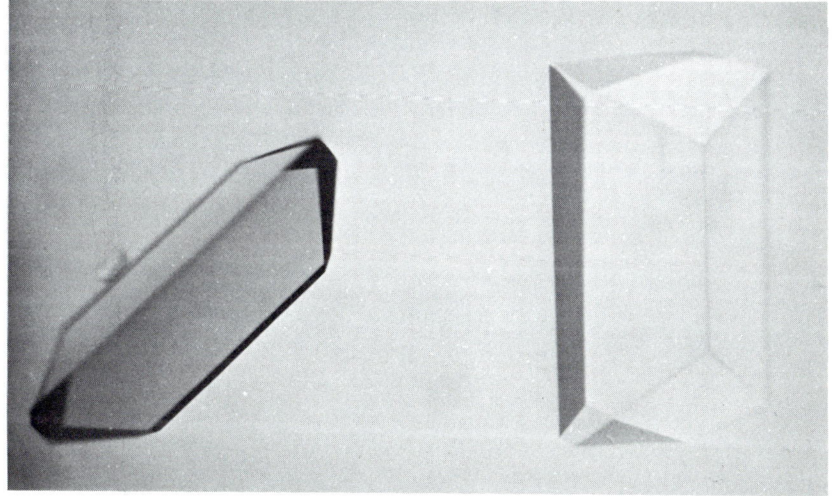

FIG. 7. Crystals of elastase grown from a 2% solution in 0.01 *M* sodium acetate–0.02 *M* sodium sulfate, pH 5.0. The crystals grew overnight and were approximately 1 mm long.

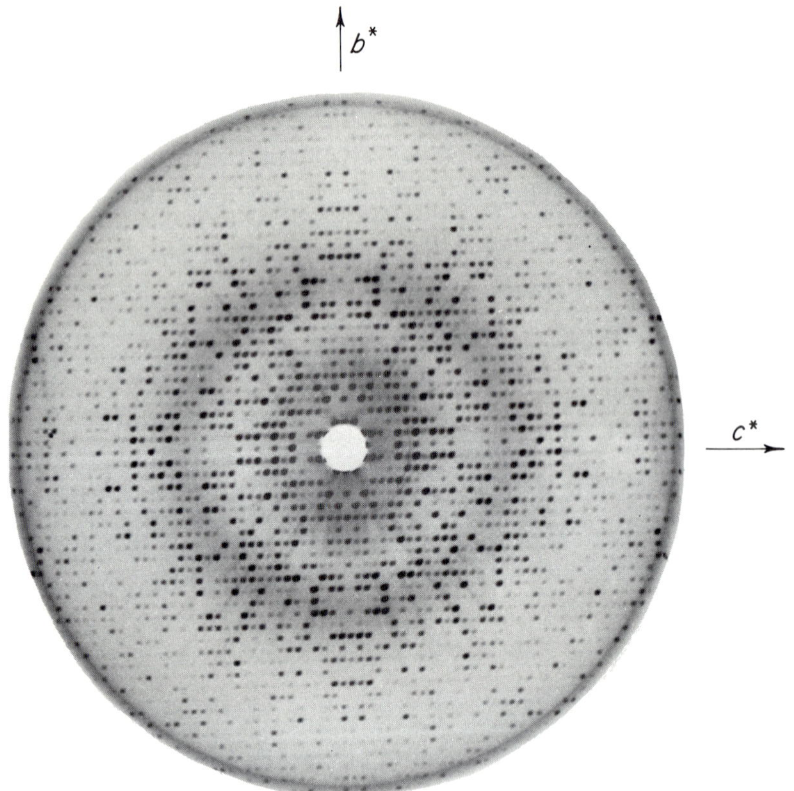

FIG. 8. A 22½% precession photograph of the 0*kl* projection of native elastase.

crystals proved very resistant to damage by X-ray irradiation, thus they seemed ideal for an X-ray crystallographic investigation.

2. *Heavy Atom Derivatives*

The search for isomorphous elastase crystals containing heavy atoms, which are necessary to determine protein structures by the isomorphous replacement technique (*95*), was made easier by utilizing experience with other proteins. Sigler *et al.* (*96*) had shown that pipsyl and tosyl fluoride reacted with Ser 195 of γ-chymotrypsin to give an excellent isomorphous pair of crystals, wherein a heavy iodine atom in the pipsyl enzyme oc-

95. D. W. Green, V. M. Ingram, and M. F. Perutz, *Proc. Roy. Soc.* **A225,** 287 (1954).

96. P. B. Sigler, H. C. W. Skinner, C. L. Coulter, J. Kallos, H. Braxton, and D. R. Davies, *Proc. Natl. Acad. Sci. U. S.* **51,** 1146 (1964).

cupied the same position as a light methyl group in the tosyl enzyme. A similar reagent, *p*-chloromercuribenzene sulfonyl (PCMBS-) fluoride had been tried with α-chymotrypsin (*97*). All three reagents were found to react stoichiometrically with elastase to give crystals which were isomorphous with those of native elastase (*63a, 98*). Therefore, PCMBS-elastase was used as the first heavy atom derivative (having a higher X-ray scattering power than the lighter pipsyl derivative), and tosyl elastase was used as the "parent" structure in all the subsequent X-ray studies in order to exploit the perfect isomorphism between it and PCMBS-elastase. Other derivatives were prepared by soaking crystals of tosyl elastase or PCMBS-elastase in uranyl nitrate solutions, which had proved useful in the determination of the structure of lysozyme (*99*). In both types of elastase crystal, uranyl ions were shown to bind to the same single site with high occupancy (*63a, 98*).

3. *Fourier Synthesis at 0.35 nm Resolution*

The positions of these heavy atoms were determined by calculating difference Patterson projections of PCMBS-elastase or uranyl tosyl elastase against tosyl elastase, and these proved immediately interpretable. The value of these derivatives for providing three-dimensional phase information was checked by calculating difference Fourier projections of one derivative using phase information derived from another. As can be seen in Fig. 9, it was obvious that these derivatives were eminently suitable (*98*).

It was decided to collect three-dimensional data to a resolution of 0.35 nm since the radiation damage was so small for these crystals that the measurement of all the reflections to this resolution could be accurately made using a single crystal and thereby eliminating problems of variable isomorphism, salt effects, heavy atom occupancies, and scaling errors inherent in the normal use of several crystals. The complete 0.35 nm data were collected from a single crystal of each of the four derivatives by an ordinate analysis procedure using a Hilger and Watts four-circle diffractometer on line to a Ferranti Argus computer, in the single month of July, 1968. After appropriate correction and scaling, the amplitude and phase data thus obtained were used to calculate the three-dimensional Fourier synthesis of tosyl elastase at 3.5 nm resolution (*63a*).

97. P. B. Sigler, B. A. Jeffery, B. W. Matthews, and D. M. Blow, *JMB* **15**, 175 (1966).

98. D. M. Shotton and H. C. Watson, *Phil. Trans. Roy. Soc. London* **B257**, 111 (1970).

99. C. C. F. Blake, D. F. Koenig, G. A. Mair, A. C. T. North, D. C. Phillips, and V. R. Sarma, *Nature* **206**, 757 (1965).

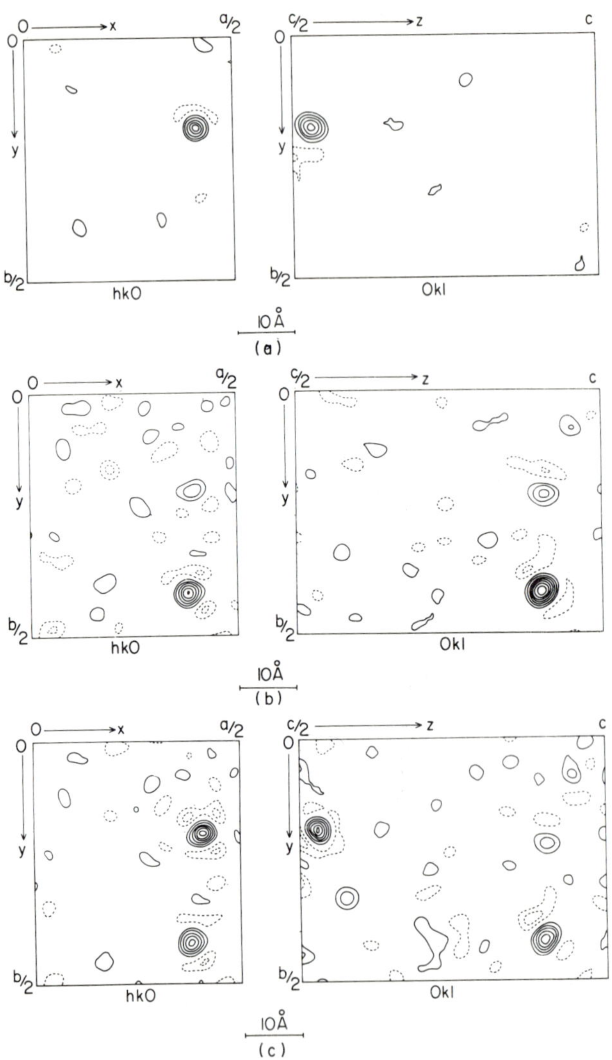

FIG. 9. *hk*0 and 0*kl* difference Fourier projections calculated at a resolution of 0.35 nm, using tosyl elastase as the parent structure, (a) of PCMBS-elastase, with phases calculated using the uranyl tosyl elastase uranyl ion position, (b) of uranyl tosyl elastase with phases calculated using the PCMBS-elastase mercury atom position, and (c) of uranyl PCMBS-elastase, with phases calculated using the PCMBS-elastase mercury atom position (*98*). Zero contours omitted and negative contours dashed.

B. The Tertiary Structure of Tosyl Elastase

1. *The Electron Density Map*

The electron density map of tosyl elastase (*63a, 98, 100*) clearly showed that each elastase molecule is an oblate spheroid, approximately 5.3 by 4.0 by 3.8 nm. There are large interstices of low electron density between molecules in the unit cell and relatively few intramolecular contacts. The entire course of the single polypeptide chain could be followed unambiguously, and the position of the amino acid side chains could be readily recognized. In some cases, such as the sulfur atoms of the four disulfide bridges or the two methionine residues, these side chains could

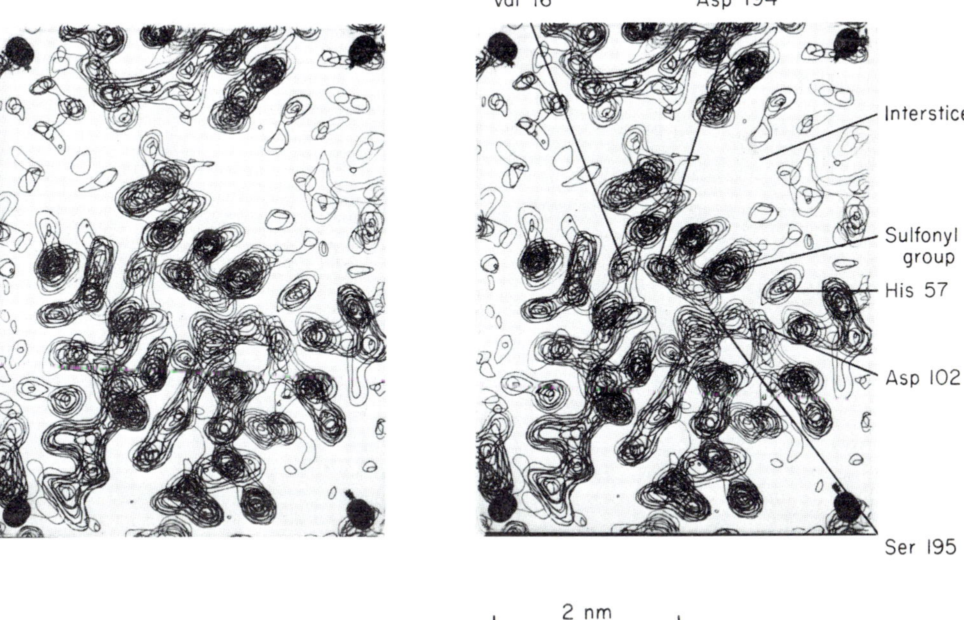

FIG. 10. A stereophotograph of the central region of a stack of seven sections from the complete three-dimensional 0.35-nm resolution electron density map of tosyl elastase showing the active center region. The positions of the large intermolecular interstice adjacent to the active center, the dense sulfonyl group of the tosyl inhibitor attached to the γ-oxygen of the active center serine residue Ser 195, and the side chains of some of the catalytically important residues Val 16, His 57, Asp 102, Asp 194, and Ser 195, whose functions are discussed in the text, are indicated.

100. D. M. Shotton and H. C. Watson, *Nature* **225**, 811 (1970).

be definitely identified, and in other cases a characteristic size and planarity indicated an aromatic ring. Glycine residues were usually distinguishable by the absence of side chain electron density, and other side chains appeared as stalks or bulges of various sizes along the main chain and were not identifiable from the electron density map alone at this resolution.

It was however possible, by using the known amino acid sequence, to build an atomic model of elastase which uniquely satisfied all the electron density in the map (*100*). Figure 10 is a stereophotograph of seven sections from the central part of the electron density map, which illustrates the clearly defined electron density of part of the molecule, showing the tosyl sulfur atom and several side chains of enzymic interest.

2. *The Model of Tosyl Elastase*

It was immediately clear that the conformation of the polypeptide chain was almost identical to that of α-chymotrypsin. Stereo views of the elastase chain are shown in Figs. 11 and 12, where one can see that as in chymotrypsin the molecule appears to be divided into two halves composed of residues 27–127 plus the C-terminal helix in the upper left hemisphere and residues 128–230 plus the N-terminal sequence (residues 16–26) in the lower right hemisphere.

A convenient diagrammatic representation of the structure is shown in Fig. 13, where one can see that each half of the molecule is composed of large antiparallel loops forming six regions of close antiparallel interactions. There are three main loops in each half of the molecule, five of which are named, in Fig. 13, according to the chymotrypsin usage, while the sixth is called the *uranyl loop* since it forms the principal binding site for the uranyl ion in the elastase uranyl heavy atom derivatives, (*63a*) which is bound in contact distance of the side chains of Val 67, Glu 70, Glu 80, and Tyr 82, and of the carbonyl oxygen of Leu 73. This diagram shows some side chain properties and illustrates the gross structural similarity observed between each half of the molecule when the chain is followed from each terminus to the center. The diagram differs slightly from that of Birktoft *et al.* (*89*) for α-chymotrypsin, which was designed predominantly to indicate hydrogen bonding pattern and which suggested a similarity between each half molecule when the chain was followed sequentially from N-terminus to C-terminus. Both enzymes can, of course, be written in either formulation.

Birktoft *et al.* (*89*) pointed out that in each half of the chymotrypsin molecule these antiparallel chains form a twisted sheet which curves around to unite with itself to form a "cylinder" around the surface of

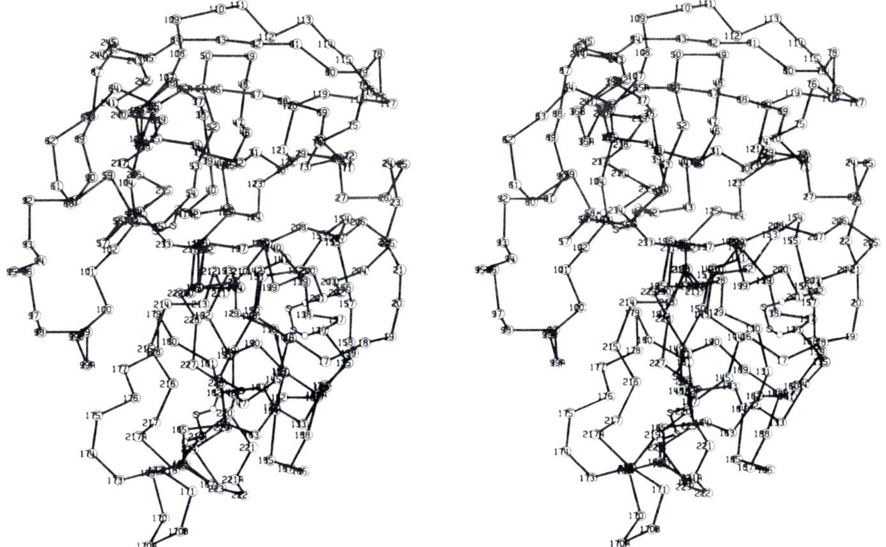

Fig. 11

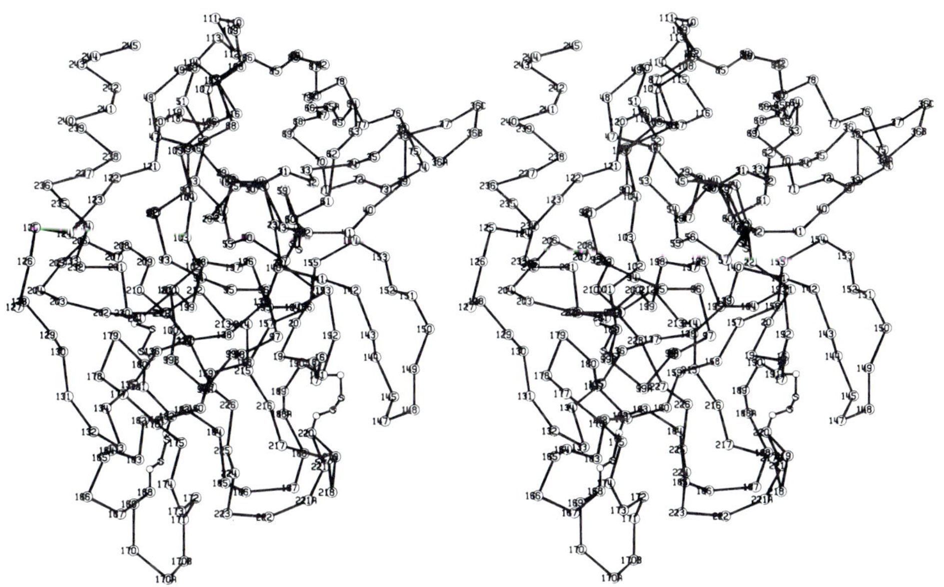

Fig. 12

Figs. 11 and 12. Two computer-prepared stereo views of the conformation of the elastase polypeptide chain, prepared from the approximate α-carbon coordinates of tosyl elastase, derived from the 0.35-nm resolution study *(100)*. Residues are numbered on the chymotrypsinogen A numbering scheme, insertions in the sequence relative to chymotrypsinogen A being numbered 36A, 36B, etc. Residue 146 is not present in elastase. For clarity all side chains are omitted except those forming the four disulfide bridges. Figure 12 is viewed from 90° to the left of Fig. 11.

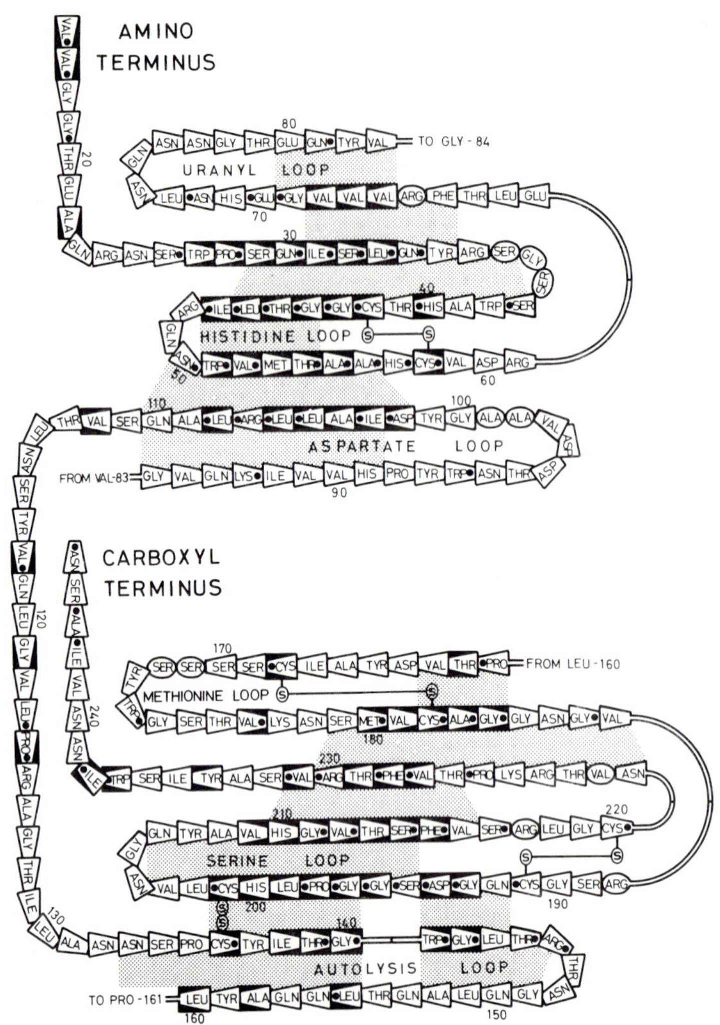

FIG. 13. A diagrammatic representation of the conformation of the elastase polypeptide chain, which is folded into two distinct halves in which looping of the chain brings neighboring regions into close antiparallel contact. All neighboring antiparallel regions in the molecule are represented as such in the diagram, those regions where chains run within hydrogen bonding distance of one another being shaded. Internal residues, judged to be inaccessible to water molecules in the atomic models of elastase and α-chymotrypsin, are boxed with heavy outlines. Residues which are homologous (identical or chemically simliar, see Fig. 4) in elastase, trypsin, chymotrypsin A and chymotrypsin B are indicated by a black dot next to the residue name. Disulfide bridges are shown. The numbering system is as for Fig. 4. Insertions are indicated by ovals and can be seen to occur externally at the ends of loops. The center of each half of the diagram represents the hydrophobic homologous

which the chains run in an oblique fashion. This cylinder can be imagined by curling the plane of Fig. 13 in such a way as to connect Val 83 to Gly 84 or Leu 160 to Pro 161, but it can be much more easily seen in the stereo views of elastase (Figs. 11, 12, and 14). Figure 12 is a stereo view looking down the axis of the N-terminal cylinder, which lies in the upper half of the diagram. One sees, in clockwise order, the "uranyl loop" (residues 62–83) sinking away to the upper right, the "N-terminal loop" (residues 30–42) protruding on the right, the "histidine loop" (residues 42–57) sinking into the molecule on the lower left, and the "aspartate loop" (residues 84–110) protruding forward on the upper left. The C-terminal cylinder is less obvious in Fig. 12 since the axis runs from left to right more or less in the plane of the paper.

A comparison of the chains of elastase and α-chymotrypsin (Fig. 14) in similar orientations illustrates the high degree of similarity between these two enzymes. The few insertions (residues 36A, B, and C; 65A; 99A and B; 170A and B; 184A; 188A; 217A; and 221A) and the deletion of residue 146 in elastase near the site of autolytic cleavage in α-chymotrypsin represent the only major differences which all occur in surface positions.

Access to the substrate binding pocket of elastase, which otherwise is very similar to that of chymotrypsin, is obstructed by the side chains of Val 216 and Thr 226. Substrates with bulky aromatic side chains are therefore sterically unable to enter this pocket, explaining the complete lack of activity of elastase toward such substrates. L-Alanyl substrates can, however, be readily fitted in similar conformation to that of *N*-formyl-L-tryptophan with chymotrypsin (*94*), the α-methyl groups making good van der Waals contact with hydrophobic groups at the mouth of the pocket and the carbonyl group of the peptide bond to be hydrolyzed lying near the γ-oxygen atom of the active center serine residue, thus explaining in atomic terms the substrate specificity of this enzyme (Fig. 15).

One consequence of the construction of this pocket is that the tosyl group of tosyl elastase is unable to fit into the substrate binding pocket as it does in tosyl α-chymotrypsin (*92, 93*). Instead it is found to protrude into solution, displacing the imidazole ring of His 57 from its native conformation by a rotation of about 180° about its Cα–Cβ bond so that it lies pointing away from the molecule, parallel to and in van

---

core of each half of the elastase molecule. Two breaks in the chain have had to be made in the diagram in order to represent the three-dimensional conformation of elastase on a plane surface. The diameter of the elastase molecule corresponds to about 10 residues length of extended polypeptide chain.

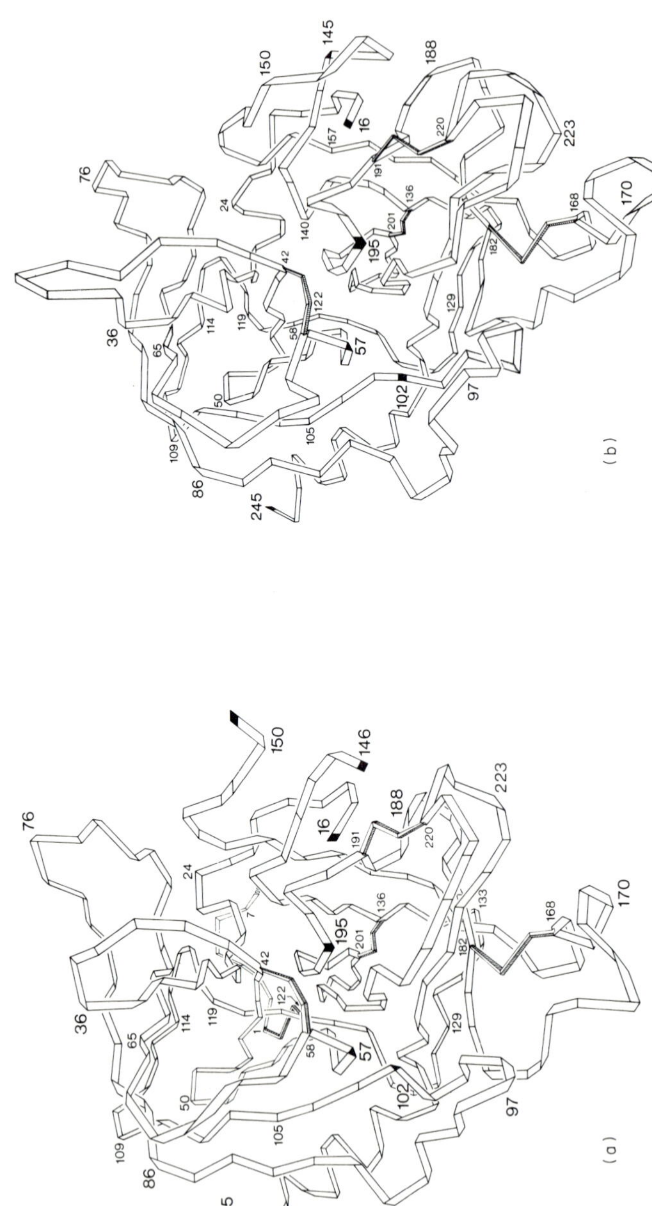

FIG. 14. Drawings of the conformations of (a) chymotrypsin and (b) elastase. Creases in the tape indicate α-carbon positions; shaded areas indicate residues in elastase which are "insertions" in the sequences shown in Fig. 4. (See also Fig. 5, Chapter 6, p. 194, and related text discussion.)

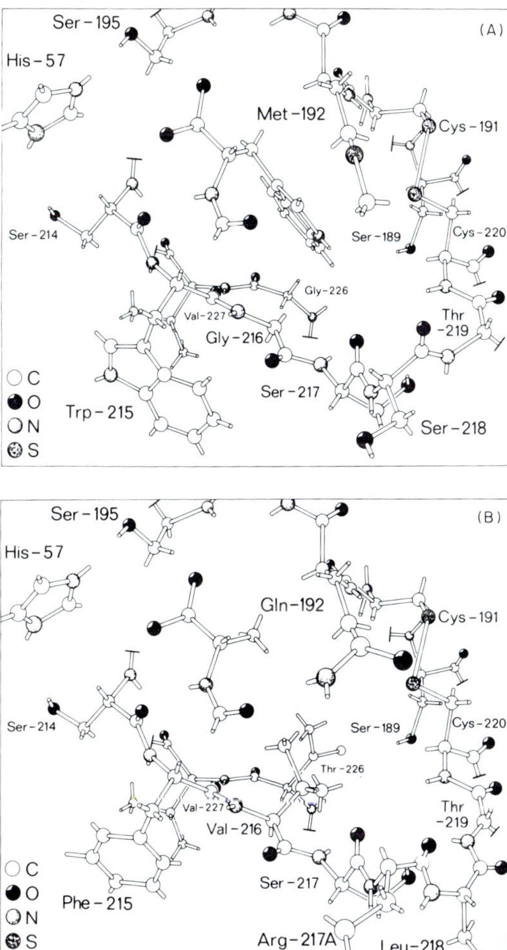

FIG. 15. Perspective drawings of the active center regions of the models of (A) chymotrypsin with bound formyl-L-tryptophan; (B) elastase with formyl-L-alanine arranged in similar conformation.

der Waals contact with the tosyl ring (*100*). A three-dimensional difference Fourier between tosyl elastase and native elastase (*101*) shows that in the native enzyme the imidazole ring of His 57 lies between Asp 102 and Ser 195, and within hydrogen bonding distance of both, and is thus able to form a "charge-relay" system exactly as in α-chymotrypsin (*85*), explaining the common catalytic mechanism possessed by these two enzymes.

101. H. C. Watson and D. M. Shotton, unpublished evidence (1970).

3. *Comparison with the Hypothetical Model*

A comparison of the hypothetical elastase model discussed previously with this model of tosyl elastase showed that the two were identical in all but minor particulars. Most impressive, perhaps, is the observation that the substrate binding pocket has exactly the structure predicted. The orientation of most internal side chains is the same as that in the hypothetical model, although the precise conformation of branched side chains cannot be established at 0.35 nm resolution. Figure 16 shows the distribution of residues between internal and external positions in elastase, from which it can be seen that all the charged groups except Asp 102 and Asp 194 lie on the surface, and the hydrophobic interior is highly homologous with other serine proteinases. The hypothetical model was wrong in only one cluster of internal side chains, involving Ser 29, Thr 45, Met 53, His 200, Val 209, His 210, and Val 212, where a collection of amino acid changes had resulted in a different stacking of the side chains without any disturbance of the main chain. Other errors in the hypothetical model were at insertions in elastase relative to chymotrypsin,

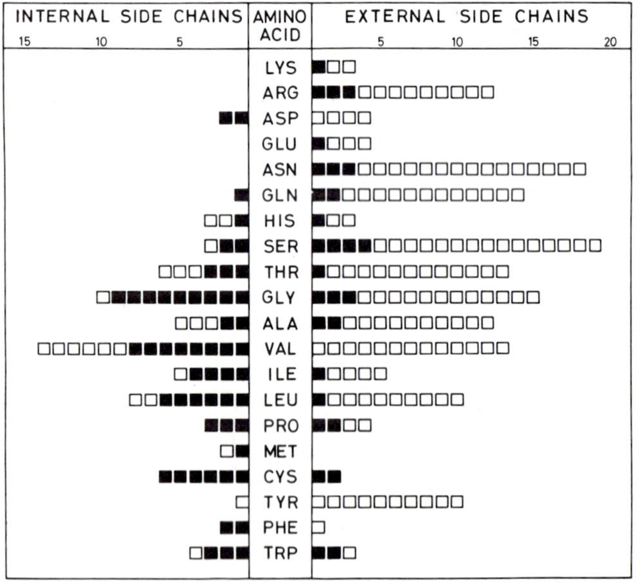

FIG. 16. The distribution between internal and external positions of the side chains of the amino acid residues in the porcine elastase molecule. (■) Residues homologous with corresponding residues in bovine trypsin, chymotrypsin A, and chymotrypsin B. (□) Residues not homologous with all three other enzymes. "Internal" and "homologous" are defined as for Fig. 13.

such as residues 36–37, where the external loop had necessarily been built by guesswork. The precise conformation of some surface side chains is also, of course, in doubt; but otherwise the hypothetical model proved to be a good representation of the true structure. We can, therefore, have faith that the trypsin model will be a good representation of that enzyme also.

4. *Conclusions from the Tertiary Structure*

The elastase structure strongly supports conclusions about the catalytic activity drawn from the tertiary structure of α-chymotrypsin. The conformation of Asp 102, His 57, and Ser 195 in the native enzyme is that necessary for the charge relay mechanism of Blow *et al.* (*85*). The structure of the substrate binding pocket explains the preference of elastase for alanyl substrates rather than amino acids with large aromatic or charged side chains. The internal ion pair between the N-terminus and Asp 194 is similar to that in chymotrypsin and suggests that proelastase and chymotrypsinogen have a similar activation mechanism. In more general terms it has demonstrated that enzymes possessing a common catalytic mechanism and exhibiting a high degree of sequence homology do have very similar tertiary structures.

C. Activity of Elastase Crystals

The elastase crystals have considerable advantages over those of chymotrypsin for further studies. The crystals are well ordered and offer an excellent prospect for investigation at 0.2 nm resolution, where precise questions about interactions between side chains should be answered. The stability of the crystals is such that they can be titrated to pH 11 without cracking, thus it may be possible to investigate any conformational changes consequent on titration of the charge-relay system to its active ionic form and on deprotonation of the Val 16–Asp 194 ion pair. Moreover, the packing of molecules in the crystal is such that the catalytic site lies facing a solvent cavity of approximately 1.8 nm diameter (unlike chymotrypsin where this site is blocked by intermolecular contacts) so the elastase crystals may prove suitable for studies of polypeptide binding.

In this context, preliminary studies of the catalytic activity of crystals of native elastase are of interest (*32*). They react stoichiometrically with tosyl fluoride, and difference Fourier projections of the product show that conformational changes are small and confined to the local environment of the tosyl group. When microcrystals of elastase (2–4 μ long) were assayed against *N*-benzoyl-L-alanine methyl ester (5 m*M*) in 1.2 *M* sodium sulfate, 0.1 *M* potassium chloride, 10 m*M* tris chloride pH 8.0 in the pH stat, the activity was at least 42% of that of a solution of elastase in the same medium. Under these conditions, diffusion was probably not rate limiting. The activity is much higher than that so far reported for microcrystals of carboxypeptidase A (0.3% that of the dissolved enzyme) (*102*), ribonuclease S (2% that of the amorphous enzyme) (*103*), or α-chymotrypsin (*104*) and gives one confidence that the elastase crystallographic structure represents that of the catalytically active species.

D. Reactivity and pK_a of the Amino Groups

1. *The Competitive Labeling Technique*

One of the major problems which remain when one has determined the tertiary structure of an enzyme and identified the catalytic groups and substrate binding site concerns the reactivity and state of ionization of functional groups. To help answer such questions, Kaplan *et al.* (*105*) devised a method of "competitive labeling" of such groups and applied it to the amino groups of elastase.

In this method, a small amount of radioactively labeled acetic anhydride is allowed to react to completion at a series of different pH with a large excess of protein in the presence of a competing nucleophile such as the amino group of phenylalanine. Figure 17 (*106*) demonstrates that each acetic anhydride molecule is faced with a set of competing nucleophiles such as the amino groups or phenolic groups of the protein, the amino group of X—the calibrating nucleophile, hydroxyl ions, water, etc. Since each nucleophile remains in large excess of the reagent throughout the reaction, the final degree of acetylation of any two groups will be in proportion to their apparent velocity constants, i.e.,

$$\frac{\%\ \text{Acetyl—group 1}}{\%\ \text{Acetyl—X}} = \frac{k_1'[\text{E}]}{k_x'[\text{X}]} = \frac{k_1[\text{E}](1 + \text{H}^+/K_x)}{k_x[\text{X}](1 + \text{H}^+/K_1)} \tag{1}$$

where [E] and [X] are the concentrations of enzyme and calibrating nucleophile, respectively; k_1' and k_x' are the apparent velocity constants;

102. F. A. Quiocho and F. M. Richards, *Biochemistry* **5,** 4062 (1966).
103. M. S. Doscher and F. M. Richards, *JBC* **238,** 2399 (1963).
104. D. M. Blow, personal communication (1969).
105. H. Kaplan, B. S. Hartley, and K. J. Stevenson, *Federation Proc.* **28,** 533 (1969).
106. B. S. Hartley, *BJ* **119,** 805 (1970).

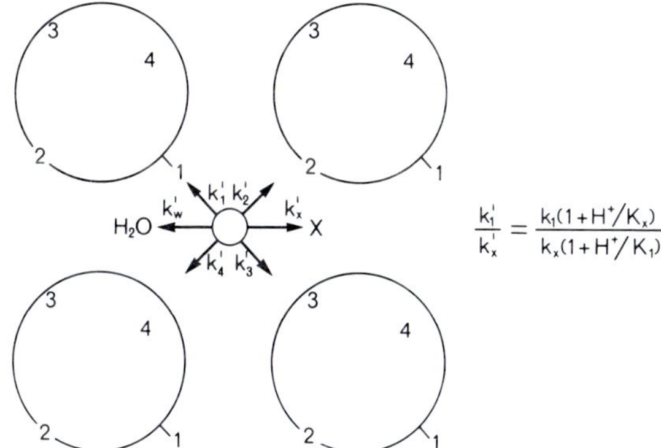

FIG. 17. The kinetics of competitive labeling. 1, 2, 3, and 4 represent amino groups of increasing "buriedness" in the protein. The reagent is surrounded by an excess of competing nucleophiles including X—an added "calibrating" nucleophile.

k_1 and k_x are the velocity constants of the unprotonated amino groups; and K_1 and K_x are the ionization constants. Knowing K_x, one can calculate the pK_a and velocity constant for each amino group if its degree of acetylation can be determined. Objections that the acetylation of one group might affect the reactivity of another are not applicable in this case since the native protein is in excess and is therefore the major reacting species.

After reaction with radioactive acetic anhydride, the mixture is reacted in 8 M urea with an excess of nonradioactive acetic anhydride in order to produce chemically homogeneous, radioactively heterogeneous acetylated products. The acetylphenylalanine is purified by ethyl acetate extraction from acid solution and high voltage paper electrophoresis. The protein is then digested with a suitable proteinase and the radioactive peptides are purified. Counting and amino acid analysis gives the specific activity of each acetyl group, and sequence analysis can identify the position of each amino group in the primary structure. The ratio of the specific activities of each peptide to that of the acetylphenylalanine gives k', which can be plotted as a function of pH to give pK_a and k for each separate amino group.

2. *Reactivity and* pK_a *of the Amino Groups of Elastase*

Kaplan *et al.* (*105*) applied this method to native elastase and isolated from a combined peptic plus tryptic plus chymotryptic digest the acetylated peptides Ac–Val–Val–Gly–Gly–Thr–Glu–Ala–Gln–Arg–Asn–Ser (residues 16–26), Val–Gln–Lys(Ac) (residues 85–87), and Lys(Ac)–Pro–

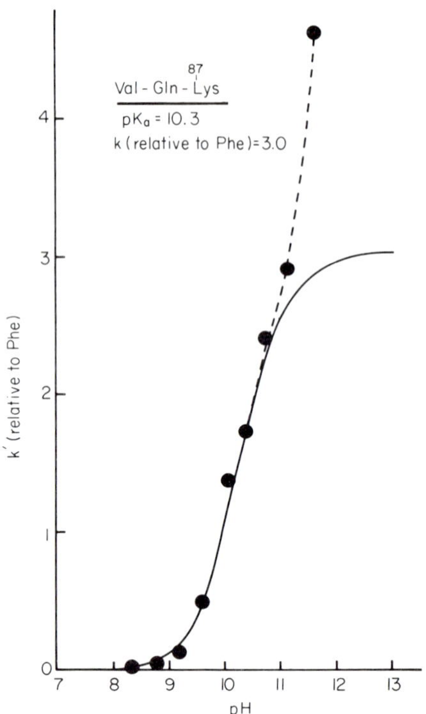

FIG. 18. Competitive labeling of Lys 87 in elastase with acetic anhydride. The full line represents the theoretical titration curve for a group with $pK_a = 10.3$. k_1' (relative to Phe) $= k_1'/k_x$ derived from Eq. (1).

Thr–Val–Phe (residues 224–229). Peptides derived from acetylated Lys 177 were difficult to purify and were not studied in detail. Plots of k'/k_{Phe} derived from the specific activities of these peptides are shown in Figs. 18–20. In each case the data between pH 8 and 10 provide a good fit to a theoretical titration curve for a single group. Above pH 10 a discontinuity is observed which appears as a sharp increase in reactivity of each group. The increased reactivity may represent an unfolding of the molecule at high pH, but it is probably premature to attempt such interpretations since errors in the competitive labeling technique at this high pH such as hydroxyl ion catalysis of the acetylation of amino groups might be involved.

The change in reactivity between pH 8 and 10 is more amenable to interpretation. Theory demands that for a series of unhindered primary amino groups reacting with acetic anhydride Bronsted plots of log k vs. pK_a for each compound should lie on a straight line. Kaplan *et al.*

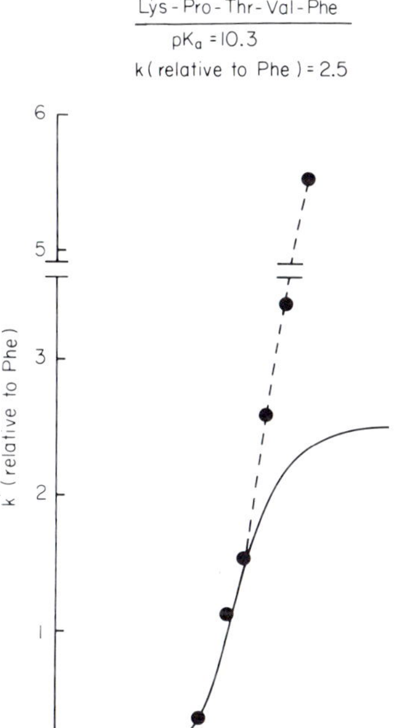

FIG. 19. Competitive labeling of Lys 224 in elastase. Details as for Fig. 18.

(*107*), by applying the competitive labeling method to a mixture of simple primary amines, have determined this slope as shown in Fig. 21. The reactivities fall satisfactorily on a line of Bronsted slope 0.46, in good agreement with results of Brouwer *et al.* (*108*) for amino acids plus acetic anhydride determined by conventional methods. The pK_a of both Lys 87 and Lys 224 in elastase is 10.3, in good agreement with the "expected" value of 10.4 (*109*) for an ϵ-amino group in water. The reactivities with acetic anhydride also lie close to the Bronsted plot for simple amines; thus, there is no doubt that the ϵ-amino groups of Lys 87 and Lys 224, which are situated in surface positions in the elastase molecule, lie free in the solvent in solutions of native elastase (see Table V).

The reactivity of the N-terminal amino group is dramatically different. The apparent pK_a of 9.7 is very much higher than that "expected"

107. H. Kaplan, K. J. Stevenson, and B. S. Hartley, unpublished evidence (1969).
108. D. M. Brouwer, M. J. Van der Vlugt, and E. Havinga, *Kon. Ned. Akad. Wetenschap., Proc.* **B61**, 141 (1958).
109. C. Tanford, *Advan. Protein Chem.* **17**, 69 (1962).

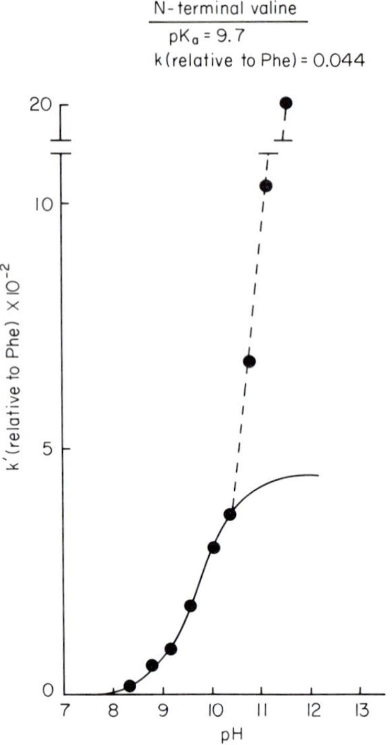

FIG. 20. Competitive labeling of the N-terminus of elastase (Val 16). Details as for Fig. 18.

for a free α-amino group—pK_a 7.6–7.9 (*109*)—and the relative reactivity is only 4% of that of a simple amine of this pK_a in solution determined from the Bronsted slope (Fig. 21). These data give us a quantitative estimate of the degree of "buriedness" of the protonated N-terminal

TABLE V
KINETIC CONSTANTS FOR THE REACTION OF ACETIC ANHYDRIDE WITH THE AMINO GROUPS OF ELASTASE

Amino group	p$K_{a(app)}$	Velocity constant for the uncharged amino group[a]
N-terminal	9.7	0.044
Lys 87	10.3	3.0
Lys 224	10.3	2.5

[a] Expressed as the ratio relative to the uncharged amino group of phenylalanine.

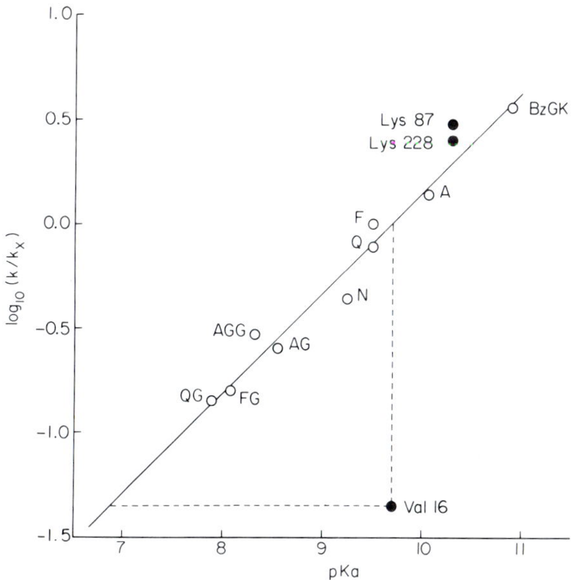

FIG. 21. Bronsted plot of the reactivities of amino groups with acetic anhydride at 10° in 0.1 M potassium chloride. k/k_x is the ratio of the velocity constant of the unprotonated group to that of unprotonated phenylalanine. QG, L-glutaminyl-glycine; FG, L-phenylalanyl-glycine; N, L-asparagine; Q, L-glutamine; F, L-phenylalanine; A, L-alanine; BzGK, N-benzoyl-glycyl-L-lysine. (●) The reactivities of amino groups of native elastase. The slope is 0.46 ± 0.03.

amino group of elastase in solution which can be interpreted in terms of the following model:

$$\begin{array}{ccc} \text{E-H} & \underset{}{\overset{K'}{\rightleftharpoons}} & \text{E}'\text{-H} \\ \upharpoonleft\downharpoonright K_a & & \upharpoonleft\downharpoonright K'_a \\ \text{E} + \text{H}^+ & \overset{K}{\rightleftharpoons} & \text{E}' + \text{H}^+ \end{array} \tag{2}$$

where E-H represents the conformation of elastase shown in the crystal structure with protonated Val 16 forming an internal ion pair with Asp 194; E′-H represents a hypothetical conformation in which Val 16 has moved from its internal cleft in order to leave the protonated amino group free in solvent, perhaps close to the position which it might occupy in the zymogen (*110*); E and E′ represent the unprotonated amino groups in these conformations. It is reasonable to assume that acetic anhydride reacts only with the E′ form, and if we assume a "normal" pK_a of 7.7 (*109*) for E′-H, the apparent pK_a of 9.7 for the system gives

110. J. Kraut, Chapter 5, this volume.

us an equilibrium constant $K' = 10^{-2}$ for "unfolding" of the protonated conformation. On this basis, the reactivity of the free amino group (0.044) is 40% of that expected from Fig. 21 (0.110) for a free amino group of pK_a 7.7; thus, K, the equilibrium constant of the unprotonated "native" structure, would be about 0.7 and the pK_a for the E-H form would be about 10.5. These figures are, of course, largely illustrative but do give us a picture of the stability of the Val 16–Asp 194 ion pair. Garel and Labouesse (*111*) have reached very similar conclusions for the behavior of Ile 16 in chymotrypsin.

3. *Reaction of the Amino Groups with Nitrous Acid*

Gertler and Hofmann (*22*) have investigated the reaction of elastase with nitrous acid at 0° in acid solutions. They found that at pH 3 the rate of inactivation, measured by the activity toward elastin or casein, was proportional to the loss of the N-terminal amino group, although up to 3 moles of tryptophan, 1 mole of tyrosine, and 0.5 mole of lysine also reacted by the time 90% inhibition was achieved. When the reaction was carried out at pH 4.0, less than 10% inhibition was observed and only 6% of the N-terminal amino group was lost, although the tryptophan, tyrosine, and lysine were destroyed in the same amounts as at pH 3.0. Gertler and Hofmann therefore concluded that a conformational transition in the enzyme takes place below pH 4, whereby the N-terminal valine is removed from a protected environment in the protein to a form in which it is freely accessible to nitrous acid. Such a conformational change must be reversible since the enzyme was fully reactive at pH 8.8 after 1 hr at 0° at pH 3.0. These experiments support the hypothesis that native elastase in solution between pH 4 and 9 is predominantly in the structure determined by X-ray crystallography with the internal ion pair of Val 16 and Asp 194 intact.

4. *The Role of Val 16 and Asp 194 in Enzymic Activity*

In acetylated δ-chymotrypsin, as discussed by Hess in Chapter 7 of this volume (*54*), deprotonation of Ile 16, with apparent pK_a 8.5, seems to be responsible for a conformational change in the enzyme which decreases the affinity for substrates such as acetylphenylalanine amide. This conformational change can be demonstrated by changes in optical rotation or circular dichroism. The N-terminal isoleucine can be specifically acetylated, and the product shows only 9% of the activity of δ-chymotrypsin toward *N*-acetyl-L-tyrosine ethyl ester but still reacts

111. J.-R. Garel and B. Labouesse, *JMB* **47**, 41 (1970).

with *p*-nitrophenyl acetate to give a "burst" of 1 mole of nitrophenol although both acylation and deacylation rates are slower than for the native enzyme (*112*). One must therefore conclude that although disruption of the internal ion pair by deprotonation or acetylation of Ile 16 disturbs the conformation of the enzyme considerably it does not completely destroy either the catalytic activity or the ability to refold into an active conformation.

The situation with elastase is rather different. Kaplan *et al.* (*107*) were unable to find any change in optical rotation of the enzyme up to pH 11, and Kaplan and Dugas (*27*) showed that the native enzyme showed no change in activity against benzoylalanine methyl ester over the range pH 8–10, although the activity was dependent on an ionizing group of pK_a 6.5, presumably that of the charge-relay system of Asp 102, His 57, and Ser 195. Moreover, reaction of elastase with excess acetic anhydride appeared to acetylate all amino groups including that of Val 16. This fully acetylated enzyme showed exactly the same activity toward benzoylalanine methyl ester as the native enzyme, and the pH dependence was identical for both, but the acetylated enzyme was only 50% active toward elastin at pH 8.8. The apparent pK_a of Val 16 determined by the competitive labeling experiment is 9.7, thus one must conclude that the enzyme is fully active toward benzoylalanine methyl ester with its N-terminal group unprotonated. Furthermore, the enzyme with Val 16 acetylated is also fully active toward *N*-benzoyl-L-alanine methyl ester, so the internal ion pair of Val 16 and Asp 194 cannot be essential for the activity of elastase against this small substrate.

These apparently arcane observations become important when we consider possible explanations for the activation of zymogens of the serine proteinases such as chymotrypsinogen. The simple and attractive hypothesis of Sigler *et al.* (*93*) that the internal ion pair is essential to orient correctly Ser 195 and His 57 is no longer entirely adequate (*110*), thus we must look to further fine details of the structures of the enzymes and their zymogens and the interactions with their differing substrates and to further exploration of the ionizing properties of the relevant groups to answer these tantalizing questions.

112. C. Ghelis, J. Labouesse, and B. Labouesse, *BBRC* **29**, 101 (1967).

11

Protein Proteinase Inhibitors—Molecular Aspects

MICHAEL LASKOWSKI, JR. • ROBERT W. SEALOCK

I. Introduction

The presence of proteinlike trypsin inhibitors in nature was known even before the turn of the century, but the beginning of productive chemical study of these substances must be marked by the pioneering work of Moses Kunitz. Kunitz and Northrop (*1*) in 1936 crystallized the basic trypsin inhibitor from bovine pancreas and demonstrated that it consists "largely of amino acids." Kunitz later (*2–4*) isolated and studied the well-known soybean trypsin inhibitor. Not only was this series of papers remarkably elegant for its time but also many of the fundamental concepts introduced by Kunitz are still retained in modern theories of proteinase–inhibitor interaction. This legacy to biochemistry includes 1:1 crystallizable complexes of enzyme and inhibitor (*1, 5*), a quantitative assay method for inhibitors (*4*), dissociation of enzyme–inhibitor complexes at low pH and subsequent reassociation as the pH is raised (*1, 5*), the concept of equilibrium between complex and enzyme and inhibitor and the first measurements of equilibrium constants for such reactions (*4*), the recognition of vast difference in rates of association and dissociation in different enzyme–inhibitor systems (*1, 4*), and the recognition that the inhibitors must be native in order to combine with enzymes (*4*). The research on proteinase inhibitors since the Kunitz period has been carried out by workers in a vast variety of fields and has spawned a huge literature. It was a multipronged attack. One prong was the isolation and characterization (in the Kunitz tradition) of a truly enormous number of inhibitors from a vast variety of plant and animal tissues, another prong was the work on possible physiological function and pharmacological and nutritional significance of the inhibitors. The most recent work (1960's) is an attempt to understand the enzyme–inhibitor interaction and specificity in molecular terms. This approach is recent since only within the last decade was enough learned about the molecular aspects of the enzymes and were sufficiently sensitive physical and chemical probes developed. It is this last aspect that is

1. M. Kunitz and J. H. Northrup, *J. Gen. Physiol.* **19,** 991 (1936).
2. M. Kunitz, *Science* **101,** 668 (1945).
3. M. Kunitz, *J. Gen. Physiol.* **29,** 149 (1946).
4. M. Kunitz, *J. Gen. Physiol.* **30,** 291 (1947).
5. M. Kunitz, *J. Gen. Physiol.* **30,** 311 (1947).

stressed in this chapter since the other two aspects have already received an extensive coverage in other reviews.

A. Other Reviews

There are several current reviews of the nutritional, pharmacological, and general aspects of proteinase inhibitors. The largest collection of references is the review by Vogel *et al.* (*6*) now available in an updated (to early 1968) version in English (*7*). Liener and Kakade (*8*) have recently published a review of inhibitors from plant sources which contains extensive tables of inhibitors and their chemical and physical properties. Feeney and Allison (*9*) have also tabulated the properties of many inhibitors and have reviewed the recent researches and viewpoints of their laboratory. Steiner and Frattali (*10*) have reviewed the preparations and properties of the Kunitz and Bowman–Birk inhibitors from soybeans. The proceedings of a 1966 symposium on pharmacological and clinical applications of inhibitors has been published (*11*). Pusztai has reviewed inhibitors from the standpoint of animal nutrition (*12*). Discussions of many topics not included herein may be found in these sources. Earlier reviews include those by Laskowski and Laskowski (*13*), Laskowski, Sr. (*14*), Green and Neurath (*15*), Desnuelle (*16*). A series on trypsin inhibitors is scheduled for a forthcoming volume of "Methods in Enzymology" (*16a, 16b*). One of us (*17*) recently wrote a short discussion of the reactive site of soybean trypsin inhibitor (Kunitz).

6. R. Vogel, I. Trautschold, and E. Werle, "Naturliche Proteinasen-Inhibitoren." Thieme, Stuttgart, 1966.

7. R. Vogel, I. Trautschold, and E. Werle, "Natural Proteinase Inhibitors." Academic Press, New York, 1968.

8. I. E. Liener and M. L. Kakade, *in* "Toxic Constituents of Plant Foodstuffs" (I. E. Liener, ed.), p. 7. Academic Press, New York, 1969.

9. R. E. Feeney and R. G. Allison, "Evolutionary Biochemistry of Proteins." Wiley, New York, 1969.

10. R. F. Steiner and V. Frattali, *J. Agr. Food. Chem.* **17,** 513 (1969).

11. E. M. Weyer, *Ann. N. Y. Acad. Sci.* **146,** 361 (1968).

12. A. Pusztai, *Nutr. Abstr. Rev.* **37,** 1 (1967).

13. M. Laskowski and M. Laskowski, Jr., *Advan. Protein Chem.* **9,** 203 (1954).

14. M. Laskowski, Sr., "Methods in Enzymology," Vol. 2, p. 36, 1955.

15. N. M. Green and H. Neurath, *in* "The Proteins" (H. Neurath and K. Bailey, eds.), 1st ed., Vol. 2, Part B, p. 1057. Academic Press, New York, 1954.

16. P. Desnuelle, "The Enzymes," Vol. 4, p. 128, 1960.

16a. B. Kassell, "Methods in Enzymology," Vol. 19, p. 839, 1970.

16b. P. J. Burck, "Methods in Enzymology," Vol. 19, p. 906, 1970.

17. M. Laskowski, Jr., *Proc. IUB/IUBS Intern. Symp. Struct.-Funct. Relationship Proteol. Enzymes, 1969* p. 89. Munksgaard, Copenhagen, 1970.

The proceedings of the First International Research Conference on Proteinase Inhibitors, held in Munich in November, 1970, are to be published (*17a*), and should be consulted for more recent information than that included in this review.

B. Definition

For the purposes of this review, the term "protein proteinase inhibitor" refers to a protein which may associate reversibly with one or more proteinases to form complexes of discreet stoichiometry in which *all* of the catalytic functions of the proteinase are *competitively* inhibited. These inhibitors usually have molecular weights in the range 5,000–60,000 with most being less than 20,000. With the exception of certain inhibitors from serum, all apparently contain cysteine in disulfide linkage, all apparently contain proline, and many contain no tryptophan. Many, primarily the ovomucoids, contain a large carbohydrate component. As a class they are highly stable proteins. The inhibited proteinases are usually endopeptidases, i.e., peptidyl-peptide hydrolases, although there are reports of protein inhibitors(s) of carboxypeptidase B [e.g., Rancour and Ryan (*18*)].

The above definition and the intended scope of this review exclude from consideration antigen–antibody interactions, the high molecular weight carrier protein of the α_2-macroglobulin fractions [reviewed by Vogel *et al.* (*7*)], and the pepsin inhibitor fragment of pepsinogen (*19, 20*). Other polymeric but nonprotein inhibitors of proteinases, e.g., the polysaccharide heparin (*21*), polyglutamic acid (*22*), and the acidic macromolecular inhibitor from *Penicillium cyclopium* (*23, 24*), are also excluded.

C. Identities and Nomenclature

The truly astonishing number of proteinase inhibitors which have been isolated or at least detected, as well as the frequent presence of several proteinase inhibitors in the same source tissue compound the difficulties

17a. *Proc. First Intern. Res. Conf. Proteinase Inhibitors,* de Gruyter, Berlin, 1971 (in press).

18. J. M. Rancour and C. A. Ryan, *ABB* **125,** 380 (1968).
19. R. M. Herriott, *J. Gen. Physiol.* **24,** 325 (1941).
20. H. VanVunakis and R. M. Herriott, *BBA* **22,** 537 (1956).
21. T. Astrup, U. Nissen, and J. Rasmussen, *ABB,* **113,** 634 (1966).
22. M. Rigbi and M. Sela, *Biochemistry* **3,** 629 (1964).
23. K. Shimada and K. Matsushima, *Agr. Biol. Chem.* (*Tokyo*) **33,** 544 (1969).
24. K. Shimada and K. Matsushima, *Agr. Biol. Chem.* (*Tokyo*) **33,** 549 (1969).

of any logical nomenclature scheme. Another aspect raising the level of confusion is that the same proteinase inhibitors are frequently present in different tissues of the same organism and that they frequently inhibit more than one enzyme. Therefore, it should be expected (and as discussed below is frequently observed) that the same inhibitor may be isolated from two different tissues characterized by inhibitory activity against different enzymes so that the identity of the two preparations escapes immediate detection.

1. *Kallikrein Inactivator and Pancreatic Inhibitors*

A protein inhibitor of trypsin, chymotrypsin, plasmin, and kallikrein from bovine parotid gland was first described in 1928 (*25*). It was purified in 1960 (*26*), and a similar substance from lungs was isolated in 1964 (*27*). Chemical and physical evidence suggested these two were identical. In 1965, Anderer and Hornle (*28, 29*) published the amino acid sequence of this "kallikrein inactivator" from lungs and by comparison of tryptic peptides established its identity with the inactivator from parotid gland (*30*). Simultaneously the groups led by Acher, Šorm, and Laskowski, Sr., were independently determining the sequence of the bovine pancreatic trypsin inhibitor of Kunitz. Anderer (*30*), on the basis of the first (incorrect) sequence of Chauvet *et al.* (*31*), suggested that these two inhibitors might also be identical. The correct sequence (*32, 33*) and disulfide pairings (*34, 35*) of the pancreatic inhibitor were published by Kassell and Laskowski in 1965. The 1966 publication by Anderer and Hornle (*36*) of the disulfide pairings in the kallikrein inactivator proved the identity. The other groups have since revised their sequences (*37–41*).

25. H. Kraut, E. K. Frey, and E. Bauer, *Z. Physiol. Chem.* **175,** 97 (1928).
26. H. Kraut, W. Korbel, W. Scholtan, and F. Schultz, *Z. Physiol. Chem.* **321,** 90 (1960).
27. H. Kraut and N. Bhargava, *Z. Physiol. Chem.* **328,** 231 (1964).
28. F. A. Anderer and S. Hornle, *Z. Naturforsch.* **20b,** 457 (1965).
29. F. A. Anderer, *Z. Naturforsch.* **20b,** 462 (1965).
30. F. A. Anderer, *Z. Naturforsch.* **20b,** 499 (1965).
31. J. Chauvet, G. Nouvel, and R. Acher, *BBA* **92,** 200 (1964).
32. B. Kassell and M. Laskowski, Sr., *BBRC* **17,** 792 (1964).
33. B. Kassell, M. Radicevic, M. J. Ansfield, and M. Laskowski, Sr., *BBRC* **18,** 255 (1965).
34. B. Kassell and M. Laskowski, Sr., *BBRC* **20,** 463 (1965).
35. B. Kassell and M. Laskowski, Sr., *Acta Biochim. Polon.* **13,** 287 (1966).
36. F. A. Anderer and S. Hornle, *JBC* **241,** 1568 (1966).
37. V. Dlouhá, P. Pospišilova, B. Meloun, and F. Šorm, *Collection Czech Chem. Commun.* **30,** 1705 (1966).
38. V. Dlouhá, P. Pospišilova, B. Meloun, and F. Šorm, *Collection Czech Chem. Commun.* **33,** 1363 (1968).

It is now known that this inhibitor is found in all tested organs of some, though not all, ruminants (*42*). It is not found in other mammals. It is an intracellular inhibitor, distinguished from the secretory (extracellular) pancreatic inhibitors of the Kazal type, found in all mammals tested (Section IV,A). The Kunitz inhibitor is also known as the basic pancreatic inhibitor and occasionally referred to by the names of commercial drug preparations of which it is the active component, e.g., Cortrykal, Iniprol, and Trasylol (*43*). In this review it is known as the pancreatic trypsin inhibitor (Kunitz).

The pancreatic trypsin inhibitor common to all mammals (and only the pancreas thereof) was first isolated from the cow by Kazal *et al.* (*44*). This inhibitor and its homologs are generally referred to in this review as pancreatic secretory inhibitors (Kazal). Other common designations in the literature are acidic, secretory, and specific inhibitor and acronyms such as APTI, SPTI, and PSTI. The designation *acidic* is a misnomer for this class of inhibitors, that from human pancreas, for example, being a slightly basic protein (*45*).

2. *Soybean Inhibitors*

The well-known soybean trypsin inhibitor (variously known as STI, SBTI, or SbI) is the one isolated by Kunitz in 1945 (*4*). Even then the presence of multiple inhibitors in soybeans was known; Bowman (*46*) isolated acetone- and alcohol-insoluble inhibitors, the latter being the Kunitz inhibitor. Birk (*47*) and Birk *et al.* (*48*) have purified the acetone-insoluble fraction and called it STIAA. In this review, this and similar preparations are known as the Bowman–Birk inhibitor after a suggestion by Frattali (*49*). Rackis and Anderson (*50*) have isolated active chromatographic fractions $STIA_1$, A_2, B_1, and B_2 from soybean whey. Frattali and Steiner (*51*) found fractions F_1, F_2, and F_3 in com-

---

39. J. Chauvet, G. Nouvel, and R. Acher, *BBA* **115,** 121 (1966).
40. J. Chauvet, G. Nouvel, and R. Acher, *BBA* **115,** 130 (1966).
41. J. Chauvet and R. Acher, *Bull. Soc. Chim. Biol.* **49,** 985 (1967).
42. E. Werle, *Z. Physiol. Chem.* **338,** 228 (1964).
43. F. Markwardt and M. Richtor, *Pharmazie* **24,** 620 (1969).
44. L. A. Kazal, D. S. Spicer, and R. A. Brahinsky, *JACS* **70,** 3034 (1948).
45. P. J. Keller and B. J. Allan, *JBC* **242,** 281 (1967).
46. D. E. Bowman, *Proc. Soc. Exptl. Biol. Med.* **63,** 547 (1946).
47. Y. Birk, *BBA* **54,** 378 (1961).
48. Y. Birk, A. Gertler, and S. Khalef, *BJ* **87,** 281 (1963).
49. V. Frattali, *JBC* **244,** 274 (1969).
50. J. J. Rackis and R. L. Anderson, *BBRC* **15,** 230 (1964).
51. V. Frattali and R. F. Steiner, *Biochemistry* **7,** 521 (1968).

mercial crude Kunitz inhibitor. Yamamoto and Ikenaka (*52*) introduced the 1.9 S inhibitor. The Kunitz inhibitor, A_2, and F_2 are apparently identical, although the recent amino acid analysis of A_2 (*53*) is not in complete agreement with that of STI. The originally found molecular weights of the Bowman–Birk and 1.9 S inhibitors were 24,000 (*48*) and 16,000 (*52*), respectively, and from these and few other data they were considered distinct. However, the amino acid analysis of the Bowman–Birk inhibitor from Lee variety soybeans (*49*) is nearly identical with that of the 1.9 S inhibitor (*52*) and requires a minimum molecular weight of 8000. Steiner and co-workers have verified this value and have shown this inhibitor to be a self-associating system (*54, 55*), thus accounting at least partially for the variable molecular weights. For these reasons the Bowman–Birk and 1.9 S inhibitors are probably identical.

Comparisons among the other fractions are more difficult. After recalculation (*53*) of the original values (*50*), the sedimentation coefficients (1.7–1.8) of A_1, B_1, and B_2 suggest molecular weights of 14,000–16,000 and their equivalent weights for trypsin inhibition are about 12,000 (*50*). Thus, these and the Bowman–Birk inhibitors may be soybean homologs of the lima bean "isoinhibitors" (*56–58*), which have molecular weights of less than 10,000, and of which at least one is known to dimerize (*58*) (apparent molecular weight 16,000, $s^{\circ}_{20,w}$ 1.9). By ultracentrifugation, molecular weights of 18,000 and 23,000 were assigned to F_1 and F_3 (*51*). By amino acid analysis, they are distinct from each other and from the Kunitz and Bowman–Birk inhibitors, but their relationship to the others is unclear (*58a*).

This considerable confusion among soybean inhibitors led Steiner and Frattali (*10*) to propose a general scheme for inhibition nomenclature which makes use of the source of the inhibitor and the nature and location of the residues comprising its reactive site. It is clear that such a scheme would have a very great advantage over the present confusion

52. M. Yamamoto and T. Ikenaka, *J. Biochem.* (*Tokyo*) **62,** 141 (1967).
53. A. C. Eldridge and W. J. Wolf, *Cereal Chem.* **46,** 470 (1969).
54. D. B. S. Millar, G. E. Willick, R. F. Steiner, and V. Fratalli, *JBC* **244,** 281 (1969).
55. J. B. Harry and R. F. Steiner, *Biochemistry* **8,** 5060 (1969).
56. B. Jirgensons, T. Ikenaka, and V. Gorguraki, *Makromol. Chem.* **39,** 149 (1960).
57. G. Jones, S. Moore, and W. H. Stein, *Biochemistry* **2,** 66 (1963).
58. R. Haynes and R. E. Feeney, *JBC* **242,** 5378 (1967).

58a. Additional complications may arise resulting from the existence of genetic variants of inhibitors in various soybean strains. A clear example of a genetic variant was discovered by Singh *et al.* (*58b*). It is unlikely, however, that this particular variant was ever present in commercial samples of the inhibitor.

58b. L. Singh, C. M. Wilson, and H. H. Hardley, *Crop. Sci.* **9,** 489 (1969).

provided that it would lead to unambiguous designation of all distinct inhibitors. Not enough reactive sites have yet been determined to predict whether the Steiner–Frattali scheme could distinguish closely related inhibitors such as isoinhibitors from lima beans.

3. *Serum Inhibitors*

The proteinase inhibitors found in sera have been reviewed by Vogel *et al.* (*7*). Rather little is known about their mode of interaction with proteolytic enzymes, and they are, therefore, little dealt with in this review. Serum inhibitors typically have higher molecular weights than the inhibitors discussed herein, and rather surprisingly the α_1-trypsin and α_1-chymotrypsin inhibitors from human serum contain no half-cystine (*59*). These two proteins thus form, or may be only a part of, a class of inhibitors to which much of the latter discussion in this review may not apply.

Habermann (*60*) and Fritz *et al.* (*61*) have shown that both bovine and human sera contain (at least) two inhibitors of swine kallikrein. In each species one inhibitor has the interesting property of combining rapidly with trypsin but only very slowly with kallikrien. After purifying this "progressive anti-kallikrein" from bovine serum, Habermann noted that it possessed properties similar to those of the acid-labile trypsin, chymotrypsin, elastase, and plasmin inhibitor crystallized from bovine serum by Wu and Laskowski, Sr. (*62*). In parallel experiments he showed that the two inhibitors are identical by starch gel electrophoresis, by their acid lability, and by comparison of their time courses for inhibition of trypsin and kallikrein. It is likely that the bovine α_1-trypsin inhibitor is also identical to this inhibitor since Fritz *et al.* (*63*) have shown by similar experiments that the α_1-trypsin and progressive kallikrein inhibitors of human serum are the same substance.

Heimburger and Haupt have also demonstrated the identity of the α_1-X-glycoprotein and the α_1-chymotrypsin inhibitor of human serum (*64*).

59. N. Heimburger, K. Heide, H. Haupt, and H. E. Schultze, *Clin. Chim. Acta* **10,** 293 (1964).
60. E. Habermann, *Ann. N. Y. Acad. Sci.* **146,** 479 (1968).
61. H. Fritz, I. Trautschold, H. Haendle, and E. Werle, *Ann. N. Y. Acad. Sci.* **146,** 400 (1968).
62. F. C. Wu and M. Laskowski, Sr., *JBC* **235,** 1680 (1960).
63. H. Fritz, B. Brey, A. Schmal, and E. Werle, *Z. Physiol. Chem.* **350,** 1551 (1969).
64. N. Heimburger and H. Haupt, *Clin. Chim. Acta* **12,** 116 (1965).

D. Crystalline Inhibitors and Complexes

Many of the speculations and questions about the interactions of enzymes and inhibitors which are developed in the following pages can hopefully be resolved, or at least be more elegantly rephrased, by the results of X-ray crystallographic studies. For this reason we have listed in Table I all examples of crystalline inhibitors and inhibitor–enzyme complexes of which we are aware. In most cases, however, the crystals are small crystals detectable by microscopy and the large ones suitable for X-ray crystallography have not yet been grown.

For obvious reasons the crystallographic structures of the complexes will be of primary interest. However, if for technical or other reasons such studies are not feasible, we wish to suggest that a structure of a modified inhibitor (i.e., one whose reactive site peptide bond is hydrolyzed) would be of considerably more interest than the structure of a virgin inhibitor. Largely on the basis of results discussed below, it ap-

TABLE I
Crystalline Inhibitors and Complexes

Inhibitor	Crystalline inhibitors	Crystalline enzyme–inhibitor complex
Bovine pancreatic (Kunitz)	(*1*)	Trypsin (*1*)
Bovine colostrum	(*65*)	Trypsin (*65*)
Porcine colostrum	(*66*)	Trypsin (*66*)
Bovine pancreatic secretory (Kazal)	(*44*)	—
Bovine serum acid labile	(*62*)	—
Soybean (Kunitz)	(*3*)	Trypsin (*5*)
Indian field bean	(*67*)	—
Double bean	(*67*)	—
Potato type I (*68*)	(*69*)	Chymotrypsin (*69*)
Ascaris chymotrypsin	(*70*)	—
Mung bean A and B	(*71, 72*)	Trypsin (*73*)[a]

[a] Both 1:1 and 2:1 trypsin–inhibitor complexes have been crystallized.

65. M. Laskowski, Jr. and M. Laskowski, Sr., *JBC* **190**, 563 (1951).
66. M. Laskowski, Sr., B. Kassell, and G. Hagerty, *BBA* **24**, 300 (1957).
67. K. Sohonie and K. S. Ambe, *Nature* **175**, 508 (1955).
68. C. A. Ryan, *Biochemistry* **5**, 1592 (1966).
69. A. K. Balls and C. A. Ryan, *JBC* **238**, 2976 (1963).
70. R. J. Peanasky and M. Laskowski, Sr., *BBA* **37**, 167 (1960).
71. H.-M. Chu and C.-W. Chi, *Sci. Sinica* (*Peking*) **14**, 1441 (1965).
72. H.-M. Chu, S.-S. Lo, M.-H. Jen, C.-W. Chi, and T.-C. Tsao, *Sci. Sinica* (*Peking*) **14**, 1454 (1965).
73. H.-M. Chu and C.-W. Chi, *Acta Biochem. Biophys. Sinica* **5**, 519 (1965).

pears that the structure of modified inhibitors is very nearly identical to that of virgin inhibitors except in the immediate region of the reactive site. The study of a modified inhibitor is therefore of greater value not only for much the same general reasons as was the study of ribonuclease-S (*73a*) but also from the standpoint of speculations about the nature of the stable complex. Several X-ray investigations of either an inhibitor or an enzyme–inhibitor complex are now in progress. The X-ray crystallographic structure determination of pancreatic trypsin inhibitor (Kunitz) has now been annouced by Huber *et al.* (*73b*) and atomic coordinates are available. The structure itself and combination of the inhibitor model with that of trypsin offers an opportunity for detailed interpretation of trypsin–inhibitor interactions. The work is too recent to be reviewed here. It is gratifying that the results offer strong support for many of the conclusions obtained by studies carried out in solution and discussed in this review.

II. General Characterization of Inhibitors and of Their Interactions with Enzymes

A. Physical Properties of Inhibitors

Various properties of inhibitors such as amino acid compositions, molecular weight, and extinction coefficient are tabulated in the reviews mentioned above and are therefore not collected here, athough instances in which new results require reinterpretation of the older data are cited in this review. There is, however, important literature which leads to a rather general description of an inhibitor protein and which has not been previously reviewed from that standpoint.

The most remarkable physical property of inhibitors as a class of proteins is their stability under the usual denaturing conditions, a fact which has considerable importance for the nutritional qualities of many bean meals. Several inhibitors, e.g., Kunitz (*1*) and Kazal (*74*) pancreatic inhibitors, bovine and porcine colostrum inhibitors (*65, 66*), various bean inhibitors (*7*), and the *Ascaris* trypsin (*75*) and chymo-

73a. H. W. Wyckoff, D. Tsernoglou, A. W. Hanson, J. R. Knox, B. Lee, and F. M. Richards, *JBC* **245**, 305 (1970).

73b. R. Huber, D. Kukla, A. Rühlmann, O. Epp, and H. Formanek, *Naturwissenshaften* **57**, 389 (1970).

74. M. Laskowski, Sr. and F. C. Wu, *JBC* **204**, 797 (1953).

75. H. B. Collier, *Can. J. Res.* **B19**, 91 (1941).

trypsin (*76*) inhibitors, remain active after exposures of several minutes to 2–3% trichloroacetic acid at temperatures up to 95°. Chicken ovomucoid is not precipitable by trichloroacetic acid and is also stable to heating in solution (*77*). The Kazal pancreatic (*44*) and Bowman–Birk soybean (*46*) inhibitors withstand treatment with up to 90% ethanol. Several of these inhibitors have been shown to be stable to denaturation by 8 *M* or 9 *M* urea solution at neutral pH and room temperature.

With the exception of the serum inhibitors cited earlier, all inhibitors whose amino acid compositions we have examined contain cystine. Lima bean inhibitor component 2 contains 14 half-cystines in its total of 76 residues (*57*), and the Bowman–Birk inhibitor from soybean has 14 half-cystines in a total of 78 residues (*49*). Most of the other inhibitors mentioned above contain 9–15% half-cystine. None contains free sulfhydryl groups. This extensive cross-linking is obviously an important source of stability for these inhibitors. The Kunitz soybean inhibitor, containing only 4 half-cystines in its 198 residues (*78, 79*), rather readily undergoes thermal denaturation in aqueous solution (*4, 80*), but it is nonetheless also stable to denaturation in 9 *M* urea (*81*) and in 6 *M* guanidine hydrochloride at neutral pH and room temperature (*82, 83*).

By measurement of decreased solubility at the isoelectric point, Kunitz (*4*) found that this inhibitor can be reversibly denatured by heating or irreversibly but slowly denatured in 0.1 *M* NaOH. He also established that it is the native form which associates with trypsin. Wu and Scheraga (*84*) further studied these denaturations by optical rotatory dispersion and ultraviolet difference spectroscopy. No transition was observable in the dispersion measurements at long wavelengths, and the data suggested that the inhibitor under all solution conditions is devoid of the usual ordered backbone structures typical of globular proteins. Nonetheless, a rather sharp pH and ionic strength dependent thermal transition is readily observed by difference spectroscopy and by optical rotation at short wavelengths (*17, 85*). The transition is fully reversible

76. N. M. Green, *BJ* **66,** 416 (1957).
77. H. Lineweaver and C. W. Murray, *JBC* **171,** 565 (1947).
78. Y. V. Wu and H. A. Scheraga, *Biochemistry* **1,** 698 (1962).
79. K. Ozawa and M. Laskowski, Jr., *JBC* **241,** 3955 (1966).
80. M. Kunitz, *J. Gen. Physiol.* **32,** 241 (1948).
81. H. Edelhoch and R. F. Steiner, *JBC* **238,** 931 (1963).
82. B. Jirgensons, M. Kawabata, and S. Capetillo, *Makromol. Chem.* **125,** 126 (1969).
83. R. W. Sealock and M. Laskowski, Jr., *Biochemistry* **8,** 3703 (1969).
84. Y. V. Wu and H. A. Scheraga, *Biochemistry* **1,** 905 (1962).
85. C. W. Niekamp and M. Laskowski, Jr., *158th Am. Chem. Soc. Meeting New York,* Abstract No. 273, Biol. Chem. Section (1969).

provided that the protein concentration and ionic strength are kept very low and the time of incubation at elevated temperature is minimized (*85*). No changes in conformation were observed at low pH and room temperature, although a significant alkaline pH transition (25°) was observed (*84*). The difference spectra accompanying both the thermal and high pH transitions are positive, corresponding to a red shift in tryptophyl and tyrosyl absorbance and thereby suggesting the burial of these residues during the transitions. It is interesting to note that the Linderstrom–Lang electrostatic factor required to fit the titration curve of the inhibitor remains constant between pH 2 and 7 but increases at higher pH, suggesting the formation of a more compact structure during the alkaline transition (*78*).

Steiner and Edelhoch (*86*) have also investigated the native state of soybean inhibitor (Kunitz) in aqueous solution as a function of solution conditions. They confirmed the alkaline transition and in addition noted an acid transition (pH 4) which is accompanied by a very small red-shifted difference spectrum, in contrast to the blue shift observed with most proteins at acid pH. At very low pH values there is an approximately 15% increase in levorotation at 365 nm, in spite of the fact that native soybean inhibitor (Kunitz) is one of the most levorotatory native proteins known.

Edelhoch and Steiner (*81*) have also characterized the state of the Kunitz inhibitor in 9 *M* urea. The difference spectrum of the inhibitor in urea at neutral pH and room temperature was very similar to that induced by 53% ethylene glycol, i.e., a red shift and linear dependence on urea concentration were observed, and no evidence of denaturation could be detected. The specific rotation of the inhibitor is essentially unchanged in 9 *M* urea from that in water at pH values between 2 and 11. At higher pH values and room temperature or at neutral pH and higher temperatures, denaturation readily occurs. Jirgensons *et al.* (*82*) have also shown the inhibitor to be stable to denaturation by 6 *M* guanidine·HCl at neutral pH, as measured by circular dichroism.

Chicken ovomucoid, another inhibitor which is quite stable in 9 *M* urea (*77*), has also been shown to undergo low pH transition. Herskovits and Laskowski, Jr. (*87*) found by solvent perturbation techniques that at pH values near 3.5 the tyrosyls of ovomucoid are equally exposed to large and small perturbants but that as the pH is raised they become accessible to only the small perturbants. This rather exceptional behavior appears to be the result of limited conformational alterations since it is accompanied by only small changes in specific optical rotation

86. R. F. Steiner and H. Edelhoch, *JBC* **238**, 925 (1963).
87. T. T. Herskovits and M. Laskowski, Jr., *JBC* **237**, 3418 (1962).

(*87, 88*). Donovan (*88*) has identified a larger transition in the acid region (pK 2.8) which gives rise to a large blue shift in ultraviolet absorption spectrum, a drastic increase in tyrosyl fluorescence, and an approximately 15% increase in intrinsic viscosity. Chromophores apparently do not become more exposed to solvent during this transition. Esterified ovomucoid gives no low pH difference spectrum and at all pH values has the fluorescence characteristics of protonated ovomucoid and is therefore probably frozen in the low pH conformation (*88*).

In spite of the occurrence of minor transitions in these inhibitors, the general description of an inhibitor as a protein which emerges from these studies is that of a protein which possesses extreme internal rigidity and considerable resistance to overall denaturation. This stability often includes resistance to general degradative action of proteolytic enzymes, most notably pepsin at pH 2 or 3 (*7*). The secondary structures of native inhibitors are responsible for this observation since inhibitors which have been previously denatured are readily hydrolyzed by most proteinases. It is interesting in this regard that pancreatic inhibitor (Kazal) and chicken ovomucoid are temporary inhibitors (Section III,J), i.e., they are slowly degraded by equimolar trypsin at neutral pH. The stability of sulfur-containing inhibitors to adverse solution conditions contrasts markedly with the acid lability of the serum inhibitors. For example, Wu and Laskowski, Sr. (*62*) noted that their serum inhibitor (the α_1-trypsin inhibitor) suffers considerable loss of inhibitory activity upon incubation at pH 4 or below.

The frequently observed exceptional conformational stability of inhibitors might suggest the presence of some conformational feature common to most inhibitors, which accounts for their stability. However, the study of the protein conformation of inhibitors by optical rotatory dispersion and circular dichroism has thus far failed to indicate the presence of any common secondary structure in inhibitors. Indeed, quite the contrary is probably the case. Jirgensons *et al.* (*56*) compared the optical rotatory dispersion in the visible spectrum of four of their lima bean inhibitor fractions, of the Kunitz soybean inhibitor, and of chicken ovomucoid. The latter two are clearly distinct from each other and from the lima bean inhibitors, of which two fractions were distinct from the two remaining fractions.

Ikeda *et al.* (*89*) have compared the same inhibitors and the 1.9 S inhibitor (Section I,C,2) by optical rotatory dispersion and circular dichroism in both the visible and ultraviolet. By both techniques the

88. J. W. Donovan, *Biochemistry* **6,** 3918 (1967); correction: *ibid.* **7,** 1252 (1968).
89. K. Ikeda, K. Hamaguchi, M. Yamamoto, and T. Ikenaka, *J. Biochem.* (*Tokyo*) **63,** 521 (1968).

same result was again obtained, i.e., that qualitatively and/or quantitatively there is little resemblance among these inhibitors. Because of the possibility of their homology discussed in Section IV,A, it is interesting to note that the 1.9 S and lima bean inhibitors are easily distinguished by these measurements. The 1.9 S, the lima bean, and the Kunitz soybean inhibitors are, however, qualitatively similar in position (210–213 nm) of the negative troughs in their optical rotatory dispersion curves. These data and the earlier data of Jirgensons *et al.* (*56*) are interpreted to mean that these inhibitors contain essentially no ordered structure. The trough in the ovomucoid curve is at 229–230 nm, suggesting the presence of helix.

Baba *et al.* (*90*) and Jirgensons *et al.* (*82*) have studied the circular dichroism of various derivatives of soybean inhibitor (Kunitz). Significant changes in the spectra were observed following partial reduction and/or carboxymethylation of the inhibitor, but after air oxidation of the reduced inhibitor the spectrum was again very similar to that of the native form. Baba *et al.* also noted minor changes upon incubation of the inhibitor at acid pH, and rather extensive changes after incubation at very alkaline pH values, in agreement with the earlier work of Steiner and Edelhoch.

Šorm and co-workers (*91*, *92*) have examined the dispersion and dichroism of pancreatic inhibitor (Kunitz) and its partially and completely reduced forms. The dispersion curve of the native inhibitor has a trough at 230 nm and suggests to these workers the presence of roughly 25% α helix in the inhibitor. Complete reduction of the inhibitor with 2-mercaptoethanol leads to reduction in the levorotation at 230 nm and elimination of the aromatic Cotton effect (*91*). Inhibitor in which only the Cys 14–Cys 38 disulfide bond has been reduced and carboxymethylated is about 20% less rotatory in the region of the aromatic Cotton effect than the unmodified inhibitor, but only insignificant changes are observed in the transitions of the peptide chromophores (*92*). Apparently the peptide conformation is essentially unaltered, in spite of this derivative's complete inactivity against trypsin. This result lends support to the hypothesis advanced in Section III,C,4 that the inactivity arises as a result of the negative charges of carboxymethyl groups so near the reactive site of the inhibitor.

90. M. Baba, K. Hamaguchi, and T. Ikenaka, *J. Biochem.* (*Tokyo*) **65,** 113 (1969).

91. D. Pospišilova, B. Meloun, I. Fric, and F. Šorm, *Collection Czech. Chem. Commun.* **32,** 4108 (1967).

92. B. Meloun, I. Fric, and F. Šorm, *Collection Czech. Chem. Commun.* **33,** 2299 (1968).

B. Special Purification Techniques

Certain characteristics common to many proteinase inhibitors can be used to circumnavigate problems in purification and laboratory manipulation. A most striking application of the extraordinary stability of some inhibitors (Section II,A) was provided when Kunitz exposed his preparation of pancreatic inhibitor to boiling 2.5% trichloroacetic acid, leaving the still intact and soluble inhibitor largely free of protein contaminants (*1*). He further used trichloroacetic acid treatment to dissociate the trypsin–inhibitor complex and recover free inhibitor from the supernatant. Similar procedures have been employed by many workers who isolated other very stable inhibitors. In cases when the enzyme–inhibitor association constant is sufficiently high, complex can be separated from excess enzyme or inhibitor by simple exclusion chromatography. A notable exception is chicken ovomucoid, long known to be unsuitable as a standard for gel filtration molecular weights (*93*). In this case, at neutral pH the free inhibitor [molecular weight 27,000 (*94*)] is nearly indistinguishable from its trypsin complex (molecular weight 52,000). At sufficiently low pH (<2), where the complex is no longer stable, trypsin denatures and precipitates, while ovomucoid remains in solution (J. Schrode, unpublished results). As a rule complexes can be dissociated and the enzyme and inhibitor readily separated by lowering of pH followed by gel exclusion chromatography, also at low pH.

Separation by gel filtration of soybean trypsin inhibitor (Kunitz) [molecular weight 22,000 (*78, 79*)] and trypsin is impossible at any pH in the usual aqueous solvents. On the other hand, in 6 *M* guanidine·HCl-0.01 *M* tris, pH 7, trypsin denatures but the inhibitor does not, and the resulting difference in molecular volumes permits separation on Sephadex G-100 in the same solvent (*83*). Native inhibitor is recovered from the appropriate fractions by normal dialysis procedures.

A separation of the dissociated components of soybean trypsin inhibitor–trypsin complex at pH 2.6 on sulfoethyl Sephadex was recently described (*94a*).

By far the most generally useful new tools for inhibitor preparation are the insoluble enzyme resins. In 1960, Kent and Slade (*95*) bound an active antibody to styrene beads; in 1961, Mitz and Summaria (*96*)

93. J. R. Whitaker, *Anal. Chem.* **35,** 1950 (1963).
94. M. B. Rhodes, N. Bennett, and R. E. Fenney, *JBC* **235,** 1686 (1960).
94a. S. E. Papaioannou and I. E. Liener, *JBC* **245,** 4931 (1970).
95. L. H. Kent and J. H. R. Slade, *BJ* **77,** 12 (1960).
96. M. A. Mitz and L. J. Summaria, *Nature* **189,** 576 (1961).

prepared insoluble trypsin and chymotrypsin on carboxymethyl cellulose and diazobenzyl cellulose supports. Since then there have been many examples of biologically active resins.

The resin which has found the greatest application for the present purpose is the polyanionic trypsin resin introduced by Katchalski and co-workers (*97*), which uses maleic anhydride–ethylene copolymers as support and 1,6-diaminohexane as cross-linking reagent. Fritz and Hochstrasser in Werle's group have used this resin to singular advantage in inhibitor isolation (*98, 99*). The resin was used to isolate the Kazal-type inhibitors from human (*100*), porcine, canine, and bovine pancreas (*100, 101*); from corn (*102*); and from seminal vesicles of mice (*100*). These workers also established that the failure of the resin to bind soybean inhibitor (Kunitz) and certain other inhibitors in high yield was not the result of steric exclusion but rather of electrostatic repulsion; inhibitors with isoelectric points below about pH 5 were not bound (*103*). Thus they modified the preparation of the resin so that both trypsin and *N,N*-dimethylethylenediamine were incorporated (*103*), the latter to electrostatically neutralize the free carboxylates of the polyanionic resin. This new resin, termed *polyamphoteric,* was found to have a high binding capacity for many more inhibitors, and was used to isolate inhibitors from wheat, rye (*104*), potatoes (*105*), peanuts (*106*), ovine pancreas (Kazal type) (*107*), and ovine lung (*108*). [The latter inhibitor is identical (*108*) with Kunitz's bovine pancreatic trypsin inhibitor,

97. Y. Levin, M. Pecht, L. Goldstein, and E. Katchalski, *Biochemistry* **3,** 1905 (1964).

98. H. Fritz, H. Schult, M. Hutzel, M. Wiedemann, and E. Werle, *Z. Physiol. Chem.* **348,** 308 (1967).

99. H. Fritz, I. Trautschold, H. Haendle, and E. Werle, *Ann. N. Y. Acad. Sci.* **146,** 400 (1968).

100. H. Fritz, I. Huller, M. Wiedemann, and E. Werle, *Z. Physiol. Chem.* **348,** 405 (1967).

101. H. Fritz, H. Schult, M. Neudecker, and E. Werle, *Angew. Chem. Intern. Ed. Engl.* **5,** 735 (1966).

102. K. Hochstrasser, M. Muss, and E. Werle, *Z. Physiol. Chem.* **348,** 1337 (1967).

103. H. Fritz, M. Gebhardt, E. Fink, W. Schramm, and E. Werle, *Z. Physiol. Chem.* **350,** 129 (1969).

104. K. Hochstrasser and E. Werle, *Z. Physiol. Chem.* **350,** 249 (1969).

105. K. Hochstrasser, E. Werle, R. Siegelmann, and S. Schwarz, *Z. Physiol. Chem.* **350,** 897 (1969).

106. K. Hochstrasser, K. Illchmann, and E. Werle, *Z. Physiol. Chem.* **350,** 929 (1969).

107. K. Hochstrasser, W. Schramm, H. Fritz, S. Schwarz, and E. Werle, *Z. Physiol. Chem.* **350,** 893 (1969).

108. H. Fritz, B. Greif, W. Schramm, K. Hochstrasser, and E. Werle, *Z. Physiol. Chem.* **351,** 139 (1970).

Section I,C,1.] The use of these resins for the preparation of chemically modified derivatives of proteinase inhibitors has also been reported (*109*).

With either resin the procedure is quite simple (*103*). Minimally pretreated extracts are combined with the resin at neutral pH, either in a column or in batches. After sufficient time for the inhibitors to bind in high yield the resin is washed several times with buffer to remove noninhibitory proteins. Final treatment with 0.2 *M* KCl or 6 *M* urea solution at pH 2 releases the inhibitors, which can then be collected in the usual fashion. The resins may be used repeatedly. Because many sources, particularly plant tissues, contain several inhibitors, the product obtained from the resin is subjected to ion exchange chromatography. A second reason for this last step is the interesting observation that the inhibitors may be released from the resin as modified inhibitors (*102, 104, 110, 111*), i.e., with the reactive site peptide bond hydrolyzed (Section III,B). The modified and virgin forms can usually be separated by this chromatography.

These same workers have also used water-insoluble resins containing bovine pancreatic inhibitor (Kunitz) to isolate kallikrein and plasmin by similar techniques (*112*).

C. Competitive Inhibition

Our definition of proteinase inhibitor states that the activity of the proteinase is competitively inhibited. A very large body of data argues in favor of competitive inhibition, although claims to the contrary are still common in the literature.

Most claims for noncompetitive inhibition are based on the use of Lineweaver–Burk plots in the standard manner. In 1953, Green and Work (*113*) pointed out that when the enzyme–inhibitor association constant K_{assoc} is much larger than $1/K_m$ for the substrate used in the assay, the substrate does not substantially perturb the enzyme–inhibitor equilibrium. Furthermore, whatever the value of K_{assoc}, if the half-time for dissociation of the enzyme–inhibitor complex (since substrate is usually added to previously prepared complex) is long compared to the assay time, the three components do not equilibrate. In both cases Lineweaver–Burk plots will indicate noncompetitive inhibition.

109. H. Fritz, M. Gebhardt, R. Meister, and H. Schult, *Z. Physiol. Chem.* **351,** 1119 (1970).

110. H. Tschesche, *Z. Physiol. Chem.* **348,** 1216 (1967).

111. H. Tschesche, *Z. Physiol. Chem.* **348,** 1653 (1967).

112. H. Fritz, B. Breg, A. Schmal, and E. Werle, *Z. Physiol. Chem.* **350,** 617 (1969).

113. N. M. Green and E. Work, *BJ* **54,** 347 (1953).

Competition between protein inhibitors, synthetic inhibitors, and substrates for the active site of trypsin has been amply demonstrated. Green (*114*) has shown that presence of very high concentrations of ester substrate can prevent association of trypsin with pancreatic and soybean inhibitors (Kunitz) and with chicken ovomucoid. Conversely, inhibitors can displace substrate from trypsin (*114*). Finkenstadt (*115*) has demonstrated that soybean inhibitor (Kunitz) and benzamidine, a specific competitive inhibitor of trypsin (*116*), compete for the enzyme. Displacement of proflavin from the active centers of trypsin (*117*) and chymotrypsin [e.g., Brandt and Hess (*118*)] upon formation of enzyme–inhibitor complex has been used (*119*) to follow the reaction by fast reaction techniques. The active site of trypsin is also protected by soybean inhibitor (Kunitz) from the specific trypsin titrant of Chase and Shaw (*120*).

The reactive site hypothesis of proteinase inhibitor action (Section III) clearly provides a compelling proof of competitive inhibition since inhibitors have been shown to be substrate analogs in which a specific enzyme-susceptible bond (reactive site) is cleaved. Particularly strong proof comes from the detailed analysis of the kinetic control dissociation of complexes made from modified inhibitors and enzyme. The resynthesis of reactive site peptide bond on such dissociation and the kinetics of the reactions involved prove that the reactive site of the inhibitor and the active site of the enzyme are in direct contact in the stable complex (Section III,F).

It is a corollary of competitive inhibition that all catalytic functions must be equally inhibited in the enzyme–inhibitor complex. On the other hand, the literature contains many reports of mild to very extreme dependence of extent of inhibition upon type of substrate used in the assay of a particular inhibitor. These observations apparently follow no particular pattern and are often difficult to rationalize. In the case of fairly weak inhibitor–enzyme associations (K_{assoc} more of the order of $1/K_m$ for a good specific substrate of the enzyme), the substrate can effectively compete for the enzyme and lessen the apparent degree of inhibition. This probably explains the fairly common observations that an inhibitor

114. N. M. Green, *JBC* **205,** 535 (1953).
115. W. R. Finkenstadt, Ph.D. Thesis, Purdue University, West Lafayette, Indiana, 1968.
116. M. Mares-Guia and E. Shaw, *JBC* **240,** 1579 (1965).
117. S. A. Bernhard and H. Gutfreund, *Proc. Natl. Acad. Sci. U. S.* **53,** 1238 (1965).
118. K. G. Brandt and G. P. Hess, *BBRC* **22,** 447 (1966).
119. G. Feinstein and R. E. Feeney, *Biochemistry* **6,** 749 (1967).
120. T. Chase, Jr. and E. Shaw, *BBRC* **29,** 508 (1967).

abolishes proteolytic (high K_m substrate) but not esterolytic (low K_m) activity of its respective enzyme. In esterolytic assays the inhibitor–substrate competition is usually exacerbated by the use of very high substrate concentrations (approximately 100–1000 times K_m). A contrary example is the observations by Rhodes *et al.* (*94*) and Bigler and Feeney (*121*) that 2–3 times more penguin ovomucoid, guinea ovomucoid, or chicken ovoinhibitor is required to inhibit chymotrypsin when casein is the substrate as when *N*-benzoyltyrosine ethyl ester is used. No such results were observed with the other ovomucoids tested nor when trypsin was the enzyme and tosylarginine methyl ester (TAME) the substrate, which facts make the observations particularly difficult to explain unless we postulate that while K_m for TAME–trypsin is more favorable than for casein–trypsin the opposite is true for the benzoyltyrosyl ethyl ester–chymotrypsin and casein–chymotrypsin interaction.

D. Inhibitor Assays

1. *Active Site Titrants*

Inhibitors are assayed by measurement of the extent of complex formation or of inhibition as a function of the inhibitor to proteinase ratio. The signals used to monitor these titrations have included sedimentation coefficient (*122*), fluorescence intensity (*123, 124*), fluorescence depolarization (*125*), UV difference spectroscopy (*124*), light scattering (*125*), sedimentation equilibrium (*126*), and the release of protons (*127*), but by far the most commonly used is enzymic activity against natural or synthetic substrates. This method employs extremely dilute enzyme solutions so that it works well for systems whose association constants are large in the pH region of optimal enzymic activity. It has the disadvantage that the concentration of active enzyme is not generally known. This fact has historically led to some confusion over the precise weight–weight combining ratios of proteinases and their inhibitors. In favorable cases, for example, trypsin and soybean inhibitor (Kunitz), unsuspected

121. J. C. Bigler and R. E. Feeney, *JBC* **246** (1971) (in press).
122. E. Sheppard and A. D. McLaren, *JACS* **75,** 2587 (1953).
123. D. Millar, K. Minzghor, and R. F. Steiner, *BBA* **65,** 153 (1962).
124. H. Edelhoch and R. F. Steiner, *JBC* **240,** 2877 (1965).
125. R. F. Steiner, *ABB* **49,** 71 (1954).
126. W. Scholtan, *Z. Physiol. Chem.* **348,** 1193 (1967).
127. J. Lebowitz and M. Laskowski, Jr., *Biochemistry* **1,** 1044 (1962); correction: *ibid.* **3,** 469 (1964).

impurities in either or both proteins (*128, 129*) and/or incorrect molecular weights resulted in cancellation of errors.

The use of reactive site titrants for proteinases eliminates these problems and is considerably more convenient. To be acceptable a reagent need only react rapidly in the pH region where enzyme–inhibitor association takes place. Such a reagent is *p*-nitrophenyl-*p'*-guanidobenzoate (NPGB) introduced by Chase and Shaw (*120*) for analysis of trypsin solutions. Typical titrations of trypsin with soybean inhibitor (Kunitz) and chicken ovomucoid are shown in Fig. 1. For each point 1 ml of Veronal buffer, pH 8.3, 100 μl of trypsin solution (5×10^{-5} *M* in 0.001 *M* HCl, 0.02 *M* $CaCl_2$), and varying amounts of inhibitor solution are combined in a cuvette and the base line against buffer is determined at 410 mm. Because soybean inhibitor and trypsin combine rapidly, the reagent

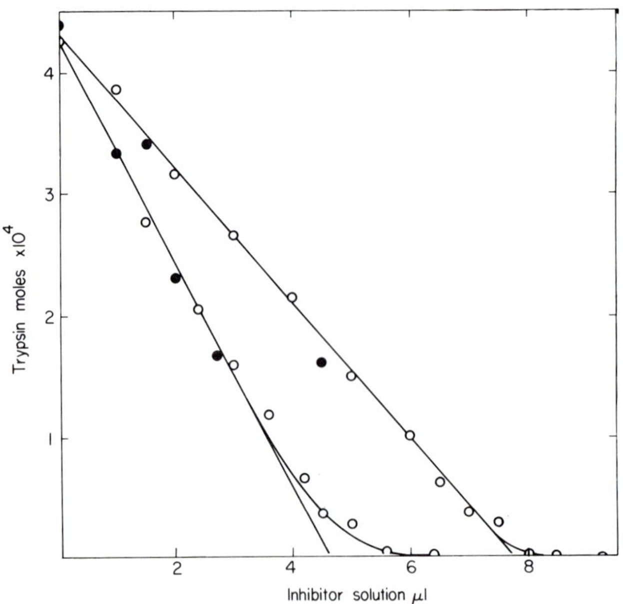

FIG. 1. Titrations of 4.32×10^{-9} mole of trypsin with 5.48×10^{-4} *M* soybean inhibitor (Kunitz) (upper curve) and 9.25×10^{-4} *M* chicken ovomucoid (lower curve). Free trypsin was analyzed by measuring the burst upon reaction with *p*-nitrophenyl-*p'*-guanidobenzoate (*120*) as described in text. Solid points were done after the others were completed.

128. M. L. Bender, M. L. Begue-Canton, R. L. Blakeley, L. J. Brubacher, J. Feder, C. R. Gunter, F. J. Kezdy, J. V. Killheffer, Jr., T. H. Marshall, C. G. Miller, R. W. Roeske, and J. K. Stoops, *JACS* **88,** 5890 (1966); correction: *ibid.* **89,** 2241 (1967).

129. A. C. Eldridge, R. L. Anderson, and W. J. Wolf, *ABB* **115,** 495 (1966).

can be added immediately and the burst measured (however, see below). The amount of free trypsin is then determined according to Chase and Shaw. At the end of the titration, initial points are repeated to insure that tryptic hydrolysis is negligible. At the time of this writing, several manuscripts which describe the use of the Shaw reagent for this purpose are believed to be in preparation.

Trypsin and soybean trypsin inhibitor (Kunitz) combine essentially stoichiometrically at pH 8.3, and the equivalence point can be determined to within about 5%. It is assumed that any inactive trypsin present in the enzyme preparation is incapable of combining with the inhibitor. In less favorable cases (K_{assoc} not so large) the inhibition should be stoichiometric to at least 50% in order for the extrapolation to be justified. At the fairly high concentrations (10^{-5}–10^{-6} M) employed in these assays, this condition requires that K_{assoc} be $\sim 10^{6}$–10^{7} M^{-1}. Thus, many trypsin inhibitors which combine with trypsin too weakly to be quantitatively assayed by the enzymic technique can be assayed by the use of NPGB, although at the expense of larger amounts of material.

It is assumed in these assays, as well as in some assays based on enzymic hydrolysis of simple substrates, that the dissociation of the trypsin–inhibitor complex is very slow and, therefore, that only trypsin which is free when the reagents are mixed produces a signal. If the complex dissociates relatively rapidly, then upon addition of active site titrant the initial burst due to free trypsin may be followed by further slow increase in signal due to the reaction of titrant with trypsin released from the complex. In such cases the rate of dissociation of the complex may be obtained from the slow part of the reaction (*130*).

It should be noted that for the analysis of Fig. 1, 1:1 combination (on a molar basis) of trypsin and soybean inhibitor (Kunitz) or chicken ovomucoid was assumed. In other cases, comparisons of the equivalent weight of an inhibitor with its independently determined molecular weight have led to two important findings. By titration data it was first determined that chicken ovoinhibitor has two independent sites for trypsin and two for chymotrypsin (*131*), that duck (*94*) and emu (*132*) ovomucoids have two sites for trypsin, and that potato type I inhibitor (*68*) apparently has up to four sites for chymotrypsin (*69, 133*). Corroborating evidence has been reported for ovoinhibitor, duck ovomucoid (including gel electrophoresis of the various complexes), and the potato inhibitor (see Section IV,B,2). In the same way it was established that

130. J. C. Zahnley and J. G. Davis, *Biochemistry* **9,** 1428 (1970).
131. Y. Tomimatsu, J. J. Clary, and J. J. Bartulovich, *ABB* **115,** 536 (1966).
132. D. T. Osuga and R. E. Feeney, *ABB* **124,** 560 (1968).
133. C. A. Ryan and A. K. Balls, *Proc. Natl. Acad. Sci. U. S.* **48,** 1839 (1962).

several inhibitors from plants exist in more concentrated solutions as oligomers of the monomeric inhibitory unit. Examples are the Bowman–Birk inhibitor from soybeans (*54, 55*), several lima bean inhibitors (*58*) which are probably homologous to the Bowman–Birk (*49*), and several others from grains and potatoes (*134*). These inhibitor aggregates apparently dissociate upon reaction with trypsin to form complexes composed from one molecule of inhibitor and one of trypsin. On the other hand, potato I, which has been clearly shown to be a tetramer (*134a*) of identical or similar subunits, combines with four molecules of chymotrypsin without dissociation.

2. *Competitive Enzyme Assays*

The association of inactive derivatives of proteolytic enzymes with protein inhibitors can be monitored by a variation of the enzymic titration known as a competitive enzyme assay (*135*). In the simplest form of the experiment, constant and equivalent amounts of active enzyme are titrated with varying quantities of the derivative. Enzymic activity appears in the mixture to the extent that the derivative competes with the active enzyme for the inhibitor. By this means (and by disc gel electrophoresis), Feinstein and Feeney (*135*) have demonstrated the association of anhydro- and TPCK-chymotrypsin with various ovomucoids and potato inhibitor, and of TLCK-trypsin with chicken and turkey ovomucoids. While the competitive method appears to lead to correct results it is quite indirect and therefore subject to errors of many indirect methods. We prefer measurement of direct signals due to the association such as increase in molecular weight and spectral change.

3. *Errors in Assays*

a. Poorly Characterized Assays. There are many inhibitor assay procedures currently in use which as a rule give at least qualitatively correct results. However, several possible ways exist by which misleading results can be obtained. Since these difficulties most often affect determination of an inhibitor's specificity pattern (Section IV,B), they are discussed in that context.

Included among the various procedures are the biological or quasi-biological methods which often suffer from lack of an obvious and direct cause and effect relationship. Probably the most commonly mislead-

134. K. Hochstrasser, K. Illchmann, and E. Werle, *Z. Physiol. Chem.* **350,** 655 (1969).

134a. J. C. Melville and C. A. Ryan, *ABB* **138,** 700 (1970).

135. G. Feinstein and R. E. Feeney, *JBC* **241,** 5183 (1966).

ing example is the blood clotting assay for thrombin inhibition by trypsin inhibitors. It is seldom, if ever, established that it is the thrombin and not some other trypsinlike enzyme in the clotting cascade which is being inhibited. The use of pure thrombin acting on specific synthetic thrombin substrates [tosyl-L-arginine methyl ester (*136*)] can settle the question, because from the remarks made in Section II,C it is clear that if an inhibitor does not inhibit the esterolytic activity of thrombin, it will not inhibit its clotting activity. Much of the thrombin inhibition data need to be reconsidered on this basis, and it is very likely that the recent listing by Feeney *et al.* (*137*) of 16 inhibitors, once thought to be active against thrombin, as in fact inactive, is entirely correct.

b. Impure Inhibitor Preparations. The multiplicity of inhibitors in most source tissues and the often encountered difficulties in their separation commonly lead to the use of impure inhibitor preparations which can easily give misleading inhibition data. If a preparation of an inhibitor which is inactive against a particular enzyme contains a contaminating inhibitor which is strongly active against this enzyme, apparent weak inhibition by the primary inhibitor will be the result if care is not exercised in the analysis of the inhibition data. Several examples exist in the literature in which inhibitor specificities were originally reported as far broader than those which have now been definitely established on much better inhibitor preparations.

In a remarkably clear example, Green (*76*) nicely demonstrated how such errors may be avoided at the outset. By careful titration assays he showed the apparent equivalent weight of his *Ascaris lumbricoides* trypsin inhibitor for chymotrypsin inhibition was 50% larger than that for trypsin inhibition. He proposed the presence of two inhibitors, each specific for one of these enzymes. The proposal was later confirmed by isolation of the chymotrypsin inhibitor (*70, 138, 139*).

c. Impure Enzyme Preparations. Many proteolytic enzyme preparations, quite frequently those of major biological interest, are only partially purified and may contain several enzymes. When such enzyme mixtures are exposed to proteinase inhibitors, confusing results are often obtained. Only part of the activity of such a mixture may be inhibited if the assay is carried out with some general proteinase substrate such as casein. If sufficient care in the interpretation of the inhibition curve is not exercised, strong inhibition of one enzymic component of the mix-

136. S. Sherry and W. Troll, *JBC* **208,** 95 (1954).
137. R. E. Feeney, G. E. Means, and J. C. Bigler, *JBC* **244,** 1957 (1969).
138. R. J. Peanasky and M. M. Szucs, *JBC* **239,** 2525 (1964).
139. F. H. Rola and J. Pudles, *ABB* **113,** 134 (1966).

ture may well be considered to be weak inhibition of the whole preparation. On the other hand, if specific substrates are used and only one component of the enzyme mixture is inhibited, the answer obtained may be totally dependent upon the choice of substrate. Some authors using one substrate may well see strong inhibition, while others employing another substrate see no inhibition at all. Obviously, such results are quite useful in recognizing that an enzyme preparation is a mixture.

Pronase, a proteolytic enzyme preparation from *Streptomyces griseus*, is known to be a mixture of several enzymes. Bowman–Birk soybean inhibitor inhibits about 30% of its activity toward casein and 100% of its activity toward benzoylarginine ethyl ester (*140*). Chicken ovoinhibitor inhibits both the trypsinlike activity of pronase toward benzoylarginine ethyl ester and the chymotrypsinlike activity toward acetyltyrosine ethyl ester, but the chymotrypsinlike activity is inhibited far more strongly than the trypsinlike activity (*141*). Trop and Birk (*142*) have shown that the caseinolytic activity of the isolated (*142, 143*) trypsinlike component of pronase can be inhibited up to 80–100% (depending upon the inhibitor) by soybean and pancreatic inhibitors (Kunitz), by lima bean and Bowman–Birk soybean inhibitors, and by chicken ovomucoid.

The inhibition of elastase, which was very difficult to purify, is even more perplexing. Proteolytic activity toward casein of older elastase preparations was relatively strongly inhibited by soybean trypsin inhibitor (Kunitz), but only moderately weak inhibition of hydrolysis of elastin was observed (*144*). These observations would suggest that the contaminating enzymes in the elastase preparations were inhibited but that elastase itself was not inhibited at all. More recent results on newer and presumably purer elastase preparations indicate that elastase itself is weakly inhibited by soybean trypsin inhibitor (Kunitz) ($K_{assoc} \sim 6 \times 10^5 M^{-1}$) but that these preparations also contain small amount of unusual proteolytic enzymes—highly specific for *N*-carbobenzoxytyrosyl *p*-nitrophenyl esters—which are moderately strongly inhibited by Kunitz inhibitor ($K_{assoc} \sim 3 \times 10^7 M^{-1}$) (*145, 146*). In addition, Pütter and

140. Y. Birk, *Ann. N. Y. Acad. Sci.* **146,** 388 (1968).
141. J. G. Davis, J. C. Zahnley, and J. W. Donovan, *Biochemistry* **8,** 2044 (1969).
142. M. Trop and Y. Birk, *BJ* **116,** 19 (1970).
143. Y. Narahashi and J. Fukunaga, *J. Biochem.* (*Tokyo*) **66,** 743 (1969).
144. R. L. Walford and B. Kickhöfen, *ABB* **98,** 191 (1962).
145. T. H. Marshall, J. R. Whitaker, and M. L. Bender, *Biochemistry* **8,** 4665 (1969).
146. T. H. Marshall, J. R. Whitaker, and M. L. Bender, *Biochemistry* **8,** 4671 (1969).

Schmidt-Kastner (*147*) found that the slower hydrolysis of elastin by elastase in the presence of pancreatic trypsin inhibitor (Kunitz) results from strong interaction between the inhibitor and substrate and not as would be expected between the inhibitor and enzyme. Such substrate–inhibitor interactions introduce another possible complication in assays on natural substrates.

In most detailed kinetic and thermodynamic studies of enzyme–inhibitor interaction it will be necessary to define "pure enzymes" even more rigorously. Several proteolytic enzymes are mixtures of inactive material and of two or more active partial autolysis products. Aside from the classic case of such products in chymotrypsin, Schroeder and Shaw (*148*) recently isolated bovine β-trypsin (single chain) and bovine α-trypsin (Lys 131–Ser 132 bond cleaved). These two hydrolyze substrates and combine with inhibitors at somewhat different rates. In order to simplify interpretation of kinetic data (Section III,E,1) most of the very recent work in our laboratory is done only with purified β-trypsin.

d. Kinetic Inactivity. It is almost always tacitly assumed in the literature that inhibition or titration curves such as in Fig. 1 describe equilibrium data. Typically the enzyme and inhibitor are preincubated for "a reasonable period of time" such as 5 min, and seldom, if ever, have detailed kinetic studies been done to ensure that this period of time is sufficient. Thus, claims that an inhibitor is inactive against a particular enzyme must be viewed with caution, particularly if the enzyme in question is otherwise similar to one which is inhibited.

The above view is dramatically supported by an example from the literature. The α_1-trypsin inhibitor reacts essentially instantaneously with trypsin, yet its reaction with kallikrein under typical assay conditions is characterized by a half-life measured in hours (*60, 63*). Conventional laboratory practice (a 3–5 min preincubation) would detect essentially no inhibition of kallikrein, and it is by no means certain that this large variation in rates of reaction is an isolated case. And even considerably shorter half-lives can cause significant error, as Green (*149*) has discussed with respect to measurement of the equilibrium constant for pancreatic trypsin inhibitor (Kunitz) and trypsin association (half-life under typical conditions, 1 min).

The necessary cautions implied above must also be exercised in the study of chemically or enzymically modified inhibitors. It was originally

147. J. Pütter and G. Schmidt-Kastner, *BBA* **127,** 538 (1966).
148. D. D. Schroeder and E. Shaw, *JBC* **243,** 2943 (1968).
149. N. M. Green, *BJ* **66,** 407 (1957).

thought (*150*) on the basis of conventional assay procedure, that chymotrypsin-modified (see Section III,H) Bowman–Birk soybean inhibitor is inactive against chymotrypsin. Frattali and Steiner (*151*) later elegantly proved that it is indeed active but with a half-time for association (under the conditions employed) of 10 hr. In another example, Haynes and Feeney (*152*) have shown that guanidinated lima bean inhibitor associates with trypsin with a half-life only 10% of that for the native inhibitor. Oppositely directed variations by orders of magnitude in association rates can also be expected upon modification of inhibitors, and this must be considered before a chemically modified inhibitor is pronounced thermodynamically inactive.

E. Association Constants

1. *Physicochemical Methods*

The application of thermodynamics to the study of proteinase–inhibitor interaction will depend in large part on measurements of the free energy of association over a wide range of conditions, most notably pH. The usual physicochemical methods have limited value for this purpose because of their twin requirements of fairly high protein concentration and the simultaneous presence of free inhibitor, free enzyme, and complex, all in significant quantities. With these requirements the usual range of measurable association constants is approximately $10^2 M^{-1}$ to $10^6 M^{-1}$. These values typically characterize proteinase–inhibitor associations only at pH values well below those of enzymic activity, i.e., outside the usual range of interest. Furthermore, with one exception, no method of this type has been used to span more than three orders of magnitude in K_{assoc}. The exception is the gel filtration method of Fritz *et al.* (*153*), who enzymically measured very small concentrations of trypsin and kallikrein in eluted fractions. Their data span nearly seven decades, but they appear to understate K_{assoc} when it is very high ($10^{10} M^{-1}$).

The known measurements by physical methods are summarized in Table II. The older data for the trypsin-STI system show considerable scatter, only in part because they were measured under slightly different conditions. Millar *et al.* (*123*) have noted that the experimental error of measurements from that laboratory was necessarily rather large, and

150. Y. Birk, A. Gertler, and S. Khalef, *BBA* **147,** 402 (1967).
151. V. Frattali and R. F. Steiner, *BBRC* **34,** 480 (1969).
152. R. Haynes and R. E. Feeney, *Biochemistry* **7,** 2879 (1968).
153. H. Fritz, I. Trautschold, and E. Werle, *Z. Physiol. Chem.* **342,** 253 (1965).

TABLE II
TRYPSIN- AND KALLIKREIN-INHIBITOR ASSOCIATION CONSTANTS

Enzyme	Inhibitor	Method	pH	$K_{assoc}(M^{-1})$	Ref.
Trypsin	Soybean (Kunitz)	Sedimentation velocity	3.1	6×10^3	*(122)*[a]
			3.3	5×10^4	
			3.55	2×10^4	*(83)*[b]
			3.77	1.5×10^5	
		Light scattering	3.40	2×10^3	*(125)*
			3.89	1×10^4	
			4.24	2×10^4	
		Fluorescence depolarization	4.05	5×10^4	*(125)*[a,d]
			4.45	5×10^5	
		Fluorescence intensity	3.8	2×10^4	*(123)*[d]
			3.9	4×10^4	
			4.0	9×10^4	
		Gel filtration	4.25	9×10^5	*(154, 155)*
			4.50	4×10^6	
			4.75	2×10^7	
	Modified[c] soybean (Kunitz)	Light scattering	2.95	2×10^2	*(156)*
			3.37	1×10^3	
			3.55	1×10^4	
			3.78	5×10^4	
			3.98	9×10^4	
	Lys 64–soybean (Kunitz)	Sedimentation velocity	3.72	2×10^4	*(83)*[b]
			3.78	1.5×10^5	
	Pancreatic (Kunitz)	Gel filtration	2.0	3×10^3	*(153)*
			3.0	4×10^5	
			4.0	4×10^8	
	Modified[c] ovomucoid	Light scattering	3.02	1×10^2	*(156)*
			3.33	6×10^2	
			3.51	2×10^3	
			3.62	5×10^3	
			3.74	1×10^4	
			3.87	2×10^4	
			4.12	4×10^4	
			4.35	1×10^5	
			4.90	8×10^5	
Kallikrein	Pancreatic (Kunitz)	Gel filtration	5.0	5×10^4	*(153)*
			6.0	3×10^5	
			7.8	8×10^7	

[a] Calculated *(127)* from the original data.
[b] Calculated by us from the original data *(83)*.
[c] For definition of modified inhibitor see Section III,B.
[d] Determined with dansyl conjugates of the inhibitor.

the sedimentation data of Sheppard and McLaren *(122)* and of Sealock and Laskowski, Jr. *(83)* are not well suited to this purpose. Furthermore, the third and higher order of pH dependence of log K_{assoc} considerably magnifies the effect of errors in pH measurement. On the other hand, the gel filtration values of Nichol and Winzor (*154*) as recalculated by Gilbert (*155*) and the light scattering values of McKie (*156*) agree excellently with the potentiometrically determined values (see Section II,E,3). With the exception of the two high values, those determined for pancreatic inhibitor (Kunitz) and trypsin by gel filtration (*153*) are in good agreement with enzymically determined values of Green.

2. *Enzymic Titration Methods*

Association constants can be measured by the extent of deviation from stoichiometric inhibition in the region of the equivalence point of a titration, as done by Green and Work (*113, 149*) for pancreatic trypsin inhibitor (Kunitz)-trypsin association. For a titration to be suitable, the extent of inhibition at the equivalence point should be greater than about 0.7, which is to say that the product $K_{assoc}\ C$, where C is the initial proteinase concentration, should be greater than about eight. The NPGB titration (Section II,D,1), using at the highest about $10^{-5}\ M$ trypsin, can in principle be used to measure K_{assoc} as low as $10^{6}\ M^{-1}$. Enzymic titrations ($C < 10^{-7}\ M$) are useful when $K_{assoc} \geqslant 10^{8}\ M^{-1}$, a range commonly found in proteinase–inhibitor systems. Thus the bulk of equilibrium constants for A,B type protein–protein associations found in the literature comes from such titrations. A representative list is shown in Table III. They are all about $10^{8}\ M^{-1}$ or higher and are restricted to a narrow range of pH. The chief exceptions are the low pH values for pancreatic inhibitor (Kunitz) and trypsin measured by Green (*149*). He allowed equimolar enzyme and inhibitor to equilibrate at the desired pH, then plunged aliquots into buffer at pH 7.8. The unusually slow rates of association and dissociation (Section III,E) prevented equilibration at the higher pH. These values are in good agreement with those found later by gel filtration (*153*).

In practice it is very difficult to assess the reliability of the values in Table III (*62, 70, 76, 114, 138, 149, 153, 157–162*). In no case was the

154. L. W. Nichol and D. J. Winzor, *BBA* **94,** 591 (1965).
155. G. A. Gilbert, *Nature* **210,** 299 (1966).
156. J. E. McKie, Jr., M.S. Thesis, Purdue University, West Lafayette, Indiana, 1966.
157. J. Pütter, *Z. Physiol. Chem.* **348,** 1197 (1967).
158. J. S. Ram, L. Terminiello, M. Bier, and F. F. Nord, *ABB* **52,** 451 (1954).
159. J. C. Bigler and R. E. Feeney, *JBC* **246,** (1971) (in press).
160. C. J. Martin, *JBC* **237,** 2099 (1962).
161. H. Fritz, H. Schult, R. Meister, and E. Werle, *Z. Physiol. Chem.* **350,** 1531 (1969).
162. K. Fossum and J. R. Whitaker, *ABB* **125,** 367 (1968).

TABLE III
ENZYMICALLY DETERMINED ENZYME–INHIBITOR ASSOCIATION CONSTANTS

Enzyme	Inhibitor	pH	$K_{assoc}(M^{-1})$	Ref.
Trypsin	Pancreatic (Kunitz)	2.6	7×10^4	*(149)*
		2.8	2×10^5	
		2.95	3×10^5	
		3.3	7×10^6	
		3.8	4×10^7	
		7.8	10^{11}–10^{12}	
		7.8	6×10^{10}	*(157)*
	Soybean (Kunitz)	7.8	5×10^9	*(114)*
	Chicken ovomucoid	7.8	2×10^8	*(114)*
		8.0	7×10^7	*(158)*
	Penguin ovomucoid	8.0	1×10^7	*(159)*
	Ascaris trypsin inhibitor	7.8	1×10^9	*(76)*
	Bovine acid-labile inhibitor	7.9	5×10^9	*(62)*
	Sheep serum inhibitor	8.0	2×10^8	*(160)*
α-Chymotrypsin	Pancreatic (Kunitz)	7.8	1×10^7	*(161)*
	Penguin ovomucoid	8.0	3×10^5	*(159)*
	Ascaris chymotrypsin inhibitor	7.0	2×10^8	*(138)*
		5.0	3×10^7	*(70)*
		7.1	9×10^7	
		9.0	4×10^7	
	Sheep serum inhibitor	8.0	5×10^8	*(160)*
Chymotrypsin B	*Ascaris* chymotrypsin inhibitor	7.0	3×10^7	*(70)*
Kallikrein	Pancreatic (Kunitz)	8.0	1×10^9	*(161)*
		8.0	1×10^8	*(153)*
Plasmin	Pancreatic (Kunitz)	7.8	3×10^8	*(161)*
Ficin	Egg white inhibitor	7.0	1×10^9	*(162)*
Subtilisin	Penguin ovomucoid	8.0	6×10^8	*(159)*

concentration of active enzyme (C) known with any certainty. When $K_{assoc}\ C$ is large, both the curvature of the plot and the equivalence point must be measured with considerable precision. Small absolute errors in either become large errors in K_{assoc}. When $K_{assoc}\ C < 10$ determination of the equivalence point is often difficult. Also, Green (*149*) pointed out that his initial measurements on pancreatic trypsin inhibitor (Kunitz) at pH 7.8 were in error because he failed to wait long enough for complete reaction. For these and similar reasons the values in Table III are likely at best to be reliable only to an order of magnitude.

3. *The Potentiometric Method*

The association of one mole each of trypsin and of soybean trypsin inhibitor to form one mole of complex is accompanied by the release of an average number $\bar{q}$ of protons. The value of $\bar{q}$ of course varies with pH.

The association constant and its relationship to $\bar{q}$ are defined by three equations, the third being the integral form of the second.

$$K_{\text{app}} = \frac{(\text{C})}{(\text{T})(\text{I})} \tag{1}$$

$$\frac{d \log_{10} K_{\text{app}}}{d\text{pH}} = \bar{q} \tag{2}$$

$$\log_{10} (K_{\text{app}})_{\text{pH}_2} - \log_{10} (K_{\text{app}})_{\text{pH}_1} = \int_{\text{pH}_1}^{\text{pH}_2} \bar{q}(\text{pH})\, d\text{pH} \tag{3}$$

Here (C), (T), and (I) are the concentrations of complex, trypsin, and inhibitor, respectively. The potentiometric method (*127*) utilizes Eq. (3), which shows that if $\bar{q}$(pH) can be determined over a range of pH and if K_{assoc} is known at a single pH within that range, then K_{assoc} can be found throughout the range by graphical integration.

The measurement of $\bar{q}$ is carried out by titration of a quantity of trypsin with small additions of inhibitor in a suitably sensitive pH meter. The small change in pH after each addition is monitored on a recorder then neutralized by the addition of standard base. In this way the total number of protons released as a function of inhibitor added was measured over the pH range 3.75–5.75. The analysis of these data allowed calculation of $\bar{q}$(pH) and a reference association constant (at 4.00). The values of K_{assoc} were determined over the whole range by application of Eq. (3).

Table IV lists the values of $\bar{q}$ and log K_{assoc} so obtained. Figure 2 is a plot of these and other equilibrium data (Tables II and III) for association of trypsin with soybean inhibitor (Kunitz), pancreatic inhibitor

TABLE IV

pH Dependence of Soybean Trypsin Inhibitor–Trypsin Association[a,b]

pH	$\bar{q}$	log K_{app}
3.75	3.35	4.565
4.00	2.92	5.349
4.25	2.53	6.030
4.50	2.34	6.639
4.75	2.00	7.182
5.00	1.80	7.657
5.25	1.50	8.070
5.50	1.33	8.424
5.75	1.13	8.732
8.30	0	9.832

[a] In 0.50 M KCl, 0.05 M $CaCl_2$ at 20 ± 0.02°.

[b] From Lebowitz and Laskowski, Jr. (*127*).

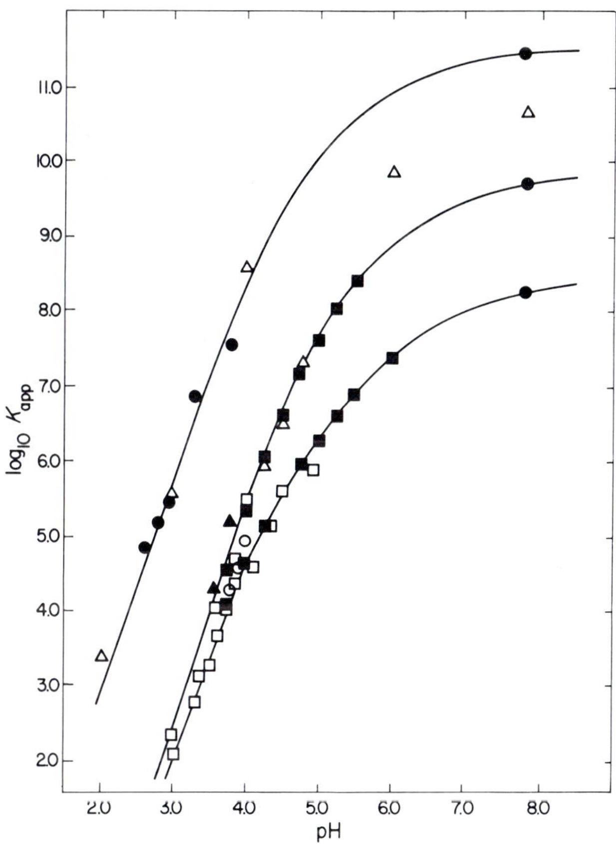

FIG. 2. pH Dependence of the logarithm of the association constant of bovine trypsin with pancreatic inhibitor (Kunitz) (top curve), soybean inhibitor (Kunitz) (middle curve), and chicken ovomucoid (bottom curve) obtained by (■) potentiometric method, (□) light scattering, (△) gel exclusion chromatography, (▲) sedimentation velocity, (●) residual enzymic activity at equivalence, and (○) change in fluorescence intensity of dansyl derivatives. For literature references see Tables II and III.

(Kunitz), and chicken ovomucoid. The agreement of the values determined by other means and the potentiometric values, as well as the broad range of measurement possible with the potentiometric method, is clear. Extension into the pH range 6–8 requires only the measurement of $\bar{q}$, but autolysis of trypsin and the need for CO_2-free solutions makes the measurements somewhat more difficult. Particularly striking in Fig. 2 is the enormous change (about 10^7-fold) in K_{app} between pH 3 and 6 and the fairly minor change (about 10-fold) between pH 6 and 8.

The similarity between the plots in Fig. 2 for the pancreatic inhibitor (Kunitz)-trypsin and soybean inhibitor (Kunitz)-trypsin associations is striking. Application of Eq. (2) to these curves shows that a single function $\bar{q}$(pH) very well describes the pH dependence of both associations. Particularly in view of the large difference in charge between the inhibitors in the region of pH 5, this observation very strongly suggests that identical (i.e., tryptic) sites (four sites, Table IV) must be deprotonated in both complexes. This contention is strengthed by the findings that several trypsin inhibitors do not lose activity when their free carboxyls are extensively modified (see Section III,I). At first glance, ovomucoid (Fig. 2) cannot be included in the generalization. However, ovomucoid is known to undergo a transition at low pH which is superimposed on the association with trypsin and makes precise analysis difficult (Section II,A).

It has only recently been recognized that the complexes coexist in equilibrium with trypsin and with both virgin and modified inhibitors (Section III,D). Thus two, rather than one, equilibrium constants can be defined. It is not clear whether all of the equilibrium constants discussed above refer to an equilibrium between complex, trypsin, and virgin inhibitor or complex, trypsin, and the equilibrium mixture of virgin and modified inhibitors. The answer depends on whether a sufficient time for the virgin $\rightleftharpoons$ modified inhibitor equilibration has been allowed before measurements are taken.

F. Changes in Conformation-Sensitive Parameters on Complex Formation

Measurements of changes in parameters which are sensitive to protein conformation have been intended to elucidate the nature and extent of conformational changes in either or both proteins which may accompany complex formation and to determine the nature and extent of the contact area between enzyme and protein inhibitor. In addition, an important practical reason for such measurements is that they provide more convenient signals for monitoring of enzyme–inhibitor interaction—particularly for kinetic studies—than do the more common methods such as disc gel electrophoresis and measurements of loss of enzymic activity, of molecular weight change, and of proton release.

Steiner (*125*) studied the effect of complex formation with trypsin upon the polarization of fluorescence of a dansyl (1-dimethylamino-naphthalene-5-sulfonyl) derivative of soybean trypsin inhibitor

(Kunitz). He found that the rotational relaxation time of the complex is almost twice as great as that of the free inhibitor. This result suggests that the hydrodynamic volume of the complex is almost double that of the inhibitor (a result consistent with doubling of the molecular weight) and that there is essentially no rotation around the contact area in the complex. Incidentally, the fluorescence depolarization data can be used to evaluate an equilibrium constant for complex formation (*127*) over a limited pH range.

Changes in optical rotation on complex formation have been studied by several workers. Such results, however, are subject to considerable error since they were not obtained differentially but by subtraction of the optical rotatory results for free enzyme and inhibitor from those of the complex. The lack of really pure enzyme and the autolysis of the enzyme introduce considerable uncertainty.

Kunitz (*5*) found almost no difference in specific rotation of a complex of trypsin and his soybean inhibitor and the mean of the specific rotations of the two components at the sodium D line. Wu and Scheraga (*84*) detected a 15% decrease in levorotation at this wavelength on complex formation but felt that the difference was within experimental error. More extensive measurements were made by Edelhoch and Steiner (*124*). Upon formation of a complex between trypsin and either soybean inhibitor (Kunitz) or chicken ovomucoid there was a small but detectable increase in levorotation at 325 nm. There was also an increase in levorotation at this wavelength when complex was formed with lima bean inhibitor and with pancreatic inhibitor (Kunitz), but the magnitude was within experimental error. The circular dichroism spectra of the α-chymotrypsin–pancreatic inhibitor (Kunitz) complex and of the free components were studied by Meloun *et al.* (*163*). Small differences were found in the 270–300 nm and in the 220–240 nm region and assigned to a change in environment of tryptophyl residue(s) in the enzyme (the inhibitor contains no tryptophan). Since there was no change at shorter wavelengths the authors conclude that the peptide backbone of the proteins is not greatly affected by the combination. Similarly, little change in the amide circular dichroism, but a change in aromatic circular dichroism, has been detected on combination of Bowman–Birk soybean inhibitor with α-chymotrypsin. On the other hand, combination of this double-headed inhibitor (Section IV,B,2) with trypsin led to a change in amide but not in aromatic circular dichroism (*10*).

163. B. Meloun, I. Fric, and F. Šorm, *Collection Czech. Chem. Commun.* **34**, 3127 (1969).

Difference spectra between complexes of chicken ovomucoid (*124, 164*), pancreatic inhibitor (Kunitz), lima bean (*124*), and Bowman–Birk soybean inhibitor (*10*) with trypsin versus the free components in tandem double cells show small but definite red shift of tyrosyl and tryptophyl absorptions usually assigned to transfer of these residues from aqueous to hydrophobic environment. Since none of these inhibitors contains tryptophan, the environment of tryptophyl residue(s) in trypsin must be affected. It is likely also to be the trypsin tyrosyls which are affected, at least in the complex with pancreatic inhibitor (Kunitz), in view of the equal accessibility of tyrosyls of this inhibitor to nitration in the complex and in the free inhibitor (*163*) (Section III,I,1).

Difference spectra in the tyrosyl-tryptophyl region are also obtained between the complex of soybean inhibitor (Kunitz) and trypsin and the free proteins (*124, 164*). The shape of these spectra is, however, rather unusual and it is somewhat difficult to decide clearly whether a red or a blue shift is seen. There are two possible important reasons for this difference in shape. Of the inhibitors studied only soybean inhibitor contains tryptophan and further a tyrosyl (Tyr 63) is immediately adjacent to its Arg 64–Ile 65 reactive site (Section III,B). No aromatic residue is immediately adjacent to the reactive site of pancreatic inhibitor (Kunitz). The known sequences of the reactive sites of lima bean, Bowman–Birk soybean inhibitor and of chicken ovomucoid are not sufficiently complete to settle this question.

It is of considerable interest to compare the trypsin–inhibitor difference spectra with trypsin–small substrate difference spectra. Such a comparison was made between trypsin-chicken ovomucoid (pH 5.0) and trypsin-tosyl-L-arginine methyl ester (pH 3.0) (*164*). The difference spectra with ovomucoid are 2–3 times larger, but the peak positions and approximate shapes are similar although not identical.

Solvent perturbation difference spectrum of pancreatic inhibitor (Kunitz)–trypsin complex with 30% propylene glycol shows a somewhat smaller exposure of chromophores (tyrosyls and tryptophyls) than the sum of the solvent perturbation difference spectra of the free proteins (*124*).

Tryptophan fluorescence of trypsin is sharply quenched on complex formation with pancreatic inhibitor (Kunitz) (*124*); it is enhanced on interaction with chicken ovomucoid (*124*) and with Bowman–Birk soybean inhibitor (*10*). There is also an enhancement of tryptophyl fluorescence on trypsin–soybean inhibitor complex formation (both proteins

164. P. Benmouyal and C. G. Trowbridge, *ABB* **115,** 67 (1966).

contain tryptophan) (*124*). In all cases the maximum in the fluorescence emission is shifted by about 5 nm to shorter wavelength, a conclusion consistent with transfer of tryptophyl residue(s) to a less polar medium and in agreement with difference spectral results.

The apparent burial of tryptophyl residues of trypsin observed in all of the experiments described above has been further studied by comparing the susceptibility of trypsin to *N*-bromosuccinimide oxidation of tryptophyls to that of the complex. Spande *et al.* (*165*), Steiner (*166*), and Spande and Witkop (*167*) found that invariably fewer tryptophyls were oxidizable in the complex than in the free enzyme. The number protected varied depending on the inhibitor involved, pH of oxidation, etc., but it occasionally rose as high as two or three. The marked dependence of the extent of oxidation of the free proteins on pH (in a region where no conformational transitions are known to occur) suggests that oxidizability is an extremely sensitive parameter, and, therefore, the effect on some of the tryptophyls may be quite small.

Steiner (*166*) has also shown by iodination that perhaps 2–3 tyrosyls are involved in the formation of complexes of trypsin and soybean and pancreatic inhibitors (Kunitz), lima bean inhibitor, or chicken ovomucoid. No attempt was made to determine whether the protected tyrosyls were in the inhibitor or in trypsin or both.

All of the evidence is clearly consistent with the involvement of one tryptophyl residue of trypsin in the contact area between trypsin and all of the inhibitors. Additional changes in tryptophyl properties may arise from minor conformational changes in the trypsin molecule. To test for such conformational changes, Edelhoch and Steiner (*124*) devised a new probe by making a dansyl derivative of trypsin–pancreatic trypsin inhibitor (Kunitz) complex, thus insuring that the dansyl substitution took place away from the contact area. The dansyl trypsin was then isolated by dissociating the complex at low pH and separation of the components by gel filtration (Section II,B). The direct fluorescence of the dansyl groups is quenched by about 10% by complex formation with pancreatic inhibitor, indicating a conformational change in the enzyme. When the dansyl fluorescence is excited at 290 nm (the events occur in the sequence tryptophyl absorption, energy transfer to dansyls, and fluorescence from dansyl groups) the dansyl fluorescence is quenched by 30% on complex formation, but this may simply be a consequence of change in tryptophyl environment.

165. T. F. Spande, N. M. Green, and B. Witkop, *Biochemistry* **5,** 1926 (1966).
166. R. F. Steiner, *Biochemistry* **5,** 1964 (1966).
167. T. F. Spande and B. Witkop, *BBRC* **21,** 131 (1965).

In summary, evidence has probably been adduced for small conformational changes in the enzyme (and less likely in the inhibitors) on complex formation. Some authors differ and say that no conformational change is observed. Even if a conformational change has been established it has not been well characterized. Therefore, the really important comparison between this change and changes which occur in enzymes on substrate binding, acylation, ionization of critical groups, etc., cannot be made at present and probably must await X-ray crystallography.

III. The Reactive Site Model

A. Overshoot of Complex

The potentiometric experiments discussed in Section II,E,3 produced, aside from the determination of the equilibrium constants, an important and surprising observation (*127*). When trypsin and soybean trypsin inhibitor (Kunitz) solutions are mixed under conditions where K_{assoc} $C \gg 1$ (high pH values and high protein concentration) and thus essentially all of the stoichiometrically possible complex forms (and remains at equilibrium), the proton release on mixing is fast and monotonic. On

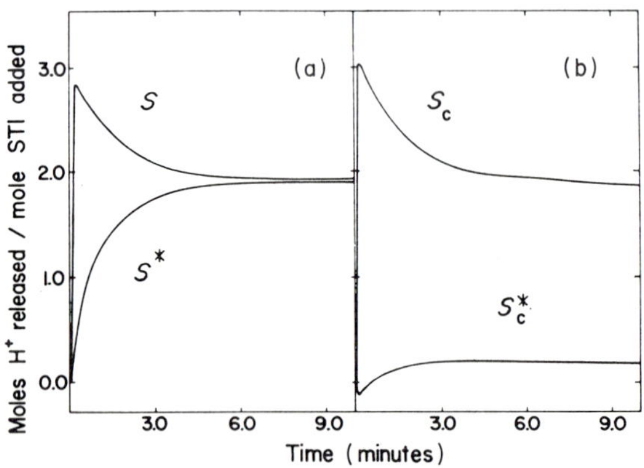

FIG. 3. Hydrogen ion release as a function of time upon addition of soybean trypsin inhibitor (Kunitz) to trypsin at pH 3.75. (a) Upper curve, virgin inhibitor; lower curve, inhibitor modified by incubation with 1 mole-% trypsin at pH 3.75, 24 hr. (b) Virgin (upper curve) and modified (lower) inhibitor after treatment with 1 mole-% carboxypeptidase B at pH 7.5, 8 hr (*168*). At this pH, 3.35 protons are released per mole of complex formed (Table IV).

the other hand, upon mixing of trypsin and inhibitor solutions under the conditions where $K_{assoc}\ C \sim 1$ [lower pH values, lower protein concentration, and presence of benzamidine which effectively lowers K_{assoc} (*115*)], initial fast proton release is followed by a slow partial uptake of the released protons [curve labeled S in Fig. 3a (*168*)]. It is, therefore, clear that at least two different reactions follow mixing: the first, responsible for proton release, is the association; the second is temporarily unidentified. However, the most facile explanation—that the slow proton uptake is due to an isomerization of the newly formed complex into an equilibrium form—was immediately ruled out (*127*). In the potentiometric experiments small, equal aliquots of inhibitor were added to a large quantity of trypsin. It was found that the overshoot observed became more significant with each added aliquot. Were the overshoot due to isomerization of the complex the opposite result should be expected since (because of equilibrium considerations) more complex forms on initial additions than on later ones. The only interpretation that is consistent with all of the data is that the slow proton uptake is due to dissociation of the complex, i.e., that upon mixing more than equilibrium amount of complex forms and then dissociates to equilibrium concentration. The proton release overshoot is thus due to an overshoot in the amount of complex formed (*127*).

Such a result, however, is inconsistent with any reaction scheme in which reagents are in equilibrium with a complex in homogeneous solution (this conclusion is independent of the detailed reaction mechanism or of the number and kind of kinetic intermediates postulated). Thus the result shows that either the solutions under study are not homogeneous or that the encounter between reagents leads to the modification of one or more of the reagents; therefore, the trypsin or inhibitor or both arising from complex dissociation are not necessarily the same as the original molecules which have entered into the complex. In view of the overwhelming evidence that the reaction between trypsin and inhibitors is reversible we originally adopted the first—inhomogeneous solution—explanation (*127*). It was, however, shown later (*168*) that this conclusion is incorrect and that modification of reagents is the correct explanation.

In order to find out whether it is the trypsin or the inhibitor or both which are modified, it is only necessary to reverse the order of the aliquot additions, i.e., to add aliquots of trypsin to a large quantity of the inhibitor. In this experiment the overshoot occurs only once upon the

168. W. R. Finkenstadt and M. Laskowski, Jr., *JBC* **240,** PC962 (1965).

addition of the first aliquot of trypsin. On subsequent additions only smooth, slow, monotonic proton release occurs (*115*). This result shows that it is the inhibitor that is modified by trypsin since the first trypsin addition accomplishes all of the modification, and subsequent additions cause only simple complex formation from trypsin and the equilibrium mixture of virgin and modified inhibitor.

All of the results described thus far in this section are consistent with the mechanism (*168, 169*):

$$T + I \underset{k_{-1}}{\overset{k_1}{\rightleftharpoons}} C \underset{k_{-2}}{\overset{k_2}{\leftrightharpoons}} T + I^* \tag{4}$$

For such a mechanism it has been shown that an overshoot in C will occur if and only if $k_1 > k_{-2}$ (*170*). Since the mechanism is symmetric this means that if mixing of $T + I$ leads to an overshoot in C, mixing of $T + I^*$ cannot give rise to an overshoot in C.

Since trypsin modifies the inhibitor, modified inhibitor solution can be prepared by incubating the inhibitor with catalytic quantities of trypsin for long periods of time (*168*) [the time required is a function of pH—for the soybean trypsin inhibitor (Kunitz)–bovine trypsin system optimal pH is about 3.5 (*115*)]. When such modified inhibitor (more properly an equilibrium mixture of virgin and modified inhibitors) is allowed to react with an excess of trypsin under conditions where virgin inhibitor would have caused an appreciable overshoot, monotonic (no overshoot), slow proton release is observed (Fig. 3, S*). This result strongly confirms the validity of Eq. (4) since (1) as predicted by it no overshoot is observed and (2) as predicted by it $k_1 > k_{-2}$.

B. Detection of Reactive Sites

Once it became clear that the inhibitor may be modified by the encounter with trypsin, it was natural to assume that the modification was a cleavage of an Arg-X or a Lys-X peptide bond. Treatment of modified (Fig. 3, S_C^*) but not of virgin (Fig. 3, S_C) inhibitor with carboxypeptidase B led to virtually complete loss of activity (*168*), thus supporting the surmise about the Arg-X or Lys-X bond and further suggesting

169. This is not in fact the simplest possible mechanism which will explain all of the known data. A somewhat more complex, although certainly too simple, mechanism is described in Section III,E,1.

170. C. Walter, "Steady State Applications in Enzyme Kinetics," pp. 47–51. Ronald Press, New York, 1965.

the possibility that this bond was the reactive site (*171*) of soybean trypsin inhibitor (Kunitz). The newly formed COOH terminal in modified inhibitor was shown to be arginine and the newly formed NH_2-terminal isoleucine (*79*). Reduction and carboxymethylation of modified inhibitor produced two fragments. The smaller one, composed of 64 amino acids (one of which is carboxymethylcysteine), has the original NH_2-terminal aspartic acid of the virgin inhibitor and the newly formed COOH terminal arginine. The large fragment, composed of 134 amino acids (including three carboxymethylcysteine residues), has an NH_2-terminal isoleucine. Reduction and carboxymethylation of virgin inhibitor gave rise to only a single fragment having an NH_2-terminal aspartic acid and containing 198 residues (precisely accounting for the sum of the two fragments of modified inhibitor) (*79*). The chemical events occurring on modification by trypsin of soybean trypsin inhibitor (Kunitz) and on its subsequent treatment with carboxypeptidase B are summarized in Fig. 4.

Similar splitting of a single, specific Arg-X or Lys-X bond upon exposure to catalytic quantities of trypsin has been found in most, but not all, protein trypsin inhibitors on which the experiment was attempted. The successful examples are included in Table V. We have postulated that this is a general result and that the few cases in which experiments

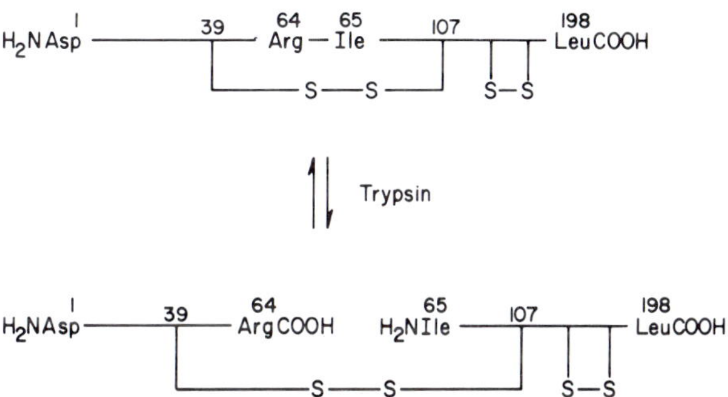

FIG. 4. Chemical event occurring on conversion of virgin to modified soybean trypsin inhibitor (*79*). Assignments of positions of residues shown are also discussed in Section IV,A.

171. The phrase *reactive site* is used to denote the part of the inhibitor molecule which comes in direct contact with the active site of the enzyme in the stable enzyme–inhibitor complex. Free inhibitor whose reactive site peptide bond is intact is called *virgin;* the inhibitor in which this bond is hydrolyzed is called *modified.* This nomenclature is unfortunate since the term *modified* is not sufficiently descriptive. However, there is as yet no systematic nomenclature for native proteins with one or more specific peptide bonds hydrolyzed.

TABLE V
Reactive Site Sequences of Trypsin Inhibitors

Inhibitor	Reactive site sequence[a]	Identified by[b]			
				Inactivation with	
		Cleavage	Homology	Lysine reagents	Arginine reagents
Soybean (Kunitz)	. . .Cys. . .SerProSerTyrArg^{64}-IleArgPheIle. . .Cys[c]. . . (*79, 172, 172a*)	+ (*79, 172, 173*)	—	− (*179, 180*)	+ (*179, 181*)
(Lys 64)-soybean (Kunitz)	. . .Cys. . .SerProSerTyrLys^{64}-IleArgPheIle. . .Cys. . . (*83*)	+ (*83*)	—	—	—
Bovine, ovine pancreatic (Kunitz)	. . .TyrThrGlyProCysLys^{15}-AlaArgIleIle. . .Cys. . . (*34, 108*)	+ (*182*)[d]	—	+ (*34, 179, 183, 188–192*)	—
Colostrum	. . .AlaArgGlyProCysLys^{18}-AlaAlaLeuLeu. . .Cys. . . (*176*)	—	+[e]	+ (*180*)	—
Bovine, ovine pancreatic (Kazal)	. . .ValAsnGlyCysProArg^{18}-IleTyrAsnPro. . .Cys. . . (*107, 174, 174a*)	+ (*184*)	+[f]	− (*179*)	+ (*179*)
Porcine I, pancreatic (Kazal)	. . .ValSerGlyCysProLys^{18}-IleTyrAsnPro. . .Cys. . . (*175*)	+ (*110*)	+[g]	+ (*179*)	—
Ovomucoid, chicken	Arg-Ala (*79*)	+ (*79, 173*)	—	− (*180, 185, 186*)	+ (*179, 181*)
Corn	. . .Cys. . .ThrGlyIleProGlyArg^{25}-LeuProProLeu. . .Cys. . . (*102, 175b*)	+ (*102*)	—	− (*179*)	—
Rye	. . .Cys. . .Arg-Ala. . .Cys. . . (*104*)	+ (*104*)	—	− (*179*)	+ (*179*)
Wheat	. . .Cys. . .Arg-Ala. . .Cys. . . (*104*)	+ (*104*)	—	− (*179*)	+ (*179*)
Peanut	. . .Cys. . .Arg-Ala. . .Cys. . . (*106*)	+ (*106*)	—	− (*179*)	+ (*179*)
Lima bean	Lys-X (*177*)	+ (*177*)	—	+ (*58, 179, 180, 187*)	− (*181*)
Soybean (Bowman–Birk)	Lys-X (*178*)	+ (*178*)	—	+ (*9*)	—

[a] The dash represents the reactive site peptide bond. The numbers give the residue locations in the primary sequence, if known.

[b] Symbols: (—) column does not apply or experiment was not performed; (+) reactive site identified by cleavage or homology and inhibitor is inactivated by lysine or arginine reagents, and (−) inhibitor not inactivated by these reagents.

[c] T. Ikenaka (*172*) tentatively locates the reactive site at Arg 63–Ile. For consistency we are continuing to use the earlier designation (*79*).

[d] Cleavage could be demonstrated only after reduction and carboxamidomethylation of the Cys 14–Cys 38 disulfide bond.

[e] By homology with bovine pancreatic inhibitor (Kunitz) (Section IV,A).

[f] The ovine inhibitor differs by one residue, at position 32, from the bovine inhibitor (*107*).

[g] By homology to bovine pancreatic inhibitor (Kazal) (Section IV,A).

have failed thus far to detect modified inhibitor can be rationalized (Section III,L,2).

It should of course be clear that simply the observation of cleavage of a specific bond by catalytic quantities of trypsin does not prove that this bond is the reactive site. An extensive proof can be provided on a variety of bases that Arg 64–Ile 65 bond in soybean trypsin inhibitor (Kunitz) is the reactive site *(79, 83, 168, 193)*. The proof is less complete for other inhibitors in Table V but nonetheless quite convincing. A simple part of such a proof is loss of inhibitory activity upon removal of newly formed COOH terminal arginine or lysine from the modified inhibitor by carboxypeptidase *(58, 150, 168, 177, 184)*. However, lack of inactivation of modified inhibitor by prolonged incubation with carboxypeptidase B does not prove that the bond cleaved is not at the reactive site since in

---

172. T. Ikenaka, T. Koide, and S. Tsunazawa, *Symp. Protein Struct., Tokyo, 1968.*

172a. T. Ikenaka, S. Tsunazawa, and T. Koide, *J. Biochem.* **69,** (1971) (in press).

173. R. Haynes and R. E. Feeney, *BBA* **159,** 209 (1968).

174. L. J. Greene and J. S. Giordano, Jr., *JBC* **244,** 285 (1969).

174a. L. J. Greene and D. C. Bartelt, *JBC* **244,** 2646 (1969).

175. H. Tschesche, E. Wachter, S. Kupfer, and K. Niedermeier, *Z. Physiol. Chem.* **350,** 1247 (1969).

175a. H. Tschesche and E. Wachter, *Eur. J. Biochem.* **16,** 187 (1970).

175b. K. Hochstrasser, K. Illchman, and E. Werle, *Z. Physiol. Chem.* **351,** 721 (1970).

176. D. Čechova, V. Švestkova, B. Keil, and F. Šorm, *FEBS Letters* **4,** 155 (1969).

177. F. C. Stevens, *Proc. Can. Fed. Biol. Soc.* **12,** 16 (1969).

178. Y. Birk, "Methods in Enzymology," 1970 (cited by B. Kassell) (in press).

179. H. Fritz, E. Fink, M. Gebhardt, K. Hochstrasser, and E. Werle, *Z. Physiol. Chem.* **350,** 933 (1969).

180. R. Haynes, D. T. Osuga, and R. E. Feeney, *Biochemistry* **6,** 541 (1967).

181. W.-H. Liu, G. Feinstein, D. T. Osuga, R. Haynes, and R. E. Feeney, *Biochemistry* **7,** 2886 (1968).

182. L. F. Kress and M. Laskowski, Sr., *JBC* **243,** 3548 (1968).

183. N. Sevilla and M. Rigbi, *Israel J. Chem.* **7,** 130p (1969).

184. M. Rigbi and L. J. Greene, *JBC* **243,** 5457 (1968).

185. F. C. Stevens and R. E. Feeney, *Biochemistry* **2,** 1346 (1963).

186. H. Fraenkel-Conrat, R. S. Bean, and H. Lineweaver, *JBC* **177,** 385 (1949).

187. H. Fraenkel-Conrat, R. S. Bean, E. D. Ducay, and H. S. Olcott, *ABB* **37,** 393 (1952).

188. J. Chauvet and R. Acher, *JBC* **242,** 4274 (1967).

189. R. Acher, J. Chauvet, R. Arnon, and M. Sela, *European J. Biochem.* **3,** 476 (1968).

190. J. Chauvet and R. Acher, *BBRC* **27,** 230 (1967).

191. F. A. Anderer and S. Hornle, *Ann. N. Y. Acad. Sci.* **146,** 381 (1968).

192. R. Avineri, G. Blauer, and M. Rigbi, *Israel J. Chem.* **1,** 194 (1963).

193. W. R. Finkenstadt and M. Laskowski, Jr., *JBC* **242,** 771 (1967).

some cases carboxypeptidase B may fail to remove the COOH terminal Arg or Lys due to an unfavorable protein conformation or an unfavorable sequence of preceding residues. The frequent presence of –Pro–Lys or –Pro–Arg sequences at the reactive site (Table V) increases the probability of such a failure.

On the basis of their results with soybean trypsin inhibitor (Kunitz) and with chicken ovomucoid and of the information about chemical modification of various inhibitors, Ozawa and Laskowski, Jr. (*79*) postulated that some trypsin inhibitors have a trypsin susceptible Arg-X bond in their reactive site while others have a Lys-X bond. This postulate gave rise to an elegant method of distinguishing between lysyl and arginyl inhibitors, pioneered by Liu *et al.* (*181*) and extensively applied by Feeney's group (*9, 152, 180, 181*) and by Fritz *et al.* (*179*). Since trypsin does not attack peptide bonds of the Lys-X type in which the ε-amino group of lysine is suitably (*194*) modified (e.g., acetylated, succinylated, maleylated, deaminated, or reacted with trinitrobenzene sulfonic acid), such modifications of lysyl trypsin inhibitors should destroy their inhibitory activity. On the other hand (and assuming that highly reactive lysyl residues in positions other than the reactive site play no significant role in the inhibition process), the arginyl inhibitors should retain their inhibitory activity even after extensive lysyl modification. Since modification of arginyls in Arg-X bonds by such reagents as 1,2-cyclohexadione (*181*) and 2,3-butadione (*179*) renders these bonds resistant to trypsin, similar conclusions should apply to arginyl inhibitors. It is experimentally found that virtually all inhibitors tested can be cleanly divided into two classes: (1) lysyl inhibitors which are rapidly inactivated by all lysyl substitutions which make Lys-X bonds resistant to tryptic attack and are only slowly, if at all, inactivated by arginine modification; and (2) arginyl inhibitors which are relatively easily inactivated by arginyl modifications but are virtually resistant to lysyl modifications.

All inhibitors which to date have been classified by reactive site cleavage and/or by chemical modifications are included in Table VI. Chemical modification appears at present to be a simpler technique for classification of trypsin inhibitors than does enzymic modification with its necessary search for appropriate conditions, and therefore many more inhibitors have been classified by the former method. It is reassuring to note in Table V no instance in which the two methods have given conflicting results.

194. The question of what are suitable modifications is still hotly argued [e.g., Gorecki and Shalitin (*195*)] and is quite relevant to the discussion about trypsin inhibitors (see Section III,L,1).

195. M. Gorecki and Y. Shalitin, *BBRC* **29**, 189 (1967).

TABLE VI
CLASSIFICATION OF TRYPSIN INHIBITORS BY REACTIVE SITE

Lysine inhibitors		Arginine inhibitors	
Ovomucoid, turkey	(*152, 180, 185*)	Ovomucoid, chicken	(Table V)
Ovomucoid, cassowary	(*180, 181*)	Ovoinhibitor, chicken	(*9*)
Ovomucoid, penguin	(*180, 181*)	Soybean (Kunitz)	(Table V)
Ovomucoid, duck	(*180, 181*)	*Ascaris*	(*196*)
Ovomucoid, ostrich	(*181*)	Pancreatic (Kazal) bovine	(Table V)
Ovomucoid, rhea	(*181*)	Pancreatic (Kazal) ovine	(Table V)
Ovomucoid, emu	(*181*)		
Colostrum, bovine	(Table V)		
Pancreatic (Kunitz)	(Table V)	Submandibular gland, canine	(*179*)
Lima bean	(Table V)	Wheat	(Table V)
Soybean (Bowman–Birk)	(Table V)	Corn	(Table V)
Pancreatic (Kazal), porcine I, II	(Table V)	Rye	(Table V)
Pancreatic (Kazal), canine	(*179*)	Potato I	(*179*)
Pancreatic (Kazal), feline	(*179*)	Peanut	(Table V)
White bean (*Phaseolus vulgaris*)	(*179*)	Inter-α, human serum	(*179*)

The chemical modification technique can be further refined to answer more detailed questions. An example is the work of Haynes *et al.* (*180*) on the comparison of the rate of inactivation with the rate of amino group substitution by trinitrobenzene sulfonic acid. For several lysyl inhibitors studied it was shown that the rate of inactivation parallels the rate of substitution of rapidly reacting lysyl residues. This is very much in line with expectation since in order to be clearly recognized by trypsin the reactive site must be quite exposed. Further, the application of the method of Koshland *et al.* (*197*) and Ray *et al.* (*198*) to such data allows one to show how many groups are required for activity. As is seen in Fig. 5, turkey ovomucoid is inactivated by the substitution of a single highly reactive ϵ-amino group (*181*).

Another valuable extension of the chemical modification technique allowed Chauvet and Acher (*183*) to determine which lysyl residue was at the reactive site of pancreatic trypsin inhibitor (Kunitz). Polyalanylation of all four lysyls in this molecule renders the inhibitor in-

196. J. Pudles, F. H. Rola, and A. K. Matida, *ABB* **120,** 594 (1967).

197. D. E. Koshland, Jr., W. J. Ray, Jr., and M. J. Erwin, *Federation Proc.* **17,** 1145 (1958).

198. W. J. Ray, H. G. Lathan, M. Katsoulis, and D. E. Koshland, Jr., *JACS* **82,** 4743 (1960).

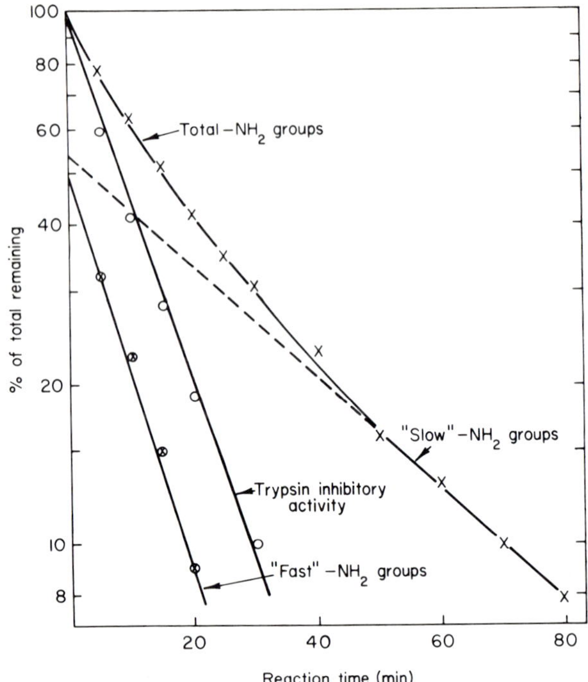

FIG. 5. Semilogarithmic plot for loss of amino groups and loss of trypsin inhibitory activity in turkey ovomucoid by modification with trinitrobenzenesulfonic acid (*180*).

active. On the other hand, after polyalanylation of the trypsin-inhibitor complex and subsequent dissociation of the complex and isolation of the inhibitor, the inhibitor is still active and only three of the four possible lysyls are polyalanylated. The free Lys 15, is, therefore, at the reactive site. This study was further extended by Fritz *et al.* (*161*) who maleylated the trypsin–pancreatic inhibitor (Kunitz) complex. Upon dissociation they obtained the inhibitor with its reactive site Lys 15 unmodified and the other lysyls and the α-amino group maleylated. This active derivative was then used to make several specific modifications of Lys 15 only and to study the associations of such chemically modified inhibitors with various enzymes.

C. GENERAL PROPERTIES OF THE REACTIVE SITES

The data of Table V and additional literature data are sufficient to allow some limited conclusions to be drawn with some confidence.

1. *Enzyme-Susceptible Bond in the Reactive Site*

All of the evidence appears to suggest that an enzyme-susceptible bond is required. If we restrict consideration to natural amino acids then the reactive site of trypsin inhibitors must be either Lys-X or Arg-X. All of the data of Table V and VI are in accord with this view. Analogous reasoning leads one to believe that chymotryptic reactive sites should be of the Trp-X, Tyr-X, Phe-X, or Leu-X type. The only case so far elucidated—the chymotryptic site of lima bean inhibitor—is, in fact, Leu–Ser (*199*). In all cases where the newly formed COOH terminals of the trypsin-modified inhibitors have been removed, inhibitory activity was lost (*150, 168, 173, 177, 184*). In all chemical substitutions which clearly remove the tryptic susceptibility of the Lys-X bonds the inhibitory activity of lysyl trypsin inhibitors was lost, and the corresponding conclusion holds for arginyl inhibitors. [There is room for argument (see Section III,L,1) whether some modifications such as conversion of lysyl to homoarginyls, which occasionally do not lead to loss of inhibitory activity, do, in fact, remove the tryptic susceptibility.]

Discussion of specificity (Section IV,B), of enzymic replacement in soybean inhibitor (Section III,C) and of homology of secretory inhibitors (Section IV,A) shows that most or all trypsin inhibitors are capable of retaining their activity and specificity on Lys → Arg replacement.

It is not yet known whether a Lys → Leu replacement, for example, would convert a tryptic to a chymotryptic reactive site or vice versa. This need not necessarily be true in view of the additional possible requirements of chymotryptic sites.

2. *Nature of the NH_2-Terminal Residue*

Inspection of Table V shows that the NH_2-terminal residue in the cleaved reactive site is Ile, Leu, or Ala. These residues are all aliphatic and hydrophobic. Appearance of these residues in reactive sites is closely reminiscent of the limited number of similar closely related residues (Ile, Val, and Ala) which are X residues in bonds cleaved by trypsin during activation of a variety of zymogens. Although chance appearance of only these residues in the reactive sites (Table V) is not probable, it is not excluded.

No experiments have been carried out so far on removal of the NH_2-terminal residue from the reactive sites of modified inhibitors. Limited results have been reported on chemical substitution of these residues. Haynes and Feeney (*173*) observed rapid loss in inhibitory activity of modified but

199. J. Krahn and F. C. Stevens, *Biochemistry* **9**, 2646 (1970).

not of virgin soybean trypsin inhibitor (Kunitz) upon treatment with trinitrobenzenesulfonic acid and suggested that the free α-NH_2 group of modified inhibitor is required for activity. Modified chicken ovomucoid lost activity only slowly on exposure to trinitrobenzenesulfonic acid. Since under these conditions α-NH_2 groups are expected to react rapidly, Haynes and Feeney (*173*) concluded that the reactive site α-NH_2 of modified ovomucoid is not essential. Fritz *et al.* (*179*) maleylated several modified arginyl inhibitors (corn, rye, and wheat) and found no loss in inhibitory activity. They surmised that α-NH_2 groups in modified inhibitors are never required and that the loss of inhibitory activity of modified soybean trypsin inhibitor (Kunitz) on blocking of its α-NH_2 group with trinitrobenzenesulfonic acid was a consequence of large steric bulk of the blocking group.

Recent results from our laboratory (D. Kowalski, unpublished) indicate that specific maleylation, citraconylation, and carbamylation of the newly formed NH_2 terminal of either modified soybean trypsin inhibitor (Kunitz) or modified chicken ovomucoid lead to complete loss of inhibitory activity. We presume that any blockage of the reactive site NH_2 terminal in any modified inhibitor will lead to such a loss. The earlier results (*179*) can be explained since the inhibitors were presumed to be completely modified but in fact contained sizeable amounts of virgin material.

3. *The Reactive Site Is in a Disulfide Loop*

It seems clear that if both virgin and modified inhibitors are to be active, then the two peptide chains of the modified inhibitor must be strongly held together since dissociation would destroy the modified inhibitor. The interaction between the two fragments can be due to noncovalent interaction [as is the case in S peptide–S protein interaction of pancreatic ribonuclease (*200*) or of the two fragments of staphylococcal nuclease (*201*)] or from covalent linkages (and noncovalent interactions as well). The disulfide bond is the most obvious covalent bond. The reactive site of soybean trypsin inhibitor (Kunitz) is contained in a single disulfide loop (Fig. 4) (*79*), and the reactive site of bovine pancreatic trypsin inhibitor (Kunitz) is contained in two disulfide loops (*34, 36*). While unambiguous proof is lacking, it is likely that all of the other reactive sites are held by one or more disulfide loops since there are half-cystine residues in both fragments produced by the reactive site cleavage, and the balance of the evidence strongly favors this impression.

200. F. M. Richards, *Proc. Natl. Acad. Sci. U. S.* **44,** 162 (1958).
201. H. Taniuchi and C. B. Anfinsen, *JBC* **243,** 4778 (1968).

With the exception of some serum inhibitors *(59)*, all protein proteinase inhibitors whose amino acid analyses we could locate contain cystine (and no free sulfhydryls). In all cases where the experiment has been attempted, reduction or oxidation of all the disulfide bridges leads to a complete loss of inhibitory activity and in many cases, but not in all, activity is restored on reoxidation of reduced disulfides (Table VII). In those cases where activity was not restored by reoxidation no extensive attempts to improve on the original possibly incorrect disulfide pairings were made. In two cases selective reduction of disulfides was achieved. In bovine pancreatic trypsin inhibitor (Kunitz), the Cys 14–Cys 38 disulfide can be selectively reduced with sodium borohydride *(211)*, 2-mercaptoethanol *(92)*, or dithioerythritol *(212)* and the inhibitor remains active provided that nonnegatively charged substituents are used to block the sulfhydryls *(213)*. The selectivity of these reductions is easily explained since the X-ray crystallographic structure shows that

TABLE VII
REDUCTION AND REOXIDATION OF DISULFIDES IN INHIBITORS[a]

Inhibitor	Recovery upon reoxidation[b]	Reference
Pancreatic inhibitor (Kunitz)	+	*(36, 91, 202, 203)*
Soybean inhibitor (Kunitz)	+	*(82, 204–206)*
Chicken, turkey, and golden pheasant ovomucoids	+	*(207, 208)*
Ascaris chymotrypsin inhibitor	+	*(139)*
Ascaris trypsin inhibitor	+	*(209)*
Soybean inhibitor (Bowman–Birk)	−	*(140)*
Kidney bean inhibitor	−	*(210)*
Lima bean inhibitor	+	*(57)*
Mung bean inhibitor	+	*(71)*

[a] In all cases the fully reduced inhibitors possess no inhibitory activity.
[b] Complete or partial reactivation occurred (+), or did not (−), on air reoxidation.

202. R. Avineri-Goldman, I. Snir, G. Blauer, and M. Rigbi, *ABB* **121,** 107 (1967).
203. J. Chauvet and R. Acher, *Bull. Soc. Chim. Biol.* **48,** 1284 (1966).
204. R. F. Steiner, *Nature* **233,** 579 (1964).
205. R. F. Steiner, *BBA* **100,** 111 (1965).
206. R. F. Steiner, *JBC* **240,** 4648 (1965).
207. M. M. Simlot, F. C. Stevens, and R. E. Feeney, *BBA* **130,** 549 (1966).
208. L. B. Sjoberb and R. E. Feeney, *BBA* **168,** 79 (1968).
209. J. Pudles, F. H. Rola, and A. K. Matida, *ABB* **120,** 594 (1967).
210. A. Pusztai, *European J. Biochem.* **5,** 252 (1968).
211. L. F. Kress and M. Laskowski, Sr., *JBC* **242,** 4925 (1967).
212. S. K. Liu and J. Meienhofer, *BBRC* **31,** 467 (1968).
213. L. F. Kress, K. A. Wilson, and M. Laskowski, Sr., *JBC* **243,** 1758 (1968).

the Cys 14–Cys 38 disulfide is highly exposed while the remaining two are buried (*73b*).

It should be noted, however, that the Lys 15–Ala 16 reactive site is still within the Cys 5–Cys 55 disulfide loop. Selective reduction of the small disulfide loop in soybean trypsin inhibitor (Kunitz) (Fig. 4) by sodium borohydride was achieved by DiBella and Liener (*214*) without loss of inhibitory activity. Since complete reduction abolishes inhibitory activity (*204*), the reduction of the large disulfide loop in which the reactive site is contained appears to be responsible for this loss. However, the interpretation of all of the disulfide reduction results where activity is lost is not straightforward since loss of conformational stability is generally coupled with disulfide reduction; therefore, it is not clear whether the loss of inhibitory activity is a direct consequence of disulfide reduction or an indirect one due to conformational change. In case of proteinase inhibitors the problem is compounded by the fact that denatured inhibitors become proteinase substrates at a variety of sites, and thus their lack of inhibitory activity may in part be caused by proteolytic cleavage.

Bovine and human α_1-serum trypsin and chymotrypsin inhibitors appear to be exceptions to the statements above since they contain no cystine or cysteine (*59*). Nothing is yet known about the possible virgin $\rightleftharpoons$ modified conversion of these inhibitors. It may be worth noting that α_1-serum inhibitors have a high carbohydrate content (*59*) and that, therefore, a remote possibility exists that the carbohydrate functions in these inhibitors as a covalent cross-link.

4. *Absence of Negative Charge*

It is part of the general lore of studies with trypsin that the presence of negatively charged groups near a trypsin-susceptible bond makes this bond more resistant to attack by trypsin. This contention was recently confirmed in studies on model peptides by Abita *et al.* (*215*) who showed that the consequence of introduction of Asp residues closely preceding a Lys-X or Arg-X bond is an increase in K_m and a dramatic decrease in k_{cat} for tryptic hydrolysis. Since trypsin inhibitors appear to require an unusually trypsin-susceptible bond, the total absence of Asp and Glu residues in the vicinity of all tryptic reactive sites whose sequence is known is not surprising (see Table V and Section IV,A). The

214. F. P. DiBella and I. E. Liener, *JBC* **244,** 2824 (1969).
215. J. P. Abita, M. Delaage, M. Lazdunski, and J. Savrda, *European J. Biochem.* **8,** 314 (1969).

same conclusion holds for the sequence of the only known chymotryptic reactive site—that of lima bean inhibitor (*199*). The deleterious effect of neighboring negative charges in the vicinity of tryptic reactive sites was subjected to a direct test by Kress *et al.* (*213*). These workers found that the carboxymethyl derivative of bovine pancreatic trypsin inhibitor (Kunitz) in which Cys 14–Cys 38 disulfide was selectively reduced (negative charge of *S*-carboxymethyl-Cys 14 adjoins the Lys 15–Ala 16 reactive site) was inactive while the carboxamidoethyl derivative was active.

5. *The Role of Proline*

Several lines of evidence argue that the reactive sites have a rather unusual conformation, and prolyl residues are frequently found or postulated in regions of proteins containing unusual conformational features. It is therefore of interest that proline is present in all of the proteinase inhibitors whose amino acid analyses we have been able to examine. Since proline is a relatively rare amino acid and since several other amino acids are frequently or occasionally absent from many inhibitors, this may be more than coincidence. Inspection of Table V shows that in all trypsin inhibitors in which a sufficient part of the sequence was determined a prolyl residue occurs near the reactive site on the NH_2-terminal side of the Lys-X or Arg-X bond, but its specific location is not invariant. Further, the NH_2-terminal fragments of corn (*102*), wheat, and rye (*104*) inhibitors all contain proline. Thus, presence of prolyl residues near the reactive sites and on the NH_2-terminal side thereof is consistent with, although not proved by, the data. On the other hand, there is no prolyl near the reactive site on the NH_2-terminal side of the Leu–Ser chymotryptic reactive site of lima bean inhibitor although there is a prolyl nearby on the COOH terminal side of this bond (*199*). The common presence but lack of an invariant position of the prolyl residues near the reactive sites suggests that the prolyl residues are not absolutely essential but that they contribute to the required rigidity of the reactive sites as evidenced by low values of K_{hyd} (Section III,D). Similarly, the presence of prolyls in the remainder of the inhibitor molecules probably is partly responsible for the rigidity and resistance to denaturation of these molecules.

D. Equilibria of Hydrolysis of Reactive Site Bonds

The arguments developed in previous sections suggest that the virgin inhibitor, the modified inhibitor and trypsin are in equilibrium with the trypsin–inhibitor complex [Eq. (4)]. Since we have two equilibria, two equilibrium constants should be defined, for example,

$$(K_{\text{assoc}})_{\text{virgin}} = \frac{(\text{C})}{(\text{T})(\text{I})} \tag{5}$$

$$(K_{\text{assoc}})_{\text{modified}} = \frac{(\text{C})}{(\text{T})(\text{I}^*)} \tag{6}$$

Combination of these two expressions leads to a new equilibrium constant for the conversion of virgin into modified inhibitor.

$$K_{\text{hyd}} = \frac{(\text{I}^*)}{(\text{I})} = \frac{(K_{\text{assoc}})_{\text{virgin}}}{(K_{\text{assoc}})_{\text{modified}}} \tag{7}$$

It is clear that K_{hyd} will be of the order of magnitude of unity if virgin and modified inhibitors have a comparable thermodynamic affinity for trypsin. This is in agreement with experience on trypsin inhibitors but is in sharp contrast with the generally held view on the magnitude of equilibrium constants for hydrolysis of peptide bonds in proteins, which are believed to be very large. Since virgin and modified inhibitors can be separated by disc gel electrophoresis at high pH on the basis of the difference in charge (the modified inhibitor is more negative because it has COO^- at the newly formed COOH terminal while its newly formed NH_2 terminal is uncharged at high pH), the matter was put to an experimental test (*216*). The conversion of virgin to modified soybean trypsin inhibitor by catalytic quantities of trypsin was monitored at pH 4.00 by disc gel electrophoresis, and it was shown that a steady state at $86 \pm 2\%$ modified and $14 \pm 2\%$ virgin was reached. This steady state persisted for a long time, and the composition of the steady state mixture was independent of the amount of trypsin used. In order to prove that this steady state was indeed an equilibrium state, pure modified inhibitor was isolated by preparative disc gel electrophoresis (*217*, *218*) and its conversion to virgin inhibitor by catalytic quantities of trypsin was monitored. Again a persistent steady state with a composition $86 \pm 2\%$ modified, $14 \pm 2\%$ virgin was attained (Fig. 6). This completes the proof of equilibrium. A series of somewhat similar, although less complete, experiments was carried out over a broad pH range, but because of the extreme slowness of the reaction the proof of attainment of equilibrium was less rigorous (*216*, *219*). The results shown in Fig. 7

216. C. W. Niekamp, H. F. Hixson, Jr., and M. Laskowski, Jr., *Biochemistry* **8**, 16 (1969).
217. S. Raymond, *Clin. Chem.* **8**, 455 (1962).
218. M. D. Melamed, *BJ* **103**, 805 (1967).

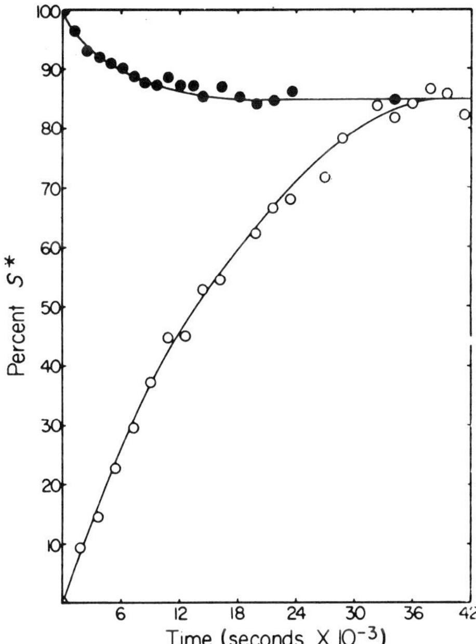

FIG. 6. Tryptic conversion of both pure virgin (S) and pure modified (S*) soybean trypsin inhibitor into the equilibrium mixture as monitored by disc gel electrophoresis. Reactions were conducted at pH 4.0 and 20°. For the S → S* conversion (○) 2 mole-% trypsin and for the S* → S conversion (●) 0.5 mole-% trypsin were added (*216*).

(*219*) describe a caternary closely similar, though not identical, to that expected for the hydrolysis of an internal peptide bond in a large peptide or in a dipeptide with blocked terminal groups (*220, 221*), given by Eq. (8), where K°_{hyd} is the minimal value of

$$K_{\text{hyd}} = K^{\circ}_{\text{hyd}}\left(1 + \frac{[\text{H}]}{K_1} + \frac{K_2}{[\text{H}]}\right) \tag{8}$$

K_{hyd}, corresponding to the hydrolysis to fully ionized products, and K_1 and K_2 are the ionization constants of the newly formed COOH and NH_2 terminals.

This result shows that we are dealing with a reasonably normal hydrol-

219. H. F. Hixson, Jr., Ph.D. Thesis, Purdue University, West Lafayette, Indiana, 1970.

220. A. Dobry, J. S. Fruton, and J. M. Sturtevant, *JBC* **195,** 149 (1952).

221. L. V. Kozlov, L. M. Ginodman, V. N. Orekhovich, and T. A. Valueva, *Biokhimiya* **31,** 315 (1966).

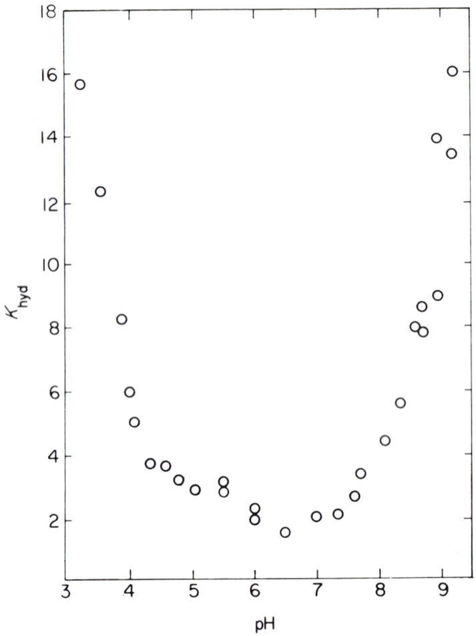

FIG. 7. The effect of pH on the equilibrium constant K_{hyd} for the reactive site cleavage in soybean trypsin inhibitor (Kunitz) at 20 ± 2° (revised preliminary data) (*219*).

ysis of a peptide bond and that this hydrolysis causes relatively small perturbations of the numerous other ionizable groups on the inhibitor molecule. Such a situation is consistent with only minimal conformational changes upon virgin → modified inhibitor conversion.

Since the equilibrium constants given in Fig. 7 are unusually small compared with the general expectation, it was of interest to test whether the low value is a direct consequence of an unusual amino acid sequence in the reactive site or whether it arises as a result of conformational restrictions on this site. To this end "melting curves" for the denaturation of virgin, modified, and des-64-arginine–modified inhibitor were obtained (*17, 85*). The most striking feature of these data is the coincidence of the curves for native and denatured states of all three proteins. This result is a strong argument for similar conformation of all three proteins in both states. Furthermore, the equilibrium constant for denaturation of modified inhibitor can be shown to be about 20 times greater than that of the virgin inhibitor at all pH values and at all temperatures investigated. The results were combined into a thermodynamic cycle (*85*),

$$\begin{array}{ccc} & (K_{\text{hyd}})_{\text{native}} = 3 & \\ \text{I} & \rightleftharpoons & \text{I}^* \\ K_{\text{den}} = 1 \; \downarrow\uparrow & & \uparrow\downarrow \; K_{\text{den}} = 20 \\ \text{D} & \rightleftharpoons & \text{D}^* \\ & (K_{\text{hyd}})_{\text{den}} = 60 & \end{array} \tag{9}$$

Thus, it can be seen that the equilibrium constant for reactive site hydrolysis is grossly dependent on conformation and the low value in the native protein is due to the conformation of the reactive site. Tentatively, it can be assumed that upon cleavage of the reactive site bond in the native protein the peptide chains largely remain in place and therefore gain but little entropy. Thus, upon denaturation of modified inhibitor the potential entropy gain on opening a ring is realized.

The question of whether all protein proteinase inhibitors must have reactive site conformations which lead to low K_{hyd} is not yet fully resolved but the balance of evidence favors generality. Rigbi and Greene (*184*) have shown that conversion of virgin-to-modified bovine pancreatic secretory trypsin inhibitor reaches a steady state plateau of about 30% modified, 70% virgin at pH 3.22. This plateau composition is independent of the amount of trypsin used. While the reverse experiment of converting modified-to-virgin inhibitor was not attempted, all the other data suggest that the steady state was the equilibrium state. The value of the equilibrium constant at pH 3.22 is about a factor of 40 smaller than that given in Fig. 7. It therefore appears that over most of the pH range virgin secretory inhibitor is thermodynamically favored over the modified one. Similar although less extensive measurements by Haynes and Feeney (*173*) and by Schrode (*17*) suggest a steady state plateau in the conversion of virgin to modified chicken ovomucoid.

A moderately strong argument can be made for small values of K_{hyd} in all inhibitors on the basis that modified inhibitors all appear to be active and, therefore, that the equilibrium constant for their combination with trypsin is not likely to be much smaller that that of the corresponding virgin inhibitors [Eq. (7)]. If this were not the case the inhibitors would slowly become used up by conversion to modified inhibitor, which was inactive. While this would offer an interesting explanation for the "temporary inhibition" phenomenon, experimental evidence seems to be against it, and temporary inhibition appears to be a consequence of cleavage of bonds other than the reactive site (see Section III,J).

A possible exception to these generalizations is the behavior of bovine pancreatic trypsin inhibitor (Kunitz) in which the Cys 14–Cys 38 disulfide bridge was selectively reduced and the resultant cysteine residues converted to carboxamidomethylcysteines. Kress and Laskowski, Sr. (*211*) have shown that the virgin form of this inhibitor is active but that the modified form (reactive site Lys 15–Ala 16 cleaved) is not. There are two possible explanations for this observation. The value of K_{hyd} may be quite high, presumably resulting from loss of the stabilization by the disulfide bridge. [The two fragments are still held together by the Cys 5–Cys 55 bridge (*34*).] On the other hand, during the activity test insufficient time may have been allowed for combination of the modified derivative with trypsin. As discussed in the next section, modified inhibitors associate considerably more slowly than do the virgin inhibitors, and even virgin pancreatic inhibitor is well known as a slowly reacting inhibitor. The chymotrypsin-modified Bowman–Birk inhibitor was originally thought to be inactive for just this reason (*150, 151*).

E. Kinetics of Interaction of Proteinases with Proteinase Inhibitors

The discovery that the conversion of virgin to modified inhibitors is closely connected with the enzyme–inhibitor association process has greatly increased the work involved in obtaining detailed kinetic understanding of the process, but it has also raised what still appears a realistic hope that once such an understanding is obtained it will shed a great deal of light on the chemistry of the interaction. Even if the mechanism such as given by Eq. (4) were correct, four rate constants rather than two as formerly believed need be determined. Many of the kinetic studies reported in the literature were undertaken before the development of the reactive site concept and before methods of preparing modified inhibitors were known. Therefore, the literature seems unbalanced by containing far more information on virgin inhibitor plus enzyme reactions than on equally interesting modified inhibitor plus enzyme reactions, and relatively little work on virgin $\rightleftharpoons$ modified conversion kinetics has yet been reported.

1. *Association of Enzyme with Inhibitor*

In cases where the association of enzyme and inhibitor is relatively slow the measurement of the rate of association can be carried out by testing for remaining enzymic activity at various times after mixing of enzyme and inhibitor in absence of substrate. Such measurements are usually done in very dilute solution to slow the reaction down. By these means Pütter (*157*) measured the rate of association of pancreatic trypsin inhibitor (Kunitz) with trypsin as a function of pH (Fig. 8) and Haynes and Feeney (*152*) of several trypsin inhibitors with trypsin at neutral pH. In all cases second-order rate constants were obtained (Table VIII), but Haynes and Feeney (*152*) remarked that the reactions

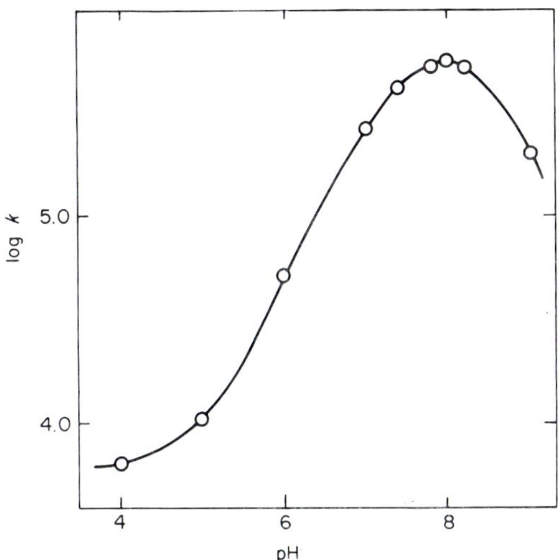

FIG. 8. pH Dependence of the logarithm of the second-order association rate constant for bovine trypsin and pancreatic trypsin inhibitor (Kunitz). Redrawn from Pütter (*157*). Rate constant in liter·mole^{-1}·sec^{-1}.

TABLE VIII
ENZYMICALLY DETERMINED ASSOCIATION RATE CONSTANTS

Inhibitor	pH	k(liter·mole^{-1}·sec^{-1})	Ref.
	Trypsin		
Pancreatic (Kunitz)	7.8	2.8×10^5	(*149*)
	7.8	3.1×10^5[a]	(*157*)
Soybean (Kunitz)	8.1	8.2×10^6	(*152*)
	5.0	5×10^5[b]	(*222*)
Modified soybean (Kunitz)	5.0	1×10^4[b]	(*222*)
Turkey ovomucoid	8.1	9.7×10^5	(*152*)
Chicken ovomucoid	8.1	2.7×10^6	(*152*)
Lima bean	8.1	1.4×10^6	(*152*)
Guanidinated lima bean	8.1	1.1×10^7	(*152*)
Ascaris trypsin inhibitor	7.8	2.1×10^6	(*76*)
	Chymotrypsin		
Ascaris chymotrypsin inhibitor	7.8	1.5×10^6	(*76*)

[a] The pH dependence of this rate constant is shown in Fig. 8.
[b] Determined by the method of Green (*114*) discussed in text.

222. H. F. Hixson, Jr., W. R. Finkenstadt, and M. Laskowski, Jr., *Federation Proc.* **26**, 276 (1967).

are not diffusion controlled and therefore probably involve at least one intermediate between the free enzyme and inhibitor and the stable complex. A considerable number of experiments obtained by this method are reported for some very slow associations such as kallikrein, plasmin, and thrombin with various inhibitors [e.g., (*60, 62, 63, 223, 224*)], but in view of complications in determining molar concentrations of these enzymes from the older literature they are not suitable for calculation of rate constants in molecular units.

Faster associations can be followed by conventional kinetic methods only if they are intentionally slowed down by addition of competing substances (substrate or small organic inhibitors). The most commonly used method (*114*) is Green's and consists of adding enzyme to an inhibitor–substrate mixture and observing the decline in the rate of substrate hydrolysis (Fig. 9). This decline is relatively small since substrate competes effectively and rapidly for the free enzyme. The

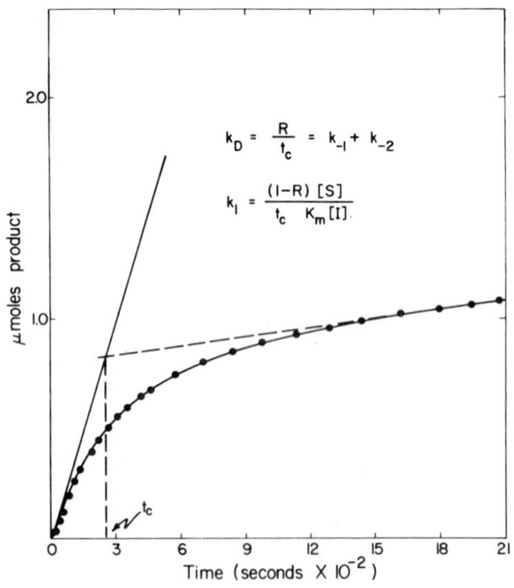

FIG. 9. Hydrolysis of *p*-tosylarginine methyl ester (TAME) by trypsin obtained by addition of trypsin to a TAME-soybean trypsin inhibitor (Kunitz) mixture. Quantitative analysis of these data and knowledge of K_m for trypsin–TAME system permits evaluation both of second-order association rate constant for trypsin and inhibitor and of the rate constant for dissociation of complex according to the simple model of Eq. (4) (*219, 222*). The quantity R is the ratio of the final rate of hydrolysis in the presence of inhibitor to the rate of hydrolysis in the absence of inhibitor (also the initial rate in the presence of inhibitor).

223. P. S. Norman, *Federation Proc.* **25,** 63 (1966).
224. W. H. Seegers, *Ann. N. Y. Acad. Sci.* **146,** 593 (1968).

method has been used by Green (*114*) to obtain relative rates of association of soybean trypsin inhibitor (Kunitz), chicken ovomucoid, and pancreatic trypsin inhibitor (Kunitz) with the result that the rates decrease in the order in which the inhibitors are listed. Simlot and Feeney (*225*) applied a similar method to the study of comparative rates of chemically modified turkey ovomucoid with trypsin and chymotrypsin.

Recently, a quantitative evaluation of the second-order rate constant (and of the dissociation rate constant and the association equilibrium constant) by this method has been proposed and applied to the soybean trypsin inhibitor (Kunitz)–trypsin system in the presence of tosylarginine methyl ester (*219, 222*). The quantitative aspects depend on a number of assumptions and on the knowledge of the K_m for the substrate (Fig. 9). While the data agree reasonably well with those obtained by stopped flow on this system, we have relatively little faith in such very indirect determinations. At best, they obscure mechanistic features which are apparent in direct measurements (see below) because they are necessarily based on an assumed mechanism. At worst, gross errors can be made by slightly incorrect treatment of data.

Upon examination of mechanism of Eq. (4) we note that either in the $T + I \rightarrow C$ step or in the $T + I^* \rightarrow C$ step or in both (depending on the nature of the complex C) a covalent bond must be synthesized or cleaved in a simple, bimolecular reaction. This is highly unlikely; therefore, it should be expected that intermediate noncovalent complexes should precede C. It was found to be possible to measure the association of bovine trypsin with soybean trypsin inhibitor (Kunitz) in a stopped flow apparatus by directly monitoring the change in absorbance at 260 nm (Section II,F) upon mixing of equimolar amounts of trypsin and of inhibitor (*226*). No indicators such as proflavin (*119*) need to be employed; thus, while the signals are small the analysis of data is relatively simple. When the experiment is carried out at low concentrations the oscilloscope traces show second-order reaction with respect to time and the half-time is, as expected, inversely proportional to the reactant concentration. However, as the reactant concentration is raised the traces acquire first-order character and the dependence on reactant concentration diminishes and then disappears [Fig. 10, (*226*)]. Similar results are obtained on mixing of trypsin and modified inhibitor. These data are consistent with the mechanism of Eq. (9a) (*226, 227*), where L and L* are loose, noncovalent complexes

225. M. M. Simlot and R. E. Feeney, *ABB* **113**, 64 (1966).
226. J. A. Luthy, M. Praissman, and M. Laskowski, Jr., *Abstr., 158th ACS Natl. Meeting, New York* BIOL No. 321 (1969).
227. H. F. Hixson, Jr. and M. Laskowski, Jr., *JBC* **245**, 2027 (1970).

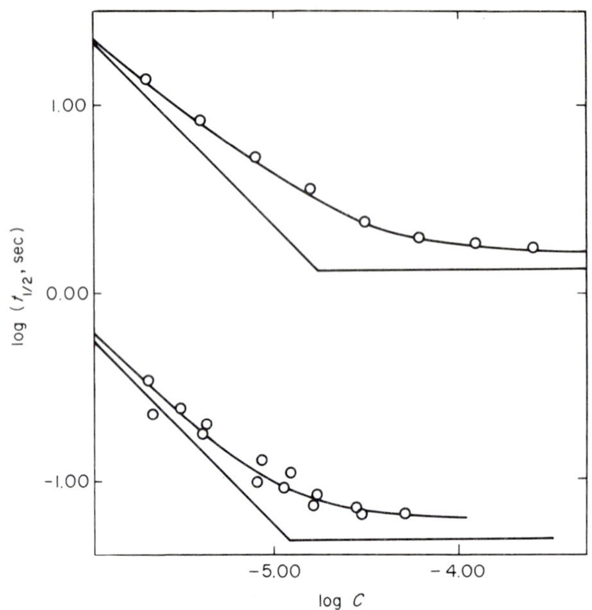

FIG. 10. Dependence of the logarithm of the half-time of change in absorbance at 260 nm observed in stopped flow spectrophotometer upon the logarithm of the concentration of reactants. Equal volumes of equimolar bovine β-trypsin (*148*) and soybean trypsin inhibitor were mixed at pH 5.6, 23°. Upper curve, modified inhibitor; lower curve, virgin inhibitor (*226*).

$$\mathrm{T} + \mathrm{I} \underset{k_{-1}}{\overset{k_1}{\rightleftharpoons}} \mathrm{L} \underset{k_{-2}}{\overset{k_2}{\rightleftharpoons}} \mathrm{C} \underset{k_{-3}}{\overset{k_3}{\rightleftharpoons}} \mathrm{L}^* \underset{k_{-4}}{\overset{k_4}{\rightleftharpoons}} \mathrm{T} + \mathrm{I}^* \tag{9a}$$

and C is the stable complex. From now on we shall assume that this is the correct mechanism although undoubtedly more detailed experiments will force incorporation of more intermediates or of branches. The mechanism is formally identical with the simplest possible mechanism for a reversible hydrolysis of a bond by any enzyme involving a covalent acyl intermediate. The difference between trypsin inhibitors and other substrates lies in the fact that for the inhibitors C is unusually stable (at neutral pH) with respect to the reactants and the products.

From the data of Fig. 10 the values of $K_{\mathrm{L}} \cong k_{-1}/k_1$ and k_2 and of $K_{\mathrm{L}}^* \cong k_4/k_{-4}$ and of k_{-3} are obtained. For soybean trypsin inhibitor (Kunitz) and bovine β-trypsin (*148*) the values of K_{L} and K_{L}^* are of the order of 10^{-5} M (*226*). Thus the loose bindings of virgin and of modified inhibitor to trypsin are similar. The value of log k_2 as a function of pH is given in Fig. 11. As can be shown by both stopped flow experiments (*226*) and by even more direct chemical experiments (see Section

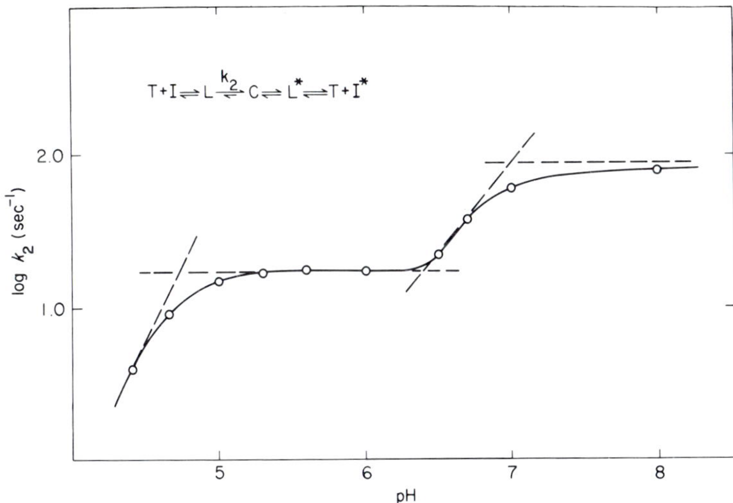

FIG. 11. Dependence of first-order rate constant, k_2 in Eq. (9a), for the conversion of L, the loose, noncovalent complex between virgin soybean inhibitor (Kunitz) and bovine β-trypsin, into the stable complex C. The data correspond to the horizontal segments of Fig. 10 (*226*).

III,F) k_{-3}, the rate constant for converting loose complex made of trypsin and modified inhibitor into the stable complex, has a very similar pH dependence to k_2. However, the second order rate constant for formation of complex from virgin inhibitor and trypsin, k_2/K_L, is about 30 times greater than the second order rate constant for formation of complex from modified inhibitor and trypsin, k_{-3}/K_L^*. The conclusion that virgin inhibitors react with enzymes faster than modified ones appears to be general (*115, 151, 177, 182, 226*).

The values of K_m and k_2 for the association are analogous to those for acylation steps and therefore to the hydrolysis of amides. The commonly found parameters for hydrolysis of amides and peptides by trypsin are of the order of $k_{cat} = 1$/sec, $K_m = 10^{-3}$ M (*228*) so that the values for association of trypsin and soybean trypsin inhibitor (Kunitz) indicate about two orders of magnitude better association and more efficient catalysis. Abita *et al.* (*215*) showed that such values can change by several orders of magnitude depending upon the sequence surrounding the cleaved bond.

It should be noted that for the purpose of comparison the old second-order association rate constants of mechanisms such as Eq. (4) correspond to k_2/K_m in the mechanism of Eq. (9a).

228. S.-S. Wang and F. H. Carpenter, *JBC* **243**, 3702 (1968).

2. *Dissociation of Complexes*

Since for a simple association

$$K_{\text{assoc}} = k_{\text{assoc}}/k_{\text{dissoc}} \tag{10}$$

an approximate value of the dissociation rate constants for the various enzyme–inhibitor complexes can be obtained by division of the second-order association rate constant (Table VIII) by the association equilibrium constant. This yields at neutral pH about 10^{-7}/sec for bovine pancreatic trypsin inhibitor (Kunitz) (half-time 12 weeks) and 10^{-3}/sec for soybean trypsin inhibitor (Kunitz) (half-life 12 min). While these are quite crude estimates it is clear that (1) the slow rate of complex dissociation is the most striking characteristic of proteinase inhibitors and (2) the rates of dissociation vary over a truly large range. The slow rate of dissociation of complexes obviously greatly increases the effectiveness of proteinase inhibitors since once a complex is formed the enzyme will continue to be inhibited for a considerable period of time even if the solution conditions are sharply changed so as to favor dissociation (e.g., great dilution or introduction of large concentration of strongly competing substrate).

Direct measurement of dissociation rate constants is complicated both by their slowness and by the difficulty of dissociating the complex under the conditions of optimal stability. This difficulty is obviated at low pH values where the association equilibrium constants are much smaller (Fig. 2) and the dissociation rate constants are larger. The soybean trypsin inhibitor (Kunitz)–trypsin complex at neutral pH was mixed in a stopped flow apparatus with acidic buffers and the rates of dissociation measured. The data are given in Fig. 12 (*226*).

Another possible expedient was suggested by Green (*114*) and applied by him to the study of comparative rates of dissociation of complexes of trypsin with chicken ovomucoid, soybean inhibitor (Kunitz), and pancreatic inhibitor (Kunitz) with the result that the dissociation rate constants decrease in the order of listing. As expected, pancreatic inhibitor–complex dissociates far more slowly than the other two. The method consists of placing the complex in an excess of low K_m competing substrate and watching the appearance of enzymic activity as the complex dissociates. If the rate constant for dissociation is of the right order of magnitude ($10^{-4}/\text{sec} < k_D < 10^{-2}/\text{sec}$) the method can be made quantitative (*219, 229, 230*). Some data obtained by such a method are also shown in Fig. 12. Data of this type are, however, difficult to interpret in view of the possible interaction of substrate with the stable complex

229. M. Laskowski, Jr. and R. W. Duran, *Federation Proc.* **25,** 752 (1966).
230. R. W. Duran, M.S. Thesis Purdue University, West Lafayette, Indiana, 1965.

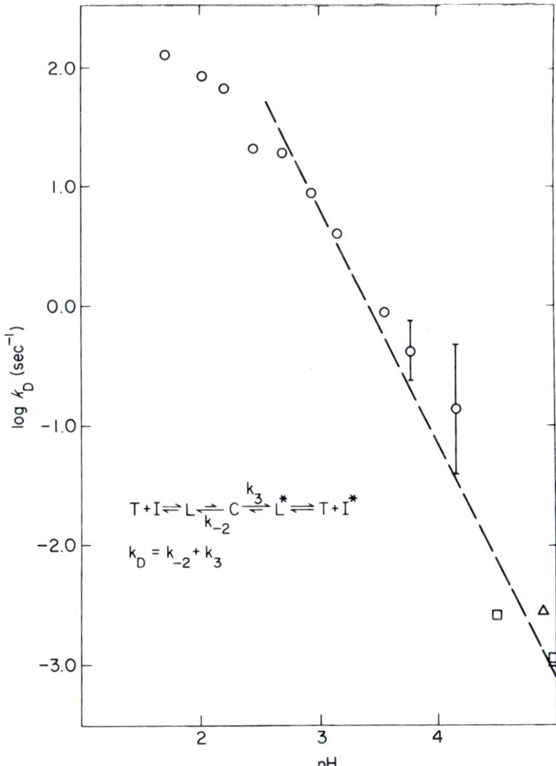

FIG. 12. First-order rate constant for the dissociation of the stable complex C of bovine β-trypsin with soybean trypsin inhibitor (Kunitz). (○) Data obtained by stopped flow (*226*). Large error bars around pH 4 arise from the necessity of correcting for incomplete dissociation of complex. (□, △) Data obtained by dissociation of complex in presence of TAME and by analysis shown in Fig. 9 (*219, 230*).

or with the loose complexes, e.g., substrate activation of dissociation (*229, 230*).

The general conclusion to be drawn from data of Fig. 12 and from other scattered data in the literature on dissociation of other complexes is that complexes dissociate very slowly at neutral pH but that the rate of dissociation increases dramatically with falling pH. The exact pH dependence of this increase is not yet elucidated for any complex.

The dissociation rate of complex [given by $k_{-2} + k_3$ of Eq. (9a)] is formally analogous to the deacylation rate in a typical acyl enzyme mechanism, but its pH dependence is precisely opposite that found for deacylation of acyl enzymes.

It may be recalled that upon complex formation (Section II,E,3) several protons are released, indicating that the pK values of several acid groups in the enzyme and/or inhibitor are greatly lowered in the complex. It appears that in order to obtain rapid dissociation some of these groups must be protonated in the complex and since their pK values are very low, high acidity is required to accomplish this.

3. *Steady State Interconversion of Virgin and Modified Inhibitors*

Another quite different set of kinetic measurements is also possible. As already pointed out, we may follow the rate of conversion of virgin → modified inhibitor as well as the reverse reaction in the presence of catalytic quantities of enzyme. An example of such data was given in Fig. 6. The appropriate analysis is simply that for one substrate–one product reversible steady state enzymic reaction (*216*). A complete study of such a system is not yet available but two conclusions, both of which are probably general for all or most proteinase-inhibitor systems, can be drawn. The pH dependence of both virgin → modified and modified → virgin inhibitor conversion is a bell-shaped plot such as given in Fig. 13 (*184*), with a pH optimum at a much lower pH value than the pH optimum for the usual activity of the relevant enzyme. Detailed steady state analysis (*230a*) shows that the catalytic rate constant for the virgin → modified inhibitor reaction is given by

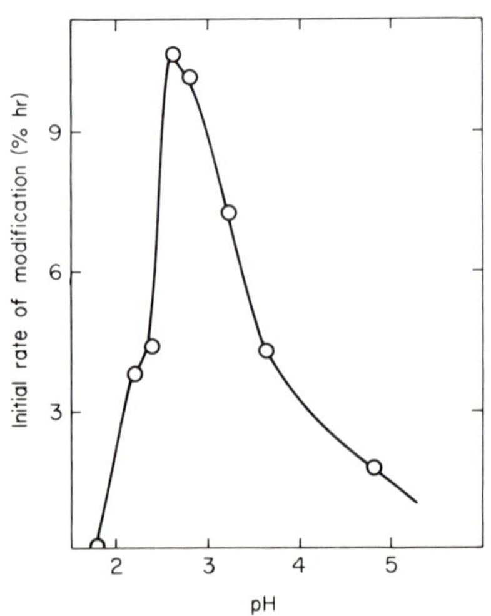

FIG. 13. Initial rate of conversion of virgin-to-modified bovine pancreatic secretory trypsin inhibitor (Kazal) by 0.7 mole-% trypsin as a function of pH. Redrawn from Rigbi and Greene (*184*).

230a. L. Peller and R. A. Alberty, *JACS* **81**, 5907 (1959).

$$k_{cat} = \frac{k_2 k_3}{k_2 + k_{-2} + k_3} \cong \frac{k_2 k_3}{k_2 + k_{-2}} \tag{10a}$$

The simplification of Eq. (10a) is possible because kinetic control dissociation of complex gives predominately virgin inhibitor (Section III,F), and therefore $k_{-2} \gg k_3$. At high pH k_2 is quite large (Fig. 11) and k_{-2} very small (Fig. 12). Thus $k_{cat} = k_3$. This accounts for the right limb of Fig. 13. As the pH is lowered k_2 decreases and k_{-2} increases and at very low pH $k_{-2} \gg k_2$, which accounts for the left limb of Fig. 13. The ratio k_2/k_{-2} is the fraction of the complex that exists as C—at low pH this fraction is very low.

The second generalization is that the modified → virgin rate constants are (at pH values higher than optimum pH) always much faster than virgin → modified rate constants. This conclusion is borne out by the data of Fig. 6 and by the studies on cocoonase–soybean trypsin inhibitor interaction (*231*). Since these rates have been assigned to dissociation of complex to virgin and to modified inhibitor, respectively, this conclusion predicts that kinetic control dissociation should always produce virgin inhibitor. As detailed in the next section, this was the case in all cases when it was tried.

F. Kinetic Control Dissociation of the Stable Complex

As already pointed out, the striking feature of the mechanisms of Eqs. (4) and (9a) is the assumption that the stable enzyme–inhibitor complex is the same substance whether it is made from virgin (reactive site peptide bond intact) or from modified (reactive site peptide bond hydrolyzed) inhibitor. As a consequence, dissociation of the complex made from either form of the inhibitor should yield the same products. Probably the most interesting dissociation reactions are kinetic control dissociations where the complex is suddenly transferred from solution conditions under which it is quite stable to solution conditions where it is very unstable. Under such circumstances the complex must dissociate, and the distribution of dissociation products (i.e., the virgin-to-modified inhibitor ratio) is governed solely by the relative values of the dissociation rate constants and not by the equilibrium constant between virgin and modified inhibitor (*193*). Since previous data suggest that the rate constant for dissociation to virgin inhibitor is generally faster than for dissociation to modified inhibitor predominantly virgin inhibitor should be expected.

231. H. F. Hixson, Jr. and M. Laskowski, Jr., *Biochemistry* **9**, 166 (1970).

Complexes made from trypsin and both virgin and modified soybean trypsin inhibitor (Kunitz) have been dissociated by either sudden pH drop (to pH 2) (*193*) or by placement in 6 *M* guanidine hydrochloride solution at neutral pH (*83*). In all cases, predominantly ($> 95\%$) virgin inhibitor was obtained even though equilibrium favors predominantly modified inhibitor. A similar experiment was performed by Frattali and Steiner (*151*) who subjected complex made from chymotrypsin-modified Bowman–Birk soybean inhibitor (*150*) and chymotrypsin and from trypsin-modified Bowman–Birk inhibitor and trypsin to kinetic control dissociation. In both cases virgin inhibitor was the predominant product even though equilibrium in both cases leads to predominantly modified inhibitors. The resynthesis of the cleaved peptide bond in modified inhibitors by kinetic control dissociation might be general for all enzyme–inhibitor interactions discussed in this chapter. This result tends to explain why such a long time elapsed between the beginning of study of enzyme–inhibitor reactions and the discovery of modified inhibitors since simple dissociation of complexes (which is frequently carried out under kinetic control conditions) leads to virgin rather than to modified inhibitor.

The kinetic control dissociation studies were further extended to probing of the mechanism of Eq. (9a) (*227*). Solutions made by mixing of concentrated modified inhibitor and trypsin were subjected to kinetic control dissociation at various short time intervals after mixing. It was assumed that the loose complex L^* of modified inhibitor and trypsin forms immediately after mixing and then converts to the stable complex C in a first-order process. Upon kinetic control dissociation, L^* should yield modified inhibitor I^*, and C should yield essentially pure virgin inhibitor I. Thus, the composition of the product obtained should smoothly convert from predominantly I^* to predominantly I in a first-order process as a function of time between mixing and dissociation. This is what was found (*227*), and further the first-order rate constant at several pH values closely agrees with the $L^* \rightarrow C$ rate constant determined by stopped flow (*226*).

The kinetic control resynthesis of the cleaved peptide bond offers the strongest available proof that the cleaved bond is the reactive site. Such a complete resynthesis of an accidentally cleaved bond would stretch the limits of credibility. Further, the rapid formation of C from L^* and the dissociation of $C \rightarrow L \rightarrow I$ prove that in the complex the reactive site of the inhibitor and the active site of trypsin are in a close association.

The rapid dissociation of the complex C to virgin rather than modified inhibitor might at first glance be taken to imply that in the complex C the peptide bond must be already intact. However, Tobias *et al.* (*232*) have shown that lactones (I) and (II) both acylate chymotrypsin and that upon kinetic control dissociation the acyl complex of lactone I and

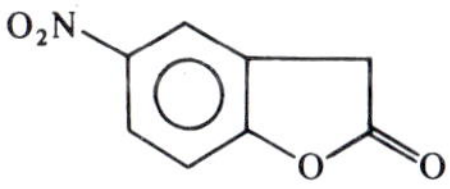

O_2N O_2N

(I) 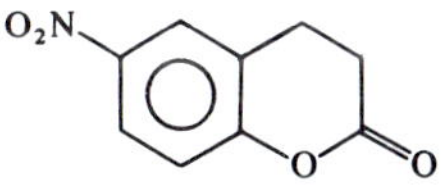 (II)

chymotrypsin yields the lactone (I) (not the hydroxy acid) while the acyl complex with lactone (II) yields the hydroxy acid. Thus, model studies show that the predominant yield of virgin inhibitor upon dissociation does not rule out an acyl complex as a stable complex C but only implies that if C is an acyl complex the α-NH_2-terminal of the reactive site must be held quite close to the acyl bond. A more extensive discussion of kinetic control dissociation has appeared (*227*).

G. Enzymic Replacement of Residues at or near the Reactive Site

The above discussion suggests that a synthetic variant of soybean trypsin inhibitor (Kunitz), in which the reactive site Arg 64 has been replaced by a lysyl should be fully able to complex with trypsin. An enzymic route to this substance was suggested by the following reactions:

$$S_C^* + \text{Lys} \underset{}{\overset{\text{C}_{\text{pase}}\text{B}}{\rightleftharpoons}} S_L^* \qquad K = 10^{-2}\ M^{-1} \tag{11}$$

$$S_L^* + T \rightleftharpoons C_L \qquad K = 10^{9}\ M^{-1} \tag{12}$$

$$S_C^* + \text{Lys} + T \rightleftharpoons C_L \qquad K = 10^{7}\ M^{-2} \tag{13}$$

The first of these reactions is the carboxypeptidase B–catalyzed addition of lysine to the Tyr 63 carboxyterminal of des-64-arginine–modified inhibitor (S_C^*) whose lack of activity against trypsin has been noted (Section III,B). The estimation of the unfavorable equilibrium constant for this reaction is discussed by Sealock and Laskowski, Jr. (*83*). The product [Lys 64]-modified inhibitor (S_L^*) should be a strong inhibitor of trypsin (T), as shown in Eq. (12). The equilibrium constant for this reaction was estimated from the data of Fig. 2 based on the assumption

232. P. Tobias, J. H. Heidema, K. W. Lo, E. T. Kaizer, and F. J. Kezdy, *JACS* **91,** 202 (1969).

(later proved correct) that the Arg → Lys substitution would not have a large effect on K_{assoc}. If the reactions of Eqs. (11) and (12) are coupled [Eq. (13)], the overall equilibrium constant for synthesis of (Lys 64)-soybean inhibitor–trypsin complex is quite favorable.

To prepare S_C^*, virgin inhibitor is incubated with catalytic quantities of trypsin at pH 3.75 [reaction (2) of Fig. 14]; this produces an equilibrium mixture of virgin (upper band in Fig. 15, gel A) and modified inhibitors (lower band, gel A). Incubation (pH 7.6) with carboxypeptidase B converts the modified inhibitor to des-64-arginine–modified inhibitor (gels B–J, lowest band), which is more negative by one charge unit than the modified inhibitor. By repetition of these steps (gels K and L), the des-arginine inhibitor is prepared in nearly quantitative yields (*216*).

In the synthetic reaction, des-arginine inhibitor ($1 \times 10^{-4}\,M$), trypsin ($2 \times 10^{-4}\,M$), carboxypeptidase B ($1 \times 10^{-4}\,M$), and $0.1\,M$ lysine were incubated at pH 6.7 for a few days [reactions (3a) and (3b), Fig. 14]. Synthesized C_L was isolated by Sephadex chromatography and subjected to kinetic control dissociation in $6\,M$ guanidine·HCl [reaction (4), Fig. 14]. The resulting product was subjected to a variety of tests to establish its identity (*83*).

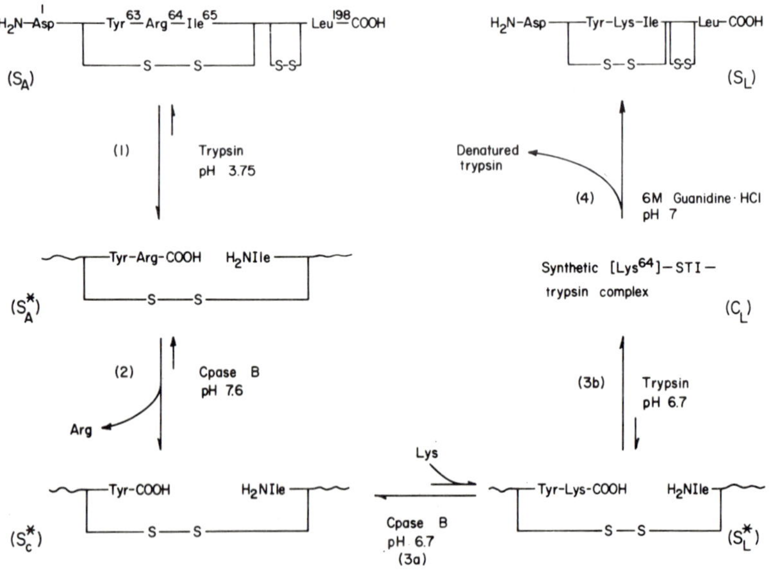

FIG. 14. The sequence of reactions by which Arg 64–soybean trypsin inhibitor (Kunitz) is converted to Lys 64–inhibitor (*83*). Assignments of residue positions are discussed in Sections III,B and IV,A.

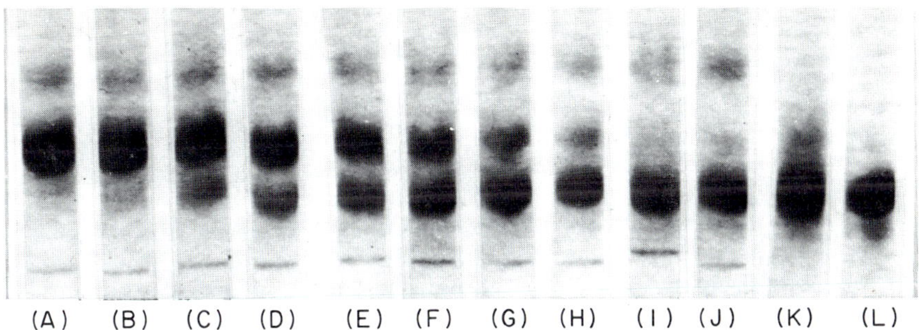

FIG. 15. Disc gel electrophoresis of aliquots of a des-Arg 64–soybean inhibitor preparation (*83*). From the top the bands are assigned to virgin, modified, and des-arginine inhibitors, respectively. (A) Initial virgin-modified equilibrium mixture (pH 3.75). (B–J) Aliquots removed at 1, 2.5, 4.5, 7.5, 11, 15, 20, 30, and 40 min, respectively, after addition of 1 mole-% carboxypeptidase B (pH 7.6). (K) Result of second tryptic equilibration at pH 3.75. (L) Result of second reaction with carboxypeptidase B.

Studies on the [Lys 64]-inhibitor have shown the relative unimportance of Arg → Lys exchange on thermodynamic properties of the inhibitors, but preliminary results indicate that several of its reactions with trypsin are slower than those of the original [Arg 64]-inhibitor (*83*). This observation may explain the apparently somewhat greater difficulty of obtaining reactive site cleavage of lysyl rather than arginyl inhibitors.

The enzymic replacement reaction offers a bright promise of selectively replacing several residues near the reactive sites of various proteinase inhibitors provided only that enzymes similar to carboxypeptidase B can be found which rapidly and selectively cleave a single peptide bond in modified inhibitors and thereby inactivate the inhibitor. A variant of the released peptide or amino acid could then be reintroduced provided that its incorporation restores inhibitory activity. Such studies should allow for a critical test of the present surmises about the reactive site sequences and a considerable extension of them.

H. SITES FOR OTHER ENZYMES

Unfortunately, rather little can be said about combining (reactive) sites of inhibitors for many enzymes other than trypsin. Reactive sites for enzymes such as plasmin, kallikrein, urokinase, and cocoonase are likely to be similar or identical to tryptic reactive sites (Section IV,B). The enzyme most studied after trypsin is chymotrypsin, which by analogy to the trypsin systems, should react with inhibitors at Trp-X,

Tyr-X, Phe-X, or Leu-X peptide bonds. The applications of chemical modifications to the identification of these sites may be rather limited since here it is even less obvious which modifications clearly remove chymotryptic susceptibility, and since phenylalanine and leucine are not easily modified. There are, however, two cases in which cleavage has been used to indicate or identify a chymotryptic reactive site.

The Bowman–Birk soybean inhibitor has been shown to inhibit trypsin and chymotrypsin independently (*140*). Additional proof that the inhibitor was not a mixture was obtained in paper electrophoresis at pH 6.9. An equimolar mixture of Bowman–Birk inhibitor and trypsin showed only a band due to complex and none due to free inhibitor. Identical results were obtained with an equimolar mixture of inhibitor and chymotrypsin (*140*).

In further studies Birk *et al.* (*150*) [work extended by Frattali and Steiner (*151*)] showed that on incubation of Bowman–Birk inhibitor with catalytic quantities of trypsin at pH 3.75 almost all (as judged by disc gel electrophoresis) of the inhibitor molecules are converted to trypsin-modified inhibitor, presumably by cleavage of the reactive site Lys-X bond (Table V). This trypsin-modified inhibitor retains its inhibitory activity toward chymotrypsin and trypsin, although like the modified Kunitz inhibitor it reacts with trypsin more slowly than virgin inhibitor. Upon treatment with carboxypeptidase B it loses its trypsin but not the chymotrypsin inhibiting activity. Similarly, on incubation with catalytic quantities of chymotrypsin at low pH almost all of the inhibitor molecules are converted to chymotrypsin-modified inhibitor, which retains its tryptic inhibitory activity but combines with chymotrypsin so slowly that it appears to be inactive as a chymotrypsin inhibitor in standard assay. Frattali and Steiner (*151*) prepared complexes of trypsin with trypsin-modified inhibitor and of chymotrypsin with chymotrypsin-modified inhibitor and subjected them to kinetic control dissociation. In both cases the inhibitor obtained was predominantly virgin.

These important experiments offer very strong proof that the reactive site model described above also applies in its entirety to (1) a second trypsin inhibitor and (2) most importantly, to chymotrypsin inhibitors, although the chymotrypsin reactive site has not been otherwise identified.

In the second example, kinetic control dissociation has not been carried out, but a chymotrypsin reactive site has been identified. Stevens (*177*) has shown that incubation of lima bean inhibitor (component 3; see Section IV,A) with catalytic quantities of chymotrypsin produces chymotrypsin-modified inhibitor which loses its inhibitory activity against chymotrypsin (but not against trypsin) on treatment with carboxypeptidase A. Further, Krahn and Stevens (*199*) determined the chymotryptic reactive site sequence in this inhibitor to be Ser–Thr–Leu↓Ser–Ile–Pro. To the best of our knowledge this is the only chymotryptic reactive site sequence heretofore determined. An interesting aspect of this sequence is the presence of three hydroxyamino acid residues surrounding the reactive site leucyl. Amino acid residues contributing the carbonyl partner of the chymotryptic reactive sites will probably all be hydrophobic, and thus some special structure must be present to insure that they are exposed rather than buried. This requirement is virtually automatically satisfied for the tryptic active site, and therefore it is quite likely that sequences surrounding the chymotryptic and tryptic reactive sites will generally differ.

I. Chemical Modifications of Inhibitors

Information about chemical modifications of inhibitors is widely scattered in the literature, and since so many proteinase inhibitors have been studied summarizing it all would be a prohibitive task. The listing here is a representative sample. Chemical modifications intended to locate or classify reactive sites are discussed in Sections III,B and III,H.

1. *Pancreatic Trypsin Inhibitor (Kunitz)*

Pancreatic trypsin inhibitor (Kunitz) was studied far more intensively than any other since it is quite small (58 residues), reasonably resistant to chemical environments which are not often tolerated by many other proteins, and, most importantly, is the only one whose amino acid sequence (Sections I,C,1 and IV,A) has been known for some time.

Kassell and Chow (*233*) have succeeded in removing the NH_2-terminal tripeptide by successive Edman degradation of guanidinated inhibitor without loss of activity. Chauvet and Acher (*188*) have shown that Lys 26, 41, and 46 are nonessential by polyalanylating the trypsin–inhibitor complex and showing that the recovered substituted inhibitor was active. These experiments were extended to maleylation by Fritz *et al.* (*161*). It should be recalled that the remaining Lys 15 is at the reactive site. Kassell (*234*) eliminated Met 52 by showing that after its oxidation with H_2O_2 the inhibitor was still active. Meloun *et al.* (*235*) showed that only Tyr 10 and 21 are nitrated with tetranitromethane without loss of either antitryptic or antichymotryptic activity. In a later study from that group (*163*) it was shown that these tyrosyls are also nitrated in the complexes of the inhibitor with either trypsin or chymotrypsin.

233. B. Kassell and R. B. Chow, *Biochemistry* **5**, 3449 (1966).
234. B. Kassell, *Biochemistry* **3**, 152 (1964).
235. B. Meloun, I. Fric, and F. Šorm, *European J. Biochem.* **4**, 112 (1968).

Sherman and Kassell (*236*) have found that three out of four tyrosyls ionize readily and reversibly and that three are iodinated—Tyr 10 and 21 to monoiodotyrosyls, Tyr 35 to diiodotyrosyl. It is, therefore, rather surprising that Tyr 35 is not nitrated. On the basis of these results (*236*) it is surmised that Tyr 23 is buried and none of the four tyrosyls participates in direct trypsin–inhibitor contact.

The role of carboxylates is not yet fully elucidated, but Chauvet and Acher (*237*) obtained derivatives in which an average of three out of five carboxyls were substituted with glycine amide and were fully active. On the basis of peptide maps of these derivatives they showed clearly that the COOH-terminal Ala 58 can be eliminated from consideration. The carboxyl of Asp 3 was eliminated by the removal experiment of Kassell and Chow (*233*). However, Avineri-Goldman *et al.* (*202*) found a dramatic decrease in inhibition of trypsin by the inhibitor methyl ester. But decrease was far more pronounced at low ionic strength than at high and quite probably arose from a large decrease in the rate of association because of greatly increased charge repulsion. While the role of the carboxylates is not yet clear, in view of our comments about avoidance of negative charge near reactive sites, we are inclined to doubt that any of them are involved in direct interaction. It is clear from the foregoing that if all of the results above are correct and if the somewhat hazardous interpretation that the remaining activity after substitution means that the residues in question are not involved in important enzyme–inhibitor contact is correct, then the search is almost over. Some doubt remains about the carboxylates and Tyr 35. To the best of our knowledge the arginyls were not substituted, but substitution of arginyls in some other lysyl inhibitors was without effect on activity (Table V). It appears plausible that the only readily modifiable residue in the inhibitor molecule whose specific substitution leads to a loss of activity is reactive site Lys 15. In the recently announced three-dimensional structure of Huber *et al.* (*73b*) all of the carboxylates of this inhibitor are quite remote from the reactive site Lys 15.

2. *Other Inhibitors*

The NH_2-terminal tetrapeptide in porcine secretory pancreatic inhibitor I (Kazal) is nonessential because inhibitor II (*175, 238*) which lacks this tetrapeptide is fully active (Section IV,A). This conclusion

236. M. P. Sherman and B. Kassell, *Biochemistry* **7**, 3634 (1968).
237. J. Chauvet and R. Acher, *FEBS Letters* **2**, 17 (1968).
238. L. J. Greene, J. J. DiCarlo, A. J. Sussman, D. C. Bartelt, and D. E. Roark, *JBC* **243**, 1804 (1968).

has been recently extended to the NH_2-terminal pentapeptide which was removed by trypsin from inhibitor I without loss of biological activity (*239*).

Lima bean inhibitor was extensively esterified at side chain carboxyls without loss of inhibitory activity (*187*). Soybean trypsin inhibitor (Kunitz) after carbodiimide-mediated reaction with glycine amide had 30 of its 35 carboxyl groups modified and retained full activity, but the association rate constant decreased about 10-fold (*219*), presumably because of unfavorable electrostatic interaction. Chicken ovomucoid loses its inhibitory activity upon esterification (*240*). This, however, is probably not a consequence of direct interaction of carboxylates with trypsin. Donovan (*88*) has shown that ovomucoid undergoes a large transition at low pH. Esterified ovomucoid appears to be locked in the low pH form and thus may well have unsatisfactory conformation for complex formation with trypsin (Section II,A). On the basis of these scattered results and especially of the work of Chauvet and Acher (*237*), we conclude that highly reactive carboxyls are probably not involved in direct specific enzyme–inhibitor contact. A corollary of this conclusion is that most of the protons released on trypsin–inhibitor complex formation are primarily released by the carboxyl groups of trypsin and not of the inhibitors (Section II,E,3).

The whole rationale of the chemical modification method as a search for lysyl and arginyl reactive sites described in Section II,B is the assumption that highly reactive lysyls and arginyls (other than the reactive site ones) can be modified without loss of inhibitory activity. The great success of the chemical modification method (see Table V) argues that this assumption is most often correct.

Steiner (*241*) carried out an extensive study of the effect of chemical substitutions on activity and on thermodynamic stability of soybean trypsin inhibitor (Kunitz). Several reagents were used and several groups modified. While the detailed results are somewhat more complex than this, the simplest way to summarize his conclusions is to say that substitution of fully exposed, readily reactive groups did not interfere with inhibitory activity either immediately after substitution or after a denaturation–renaturation cycle. On the other hand, modification of buried or partially buried groups, which reacted only after complete or partial denaturation, completely abolished activity. Thus, many groups were implicated in the maintenance of the inhibitor conformation, but

239. H. Tschesche, E. Wachter, and G. Kallup, *Z. Physiol. Chem.* **350,** 1662 (1969).
240. H. Fraenkel-Conrat, R. S. Bean, and H. Lineweaver, *JBC* **177,** 385 (1949).
241. R. F. Steiner, *ABB* **115,** 257 (1966).

none was clearly indicated as required for direct contact with the enzyme. (No attempt was made to modify the reactive site arginyl.)

Numerous chemical modifications of ovomucoids were carried out by Feeney and collaborators. Most of these studies were concerned with the search for reactive sites (Section III,B), but other conclusions can be drawn concerning the effect of change in charge on association rate constant. All of the substitutions which did not affect the reactive site but which increased the overall negative charge on the molecule led to an increase in the rate of association with α-chymotrypsin, while if the charge remained the same (amidination) the reaction proceeded at the same rate (*225*).

The effect of several modifications of enzymes, e.g., trypsin and chymotrypsin, on their association with inhibitors has been reported. We chose not to review such modifications because of the great complexity of active derivatives of enzymes. Thus, acetylated trypsin has been reported to be more active, less active, or not active at all in interaction with simple substrate. For a recent paper attempting to unravel such complexities, see Houston and Walsh (*242*). It should, however, be recalled that in Section II,F evidence was discussed which implicates that one or more of trypsin's tryptophyls and possibly some of trypsin's tyrosyls are in enzyme–inhibitor contact.

The present information on inhibitor modification stands in sharp contrast with the old literature in which an interaction of very many groups on the inhibitor with an equal amount on the enzyme was invoked (e.g., references *3*, *4*, *13*, *240*). The small extent of changes in optical properties on enzyme–inhibitor interaction suggests that the contact area between enzyme and inhibitor is quite small (Section II,F). This is further supported by low angle X-ray scattering studies on the trypsin–soybean trypsin inhibitor complex (*243*).

Ozawa and Laskowski, Jr. (*79*) remarked that in active trypsin inhibitors aside from the reactive site there must be present other specific groups which interact with the trypsin molecule and thus account for the stability of complexes and for the pH dependence of trypsin–inhibitor interaction. We are no longer certain that this conclusion is correct. Prolonged search by various workers failed to clearly identify any such additional groups, but a great many groups have been excluded from consideration. Obviously such groups may still be found. On the other hand, an alternative is that the interaction with enzymes involves only the reactive site alone and that the strong interaction is a consequence

242. L. L. Houston and K. A. Walsh, *Biochemistry* **9,** 156 (1970).

243. G. Damaschum, P. Fichtner, H.-V. Purschel, and J. B. Reich, *Acta Biol. Med. Ger.* **21,** 309 (1968).

of unusual and extremely rigid conformation of the reactive site and of its nearest neighbor residues. Whenever this local conformation is destroyed by denaturation or by substitution of groups which stabilize the native structure, proteinase inhibitors lose their activity.

One striking feature of proteinase inhibitor–proteinase interaction, *viz.*, the extremely slow rate of dissociation of a stable complex, may have an analog in a simple, specific molecule–enzyme reaction. Mares-Guia and Shaw (*244*) and Chase and Shaw (*120*) have shown that the specific active site titrant *p*-nitrophenyl-*p'*-guanidobenzoate combines with trypsin very rapidly (as some inhibitors do) to form a very stable acyl complex, which, however, deacylates extraordinarily slowly. The analogy between their system and the trypsin–soybean trypsin inhibitor reaction was already pointed out by these workers (*244*). Clearly the *p*-guanidobenzoyl group (analogues to the reactive site amino acid residue) interacts only with the specificity site of trypsin, and no additional interactions stabilize the acyl enzyme.

J. Temporary Inhibition

Complexes of trypsin with a few trypsin inhibitors do not survive indefinitely but decompose after long incubation times to free trypsin and to digested inhibitor. A particularly dramatic illustration is shown in Fig. 16 (*245*). The phenomenon of initial disappearance (upon original mixing of trypsin and inhibitor) and later reappearance (after prolonged incubation) of trypsin activity was called temporary inhibition by Laskowski, Sr. and Wu (*74*), and inhibitors whose complexes show this phenomenon are frequently referred to as *temporary inhibitors.* They are chicken ovomucoid (*246, 247*) and possibly other avian ovomucoids (but see below), all of the tested mammalian pancreatic secretory inhibitors of Kazal type (*74, 100, 245, 248, 249*), and several, but not all, derivatives of the pancreatic trypsin inhibitor (Kunitz) whose Cys 14–Cys 38 disulfide bridge was selectively cleaved and the cysteinyls modified (*213*). Unmodified pancreatic trypsin inhibitor (Kunitz) is not a temporary inhibitor (*74, 250*) and is totally resistant to proteolytic attack

244. M. Mares-Guia and E. Shaw, *JBC* **242,** 5782 (1967).

245. P. J. Burck, R. L. Hamill, E. W. Cerwinsky, and E. L. Grinnan, *Biochemistry* **6,** 3180 (1967).

246. L. Gorini and L. Audrain, *BBA* **8,** 702 (1952).

247. L. Gorini and L. Audrain, *BBA* **10,** 570 (1953).

248. H. Tschesche and H. Klein, *Z. Physiol. Chem.* **349,** 1645 (1968).

249. L. J. Greene, M. Rigbi, and D. S. Fackre, *JBC* **241,** 5610 (1966).

250. B. Kassell and M. Laskowski, Sr., *Federation Proc.* **24,** 593 (1965).

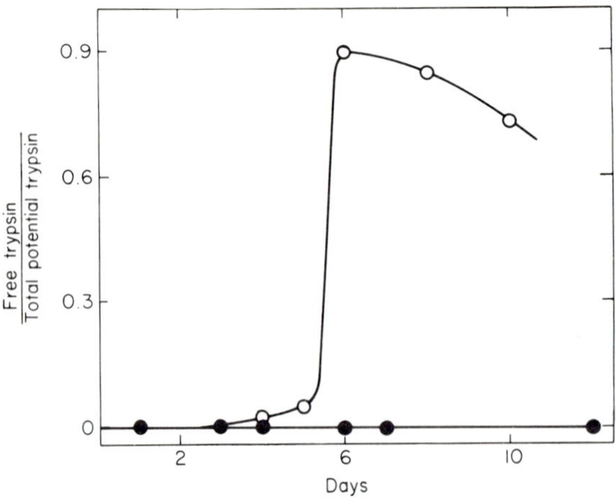

Fig. 16. Trypsin activity as function of time upon incubation of complexes of bovine trypsin with bovine pancreatic secretory trypsin inhibitor (Kazal) and bovine pancreatic inhibitor (Kunitz) at pH 8.0. Slight excess of inhibitor was initially added to impart stability. Redrawn from Burck *et al.* (*245*).

by numerous proteolytic enzymes with which its cleavage was attempted (*251*).

Kress *et al.* (*213*) made the interesting generalization that none of the presently known temporary trypsin inhibitors inhibit chymotrypsin. This observation is particularly striking because the list of temporary inhibitors includes some of the above-mentioned derivatives of the Kunitz inhibitor which do not inhibit chymotrypsin. Other derivatives which do inhibit chymotrypsin are not temporary inhibitors (*213*). An explanation of this generalization in molecular terms seems difficult if double-headed trypsin–chymotrypsin inhibitors (Section IV,B) are included. The statement predicts that the ovomucoids which do not inhibit chymotrypsin may be temporary inhibitors while those that do inhibit it cannot be. A test with turkey ovomucoid is clearly indicated. However, a more restricted version of the statement—those inhibitors which inhibit both trypsin and chymotrypsin on overlapping independent reactive sites (Section IV,B,2) may not be temporary inhibitors—agrees with the known facts and admits to a molecular explanation in terms of the additional rigidity of the reactive region (*161*) required to accommodate two independent reactive sites in close proximity.

There is considerable discussion in the literature on the question of the mechanism of inhibitor digestion in temporary inhibition. Some authors (*246–248*) prefer to think of the reaction as proteolytic break-

251. B. Kassell and M. Laskowski, Sr., *JBC* **219,** 203 (1956).

down of trypsin–inhibitor complex itself. Laskowski, Sr. and Wu (*74*), after consideration of the dependence of the rate of the reaction on addition of excess inhibitor to the complex (decrease) and of excess trypsin (increase), postulated an attack on the inhibitor in the complex by an additional trypsin molecule as in Eqs. (14) and (15):

$$\mathrm{T} + \mathrm{I} \rightleftharpoons \mathrm{TI} \tag{14}$$

$$\mathrm{TI} + \mathrm{T} \rightleftharpoons \mathrm{TIT} \rightarrow 2\ \mathrm{T} + \text{products} \tag{15}$$

However, Tschesche and Klein (*248*) found that proteolytic conversion of porcine pancreatic trypsin inhibitor (Kazal) to largely inactive product and to totally inactive fragments took place both in free solution and on insoluble trypsin resin. Since they considered it unlikely on steric grounds that an inhibitor molecule could be in contact with two trypsin molecules on the resin they preferred direct TI breakdown. It must be emphasized that breakdown of a complex of the form TI does not mean that this is the stable trypsin–inhibitor complex C, in which the active site of trypsin is close to the reactive site of the inhibitor. Obviously, trypsin occasionally comes in contact with other Lys-X and Arg-X peptide bonds in the inhibitor molecule and these can then be hydrolyzed. While the contact with other bonds is quite improbable in view of large K_{assoc} at the reactive site, the hydrolysis there may well be moderately rapid and thus account for the slow overall digestion observed in temporary inhibition.

Some authors (*9*) are impressed with the similarity between temporary inhibition and reactive site hydrolysis. Indeed, formally the analogy is rather strong; the delayed release of trypsin is an "overshoot," and the cause is proteolysis of the inhibitor. There are, however, sharp differences. The pH optimum for temporary inhibition is near neutral pH (*74, 245, 248*), while that for reactive site modification is in the acid range (Section III,E,3). The products of temporary inhibition have little or no inhibitory activity and frequently suffer multiple proteolytic cleavages (*74, 248*), while in reactive site modification one bond is selectively split and the resultant modified inhibitor has considerable inhibitory activity. Finally, while the experiment was not tried, it seems to us unlikely that the bonds split in temporary inhibition could be resynthesized by kinetic control dissociation of complexes between the cleaved molecules and trypsin. It can happen, however, that genuine reactive site modification (hydrolysis) will appear as temporary inhibition due to the extreme slowness with which some modified inhibitors act. Such a situation with the chymotryptic reactive site of the Bowman–Birk soybean inhibitor has been observed and resolved (Sections II,D,3 and III,H).

K. Speculation on the Nature of the Stable Complex

No firm statement can be made about the detailed nature of bonding in the stable complex, i.e., whether a covalent bond or only secondary forces are involved in the enzyme–inhibitor interaction. Finkenstadt and Laskowski, Jr. (*168, 193*) have postulated that the complex is an acyl complex between the newly formed COOH terminal residue of the reactive site and the active site seryl of the enzyme. This view was, however, subjected to considerable criticism and a number of difficulties in it were presented (see Section III,L). Further, all attempts to date to trap the denatured acyl complex by rapid denaturation have thus far failed—in all cases virgin inhibitor and the enzyme were obtained.

The present argument for the stable complex being an acyl enzyme rests on the general belief that an acyl enzyme must be present somewhere in the kinetic scheme for all trypsin-catalyzed hydrolyses (*252*). This does not demand that the stable complex be an acyl enzyme; however, the following should be considered:

(1) Peptide bond cleavage has been found in almost all, but not all, cases of protein inhibitor–serine esterase interaction where a serious attempt to find it was made. Thus the peptide bond hydrolysis and complex formation are intimately connected.

(2) If an acyl enzyme is a mandatory intermediate but not any of the species recognized in Eq. (9a), then it presumably has to be inserted between C and L*. Yet, since it has not been kinetically recognized, extraordinarily high rate constants for the $L^* \rightarrow$ acyl and the acyl $\rightarrow$ C conversion would be required to account for the kinetic data (*227*).

(3) The mechanism of Eq. (9a) and the pH dependence of all the rate constants (although not their magnitudes) are symmetric around the stable complex C (*226*). If C is a noncovalent complex between virgin inhibitor and the enzyme, the observed symmetry is hard to rationalize; but if C is an acyl enzyme it becomes easy to explain.

(4) The rate of complex formation from trypsin and virgin inhibitor is reasonably consistent in the pH dependence, existence of loose complex and its conversion to next (stable) complex, and in the values of the parameters K_m and k_2 with an acylation reaction of peptides and amides (*226*).

252. This general statement is frequently subject to challenge. It should be made quite clear that the acyl enzyme hypothesis for the stable complex critically depends on the assumption that an acyl enzyme complex exists somewhere in the pathway for conversion of virgin to modified inhibitor.

None of this can be viewed as compelling and we entertain some serious doubts, but on the balance we prefer the acyl model. Even if the acyl model is proved to be correct it will remain to explain the strikingly unusual rate constants of enzyme–inhibitor system compared to normal enzyme–substrate rate constants, especially the increase of rate of dissociation on lowering of pH rather than a decrease as is found in analogous deacylation reactions. To the best of our knowledge none of these problems becomes any easier if the notion of the stable complex as an acyl enzyme is rejected.

L. Objections to the Reactive Site Model

The overall mechanism of proteinase inhibitor–proteinase interactions discussed in this review has enjoyed both considerable acceptance and considerable criticism during its development. Although the views of proponents and critics alike have changed with the passing of time, two clear objections to the reactive site model appear to remain:

(1) The mechanism outlined here is correct for a few inhibitor–enzyme systems, but other mechanisms must be invoked for other known systems.

(2) The stable complex between enzyme and inhibitor involves only secondary interactions, and the inhibitor in the complex is virgin (reactive site peptide bond intact).

The most telling points adduced in favor of these propositions are summarized below.

1. *Activity of Inhibitors with Homoarginyl or ϵ-Acetamidyllysine in Their Reactive Sites*

A number of statements in the literature (*253–255*) assert that homoarginyl and ϵ-acetamidyllysine peptide bonds are not cleaved by trypsin. If these statements are correct, then it should be expected that lysyl trypsin inhibitors whose reactive sites are guanidinated or amidinated would be inactive just as are the acetylated, succinylated, deaminated, etc., derivatives if peptide bond cleavage is required for complex formation. In fact, it is known that several inhibitors do not lose their activity on guanidination and amidination.

Kassell and Chow (*233*) guanidinated and amidinated all of the lysyls

253. L. Weil and M. Telka, *ABB* **71,** 473 (1957).
254. G. S. Shields, R. L. Hill, and E. L. Smith, *JBC* **234,** 1747 (1959).
255. M. J. Hunter and M. L. Ludwig, *JACS* **84,** 3491 (1962).

of the pancreatic trypsin inhibitor (Kunitz) and found the inhibitor fully active. These results were further confirmed and extended in the literature (*161, 190*). At first the data were used to argue that this inhibitor was not a lysyl inhibitor (*233*), but the present (Section III,B) in favor of Lys 15–Ala 16 as the reactive site appears overwhelming.

Haynes and Feeney (*180*) guanidinated and amidinated the following lysyl trypsin inhibitors—turkey, cassowary, and duck ovomucoids and lima bean inhibitor. While the substitution of lysyls was not complete, convincing additional evidence was provided that the reactive site lysyls were substituted. The amidinated inhibitors possessed an extremely feeble (but detectable with casein as substrate) inhibitory activity. Among the guanidinated inhibitors, lima bean inhibitor was fully active and the ovomucoids were only weakly active. Surprisingly, guanidinated lima bean inhibitor combines with trypsin more rapidly than the unsubstituted inhibitor (Section III,E,1).

The validity of the objection rests on how absolute are the statements that trypsin does not cleave homoarginyl and ϵ-acetamidyllsyl peptide bonds and whether these rather general findings can occasionally be violated when the residues are present in the unusual amino acid sequences and/or conformations of inhibitor reactive sites.

Kitagawa and Izumiya (*256*) and Baines *et al.* (*257*) have shown that acyl esters of homoarginine are hydrolyzed by trypsin although the K_m and k_{cat} values are poorer than for corresponding lysyl and arginyl esters. More importantly, Rigbi and Elbaz (*258*) have shown that polyhomoarginine is hydrolyzed by trypsin, although less well than polylysine. This finding clearly shows that the prohibition against hydrolysis of homoarginyl peptide bonds is not absolute.

Since it has now been shown that a Lys-X reactive site bond is hydrolyzed upon incubation of lima bean inhibitor with catalytic amounts of trypsin at low pH (*177*), repeating the experiment with the very active homoarginyl-X derivative of this inhibitor may well resolve this question.

2. *Lack of Peptide Bond Cleavage in Some Inhibitors*

One of the most impressive arguments for generality of the proposed mechanism is the relative ease with which a number of specific reactive site cleavages by catalytic quantity of enzyme have been found in so many inhibitors within only 4 years since the announcement of the first

256. K. Kitagawa and N. Izumiya, *J. Biochem.* (*Tokyo*) **46,** 1159 (1959).
257. N. J. Baines, J. B. Baird, and D. T. Elmore, *BJ* **90,** 470 (1964).
258. M. Rigbi and L. Elbaz, *FEBS Symp.* **18,** No. 730 (1970) (abstr.).

modified inhibitor. However, in a few cases cleavage was not obtained and these are properly taken as objections to the generality of the proposed mechanism.

Numerous laboratories incubated pancreatic trypsin inhibitor (Kunitz) with catalytic quantities of trypsin under a variety of solvent conditions and for very long periods of time. None of them reported reactive site cleavage (*115, 182, 184*).

This result can be dealt with in a variety of ways. One is to exempt pancreatic trypsin inhibitor (Kunitz) from the general mechanism and invoke another mechanism for it. This seems to be a poor choice in view of the great similarity of the pH dependences of its association equilibrium constant (Fig. 2), association rate constant (Figs. 8 and 11), and dissociation rate constant with those of soybean trypsin inhibitor (Kunitz). Even more convincing are the findings of Kress and Laskowski, Sr. (*182*) that the Lys 15–Ala 16 bond is cleaved in an inhibitor whose Cys 14–Cys 38 disulfide bond is reduced, since Chauvet and Acher (*188*) have shown that Lys 15 is part of the reactive site of the fully intact inhibitor molecule.

We must therefore argue that failure to obtain cleavage results from either kinetic or thermodynamic difficulties. Pancreatic trypsin inhibitor–trypsin complex dissociates extraordinarily slowly, and for most trypsin inhibitors the rate of dissociation to modified inhibitor is much slower than to virgin. Thus, the rate of virgin → modified pancreatic inhibitor conversion may be so slow that even the prolonged incubations employed by various workers are too short.

The second possible explanation is that K_{hyd}, the virgin ⇌ modified equilibrium constant, is so low that no modified inhibitor can be detected at equilibrium. A finite, but reasonably large, equilibrium constant has been clearly measured for soybean trypsin inhibitor (Kunitz) (Section III,D). If the data of Rigbi and Greene indicate equilibrium as is thought, the K_{hyd} constant for pancreatic secretory inhibitor (Kazal) is much lower and in fact at most pH values favors virgin inhibitor (*184*). In soybean trypsin inhibitor (Kunitz) the half-cystine residue of the reactive site disulfide loop is quite far from the reactive site (Fig. 4) and thus does not greatly interfere with the motion of the peptide chains in the modified inhibitor. In the pancreatic (Kazal) inhibitor this half-cystine is only one residue removed from the reactive site, and the equilibrium constant is presumably much smaller. In the pancreatic (Kunitz) inhibitor the half-cystine is adjoining (Section IV,A), and therefore the equilibrium constant could be expected to be smaller yet. The aforementioned results of Kress and Laskowski are in accord with such an interpretation.

Feinstein *et al.* (*259*) have argued against the reactive site model on the basis of their findings that turkey and cassowary ovomucoids were not inactivated by consecutive treatments with trypsin and carboxypeptidase B. However, only one set of conditions for incubation of trypsin and the ovomucoids was employed, and the data of Rigbi and Greene (*184*) indicate how significantly the rate of cleavage can vary with conditions and with the inhibitor being studied. Furthermore, Feinstein *et al.* (*259*) did not establish that the cleavage of the reactive site bond did not in fact take place. The alternative remains that the bond was hydrolyzed but that an unfavorable sequence (e.g., Pro–Lys) at the new COOH terminus or other factors prevented the release of lysine by carboxypeptidase B.

3. *Recovery of Virgin Inhibitor from Inhibitor–Trypsin Fragment Complex*

Dlouhá *et al.* (*260*) subjected the trypsin–pancreatic trypsin inhibitor (Kunitz) complex to extended digestion by additional trypsin and by chymotrypsin. Owing to the exceptionally slow dissociation of this complex (Section III,E,2) and to the exceptional resistance to proteolysis of the Kunitz inhibitor, only the trypsin part of the complex was digested with no dissociation of free inhibitor. Gel exclusion chromatography led to isolation of material composed of the inhibitor and about 48% of a trypsin molecule containing *inter alia* the active site serine and histidine. This material is stable in neutral solution but dissociates irreversibly into virgin inhibitor and tryptic peptides upon performic acid oxidation.

The authors argued that since virgin inhibitor was obtained the complex could not have contained an acyl bond. On the other hand, they reported only 50% dissociation of the inhibitor–fragment complex after 3 hr in 0.01 *M* HCl, whereas native enzyme and inhibitor associate only very weakly at pH 2. The source of this stability was not established. The failure to trap denatured and acyl-linked inhibitor and trypsin during performic acid oxidation generally argues against the existence of the acyl complex. However, it seems possible that the fragment of the trypsin retains sufficient characteristics of the active site to permit kinetic control dissociation to occur if the acyl complex indeed exists. Virgin inhibitor is the expected product in such a case.

4. *Interaction of Inhibitors with Inactive Enzymes*

In proposing the acyl enzyme mechanism (*168*), Finkenstadt and Laskowski, Jr., made the statement "*only* enzymatically active trypsin

259. G. Feinstein, D. T. Osuga, and R. E. Feeney, *BBRC* **24,** 495 (1966).
260. V. Dlouhá, B. Keil, and F. Šorm, *BBRC* **31,** 66 (1968).

is capable of forming complexes with naturally occurring inhibitors" and used the statement as a part of the argument. While this statement was not contradicted by any reports available at the time of its writing, three papers contradicting it have now appeared. Those papers form probably the strongest objection to the acyl enzyme mechanism.

Green (*114*) has shown that several trypsin inhibitors do not form complexes with diisopropylphosphoryl trypsin ($K_{assoc} < 10^3\ M^{-1}$), and these observations were widely confirmed. Trypsin preparations contain not only active trypsin but also several kinds of inactive material. It was shown that only active trypsin in these preparations combines with soybean trypsin inhibitor (Kunitz) (*127*). This conclusion has been effectively generalized to all trypsin inhibitors since most authors in carrying out trypsin inhibitor assays make the assumption that only active trypsin in their preparations combines with the inhibitors and on the basis of this assumption obtain correct trypsin–inhibitor combining ratios. Estermann and McLaren (*261*) have shown that photoinactivated trypsin does not form complexes with soybean inhibitor, and Beeley and Neurath (*262*) found no complex formation when either soybean (Kunitz) or pancreatic (Kunitz) inhibitors were mixed with a bromoacetone derivative of trypsin [specific inactivation by reaction with active site histidyl (*46*)].

However, Foster and Ryan (*263*) showed that anhydrochymotrypsin combines with potato chymotrypsin inhibitor. In an extension of these studies Feinstein and Feeney (*135*) have investigated the interaction of tosylchymotrypsin, anhydrochymotrypsin, and the tosylphenylalanine chloromethyl ketone (TPCK) derivative of chymotrypsin with several chymotrypsin inhibitors and of tosyllysine chloromethyl ketone (TLCK) derivative of trypsin with trypsin inhibitors. Tosylchymotrypsin combines with none of the chymotrypsin inhibitors, but it was shown that anhydrochymotrypsin combines with turkey ovomucoid and TPCK derivative of chymotrypsin forms complexes with potato inhibitor and with turkey and golden pheasant ovomucoids. In the class of trypsin inhibitors it was shown that TLCK derivative of trypsin forms complexes with turkey and chicken ovomucoid, possibly a weak complex with penguin ovomucoid and no complexes at all with either soybean or lima bean inhibitor. The inactive trypsin–inhibitor combinations reported by Feinstein and Feeney (*135*) were sharply criticized by Pudles and Bachellerie (*264*), but the latter authors' conclusion appears to suffer from an error in calculations. In our laboratory we confirmed the interaction be-

261. E. F. Estermann and A. D. McLaren, *Photochem. Photobiol.* **1**, 109 (1962).
262. J. G. Beeley and H. Neurath, *Biochemistry* **7**, 1239 (1968).
263. R. J. Foster and C. A. Ryan, *Federation Proc.* **24**, 473 (1965).
264. J. Pudles and D. Bachellerie, *ABB* **128**, 133 (1968).

tween chicken ovomucoid and the TLCK derivative of trypsin. In all the cases reported above the interaction of inhibitors with inactive enzyme is both much slower and thermodynamically much weaker than with active enzymes.

The interaction of TLCK-trypsin derivative and chicken ovomucoid is paradoxical for a variety of reasons. First it is rather hard to understand why chicken ovomucoid combines with the TLCK derivative rather strongly, while a much stronger trypsin inhibitor—soybean inhibitor of Kunitz—does not combine at all. Even if this is rationalized on the basis of more stringent steric requirements of soybean inhibitor, it still remains to explain why neither soybean inhibitor nor pancreatic trypsin inhibitor (a particularly strong inhibitor, Fig. 2) combine with the bromoacetone derivative of trypsin, in which the same His **46** is modified as in TLCK derivative. Much more paradoxical is the fact that for combination of chicken ovomucoid with active trypsin the reactive site arginyl is absolutely required—its removal with carboxypeptidase B from modified inhibitor or blockage with arginyl blocking reagents completely removes the activity of the ovomucoid (Table V). Presumably any theory of trypsin–inhibitor interaction requires that this arginyl side chain combines with trypsin's "substrate combining site." But in TLCK derivative of trypsin the "substrate combining site" is blocked by the lysyl side chain of the reagent. This blockage, which forms the rationale of TLCK action, can be specifically proved since it was shown that both diisopropylphosphoryl trypsin and active trypsin bind benzamide but the TLCK trypsin derivative does not (*265*).

The only comment that can be added to "clear up" this paradox is that TLCK derivative of trypsin (and probably the other inactive derivatives of trysin) are particularily ill-behaved protein derivatives. Petra *et al.* (*266*) have shown that the derivative spontaneously loses tosylamide forming an unknown product. We have observed great variability from lot to lot of TLCK trypsin derivative and erratic extensive polymerization and precipitation. Whether this erratic behavior is related to complex formation with inhibitors is not known.

Another intriguing report of an interaction with inactive enzyme is the weak association of pancreatic trypsin inhibitor (Kunitz) and of bovine colostrum inhibitor with bovine trypsinogen (*266a*). On the other hand, trypsinogen does not form complexes with chicken ovomucoid or

265. E. J. East and C. G. Trowbridge, *ABB* **125,** 344 (1968).
266. P. H. Petra, W. Cohen, and E. N. Shaw, *BBRC* **21,** 612 (1965).
266a. V. Dlouhá and B. Keil, *FEBS Letters* **3,** 137 (1969).

with soybean trypsin inhibitor (Kunitz) (J. Schrode, unpublished results).

IV. Molecular Differences among Inhibitors

A. Analogies and Homologies

In the preceding discussion we have tried to bring out the possible generality of the reactive site model, the probable similarity of mechanisms of interaction of most proteinases with their protein inhibitors [Eq. (9a)], similarity of ratios of certain rate constants (e.g., Sections III,E and III,F), and some regularities in the amino acid sequences near the reactive sites (Section III,B). Insofar as these generalizations are correct, protein proteinase inhibitors are analogous.

Recently, several amino acid sequences of proteinase inhibitors have been determined. A glance at the sequences and a detailed computer analysis (*267*) yield the same conclusion—all trypsin inhibitors whose sequences are known do not form a single homologous class, but they do form several homologous subclasses. Since the number of sequences determined so far is small the number of the homologous subclasses is impossible to estimate, but at least 10 such subclasses are known if unsequenced inhibitors are grouped by amino acid compositions and similarities in physical and inhibitory properties.

Probably the most surprising finding based on the presently known sequences is the complete lack of homology between bovine pancreatic trypsin inhibitor (Kunitz) (Fig. 17) and bovine secretory trypsin inhibitor (Kazal) (Fig. 18). This lack of homology clearly emphasizes the large differences in the site of synthesis, general properties, and probable physiological function of the two inhibitors. Homologous secretory trypsin inhibitors are apparently present in all mammals and are secreted along with zymogens in the pancreatic juice (*249*). Their physiological role is presumably prevention of premature zymogen activation by inhibition of trypsin, a conclusion in accord with their narrow inhibitory specificity (Section IV,B). On the other hand, the pancreatic inhibitor (Kunitz) is present in all bovine tissues and is apparently restricted only to a limited number of ruminants; it has thus far been found in cows and sheep (*42*).

267. M. O. Dayhoff, ed., "Atlas of Protein Sequence and Structure," Vol. 4, p. D144. Natl. Biomed. Res. Found., Silver Spring, Maryland, 1969.

10

Arg-Pro-Asp-Phe-Cys-Leu-Glu-Pro-Pro-Tyr-Thr-Gly-Pro-Cys-Lys-Ala-Arg-Ile -Ile -

Phe-Gln-Thr-Pro-Pro-Asp-Leu-Cys-Gln-Pro-Pro-Gln-Ala -Arg-Gly-Pro-Cys-Lys-Ala-Ala -Leu-Leu-

20 30 40

-Arg-Tyr-Phe-Tyr-Asn-Ala-Lys-Ala-Gly-Leu-Cys-Gln-Thr-Phe-Val -Tyr-Gly-Gly-Cys-Arg-Ala-Lys-

-Arg-Tyr-Phe-Tyr-Asx-Ser-Thr-Ser-Asn-Ala -Cys-Glu-Pro-Phe-Thr-Tyr-Gly-Gly-Cys(Asx,Asx,Asx,

50

-Arg-Asn-Asn-Phe-Lys-Ser-Ala-Glu-Asp-Cys-Met-Arg-Thr-Cys-Gly-Gly-Ala

-Asx,Glx,Glx,Glx,Thr,Thr,Gly,Met,Phe)Cys-Leu -Arg-Ile -Cys-Asx-Pro-Pro-Glx-Glx-Thr-Glx-Lys-Ser

Fig. 17. The amino acid sequences of bovine pancreatic trypsin inhibitor (Kunitz) (upper) and of major component of bovine colostrum inhibitor (lower) (*268*).

It is almost equally surprising that the bovine pancreatic trypsin inhibitor (Kunitz) is homologous to the bovine colostrum inhibitor (Fig. 17) since the distribution of the Kunitz pancreatic inhibitor is so narrow, while the colostrum inhibitor, thus far found in man, cows (*65*), and pigs (*66*), is probably common to all mammals. In these two sequences there are only three partial sequences longer than two amino acids which are identical. One of these is the –Gly–Pro–Cys–Lys–Ala– sequence around the reactive site Lys 15 of pancreatic inhibitor. (The others are Arg 20–Tyr 23 and Tyr 35–Cys 38.) Of particular interest at this location is the sequence –Lys 15–Ala–Arg–Ile 18–, which differs in the colostrum inhibitor by the substitution of –Ala–Leu– for –Arg–Ile. In addition to wondering why Lys 15 and not Arg 18 is the tryptic reactive site of pancreatic inhibitor [cf. –Arg 64–Ile– in soybean inhibitor (Kunitz)], one can speculate that the amino acid penultimate to the reactive site amino terminal is unimportant. It is interesting that all the half-cystine residues can be aligned, but somewhat surprising that the distinctive –Pro–Pro sequences do not align exactly. The observations of Kassal and Chow (*233*) that removal of the three amino terminal residues in pancreatic inhibitor by sequential Edman reaction does not affect activity is supported, or explained, by the considerable differences in these regions of the two inhibitors.

The three secretory inhibitors on which extensive sequence information is available clearly form a homologous class (Fig. 18). With the exception of the Arg 18 ↔ Lys 18 substitution, the reactive site seqences are immutable. However, this substitution shows again that the choice of trypsin susceptible amino acid in the reactive site is relatively unimportant, in agreement with the results obtained with [Lys 64]-soybean inhibitor (Kunitz) (Section III,G). Once again the high mutability of the amino terminal region is expected since there exists a second inhibitor designated pig II, which lacks the four amino terminal residues of pig I but is otherwise identical (*175, 238*). A fully active form lacking the first five amino terminal residues of pig I has also been described (*239*). The occurrence of "isoinhibitors" in pig is similar to the situation in lima

---

268. The sequence of Kunitz inhibitor shown here was determined by Kassell and Laskowski, Sr. (*32–34*) and confirmed by Chauvet and Acher (*41*) and by Dlouhá *et al.* (*38*). Anderer and Hornle (*36*) also confirm it, but substitute Asn 50 for Asp 50 in bovine kallikrein inactivator (Section I,C,1). Disulfide bridges link positions 5–55, 14–38, and 30–51 (*34, 36*). The amino acid analysis of ovine organ inhibitor is identical to that of bovine; therefore, the sequences are presumed to be the same (*108*). Partial bovine colostrum inhibitor sequence was detemined by Čechová *et al.* (*176*).

```
                                       10                                        20
Bovine      Asn-Ile -Leu-Gly-Arg-Glu-Ala-Lys-Cys-Thr-Asn-Glu-Val-Asn-Gly-Cys-Pro-Arg-Ile-Tyr-Asn-Pro-Val-Cys-Gly-Thr-Asp-Gly-
  (Ovine)

Porcine I   Thr-Ser-Pro-Gln-Arg-Glu-Ala-Thr-Cys-Thr-Ser -Glu-Val-Ser -Gly-Cys-Pro-Lys-Ile-Tyr-Asn-Pro-Val-Cys-Gly-Thr-Asp-Gly-

             30                                      40                                      50
           -Val-Thr-Tyr-Ser-Asn-Glu-Cys-Leu-Leu-Cys-Met-Glu-Asn-Lys-Glu-Arg-Gln-Thr-Pro-Val-Leu-Ile-Gln-Lys-Ser-Gly-Pro-Cys
                        (Ala)

           -Ile -Thr-Tyr-Ser-Asn-Glu-Cys-Val -Leu-Cys- Ser -Glu-Asn-Lys-Lys-Arg-Gln-Thr-Pro-Val-Leu-Ile-Gln-Lys-Ser -Gly-Pro-Cys
```

FIG. 18. The amino acid sequences of bovine, ovine, and porcine pancreatic secretory inhibitors (Kazal) (*269*).

beans, and likewise their origin is unknown. We note that the difference between the pig I and pig II inhibitors is consistent with the specificity of the recently described dipeptidylpeptidase cathepsin C (*271*).

Speculation about homology on the basis of amino acid composition alone is of limited value. However, it appears that there are several classes of inhibitors whose sequences would make interesting comparisons.

Apparently all varieties of beans which have been examined contain inhibitors, many of which have similar amino acid compositions. Several groups (*56–58*) have demonstrated the presence of several similar inhibitors in lima beans. Of the four fractions isolated by Jones *et al.* (*57*) one appears to be a precursor for the others (*272*) but the exact relationship among these inhibitors is unknown. The relative amounts of the various components which can be isolated vary with the variety of lima beans (58). It is often surmised on the basis of amino acid compositions (particularly the exceptionally high content of cystine), molecular weights, tendency to associate, etc., that the Bowman–Birk inhibitor from soybeans is homologous with lima bean inhibitors. Inhibitors from several other species of beans and some inhibitors from potatoes (*105, 134*) appear to belong in this class also.

The Kunitz soybean inhibitor appears to be distinct, however, since it has a higher molecular weight (22,500) (*79*), does not associate (*10*), and contains only 4 half-cystine residues (out of a total of 198 residues). The partial sequences of several fragments are shown in Fig. 19.

As emphasized by Feeney and Allison (*9*) avian ovomucoids are probably homologous, although their specificity patterns at first glance appear very different. Because certain of the ovomucoids have multiple

---

269. The bovine sequence was determined by Greene and Bartelt (*174a*), the porcine I by Tschesche *et al.* (*175*; revised, *175a*). Porcine II inhibitor lacks the first four residues of porcine I but is otherwise identical (*175, 238*). The statement (*107*) that ovine differs from bovine only by substitution of Ala 32 for Ser 32 is based on amino acid analyses of tryptic peptides and on homology to the bovine inhibitor. Disulfide bridges link positions 9–38, 16–35, and 24–56 (*270*).

270. L. J. Greene, *Proc. First Intern. Conf. Proteinase Inhibitors,* de Gruyter, Berlin, 1971 (in press).

271. J. K. McDonald, P. X. Callahan, B. B. Zeitman, and S. Ellis, *JBC* **244,** 6199 (1969).

272. From the amino acid analyses of the component inhibitors it appears that components 1, 2, and 4 (77, 76, and 86 residues, respectively) could each be derived from component 3 (93 residues) by proteolytic cleavage, presumably leading to loss of terminal peptides, except that component 2 contains 14 Asp, and component 3 only 13. The authors (*57*) remark that the difference may be within experimental error. Fraction 6 (89 residues) of Haynes and Feeney (*58*) likewise could be a precursor of fraction 4 (84 residues).

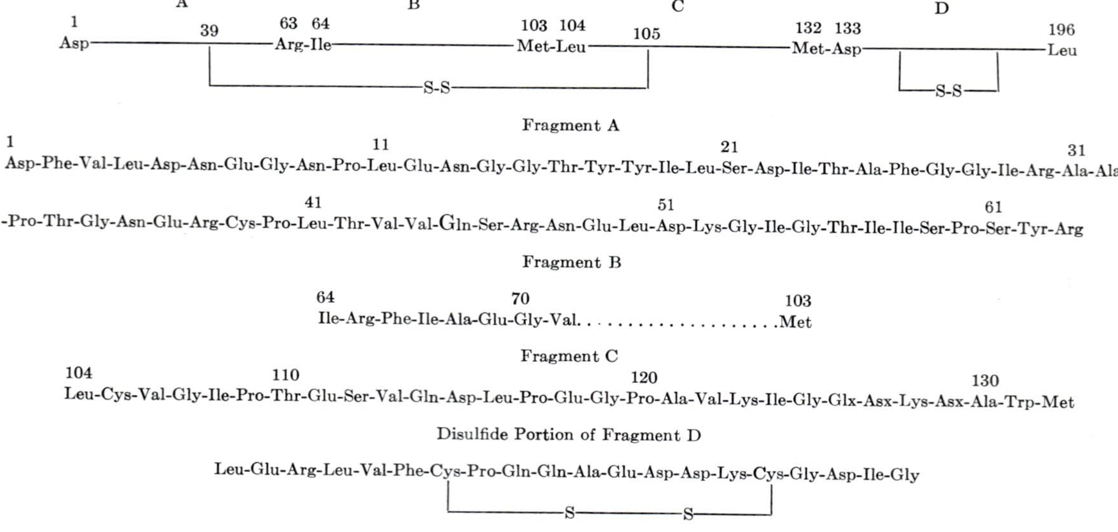

FIG. 19. Alignment of fragments produced by reactive site cleavage and by cyanogen bromide cleavages in soybean trypsin inhibitor (Kunitz) and partial sequences of some of these fragments (*273*).

and independent sites, presumably a great deal could be learned from comparisons of corresponding portions of sequences which in some ovomucoids are active and in others are not. In Section IV,B it is shown that quail ovomucoid inhibits human trypsin, presumably at its normal trypsin site, while the others tested do not. Comparison of sequences at the trypsin sites of these ovomucoids might in part explain some of the puzzling aspects of human trypsin inhibitions (Section IV,B).

It is obvious that several additional classes of inhibitors must exist. We have already emphasized the differences between serum inhibitors and other proteinase inhibitors. The inhibitor from *Ascaris* (Fig. 20) seems to represent a separate class. On the basis of differences in their molecular weights and amino acid composition, ovoinhibitors (*9*) probably form a separate class not closely homologous to ovomucoids. It is rather striking that in at least four cases two or more nonhomologous inhibitors have been isolated from the same tissue of the same organism—Kunitz and Kazal inhibitors from bovine pancreas, Kunitz and Bowman–Birk inhibitors from soybean, ovomucoid and ovoinhibitor from chicken eggs, the potato type I inhibitor (*68, 69*), and the several potato inhibitors isolated by Hochstrasser *et al.* (*105*). This fact and the presence of isoinhibitors in the same tissue appear to emphasize the ubiquity of protenase inhibitors in nature.

B. Specificity of Inhibitors toward Other Enzymes

The specificity pattern of an inhibitor, i.e., the tabulation of enzymes inhibited and not inhibited, is important for the understanding of its physiological function and as a means for classifying new enzymes. These data are also an important input into any detailed molecular theory of proteinase–inhibitor interactions.

At first glance there appears to be a large amount of specificity data already in the literature, but unfortunately the critical evaluation of

---

273. The information in this figure is based on the work of Ikenaka *et al.* (*172, 172a*) except that the sequence surrounding the small, nonessential (*214*) disulfide loop was first determined by Brown *et al.* (*274*). Ozawa and Laskowski, Jr. (*79*) find 64 amino acid residues in fragment A and 198 in the whole molecule. The assignment of COOH terminal leucine here and in Figs. 4 and 14 is based on the paper of Davie and Neurath (*275*). We are unable to confirm this result.

274. J. R. Brown, N. Lerman, and Z. Bohak, *BBRC* **23**, 561 (1966).

275. E. W. Davie and H. Neurath, *JBC* **212**, 507 (1955).

1 10 20
Glu-Ala-Glu-Lys-Cys(Asx,Glx,Glx,Pro,Gly,Trp)Thr-Lys-Gly-Gly-Cys-Glu-Thr-Cys-Gly-Cys-Ala-Glu-Lys-Ile-Val-Pro-Cys-Thr-Arg-Glu-Thr-Lys

40 50 60
-Pro-Asn-Pro-Gln-Cys-Pro-Arg-Lys-Glu-Cys-Cys-Ile-Ala-Ser-Ala-Gly-Phe-Val-Arg-Asp-Ala-Gln-Gly-Asn-Cys-Ile-Lys-Phe-Glu-Asp-Cys-Pro-Lys

FIG. 20. The amino acid sequence of a trypsin inhibitor from *Ascaris lumbricoides* var. suum (*276*). Disulfide bridges link positions 5–28, 16–44, 19–58, 21–38, and 43–64 (*276a*). The reactive site has not been established.

most of it is quite difficult. The reasons most often responsible for this fact have been discussed in Section II,D,3. The specificity tabulation by Feeney *et al.* (*137*) presented in Table IX was recently done and is probably largely correct. However, even here some of the puzzling entries may be subject to reinterpretation. It has for example been reported that human trypsin is in fact a mixture of enzymes (*277*) and some of the weak inhibitions of this preparation may result from inhibition of only some of the components. Likewise, the lima bean inhibitor used is a mixture of inhibitors (Section IV,A) and only certain components of the mixture may be responsible for weak inhibition entries. Far more

TABLE IX
SPECIFICITY PATTERNS OF INHIBITORS[a]

Inhibitor	Enzymes inhibited[b]				
	Bovine trypsin	Human trypsin	Bovine α-chymotrypsin	Human plasmin	Human thrombin
Chicken ovomucoid	++++	−	−	−	−
Quail ovomucoid	++++	++	−	−	−
Turkey ovomucoid	++++	−	+++	−	−
Tinamon ovomucoid	+	−	+++	−	−
Chicken ovoinhibitor	++++	−	+++	−	−
Pancreatic (Kunitz)	++++	++++	++	++++	−
Bovine colostrum	+++	+++	−	+++	−
Bovine (Kazal)	++	−	−	−	−
Porcine (Kazal)	++	−	−	−	−
Soybean (Kunitz)	++++	+	+	+++	−
Soybean (Bowman–Birk)	+	+	++	+	−
Lima bean	++++	++++	+++	+	−
Navy bean	+++	++	+++	+	−
Kidney bean	+++	+	+++	+	−
Pea (black-eyed)	++	++	+++	+	−
Potato type I[c]	+	−	+++	−	−

[a] Several of the entries in this table were determined earlier by other workers and were only confirmed by Feeney *et al.* (*137*). The complete listing of original references would be prohibitively long.

[b] The degree of inhibition is indicated as −, for extremely weak or inactive, and +, + +, + + +, and + + + + for varying degrees, progressing from weak to strong inhibition.

[c] Ryan (*68*).

276. W. Fraefel and R. Acher, *BBA* **154,** 615 (1968).

276a. A. Induni, Ph.D. Thesis, University of Freiburg, Switzerland, 1969. Cited in (*16a*).

277. J. Travis and R. C. Roberts, *Biochemistry* **8,** 2884 (1969), correction: *ibid.* **9,** 1048 (1970).

extensive tabulations of specificity patterns are listed in Vogel *et al.* (*7*). However, in consulting these and other specificity tables the original papers reporting the table entries should be carefully checked for possible errors.

1. *Inhibition of More Than One Enzyme Molecule*

The inhibitions of two enzymes by a single inhibitor may be competitive, cooperative, or independent. Firm evidence is lacking for cooperative inhibition of two or more enzyme molecules by one molecule of inhibitor. Independent and competitive inhibition may involve three molecular possibilities, namely,

(1) Inhibition at distinct, nonoverlapping sites
(2) Inhibition at distinct, overlapping sites
(3) Inhibition at the same reactive site

As pointed out before, case (1) would give rise to independent binding and cases (2) and (3) would give rise to competitive binding and therefore cannot be distinguished by inhibition studies alone.

2. *Inhibition at Distinct, Nonoverlapping Reactive Sites*

Wu and Laskowski, Sr. (*270*) observed that the trypsin–soybean trypsin inhibitor (Kunitz) complex still inhibits chymotrypsin B, although more weakly than the free inhibitor. Since trypsin is inhibited by soybean trypsin inhibitor (Kunitz) far more strongly than chymotrypsin B, this result is unlikely to be caused by partial displacement of the trypsin and strongly suggests that at least one (if there is more than one) of the chymotrypsin B sites on the inhibitor molecule is distinct from and does not significantly overlap the trypsin site.

Rhodes *et al.* (*94*), in an extensive study of inhibition specificity of avian ovomucoids, have shown that several of them inhibit both trypsin and chymotrypsin (Table IX) and that in those cases where both enzymes are inhibited the inhibition is independent (for example, Fig. 21). The independent inhibition was shown not only by inhibition curves but also by the observation of a mixed trypsin–chymotrypsin–duck ovomucoid complex on paper electrophoresis as a single band. Rhodes *et al.* coined the very descriptive term *multiheaded* to describe the inhibitors with several nonoverlapping independent sites.

Chicken ovoinhibitor, a glycoprotein from egg white which is quite distinct from chicken ovomucoid, has been shown to inhibit 2 moles of trypsin and 2 moles of chymotrypsin simultaneously or separately and independently (*131*). This protein apparently consists of a single poly-

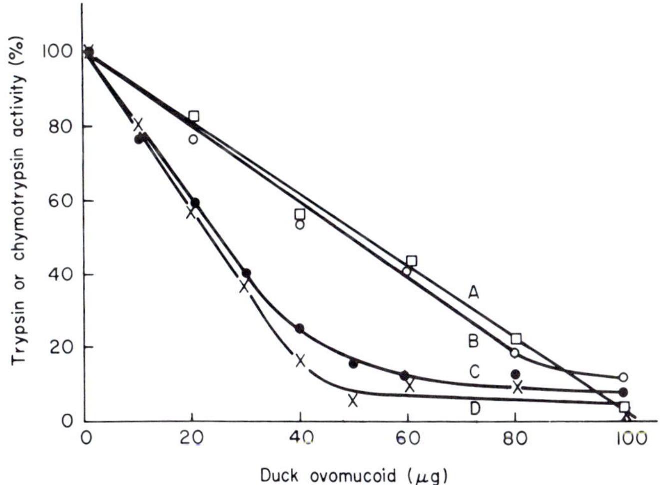

FIG. 21. Titrations of trypsin and α-chymotrypsin with duck ovomucoid in the presence and absence of saturating quantities of chymotrypsin or trypsin, respectively (*94*): A, chymotrypsin + 4 trypsin (□); B, chymotrypsin (○); C, trypsin (●); and D, trypsin + 4 chymotrypsin (×).

peptide chain with a weight-average molecular weight of 49,000 (of which about 2300–4800 daltons is due to carbohydrate) since the molecular weight is the same in mild aqueous solution and in strong denaturing and reducing media (*141*).

Ryan and Clary (*278*) have shown that lima bean inhibitor (now known to be a mixture of several closely related components—isoinhibitors) inhibits turkey trypsin and chicken chymotrypsin independently. Haynes and Feeney (*58*) observed independent inhibition of bovine trypsin and of bovine α-chymotrypsin in several of the isolated components obtained from lima bean inhibitor preparations. These authors completed the proof that at least component 6 of this mixture (and probably all of the components) were "double-headed" by showing that upon Sephadex chromatography of lima bean component 6 (*58*) in the presence of excess trypsin the chymotrypsin inhibitory activity eluted only with the highest molecular weight material, i.e., the trypsin–lima bean inhibitor complex, and that no chymotrypsin inhibitory activity was present at the previously established elution position of the free inhibitor. Similar analysis by Pusztai (*210*) shows that kidney bean inhibitor, which inhibits both trypsin and α-chymotrypsin independently, is also double-headed and not a mixture.

An interesting example of a double-headed inhibitor both of whose re-

278. C. A. Ryan and J. J. Clary, *ABB* **108**, 169 (1964).

active sites bind the same enzyme is the mung bean inhibitor (*71–73*). This inhibitor has been shown to be double-headed for trypsin not only by the demonstration that 2 moles of trypsin are inhibited by a mole of inhibitor but also by isolation of crystallizable 1:1 and 2:1 trypsin: inhibitor complexes. The 1:1 complex is a trypsin inhibitor, the 2:1 complex is not.

The most thoroughly characterized examples of double-headed trypsin–chymotrypsin inhibitors are the Bowman–Birk and lima bean inhibitors (Section III,H). The cleavage of the tryptic reactive site of Bowman–Birk inhibitor followed by release of the reactive site lysyl by carboxypeptidase B converts the double-headed inhibitor into a single-headed α-chymotrypsin inhibitor, an impressive feat of rationally changing the inhibitor specificity (*150*). Stevens (*177*) converted the lima bean inhibitor into a single-headed α-chymotrypsin inhibitor by the method described above and (in a separate experiment) into a single-headed trypsin inhibitor by cleaving the chymotryptic reactive site with chymotrypsin and eliminating it by subsequent treatment with carboxypeptidase A. A similar feat has been achieved by chemical modification. Reaction of the double-headed trypsin–chymotrypsin inhibitors turkey ovomucoid (*185*) and lima bean inhibitor (*180*) with trinitrobenzenesulfonic acid [both inhibitors are lysyl trypsin inhibitors (Table V)] abolished their tryptic inhibitory activity by substituting the reactive site lysyl residues but left their chymotrypsin inhibitory activity entirely intact. It appears that a similar specificity restriction (arginyl-modifying reagents should be used for arginyl inhibitors) could be carried out on any trypsin–chymotrypsin double-headed inhibitor.

The frequent presence of more than one reactive site (and as many as four for ovoinhibitor) on the same single chain inhibitor molecule is surprising since most proteins and enzymes have only one active site per peptide chain. The fact that so many sites occur here suggests to us (as does a good deal of other evidence) that relatively few amino acid residues are involved in forming the reactive site but that for such reactive site to be operative requires the great rigidity and stability of the whole inhibitor molecule.

3. *Overlapping Independent Reactive Sites*

If the existence and locations of multiple reactive sites in a single inhibitor were determined more by accidents of primary structure than by physiological function, it seems that overlapping reactive sites should be quite common. If multiheaded inhibitors arise largely because of a physiological necessity, overlapping sites, for the same or different enzymes, should be quite rare. At the moment discussion of these possibilities must consist mainly of speculation.

The meager evidence suggests that occurrences of overlapping sites are rare, but this may simply arise from the difficulty of detecting this situation with the techniques heretofore applied to the problem. Mutually competitive inhibition of two enzymes implies either overlapping or common sites. Cleavage would show which is the case, but cleavage in this situation has been used only once—in the study of trypsin and cocoonase inhibition by soybean trypsin inhibitor (Kunitz) (*227*).

The two most likely examples of overlapping reactive sites are the sites for inhibition of trypsin and chymotrypsin on pancreatic trypsin inhibitor (Kunitz) and on component III of Japanese radish seed (*Raphanus Sativus*) inhibitor (*279*).

Wu and Laskowski, Sr. (*280*) have shown that pancreatic trypsin inhibitor (Kunitz) inhibits trypsin and chymotrypsin in a competitive manner. That the reactive sites are different can be argued on the basis of different specificity of the two enzymes and of the frequent demonstration (see nonoverlapping sites) that in many other inhibitors the reactive sites for trypsin and chymotrypsin are distinct. Reaction of turkey ovomucoid with acetic anhydride or potassium cyanate (carbamylation) (*185*) or of lima bean inhibitor with trinitrobenzenesulfonic acid (*180*) abolished the tryptic inhibitory activities of these double-headed trypsin–chymotrypsin inhibitors by substitution of their tryptic reactive site lysyls (Table V), but left intact their chymotryptic inhibitory activities.

The only more direct evidence for different reactive sites is from chemical modifications. Sevilla and Rigbi (*183*) argued that the sites are distinct because acetylation completely removes trypsin inhibitory activity (presumably by eliminating tryptic susceptibility of Lys 15–Ala tryptic reactive site) while the antichymotryptic activity, though diminished, remains. Many chemical modifications either leave both activities unaffected or destroy both activities [e.g., maleylation of Lys 15 (*161*) or reduction of Cys 14–Cys 38 disulfide bridge followed by blockage with *S*-carboxymethylcysteine (*213*)]. Both of the results could be explained by introduction of negative charges near the reactive site for both enzymes. Several other modifications (*213*) destroy chymotryptic but not tryptic inhibitory activity. On the basis of an extensive analysis of chemical modification, Fritz *et al.* (*161*) argued that trypsin and chymotrypsin share the same reactive region on the inhibitor molecule

279. T. Ogawa, T. Higasa, and T. Hata, *Agr. Biol. Chem.* (*Tokyo*) **32,** 484 (1968).
280. F. C. Wu and M. Laskowski, Sr., *JBC* **213,** 609 (1955).

but not necessarily the same reactive site. All of the evidence is consistent with two separate overlapping reactive sites but it probably could also be reinterpreted to argue for the same site. Only the chemical identification of the chymotryptic reactive site will settle the question.

Competitive inhibition of trypsin and chymotrypsin by Japanese radish seed inhibitor was clearly demonstrated by Ogawa *et al.* (*279*). A suggestion was also made that the two enzymes are inhibited on different reactive sites, but the evidence for this is just preliminary.

4. *Inhibition at the Same Reactive Site*

Only one clearly documented example of this situation has been published so far. This is the case of the interaction of bovine trypsin and of the trypsinlike enzyme from silk moths—cocoonase—with soybean trypsin inhibitor of Kunitz. It has been shown that when the inhibitor is incubated with either enzyme the same bond, Arg 64–Ile, is cleaved (*281*) and that while the virgin inhibitor inhibits either enzyme, the des-64-arginine-modified inhibitor inhibits neither one (*231*). This is very strong although not totally convincing proof for identity of the reactive site involved. An additional experiment—kinetic control dissociation of a complex made from cocoonase and modified inhibitor—would clearly be desirable. If the experiment showed resynthesis of the Arg 64–Ile bond, as is the case with the dissociation of trypsin–inhibitor complex, we feel that the proof would be compelling.

However, in general a great many cases of inhibition of more than one enzyme by a proteinase inhibitor have led several authors to a more or less tacit assumption that the same reactive site is involved. This is especially so when competitive behavior between the enzymes was demonstrated or when the two enzymes shared the same general specificity, e.g., trypsins from different species. If one makes the reasonable assumption that the number of reactive sites in any inhibitor is quite limited, this surmise seems correct.

A strong supportive argument for the identity of trypsin and plasmin reactive sites has been presented by Fritz *et al.* (*179*). A large number of trypsin inhibitors were subjected to maleylation and to reaction with

281. The identity of the cleaved bond was demonstrated by chemical analysis of resultant fragments. However, a much simpler and more elegant proof was also used. It was shown by disc gel electrophoresis that catalytic amounts of cocoonase at pH 6 convert an equilibrium mixture of virgin and modified inhibitor obtained at pH 3 with trypsin to the pH 6 equilibrum mixture (Fig. 7) by resynthesizing the cleaved bond in the modified inhibitor. This can be the case only if trypsin and cocoonase actions are directed toward the same peptide bond in the inhibitor molecule.

2,3-butadione in order to divide them into lysyl or arginyl inhibitors. Several, but not all, of the inhibitors studied were also plasmin inhibitors—some of the lysyl type and others of the arginyl type. In all cases studied the loss of trypsin inhibitory activity upon chemical substitution was paralleled by loss of plasmin inhibitory activity. If this were a single case, it would only suggest that an inhibitor which is, say, a lysyl trypsin inhibitor is also a lysyl plasmin inhibitor but the reactive sites are not necessarily the same. In fact, however, the lack of a single exception appears to be a very strong indication of identity of plasmin and trypsin reactive sites in all those inhibitors which inhibit both enzymes.

If it is true that enzymes of similar specificities are frequently inhibited at the same reactive site, then it immediately becomes apparent that specificity is a function of both partners, since inhibitors variously select among serine proteinases such as trypsin, thrombin, plasmin, urokinase, cocoonase, and even among the same enzyme from different species (Table IX). It will be a matter of great interest to decide whether a given reactive site may have any specificity pattern whatever or whether in fact there exists only a limited number of possible patterns. Should the question of specificity patterns be resolved in the latter way, then the next question will arise of whether the specificity patterns can be related to amino acid sequences, especially to the sequences near the reactive site.

Such speculations gain partial support from the data already in the literature (e.g., data of Table IX). Homologous inhibitors have similar specificity patterns; for example, the patterns of homologous (Section IV,A) porcine and bovine secretory inhibitors (Kazal) are identical. This comparison incidentally suggests that Lys ↔ Arg interchange in the reactive site has no important effect on site specificity. The fact that neither bovine nor porcine secretory pancreatic inhibitors inhibits human trypsin is startling. Pancreatic secretory inhibitors have been isolated from many mammals including man (*45, 100*), and it is believed that they are present in all mammals and are homologous. Their physiological function is believed to be inhibition of small amounts of prematurely formed trypsin in the pancreatic gland in order to prevent premature activation of pancreatic zymogens (*249*). If this is the case then human secretory trypsin inhibitor must be able to inhibit human trypsin [it is known to inhibit bovine trypsin (*100*)] yet be homologous to the porcine and bovine inhibitors which do not. Clearly, if the facts listed above are correct, determination of the sequence of human secretory inhibitor may shed a great deal of light on the specificity question.

The general lack of inhibition of thrombin by all of the inhibitors listed in Table IX is not too surprising since it is well known that while thrombin shares with trypsin the lysine–arginine specificity it has some additional and very restrictive specificity requirements. Therefore, only very few of trypsin-susceptible bonds in proteins and large peptides are hydrolyzed by thrombin (*282*). On the other hand, this type of reasoning makes it very surprising that the specific thrombin inhibitor hirudin does not inhibit trypsin (*283*) since in general all thrombin-susceptible bonds are also hydrolyzed by trypsin.

A benzoylarginine methyl ester hydrolyzing component was recently isolated (*142, 143*) from pronase, a mixture of proteolytic enzymes from *S. griseus*. This bacterial enzyme was temporarily named *pronase trypsin* (*142*). Pronase trypsin is inhibited by all bovine trypsin inhibitors which were tested—soybean inhibitors (Kunitz) and (Bowman–Birk), lima bean inhibitor, and chicken ovomucoid. It is truly striking that chicken ovomucoid inhibits trypsin from both bacteria and cows but inhibits neither human trypsin nor human plasmin (Table IX).

In a recent paper, Bigler and Feeney (*159*) reported on the inhibition of subtilisin by several trypsin inhibitors. The salient results are that subtilisin (either Novo or BPN′) is inhibited only by those inhibitors which can also inhibit bovine α-chymotrypsin, although not all inhibitors which can inhibit α-chymotrypsin inhibit subtilisin. If the surmises that follow are correct this is an indication of specificity patterns on the chymotryptic sites. In all cases tested where both α-chymotrypsin and subtilisin were inhibited, the inhibition was competitive between these two enzymes while there was no competition between the inhibition of trypsin and either chymotrypsin or subtilisin. These results prove that subtilisin and chymotrypsin are inhibited either at the same reactive site or at two overlapping reactive sites. While no data are available to make a definite distinction between these two alternatives, the bias is clearly in favor of the same reactive site. If this is so the reactive site specificity patterns of proteinase inhibitors are indeed intriguing. In some cases general specificity (e.g., trypsin and plasmin) and this discrimination is even carried to the same enzyme of different mammalian species (e.g., bovine and human trypsins), but in other cases there is no discrimination between two enzymes which, though similar in properties, are only related by analogy rather than by homology (*284*).

A matter of considerable interest here is the inhibition of cysteinyl proteinases such as papain, chymopapain, and bromelain by proteinase inhibitors. Several inhibitors of these enzymes have been described (*7*).

282. B. Blomback, *Ann. N. Y. Acad. Sci.* **146,** 364 (1968).
283. H. Fritz, K.-H. Oppitz, M. Gebhardt, I. Oppitz, E. Werle, and R. Marx, *Z. Physiol. Chem.* **350,** 91 (1969).
284. S. A. Olaitin, R. J. DeLange, and E. L. Smith, *JBC* **243,** 5296 (1968).

Particularly clear-cut is the isolation of a specific papain inhibitor from egg white by Fossum and Whitaker (*162*). This inhibitor inhibits neither trypsin nor α-chymotrypsin. There are many reports in the literature that partially purified inhibitory extracts from various animal and plant tissues inhibit both some serine proteinases and also some cysteinyl proteinases. However, to the best of our knowledge, clear-cut cases of inhibition of trypsin and of papain by the same highly purified inhibitor are very rare. It, therefore, seems probable that these two enzymes are generally inhibited on different reactive sites. It should be mentioned that there is at present no evidence whatever bearing on the question of the inhibition mechanism of cysteinyl proteases by protein inhibitors.

Many workers concerned with the physiological role of proteinase inhibitors are concerned with the question of whether or not plant inhibitors, which appear to be specific for animal proteinases, inhibit the endogenous proteolytic enzymes of the plant (e.g., references *7, 8, 285*). Surprisingly, the balance of the evidence appears to be that they do not. However, this conclusion may simply reflect the improper choice of proteolytic enzymes since the ones which are likely to be inhibited may be present in most plant tissues as inactive enzyme–inhibitor complexes. Shain and Mayer (*285a*) found in germinating but not in dry lettuce seeds a trypsin-like enzyme. This enzyme was inhibited by a trypsin inhibitor which could be isolated from dry but not from germinating seeds. Polanowski (*286*) isolated from rye a protein which upon dissociation yields rye proteinase and a rye inhibitor. The inhibitor inhibits both the rye proteinase and bovine trypsin. Hurain, a trypsinlike enzyme from Venezuelan jubilee tree (*287*), is frequently inhibited by trypsin inhibitors (*7*).

ACKNOWLEDGMENT

The work from our laboratory reported here was supported by grants GM 10831 and GM 11812 from the Institute of General Medical Sciences, National Institutes of Health. One of us (R.W.S.) also acknowledges support as a National Science Foundation Predoctoral Fellow.

285. J. Mikola and E. M. Suolinna, *European J. Biochem.* **9,** 555 (1969).
285a. Y. Shain and A. M. Mayer, *Physiol. Plant.* **18,** 853 (1965).
286. A. Polanowski, *Acta Biochim. Polon.* **14,** 389 (1967).
287. D. S. Seidl and W. G. Jaffe, *Enzymologia* **33,** 313 (1967).

12

Cathepsins and Kinin-Forming and -Destroying Enzymes

LOWELL M. GREENBAUM

I. Introduction

The preceding edition of this treatise reviewed much of the earlier important information about cathepsins (*1*). This review will serve to give an insight into our current understanding of these complex enzymes without repeating earlier reported details. Included also is a brief discussion of kinin-forming and -destroying enzymes which will serve to direct the reader to some of the possible important functions of the cathepsins in terms of the metabolism of biologically active peptides.

1. J. S. Fruton, "The Enzymes," 2nd ed., Vol. 4, p. 233, 1960.

While there is still much unknown about catheptic enzymes it would appear that two of these enzymes, cathepsins D and E, are the major catheptic endopeptidases especially in the reticuloendothelial system. Most other cathepsins seem to have only exopeptidase activity in that they attack either the carboxyl terminus (catheptic carboxypeptidases A and B; cathepsin A) or the amino terminus (cathepsin C) of proteins. Cathepsin B, which has been reported to have endopeptidase activity in that it activates trypsinogen to trypsin (*2*) at acid pH, should be investigated further with other protein substrates to help in the understanding of its role as a protease or peptidase.

II. Catheptic Endopeptidases

A. Cathepsin D

Cathepsin D was named for a series of proteases found in beef spleen which did not hydrolyze the usual synthetic substrates acted upon by other cathepsins (*3*). The enzyme is considered to be the same as the hemoglobin-splitting enzyme prepared by Anson (*4*). The purification of this enzyme from rabbit spleen has also been carried out (*5*). The beef enzyme was found to hydrolyze the B chain of insulin at five different bonds (*3*), while the rabbit enzyme is more extensive in its actions on this substrate (*6, 7*). Its presence in a variety of white cells such as the polymorphonuclear leukocytes and lymphocytes has been noted (*5*). The activity of cathepsin D on a heptapeptide sequence of the B chain of insulin has been studied extensively (*8, 9*). As seen in Fig. 1, the hydrolysis of the peptide occurs at two bonds. Pepsin's action is similar; however, on a synthetic hexapeptide cathepsin D hydrolyzes only the Phe–Leu bond while pepsin hydrolyzes three bonds. The hexapeptide has been used as a substrate during purification

2. L. M. Greenbaum, A. Hirshkowitz, and I. Shoichet, *JBC* **234,** 2885 (1959).
3. E. Press, R. R. Porter, and J. Cebra, *BJ* **74,** 501 (1960).
4. M. L. Anson, *J. Gen. Physiol.* **23,** 695 (1940).
5. C. Lapresle and T. Webb, *BJ* **84,** 455 (1962).
6. C. Lapresle and T. Webb, *BJ* **76,** 538 (1960).
7. H. Rangel and C. Lapresle, *BBA* **128,** 372 (1966).
8. H. Keilova, K. Bláha, and B. Keil, *European J. Biochem.* **4,** 442 (1968).
9. H. Keilova, K. Bláha, and B. Keil, *Collection Czech. Chem. Commun.* **33,** 131 (1968).

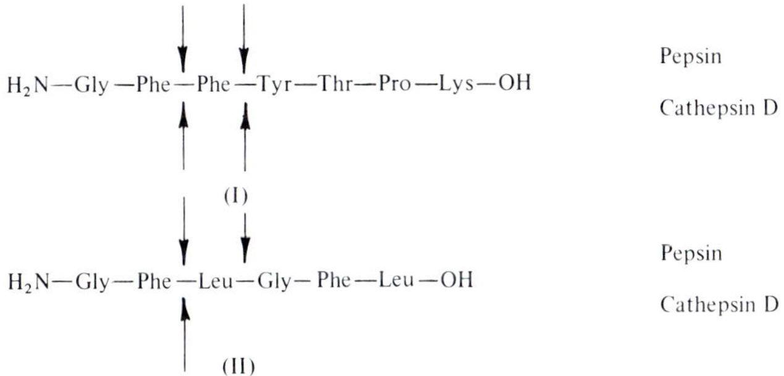

FIG. 1. Comparison of the specificity of cathepsin D and pepsin [after Keilova *et al.* (*8, 9*)].

and to indicate the effect of steric factors on the activity of the enzyme (*10*).

Properties of the enzyme: No known activators are required. No specific inhibitors have been found. Diisopropylfluorophosphate (DFP) does not inhibit the enzyme. The molecular weight has been determined as 58,000 and an amino acid analysis given (*8, 9*). Glycine has been found to be the N-terminal amino acid (*3*). The enzyme was found to be unstable at 60° for 15 min (*3*). Other properties of the enzyme may be found in Table I.

B. CATHEPSIN E

Cathepsin E is a protease in the reticuloendothelial system which may be differentiated from the D enzyme on the basis of the rate at which various protein substrates are attacked and its pH optimum (*5*) (see Table I).

While cathepsin D makes up most of the acid protease activity of spleen, cathepsin E is the more active enzyme in polymorphonuclear leukocytes, and macrophages (*5*). In this regard it may be the enzyme in these cells responsible for the formation of kinins (see below). Its action on the B chain of insulin has been studied in detail (*6, 7*).

Properties of the enzyme: The enzyme has been found to be unstable below pH 4 and above 80°. A molecular weight of 305,000 ± 40,000 has been proposed (*11*). It has also been noted that cathepsin E may be

10. H. Keilova, O. Markovic, and B. Keil, *Collection Czech. Chem. Commun.* **34,** 2154 (1969).

11. V. Turk, I. Kregar, and D. Lebez, *Enzymologia* **34,** 89 (1968).

TABLE I
SOME COMPARISONS OF CATHEPSINS D AND E

	Cathepsin D[a]	Cathepsin E
Rate of protein digestion	Hb > HSA or BSA	BSA > Hb
pH optimum, HSA	3–4.5	2.5
Activity in the reticuloendothelial system	Spleen +++ PMN + Macrophages + Lymphocytes +	Spleen + PMN +++ Macrophages + Lymphocytes ±
Molecular weight (approx.)	58,000	305,000
Inhibitors and activators	None	None

[a] Hb = acid denatured hemoglobin
HSA or BSA = acid denatured human or bovine serum albumin
PMN = polymorphonuclear leukocytes
+ = relative activity

converted into the smaller D enzyme during low temperature storage (*12*). This observation should be pursued further in that it may clarify the formation and relationships of these endopeptidases.

C. CATHEPSIN B

This enzyme which is present in most tissues is characterized by its ability to hydrolyze at acid pH the amide (or ester) bonds involving α-*N*-acylated L-arginine or L-lysine (*13*). Thus, while benzoyl-L-argininamide (BAA) or its L-lysine analog is hydrolyzed to the corresponding benzoyl-L-amino acid and ammonia, L-argininamide and L-arginine ethyl ester are not attacked. This is in contrast to trypsin which hydrolyzes these latter substrates. The enzyme requires thiol activation (see below) for activity. The enzyme has been shown to activate trypsinogen to trypsin *in vitro* and *in vivo* and thus would appear to have endopeptidase activity (*2*, *14*). At neutral pH the enzyme catalyzes transamidation reactions with benzoyl-L-argininamide and a variety of peptides (*15*).

A variation of cathepsin B known as cathepsin B′ has been separated

12. V. Turk, I. Kregar, F. Gubensek, and D. Lebez, *Enzymologia* **36,** 182 (1969).
13. L. M. Greenbaum and J. S. Fruton, *JBC* **226,** 173 (1957).
14. L. M. Greenbaum and A. Hirshkowitz, *Proc. Soc. Exptl. Biol. Med.* **107,** 74 (1961).
15. R. B. Johnston, M. J. Mycek, and J. S. Fruton, *JBC* **187,** 205 (1950).

from cathepsin B by Otto (*16*). This enzyme has a lower molecular weight than cathepsin B. It has kinin-forming activity (*16a*).

Properties of the enzyme: The molecular weight of cathepsin B has been estimated at 52,000 (*16*). Cathepsin B is activated by β-mercapto-ethylamine, cysteine, 2,3-dimercaptopropanol, and glutathione, listed in decreasing order of effectiveness. KCN does not activate the enzyme. The enzyme is completely inhibited by 0.001 *M* iodoacetamide. It is not inhibited by DFP, trypsin soybean inhibitor, or *N*-ethylmaleimide. Preparation of the enzyme from spleen or liver has been reported (*13, 16*). The pH optima of the enzyme depend upon the substrate and buffer used and varies from 4.0 to 6.5.

III. Catheptic Exopeptidases

A. Cathepsin C

Cathepsin C was first described in detail by Gutmann and Fruton (*17*). In an important series of investigations by Fruton and his collaborators, it was indicated that cathepsin C seemed to be similar to chymotrypsin since both enzymes catalyzed the splitting of substrates resembling Gly–Phe–NH_2 although cathepsin C has a thiol cofactor requirement. A good deal of information has been accumulated as to the substrates hydrolyzed as well as the requirements for transpeptidation (*1*). More recent work indicates that cathepsin C is a dipeptidyl-transferase (*18*) or dipeptidyl aminopeptidase (*19*) in that it cleaves dipeptides out of polypeptides with certain specificities from the amino-terminal position or proteins. The molecular weight of the purified beef spleen enzyme is 210,000 and can be dissociated into subunits in the presence of *p*-mercuribenzoate or sodium dodecyl sulfate (*17*). Planta and Gruber (*20*) showed that a dipeptide Ser–Tyr could be cleaved from corticotrophin by cathepsin C. An important finding was made by McDonald *et al.* (*21*) when it was discovered that cathepsin C is a Cl^-

16. V. K. Otto, *Z. Physiol. Chem.* **348,** 1449 (1969).
16a. Y. Okuda and K. Yamafuji, personal communication.
17. H. R. Gutmann and J. S. Fruton, *JBC* **174,** 851 (1948).
18. R. M. Metrione, A. G. Neves, and J. S. Fruton, *Biochemistry* **5,** 1597 (1966).
19. J. K. McDonald, B. B. Zeitman, T. J. Reilly, and S. Ellis, *JBC* **244,** 2693 (1969).
20. R. J. Planta and M. Gruber, *BBA* **89,** 503 (1964).
21. J. K. McDonald, T. J. Reilly, B. B. Zeitman, and S. Ellis, *BBRC* **24,** 771 (1966).

requiring enzyme in addition to its thiol requirement. In the presence of Cl^- and thiol groups these workers showed that cathepsin C, whether derived from rat liver or bovine spleen, could cleave five dipeptides from the amino-terminal position of β-corticotrophin in sequence (*19*). Glucagon, angiotensin II amide, and gastrin also act as substrates for this enzyme. McDonald *et al.* also found that the halide requirements for cathepsin C were: $Cl^- > Br^- > I^- > F^-$. The current favored method of preparation of cathepsin C is by the method of Metrione and Fruton (*18*).

Other properties: The relative rates at which a variety of peptides and proteins are hydrolyzed by cathepsin C as well as recent findings on its ability to polymerize peptides may be found in papers by McDonald *et al.* (*19*) and Metrione and Fruton (*18*). The presence of cathepsin C in lysosomes has been noted (*22*).

B. Catheptic Carboxypeptidases A and B

Fruton *et al.* (*23*) first demonstrated the presence of a cysteine-activated enzyme in spleen which hydrolyzed CBZ-glycyl-L-phenylalanine (where CBZ stands for carbobenzoxy), a substrate now recognized to be hydrolyzed by pancreatic carboxypeptidase A. Fruton *et al.* termed this enzyme cathepsin IV. Greenbaum and Sherman (*24*) investigated this and related activities in detail after 100 times purification from beef spleen. They demonstrated that the cathepsin IV activity was indeed a catheptic carboxypeptidase "A" type of enzyme in that it hydrolyzed a series of synthetic peptide substrates at the COOH-terminal positions containing aromatic or neutral amino acids at acid pH. A second activity was also noted which was termed catheptic carboxypeptidase B. This activity, which is also optimum at acid pH, requires a much higher concentration of thiol and has a slightly higher pH optimum than does the catheptic carboxypeptidase A enzyme.

The nonapeptide, bradykinin, has been used as a substrate of known sequence for these enzymes, and its use has confirmed the specificity determined by smaller synthetic substrates (*25*). It should be noted that catheptic carboxypeptidases A and B, although considerably purified, have not as yet been purified to homogeneity.

22. J. M. W. Bouma and M. Gruber, *BBA* **113,** 350 (1966).
23. J. S. Fruton, G. W. Irving, and M. Bergmann, *JBC* **138,** 249 (1941); **141** 763 (1941).
24. L. M. Greenbaum and R. Sherman, *JBC* **237,** 1082 (1962).
25. L. M. Greenbaum and K. Yamafuji, *Life Sci.* **4,** 657 (1965); *Brit. J. Pharmacol.* **27,** 230 (1966).

C. Cathepsin A

Cathepsin A is an intracellular protease which, like catheptic carboxypeptidase A, is an enzyme that originally was found to hydrolyze CBZ-Glu-L-tyrosine and is now believed to be an exopeptidase (*26*). Unlike catheptic carboxypeptidase A, however, it has no requirement for thiols.

The enzyme obtained from muscle and spleen has been shown to attack glucagon at the carboxyl-terminal position to remove the five terminal amino acids (*26*). In the absence of catheptic endopeptidases such as cathepsin D, little or no proteolysis occurs with substrates such as hemoglobin. These results and the finding that catheptic endopeptidases and cathepsin A are present in rat-kidney lysosomes (*27, 28*) have led to the suggestion that cathepsin A acts synergistically with cathepsin D in the hydrolysis of protein substrates.

Properties of the enzyme: A summary of these may be found in papers by Iodice *et al.* (*26, 27*) and in Table II. The enzyme from chicken muscle is very stable when stored frozen. CBZ-Gly-L-phenylalanine is considered to be the best synthetic substrate.

TABLE II
Summary of Catheptic Carboxypeptidases

	Catheptic carboxypeptidase A	Catheptic carboxypeptidase B	Cathepsin A
Typical substrate	CBZ-Glu-L-tyrosine	Hippuryl-L-arginine	CBZ-Gly-L-phenylalanine
pH optima	3.7	4.0	5.0
Cofactor	0.008 *M* cysteine (max)	0.04 *M* cysteine (max)	None
Approx. K_m	2.2×10^{-3} *M*		
Inhibitors	Iodoacetic acid (0.001 *M*) Zn^{2+} Indole-3-acetic acid (0.01 *M*) Indole-3-proprionic acid (0.01*M*)	Iodoacetic acid	None
Distribution	Probably found in most tissues although only spleen and liver have been used as sources for purification of catheptic carboxypeptidases A and B. Cathepsin A is best purified from muscle		

26. A. A. Iodice, *ABB* **121,** 241 (1967).
27. A. A. Iodice, V. Leong, and I. M. Weinstock, *ABB* **117,** 477 (1966).
28. S. Shibko and A. L. Tappel, *BJ* **95,** 731 (1965).

A survey of acid-carboxypeptidase activity in human tissues has been made (*29*).

IV. Kinin-Forming Enzymes (Kininogenases)

A. Kallikreins

These are a group of enzymes present in activatable forms in plasma and various glandular organs such as pancreas and salivary glands as well as in lymph and urine. The major action of these enzymes is to liberate kinins such as bradykinin (Arg–Pro–Pro–Gly–Phe–Ser–Pro–Phe–Arg) or kallidin (Lys-bradykinin) from α-2-globulins in plasma and lymph known as kininogens. The kinins are potent biologically active peptides. The term kallikrein is derived from the Greek *kallikreis* meaning pancreas, the organ from which the enzyme was originally extracted (*30*).

Properties of the enzymes: Most kallikreins have little protease activity on substrates such as casein but have esterase activity on esters of benzoyl-L-arginine such as BAEE or BAME or on TAME. They do not hydrolyze the amides of these substrates (*31*). Kallikreins from different sources differ in their properties. The reader is referred to the excellent review by Schachter (*32*) for further details of these differences. The molecular weights of pure kallikreins from pancreas, urine, and the submaxillary gland range from 33,000 to 36,000. Plasma kallikrein has a molecular weight of 97,000. Trypsin inhibitors such as soybean, lima bean, bovine pancreatic (Kunitz), and bovine lung (commercially known as Trasylol) have inhibitory actions on the kallikreins (see Chapter II by Laskowski and Sealock, this volume). The bovine lung inhibitor, which is the same as the Kunitz inhibitors, inhibits to some extent almost all kallikreins. It is, however, not particularly potent on plasma kallikrein. Soybean inhibitor, however, is very active against the plasma enzyme (*33, 34*).

Urinary, pancreatic and salivary kallikreins have been purified and their amino acid composition and molecular weight recorded (*34, 35*).

29. U. Stein and D. Platt, *Klin. Wochschr.* **46,** 1145 (1968).
30. H. Kraut, E. K. Frey, and E. Bauer, *Z. Physiol. Chem.* **175,** 97 (1928).
31. M. E. Webster and J. B. Pierce, *Proc. Soc. Exptl. Biol. Med.* **107,** 186 (1961).
32. M. Schachter, *Physiol. Rev.* **49,** 509 (1969).
33. N. Back and R. Steger, *Federation Proc.* **27,** 96 (1968).
34. E. K. Frey, H. Kraut, and E. Werle, "Das Kallikrein, Kinin System und Seine Inhibitoren." Enke, Stuttgart, 1968.
35. H. Moriya, A. Kato, and H. Fukushima, *Biochem. Pharmacol.* **18,** 549 (1969).

Plasma kallikrein has been highly purified (*36, 37*). It is claimed to exist in at least three different forms (*37*). A study on agents such as Hageman factor that activate the precursor of plasma kallikrein has been carried out (*38, 39*).

B. Catheptic Kininogenases

Studies have shown that tissues such as spleen and white cells contain kinin-forming enzymes (*25, 40*) which are active at about pH 4. In view of the knowledge that cathepsins D and E are contained in white cells and in spleen (Table I) it is reasonable to assume that these enzymes are responsible for the kinin-forming activity. Since white cells are believed to provide "extra-plasma" sources of kinins at inflammatory sites (*40*), the role of these cathepsins in inflammation becomes most important.

V. Kinin-Destroying Enzymes (Kininases)

Plasma and tissues contain enzymes which by virtue of their peptidase action destroy the activity of bradykinin and related kinins. In plasma, two enzymes known as kininase I and kininase II inactivate bradykinin by cleaving off the terminal-carboxyl amino acids (*41*). These enzymes may be inhibited by 8-hydroxyquinoline and phenylalanyl-arginine, respectively. In addition to these, catheptic carboxypeptidases from spleen acting at acid pH can inactivate bradykinin (*25*). Kininases whose mechanism of action are still uncertain are present in white cells and act at neutrality (*40*). The lung is now known to be a major source of kinin inactivation (*42*). Studies on kininase in this tissue will be most valuable (*43*). Recently, peptides have been isolated from snake venoms which potentiate kinin response *in vivo* by blocking lung kininase activity (*44, 45*).

36. E. Habermann and W. Klett, *Biochem. Z.* **346**, 133 (1966).
37. R. W. Colman, L. Mattler, and S. Sherry, *J. Clin. Invest.* **48**, 11 (1969).
38. S. Nagasawa, H. Takahashi, M. Koida, T. Suzuki, and J. G. Schoenmakers, *BBRC* **37**, 180 (1969).
39. E. Erdös, *Advan. Pharmacol.* **4**, 1 (1966).
40. L. M. Greenbaum, R. Freer, J. Chang, G. Semente, and K. Yamafuji, *Brit. J. Pharmacol.* **36**, 623 (1969).
41. H. Y. T. Yang and E. Erdös, *Nature* **215**, 1402 (1967).
42. J. R. Vane, *Brit. J. Pharmacol.* **35**, 209 (1969).
43. J. M. Stewart, J. Roblero, and J. W. Ryan, *Abstr. 4th Intern. Congr. Pharmacol., Basel, 1969,* p. 214. Benno Schwabe, Basel, 1970.
44. S. H. Ferreira, D. C. Bartelt, and L. J. Greene, *Biochemistry* **9**, 2583 (1970).
45. H. Kato and T. Suzuki, *Biochemistry* (1970) (in press).

13

Papain, X-Ray Structure

J. DRENTH • J. N. JANSONIUS • R. KOEKOEK •
B. G. WOLTHERS

I. Introduction

Papain was the first recognized member of the class of proteolytic enzymes that need a free sulfhydryl group for activity. Other enzymes belonging to this group are, e.g., chymopapain, which can also be isolated from papaya latex, ficin from the fig tree, and bromelain from pineapple. The enzymes from animal tissue, called *cathepsins,* belong to this group, as well as clostridio-peptidase B and streptococcal proteinase from bacteria. These enzymes usually need an activator for activity which has the function of releasing the blocked SH group. The most usual activators are cyanide, cysteine, or glutathione.

Although papain can now be prepared in an extremely pure form (*1*),

1. S. Blumberg, I. Schechter, and A. Berger, *Proc. 39th Meeting Israel Chem. Soc.* p. 125 (1969).

most investigations on papain have been performed with material prepared according to Kimmel and Smith (*2*). In the Kimmel and Smith preparation about half of the molecules have a nonactivatable sulfhydryl group (papain-S-X) which cannot be reduced to free SH by the usual activators. The remaining molecules are in the form papain-S-S-Cys, as proposed by Sluyterman (*3*) and by Glazer and Smith (*4*). Klein and Kirsch gave a direct proof of the presence of a mixed disulfide between papain and cysteine (papain-S-S-Cys), by the isolation of ^{14}C-labeled 2-iminothiazolidine-4-carboxylic acid from cyanide-activated papain (*5*). This compound is formed by a cyclization reaction between cysteine and cyanide ion. In freshly prepared nonactivated papain the free SH content varies around a mean value of 0.30 ± 0.15 mole SH/mole of protein, but on aging the free SH content rapidly decreases to zero. The Kimmel and Smith preparation has been used by Drenth *et al.* (*6*) for the elucidation of the complete three-dimensional structure of the papain molecule by X-ray crystal structure analysis. They applied the isomorphous replacement technique for obtaining the phases of the diffracted X-ray beams. Their X-ray work not only gave the folding of the main chain and the position of all the side chains but also contributed to the final solution of the amino acid sequence problem. The sequence is shown in Fig. 1. The single polypeptide chain in the molecule contains 212 residues, and from its composition a molecular weight of 23,350 can be calculated.

II. Crystallization

For an accurate structure determination by X-ray diffraction, single crystals with dimensions of 0.2–0.5 mm are required. The periodic repetition of the unit cells leads to a pattern of diffracted X-ray beams. From the diffraction directions, the symmetry in the pattern, and the extinction of certain beams, the dimensions and the symmetry of the unit cell can be derived. To calculate the electron density in the unit cell one needs the intensities of the diffracted beams and their phase relationship.

Kimmel and Smith [in ref. (*7*)] observed that papain could be crystallized easily, and later on various crystal forms were found (Table I).

2. J. R. Kimmel and E. L. Smith, *Biochem. Prep.* **6,** 61 (1958).
3. L. A. AE. Sluyterman, *BBA* **139,** 430 (1967).
4. A. N. Glazer and E. L. Smith, *JBC* **240,** 201 (1965).
5. I. B. Klein and J. F. Kirsch, *BBRC* **34,** 575 (1969).
6. J. Drenth, J. N. Jansonius, R. Koekoek, H. M. Swen, and B. G. Wolthers, *Nature* **218,** 929 (1968).
7. A. N. Glazer and E. L. Smith, Chapter 14, this volume.

TABLE I
CRYSTAL MODIFICATIONS OF PAPAIN

Crystal form	Space group	a Å	b Å	c Å	β deg	Molecules/ cell	Volume/ molecule
A	$P2_1$	56	50	42	96.2	2	58,500
B	$C2$	101	51	53	116.5	4	61,800
C	$P2_12_12_1$	45.0	104.3	50.8		4	59,600
D	$P2_12_12_1$	42	96	50		4	50,400
E	$P2_1$	65.4	50.7	31.5	97.1	2	51,000
S	$C222_1$	44	66	124		8	45,000

Crystals of the A form were obtained by Drenth and Jansonius (*8*) from a solution of mercuripapain in 70% ethanol. The crystals appeared as very thin laths and therefore were not suitable for an extensive X-ray study. The modifications B, D, and E were obtained only occasionally (*9*). Crystals of form D, appearing as regular octahedra, were grown from a solution containing only 45% methanol. The form S has been reported by Stockell (*10*) and was obtained from an aqueous solution. The crystals of form C were most suitable for an X-ray diffraction analysis because of their size—a few tenths of a millimeter—and because they were frequently found in the crystal preparations. They were grown from a methanol–water mixture 2:1 by volume (62% methanol) and appear as well-developed needles.

Since the structure of the C-form crystals was determined, one might wonder whether or not the molecular structure of the enzyme would be the same in this medium as in water. Even more so because in 62% methanol the enzyme is inactive and cannot be activated. Drenth *et al.* (*11*) showed, however, that this lack of activity was not owing to a profound change in conformation. They could crystallize papain from 15 vol-% dimethylsulfoxide, pH 6.0 containing 0.2 *M* ammonium sulfate. In this medium the enzyme does show activity. The crystals were identical with the previously found B crystals grown from a methanol-rich medium. Thus, the medium did not affect the structure of the B-type crystals and the authors assumed the same to be true for modification C. This was supported by activity measurements on a C-crystal suspension in a water medium (*12*). These crystals were cross-linked to avoid dis-

8. J. Drenth and J. N. Jansonius, *Nature* **184,** 1718 (1959).
9. R. Koekoek, Thesis, Groningen, 1969.
10. A. Stockell, *JMB* **3,** 110 (1961).
11. J. Drenth, W. G. J. Hol, J. W. E. Visser, and L. A. AE. Sluyterman, *JMB* **34,** 369 (1968).
12. L. A. AE. Sluyterman and M. J. M. de Graaf, *BBA* **171,** 277 (1969).

integration. Solutions of papain in water and in 62% methanol gave identical ORD curves in the UV spectral region with a small Cotton effect at 290 nm which must result from the side chains of the aromatic amino acids (*11*, *13*). From all the evidence, we can conclude that the enzyme has the same structure in water and in a methanol-rich medium.

III. Heavy Atom Derivatives

Bokhoven *et al.* (*14*) suggested the use of the isomorphous replacement method to determine the phase relationship of the diffracted X-ray beams for noncentrosymmetric structures. Green *et al.* (*15*) were the first to apply the method in a protein structure determination. This requires the attachment of heavy atoms to the protein structure without a distortion of that structure. The heavy atom derivatives of papain were prepared in either of two ways:

(1) By soaking crystals of the native protein in a solution of 62% methanol (at pH 9.3) containing the heavy atom reagent in a concentration of 0.001–0.002 *M*.

(2) By reaction of the native protein in solution with the heavy atom reagent and subsequent crystallization.

Table II summarizes the various isomorphous derivatives used by

TABLE II
HEAVY ATOM DERIVATIVES USED IN DIFFERENT STAGES OF THE PAPAIN STRUCTURE DETERMINATION

Derivative	Heavy atom group[a]	Binding site
(I)	PCMB	His 81; His 159
(II)	PCMS	His 81; His 159
(III)	PCMA	His 81; His 159; Lys 10
(IV)	(Papain-S)-HgCl	Cys 25
(V)	(Papain-S)-HgCl + $HgCl_2$	Cys 25; Asn 194
(VI)	Na_2PtCl_6	Ile 1; His 159
(VII)	K_2HgI_4	Ile 1; His 159
(VIII)	Na_3IrCl_6	Ile 1; His 159
(IX)	$HgCl_2$	His 159; Asn 194

[a] Abbreviations are as follows: PCMB, *p*-chloromercuribenzoate; PCMS, *p*-chloromercurisulfonate; and PCMA, *p*-chloromercurianiline.

13. S. S. Husain and G. Lowe, *BJ* **110,** 53 (1968).
14. C. Bokhoven, J. C. Schoone, and J. M. Bijvoet, *Acta Cryst.* **4,** 275 (1951).
15. D. W. Green, V. M. Ingram, and M. F. Perutz, *Proc. Roy. Soc.* **A225,** 287 (1954).

Drenth and co-workers in their X-ray study. In the low resolution study of papain (*16–18*), PCMB (I), Na_2PtCl_6 (VI), K_2HgI_4 (VII), Na_3IrCl_6 (VIII), and $HgCl_2$ (IX) were used for the phase determination. *p*-Chloromercuribenzoate gave the best results and was also perfectly satisfying in the structure determination with a resolution of 2.8 Å (*6*). Here also two other aromatic mercurials were used besides PCMB: the corresponding sulfonate [PCMS (II)] and aniline [PCMA (III)]. The sites occupied by PCMS and PCMA were very much the same as those occupied by PCMB, but nevertheless these derivatives gave additional information: PCMS because it contains a relatively heavy sulfur atom and PCMA because it also occupies a third site.

The active site SH group (Cys 25) was not easily used as a site for attachment of a heavy atom. In a part of the papain molecules in the crystals this group is irreversibly blocked (papain-S-X) while in the other part it has formed a disulfide bridge with cysteine to form papain-S-S-Cys. Wolthers (*18*), in an attempt to locate the active site sulfhydryl group, has achieved substitution of the active SH group by adding 0.001 *M* ethylenediaminetetraacetate (EDTA) to a freshly prepared papain solution prior to the addition of the heavy atom reagent. In this way several specific SH reagents could be combined with the active site SH group (Cys 25), e.g., $HgCl_2$ yielding *S*-chloromercuripapain. This latter compound could be crystallized from 62% methanol by the normal procedure. The crystals obtained were exactly isomorphous with the native protein crystals of the C form.

In spite of the SH group being blocked in ordinary papain crystals, very useful heavy atom derivatives could be obtained by soaking these crystals in solutions of PCMB, $HgCl_2$, and other SH reagents. For instance, with PCMB a stable compound was prepared in which, however, the predominant heavy atom position is far away from the active site. Also, when papain crystals were soaked in a $HgCl_2$ solution, a suitable derivative was obtained [(IX) in Table II]. This derivative however was extremely sensitive to X-irradiation and has been used only in the 4.5 Å work. The main occupancy site of this derivative (IX) is very close to but significantly different from the mercury position found in *S*-chloromercuripapain and coincides with a second, minor binding site of PCMB. The group responsible for this binding site turned out to be the imidazole ring of His 159 situated in the active center close to the sulfhydryl group (Fig. 4). The distance between the nitrogen atom N-1

16. J. N. Jansonius, Thesis, Groningen, 1967.

17. J. Drenth, J. N. Jansonius, R. Koekoek, J. Marrink, J. Munnik, and B. G. Wolthers, *JMB* **5**, 398 (1962).

18. J. Drenth, J. N. Jansonius, and B. G. Wolthers, *JMB* **24**, 449 (1967).

of His 159 and the mercury atom connected to it in derivatives (I), (II), (III), and (IX) is 2.5 Å, whereas this mercury atom is 3.4 Å away from the sulfur atom of Cys 25 and therefore not covalently connected to this sulfur atom. On the other hand, in *S*-chloromercuripapain (IV) the mercury position is only 2 Å away from the sulfur atom. This clearly indicates that in this compound the Hg atom is covalently bound to sulfur while the Hg atom in derivatives (I), (II), (III), and (IX) is not. Another stable derivative (V) was prepared by soaking crystals of *S*-chloromercuripapain in $HgCl_2$ solution.

IV. The Three-Dimensional Structure

A. The Electron Density Map

The heavy atom derivatives mentioned in Section III were used in the isomorphous replacement method to determine the phases of the diffracted waves. The general procedure has been described extensively by Phillips (*19*). The method relies on the changes in the X-ray diffraction pattern caused by the heavy atoms. Because these changes are relatively small the intensities should be measured with great accuracy.

Low resolution data on papain were collected on film and measured with an automatic integrating densitometer (*20*). The linear diffractometer, designed by Arndt and Phillips especially for protein crystallography (*21–23*), was used in the high resolution study. With the three-channel version of this instrument the data for 9525 reflections were collected. The anomalous scattering of the mercury atoms was also used in the phase determination (*24, 25*). This improved the phases significantly. Moreover, the anomalous scattering allowed the correct choice between the two enantiomorphs to be made. From the amplitudes of the diffracted waves and their phases the electron density distribution in the crystal structure could be calculated by a Fourier summation. Contour lines were drawn at intervals of 0.25 e/Å^3. The highest density of the map was found at the sulfur atom of Cys 25 (2.0 e/Å^3). The three di-

19. D. C. Phillips, *Advan. Struct. Res. Diffraction Methods* **2,** 76 (1966).
20. J. Drenth, D. Kloosterman, J. van der Woude, H. C. Croon, and L. C. M. van Zwet, *J. Sci. Instr.* **42,** 222 (1965).
21. U. W. Arndt and D. C. Phillips, *Acta Cryst.* **14,** 807 (1961).
22. D. C. Phillips, *J. Sci. Instr.* **41,** 123 (1964).
23. U. W. Arndt, A. C. T. North, and D. C. Phillips, *J. Sci. Instr.* **41,** 421 (1964).
24. A. C. T. North, *Acta Cryst.* **18,** 212 (1965).
25. B. W. Matthews, *Acta Cryst.* **20,** 82 (1966).

sulfide bridges show up as broad maxima with a highest density of 1.5 e/Å^3.

It appeared to be possible to follow the backbone of the molecule as one continuous chain of density with a height between 1.0 and 1.3 e/Å^3, paying no attention at all to the amino acid sequence. The majority of the carbonyl groups of the peptide bonds bulge out of the main chain and are easily recognizable. Their position and the position of the side groups in the electron density map enabled the construction of a molecular model. This model could be checked in various ways. First of all it appeared that the heavy atoms were in acceptable positions near binding sites at the surface of the molecule. The mercury atom in compound (IV), is at covalent distance of the sulfur atom of Cys 25. Moreover, Wolthers (*26*) could label Cys 25 by reaction of papain with *N*-tosyl-lysyl-chloroketone. This reagent showed up as additional electron density in the active site touching the density of the sulfur atom of Cys 25. Furthermore, the mercury atom at the main binding site for the aromatic mercurials (I), (II), and (III) was found close to the imidazole ring of His 81.

Drenth *et al.* (*6*) could also label the α-amino group of Ile 1 by reaction with phenyl isothiocyanate and its *p*-iodo derivative. The positions of these labels in the crystal structure were easily calculated by the difference-Fourier technique in which the intensity changes caused by the labels are used together with the phases of the waves diffracted by the protein. When this reaction was performed at pH 8 only the α-amino group was labeled. At pH 9.3 additional labeling of Lys 211, Lys 100, and Lys 190 occurred. In the same way the iodine positions in KI_3-treated papain crystals could be located. This reaction caused a disintegration of the protein crystals, which could, however, be prevented by diminishing the solubility of papain. This was done by cross-linking the crystals with glutaraldehyde or by reacting them with *p*-iodophenyl isothiocyanate. Four iodine binding sites could be identified as tyrosines, which are fully exposed at the molecular surface: Tyr 61, Tyr 78, Tyr 82, and Tyr 123. Three additional iodine positions could not be interpreted.

B. Description of the Structure

The problems encountered in the chemical sequence studies on papain are dealt with by Glazer and Smith in this volume (*7*). We shall restrict ourselves here to the contribution made by the X-ray structure determi-

26. B. G. Wolthers, *FEBS Letters* **2**, 143 (1969).

nation. Although the main chain could be traced easily in the electron density map and the position of the side chains can be seen without any ambiguity, a unique identification of the side chains is not possible at a resolution of 2.8 Å. However, the disulfide bridges show up clearly as high density connections of main chain parts. The tryptophan residues are also outstanding features because of the size and shape of their side chain. On the other hand, tyrosines, phenylalanines, and histidines all have a bulky and rather flat side chain but cannot always be distinguished from each other. The absence of any side chain density indicates that a glycine is present. Using the preliminary sequence published by Light *et al.* (*27*) as a guide, the electron density map of papain could be interpreted completely. Drenth *et al.* (*6*) were able to show that the correct sequence differed in two main respects from the preliminary chemical sequence:

(1) A peptide of 13 residues had to be inserted in the preliminary sequence behind Phe 28 instead of the two Ile residues.

(2) A peptide, 39 residues long, from Thr 138 up to and including Tyr 176 in the preliminary sequence had to be transposed to a position immediately following the inserted 13-residue peptide.

Some other corrections of the preliminary sequence were suggested by the electron density map, e.g., the addition of an extra Val residue at position 130. Husain and Lowe (*28*) isolated the missing peptide behind Phe 128 and determined its amino acid sequence. This sequence was in perfect agreement with the size and shape of the side chains observed in the electron density map.

The final amino acid sequence which comprises 212 residues is shown in Fig. 1. The chain is folded three dimensionally into an ellipsoidal particle with rough dimensions $50 \times 37 \times 37$ Å (Fig. 2). The hydrophilic residues are generally at the surface in contact with the surrounding liquid. The majority of the hydrophobic side chains are internal. However, some hydrophobic residues are distinctly at the surface, e.g., Val 91, Val 150, and Ile 148. They are probably responsible for the limited solubility of papain in water. The side chain of Glu 50 is completely internal. Its carboxyl group forms an internal ion pair with Arg 83. Intramolecular arginine–carboxylate interactions at the surface of the molecule are observed between Arg 8 and Asp 6, Arg 41 and Asn 212 (C-terminal), Arg 96 and Glu 89, and Arg 188 and Glu 183.

A characteristic feature of the papain molecule is its binuclear nature. It is constructed around two hydrophobic cores. Apparently such a core

27. A. Light, R. Frater, J. R. Kimmel, and E. L. Smith, *Proc. Natl. Acad. Sci. U. S.* **52,** 1276 (1964).

28. S. S. Husain and G. Lowe, *BJ* **114,** 279 (1969).

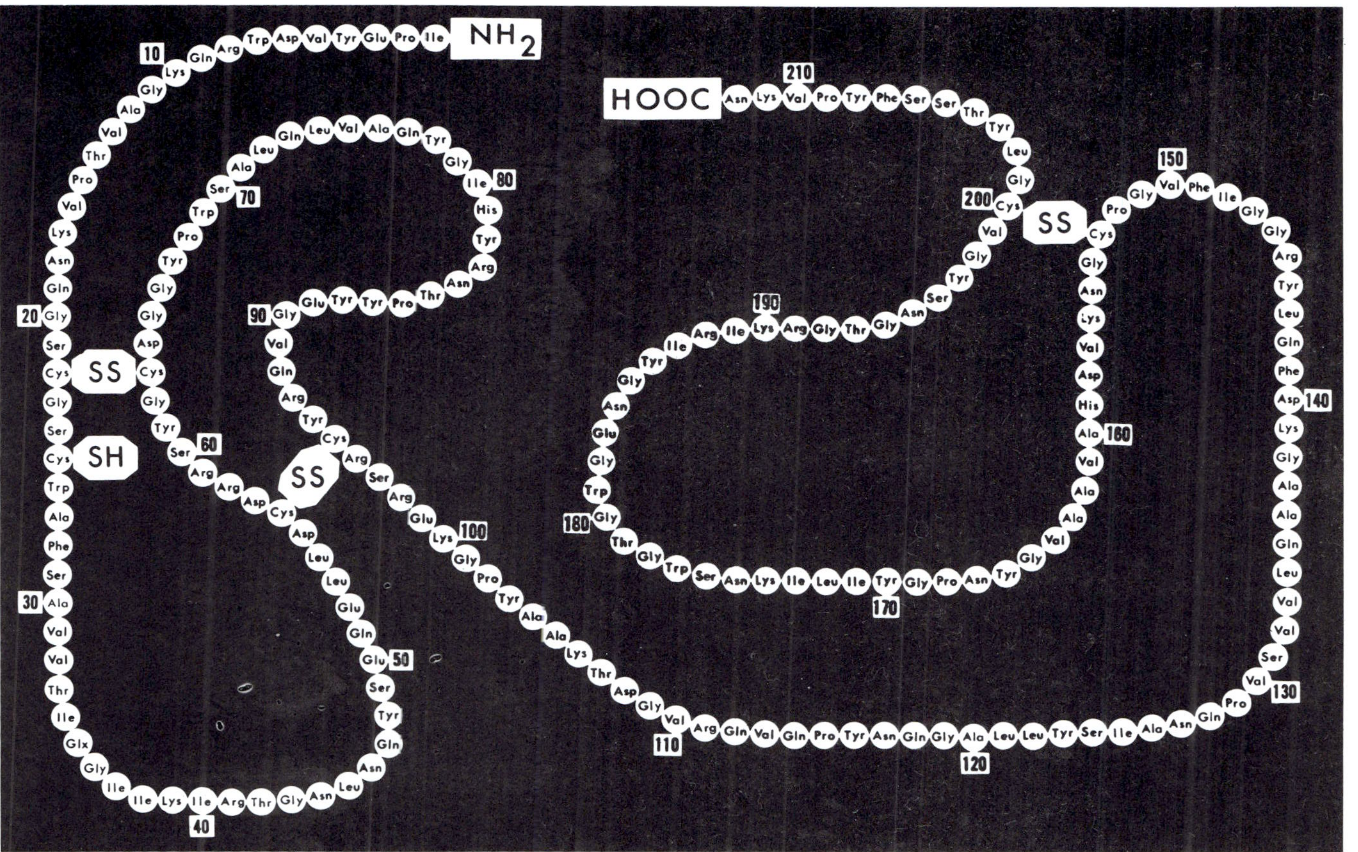

FIG. 1. The amino acid sequence of papain.

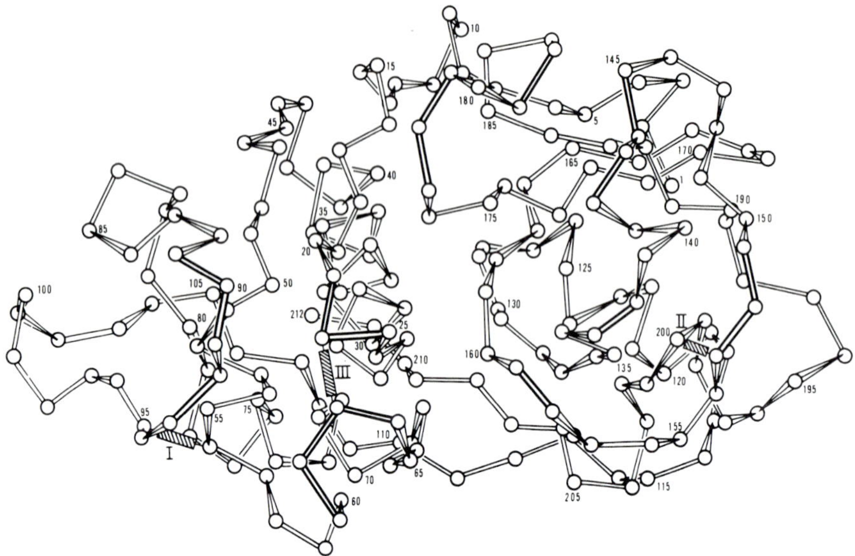

FIG. 2. The main chain of papain.

has a finite size. Once it has been surrounded completely while there is more chain to be folded another core is formed. So far we do not know the actual folding process of the chain after its synthesis at the ribosome. The chain in the papain molecule could have been folded from its N-terminal end, the actual folding process starting after residue 23 when the first α helix is formed (see Table III). The folding continues with the formation of α helices 2 and 3 and the complete surrounding of the first hydrophobic core. This part of the molecule is completed by the attachment of a mainly hydrophilic loop (residues 81–110). Now the chain crosses over to the second part of the molecule. This part has only 19 residues in α-helical conformation compared with 41 residues in the first part. However, in addition to this small α-helix content the second

TABLE III
PARTS OF THE PAPAIN POLYPEPTIDE CHAIN IN α-HELIX CONFORMATION

α-Helix number	From	To and inclusive
1	Ser 24	Gly 43
2	Ser 49	Asp 57
3	Tyr 67	Tyr 78
4	Asn 117	Gln 128
5	Gly 137	Tyr 143

part of the molecule contains a distorted pleated sheet structure to which about 30 residues contribute (Fig. 3). Between residues 165 and 172 the chain turns back along itself, forming a nearly ideal antiparallel β structure. The pleated sheet structure (Fig. 3) is bent appreciably, thereby acting as part of the wall of the hydrophobic core. This core is even more pronounced hydrophobic than the core in the first part of the molecule. It is actually a hole with a diameter of about 3 Å surrounded by 16 methyl groups.

Both terminals of the chain seem to have a similar function. They assist in the connection of the two molecular halves. The amino terminal attaches itself to the second half of the molecule by hydrogen bonds and by hydrophobic interactions between the side chain of Val 5 and the second hydrophobic core. The peptide bond between residues 211 and 212 in the carboxyl terminus is hydrogen bonded to the Arg 41 and Asp 108 side chains. Moreover, the ion pair between Arg 41 and the C-termi-

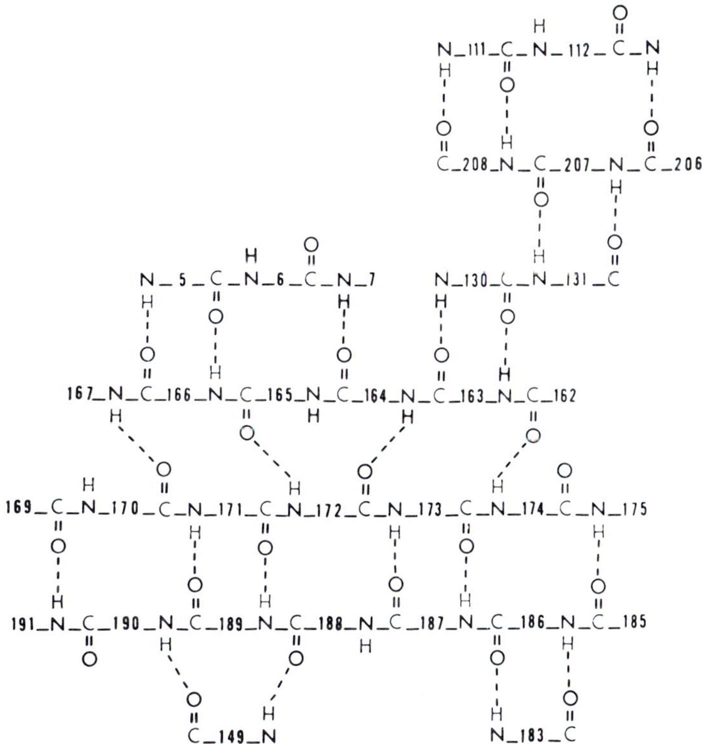

FIG. 3. Residues involved in the twisted pleated sheet structure occurring in papain.

nal carboxyl groups keeps the end of the chain attached to the first half of the molecule.

The electron density map shows the presence of several small molecules at the surface or near the surface inside the enzyme molecule. Because the medium from which the protein has been crystallized is a mixture of water and methanol these density peaks represent water or methanol molecules. They are hydrogen bonded to the protein structure.

V. The Active Site

A. The Geometry of the Active Site Region

The interaction between enzyme and substrate occurs at the surface of the papain molecule in a groove which is situated between the two parts of the molecule. The side chain of Cys 25 is found in the groove. If the SH group is not free but blocked by heavy atoms, by alkylating reagents or by disulfide formation, the activity disappears completely. Close to the sulfhydryl group is the imidazole ring of His 159 (Fig. 4). It has never been proved that this residue is actively involved in the catalytic mechanism. In ester hydrolysis catalyzed by 4-mercapto-methyl-imidazole and by 4-2-mercaptoethyl-imidazole, Schneider (*29*) found a concerted action of the sulfhydryl group and the imidazole ring. Also, in the serine proteinases, an imidazole ring is present next to the essential OH (*30, 31*). The imidazole ring of His 159 in papain is hydrogen bonded to the side chain of Asn 175. This hydrogen bond prevents any movement of the imidazole ring which is partly hidden behind the indole ring of Trp 177. Husain and Lowe (*32*) proved in a chemical way that the SH group of Cys 25 and the imidazole ring of His 159 are indeed close together.

At a distance of 6.7 Å from the imidazole ring is the side chain carboxyl group of Asp 158. Its negative charge will certainly promote the removal of the negatively charged product which is formed in the decomposition of the acyl enzyme, i.e., in the deacylation step of the catalytic reaction. A positive charge on the imidazole ring will keep the side chain of Asp 158 directed to the imidazole ring as is shown in Fig. 4. This will leave more space for the substrate in the active site groove.

29. F. Schneider, *Z. Physiol. Chem.* **348**, 1034 (1967).
30. D. M. Blow, Chapter 6, this volume.
31. J. Kraut, Chapter 15, this volume.
32. S. S. Husain and G. Lowe, *BJ* **108**, 855 (1968).

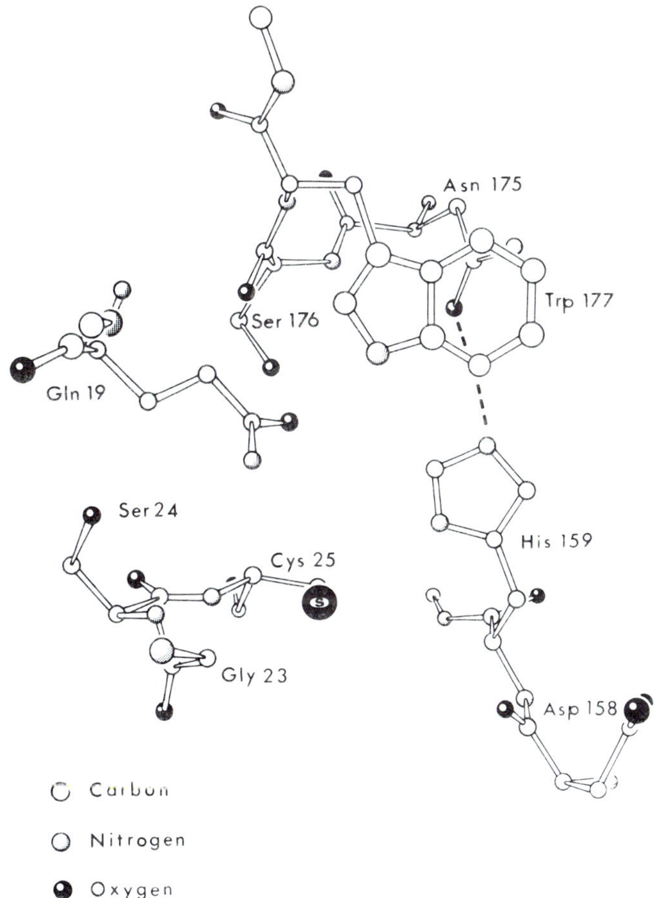

FIG. 4. The active site region of papain near the essential SH group.

When the electrostatic interaction is lost, and this will certainly be true at pH values below the pK of the carboxyl group, the side chain of Asp 158 is free to move around in the solvent. Although this will not block the active site it probably does change the mode of binding of the substrate chain. Residue 23 can only be glycine because any side chain bigger than H would seriously block the active site groove.

Opposite to Asp 158 is the position of Asp 64 which has its side chain at a distance of 10 Å from the sulfur atom of Cys 25 and at 14 Å from the imidazole ring of His 159. Though this is too far to show a significant interaction with the imidazole group this Asp residue will certainly decrease the affinity of the active site region for negatively charged ions as does Asp 158 and increase the affinity for Lys and Arg (*32a*). The active site groove is further lined by the side chains of some aromatic residues (Tyr 61, Tyr 67, and Trp 69) contributed by the first part of the molecule. The two tyrosine residues are connected by a hydrogen bond between their OH groups. The last part of the polypeptide chain contributes the side chain of Phe 207 to the wall of the groove. The side chains of Ser 205, Val 133, and Val 157, together with those of Asp 158 and His 159, form the opposite wall of the groove.

B. Comparison with Other Proteolytic Enzymes

Papain is a member of a family of proteolytic SH enzymes. The related enzymes have not been studied to such a great extent as papain. Only parts of the sequences of ficin (*33*) and stem-bromelain (*34*), especially near the essential cysteine residue, have been determined and they show a striking similarity with the corresponding part of the papain sequence (see Fig. 5). This suggests that the same folding of the main chain is present in all three enzymes. A replacement of serine at position 24 in papain by alanine in stem-bromelain would not interfere with the rest of the structure. Any side chain in position 23 would seriously block the active site region and therefore only glycine can occupy this position and it does so in all three enzymes. Chymopapain is another proteolytic sulfhydryl enzyme from papaya latex. Although its sequence (*35*) shows only a poor similarity with the papain sequence, it also has the glycine residue two positions to the left of the essential cysteine. The same is true for streptococcal proteinase (*36*). This bacterial enzyme has the same kind of specificity toward substrates as papain, suggesting a similar active site groove for both enzymes. And again glycine is found in its "space providing" position. Streptococcal proteinase also has a glutamine residue in the same position as papain (residue 19), ficin, and bromelain. The side chain of this residue is shown in Fig. 4. Its function is not yet known, but it could well take part in the binding of the substrate.

Although the folding of the main chain is completely different in papain and in the serine proteinases chymotrypsin and subtilisin, these

32a. *Note added in proof:* Husain and Lowe (*32b*) obtained evidence that residue 64 is Asn instead of Asp.

32b. S. S. Husain and G. Lowe, *BJ* **116,** 689 (1970).

33. R. C. Wong and I. E. Liener, *BBRC* **17,** 470 (1964).

34. S. S. Husain and G. Lowe, *Chem. Commun.* p. 1387 (1968).

35. J. N. Tsunoda and K. T. Yasunobu, *JBC* **241,** 4610 (1966).

36. T. Liu and S. D. Elliott, Chapter 18, this volume.

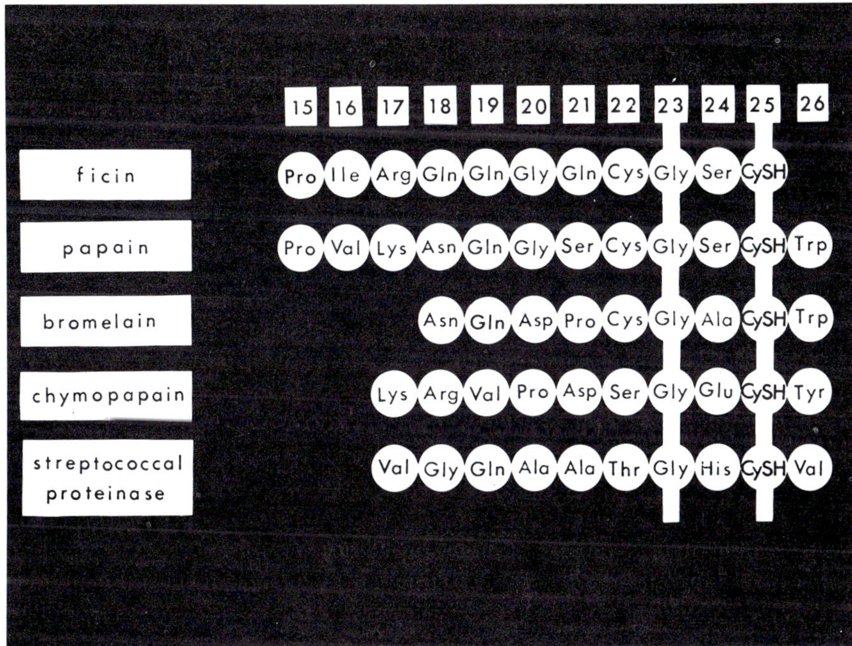

FIG. 5. Amino acid sequences near the essential cysteine in some proteolytic SH enzymes.

proteolytic enzymes have interesting common features. First of all they share with all other proteolytic SH and OH enzymes the property of having a glycine residue two positions to the left of the essential cysteine or serine (*37*). Even more surprising is the striking similarity which these proteolytic enzymes show in the arrangement of the groups responsible for the substrate splitting. In the serine enzymes this arrangement is OH . . . Im . . . Asp where Im is an imidazole ring and the ". . ." means hydrogen bond. In papain the combination SH . . . Im . . . Asn exists. It has been suggested that the nucleophilicity of the oxygen atom of the hydroxyl group in the essential serine is increased by the "charge relay" system Im . . . Asp . Apparently sulfur, which is a stronger nucleophile than oxygen, does not need an aspartic acid residue but its nucleophilicity is sufficiently increased by Im . . . Asn . The various aspects of the enzymic mechanism will not be treated further in this chapter, but are by Glazer and Smith, Chapter 14, this volume.

37. M. O. Dayhoff, "Atlas of Protein Sequence and Structure 1969," p. 49. Natl. Biomed. Res. Found., Silver Spring, Maryland.

14

Papain and Other Plant Sulfhydryl Proteolytic Enzymes

A. N. GLAZER • EMIL L. SMITH

I. Introduction

Papain is a major protein constituent of the latex and the melonlike green fruit of the small softwood tree *Carica papaya.* This tree is native to South America and is cultivated in a number of tropical countries, southern Florida, and southwestern United States. The term *papain* was introduced by Wurtz and Bouchut (*1*) to describe a proteolytic principle in papaya latex. This term is currently applied both to the crude dried latex, available commercially, as well as to the crystalline proteolytic enzyme. Much of the activity originally contained in the latex is lost in the drying and grinding, and the activity of the commercial preparations is variable (*2, 3*). A complete resume of the early history of the enzyme was presented by Hwang and Ivy (*4*), and more recent literature reviewed by Kimmel and Smith (*5*).

The most recent review devoted to papain appeared in the preceding edition of this series in 1960 (*6*). Since that time, studies of the amino acid sequence of this enzyme have been completed, its three-dimensional structure elucidated by X-ray crystallography, some understanding achieved of the nature of the interactions between papain and a variety of substrates including polypeptides, and considerable progress made toward an understanding of its mechanism of action.

A. Preparation and Crystallization

Papain is an early constituent of the juice of the papaya fruit and reaches its maximal concentration early in the development of the fruit (*7*). Procedures for the purification of papain from wet papaya latex and its isolation in a crystalline state were developed by Balls *et al.* (*8, 9*). The applicability of the procedure of Balls and Lineweaver (*9*) was limited by the scarcity and high cost of fresh papaya latex. Recently, purification of papain directly from fresh latex by chromatography on

1. A. Wurtz and E. Bouchut, *Compt. Rend.* **89,** 425 (1879).
2. A. K. Balls, R. R. Thompson, and W. W. Jones, *Ind. Eng. Chem.* **32,** 1144 (1940).
3. A. K. Balls, H. Lineweaver, and S. Schwimmer, *Ind. Eng. Chem.* **32,** 1277 (1940).
4. K. Hwang and A. C. Ivy, *Ann. N. Y. Acad. Sci.* **54,** 161 (1951).
5. J. R. Kimmel and E. L. Smith, *Advan. Enzymol.* **19,** 267 (1957).
6. E. L. Smith and J. R. Kimmel, "The Enzymes," 2nd ed., Vol. 4, p. 133, 1960.
7. G. S. Skelton, *Phytochemistry* **8,** 57 (1969).
8. A. K. Balls, H. Lineweaver, and R. R. Thompson, *Science* **86,** 379 (1937).
9. A. K. Balls and H. Lineweaver, *JBC* **130,** 669 (1939).

hydroxylapatite has been described (*10*).

In 1954, Kimmel and Smith (*11*) described the preparation of crystalline papain from commercially available dried latex. This procedure has been widely adopted, and virtually all recent studies on papain have been performed on protein prepared by this method. Since the details of the method of isolation are relevant to subsequent discussion of the activation of papain, an outline of the procedure is given here.

The dried latex is ground with sand in presence of 0.04 *M* cysteine at pH 5.7. The filtered extract is adjusted to pH 9.0 and insoluble material removed by filtration. The enzyme is precipitated first with ammonium sulfate and subsequently with sodium chloride. Crystallization is performed in 0.02 *M* cysteine solution of pH 6.5 at 4°. The crystals are redissolved in distilled water and recrystallization initiated by the addition of sodium chloride. Mercuripapain is prepared by dissolving twice crystallized papain at room temperature in 70% ethanol containing 0.001 *M* $HgCl_2$. About 90% of the active enzyme crystallizes from solution in 3–4 days at 4° in the form of heavy crystals (*11, 12*).

Papain prepared by the above procedure (*11*) may be further purified by affinity chromatography on a column of an inhibitor, Gly–Gly–Tyr(Bzl)–Arg, covalently linked to agarose resin. On activation, the enzyme exhibits about twice the specific activity of the original preparation and contains one mole of SH per mole of protein (*12a*).

II. Physical Properties

A. Hydrodynamic Properties

The hydrodynamic properties of papain have been comprehensively discussed in earlier reviews (*5, 6*), and the principal physical constants are listed in Table I.

B. Spectrophotometric and Fluorescence Properties

Papain contains no chromophoric groups other than its constituent amino acids. It is rich in tyrosine (19 residues) and tryptophan (5 residues).

The ionization behavior of the phenolic groups of papain is complex.

10. G. S. Skelton, *J. Chromatog.* **35**, 283 (1968).
11. J. R. Kimmel and E. L. Smith, *JBC* **207**, 515 (1954).
12. J. R. Kimmel and E. L. Smith, *Biochem. Prep.* **6**, 61 (1958).
12a. S. Blumberg, I. Schechter, and A. Berger, *European J. Biochem.* **15**, 97 (1970).

TABLE I
PHYSICAL PROPERTIES OF PAPAIN[a]

$s_{20,w}$	2.42 S
$D_{20,w}$ (10^{-7} cm^2 sec^{-1})	10.23
$\bar{v}$ (ml/g)	0.723[b]
Frictional ratio f/f_0	1.16
Isoelectric point pI	8.75
Absorbancy $A^{1\%}_{1\,cm}$ at 278 nm	25.0[c]
Molecular weight (S, D)	21,000
Molecular weight (sedimentation equilibrium)	23,700[d]
Molecular weight (amino acid sequence)	23,406[b]

[a] From Kimmel and Smith (*6*).
[b] From (*41*).
[c] From Glazer and Smith (*13*).
[d] From (*12b*).

As found for a number of other proteins, the tyrosine residues fall into groups of widely differing p*K*. Spectrophotometric titration indicates that 13 of the 19 tyrosine residues titrate in the pH range 8.5–12.0, in a normal reversible manner, with a pK_{int} of 9.8 (*13*). Four additional groups ionize slowly in the pH range 12–13. The ionization of the remaining two residues requires exposure to pH 14 (*13*). The titration curve for the phenolic groups ionizing above pH 12 is irreversible. Further, papain exposed to pH $>$ 12 aggregates and precipitates at pH $<$ 9.5, indicating gross changes in the conformation of the molecule.

The fluorescence properties of papain have received considerable attention (*14–16a*). The emission spectrum of papain, obtained by excitation at 295 nm, originates almost entirely from the tryptophan residues (*15, 16*). The fluorescence emission of unactivated papain increases with pH in the range 5–8, and the pH-fluorescence profile can be fitted over this range with a curve calculated for the titration of a group of apparent pK'_a of 6.6 at 25° (*14, 15*). The observed profile is very similar to those obtained with model compounds displaying intramolecular histidine–tryptophan interaction (*14*). Activation of papain is accompanied

12b. D. M. Brown, unpublished results (1968).

13. A. N. Glazer and E. L. Smith, *JBC* **236,** 2948 (1961). The values given in the text have been recalculated on the basis of a molecular weight of 23,400 for papain.

14. M. Shinitzky and R. Goldman, *European J. Biochem.* **3,** 139 (1967).

15. A. O. Barel and A. N. Glazer, *JBC* **244,** 268 (1969).

16. I. Weinryb and R. F. Steiner, *Biochemistry* **9,** 135 (1970).

16a. L. A. AE. Sluyterman and M. J. M. De Graaf, *BBA* **200,** 595 (1970). The results presented in this paper are particularly noteworthy since they were obtained with mercaptopapain separated from nonmercaptopapain on an agarose mercurial column (*16b*).

16b. L. A. AE. Sluyterman and J. Wijdenes, *BBA* **200,** 593 (1970).

by changes in the intensity of the fluorescence emission and in the pH-fluorescence profile *(14–16)*. These changes can be accounted for, in part, by perturbation of tryptophan chromophore(s) by the newly generated thiol group *(15)*. It is not clear whether the pH-dependent fluorescence changes reflect solely the ionization of the histidine residue in the area of the active site or whether they include a significant contribution dependent on the ionization of the other histidine residue in the protein. Indeed, Sluyterman and De Graaf *(16a)* find that a group of pK 6.7 is found exclusively in the pH–fluorescence profile of papain which cannot be activated. It is not seen in the profiles obtained with inactive, but activatable papain, or with active papain.

C. Stability

Papain shows an unusually high stability at elevated temperatures at near-neutral pH *(4, 17)*. At acid pH values (< 4), papain is rapidly and irreversibly inactivated at elevated temperatures *(17)*. At strongly acid pH (< 2), this irreversible inactivation is extremely rapid, even at 25°, and is associated with spectral changes indicative of the unfolding of the molecule *(18)*.

Lineweaver and Schwimmer *(17)* showed that papain remains fully active after prolonged exposure to 9 *M* urea solution. The activity of the papain was assayed after dilution of aliquots of the enzyme–urea incubation mixture into a solution of protein substrate *(17)*. These results have been confirmed and extended in later studies of Gundlach and Turba *(19)* and Sluyterman *(20)*, who showed that papain displays high activity for many hours against protein *(19)* and synthetic *(20)* substrates in concentrated urea solutions (up to 8 *M*). The inactivation of papain in solutions of low urea concentration, reported by Hill *et al.* *(21)*, appears to be a consequence of the reaction of the thiol group at the active site with cyanate arising from the decomposition of urea *(22)*. Adequate removal of metal ions from urea and guanidine solutions also appears to be of some importance *(19)*.

Papain shows a remarkable stability in concentrated urea solutions. In 8 *M* urea at near-neutral pH there are no significant changes in optical rotatory properties of the protein *(20)*, and the decrease in activity can be accounted for largely by the competition of urea for the substrate-binding site *(20)*.

17. H. Lineweaver and S. Schwimmer, *Enzymologia* **10,** 81 (1941).
18. A. N. Glazer and E. L. Smith, *JBC* **235,** PC43 (1960).
19. G. Gundlach and F. Turba, *Biochem. Z.* **342,** 303 (1965).
20. L. A. AE. Sluyterman, *BBA* **139,** 418 (1967).
21. R. L. Hill, H. C. Schwartz, and E. L. Smith, *JBC* **234** 572 (1959).
22. L. A. AE. Sluyterman, *BBA* **135,** 439 (1967).

The conformation of native papain appears to be little affected by high concentrations of organic solvents. As mentioned above, one procedure for the crystallization of papain uses 70% (v/v) ethanol at 4° as solvent (*12*); and, more recently, Drenth *et al.* (*23*) have obtained isomorphous papain crystals from aqueous solution, 60% (v/v) aqueous methanol, or 15% (v/v) dimethylsulfoxide [see also Drenth (*24*)]. Spectroscopic studies likewise indicate little change in conformation at near-neutral pH in methanol (0–50%, v/v), ethylene glycol (0–40%, v/v), and dioxane (0–30%, v/v) (*15*).

D. Immunochemical Studies

Early studies on crystalline papain demonstrated that the enzyme behaved as a single antigen and that antigenic impurities could not be detected (*6, 25*).

Arnon and Shapira (*26, 27*) isolated antipapain antibodies either by specific precipitation with the homologous antigen, papain, followed by acid dissociation and the removal of the antigen by gel filtration, or by adsorption on an insoluble immunoadsorbent (papain reacted with bromoacetyl-cellulose), followed by elution of the antibodies with 0.1 *M* acetic acid. Antibodies obtained in this manner inhibited the papain-catalyzed hydrolysis of casein up to a maximum of 85% and that of benzoyl-L-arginine ethyl ester (BAEE) up to 60%. In the course of these studies, it was demonstrated that the antipapain serum cross-reacted with chymopapain. Indeed chymopapain precipitated 32% of the antipapain antibodies in the serum (*26*). Advantage was taken of this observation to isolate the cross-reacting fraction of antibodies from the antipapain serum by adsorption on an insoluble chymopapain immunoadsorbent (*26*). These antibodies were efficient inhibitors of the action of papain and chymopapain on both large and small substrates. Benzoyl-L-arginine ethyl ester hydrolysis was inhibited to an extent of 92% for each of the proteases (*26*). The fraction of the antipapain antibodies which failed to cross-react with chymopapain was essentially non-inhibitory toward papain. These results led to the conclusion that papain and chymopapain share common antigenic determinants and that at least one of these is located close to the active site on both enzymes (*28*).

23. J. Drenth, W. G. J. Hol, J. W. E. Visser, and L. A. AE. Sluyterman, *JMB* **34**, 369 (1968).
24. J. Drenth, Chapter 13, this volume.
25. R. Arnon, *Immunochemistry* **2**, 107 (1965).
26. R. Arnon and E. Shapira, *Biochemistry* **6**, 3942 (1967).
27. E. Shapira and R. Arnon, *Biochemistry* **6**, 3951 (1967).
28. R. Arnon and E. Shapira, *Biochemistry* **7**, 4196 (1968).

III. Primary Structure

A. Amino Acid Composition

Papain is a simple protein containing only amino acids and devoid of carbohydrate (*6, 29*). All of the usual amino acids are present with the exception of methionine. Inasmuch as the complete amino acid sequence is now known, we present in Table II the composition, as derived from the sequence (Section III,B).

The sum of aspartic and glutamic acid residues, 14, is considerably less than the sum of lysine and arginine residues, 22, in accord with the isoelectric point of 8.75 at 0.1 ionic strength, which is in the titration range of the ϵ-amino groups of lysine. Thus, at the pI value only 30 groups, or approximately 14% of the residues, are in ionic form. It should be noted that papain has the properties of a globulin and is only sparingly soluble in water or at low ionic strengths. This may be related both to the relative paucity of hydrophilic groups and the abundance of hydrophobic residues, although most of the latter are in the interior of the protein.

B. Amino Acid Sequence

Most of the work on the sequence of this protein was performed by Smith and his co-workers. The studies were initiated with the demon-

TABLE II
Amino Acid Composition of Papain[a]

Amino acid	Number	Amino acid	Number
Lysine	10	Glycine	28
Histidine	2	Alanine	14
Arginine	12	Valine	18
Aspartic acid	6	Isoleucine	12
Asparagine	13	Leucine	11
Glutamic acid	8	Tyrosine	19
Glutamine	12	Phenylalanine	4
Threonine	8	Tryptophan	5
Serine	13	Cysteine	1
Proline	10	Cystine (half)	6
		Total	212

[a] These values are based on the sequence shown in Fig. 1.

29. C. Nolan and E. L. Smith, *Proc. Soc. Exptl. Biol. Med.* **105,** 287 (1960).

stration by Thompson (*30*) that papain consists of a single peptide chain with the NH_2-terminal sequence Ile–Pro–Glx. Later, Light (*31*) identified the COOH-terminal sequence as –Lys–Asn COOH.

For brevity, evidence for the complete sequence will not be described. It is sufficient to note that studies were performed on the protein oxidized with performic acid, on carboxymethylated protein, and on the denatured protein. Partial acid hydrolysis was used to obtain cysteic acid peptides (*32*) as well as certain other sequences (*33*). Peptides were isolated from tryptic (*34, 35*), chymotryptic (*36–38*), and peptic digests (*39*). From these peptides, Light *et al.* (*39*) proposed in 1964 a tentative linear sequence of the protein although it was recognized that certain gaps still had not been filled by overlapping peptides. At the same time, these investigators (*39*) reported the isolation of peptides which allowed identification of the active thiol group as Cys 25, labeled with ^{14}C-iodoacetate, and the assignment of the six half-cystine residues in three disulfide bonds.

On the basis of this tentative sequence, Drenth and his associates (*40*) proceeded with a study of the crystalline enzyme by X-ray diffraction methods. From their work it was evident that an interior section of the sequence was transposed, that 11 residues were missing in the gap near the active cysteine residue, that two residues were missing in the other indicated gap, and that a few additional minor discrepancies existed between the X-ray data and the chemically derived sequence.

To complete the chemical sequence, further studies (*41*) were performed by using papain labeled at the active site with ^{14}C-iodoacetate and carboxymethylated at the six half-cystine residues with unlabeled iodoacetate. The protein was then maleylated and hydrolyzed with trypsin. Peptides were then isolated which permitted completion of the missing portions of the sequence and confirmation of the findings of the

30. E. O. P. Thompson, *JBC* **207,** 563 (1954).
31. A. Light, *JBC* **237,** 2535 (1962).
32. J. R. Kimmel, E. O. P. Thompson, and E. L. Smith, *JBC* **217,** 151 (1955).
33. M. A. McDowall and E. L. Smith, *JBC* **240,** 281 (1965).
34. J. R. Kimmel, G. K. Kato, A. C. M. Paiva, and E. L. Smith, *JBC* **237,** 2525 (1962).
35. J. R. Kimmel, H. J. Rogers, and E. L. Smith, *JBC* **240,** 266 (1965).
36. A. Light and E. L. Smith, *JBC* **235,** 3144 and 3151 (1960).
37. A. Light, A. N. Glazer, and E. L. Smith, *JBC* **235,** 3159 (1960).
38. A. Light and E. L. Smith, *JBC* **237,** 2537 (1962).
39. A. Light, R. Frater, J. R. Kimmel, and E. L. Smith, *Proc. Natl. Acad. Sci. U. S.* **52,** 1276 (1964).
40. J. Drenth, J. N. Jansonius, R. Koekoek, H. M. Swen, and B. G. Wolthers, *Nature* **218,** 929 (1968).
41. R. Mitchel, I. M. Chaiken, and E. L. Smith, *JBC* **245,** 3485 (1970).

X-ray analysis in other segments of the sequence. Independent studies of the missing 11 residues in one gap were also reported by Husain and Lowe (*42*), as well as the identification of residue 64 as asparagine (*42*).

The complete sequence of the 212 residues of papain is shown in Fig. 1. The disulfide bonds are present between Cys 22 and Cys 63, Cys 56 and Cys 95, and Cys 153 and Cys 200. Present information concerning the structure from X-ray analysis and the conformation of papain is discussed by Drenth *et al.* (Chapter 13, this volume) and in relation to the mechanism of action and the specificity elsewhere in this chapter.

In some early studies of the treatment of papain with leucine aminopeptidase (*43, 44*) and of papain undergoing autolysis (*45*), it was found that considerable proteolysis occurred without loss of papain activity after appropriate activation. It was subsequently found (*46*) that the aminopeptidase contained some endopeptidase activity and that what had been assumed to be degradation from the NH_2-terminal end was in reality hydrolysis at internal bonds. Inasmuch as no potential papain activity was lost, either in these experiments or during autolysis, it appears likely that the degradation of unactivatable papain was occurring. As discussed later (Section IV,B), papain as prepared from dried latex does not manifest one full equivalent of SH per molecule, i.e., a portion of the papain which is inactive will co-crystallize with native papain or with mercuripapain. Presently, there is no evidence to indicate that any significant portion of the papain molecule is dispensable.

IV. Studies on the Active Site of Papain

A. Half-Cystine Content of Unactivated Papain

Studies on the elementary composition of papain indicated a sulfur content of 1.2% (*11*). This represents 8.8 g atoms of sulfur per 23,400 g of papain. Since papain contains no methionine, it would be expected that all of the sulfur was present as half-cystine. Early analyses of the protein after performic acid oxidation and acid hydrolysis showed the presence of only six residues of cysteic acid (*32, 47, 48*). The discrepancy

42. S. S. Husain and G. Lowe, *BJ* **114,** 279 (1969); **116,** 689 (1970).
43. R. L. Hill and E. L. Smith, *JBC* **231,** 117 (1958).
44. R. L. Hill and E. L. Smith, *JBC* **235,** 2332 (1960).
45. B. J. Finkle and E. L. Smith, *JBC* **230,** 669 (1958).
46. R. Frater, A. Light, and E. L. Smith, *JBC* **240,** 253 (1965).
47. E. L. Smith, A. Stockell, and J. R. Kimmel, *JBC* **207,** 551 (1954).
48. A. K. Balls and H. Lineweaver, *JBC* **130,** 669 (1939).

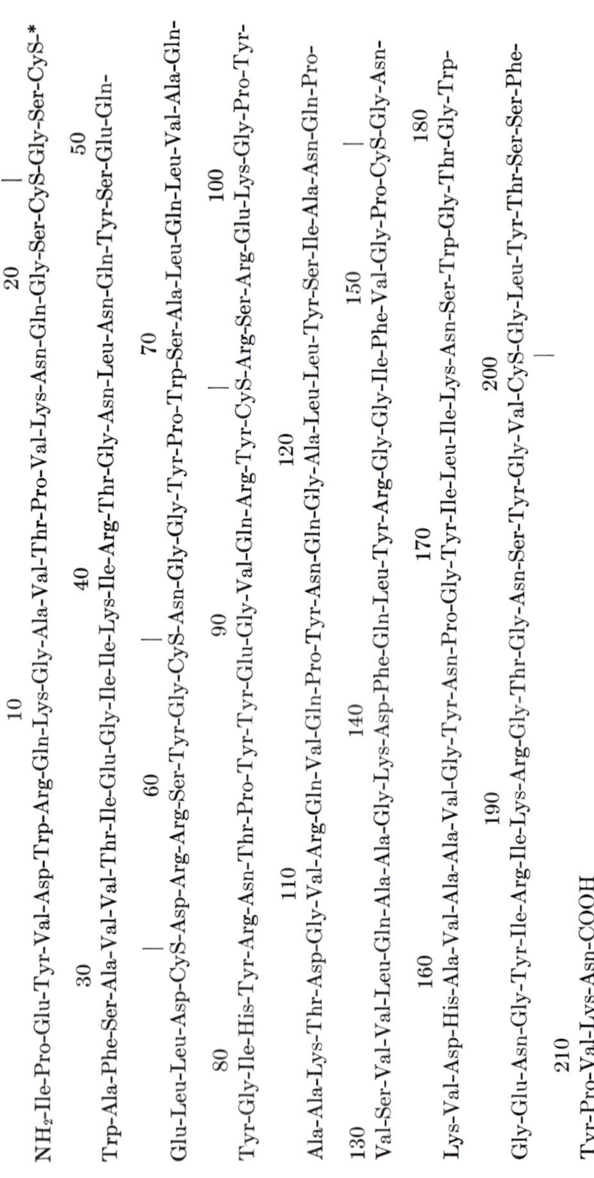

Fig. 1. Amino acid sequence of papain.

between the cysteic acid recovery and the sulfur content of the protein gave rise to a variety of speculations regarding the nature of the moieties containing the remaining S atoms (*6*). However, redetermination of the cysteic acid content of the protein by improved performic acid oxidation and analytical procedures gave a content varying from 7 to 8 residues per molecule of oxidized papain (*49, 50*). A variable contribution ranging from 0.04 to 0.51 residues has been demonstrated to derive from the mixed disulfide between the thiol group of the enzyme and cysteine (*50, 51*). This is more fully discussed below. Argentimetric amperometric titrations on papain in the presence and absence of sulfite and denaturing agents showed that the protein contained three disulfide bridges and one thiol group (*50*). These results are in agreement with subsequent structural studies on papain (see Section III). The above findings account for 7.5–8 atoms of sulfur per molecule of papain. The remaining discrepancy of about one S atom between the elementary analysis and the half-cystine content most probably represents experimental error in the former, although contamination by some sulfur-containing compound of the latex has not been excluded.

B. Activation of Papain

A comprehensive review of the literature on this subject up to 1960 has been presented earlier (*6*) and will not be repeated here.

Papain, as prepared by the method of Kimmel and Smith (*11, 12*), contains a variable sulfhydryl titer varying between 0.1 and 0.5 mole —SH per mole of enzyme (*50, 52*). Finkle and Smith (*45*) showed that in the absence of added activators a linear correlation existed between the catalytic activity of papain and its thiol content. Elicitation of the maximal hydrolytic activity of papain requires activation of the enzyme. Papain is activated by a variety of thiol compounds such as cysteine and glutathione (*5*) as well as by reducing agents such as sodium borohydride (*50*). Maximal activity is obtained with reducing agents which can also act as chelators, e.g., 2,3-dimercaptopropanol (*11*). Where this is not the case, addition of ethylenediaminetetraacetate (EDTA) is required for maximal activity (*11*). The role of the chelating agent appears to be restricted to the removal of traces of heavy metal ions which might otherwise bind to the essential thiol group of the enzyme.

49. A. N. Glazer, *Federation Proc.* **20,** 380 (1961).
50. A. N. Glazer and E. L. Smith, *JBC* **240,** 201 (1965).
51. I. B. Klein and J. F. Kirsch, *BBRC* **34,** 575 (1969).
52. T. Sanner and A. Pihl, *JBC* **238,** 165 (1963).

The causes of the variability in the thiol titer of unactivated papain and the details of the mechanism of activation have been examined by a number of investigators (*45, 50, 51, 53*). The combined results of several studies strongly suggest that the active thiol group in unactivated papain is present partly in the form of a mixed disulfide with cysteine, partly in the form of a sulfenic acid, and partly in higher oxidation state (e.g., sulfinic). The activation process consists of the conversion of the first two of these forms to the sulfhydryl.

The best documented form of inactive papain, prepared by the method of Kimmel and Smith (*12*) (see Section I,A), is that containing a mixed disulfide between the thiol group at the active site of the enzyme and cysteine (*50, 51*). The thiol group at the active site of papain readily participates in disulfide interchange reactions. Sluyterman (*53*) showed that activated papain was reversibly inactivated in the presence of air and low concentrations of cysteine and that this inactivation could be attributed directly to the formation of a mixed disulfide between the protein and cysteine. The kinetics of reactivation of this derivative and the activation kinetics for untreated unactivated papain were indistinguishable (*53*). A quantitative interchange reaction of the thiol group of active papain with cystamine had been demonstrated earlier by Sanner and Pihl (*52*). It should be noted that different batches of papain may contain different amounts of the mixed disulfide depending on the precise details of the preparative procedure such as the number of recrystallizations in absence of cysteine. Thus, Glazer and Smith (*50*) obtained a maximum of 0.28 mole of free cysteic acid after oxidation of exhaustively dialyzed preparations of unactivated papain, which yielded a thiol titer of 0.5–0.6 on activation, whereas more recently Klein and Kirsch (*51*) obtained about 0.5 mole of cysteic acid per mole of papain on preparations which gave a thiol titer of 0.4–0.5 mole on activation. Klein and Kirsch (*51*) further showed that cyanide activation of papain produces free 2-iminothiazolidine-4-carboxylic acid and a stoichiometric amount of free —SH on the protein. The activation by cyanide can therefore be represented by the scheme (*51*):

$$\text{Papain}-S-S(\leftarrow CN^-)-H_2C-CH(\overset{+}{N}H_3)(COO^-) \longrightarrow \text{Papain}-SH + S(C{\equiv}N)-H_2C-CH(\overset{+}{N}H_3)(COO^-) \longrightarrow \text{2-iminothiazolidine: } S-C(\overset{+}{\cdots}NH_2){\cdots}NH-CH(COO^-)-CH_2-S$$

There are indirect indications that some of the thiol in unactivated

53. L. A. AE. Sluyterman, *BBA* **139,** 430 (1967).

papain has been transformed to a sulfenic acid. Although aliphatic sulfenic acids have not been isolated in aqueous solution, evidence has accumulated that oxidative reactions lead to the formation of stable sulfenic acids (or sulfenyl halides) in a number of proteins (*54*). Papain activated with $NaBH_4$, and then subjected to oxidation by low levels of hydrogen peroxide, can be fully reactivated by thiols (*50*). No cysteine is available for mixed disulfide formation under these conditions, and dimerization of the protein has been excluded (*50*). Conversion of the thiol to the sulfenic acid would appear probable under the above conditions. Further, Sluyterman (*53*) has shown that the kinetics of reactivation of papain inactivated by peroxide are different from those observed for the reduction of papain–S–S–cysteine.

In virtually all papain preparations, the thiol content is less than one per molecule of enzyme after activation (*45*). This suggests that some portion of the thiol group at the active site has undergone oxidation to a higher level, e.g., sulfinic. This appears quite possible when the methods of collection and storage of the latex are taken into consideration (*4*).

On the basis of the inhibition of papain by carbonyl reagents (*55–62*), it had been proposed that an aldehyde moiety is present on papain and that the inactive form of the enzyme is that in which a thiohemiacetal is formed between the thiol group at the active site and this aldehyde (*61, 62*). The extent of inhibition by carbonyl reagents was found to be dependent on the nature of the activator. Thus, papain activated by cyanide was extensively inhibited by phenylhydrazine, whereas cysteine-activated papain showed considerably less inhibition under the same conditions (*55, 56, 60*). Klein and Kirsch (*63*) showed that the treatment of papain with carbonyl reagents such as phenylhydrazine in the presence of activators such as cysteine or $NaBH_4$ did not lead to inhibition. Treatment of papain with such reagents after removal of the activators did lead to inactivation, but this could be directly attributed to oxidation of the thiol group at the active site (*63*). Likewise, they showed that inhibition of cyanide-activated papain by carbonyl reagents is due to an oxidative coupling of cyanide to the enzyme resulting in the conver-

54. A. N. Glazer, *Ann. Rev. Biochem.* **39,** 101 (1970).
55. M. Bergmann and W. F. Ross, *JBC* **111,** 659 (1935).
56. M. Bergmann and W. F. Ross, *JBC* **114,** 717 (1936).
57. S. Maeda, *Bull. Chem. Soc. Japan* **12,** 319 (1937).
58. S. Okumura, *Bull. Chem. Soc. Japan* **13,** 534 (1938).
59. S. Okumura, *Bull. Chem. Soc. Japan* **14,** 161 (1939).
60. T. Masuda, *J. Biochem. (Tokyo)*, **46,** 1489 (1959).
61. K. Morihara, K. Monna, and S. Akabori, *Proc. Japan Acad.* **41,** 828 (1965).
62. K. Morihara, *J. Biochem. (Tokyo)* **62,** 250 (1967).
63. I. B. Klein and J. F. Kirsch, *JBC* **244,** 5928 (1969).

sion of the cysteine residue at the active site to a β-thiocyanatoalanine (*63*). These experiments appear to dispose decisively of the possibility of aldehyde group involvement in the active site of papain.

Phosphorothioate has been reported to be an effective activator of papain at very low activator-to-enzyme ratios (*64*). This is not surprising since it is an effective reducing agent capable of cleaving disulfide bonds in proteins (*65*). Neumann *et al.* (*64*) reported the stoichiometric formation of a covalent papain–phosphorothioate complex and release of a free thiol group on the enzyme on activation. These findings are difficult to reconcile with the data on unactivated papain presented above. Indeed, more recent studies show that activation by phosphorothioate, at activator enzyme ratios of 1:1, is not associated with binding of the activator to the protein but does result in the appearance of about 0.3 mole of SH per mole of enzyme (*63*).

C. Location of the Active Site Thiol Group

Finkle and Smith (*45*) demonstrated that inhibition of papain by iodoacetamide resulted from the conversion of a single cysteine residue to the corresponding alkylated derivative. Sequence analysis on papain, inactivated by treatment with ^{14}C-iodoacetic acid, demonstrated that the radioactive label was associated with a single residue in the protein, cysteine-25 (*39*). It should be emphasized that in studies on the inhibition of activated papain with bromo- or iodoacetic acid no carboxymethyl derivatives of histidine or lysine were seen on amino acid analysis of the inactivated protein (*66*). This conflicts with the report of Yu-Kum and Chen-Lu (*67*) that treatment of papain in the presence of cysteine with bromoacetate leads to the carboxymethylation of a histidine residue.

More recently, Husain and Lowe (*68*) showed that a reactive substrate analog, *p*-$CH_3 \cdot C_6H_4 \cdot SO_2 \cdot NH \cdot CH_2 \cdot CO \cdot CH_2Cl$, inactivated papain by alkylating the thiol residue at the active site and isolated a peptide from the chymotryptic digest of the protein which, from its composition and by comparison with the known amino acid sequence (*39*), was identified as containing cysteine-25.

64. H. Neumann, M. Shinitzky, and R. A. Smith, *Biochemistry* **6,** 1421 (1967).
65. H. Neumann, R. Goldberger, and M. Sela, *JBC* **239,** 1536 (1964).
66. A. Light, *BBRC* **17,** 781 (1964).
67. S. Yu-Kum and T. Chen-Lu, *Sci. Sinica* (*Peking*) **12,** 1845 (1963).
68. S. S. Husain and G. Lowe, *Chem. Commun.* p. 345 (1965).

D. Chemical Modifications at the Active Site of Papain

The sulfhydryl group at the active site can be specifically alkylated by phenylmethanesulfonyl fluoride (*69*), the chloromethyl ketones derived from *N*-tosyl-L-lysine (TLCK) (*69–71*), *N*-tosyl-L-phenylalanine (TPCK) (*69–72*), and *N*-tosyl-L-glycine (*68*), dibromoacetone (*73*), iodo-, bromo-, and chloroacetic acids and amides (*45, 66, 74–76*), as well as both antipodes of α-iodopropionic acid and amide (*77*).

The pH profile of such alkylation reactions has been studied for a number of the above compounds. The pH rate profiles for TPCK (*69, 72*) and the α-haloamides (*75, 77*) are sigmoid, showing an increase in rate which is dependent on an ionizable basic group with a pK_a in the range of pH 8.2–8.6 (25°), depending on the particular alkylating agent studied. The observed results are similar to those obtained with model sulfhydryl compounds, and it is generally agreed that the observed pH profile represents the dependence of the alkylation rate on the concentration of the thiolate ion and that the pK_a obtained from the pH-rate profile is that of the sulfhydryl group at the active site. An independent indication that the pK_a of this group may indeed be about 8.5 is suggested by a comparison of the pH dependence of the fluorescence of unactivated and active papain (*15, 16a*).

The pH-rate profile for the alkylation of papain by negatively charged alkylating agents, the halo acids, is very complex and differs radically from that described above. The pH-rate profile for chloroacetic acid, typical of those obtained with other haloacids, is compared with that for chloroacetamide in Fig. 2 (*76*). The profiles obtained with the haloacids (*75, 77*) are strikingly similar to the k_{cat}/K_m(app)–pH-rate profiles for the papain-catalyzed hydrolysis of specific substrates. Analysis of such kinetic data has been attempted by Wallenfels and Eisele (*77*) for the reaction of papain with the D(+) and L(−) antipodes of α-iodopropionic acid and by Chaiken and Smith (*76*) for chloroacetic acid. It is suggested that the alkylating agent is oriented at the catalytic site through repulsive interaction with a carboxylate group and attractive interaction

69. J. R. Whitaker and J. Perez-Villaseñor, *ABB* **124,** 70 (1968).
70. E. Shaw, M. M. Guia, and W. Cohen, *Biochemistry* **4,** 2219 (1965).
71. B. C. Wolthers, *FEBS Letters* **2,** 143 (1969).
72. M. L. Bender and L. J. Brubacher, *JACS* **88,** 5880 (1966).
73. S. S. Husain and G. Lowe, *Chem. Commun.* p. 310 (1968).
74. L. A. AE. Sluyterman, *BBA* **151,** 178 (1968).
75. I. M. Chaiken and E. L. Smith, *JBC* **244,** 5087 (1969).
76. I. M. Chaiken and E. L. Smith, *JBC* **244,** 5095 (1969).
77. K. Wallenfels and B. Eisele, *European J. Biochem.* **3,** 267 (1968).

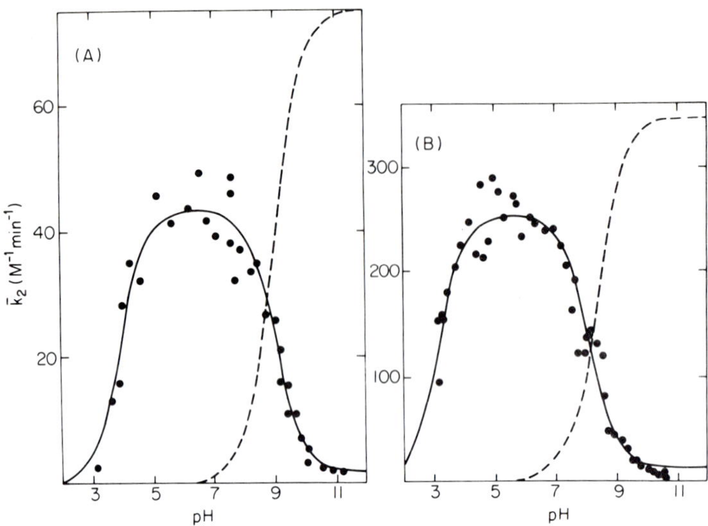

FIG. 2. Variation of apparent second-order rate constants with pH for the inhibition of activated papain by chloroacetic acid, $\Gamma/2 = 0.07$. (A) 4.5°, at a chloroacetic acid concentration of 0.667 mM, and (B) 30.5°, at a chloroacetic acid concentration of 0.167 mM. The broken lines in (A) and (B) are the pH-$\bar{k}_2$ curves which describe the reaction of chloroacetamide with activated papain at $\Gamma/2 = 0.07$ at 4.5° and 30.5°, respectively (*75, 76*).

with an imidazolium group (*76, 77*). The assignment of pK's for these groups and the interpretation of the data are complicated on the high pH side by the ionization of the thiol group and on the acid side by the protonation of the alkylating agent.

At equimolar concentrations of reagent and protein, 1,3-dibromoacetone inactivates papain rapidly (*73, 78*). This bifunctional reagent cross-links Cys 25 through the thiol group to the N-1 of the imidazole ring of His 159 (*79*). Presumably, the thiol group reacts first, followed by a slower intramolecular alkylation of the histidine residue.

V. Other Chemical Modification Reactions

A. REACTION WITH DIISOPROPYLFLUOROPHOSPHATE

Early reports of the inhibition of papain by diisopropylfluorophosphate (DFP) (*80, 81*) were subsequently shown to be erroneous (*82*).

78. S. S. Husain and G. Lowe, *BJ* **108,** 855 (1968).

The inhibition resulted from an inhibitor present in many commercial DFP preparations, which reacts with the sulfhydryl group of papain (*83, 84*) as well as of a number of other SH proteases. However, treatment of papain (or mercuripapain) with DFP does result in alkylphosphorylation of the enzyme, with the incorporation of a maximum of one mole of phosphate per mole of protein (*82, 85*). This stoichiometric reaction results in the formation of an *O*-diisopropylphosphoryl derivative of Tyr 123, but it is not associated with any loss in enzymic activity (*85*).

B. Partial Reduction of Papain

Shapira and Arnon (*86*) showed that under suitable conditions (0.32 M β-mercaptoethanol in 8 M urea at pH 8.2) one of the three disulfide bridges of papain, that linking Cys 56 to Cys 95, could be reduced selectively. After removal of the denaturing and reducing agents and addition of $HgCl_2$, the derivative could be crystallized (*87*). This crystallizable derivative contained two firmly bound atoms of Hg per molecule of protein. Presumably, one Hg atom is bound to the sulfhydryl group at the active site of papain, while the other cross-links the sulfhydryl groups arising from the reduction of the Cys 56 to Cys 95 S–S bond, forming a bridge, S–Hg–S (*87*). The latter "internal" Hg atom is retained in the molecule under the conditions used for the activation of mercuripapain (0.005 M cysteine and 0.002 M EDTA at pH 6). The derivative activated in this manner shows a similar V_{max}, with BAEE as substrate, as native papain, but a somewhat higher K_m value (*87*). From examination of the three-dimensional structure of the papain molecule, it appears likely that the conversion of the Cys 56 to Cys 95 S–S bond to a S–Hg–S bridge does not produce major changes in the overall conformation of the molecule.

C. Water-Insoluble Derivatives of Papain

Both the α-amino and the ϵ-amino groups of papain can be modified without loss of biological activity (*88, 89*). This property and the high

79. S. S. Husain and G. Lowe, *BJ* **108,** 861 (1968).
80. R. M. Heinicke and R. Mori, *Science* **129,** 1678 (1959).
81. T. Masuda, *J. Biochem.* (*Tokyo*) **46,** 1569 (1959).
82. T. Murachi, *BBA* **71,** 239 (1963).
83. N. R. Gould and I. E. Liener, *Biochemistry* **4,** 90 (1965).
84. T. Murachi and M. Yasui, *Biochemistry* **4,** 2275 (1965).
85. I. M. Chaiken and E. L. Smith, *JBC* **244,** 4247 (1969).

tyrosine content of papain facilitate coupling of the enzyme, through the amino and phenolic side chains, to a variety of water-insoluble materials with retention of most of the catalytic activity (*90, 91*).

Water-insoluble papain derivatives have been obtained by coupling papain to a water-insoluble salt derived from a copolymer of *p*-aminophenylalanine and leucine (*92, 93*), linking to insoluble preparations of collagen through bisdiazobenzidine-2,2′-disulfonic acid (*93*), coupling covalently to porous silica glass with a silane coupling agent (*94*), and insolubilization of the enzyme itself with bisdiazobenzidine (*93*) or glutaraldehyde (*95*). The properties of papain, trapped in a collodion matrix and then cross-linked through bisdiazobenzidine-3,3′-disulfonic acid, have been studied extensively as a model for a membrane-bound enzyme (*96, 97*).

Water-insoluble papain preparations can be stored for prolonged periods of time with little loss in activity and retain activity after removal of activators (*90*).

VI. Enzymic Properties

A. Methods of Assay

The amount of active enzyme in papain solutions can be determined by a number of methods. Finkle and Smith (*45*) showed that the sulfhydryl titer of papain (determined after removal of activators on an ion exchange column) was a direct measure of the amount of active enzyme present. Activator-free papain may also be readily obtained by shaking

---

86. E. Shapira and R. Arnon, *JBC* **244**, 1026 (1969).
87. R. Arnon and E. Shapira, *JBC* **244**, 1033 (1969).
88. G. S. Shields, R. L. Hill, and E. L. Smith, *JBC* **234**, 1747 (1959).
89. L. A. AE. Sluyterman, J. Wijdenes, and B. G. Wolthers, *BBA* **173**, 392 (1969).
90. I. H. Silman and E. Katchalski, *Ann. Rev. Biochem.* **35**, 873 (1966).
91. G. Manecke and G. Günzel, *Naturwissenschaften* **54**, 647 (1967).
92. J. J. Cebra, D. Givol, I. H. Silman, and E. Katchalski, *JBC* **236**, 1720 (1961).
93. I. H. Silman, M. Albu-Weissenberg, and E. Katchalski, *Biopolymers,* **4**, 441 (1966).
94. H. H. Weetall, *Science* **166**, 615 (1969).
95. E. F. Jansen and A. C. Olson, *ABB* **129**, 221 (1969).
96. R. Goldman, I. H. Silman, S. R. Caplan, O. Kedem, and E. Katchalski, *Science* **150**, 758 (1965).
97. R. Goldman, O. Kedem, I. H. Silman, S. R. Caplan, and E. Katchalski, *Biochemistry* **7**, 486 (1968).

a solution of the enzyme with *p*-thiocresol (*98*) or by reduction with sodium borohydride (*50*). "Normality" titrations on papain have been performed at low pH with the *p*-nitrophenyl esters of *N*-carbobenzoxy-L-tyrosine and *N*-acetyl-D,L-tryptophan as titrants (*72*). 2-Phenyl-4,4-dimethyl-oxazolin-5-one has been used as active site titrant for papain at pH 6.0 in the presence of cysteine (*99*).

Recently, Brocklehurst and Little (*99a*) have described an active site titration procedure based on the reaction of papain with 2,2′-dipyridyl-disulfide.

Papain is routinely assayed in the presence of freshly prepared 0.005 *M* cysteine and 0.001 *M* EDTA (*11*). At near-neutral pH, maximal activity of the enzyme is obtained within 2 min of addition to the cysteine–EDTA mixture. The observation that k_{cat} for the papain-catalyzed hydrolysis of *N*-α-benzoyl-L-argininamide (BAA) increases with increasing cysteine concentration (*52*) has not been confirmed in later studies (*100*). In much of the earlier work, assays were performed at pH 5.5 at 39°, in the presence of 0.005 *M* cysteine and 0.001 *M* EDTA, at a substrate concentration of 0.05 *M* (*5*, *6*), and the specific activity (C_1) expressed as K_1 (first-order rate constant calculated in decimal logarithms) per milliliter of protein N per milliliter of reaction mixture. Both titrimetric (*11*, *101*) and spectrophotometric (*100*) procedures have been described for the assay of papain on amide substrates such as BAA and on ester substrates such as BAEE or *N*-tosyl-L-arginine methyl ester.

B. Specificity

1. *Hydrolysis of Peptide and Amide Bonds*

Studies of the specificity of papain on a variety of synthetic substrates as well as natural polypeptides have been extensively reviewed by Kimmel and Smith (*5*), Smith and Kimmel (*6*), and Hill (*102*). Consequently, only a brief review is given here. Papain hydrolyzes the amides of α-amino-substituted arginine, lysine, glutamine, histidine, glycine, and tyrosine (*6*, *102*). The amides of α-*N*-benzoyl-L-arginine and lysine are the most susceptible (Table III). Peptide bonds involving the α-carboxyl group of glutamic acid are split much more readily by papain at pH 4

98. M. Soejima and K. Shimura, *J. Biochem.* (*Tokyo*) **49,** 260 (1961).
99. J. de Jersey, M. T. C. Runnegar, and B. Zerner, *BBRC* **25,** 383 (1966).
99a. K. Brocklehurst and G. Little, *FEBS Letters* **9,** 113 (1970).
100. J. R. Whitaker and M. L. Bender, *JACS* **87,** 2728 (1965).
101. E. L. Smith and M. J. Parker, *JBC* **233,** 1387 (1958).
102. R. L. Hill, *Advan. Protein Chem.* **20,** 37 (1965).

TABLE III
Action of Papain on Synthetic Substrates[a,b]

Substrate	pH	C_1
α-Benzoyl-L-argininamide	5.0–7.5	1.20
α-Benzoyl-L-lysinamide	5.2	0.87
Carbobenzoxy-L-glutamic acid diamide	7.35	0.24[d]
Carbobenzoxy-L-histidinamide	5.6	0.11
Carbobenzoxy-L-leucinamide[c]	6.5	0.021
Benzoylglycinamide[c]	4.5–6.85	0.014
Acetyl-L-tyrosinamide[c]	5.9	0.003

[a] The substrates at 0.05 *M* were tested in the presence of 0.005 *M* cysteine and 0.001 *M* EDTA at 39°.

[b] From Kimmel and Smith (*5*).

[c] These compounds were only partially dissolved at the start of the assay.

[d] The observed value is the extrapolated initial constant since decreasing values of *C* were observed.

than at pH 7, indicating that the state of ionization of the γ-carboxyl group influences the catalysis (*6, 102*). On prolonged hydrolysis of peptides by papain (*102*), bonds involving not only the above-mentioned residues but also those of a number of others are split as well. This is illustrated in Table IV, compiled by Hill (*102*), which lists the products isolated from papain digests of peptides obtained from the α and β chains of human hemoglobin (*103, 104*).

The wide variety of peptide bonds cleaved by papain may be superficially interpreted as indicating a low degree of specificity of the enzyme. However, the studies of Schechter and Berger (*105, 106*) have demonstrated that the factors governing the specificity of papain cannot be discerned from examination of its action on small synthetic substrates or on small peptides chosen at random. These studies were directed at the examination of the premise that endopeptidases have large active sites which extend over several amino acid residues of the substrate, i.e. a proteolytic enzyme is capable of "recognizing" a large portion of a peptide chain (*105*). From a study of the action of papain on 40 diastereoisomeric peptides of alanine, ranging in length from Ala_2 to Ala_6, Schechter and Berger (*105*) concluded that papain has an active site extending over about 25 Å, which can be divided into seven "subsites" each accommodating one amino acid residue of the substrate. The sub-

103. W. Konigsberg and R. J. Hill, *JBC* **237,** 2547 (1962).
104. J. Goldstein, W. Konigsberg, and R. J. Hill, *JBC* **238,** 2016 (1963).
105. I. Schechter and A. Berger, *BBRC* **27,** 157 (1967).
106. I. Schechter and A. Berger, *BBRC* **32,** 898 (1968).

TABLE IV

HYDROLYSIS OF CERTAIN PEPTIDES FROM α AND β CHAINS OF HUMAN HEMOGLOBIN BY PAPAIN[a,b]

Peptide	Peptide sequences	Yield (%)
A	Asp-Leu-Ser-His-Gly-Ser-Ala	
	↑ ↑	
Papain 1A	Asp-Leu-Ser	95
Papain 2A	Ser-Ala	70
Papain 3A	His-Gly	90
B	Val-Asp-Pro-Val-Asn-Phe-Lys	
	↑	
Papain 1B	Val-Asp-Pro-Val-Asn	75
Papain 2B	Phe-Lys	95
C	Ser-Ala-Leu-Ser-Asp-Leu-His-Ala-His-Lys	
	↑ ↑ ↑	
Papain 1C	Leu-Ser	25
Papain 2C	Ser-Ala	55
Papain 3C	Asp-Leu-His	50
Papain 4C	Leu-Ser-Asp-Leu-His	45
Papain 5C	Ala-His-Lys	95
D	Gly-Ala-Glu-Ala-Leu-Glu-Arg	
	↑ ↑ ↑ ↑ ↑ ↑	
Papain 1D	Arg	Not measured
Papain 2D	Ala-Glu-Ala	20
Papain 3D	Ala- or Ala	20
Papain 4D	Gly	20
Papain 5D	Leu-Glu-Arg	35
Papain 6D	Ala-Glu-Ala-Leu	45
E	Val-Ala-His-Val-Asp-Asp-Met-Pro-Asn-Ala-Leu	
	↑ ↑ ↑ ↑ ↑	
Papain 1E	Ala-Leu	20
Papain 2E	Val-Ala	75
Papain 3E	Asp-Met-Pro-Asn-Ala	70
Papain 4E	His-Val-Asp	45
Papain 5E	His-Val-Asp-Asp	55
F	Val-His-Leu-Thr-Pro-Glu-Glu-Lys	
	↑ ↑ ↑ ↑	
Papain 1F	Glu	36
Papain 2F	Leu	33
Papain 3F	Leu-Thr-Pro-Glu	34
Papain 4F	Thr-Pro-Glu	20
Papain 5F	Glu-Lys	40
Papain 6F	Lys	42
Papain 7F	Val-His	82
G	Val-Val-Ala-Gly-Val-Ala-Asn-Ala	
	↑ ↑ ↑	
Papain 1G	Gly	95
Papain 2G	Asn-Ala	75
Papain 3G	Val-Ala	65
Papain 4G	Val-Val-Ala	70

TABLE IV (*Continued*)

Peptide	Peptide sequences	Yield (%)
H	His-Val-Asp-Pro-Glu-Asn-Phe-Arg ↑ ↑	
Papain 1H	Asn	30
Papain 2H	His-Val-Asp-Pro-Glu	40
Papain 3H	His-Val-Asp-Pro-Glu-Asn	30
Papain 4H	Phe-Arg	80
I	Thr-Tyr-Phe-Pro-His-Phe ↑	
Papain 1I	Phe-Pro-His-Phe	95
Papain 2I	Thr-Tyr	80

[a] Hydrolyses were performed at 37°–40°C for 15–18 hr at pH 5.5 with papain concentrations of 0.01–0.05%.

[b] Tabulated by Hill (*102*), from the data of Konigsberg and Hill (*103, 104*).

sites are located on both sides of the catalytic site, four on the one side and three on the other, as illustrated in the accompanying diagram, where C denotes the catalytic site and point of cleavage of the substrate.

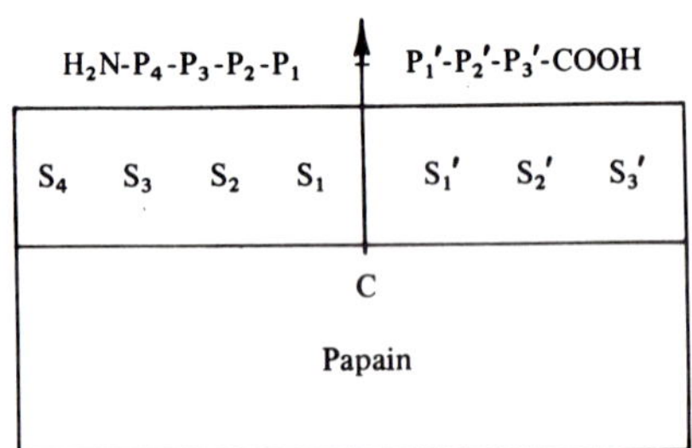

It was further found that one of the subsites, S_2, specifically interacts with a phenylalanine side chain (*106*). Thus, the presence of Phe in position three or further removed from the C-terminal of the substrate enhances the susceptibility of a peptide to hydrolysis with cleavage being at the peptide bond one residue removed from the Phe toward the C-terminal, as shown below.

Papain
↓
H_2N-Ala-Phe-Ala-Lys-Ala-$CONH_2$

The directing influence on the point of cleavage by the phenylalanine residue is sufficient to prevent the expected cleavage of the above peptide at the lysine residue (*106*). As expected from these results, peptides containing Phe as the second residue from C-terminal such as Ala–Ala–Phe–Ala or Ala–Ala–Phe–Lys are strong inhibitors (*106*). Subsite S_1

does show preference for Lys over Ala, as may be expected from studies on small synthetic substrates (*6*). Examination of the points of cleavage of the B chain of oxidized insulin (*107*) and of other peptides (see Table IV, for example) indicates that Val and Leu probably also interact strongly with subsite S_2 (*106*). Independent evidence for the existence of subsites S'_1, S'_2, and S'_3 comes from the studies of the deacylation of cinnamoyl papain by polyglycinamides of varying chain length (*108*). Brubacher and Bender (*108*) showed that the rate of deacylation increased markedly with chain length up to triglycinamide but that tetraglycinamide was only slightly more effective than triglycinamide.

A review of the progress in the mapping of the active site of papain with various substrates and competitive inhibitors has appeared recently (*108a*).

Undoubtedly, future careful studies of the specificity of each of the subsites, coupled with X-ray crystallographic studies of enzyme–inhibitor complexes, should provide a rational basis for predicting the specificity of papain on polypeptide substrates, at least with respect to the major points of cleavage.

Because of its broad specificity, papain has not been commonly used for the hydrolysis of proteins and very long polypeptides for sequence work; however, it has been exceedingly valuable for the hydrolysis of peptides of moderate size, particularly when these do not contain bonds of the type susceptible to the widely employed trypsin, chymotrypsin, etc.

Some examples of the utilization of papain on peptides may be found in the following: hemoglobin (*103, 104*), cytochromes c (*109, 110*), and subtilisin BPN′ (*111–113*).

2. *Esterase and Thiolesterase Action*

Papain is an effective catalyst of the hydrolysis of a variety of ester and thiolester (*6, 114, 115*) substrates. Although esters of *N*-acylamino acids, particularly those with an aromatic acyl group, are good sub-

107. J. T. Johansen and M. Ottesen, *Compt. Rend. Trav. Lab. Carlsberg* **36,** 265 (1968).

108. L. J. Brubacher and M. L. Bender, *BBRC* **27,** 176 (1967).

108a. A. Berger and I. Schechter, *Phil. Trans. Roy. Soc. London* **B257,** 249 (1970).

109. E. Margoliash, *JBC* **237,** 2161 (1962).

110. F. C. Stevens, A. N. Glazer, and E. L. Smith, *JBC* **242,** 2764 (1967).

111. C. B. Kasper and E. L. Smith, *JBC* **241,** 3754 (1966).

112. C. B. Kasper and E. L. Smith, *JBC* **241,** 3771 (1966).

113. F. S. Markland, G. Kreil, B. Ribadeau-Dumas, and E. L. Smith, *JBC* **241,** 4642 (1966).

114. A. N. Glazer, *JBC* **241,** 3811 (1966).

115. R. M. Metrione and R. B. Johnston, *Biochemistry* **3,** 482 (1964).

strates of papain, amino acid esters are not hydrolyzed at a measurable rate. The pH-activity curve for esterase action is similar to that obtained for amide substrates and is fully discussed below (see Section VI,C).

3. *Transamidation and Transesterification Reactions*

The synthetic reactions of papain and the hydrolysis of "co-substrates" are dealt with comprehensively in earlier reviews (*5, 6*). In brief, papain catalyzes transamidation reactions of the type

$$\text{Carbobenzoxyglycinamide} + H_2N\text{-}R \underset{}{\overset{\text{Papain}}{\rightleftharpoons}} \text{Carbobenzoxyglycyl-NH-R} + NH_3$$

where the acceptor amine can be a peptide (e.g., L-Leu–Gly) (*116*), an amino acid ester (e.g., glycine ethylester) (*117*), ammonia (*118*), hydroxylamine (*119*), or any of a wide variety of other acceptors (*6, 114, 120*). The transfer reaction proceeds more rapidly at pH 8 than at pH 5, suggesting that the attacking species is the free amino group.

Papain also catalyzes transesterification reactions (*114, 121*). As in the case of transamidation, the ratio of alcoholysis to hydrolysis varies with the substrate and alcohol under study (*114*). This point is examined further below.

4. *Binding of Nucleophiles*

Fruton and co-workers (*116, 122–124*) carried out extensive studies on papain-catalyzed transamidation reactions. They demonstrated that papain exhibits specificity toward the nucleophile (*116*). Thus, in an equimolar mixture of carbobenzoxyglycinamide, Gly–Gly, and L-Leu-Gly at pH 7.3, L-Leu–Gly inhibited the reaction leading to carbobenzoxytriglycine by 75% while the reaction leading to carbobenzoxy-Gly-L-Leu–Gly proceeded at the same rate as in the absence of Gly–Gly (*116*). Similar observations have been made with glycylglycine esters as substrates (*114*). A dramatic enhancement of the rate of deacylation of *trans*-cinnamoyl papain by certain nucleophiles has been observed (*108, 125*). Thus, L-tryptophanamide has a second-order rate constant more

116. M. J. Mycek and J. S. Fruton, *JBC* **226,** 165 (1967).
117. S. Yu-Kun and T. Chen-Lu, *Sci. Sinica* (*Peking*) **12,** 201 (1963).
118. J. S. Fruton, *Yale J. Biol. Med.* **22,** 263 (1950).
119. J. Durell and J. S. Fruton, *JBC,* **207,** 487 (1954).
120. R. P. Carty and D. M. Kirschenbaum, *BBA* **85,** 446 (1964); **110,** 339 (1965).
121. A. W. Lake and G. Lowe, *BJ* **101,** 402 (1966).
122. R. B. Johnston, M. J. Mycek, and J. S. Fruton, *JBC* **185,** 629 (1950).
123. J. S. Fruton, R B. Johnston, and M. Fried, *JBC* **190,** 39 (1951).
124. Y. P. Dowmont and J. S. Fruton, *JBC* **197,** 271 (1952).
125. A. L. Fink and M. L. Bender, *Biochemistry* **8,** 5109 (1969).

than a millionfold greater than water, whereas glycinamide, which has essentially the same pK_a, is a poorer nucleophile by a factor of over 100 (*108, 125*). Brubacher and Bender (*126*) showed that the second-order rate constants for the reaction of nucleophiles with *trans*-cinnamoyl papain were dependent on the structure of the nucleophile but were not correlated with the basicity of the nucleophile. The above results clearly show that the nucleophiles are bound to the enzyme in a highly stereospecific manner. An investigation of *n*-alkyl alcohols as nucleophiles showed that simple partition of the acyl enzyme between water and alcohols occurred only for methanol and possibly for ethanol (*125*). For higher members of the series, the results could be satisfactorily explained by a scheme in which the alcohol binds in both the area of substrate binding and in an area overlapping that of the substrate leaving group (*125*). Ternary complex formation of enzyme–substrate–nucleophile was clearly demonstrated by these (*125*), as well as earlier, studies (*114*).

It is obvious, in the light of the above discussion, that the nature of the substrate leaving group can make a significant contribution to the kinetic constants in papain-catalyzed hydrolyses.

In terms of the stereochemistry of the binding site of papain, as delineated by Schechter and Berger (*105*), the acceptor efficiency of the various nucleophiles reflects the interaction of these molecules with the S_1', S_2', and S_3' subsites. It should, therefore, be possible to use transfer reactions to study the specificity of these subsites (*106*).

C. Kinetic Studies

1. *Effect of pH, Ionic Strength, and Temperature*

The stability of papain is sufficiently high to permit kinetic studies over the pH range 2.8–10.8 (*127*). Below pH 2.0, papain is rapidly and irreversibly inactivated. The enzyme is sufficiently thermostable in the presence of substrates to allow kinetic measurements over the temperature range from 5° to 66° (*127*). Stockell and Smith (*127*) showed that reaction velocity increased with ionic strength at low ionic strength and was independent of ionic strength between 0.05 and 0.3.

2. *Acyl-Enzyme Intermediate*

Based solely on the then known essentiality of a sulfhydryl group for the activity of papain, and the ease of hydrolysis of thiolesters, Weisz

126. L. J. Brubacher and M. L. Bender, *JACS* **88**, 5871 (1966).
127. A. Stockell and E. L. Smith, *JBC* **227**, 1 (1957).

(*128*), in a long overlooked paper, proposed in 1937 that papain-catalyzed hydrolysis of peptides was a two-step process involving the formation of an acyl thiolenzyme and its subsequent hydrolysis; however, this inspired guess attracted little contemporary notice. In 1955, on the basis of chemical and kinetic evidence. Smith and his co-workers (*129*) proposed independently essentially the same scheme for papain-catalyzed reactions

$$\text{E—SH} + \text{R—}\overset{\overset{\displaystyle\text{O}}{\|}}{\text{C}}\text{—X} \underset{k_{-1}}{\overset{k_1}{\rightleftharpoons}} \text{E—S—}\overset{\overset{\displaystyle\text{O}}{\|}}{\text{C}}\text{—R} \overset{k_0}{\rightarrow} \text{E—SH} + \text{RCOOH}$$
$$+ \text{HX}$$

A similar scheme had also been proposed by Racker (*130*) for glyoxalase. In an attempt to identify groups on the enzyme involved in the catalysis as well as to probe at the mechanism, Smith and co-workers (*6, 101, 127, 131*) determined the kinetic constants, K_m and k_0, as a function of pH and temperature, for the papain-catalyzed hydrolysis of a number of synthetic *N*-acylamino acid ester and amide substrates. Basic, neutral and acidic substrates were examined, and the data analyzed according to the usual Michaelis–Menten equation:

$$\text{E} + \text{S} \underset{k_{-1}}{\overset{k_1}{\rightleftharpoons}} \text{ES} \overset{k_0}{\rightarrow} \text{E} + \text{P}$$

$$K_m = \frac{k_{-1} + k_0}{k_1}$$

and the assumption that $k_0 \gg k_{-1}$, and hence that $k_1 = k_0/K_m$ (*127*).

Similar results were obtained for the various substrates studied (Table V) (*100, 101, 127, 131–134*). For a typical substrate such as BAEE the pH-rate profile for the acylation step (i.e., k_1 vs. pH) was shown to be bell-shaped and could be interpreted to indicate the involvement of two groups of apparent pK_a values of 4.3 and 8.2 at 25° (*101*). The group with apparent pK_a of 4.3 showed a negligible heat of ionization and was proposed to be a carboxylate ion. The group with pK_a of 8.2 showed a

128. J. Weisz, *Chem. Ind.* (*London*) **15,** 685 (1937).

129. E. L. Smith, B. J. Finkle, and A. Stockell, *Discussions Faraday Soc.* **20,** 96 (1955).

130. E. Racker, *in* "The Mechanism of Enzyme Action" (W. D. McElroy and B. Glass, eds.), p. 464. Johns Hopkins Press, Baltimore, Mayland, 1954.

131. E. L. Smith, V. J. Chavré, and M. J. Parker, *JBC* **230,** 283 (1958).

132. L. A. AE. Sluyterman, *BBA* **85,** 305 (1964).

133. D. C. Williams and J. R. Whitaker, *Biochemistry* **6,** 3711 (1967).

134. W. Cohen and P. H. Petra, *Biochemistry* **6,** 1047 (1967). An incorrect value of 4.16 was used for the p*K* of benzoyl-L-citrulline in this study instead of the correct p*K* of 3.57 for this compound (*133*).

TABLE V

KINETICS CONSTANTS FOR PAPAIN-CATALYZED HYDROLYSIS OF SOME *N*-ACYLAMINO ESTER AND AMIDE SUBSTRATES

Substrate	k_2/K_s: $[k_2/K_s]$ lim (M^{-1} sec^{-1})	k_2/K_s: pK_1	k_2/K_s: pK_2	K_m(app) (M)	k_{cat}(lim) (sec^{-1})	Temp (°C)	Reference
α-*N*-Benzoyl-L-arginine ethyl ester	390	4.30	8.20	0.023	9	25	*101*
	660	4.30	8.02	0.018	12	37	*101*
	1190	4.29	8.49	0.014	16.1	25	*100*
α-*N*-Benzoyl-L-argininamide	280	3.9	8.2	0.040	11.0	38	*127*
	268	4.24	8.35	0.032	8.7	25	*100*
N-Benzoylglycine ethyl ester	147	4.1	8.0	0.021	3.1	40	*132*
N-Benzoylglycinamide	3.8	3.9	8.2	0.16	0.6[a]	38	*131*
N-Carbobenzoxyglycine *p*-nitrophenyl ester[b]	6.35×10^5	4.23	—	8×10^{-6}	5.20	25	*133*
α-*N*-Benzoyl-L-citrulline methyl ester	1808	4.23	8.50	0.010	18.6	25	*133,134*
α-*N*-Carbobenzoxy-L-histidinamide	31	4.3	8.0	0.02[c]	4.0[d]	38	*131*
N-Carbobenzoxyglycylglycine	0.26[e]	—	8.0	0.32[e]	0.083[e]	38	*131*

[a] At pH 5.2, and is not maximal.
[b] In the presence of 1.6% acetonitrile, by volume.
[c] Minimum value reported at pH 5.2.
[d] At pH 6.7, and probably not maximal.
[e] At pH 7.22.

ΔH_{ion} of 5.1 kcal/mole at 0° and was identified as a thiol group, both on the basis of these data and the known requirement of a thiol group for papain activity (*101*). The kinetic studies also led to the striking observation that the V_{max} values for the hydrolysis of BAEE and BAA by papain were nearly equal (*101*). This was in sharp contrast to the known large differences in the rates of nonenzymic hydrolysis of esters and amides and to the large differences in rates of hydrolysis for related amide and ester substrates for such enzymes as trypsin, chymotryspin, and carboxypeptidase A (*135*). This observation coupled with the chemical evidence for a thiol group at the active site of papain, led Smith (*136*) to propose that papain hydrolysis of esters and amides proceeded through a common acyl-thiolenzyme intermediate. This hypothesis has been amply verified by the spectrophotometric observation of *N-trans*-cinnamoyl papain (*137*) and thionohippuryl papain (*138*) discussed below.

A reexamination of the papain-catalyzed hydrolysis of BAEE and BAA (*100*), indicated that the minimal kinetic scheme for the hydrolysis of these substrates required a three-step, rather than a two-step mechanism:

$$E + S \underset{k_{-1}}{\overset{k_1}{\rightleftharpoons}} ES \xrightarrow{k_2} ES' + P_1 \xrightarrow{k_3} E + P_2$$

where ES is the Michaelis–Menten complex and ES′ is the acyl-enzyme intermediate. The constants in this equation are related to those of the usual Michaelis–Menten equation

$$E + S \overset{K_m(\text{app})}{\rightleftharpoons} ES \xrightarrow{k_{cat}} E + P_1 + P_2$$

by the relations

$$K_m(\text{app}) = (k_{-1} + k_2)k_3/k_1(k_2 + k_3)$$
$$k_{cat} = k_2k_3/(k_2 + k_3)$$

If it is assumed that $k_{-1} \gg k_2$, then it follows that

$$K_m(\text{app}) = [k_3/(k_2 + k_3)]K_s$$
$$k_{cat}/K_m(\text{app}) = k_2/K_s$$

where $K_s = k_{-1}/k_1$.

The limiting k_{cat} values for papain-catalyzed hydrolyses of BAEE and BAA obtained by Whitaker and Bender (*100*) were 16.1 sec^{-1} and 8.7

135. H. Neurath and G. W. Schwert, *Chem. Rev.* **46,** 69 (1950).
136. E. L. Smith, *JBC* **233,** 1392 (1958).
137. M. L. Bender and L. J. Brubacher, *JACS* **86,** 5333 (1964).
138. G. Lowe and A. Williams, *Proc. Chem. Soc.* p. 140 (1964).

sec^{-1}, respectively, at 25°, in reasonable agreement with the earlier data of Smith and co-workers (Table V). However, analysis of the individual rate constants for these hydrolyses (Table VI) indicated that for both BAEE and BAA, k_{cat} is determined by both k_2 and k_3, the predominantly rate-limiting step for the ester hydrolysis being k_3 (deacylation) while it is k_2 (acylation) for the amide (*100*). Thus the similarity in the values of k_{cat} for BAEE and BAA, does not reflect a common rate limiting deacylation step for these substrates. Whitaker and Bender (*100*) reached the same conclusions as to the apparent pK values and probable identity of the groups involved in the acylation and deacylation steps as had been previously proposed by Smith and his co-workers (*6*). It may be added that the significant inhibition by the product, benzoyl-L-arginine, particularly at low pH (*100, 101, 139*), as well as its degree of ionization as a function of pH (*100, 139*), were taken into account in these studies.

The large contribution of both k_2 and k_3 to k_{cat} for BAEE and BAA weakens the arguments based on k_{cat} values for a common intermediate for these two substrates whose rate of hydrolysis is rate limiting. Studies on other substrates, however, provide much more convincing evidence for a common intermediate. Kirsch and Igelström (*140*) showed that the k_{cat} values for the papain-catalyzed hydrolysis of *p*-, *m*-, and *o*-nitrophenyl and ethyl esters of carbobenzoxyglycine were very similar (Table VII), in contrast to the rate constants for alkaline hydrolysis which varied from 40.6 M^{-1} min^{-1} for the ethyl ester to 6900 M^{-1} min^{-1} for the *p*-nitrophenyl ester (Table VII). The rates of reaction of the esters with mercaptoethanol were found to be proportional to the concentration of the anionic species and also varied extensively from 19,000 M^{-1}

TABLE VI

KINETIC CONSTANTS OF PAPAIN-CATALYZED HYDROLYSES OF α-*N*-BENZOYL-L-ARGININE ETHYL ESTER AND α-*N*-BENZOYL-L-ARGININAMIDE AT 25°[a]

Substrate	k_2(lim) (sec^{-1})	k_3(lim) (sec^{-1})	$K_s \times 10^2$ (*M*)	Acylation pK_1	Acylation pK_2	Deacylation pK_1
BAEE	64.9 ± 13.9	20.2 ± 1.7	5.45 ± 1.17	4.29	8.49	3.91
BAA	9.70 ± 2.09	28.7 ± 25.1	3.62 ± 0.78	4.24	8.35	3.91

[a] From Whitaker and Bender (*100*), which see for the methods and assumptions involved in calculating these constants.

139. L. A. AE. Sluyterman, *BBA* **85**, 316 (1964).
140. J. F. Kirsch and M. Igelström, *Biochemistry* **5**, 783 (1966).

TABLE VII
KINETICS OF THE PAPAIN-CATALYZED HYDROLYSES OF ESTERS OF CARBOBENZOXYGLYCINE AND THE RATE CONSTANTS FOR HYDROLYSIS BY HYDROXIDE ION[a]

	Papain[b]		Hydroxide ion[c]
Esters	K_m ($M \times 10^5$)	k_{cat} (sec^{-1})	$k_{OH} = k_{obsd}/[OH^-]$ (M^{-1} min^{-1})
p-Nitrophenyl	0.93 ± 0.10	2.73 ± 0.08	6900
m-Nitrophenyl	1.89 ± 0.17	2.18 ± 0.06	4050
o-Nitrophenyl	15.2 ± 2.3	2.14 ± 0.15	3680
Phenyl	10.7 ± 0.9	2.45 ± 0.06	728
Ethyl[d]	514 ± 74	1.96 ± 0.14	40.6

[a] From Kirsch and Igelström (*140*).
[b] In 0.02 *M* sodium phosphate buffer, 0.001 *M* EDTA, 3.3 × 10^{-4} *M* cysteine, 6.7% (v/v) acetonitrile, pH 6.8, 25°.
[c] In 0.05–0.1 *M* triethylamine buffers, containing 1–2% (v/v) acetonitrile, at an ionic strength of 1.0 and 25°. There was no appreciable buffer-catalyzed or neutral hydrolysis under these conditions as k_{obsd} extrapolated to zero at $[OH^-] = 0$ in all cases.
[d] 6.0% (v/v) acetonitrile.

min^{-1} for the *p*-nitrophenyl to 531 M^{-1} min^{-1} for the phenyl ester at 25° (*140*). In a similar study, Lowe and Williams (*141*) showed that k_{cat} for the papain-catalyzed hydrolysis of a number of alkyl and aryl esters of hippuric acid was 2–3 sec^{-1} and concluded that it most probably represents the deacylation of a common intermediate, hippuryl papain. Similar results have been obtained, showing near identity of the k_{cat} values, for a series of *N*-methanesulfonylglycine esters (*142*).

The three-step kinetic scheme presented above for papain-catalyzed hydrolyses can be expanded to include transfer reactions to alcohols.

$$E + S \underset{k_1}{\overset{k_1}{\rightleftharpoons}} ES \xrightarrow{k_2} \begin{matrix} ES' \\ + \\ P_1 \end{matrix} \begin{matrix} \xrightarrow{k_3\,[H_2O]} E + P_2 \\ \\ \xrightarrow{k_4\,[ROH]} E + P_3 \end{matrix}$$

The values of K_m and k_{cat} derived from this scheme using the steady state assumption are

141. G. Lowe and A. Williams, *BJ* **96**, 199 (1965).
142. E. C. Lucas and A. Williams, *Biochemistry* **8**, 5125 (1969).

$$K_m = \frac{(k_3[H_2O] + k_4[ROH])K_s}{k_2 + k_3[H_2O] + k_4[ROH]}$$

$$k_{cat}(\text{hydrolysis}) = \frac{k_2k_3[H_2O]}{k_2 + k_3[H_2O] + k_4[ROH]}$$

$$k_{cat}(\text{alcoholysis}) = \frac{k_2k_4[ROH]}{k_2 + k_3[H_2O] + k_4[ROH]}$$

If an acyl-enzyme intermediate is formed in each case, then $k_{alcoholysis}/k_{hydrolysis}$ should be independent of the starting substrate. Henry and Kirsch (*143*) demonstrated that this criterion was satisfied for the ratio $k_{ethanol}/k_{water}$ for ethyl hippurate and *p*-nitrophenylhippurate for papain-catalyzed hydrolysis and ethanolysis of these substrates in water–ethanol mixtures.

The hydrolysis of ethanol [carbonyl-^{18}O] hippurate by papain proceeds without oxygen-18 exchange (*144*). This finding is also consistent with the formation of an acyl-enzyme intermediate rather than a tetrahedral intermediate (*145*), although the results could also be compatible with the formation of an enzyme–substrate complex so sterically hindered as to preclude oxygen-18 exchange (*144*).

The acyl-enzyme hypothesis is also consistent with the papain-catalyzed exchange of ^{18}O between the carbonyl position of carbobenzoxy-amino acids and water (*146*).

The most convincing evidence for the acyl-thiolenzyme intermediate in certain papain-catalyzed reactions comes from studies on the reaction of *O*-methylthiohippurate (*147*) and *N*-*trans*-cinnamoylimidazole (*126*) with the enzyme, as well as from the "normality" titrations discussed above (*72, 99*). Lowe and Williams (*147*) showed that *O*-methylthiohippurate was a specific substrate for papain and the ultraviolet absorption properties of the acyl-enzyme intermediate obtained with this compound

$$C_6H_5CO{\cdot}NH{\cdot}CH_2{\cdot}CS{\cdot}O{\cdot}CH_3 + HS\text{-papain} \rightarrow C_6H_5{\cdot}CO{\cdot}NH{\cdot}CH_2{\cdot}CS{\cdot}S\text{-papain} + CH_3OH$$

were those expected of a dithioester (Table VIII) and differed significantly from those of the corresponding thiono-esters and thioamides (*147, 148*). Under suitable conditions, *N*-*trans*-cinnamoyl papain is suffi-

143. A. C. Henry and J. F. Kirsch, *Biochemistry* **6,** 3536 (1967).
144. J. F. Kirsch and E. Katchalski, *Biochemistry* **4,** 884 (1965).
145. M. L. Bender, *Chem. Rev.* **60,** 53 (1960).
146. V. Grisaro and N. Sharon, *BBA* **89,** 152 (1964).
147. G. Lowe and A. Williams, *BJ* **96,** 189 (1965).
148. K. Brocklehurst, E. M. Crook, and C. W. Wharton, *Chem. Commun.* p. 1185 (1967).

TABLE VIII

ULTRAVIOLET ABSORPTION PROPERTIES OF THIONO-ACYL DERIVATIVES[a]

Chromophore	Solvent	λ_{max} (Å)	$\log_{10} \epsilon$	λ_{max} (Å)	$\log_{10} \epsilon$
—CS·OR	Ethanol	2300	4.26	—	—
—CS·O$^-$	Water	2490	2.75	—	—
—CS·NH_2	Ether	2680	4.05	3580	1.25
—CS·N(imidazole)	—	2500	4.00	2910	4.00
—CS·SR	Hexane	3050	4.08	4600	1.25
—CS·SR	Water	3050	—	—	—
Papain—CS·R	Water	3130	4.2		
Ficin—CS·R	Water	3150	3.7		
Bromelain—CS·R[b]	Water	3160			

[a] Data from Lowe and Williams (*147*).
[b] Data from Brocklehurst *et al.* (*148*).

ciently stable to permit examination of its properties (*126*). The spectral properties of this acyl enzyme (Table IX) are compatible with those of *S-trans*-cinnamoyl derivatives provided allowance is made for a red shift in the absorption band of the chromophore in the native acyl enzyme (*126*). Methylamine is 670-fold more reactive toward *trans*-cinnamoyl-papain than toward *trans*-cinnamoyl-α-chymotrypsin (*126*). This is consistent with the known greater susceptibility of thiolesters than oxygen esters to attack by nitrogen nucleophiles.

TABLE IX

ULTRAVIOLET ABSORPTION PROPERTIES OF CINNAMOYL ESTERS[a]

trans-Cinnamoyl	λ_{max} (nm)	ϵ_{max}
α-Chymotrypsin	292	17,700
Trypsin	296	19,400
Subtilisin	389	21,000
N-Acetylserinamide[b]	281.5	24,300
Thiolsubtilisin[c]	310	—
Papain	326	26,500
Cysteine[d]	306	22,600

[a] Data from (*148a*).
[b] *O*-Cinnamoyl.
[c] From (*148b*).
[d] *S*-Cinnamoyl.

148a. M. L. Bender and F. J. Kézdy, *Ann. Rev. Biochem.* **34,** 49 (1965).
148b. L. Polgar and M. L. Bender, *JACS* **88,** 3153 (1966).

3. *Deacylation*

The dependence of the deacylation step on a group of pK_a of 3.3–4.7 has been demonstrated for a large number of specific substrates, both charged and uncharged, as well as for *N-trans*-cinnamoyl papain (Table X). Hippurate esters behave in an unusual manner in that k_3 (deacylation) appears to be essentially independent of pH (*132, 142, 144*). It has been suggested that this phenomenon may be explained by the participation of the amide oxygen of the hippuryl papain (competing with the normal process involving the group with an acid pK_a) to form an oxazolone intermediate by expelling the thiol group of papain (*142*).

A large D_2O solvent isotope effect has been observed in the deacylation of α-*N*-benzoyl-L-argininyl papain [$k_{3,H_2O}/k_{3,D_2O} = 2.75$ (*100*)] and of *N-trans*-cinnamoyl papain [$k_{3,H_2O}/k_{3,D_2O} = 3.35$ (*126*)] indicating that the basic group acts as a general base catalyst and not as a nucleophilic catalyst in the deacylation (*126*).

The identification of the general base catalyst with a carboxylate ion was based on the pK_a of ∼4 and the lack of dependence of this pK_a on temperature (*6, 100*). More recently, it has been proposed on kinetic

TABLE X

KINETIC CONSTANTS OF DEACYLATION OF VARIOUS ACYL PAPAINS[a]

Substrate	pH range	Temp (°C)	k_3 (lim) (sec^{-1})	pK_1	Reference
α-*N*-Benzoyl-L-arginine ethyl ester	3.5–9.6	25	20.2	3.91	*100*
ε-*N*-Formyl-carbobenzoxy-L-lysine-*p*-nitrophenyl ester	3.7–5.4	25	32.0	3.96	*72*
Carbobenzoxy-L-lysine-*p*-nitrophenyl ester	3.2–8.3	25	45.9	3.33	*72*
Carbobenzoxy-L-lysine benzyl ester	3.15–9.4	25	45.0	3.30	*72*
α-*N*-Acetyl-L-tryptophan *p*-nitrophenyl ester	3.6–5.9	25	3.90	4.70	*72*
Benzoylglycine ethyl ester	3.7–8.4	40	3.1	pH independent	*132*
	4.2–8.7	38	3.3	pH independent	*144*
Benzoylglycine methyl ester	3.6–8.5	35	4.02	pH independent	*142*
Acetylglycine benzyl ester	4.0–8.5	35	2.76	4.14	*142*
N-Methanesulfonylglycine benzyl ester	3.4–8.95	35	13.3	4.26	*142*
N-trans-Cinnamoyl papain	3.4–12.7	25	0.00368	4.69	*72*

[a] For exact experimental conditions and calculations of k_3 values see original papers.

grounds that the group involved is in actuality an imidazole (*141, 142*). The presence of an imidazole ring within 5 Å of the active site thiol group is demonstrated by the ability to bridge the thiol group to a histidine residue with the bifunctional reagent dibromoacetone (*149, 150*). The position of the histidine inferred from this experiment is confirmed by the crystallographic studies on papain (*40*). Further, the rate of intramolecular catalysis of hydrolysis of *S*-hippurylthioglycollate (I) and *S*-ethylmonosuccinate (II) is 10^5- to 10^6-fold slower than the deacylation rates of hippurylpapain and α-*N*-benzoyl-L-argininyl papain (*151*), whereas the rate of intramolecular imidazole-catalyzed hydrolysis of *n*-propyl α-(4-imidazolyl) thiobutyrate (III) (*152*) is only an order of magnitude slower than the enzymic deacylation rate constant for reasonably good substrates (*151*). On the basis of such data, Lowe and Williams (*151*) argued for the participation of an imidazole rather than a carboxylate group as a nucleophile in the deacylation reaction.

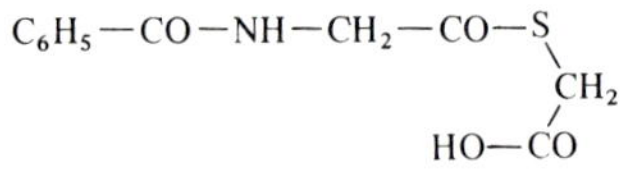

$$C_6H_5-CO-NH-CH_2-CO-S-CH_2-CO-OH$$

(I)

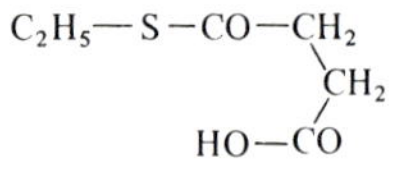

$$C_2H_5-S-CO-CH_2-CH_2-CO-OH$$

(II)

$$\text{(4-imidazolyl)}-CH_2-CH_2-CH_2-CO-S-CH_2-CH_2-CH_3$$

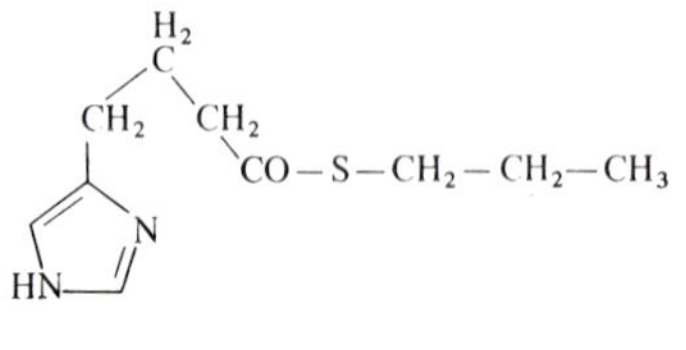

(III)

However, as pointed out by Bender and Brubacher (*72*), the applicability of the data from such model reactions is doubtful since these are examples of intramolecular nucleophilic catalysis by the carboxylate group and reflect the large difference in catalytic efficiency of carboxylate and imidazole as nucleophiles (*153*), whereas from the deuterium isotope effect it can be concluded that the deacylation of the acyl enzyme is

149. S. S. Husain and G. Lowe, *BJ* **108,** 855 (1968).
150. S. S. Husain and G. Lowe, *BJ* **108,** 861 (1968).
151. G. Lowe and A. Williams, *BJ* **96,** 194 (1965).
152. T. C. Bruice, *JACS* **81,** 5444 (1959).
153. T. C. Bruice and R. Lapinski, *JACS* **80,** 2265 (1958).

general base catalyzed (*126*). In general base catalyzed reactions, carboxylate ion is only about 100-fold less effective than imidazole (*72*).

Allen and Lowe [cited in Lowe (*153a*)] have shown that the apparent pK for the deacylation of cinnamoyl papain shifts from 4.65 in water to 4.15 in 20% dioxane–water. The solvent effect is compatible with the assignment of this pK to a positively charged acid, i.e., imidazole. An increase in pK with decrease in dielectric constant would be expected for a carboxylic acid (*153a*).

The identification of the general base catalyst with imidazole appears to be consistent with the X-ray crystallographic data [see Drenth (*24*)]. The active site of the molecule appears to consist of a cysteine residue (Cys 25) and a histidine residue (His 159). The sulfur atom of the cysteine residue and the imidazole ring of the histidine are about 4 Å apart (*24*). The imidazole ring may be hydrogen bonded to the side chain carbonyl of an asparagine residue (Asn 175) (*24*). The carboxyl group in the active site region (Asp 158) is 6.7 Å away from the imidazole ring of His 59, and its location appears to exclude it from direct participation in the catalysis, although its participation in enzyme–substrate interactions is probable (*24*).

If indeed an imidazole residue is responsible for the acid pK in the pH vs. k_{cat}/K_m(app) profile, then the absence of an observable temperature dependence of this pK is puzzling. It should be noted also that high resolution crystallographic studies have not as yet been performed on active papain, and it is possible that the side chain arrangement observed in the various derivatives examined does not reflect the exact location of the atoms in the catalytic site of the active enzyme.

4. *pH Dependence of Individual Kinetic Constants*

The exact knowledge of the dependence of the acylation rate constant (k_2) and the apparent binding constant (K_s) for papain-catalyzed hydrolyses is necessary to permit interpretation of the mechanism of action of this enzyme.

Attempts to measure k_2 and K_s directly by stopped flow spectrophotometry, with nitrophenyl esters of carbobenzoxyglycine as substrates, were unsuccessful because of the very high K_s values for these substrates (*154*). Sluyterman (*155*) studied the binding of BAEE to inactive papain and showed that the binding constant for this ES complex approached closely the K_m value obtained with the active enzyme under similar con-

153a. G. Lowe, *Phil. Trans. Roy. Soc. London* **B257**, 237 (1970).
154. C. D. Hubbard and J. F. Kirsch, *Biochemistry* **7**, 2569 (1968).
155. L. A. AE. Sluyterman, *BBA* **113**, 577 (1966).

ditions. This agreement could well be fortuitous. Sluyterman (*156*) also showed that the rate of reaction of chloroacetic acid with the active site thiol group of papain was independent of BAEE concentration and concluded that for this substrate $k_2 \ll k_3$, and hence $K_m = (k_{-1} + k_2/k_1) = K_s$ (*156*).

The recent kinetic data of Lucas and Williams (*142*) obtained with a series of esters for which deacylation was shown to be rate limiting could be successfully analyzed on the basis of the equation

$$k_{cat} = k_3 - k_3K_m/K_s$$

(which may be readily derived from the three-step kinetic model, see Section VI,C,2). If k_3 and K_s are independent of pH in the range studied, then k_{cat} will be linear in K_m with an intercept on the k_{cat} axis of k_3, and of K_s on the K_m axis. The linear relationship was found to hold for the substrates studied over the pH range of 3.6–8.5 (*142*).

The value of k_2 at each pH was then calculated from the equation

$$k_{cat}/K_m = k_1k_2/(k_{-1} + k_2) = k_2/K_s$$

The pH dependence of k_2 (acylation) was found to be bell-shaped with pK_1 ~4.5 and pK_2 ~8.5 (*142*). Lucas and Williams (*142*) concluded that K_s closely approached the true binding constant (i.e., $k_{-1} \gg k_2$) since it showed a lack of pH dependence in the region where k_2 varied significantly. A similar conclusion had been reached earlier by Whitaker and Bender (*100*) on the basis of less complete kinetic data on BAEE, BAA, and *N*-*trans*-cinnamoylimidazole hydrolysis by papain. The alternative interpretation that k_2, k_1, and k_{-1} all have similar bell-shaped profiles was considered unlikely (*142*). The view that K_s is indeed a binding constant is strengthened by the finding that uncharged reversible inhibitors, structurally related to substrates, such as benzamidoacetonitrile and acetamidoacetonitrile have pH-independent competitive inhibition constants (*142*). Lucas and Williams (*142*) have proposed that the imidazole group of His 159 gives rise to the acid pK_a in both acylation and deacylation by acting as a general base abstracting a proton from a thiol and a nucleophile, respectively. The abnormally low pK_a of this group is attributed to a hydrogen bond between the imidazole ring and Asn 175 (see above). It is generally agreed that the base pK_a in the acylation step results from the ionization of a thiol group which reacts in its unionized form in the catalytically active enzyme. Thus, Lucas and Williams (*142*) proposed that the catalytically active form of papain is that in which the imidazole is unprotonated and the thiol group is in the unionized form. Assuming a pK_a of 4.5, for the histidine residue,

156. L. A. AE. Sluyterman, *BBA* **151,** 178 (1968).

it can be calculated on the basis of the proposed mechanism that the proportion of active papain is one-hundredth the total available enzyme (*142*). This mechanism appears compatible with the kinetic and chemical data discussed above. There appears to be no compelling chemical or kinetic evidence for the existence of a thiol-imidazole hydrogen bond.

VII. Other Proteolytic Enzymes of Papaya Latex

Examination of the water-soluble proteins of commercial, dried papaya latex by chromatography on CM-cellulose shows the presence of four components possessing proteolytic activity. Three of these correspond to papain, chymopapain, and papaya peptidase A, and constitute 5, 27, and 18%, respectively, of the soluble latex protein (*157*). Little information is available on the fourth component, representing 14% of the soluble protein, beyond the fact that like the other three it appears to be a sulfhydryl enzyme (*157*).

A. Chymopapains

Chymopapain was isolated from papaya latex by Jansen and Balls (*158*). Like papain, it is a thiolenzyme, requiring activation by cysteine or cyanide, and is inhibited by sulfhydryl reagents (*159, 160*). The substrate specificity of papain and chymopapain shows a considerable degree of similarity (*107, 159, 161*). In contrast to papain, chymopapain is very stable at acid pH values (*158, 162*).

Smith and Kimmel (*6*) reported that chymopapain, prepared according to Jansen and Balls (*158*), was electrophoretically heterogeneous. Indeed, the chymopapain fraction can be separated into several major components by ion exchange chromatography (*162*). Two of these components, chymopapain A (*159*) and chymopapain B (*162*) have been obtained in crystalline and apparently homogeneous form. These two proteases appear to differ from each other with respect to the N-terminal residues, amino acid compositions, specific activities, and sulfhydryl con-

157. P. Schack, *Compt. Rend. Trav. Lab. Carlsberg* **36**, 67 (1967).
158. E. F. Jansen and A. K. Balls, *JBC* **137**, 459 (1941).
159. M. Ebata and K. T. Yasunobu, *JBC* **237**, 1086 (1962).
160. T. Cayle, L. T. Saletan, and B. Lopez-Ramos, *Wallerstein Lab. Commun.* **27**, Nos. 93/94, 87 (1964).
161. M. Ebata and Y. Takahashi, *BBA* **118**, 201 (1966).
162. D. K. Kunimitsu and K. T. Yasunobu, *BBA* **139**, 405 (1967).

tents, and they do not appear to arise from each other or from a common precursor by proteolysis (*162*). The molecular weights of chymopapains A and B, approximately 35,000, and their isoelectric points, 10.0 and 10.4 (*159, 162*), are considerably higher than those of papain (see Table I).

Pepsin digestion of chymopapain B, inactivated by reaction with *N*-(4-dimethylamino-3,5-dinitrophenyl) maleimide, showed that the label was located predominantly in one peptide, having the sequence Lys–Val–Pro–Asp–Ser–Gly–Glu–Cys–Tyr, with the derivative on the cysteine residue (*163*). This sequence shows little resemblance to those about the essential thiol group of papain and ficin.

It has been claimed that the activation of chymopapain is associated with a conformational change (*164*). As discussed above, this does not appear to be the case for papain.

B. Papaya Peptidase A

This sulfhydryl protease differs from papain and chymopapain in amino acid composition and activity toward protein substrates (*157*). The enzyme shows esterase activity toward *p*-toluenesulfonyl-L-arginine methyl ester (*157*). It appears to be less active than papain toward denatured proteins. The major points of cleavage of the B chain of oxidized insulin by papain and papaya peptidase A were the same, indicating a very similar specificity for these two enzymes (*107*). From a consideration of the amino acid composition of the enzyme, its isoelectric point, and approximate molecular weight, Schack (*157*) concluded that this enzyme did not arise by degradation of papain or chymopapain.

VIII. Ficin

The name *ficin* has been used to describe both the crude dried latex from different species of the genus *Ficus* and the crystalline sulfhydryl protease(s) prepared from the latex. The genus *Ficus* contains approximately 2000 species of tropical and subtropical trees, and it is thus one of the largest in the Moraceae family. Williams *et al.* (*165*) reported that only 13 of 46 species of *Ficus* examined contained appreciable proteo-

163. J. N. Tsunoda and K. T. Yasunobu, *JBC* **241,** 4610 (1966).

164. M. Ebata, J. N. Tsunoda, and K. T. Yasunobu, *BBRC* **22,** 455 (1966).

165. D. C. Williams, V. C. Sgarbieri, and J. R. Whitaker, *Plant Physiol.* **43,** 1083 (1968).

lytic activity and concluded that high proteolytic activity in the latex was not a distinguishing characteristic of the genus. Most of the work on *Ficus* proteases has been concerned with enzymes purified from the commercially available dried latex of *F. glabrata,* which is rich in proteolytic enzymes (*165–167*). Whitaker and his co-workers have reported that the latex of many varieties of *F. carica* (*165, 166, 168–170*) and *F. glabrata* (*165–167*) contains a number of sulfhydryl endopeptidases, as many as ten in some varieties. It appears that the various endopeptidases from a single variety of tree are virtually indistinguishable in terms of kinetics and specificity (*167, 171*), have similar molecular weights, and show considerable similarity in primary structure, as judged by peptide mapping (*167, 171*). However, the enzymes exhibit clear differences in amino acid composition, peptide maps, and chromatographic and electrophoretic properties (*171*). On the basis of the evidence available at present, it appears highly improbable that the multiplicity of the *Ficus* endopeptidases is a consequence of cleavage of a common precursor or of autodigestion (*170, 171*).

An endopeptidase, first crystallized from the clarified latex of *Ficus* spp. by Walti (*172*), and the salt-precipitated preparation of Hammond and Gutfreund (*173*) have been examined for amino acid composition (*174*), amino-terminal residues (*174*), physical properties (*175*), and amino acid sequences about the sulfhydryl groups (*174, 176*). The best characterized preparation of ficin has been described by Englund *et al.* (*177*). This enzyme, purified by salt precipitation and ion exchange chromatography from an extract of dried *F. glabrata* latex, appeared homogeneous by a variety of physical and chemical criteria, and some of its physical properties are listed in Table XI and its amino acid com-

166. V. C. Sgaribieri, S. M. Gupte, D. E. Kramer, and J. R. Whitaker, *JBC* **239,** 2170 (1964).

167. D. C. Williams and J. R. Whitaker, *Plant Physiol.* **44,** 1574 (1969).

168. D. E. Kramer and J. R. Whitaker, *JBC* **239,** 2178 (1964).

169. D. E. Kramer and J. R. Whitaker, *Plant Physiol.* **44,** 1566 (1969).

170. D. E. Kramer and J. R. Whitaker, *Plant Physiol.* **44,** 1560 (1969).

171. I. K. Jones and A. N. Glazer, *JBC* **245,** 2765 (1970).

172. A. Walti, *JACS* **60,** 493 (1938).

173. B. R. Hammond and H. Gutfreund, *BJ* **72,** 349 (1959). The preparations of Walti (*172*) and of Hammond and Gutfreund were shown to be heterogeneous by electrophoresis (*6*) or chromatography on CM-cellulose (*166*).

174. R. M. Metrione, R. B. Johnston, and R. Seng, *ABB* **122,** 137 (1967).

175. I. E. Liener, *BBA* **53,** 332 (1961).

176. R. C. Wong and I. E. Liener, *BBRC* **17,** 470 (1964).

177. P. T. Englund, T. P. King, L. C. Craig, and A. Walti, *Biochemistry* **7,** 163 (1968).

TABLE XI
PHYSICAL PROPERTIES OF FICIN[a]

Molecular weight (amino acid composition)	23,800
Molecular weight (equilibrium ultracentrifugation)	25,500
$\bar{v}$ (ml/g)	0.725
Absorbancy $A_{1\text{ cm}}^{1\%}$ at 280 nm	21.0
$-[\alpha]_{546}$ (deg)	44.1
λ_c (nm)	267
a_0 (deg)	−200
b_0 (deg)	−166

[a] From Englund *et al.* (*177*).

position in Table XII (*162, 177–179*). It has recently been reported that ficin is a glycoprotein (*180*).

Studies of the kinetic behavior of ficin up to 1960 have been reviewed by Smith and Kimmel (*6*) and the striking similarity in the catalytic properties of this enzyme and papain emphasized by these authors.

As for papain, k_{cat}/K_m(app) values for the ficin-catalyzed hydrolysis of BAEE and BAA are dependent on prototropic groups of apparent pK_a values of 4.4 and 8.4 (*173, 181–183*). The apparent pK_a values for k_{cat}/K_m(app) are essentially the same for most substrates (*173*). The deacylation step appears to depend on a group with a pK_a of about 3.5 (*173*). There are differences in the kinetics, which indicate that the various individual rate constants contribute differently to the observed kinetic parameters for the hydrolysis of the same substrate by papain and ficin (*184*). In addition, both similarities and differences are seen in the pattern of cleavage of the B chain of oxidized insulin by ficin (*177*) and papain (*107*).

A thiol group is known to participate in the catalytic activity of ficin (*176, 181, 185*), and, as in the case of papain, the kinetics of alkylation of this thiol group by iodo- and chloroacetamide depend upon a basic group of pK_a 8.6 (*185*). Further, the hydrolysis of *O*-methylthiohippurate by ficin proceeds through an intermediate with an absorption maximum at 315 nm (*138, 147*) similar to that of the dithioester deriva-

178. S. Ota, S. Moore, and W. H. Stein, *Biochemistry* **3,** 180 (1964).
179. T. Murachi, *Biochemistry* **3,** 932 (1964).
180. B. Friedenson, *Federation Proc.* **29,** 879 (1970) (abstr.).
181. S. A. Bernhard and H. Gutfreund, *BJ* **63,** 61 (1956).
182. J. R. Whitaker, *Biochemistry* **8,** 1896 (1969).
183. D. E. Kramer and J. R. Whitaker, *Plant Physiol.* **44,** 609 (1969).
184. J. R. Whitaker, *Biochemistry* **8,** 4591 (1969).
185. M. R. Hollaway, A. P. Mathias, and B. R. Rabin, *BBA* **92,** 111 (1964).

TABLE XII
AMINO ACID COMPOSITION OF CHYMOPAPAIN B, FICIN, AND STEM AND FRUIT BROMELAINS

Amino acid	Ficin[a]	Chymopapain B[b]	Stem bromelain[c]	Fruit bromelain[c,d]
Aspartic acid	17	27	29	29.8
Glutamic acid	25	29	23	23.2
Glycine	28	39	35	32.6
Alanine	20	19	35	23.8
Valine	18	25	22	19.8
Leucine	15	15	10	10.0
Isoleucine	7	12	21	16.4
Serine	14	21	28	32.2
Threonine	8	16	14	13.5
Half-cystine	8	11	10	10.0
Methionine	5	1	5	6.0
Proline	11	14	14	11.6
Phenylalanine	5	7	9	7.6
Tyrosine	15	20	21	22.4
Tryptophan	6	6	8	5.6
Histidine	1	5	2	1.4
Lysine	5	25	23	7.8
Arginine	10	10	12	8.6
Amide ammonia	25	30	42	43.0
Glucosamine	—	—	6	<0.2
Carbohydrate (%)			1.46	3.2
Molecular weight	25,000	33,500	33,730	—

[a] Data from Englund *et al.* (*177*).
[b] Amino acid analysis of chymopapain fraction IV (*162*) taken to the nearest integer.
[c] Data from Ota *et al.* (*178*). A slightly different analysis has been reported by Murachi (*179*).
[d] For comparison with stem bromelain, the data for fruit bromelain are expressed as mole ratios with leucine set at ten. Similar amino acid analyses were obtained with bromelain prepared from green and ripe fruit.

tive obtained with papain and this substrate (see Section VI,C). Degradation of ficin, inactivated with ^{14}C-iodoacetamide, results in the isolation of a single labeled peptide, and the sequence about the cysteine residue in this peptide (*176*) bears a close relationship to that about the essential cysteine residue of papain. As in papain, 1,3-dibromoacetone cross-links the essential thiol of ficin to a histidine residue (*186*).

A comparison of the amino acid sequences around the histidine residues cross-linked to the thiol group at the active site by 1,3-dibromoacetone in

186. S. S. Husain and G. Lowe, *BJ* **110**, 53 (1968).

ficin (*186a*) with the corresponding sequences from papain (*39, 149, 150*) and stem bromelain (*186b*) is as follows:

Ficin	Thr-Gly-Pro-Cys-Gly-Thr-Ser-Leu-Asp-*His*-Ala-Val-Ala-Leu
Papain	Val-Gly-Pro-Cys-Gly-Asn-Lys-Val-Asp-*His*-Ala-Val-Ala-Ala-Val-Gly-Tyr
Stem bromelain	*His*-Ala-Val-Thr-Ala-Ile-Gly-Tyr

These analogies indicate that papain and ficin share a common catalytic mechanism and that differences in the kinetics reflect differences in the stereochemistry of the binding sites in these two enzymes.

Both papain and ficin are irreversibly inactivated in acid solution, but in contrast to papain, ficin is inactive in 8 *M* urea (*177*).

A protein inhibitor of papain and ficin, which inhibits bromelain only slightly and does not inhibit trypsin or chymotrypsin, has been purified from chicken egg white (*187*).

IX. Bromelain

A large number of varieties of *Ananas comosus* (L.) Merr., the pineapple, as well as many other species of the family Bromeliaceae, contain appreciable quantities of proteolytic enzymes. Proteolytic activity in the juice of the pineapple plant was recognized during the nineteenth century (*188*), and, more recently, it was shown that the juice of the stem of the pineapple plant is a rich source of proteolytic enzymes (*189*). Proteases isolated from any member of the Bromeliaceae are called bromelains (*190*). For a given protease, the term is made specific by prefixing the binomial plant name and the organ from which the enzyme was obtained (*191*). Thus the complete name for stem bromelain is *Ananas comosus* var. Cayenne stem bromelain (*189*).

The number of proteolytic enzymes present in pineapple stem bromelain has not as yet been definitively ascertained. Heinicke and Gortner (*189*) showed that crude bromelain could be separated electrophoretically at pH 6.5 into four distinct components with proteolytic activity. Several proteolytically active components have also been separated from commercial, powdered stem bromelain by chromatography on cation ex-

186a. S. S. Husain and G. Lowe, *BJ* **117,** 333 (1970).
186b. S. S. Husain and G. Lowe, *BJ* **117,** 341 (1970).
187. K. Fossum and J. R. Whitaker, *ABB* **125,** 367 (1968).
188. R. H. Chittenden, *Trans. Conn. Acad. Arts Sci.* **8,** 281 (1892).
189. R. M. Heinicke and W. A. Gortner, *Econ. Botany* **11,** 225 (1957).
190. R. M. Heinicke, *Science* **118,** 753 (1953).
191. D. M. Greenberg and T. Winnick, *JBC* **135,** 761 (1940).

change resins at pH 6.1 (*191–195*). The preparation of a homogeneous stem bromelain, accounting for a major part of the total proteolytic activity, has been reported by Ota *et al.* (*178*) and Murachi *et al.* (*196*). The amino acid composition of the enzyme is given in Table XII and its physical properties summarized in Table XIII. However, examination of the preparation of Ota *et al.* (*178*) by end group analysis (*178*), acrylamide gel electrophoresis (*195*), and ion exchange chromatography (*193, 195*) indicates heterogeneity. It is not clear whether this heterogeneity arises through autodigestion of the enzyme (*197*), perhaps even before extraction of the juice from the stem, or whether a number of distinct but closely related enzymes are present (*194*). Neither possibility can be ruled out on the basis of the available data, and it is possible that both contribute to the observed heterogeneity. Ota *et al.* (*178*) have also purified stem bromelain from both green and ripe pineapple fruit. No difference in the enzyme from these two sources was detected. However, the fruit enzyme contained much less lysine, arginine, and histidine than the stem enzyme (Table XII). The isoelectric point of the stem enzyme is at pH 9.5, whereas that of the fruit enzyme is considerably lower (*178*). Further, the fruit enzyme contains no glucosamine, which is present in

TABLE XIII
PHYSICAL PROPERTIES OF STEM BROMELAIN[a]

$s_{20,w}$	2.73 S
$D^{\circ}_{20,w}$ (10^{-7} cm^2 sec^{-1})	7.77
$\bar{v}$ (ml/g)	0.743
Intrinsic viscosity $[\eta]$ (d/g)	0.039
Frictional ratio f/f_0	1.26
Isoelectric point pI	9.55
Absorbancy $A^{1\%}_{1\,cm}$ at 280 nm	19.0
Molecular weight (S,D)	33,200[b]
Molecular weight (S,η)	32,100
Molecular weight (Archibald)	33,500
$-[\alpha]_{546}$(deg)	43.1

[a] From Murachi *et al.* (*196*).

[b] Ota *et al.* (*178*) calculated a molecular weight of 35,730 based on a best fit to the amino acid composition. Feinstein and Whitaker (*194*) have reported molecular weights of 19,000–20,000, as determined by the Archibald method and gel filtration, for their preparations of stem bromelains. The reason for this discrepancy is not clear.

192. T. Murachi and H. Neurath, *JBC* **235,** 99 (1960).
193. M. El-Gharbawi and J. R. Whitaker, *Biochemistry* **2,** 478 (1963).
194. G. Feinstein and J. R. Whitaker, *Biochemistry* **3,** 1050 (1964).
195. J. Scocca and Y. C. Lee, *JBC* **244,** 4852 (1969).
196. T. Murachi, M. Yasui, and Y. Yasuda, *Biochemistry* **3,** 48 (1964).
197. S. Ota, *J. Biochem.* (*Tokyo*) **63,** 494 (1968).

the stem enzyme, but has a higher total carbohydrate content. Finally, the fruit enzyme is considerably more active toward both casein and BAA than the stem enzyme (*178*).

Stem bromelain is a glycoprotein containing 3 moles of mannose, 1 of fucose, 1 of xylose, and 2 of *N*-acetylglucosamine per mole (molecular weight 33,000) (*195, 198, 199*). Electrophoretically distinct components of the enzyme appear to contain the same oligosaccharide (*195*). The carbohydrate appears to be present in a single oligosaccharide moiety, which has been the subject of considerable study (*195, 198, 199*). The following structure has been proposed for the carbohydrate moiety (*199*):

α-D-Man(1 → 2)-α-D-Man-(1 → 2 or 6)-[α-L-Fuc-(1 → 6 or 2)]-α-D-Man-
(β-D-Xyl)-β-D-GlcNAc-(1 → 3 or 4)-β-D-GlcNac-(1 → β-NH_2-N-of Asn)-peptide

The sequence of the polypeptide chain around the point of attachment of the carbohydrate moiety has been shown to be (*199–201*):

```
                          CHO
                           |
Ala-Arg-Val-Pro-Arg-Asn-Asn-Glu-Ser-Ser-Met
```

The specificity of stem bromelain has been examined on a number of substrates. Although the enzyme shows similarities in its specificity to that exhibited by papain, significant differences have also been observed. Murachi and Neurath (*192*) showed that stem bromelain cleaved the Arg–Ala and Ala–Glu bonds of glucagon but left the Arg–Arg and Lys–Tyr bonds intact at pH 8.0. Autodigestion of bromelain at pH 4.6 and elevated temperature results in cleavage not only at basic amino acid residues but also at glycine, alanine, and serine residues, and to a lesser extent at other sites as well (*197*). These results are consistent with studies on synthetic substrates. At pH 6, bromelain hydrolyzed the ethyl esters of benzoyl-L-arginine and benzoyl-L-glycine, as well as those of L-phenylalanine, L-tyrosine, L-leucine, L-lysine, and glycine (*202*). In contrast to papain and ficin, bromelain shows a 140-fold difference in the value of k_{cat} for the hydrolysis of BAEE and BAA (*148, 202*).

Like the other plant proteases discussed in this review, bromelain is a sulfhydryl enzyme requiring activation by cysteine or cyanide for at-

198. T. Murachi, A. Suzuki, and N. Takahashi, *Biochemistry* **6**, 3730 (1967).

199. Y. Yasuda, N. Takahashi, and T. Murachi, *Biochemistry* **9**, 25 (1970). Scocca and Lee (*195*) reported that the oligosaccharide consists of D-glucoamine, D-mannose, D-xylose, and L-fucose in ratios of 2:2:1:1.

200. N. Takahashi, Y. Yasuda, M. Kuzuya, and T. Murachi, *J. Biochem.* (*Tokyo*) [illegible]969).

201. K. Kito and T. Murachi, *J. Chromatog.* **44**, 205 (1969).

202. T. Inagami and T. Murachi, *Biochemistry* **2**, 1439 (1963).

tainment of maximal activity (*192*) and is inhibited reversibly at very low concentrations of $HgCl_2$ (*192*), and irreversibly by iodoacetamide (*193*). The sequence around the essential sulfhydryl group of bromelain has been determined, and it shows significant similarity to the sequences about the essential thiol groups of papain and ficin (*203, 204*). A comparison of these sequences is given by Smith (*205*).

Several lines of evidence indicate that bromelain shares a common mechanism of action with ficin and papain. Bromelain is totally inactivated on alkylation of the essential thiol group with iodoacetamide (*193*) or with the chloromethyl ketones derived from *N*-tosyl-L-phenylalanine and *N*-tosyl-L-lysine (*205*). Hydrolysis of methylthionohippurate by stem bromelain proceeds through a thionohippuryl-enzyme intermediate, which exhibits the spectral characteristics of an acyl thiolester (*148*). 1,3-Dibromoacetone cross-links the active site thiol residue of bromelain with the N-1 of the imidazole of a histidine residue (*186*), showing that in this enzyme, as in papain and ficin, the essential thiol group is within 5 Å of the side chain of a histidine residue. The histidine residue is present in a strikingly similar sequence in all three enzymes (see Section VIII).

X. Concluding Remarks

It is clear that the mechanism of action of papain, ficin, chymopapain, and bromelain is very similar. While the sequences near the essential thiol groups in these enzymes display varying degrees of homology (*205*), judgment as to whether all of these enzymes arose from a common evolutionary precursor has to await more extensive information on their amino acid sequences. These enzymes are all activated by sulfhydryl compounds and cyanide, and inactivated by mild oxidizing agents. Many proteolytic enzymes from a wide variety of plants share these properties (*206*). These include mexicain (from the leaves or fruit of *Pileus mexicanus* (*207*), pinguinain (from the fruit of *Bromelia pinguin*) (*208*), asclepain (from the roots or latex of the milkweed, *Asclepia speciosa*

203. L. P. Chao and I. E. Liener, *BBRC* **27**, 100 (1967).
204. S. S. Husain and G. Lowe, *Chem. Commun.* p. 1387 (1968).
205. E. L. Smith, "The Enzymes," 3rd ed., Vol. 1, p. 267, 1970.
206. D. M. Greenberg and T. Winnick, *Ann. Rev. Biochem.* **14,** 31 (1945).
207. M. Castañeda-Agulló, A. Hernandez, F. Loaeza, and W. Salazar, *JBC* **159,** 751 (1945).
208. C. F. Asenjo and M. Capella de Fernandez, *Science* **95,** 48 (1942).

(*209*), tabernamontanain (from the sap or fruit of *Tabernamontana grandiflora* (*210*), and euphorbain (from the latex of *Euphorbia lathyris*) (*211*). These enzymes show a wide range of pH optima, from pH 8.5 to pH 3 (*212*). More extensive studies of their mechanism of action and chemistry of their active sites would be of considerable interest.

A crystalline anionic proteinase from the ripe fruit of the Chinese gooseberry (*Actidinia chinensis*) has been shown to require a SH group for its catalytic activity (*212a, 212b*). It is similar to papain with respect to activation behavior, inhibition by SH reagents and pH–velocity profile for the hydrolysis of BAEE (*212b*).

It should be noted, however, that not all plant proteases are sulfhydryl enzymes. Thus, solanain (from the berries of the horsenettle, *Solanum eleagnifolium* (*212*), hurain (from the sap of *Hura crepitans*) (*213*), and arachain (from the cotyledon and embryo of the peanut) (*214*) do not require activation by sulfhydryl reagents and are unaffected by mild oxidants (*215*). Nothing is known of their mechanism of action.

The role of the multiple sulfhydryl proteases of similar activity and specificity in a number of plants poses an interesting problem. Presumably, these enzymes protect the plant against invasion by fungi and insects (*216*). The finding of enzymes with high chitinase activity in the latex of *C. papaya* (*216, 217*) and *F. glabrata* (*218*) is consistent with such a speculation. However, as pointed out in connection with ficin, many closely related species of plants vary enormously in their content of proteolytic enzymes. It is therefore difficult at present to assign an unambiguous role to the plant sulfhydryl proteases.

209. D. C. Carpenter and F. E. Lovelace, *JACS* **65**, 2364 (1943).
210. W. G. Jaffe, *Rev. Brasil. Biol.* **3**, 149 (1943).
211. F. G. Lennox and W. J. Ellis, *BJ* **39**, 465 (1945).
212. D. M. Greenberg and T. Winnick, *JBC* **135**, 761 (1940).
212a. A. C. Arcus, *BBA* **33**, 242 (1959).
212b. M. A. McDowall, *European J. Biochem.* **14**, 214 (1970).
213. W. G. Jaffe, *JBC* **149**, 1 (1943).
214. G. W. Irving, Jr. and T. D. Fontaine, *ABB* **6**, 351 (1945).
215. T. Winnick and D. M. Greenberg, *Science* **102**, 648 (1945).
216. E. L. Smith, J. R. Kimmel, D. M. Brown, and E. O. P. Thompson, *JBC* **215**, 67 (1955).
217. J. B. Howard and A. N. Glazer, *JBC* **242**, 5715 (1967).
218. A. N. Glazer, A. O. Barel, J. B. Howard, and D. M. Brown, *JBC* **244**, 3583 (1969).

15

Subtilisin: X-Ray Structure

J. KRAUT

I. Background

The term *subtilisin* is now generally employed to designate the group of alkaline serine proteases (i.e., proteases having a peculiarly reactive serine at their catalytic site) produced by various species of bacilli. Subtilisin is found in the culture filtrates of these organisms together with a host of other proteases, esterases, amylases, and nucleases. The nomenclature, distribution in nature, enzymic properties, and amino acid sequences of the subtilisins are reviewed by Markland and Smith in the next chapter of this volume. This chapter will therefore confine itself to summarizing what is known at present about the three-dimensional structure, as determined by X-ray crystallography, of one member of the group, for which the most common designation is subtilisin BPN′. Other names that have been applied to identically the same enzyme are Nagarse and subtilopeptidase C (*1*). It is obtainable from the Nagase Company, Osaka, as a crude powder, contaminated with what are presumed to be autocatalytic breakdown products, colored materials, and, depending upon the particular batch, other proteases and/or esterases. Subtilisin Novo designates the *Bacillus* alkaline protease commercially available

1. J. Drenth and W. G. J. Hol, *JMB* **28**, 543 (1967).

from Novo Industries, Copenhagen. It has also been referred to as subtilopeptidase B (*1*), although the purified Novo enzyme appears to be identical with subtilisin BPN′ in all respects. Whether or not the strains of *Bacillus* from which subtilisin BPN′ and subtilisin Novo are obtained are also identical is not known, since the enzymes are proprietary items of commercial importance, and hence information regarding their production is not available. Welker and Campbell (*2*) have shown that the strain of *Bacillus* which makes subtilisin BPN′ is not a variety of *B. subtilis,* as was formerly supposed, but is in fact a distinct species they have designated *B. amyloliquefaciens,* since it also produces unusually large quantities of amylase. Subtilisin Carlsberg designates a closely related alkaline protease, differing from the BPN′ and Novo enzymes in 84 amino acid replacements and one deletion (see the following chapter by Markland and Smith in this volume). It is also commercially available from Novo Industries, and presumably it is produced by a strain of *B. subtilis.*

Throughout the remainder of this chapter, the term *subtilisin* will be used to designate the BPN′ or Novo enzyme, except where otherwise noted. Although the properties of subtilisin are reviewed in detail in the next chapter, for our present purposes it may be worthwhile briefly to remind the reader of the general similarity in catalytic properties between subtilisin and the more familiar serine proteases, as exemplified by vertebrate chymotrypsin and trypsin. The reason is that there turns out to be a fascinating similarity, in certain respects, between the three-dimensional arrangement of several side chains in the neighborhood of the catalytic site serines of the two classes of proteases. This is true despite the total absence of any discernible relationship between their overall amino acid sequences or their overall molecular conformations. In fact, at present, it may be said that the most important result to come out of the X-ray structure determination of subtilisin is the tentative conclusion that the chymotrypsinlike and subtilisinlike enzymes evolved independently to converge upon the same, or very similar, molecular machinery functioning in much the same way.

To begin with, both subtilisin and the chymotrypsinlike group of enzymes contain a single peculiarly reactive serine residue that can be specifically phosphorylated by diisopropylfluorophosphate (*3*), sulfonylated by phenylmethanesulfonyl fluoride (*4, 5*), or acylated by a variety of reagents. Furthermore, it has been found that both α-chymotrypsin

2. N. E. Welker and L. L. Campbell, *J. Bacteriol.* **94,** 1124 (1967).
3. F. Sanger and D. C. Shaw, *Nature* **187,** 872 (1960).
4. K. E. Neet and D. E. Koshland, *Proc. Natl. Acad. Sci. U. S.* **56,** 1606 (1966).
5. L. Polgar and M. L. Bender, *Biochemistry* **6,** 610 (1967).

and subtilisin show the same characteristic shift in the absorption maximum of the acylating chromophoric group when reacted with certain β-arylacryloylimidazoles *(6)*, and parallel effects are observed in the respective protonic equilibria of the two enzymes when they are acylated with *N*-β-(3-indole)acryloylimidazole *(7)*.

Another similarity between the enzymic properties of subtilisin and the chymotrypsin group is the participation of a histidine residue in the catalytic mechanism. For example, k_{cat} for hydrolysis of *p*-nitrophenylacetate by subtilisin depends upon a group with p*K* about 7.2, presumed to be a histidine *(5)*. The pH-velocity profile and its temperature dependence for hydrolysis of *N*-benzoyl-L-arginine ethyl ester by subtilisin Carlsberg are consistent with requirement of a nonprotonated imidazole at the catalytic site *(8)*. Various phenyl arsonates have been found to inhibit both chymotrypsin and subtilisin, and inhibitor binding for both enzymes requires protonation of a group with pK' of 7.15. Again, the group is presumed to be a histidine side chain *(9)*. Finally, Shaw and Ruscica *(10)* show that alkylation of a single histidine residue in subtilisin with benzyloxycarbonyl-L-phenylalanine bromomethyl ketone causes loss of hydrolytic activity.

With respect to substrate specificity, subtilisin's requirements are less stringent than the chymotrypsin group, but, nevertheless, the best small molecule substrates for chymotrypsin like *N*-benzoyl-L-tyrosine ethyl ester or *N*-acetyl-L-tyrosine ethyl ester are also good subtilisin substrates *(11, 12)*. In addition, certain aromatic compounds, like indole, phenol, and hydrocinnamate are competitive inhibitors for both subtilisin and chymotrypsin; and both enzymes hydrolyze amide substrates several orders of magnitude more slowly than esters *(8)*.

Despite these similarities, there are of course numerous important differences between subtilisin and the chymotrypsin group of enzymes. Most obvious and general is the broader specificity of subtilisin, already alluded to. A particular manifestation of this phenomenon is the fact that the substratelike chloromethyl ketones, which are effective alkylating reagents for the catalytic-site histidine of chymotrypsin or trypsin, do not react with subtilisin *(10)*. The same applies to the reagent methyl *p*-nitrobenzenesulfonate, which methylates the catalytic-site histidine

6. S. A. Bernhard, S. J. Lau, and H. Noller, *Biochemistry* **4,** 1108 (1965).
7. J. Keizer and S. A. Bernhard, *Biochemistry* **5,** 4127 (1966).
8. A. N. Glazer, *JBC* **242,** 433 (1967).
9. A. N. Glazer, *JBC* **243,** 3693 (1968).
10. E. Shaw and J. Ruscica, *JBC* **243,** 6312 (1968).
11. J. T. Johansen, M. Ottesen, and I. Svendsen, *BBA* **139,** 211 (1967).
12. O. Barel and A. N. Glazer, *JBC* **243,** 1344 (1968).

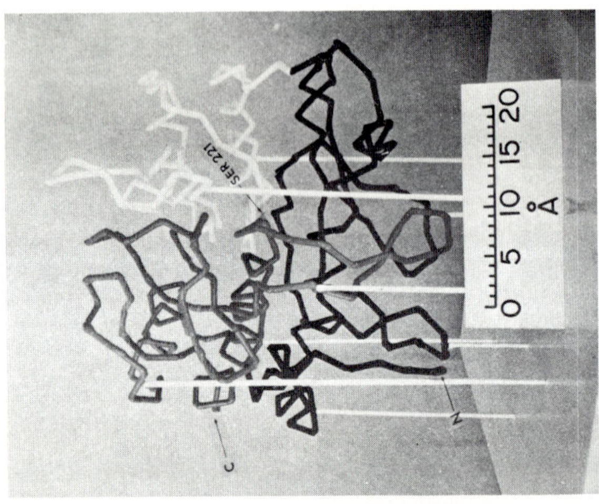

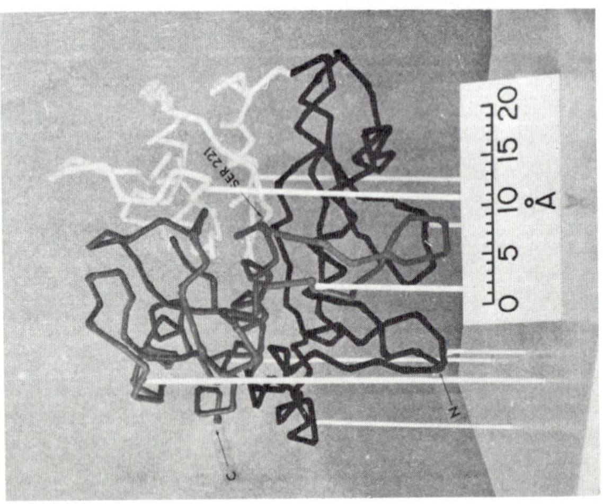

Fig. 1.

in chymotrypsin but not in subtilisin (*13*). Then too, proflavine, which is a good competitive inhibitor for chymotrypsin, is ineffective against subtilisin (*8*), while conversely 4-(4′-aminophenylazo)-phenylarsonic acid is a potent subtilisin inhibitor but does not interact with chymotrypsin (*9*). Finally, Morihara *et al.* (*14*) have obtained evidence from studying polypeptide substrates that subtilisin and chymotrypsin display quite different subsite specificities, i.e., differing hydrolysis rates due to the effects of side chains two and three residues removed in either direction from the peptide bond being split. In conclusion, then, the overall picture that emerges when comparing subtilisin with the chymotrypsin group of serine proteases is that subtilisin is rather poor as an esterase because of its relatively high K_m for simple ester substrates, but it is quite effective as a protease, probably because of its lower specificity (*15*).

Almost all the material contained in this chapter is taken from two papers (*16, 17*) describing preliminary results of the X-ray crystal structure determination of phenylmethanesulfonyl (PMS) subtilisin BPN′ at 2.5 Å resolution by the method of multiple isomorphous replacement. Wright *et al.* (*16*) should be consulted for technical details of the X-ray work, which will be omitted from this review. The results are preliminary in the sense that the models have not yet been fitted to the electron density map with the full precision now made possible by the introduction of the half-silvered mirror technique of Richards (*18*), nor has the electron density map itself been refined by taking advantage of anomalous dispersion effects, although the latter had been measured in the course of collecting the original X-ray intensity data. For these reasons

13. Y. Nakagawa and M. L. Bender, *Biochemistry* **9,** 259 (1970).
14. K. Morihara, T. Oka, and H. Tsuzuki, *BBRC* **35,** 210 (1969).
15. R. L. Hill, *Advan. Protein Chem.* **20,** 37 (1965).
16. C. S. Wright, R. A. Alden, and J. Kraut, *Nature* **221,** 235 (1969).
17. R. A. Alden, C. S. Wright, and J. Kraut, *Phil. Trans. Roy. Soc. London* **B257,** 119 (1970).
18. F. M. Richards, *JMB* **37,** 225 (1968).

FIG. 1. Stereoscopic view of the backbone chain of subtilisin BPN′ as seen from an angle of about 45° to the left of the axis of the central helix F. Each segment of rod represents a peptide unit joining adjacent Cα atoms. The amino-terminal and carboxy-terminal ends of the chain are indicated by N and C, respectively. The location of catalytic site Ser 221 is also shown. Folding of the single polypeptide chain into three distinct parts is illustrated by black, white, and gray shadings.

a list of atom coordinates has not yet been published, and some details of the present model of subtilisin BPN′ will undoubtedly have to be modified as improvements are made in the future. Nevertheless, it is unlikely that any of its major features will be changed.

II. General Description of the Molecule

The overall molecular shape of subtilisin BPN′ is approximately spherical with a diameter of about 42 Å. Two stereoscopic views of the backbone chain are shown in Figs. 1 and 2. Although there is some flattening at the active site region, it is not especially obvious by comparison with other surface irregularities, and no pronounced cleft is observed. The molecule contains eight segments of right-handed α helix, ranging in length from 6 to 16 residues. These helical regions are shown as wavy sections of sequence in Fig. 3. The helixes have been designated (A) Pro 5–Gln 10, (B) Pro 14–Gly 20, (C) His 64–Ala 73, (D) Gln 103–Asn 117, (E) Ser 132–Ser 145, (F) Ala 223–His 238, (G) Thr 242–Asn 252, and (H) Asn 269–Gln 275. Many distortions from the standard α_I-helix conformation are observed, however. The criterion for including a particular residue within a helix was that it contributed an N–H group or a C=O group, or both, to form a helix hydrogen bond. By this criterion, 31% of the residues in the molecule lie in helical segments.

The core of the subtilisin molecule is composed largely of packed nonpolar side chains, as has also been observed for all other known protein structures. Side chains projecting into the surrounding medium or lying in surface crevices may be either polar or nonpolar.

An unusual feature of the conformation of subtilisin is the high degree to which backbone chain segments tend to lie parallel to one another. Except for helix G, all the helical segments are approximately parallel, within ±15°, to a common direction, and indeed have the same N-terminal to C-terminal sense. The longest helix, F, runs roughly through the center of the molecule.

Another notable feature of the chain conformation is a twisted parallel-chain pleated sheet composed of five segments of extended chain Val 148–Ala 152, Asp 120–Met 124, Val 28–Asp 32, Ala 89–Lys 94, and Ala 45–Met 50 running from top to bottom. The angle of twist between the top and bottom segment is approximately 45°. The side chains of this pleated sheet together with those of the central helix form a large part of the nonpolar core of the molecule.

Although the subtilisin molecule consists of a single continuous poly-

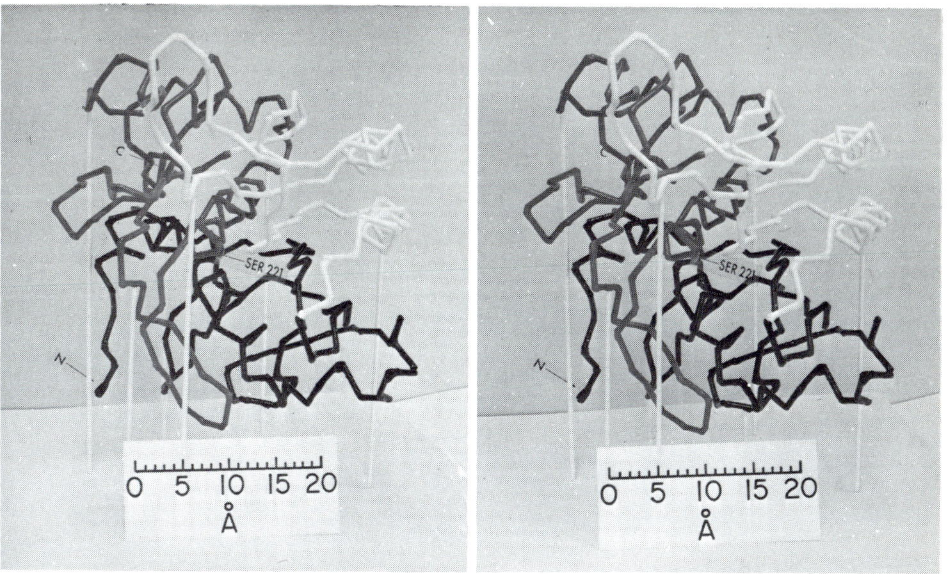

FIG. 2. Stereoscopic view of the backbone chain of subtilisin BPN′ as seen from an angle of about 30° to the right of the axis of the central helix F. See description accompanying Fig. 1.

peptide chain, the backbone chain model shown in Figs. 1 and 2 has been painted in three colors. In the photographs segment Ala 1–Gly 100 appears as black, Gly 100–Ala 176 as white, and Ala 176–Gln 275 (the C-terminus) as gray. It is immediately obvious that the backbone chain has folded into three distinct and potentially separable pieces. Most interestingly, the catalytic site occurs at one of the two places on the molecular surface where the three pieces come together.

It should be cautioned, before concluding this section, that certain portions of the electron density map are still not very clear; thus, the precise molecular conformation in these regions must still be regarded as tentative: Val 51–Thr 55, Asp 61–Ser 63, Asp 98–Gly 102, Ala 115, Gly 131, and His 238–Trp 241.

III. The Catalytic Site

For technical reasons, the parent protein crystal whose phases were determined by isomorphous replacement, and for which the electron density map was interpreted to give the picture of the molecule described in Section II, was in fact not active subtilisin BPN′ but the

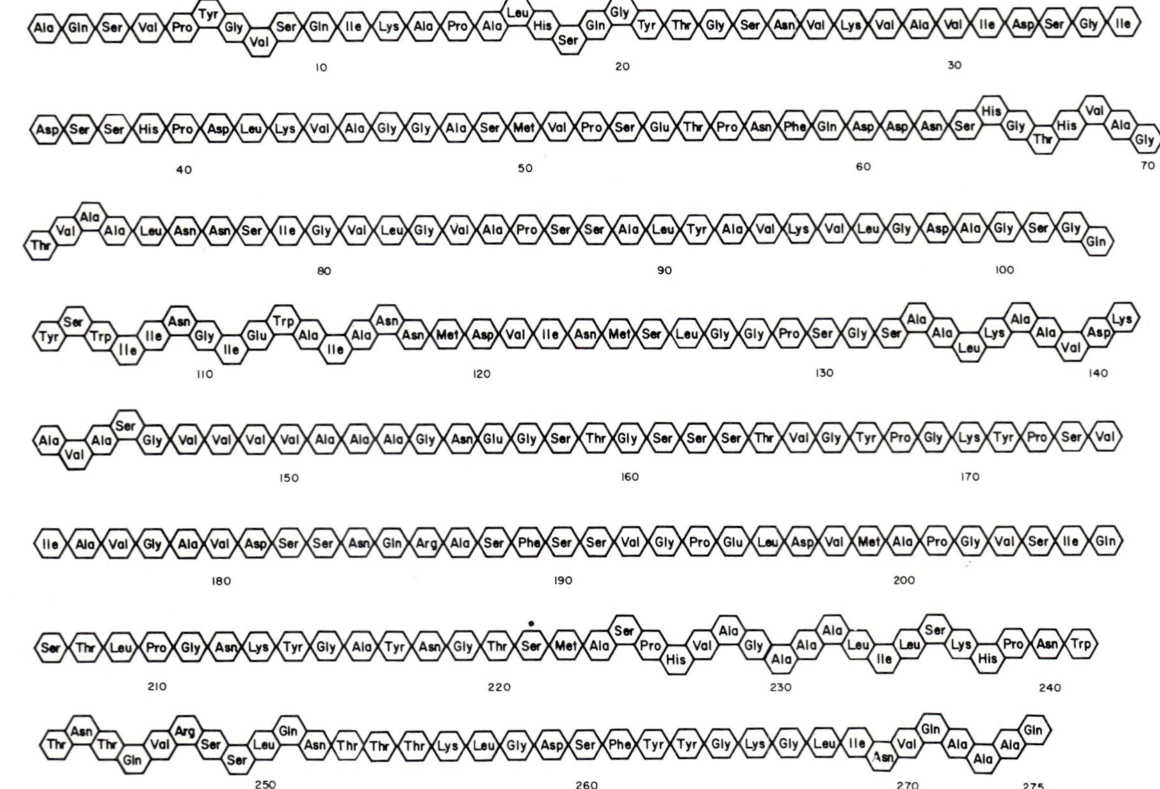

FIG. 3. Amino acid sequence of subtilisin BPN′ according to Markland and Smith (*19*). Segments of α helix are indicated by wavy portions.

PMS-inhibited enzyme. However, crystals of the active enzyme itself are isomorphous with the inhibited crystals, and so it was possible to calculate a difference electron density map between the two. The difference map clearly showed the PMS group and thus confirmed the location of reactive Ser 221. It also revealed a 4-Å movement of the side chain of His 64 upon binding of the PMS group to Ser 221, as well as a slight (about 1 Å) movement of the side chain of Met 222. With the aid of this information the region of the catalytic site as it appears in the active enzyme could be reconstructed.

Figure 4 is a stereoscopic view of the entire skeletal model for the active enzyme molecule, including all side chains. A length of knitting yarn has been threaded through the model to assist in following the course of the polypeptide backbone. The viewing direction in Fig. 4 is toward the catalytic site and roughly perpendicular to the plane of the imidazole ring of His 64 as it is oriented in the active enzyme molecule. This is the same view as in Fig. 2. A stereoscopic close-up of the catalytic site region is shown in Fig. 5. In effect, Fig. 5 is a "zoom-in" toward the catalytic site from the vantage point of Fig. 4. Similarly, Fig. 6 is a close-up of the same region as it appears in the PMS-inhibited enzyme.

From Fig. 5 it will be apparent that the side chains of several residues in the vicinity of the catalytic site are in positions that strongly suggest they are connected by a rather complex hydrogen bond network. Attention will now be focused on those features of the hydrogen bond network that are thought to be common to the active conformation of both subtilisin BPN′ and α-chymotrypsin (*20*). They appear as dotted connectors in Fig. 5.

(1) A hydrogen bond from the OH group of reactive Ser 221 to Nϵ2 of His 64 [atom numbering system is that proposed by Edsall *et al.* (*21*)]. The Oγ to Nϵ2 distance is 3.5 Å—rather long for a standard hydrogen bond but probably within allowable tolerances for the present unrefined 2.5 Å model.

(2) A hydrogen bond from Nδ1 of His 64 to the buried side chain Oδ2 of Asp 32. The N to O distance is 2.2 Å.

(3) A second hydrogen bond to Oδ2 of Asp 32 from the buried side chain OH group of Ser 33. The O to O distance is 3.1 Å.

The five atoms involved in this part of the hydrogen bond network all lie approximately within the plane of the imidazole ring. Their mean

---

19. F. S. Markland and E. L. Smith, *JBC* **242,** 5198 (1967).
20. D. M. Blow, J. J. Birktoft, and B. S. Hartley, *Nature* **221,** 337 (1969).
21. J. T. Edsall, P. J. Flory, J. C. Kendrew, A. M. Liquori, G. Némethy, and G. N. Ramachandran, *JMB* **15,** 399 (1966).

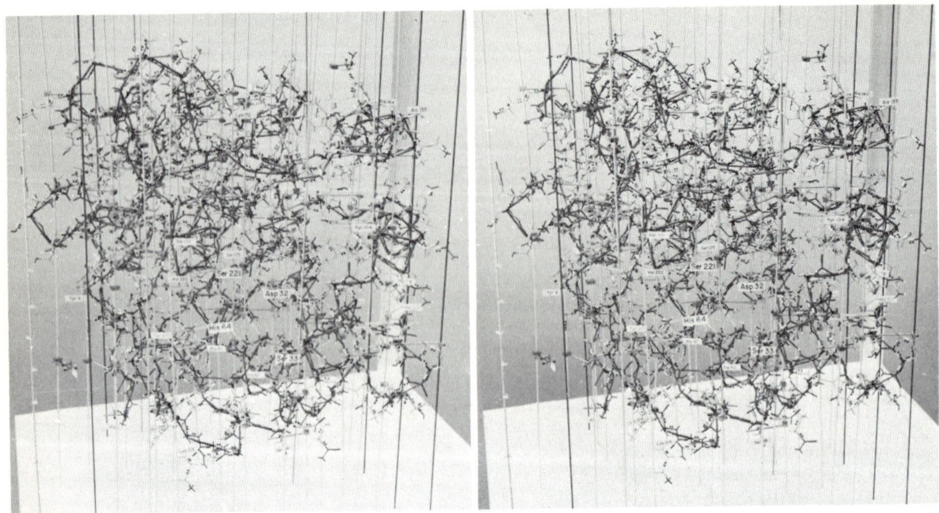

FIG. 4. Stereoscopic view of subtilisin BPN′ skeletal model, looking directly toward the active site. Viewing direction is approximately perpendicular to the imidazole ring of His 64 in the active conformation. The same viewing direction is maintained in Figs. 5 and 6.

deviation from a least squares plane is 0.1 Å, and the greatest distance from this plane is 0.2 Å, for Oγ of Ser 33. The same hydrogen bond system in α-chymotrypsin involves the side chains of Ser 195, His 57, Asp 102, and Ser 214. The close similarity in this regard between α-chymotrypsin and subtilisin BPN′ strongly suggest some fundamental role for this hydrogen bond network in the mechanism of serine protease activity.

The mechanism that has been proposed for chymotrypsin (*20, 22*) employs a "charge relay system" involving hydrogen bonds between the side chains of the reactive Ser 195, His 57, and the buried side chain of Asp 102. It is suggested that the hydrophobic environment of Asp 102 enhances polarization of the charge relay system and makes the reactive Ser 195 oxygen strongly nucleophilic, as required by the proposed mechanism. This hydrogen bond network corresponds very neatly with hydrogen bonds (1) and (2) in subtilisin, mentioned above. However, no role has been proposed for hydrogen bond (3) to the "second serine." That this residue may be important to the functioning of serine proteases in general is suggested by the fact that it exists both in chymotrypsin and subtilisin, and the suggestion is further supported by the observation

22. T. A. Steitz, R. Henderson, and D. M. Blow, *JMB* **46**, 337 (1969).

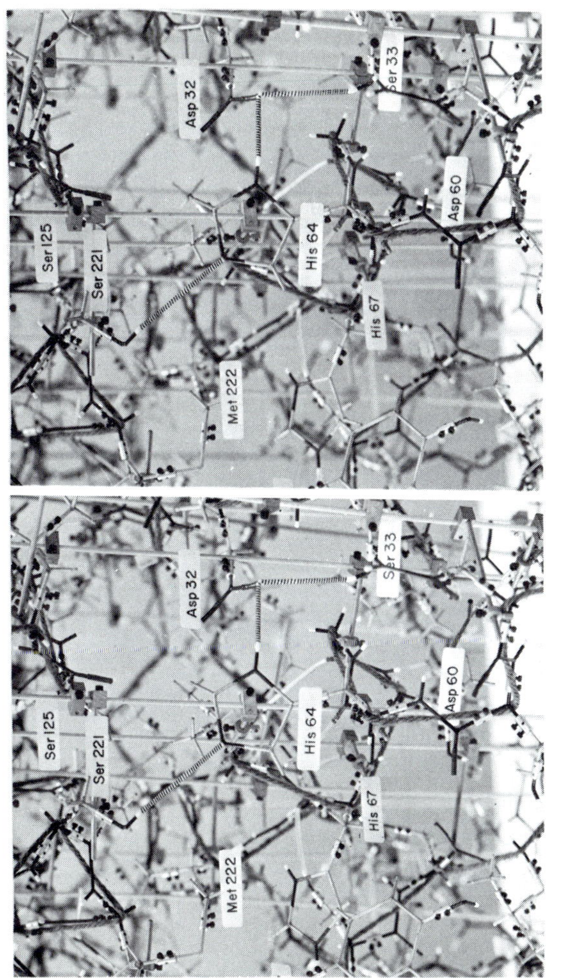

FIG. 5. Close-up of the active site region as it appears in the active enzyme. Hydrogen bonds analogous to those found in α-chymotrypsin are emphasized by insertion of dotted connectors.

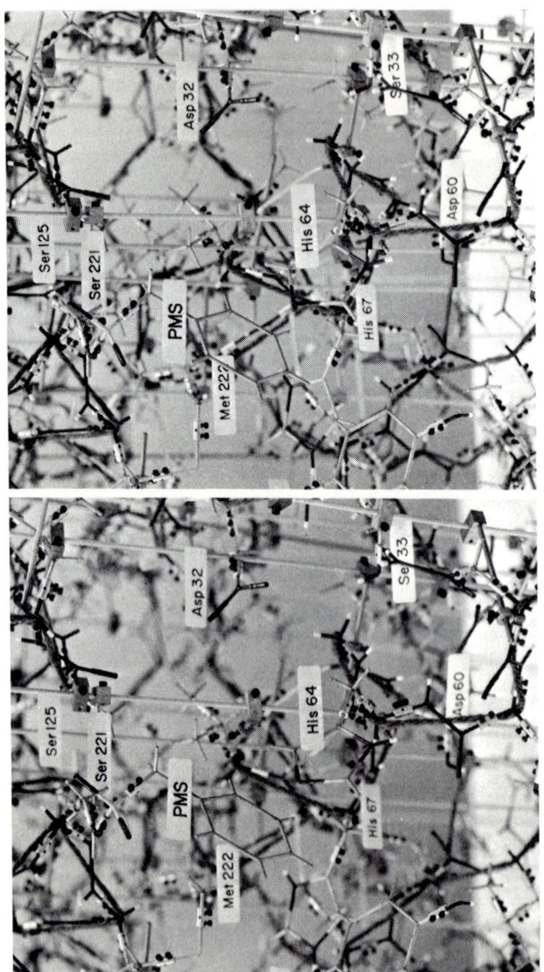

FIG. 6. Close-up of active site region with phenylmethanesulfonyl inhibitor group covalently bonded to Ser 221. Note movement of His 64 side chain.

that a serine residue is also found at the corresponding location in the amino acid sequences of five related enzymes: bovine chymotrypsinogen B, bovine trypsinogen, pig elastase (*23*), bovine thrombin (*24*), and dogfish trypsinogen (*25*). In addition, subtilisin Carlsberg (*26*) contains all components of the catalytic site hydrogen bond network, except that threonine is substituted for the second serine at position 33. From inspection of the subtilisin model, it appears that hydrogen bond (3) could be retained by utilizing the threonine side chain OH group.

Three possible roles for the second serine hydrogen bond suggest themselves. Based upon the proposal by Steitz *et al.* (*22*) that the carbonyl oxygen of Ser 214 in chymotrypsin is involved in substrate binding, one possible role for the hydrogen bond in question is that it functions to maintain the carbonyl group of this residue in the proper orientation. However, in subtilisin the corresponding carbonyl group of Ser 33 and, in fact, all the immediately neighboring backbone carbonyl groups are pointed in the wrong direction to serve a substrate binding function. The second possibility is that the second serine hydrogen bond holds the buried aspartate side chain in the correct orientation, and the third is that it is directly involved somehow in the charge relay system. However, it would appear to be very difficult to distinguish between the last two possibilities experimentally.

As mentioned in Section I, the great sterochemical similarity between subtilisin and chymotrypsin at their respective catalytic sites is highly gratifying in view of their similar enzymic activities. It is worthwhile stressing, however, that the similarity extends only to what might be termed the *connectivity* or perhaps *topology* of the hydrogen bond network. The actual geometrical arrangement of the components is different, as is the sequential order of the participating side chains at the catalytic sites of the two enzymes. This is immediately clear when it is noted that their order is Asp 32, Ser 33, His 64, Ser* 221 in subtilisin, and His 57, Asp 102, Ser* 195, Ser 214 in chymotrypsin (where the reactive serine

23. M. O. Dayhoff, ed., "Atlas of Protein Sequence and Structure," Vol. 4, p. D224. Natl. Biomed. Res. Found., Silver Spring, Maryland, 1969.

24. S. Magnusson, *in* "Structure–Function Relationships of Proteolytic Enzymes" (P. Desnuelle, H. Neurath, and M. Ottesen, eds.), p. 138. Munksgaard, Copenhagen, 1970.

25. R. A. Bradshaw, H. Neurath, R. W. Tye, K. A. Walsh, and W. P. Winter, *Nature* **226**, 237 (1970).

26. E. L. Smith, R. J. De Lange, W. H. Evans, M. Landon, and F. S. Markland, *JBC* **243**, 2184 (1968).

is denoted by Ser* in each case). This fact, in addition to the total lack of similarity in overall three-dimensional geometry between the two molecules, invites the conclusion that they represent a case of independent convergent evolution at the molecular level. It will be interesting to see whether future research can reveal exactly how the details of the two molecular geometries account for their comparative enzymic behavior.

IV. Comparison with Subtilisin Carlsberg

Subtilisin Carlsberg (*26*) differs from BPN′ at 84 residues, plus a deletion at residue 56. Based on the generally accepted assumption that overall three-dimensional folding is predominantly determined by the interior residues, and on the observed spatial distribution of the side chains that differ between Carlsberg and BPN′, it may be inferred that the two species of alkaline protease have very similar three-dimensional structures. All differences, with one exception, occur in exterior chain segments. The exception is Ile 31, which changes to a leucine in the Carlsberg enzyme. Furthermore, the side chain of 75 out of the 84 substituted residues are pointing outward, and these 75 changes appear to be more-or-less random with respect to residue type. That is to say, 44 are conservative substitutions, 15 go from polar to nonpolar, and 16 go the other way around. On the other hand, of the 9 inward-pointing side chain substitutions, 8 are conservative and only one is nonconservative, i.e., Ser 88 becomes a valine in the Carlsberg enzyme. The single deletion at Pro 56 occurs in an outside loop of the backbone chain.

16

Subtilisins: Primary Structure, Chemical and Physical Properties

FRANCIS S. MARKLAND, JR. • EMIL L. SMITH

I. Introduction

Since the last review of this subject appeared in this treatise (*1*) considerable progress has been made in our understanding of both the structure and properties of bacterial proteinases. This chapter will review only the subtilisins: the diisopropylfluorophosphate (DFP)-sensitive, extracellular, alkaline proteinases from *Bacillus subtilis* and other species of *Bacillus*. Since most of the recent work has been on the subtilisins, major emphasis will be placed on these enzymes with mention of other alkaline proteinases as applicable. A separate chapter (*2*) reviews the X-ray structure of subtilisin BPN′. Other bacterial proteinases are discussed separately (*3*).

II. Historical Background and Development

The extracellular alkaline proteinase, subtilisin, was discovered by Linderstrøm-Lang and Ottesen (*4*) during an investigation of the conversion of ovalbumin to plakalbumin. Subsequent work in the Carlsberg laboratory led to the isolation and crystallization of this enzyme, subtilisin Carlsberg (*5–7*). Later, another type of subtilisin, BPN′, was isolated and crystallized (*8*). Since the earlier review by Hagihara (*1*) described the preparation of both subtilisins Carlsberg and BPN′ only the isolation and characterization of other subtilisins are included here.

1. B. Hagihara, "The Enzymes," 2nd ed., Vol. 4, p. 193, 1960.
2. J. Kraut, "The Enzymes," 3rd ed., Vol. III, Chapter 15, 1971.
3. H. Matsubara and J. Feder, "The Enzymes," 3rd ed., Vol. III, Chapter 20, 1971.
4. K. Linderstrøm-Lang and M. Ottesen, *Nature* **159**, 807 (1947).
5. A. V. Güntelberg and M. Ottesen, *Nature* **170**, 802 (1952).
6. A. V. Güntelberg, *Compt. Rend. Trav. Lab. Carlsberg, Ser. Chim.* **29**, 27 (1954).
7. A. V. Güntelberg and M. Ottesen, *Compt. Rend. Trav. Lab. Carlsberg, Ser. Chim.* **29**, 36 (1954).
8. B. Hagihara, H. Matsubara, M. Nakai, and K. Okunuki, *J. Biochem.* (*Tokyo*) **45**, 185 (1958).

In 1960, Ottesen and Spector (*9*) reported on the characteristics of yet another extracellular alkaline bacterial proteinase, designated subtilisin Novo. Although it converted ovalbumin to plakalbumin like subtilisin Carlsberg, it exhibited different chromatographic and kinetic properties.

Subtilisin Amylosacchariticus was purified by Tsuru *et al.* (*10*) by a combination of ammonium sulfate precipitation; chromatography on Duolite A-2, CM-cellulose, and DEAE-Sephadex; and crystallization from tris·HCl-calcium acetate buffer at pH 8.5. Little difference was noted in the enzymic properties of this alkaline proteinase and other subtilisins. This proteinase as well as subtilisins BPN′ (*11, 12*), Carlsberg (*7*), and Novo (*9*), all have esterase activity and are inhibited by DFP.

Keay and Moser (*13*) have prepared additional alkaline proteinases from *B. subtilis* (NRRL B3411), *B. licheniformis,* and *B. pumilis* by acetone precipitation, batch treatment with DEAE-cellulose, chromatography on Duolite C-10 ion exchange resin, reprecipitation, dialysis, and lyophilization. They showed that these alkaline proteinases can be classified into two groups based on their amino acid compositions, serological cross-reactions, and ratio of esterase to proteinase activity. Group A includes subtilisin Carlsberg and the enzymes from *B. licheniformis* and *B. pumilis,* whereas group B includes subtilisins Novo, BPN′, Amylosacchariticus and NRRL B3411. They suggested that the name Bacillopeptidase be given to these enzymes. Inasmuch as the name subtilisin is well established, we shall follow the binomial system already in wide use by maintaining the generic term subtilisin followed by the name of the specific variety of enzyme under study (*14*).

Welker and Campbell (*15*) have indicated that the organism from which subtilisin BPN′ is derived should be called *Bacillus amyloliquefaciens* and is a different species from *B. subtilis.* These authors concluded that the relatively large differences reported by Smith *et al.* (*16*) in the sequences of subtilisins BPN′ and Carlsberg were due to the fact that they were derived from different species of *Bacillus* and not different strains of the same species as originally believed.

9. M. Ottesen and A. Spector, *Compt. Rend. Trav. Lab. Carlsberg* **32,** 63 (1960).

10. D. Tsuru, H. Kira, T. Yamamoto, and J. Fukumoto, *Agr. Biol. Chem.* (*Tokyo*) **30,** 1261 (1966).

11. H. Matsubara, B. Hagihara, M. Nakai, T. Komaki, T. Yonetani, and K. Okunuki, *J. Biochem.* (*Tokyo*) **45,** 251 (1958).

12. H. Matsubara and S. Nishimura, *J. Biochem.* (*Tokyo*) **45,** 503 (1958).

13. L. Keay and P. W. Moser, *BBRC* **34,** 600 (1969); L. Keay, P. W. Moser, and B. S. Wildi, *Biotechnol. Bioeng.* **12,** 213 (1970).

14. S. A. Olaitan, R. J. DeLange, and E. L. Smith, *JBC* **243,** 5296 (1968).

15. N. E. Welker and L. L. Campbell, *J. Bacteriol.* **94,** 1124 and 1131 (1967).

16. E. L. Smith, F. S. Markland, C. B. Kasper, R. J. DeLange, M. Landon, and W. H. Evans, *JBC* **241,** 5974 (1966).

III. Physical, Chemical, and Stability Properties of the Subtilisins

A. Subtilisin Carlsberg

Güntelberg and Ottesen (*7*) reported that subtilisin Carlsberg had an isoelectric point of 9.4 and a typical protein ultraviolet absorption spectrum. During electrophoresis experiments it was found that the protein was more stable at acid pH (pH 5.3 or 6.5) than at alkaline pH (pH 8.1 and 9.5) where autolysis of the enzyme occurs. In the alkaline range the enzyme appears to be more stable than trypsin or chymotrypsin. Furthermore, the proteinase is stable for several months in the lyophilized state, in dialyzed solution kept frozen at −10°C, or in glycerol solution at room temperature; the activity is rapidly lost at pH values below 5. The sedimentation coefficient was reported to be $s_{20,w} = 2.85$ S (*7*), and the amino-terminal sequence of the diisopropylphosphoryl (DIP) enzyme as NH_2–Ala–Glx (*17*). Table I presents some of the

TABLE I
Comparison of Physicochemical Properties of the Subtilisins

Property	BPN′ (*18–20*)	Carlsberg (*7, 17, 21, 22*)	Novo (*9, 14*)	Amylosac-chariticus (*10, 23, 24*)
$E^{1\%}_{280\ nm}$	11.7	8.6	11.7	11.9
Partial specific volume[a]	0.731	0.725	0.731	0.722
Molecular weight				
Sedimentation equilibrium	26,000	27,400[b]	—	28,000
Amino acid sequence	27,537	27,277	27,537	27,671
$s_{20,w}$	2.77	2.85	—	2.71
pI	7.8	9.4	9.1	7.8
f/f_0	1.18	—	—	1.04
NH_2-terminal residues	Ala	Ala	Ala	Ala
Elemental analysis (%)[c] N	17.3	17.05	17.3	17.01
Elemental analysis (%)[c] S	0.58	0.59	0.58	0.46

[a] Calculated from the amino acid composition.
[b] Phosphorous content of DIP enzyme.
[c] Calculated from the amino acid sequence.

17. M. Ottesen and C. G. Schellman, *Compt. Rend. Trav. Lab. Carlsberg, Ser. Chim.* **30**, 157 (1957).
18. H. Matsubara, C. B. Kasper, D. M. Brown, and E. L. Smith, *JBC* **240**, 1125 (1965).

TABLE II
AMINO ACID COMPOSITIONS OF SUBTILISINS

Amino acid	BPN′[a] (*19*)	Carlsberg[a] (*21*)	Novo[b] (*14*)	Amylosacchariticus[c] (*24*)
Lysine	11	9	11	8
Histidine	6	5	6	6
Arginine	2	4	2	4
Tryptophan	3	1	3	3
Aspartic acid	11	9	11	26
Asparagine	17	19	17	
Threonine	13	19	13	16
Serine	37	32	37	40
Glutamic acid	4	5	4	15
Glutamine	11	7	11	
Proline	14	9	14	13
Glycine	33	35	33	33
Alanine	37	41	37	36
Valine	30	31	30	25
Methionine	5	5	5	4
Isoleucine	13	10	13	16
Leucine	15	16	15	15
Tyrosine	10	13	10	12
Phenylalanine	3	4	3	3
Cysteine, cystine	0	0	0	0
Total	275	274	275	275

[a] From the amino acid sequence.

[b] From analysis of tryptic peptides; the compositions are the same as the tryptic peptides from subtilisin BPN′.

[c] From analysis of tryptic and cyanogen bromide peptides.

physicochemical properties of subtilisin Carlsberg. The amino acid composition is given in Table II (*21, 22*).

B. Subtilisin BPN′

Subtilisin BPN′ was shown to have essentially the same stability characteristics as subtilisin Carlsberg and also to be stabilized by Ca^{2+}

19. F. S. Markland and E. L. Smith, *JBC* **242,** 5198 (1967).
20. C. B. Kasper, H. Matsubara, and E. L. Smith, *JBC* **240,** 1131 (1965).
21. R. J. DeLange and E. L. Smith, *JBC* **243,** 2134 (1968).
22. G. Johansen and M. Ottesen, *Compt. Rend. Trav. Lab. Carlsberg* **34,** 199 (1964).
23. D. Tsuru, H. Kira, T. Yamamoto, and J. Fukumoto, *Agr. Biol. Chem.* (*Tokyo*) **31,** 330 (1967).
24. F. S. Markland, M. Kurihara, and E. L. Smith, unpublished studies (1970).

and other salts (*11*). The proteinase was shown to be a single polypeptide chain of molecular weight 27,600 having an isoelectric point of 7.80 (*18*). The amino acid composition is given in Table II (*19, 22*) and the amino- and carboxyl-terminal sequences of the DIP enzyme were identified as NH_2--Ala–Glx and Ala–Gln–COOH, respectively (*20*). These results agree with earlier reports by the Japanese workers who showed that the enzyme had a molecular weight of about 30,500 with alanine as the amino-terminal residue (*25*).

C. Subtilisin Novo

Subtilisin Novo was shown to be similar in physicochemical properties to subtilisin Carlsberg (*9*). The isoelectric point of Novo was, however, somewhat lower than that of subtilisin Carlsberg. The stability characteristics of subtilisin Novo are similar to the other subtilisins and all are extremely acid labile. However, the remarkable resistance of these proteinases to a variety of denaturing agents such as 6 *M* urea and 50% ethanol has been noted (*26*), and these enzymes are stable as well to a variety of detergents such as sodium dodecyl sulfate, and sodium tripolyphosphate (*27*).

D. Subtilisin Amylosacchariticus

Tsuru and his co-workers (*10*) reported that subtilisin Amylosacchariticus was much less soluble in neutral solution than the other subtilisins, and could be crystallized during dialysis against weak salt solutions between pH 6.5 and 8.5 if the enzyme concentration was higher than 0.5%. Otherwise the stability properties of this enzyme were similar to those already reported for the other subtilisins. The sedimentation coefficient was determined to be 2.89 S and from approach to sedimentation equilibrium, the molecular weight was reported to be 22,700. End group analyses showed alanine to be the amino-terminal residue (*23*).

Reinvestigation (*24*) of the molecular weight by approach to sedimentation equilibrium gave a value of 28,000 for the DIP derivative indicating that autolysis may have resulted in the low molecular weight obtained by the Japanese workers (*23*). The isoelectric point is at pH 7.8

25. H. Matsubara and S. Nishimura, *J. Biochem.* (*Tokyo*) **45,** 413 (1958).
26. A. Gounaris and M. Ottesen, *Compt. Rend. Trav. Lab. Carlsberg* **35,** 37 (1965).
27. D. Tsuru, *Kagaku To Kogyo* (*Osaka*) **43,** 199 (1969); cited in CA **71,** 79183j (1969).

(identical to that of subtilisin BPN′), and the amino acid composition (*24*) is more like that of BPN′ than of Carlsberg in agreement with the classification of Keay and Moser (*13*).

Tables I and II present the physicochemical properties and amino acid compositions, respectively, of the various subtilisins.

IV. Primary Structure of the Subtilisins

A. General Comparison

Although many of the physical properties of the subtilisins are similar (see Table I) their amino acid compositions show considerable differences [see Table II (*14, 19, 21, 22, 24*)]. Peptide mapping indicated definite similarity between subtilisin BPN′ and Novo but also showed that they were different from subtilisin Carlsberg (*28*). This was substantiated by the amino acid analyses.

A feature of the amino acid compositions of the subtilisins is the complete absence of cysteine or cystine in contrast to the high content of disulfide bridges in the several pancreatic proteinases (*29–34*).

B. Subtilisin BPN′

The sequence of subtilisin BPN′ was deduced from studies of the tryptic (*35*), chymotryptic (*36*), peptic (*37*), and cyanogen bromide digests (*19*). The complete sequence is shown in Fig. 1.

The protein consists of a single polypeptide chain of 275 residues devoid of any disulfide bridges. There is no apparent homology with the sequences of the pancreatic proteinases (*16*). The serine residue reactive

28. J. A. Hunt and M. Ottesen, *BBA* **48,** 411 (1961).
29. O. Mikeš, V. Holeyšovský, J. Tomášek, and F. Šorm, *BBRC* **24,** 346 (1966).
30. K. A. Walsh and H. Neurath, *Proc. Natl. Acad. Sci. U. S.* **52,** 884 (1964).
31. J. R. Brown and B. S. Hartley, *BJ* **101,** 214 (1966).
32. B. S. Hartley, J. R. Brown, D. L. Kauffman, and L. B. Smillie, *Nature* **207,** 1157 (1965).
33. B. Meloun, V. Kostka, J. Vaněček, I. Kluh, and F. Šorm, *Collection Czech. Chem. Commun.* **31,** 321 (1966).
34. D. M. Shotton and B. S. Hartley, *Nature* **225,** 802 (1970).
35. C. B. Kasper and E. L. Smith, *JBC* **241,** 3754 and 3771 (1966).
36. F. S. Markland, G. Kreil, B. Ribadeau-Dumas, and E. L. Smith, *JBC* **241,** 4642 (1966).
37. F. S. Markland, B. Ribadeau-Dumas, and E. L. Smith, *JBC* **242,** 5174 (1967).

Thr Ile Pro Leu Asp Lys Val Gln Ala Phe Lys Ala
NH_2-Ala-Gln-Ser-Val-Pro-Tyr-Gly-Val-Ser-Gln-Ile-Lys-Ala-Pro-Ala-Leu-His-Ser-Gln-Gly-Tyr-Thr-Gly-Ser-Asn-Val-Lys-Val-Ala-Val-
10 20 30

Leu Thr Gln Ala Asn Val Phe Ala Gly Ala — Tyr Asn Thr
Ile-Asp-Ser-Gly-Ile-Asp-Ser-Ser-His-Pro-Asp-Leu-Lys-Val-Ala-Gly-Gly-Ala-Ser-Met-Val-Pro-Ser-Glu-Thr-Pro-Asn-Phe-Gln-Asp-
40 50 60

Gly Gly Asp Thr Thr Val Ser
Asp-Asn-Ser-His-Gly-Thr-His-Val-Ala-Gly-Thr-Val-Ala-Ala-Leu-Asn-Asn-Ser-Ile-Gly-Val-Leu-Gly-Val-Ala-Pro-Ser-Ser-Ala-Leu-
70 80 90

Asn Ser Ser Ser Gly Val Ser Thr Thr Gly
Tyr-Ala-Val-Lys-Val-Leu-Gly-Asp-Ala-Gly-Ser-Gly-Gln-Tyr-Ser-Trp-Ile-Ile-Asn-Gly-Ile-Glu-Trp-Ala-Ile-Ala-Asn-Asn-Met-Asp-
100 110 120

Ala Thr Met Gln Asn Tyr Arg
Val-Ile-Asn-Met-Ser-Leu-Gly-Gly-Pro-Ser-Gly-Ser-Ala-Ala-Leu-Lys-Ala-Ala-Val-Asp-Lys-Ala-Val-Ala-Ser-Gly-Val-Val-Val-Val-
130 140 150

Ser Asn Ser Thr Asn Ile Ala Asp
Ala-Ala-Ala-Gly-Asn-Glu-Gly-Ser-Thr-Gly-Ser-Ser-Ser-Thr-Val-Gly-Tyr-Pro-Gly-Lys-Tyr-Pro-Ser-Val-Ile-Ala-Val-Gly-Ala-Val-
160 170 180

Asn Ser Asn Ala Glu Ala Gly Val Tyr Tyr
Asp-Ser-Ser-Asn-Gln-Arg-Ala-Ser-Phe-Ser-Ser-Val-Gly-Pro-Glu-Leu-Asp-Val-Met-Ala-Pro-Gly-Val-Ser-Ile-Gln-Ser-Thr-Leu-Pro-
190 200 210

Thr Thr Ala Thr Leu
Gly-Asn-Lys-Tyr-Gly-Ala-Tyr-Asn-Gly-Thr-Ser*-Met-Ala-Ser-Pro-His-Val-Ala-Gly-Ala-Ala-Ala-Leu-Ile-Leu-Ser-Lys-His-Pro-Asn-
220 230 240

Leu Ser Ala Ser Asn Arg Ser Ser Ala Tyr Ser
Trp-Thr-Asn-Thr-Gln-Val-Arg-Ser-Ser-Leu-Gln-Asn-Thr-Thr-Thr-Lys-Leu-Gly-Asp-Ser-Phe-Tyr-Tyr-Gly-Lys-Gly-Leu-Ile-Asn-Val-
250 260 270

Glu
Gln-Ala-Ala-Ala-GlnCOOH
275

FIG. 1. Comparison of amino acid sequence of subtilisins BPN′ and Carlsberg (*42*). The continuous sequence is that of subtilisin BPN′. The residues which differ in subtilisin Carlsberg are given above the corresponding residue. The dash at residue **56** indicates that the corresponding residue is lacking in the Carlsberg enzyme. The active Ser **221** is denoted by asterisk.

with DFP is at position **221** and the sequence around that serine (Thr–Ser*–Met–Ala), as originally pointed out by Sanger and Shaw (*38*), is different from that around the reactive serine in the mammalian pancreatic proteinases (see Section V,A).

There are several sequences in the polypeptide chain which appear to be repeated. These may have some relationship to the evolutionary history of the subtilisins and will be discussed in Section IV,F.

Although the possible significance is presently unknown there are many di-, tri-, and, in one case, tetrapeptide (residue **147** through **150**, Fig. 1) repetitions of the same residue. The residues appearing in these repeating sequences are Ala (six Ala–Ala sequences, three of which are Ala–Ala–Ala), Val (one Val–Val–Val–Val sequence), Ser (six Ser–Ser sequences, including one Ser–Ser–Ser), Thr (one Thr–Thr–Thr sequence), Gly (two Gly–Gly sequences), Ile (one Ile–Ile sequence), and

38. F. Sanger and D. C. Shaw, *Nature* **187**, 872 (1960).

Tyr (one Tyr–Tyr sequence). Additionally, there are many Ser–Thr or Thr–Ser sequences.

C. Subtilisin Novo

The chromatographic behavior and compositions of the 14 expected tryptic peptides from subtilisin Novo have been reported (*14*). Since these were identical in all respects to those from subtilisin BPN′, it was concluded that the two enzymes are identical. Furthermore, comparison of various enzymic and chemical properties all indicated the identity of the Novo and BPN′ enzymes (*22, 28, 39–41*).

D. Subtilisin Carlsberg

The complete sequence of subtilisin Carlsberg (*42*) was determined from the tryptic (*21, 43*) and chymotryptic peptides (*44, 45*) (see Fig. 1). The protein contains 274 residues in a single polypeptide chain and differs from subtilisin BPN′ in 85 positions including one deletion. Although the total base composition of subtilisin BPN′ (11 lysine and 2 arginine residues) and Carlsberg (9 lysine and 4 arginine residues) are identical, the base substitutions are not in the same positions in the sequence; these differences in base positions contribute to the radically different tryptic peptides of the two enzymes (*42*).

E. Subtilisin Amylosacchariticus

The partial sequence of subtilisin Amylosacchariticus (*24*) is shown in Fig. 2 in comparison with the complete sequence of subtilisin BPN′. Tryptic hydrolysis and cyanogen bromide cleavage have yielded fragments which account for the approximately 275 residues in the protein. The part of the sequence already determined shows great similarity to that of subtilisin BPN′, although there are some features which resemble subtilisin Carlsberg.

39. A. N. Glazer, *JBC* **242,** 433 (1967).
40. A. O. Barel and A. N. Glazer, *JBC* **243,** 1344 (1968).
41. J. Drenth and W. G. J. Hol, *JMB* **28,** 543 (1967).
42. E. L. Smith, R. J. DeLange, W. H. Evans, M. Landon, and F. S. Markland, *JBC* **243,** 2184 (1968).
43. R. J. DeLange and E. L. Smith, *JBC* **243,** 2143 (1968).
44. M. Landon, W. H. Evans, and E. L. Smith, *JBC* **243,** 2165 (1968).
45. W. H. Evans, M. Landon, and E. L. Smith, *JBC* **243,** 2172 (1968).

NH$_2$-Ala(Gln,Ser,Val,Pro)Tyr-Gly(**Ile**,Ser,Gln,Ile)Lys(Ala,Pro,Ala,Leu,His)(Ser,Gln,Gly,Tyr)(Thr,Gly,Ser,Asn,Val)Lys(Val,Ala,Val,
NH$_2$-Ala-Gln-Ser-Val-Pro-Tyr-Gly-**Val**-Ser-Gln-Ile-Lys-Ala-Pro-Ala-Leu-His-Ser-Gln-Gly-Tyr-Thr-Gly-Ser-Asn-Val-Lys-Val-Ala-Val-
10 20 30

Ile, Asp, Ser, Gly, Ile, Asp, Ser, Ser, His, Pro, Asp, Leu) **Arg** (Val, Ala, Gly, Gly, Ala, Ser, **Ser**, Val, Pro, Ser, Glu, Thr, Pro, **Tyr**, Phe, Gln, Asp
Ile-Asp-Ser-Gly-Ile-Asp-Ser-Ser-His-Pro-Asp-Leu-**Lys**-Val-Ala-Gly-Gly-Ala-Ser-**Met**-Val-Pro-Ser-Glu-Thr-Pro-**Asn**-Phe-Gln-Asp-
40 50 60

Gly, Asn, Ser, His*, Gly, Thr, His, **Ile**, Ala, Gly, Thr, Val, Ala, Ala, Leu, Asn,Asn,Ser,Ile,Gly,Val,Leu,Gly,**Ser**,Ala,Pro,Ser,Ser,Ala,Leu,
Asp-Asn-Ser,His*-Gly-Thr-His-**Val**-Ala-Gly-Thr-Val-Ala-Ala-Leu-Asn-Asn-Ser-Ile-Gly-Val-Leu-Gly-**Val**-Ala-Pro-Ser-Ser-Ala-Leu-
70 80 90

Tyr) (Ala, Val) Lys (Val,Leu,**Thr**,Asp,**Ile**,Gly,Ser,Gly,Gln,Tyr,Ser,Trp,Ile)(Ile,Asn,Gly,Ile,Glu)Trp(Ala,Ile,**Ser**,Asn,Asn)Met(Asp,
Tyr-Ala-Val-Lys-Val-Leu-**Gly**-Asp-**Ala**-Gly-Ser-Gly-Gln-Tyr-Ser-Trp-Ile-Ile-Asn-Gly-Ile-Glu-Trp-Ala-Ile-**Ala**-Asn-Asn-Met-Asp-
100 110 120

Val, Ile, Asn) Met (Ser, Leu, Gly, Gly, Pro,Ser,Gly,Ser,**Thr**,Ala,Leu)Lys-**Thr-Val**-Val-Asp-Lys(Ala,Val,Ala,Ser,Gly,Val,Val,Val,**Ile**,
Val-Ile-Asn-Met-Ser-Leu-Gly-Gly-Pro-Ser-Gly-Ser-**Ala**-Ala-Leu-Lys-**Ala-Ala**-Val-Asp-Lys-Ala-Val-Ala-Ser-Gly-Val-Val-Val-**Val**-
130 140 150

Ala, Ala, Ala, Gly, Asn, Glu, Gly, Ser, **Ser**, Gly, Ser, Ser, Ser,Thr,Val,Gly,Tyr,Pro,**Ala**)Lys(Tyr,Pro,Ser,Val,Ile,Ala,**Thr**,Gly,Ala,Val,
Ala-Ala-Ala-Gly-Asn-Glu-Gly-Ser-**Thr**-Gly-Ser-Ser-Ser-Thr-Val-Gly-Tyr-Pro-**Gly**-Lys-Tyr-Pro-Ser-Val-Ile-Ala-**Val**-Gly-Ala-Val-
160 170 180

Asp, Ser, Ser,Asn,Gln)Arg(Ala,Ser,Phe,Ser,Ser,**Ser**,Gly,**Ala**,Glu,Leu,Asp,Val)Met(Ala,Pro,Gly)(Val,Ser)Ile-Gln-Ser-Thr-Leu-Pro-
Asp-Ser-Ser-Asn-Gln-Arg-Ala-Ser-Phe-Ser-Ser-**Val**-Gly-**Pro**-Glu-Leu-Asp-Val-Met-Ala-Pro-Gly-Val-Ser-Ile-Gln-Ser-Thr-Leu-Pro-
190 200 210

Gly-**Gly-Thr**-Tyr-Gly-Ala-Tyr(Asn,Gly,Thr,Ser*)Met(Ala,**Thr**,Pro,His)(Val,Ala,Gly,Ala,Ala,Ala,Leu)(Ile,Leu,Ser)Lys(His,Pro,Asn,
Gly-**Asn-Lys**-Tyr-Gly-Ala-Tyr-Asn-Gly-Thr-Ser*-Met-Ala-**Ser**-Pro-His-Val-Ala-Gly-Ala-Ala-Ala-Leu-Ile-Leu-Ser-Lys-His-Pro-Asn-
220 230 240

Trp)(Thr,Asn)(**Ala**,Gln,Val)Arg(**Asn**,**Arg**)(Leu,Gln,**Ser**,Thr,**Ala**,Thr,**Tyr**)(Leu,Gly,Asp)(Ser,Phe,Tyr,Tyr,Gly)Lys(Gly,Leu,Ile,Asn,Val
Trp-Thr-Asn-**Thr**-Gln-Val-Arg-**Ser-Ser**-Leu-Gln-**Asn**-Thr-**Thr**-Thr-**Lys**-Leu-Gly-Asp-Ser-Phe-Tyr-Tyr-Gly-Lys-Gly-Leu-Ile-Asn-Val-
250 260 270

Gln)Ala-Ala-Ala-GlnCOOH
Gln-Ala-Ala-Ala-GlnCOOH
275

FIG. 2. Comparison of amino acid sequence of subtilisins BPN′ and Amylosacchariticus (*24*). The lower sequence is that of BPN′. Residues in subtilisin Amylosacchariticus whose sequences have not been determined are enclosed in parentheses. However, due to the extensive homology the arrangement of many of the residues within the parentheses is probably correct. The active His 64 and Ser 221 are denoted by asterisks. Apparent differences are indicated in boldface. The two enzymes apparently differ by at least 28 residues.

There are only 4 methionine and 12 basic residues in subtilisin Amylosacchariticus as compared to 5 and 13, respectively, in the other subtilisins. The 4 methionine residues are in the same positions as the 4 constant methionine residues in subtilisins BPN′ and Carlsberg (residues 119, 124, 199, and 222; Fig. 1). The missing base appears to be at residue 213 where lysine is replaced by threonine in subtilisin Carlsberg. The other bases are in the same positions as in subtilisin BPN′ except for residue 249 where there appears to be an arginine for serine replacement as in subtilisin Carlsberg and residue 256 where lysine is replaced by an unknown residue.

One interesting substitution occurs in the region around the reactive Ser 221 where the sequence is the same in subtilisin BPN′ and Carlsberg. This change in subtilisin Amylosacchariticus involves the substitution of threonine for serine at residue 224.

F. Comparison of Sequences

In comparing the sequences of subtilisins BPN′ and Carlsberg and the partial sequence of subtilisin Amylosacchariticus, certain generalizations can be made. Although the subtilisins resemble the pancreatic proteinases in many properties, e.g., inhibition by DFP (*12*, *17*), involvement of a histidine residue (*46*), and pH optima, there appears to be no similarity in their amino acid sequences. The absence of disulfide bonds and the high content of glycine, alanine, serine, and valine in the subtilisins is particularly noteworthy. Furthermore, the sequences around the reactive histidine and serine residues, which will be discussed in a later section, do not resemble those in the vertebrate enzymes; and the positioning and sequence around the aspartic acid residue involved in the active site of the subtilisins (*47*) are completely different from these in the mammalian proteinases (*48*) (Table III). Thus, the animal and bacterial proteinases would seem to have evolved independently, although a similar mechanism of action seems to have been the end result [also, see Smith (*49*)].

Of the 275 residues in the polypeptide chains there are 84 differences between subtilisins BPN′ and Carlsberg plus a deletion of one residue in subtilisin Carlsberg (as indicated by the dash at residue 56 in Fig. 1). This position has been chosen in order to minimize the number of muta-

46. F. S. Markland, E. Shaw, and E. L. Smith, *Proc. Natl. Acad. Sci. U. S.* **61,** 1440 (1968).
47. C. S. Wright, R. A. Alden, and J. Kraut, *Nature* **221,** 235 (1969).
48. D. M. Blow, J. J. Birktoft, and B. S. Hartley, *Nature* **221,** 337 (1969).
49. E. L. Smith, "The Enzymes," 3rd ed., Vol. I, Chapter 6, 1970.

TABLE III
SEQUENCES AROUND THE ASPARTYL RESIDUES IN ACTIVE SITES[a]

Enzyme	Sequence[b]
Chymotrypsin A (bovine)	Ser-Leu-Thr-Ile-Asn-Asn-Asp*-Ile-Thr-Leu-Leu-Lys 96 100 102 105 107
Chymotrypsin B (bovine)	Ile-Leu-Thr-Val-Arg-Asn-Asp*-Ile-Thr-Leu-Leu-Lys 96 100 102 105 107
Trypsin (bovine)	Ser-Asn-Thr-Leu-Asn-Asn-Asp*-Ile-Met-Leu-Ile-Lys 96 100 102 105 107
Thrombin (bovine)	-Glx-Asx-Leu-Asp-Arg-Asp*-Ile-Ala-Leu-Leu-Lys 97 100 102 105 107
Elastase (porcine)	Asp-Val-Ala-Ala-Gly-Tyr-Asp*-Ile-Ala-Leu-Leu-Arg 98 99 99A 99B 100 102 105 107
Subtilisin BPN′	Val-Ala-Val-Ile-Asp*-Ser-Gly-Ile-Asp 28 30 32 35 36
Subtilisin Carlsberg	Val-Ala-Val-Leu-Asp*-Thr-Gly-Ile-Gln 28 30 32 35 36
Subtilisin Amylosacchariticus	(Val,Ala,Val,Ile,Asp*,Ser,Gly,Ile,Asp) 28 30 32 35 36

[a] Residue numbers are below the sequence using the numbering of chymotrypsinogen A for the mammalian proteinases. The active aspartic acid residue is marked by an asterisk.

[b] For identification in mammalian enzymes, see Hartley *et al.* (*34, 48*); in the subtilisins, see Smith *et al.* (*42*), Wright *et al.* (*47*), and Fig. 2.

tional events in the nucleotide codons in this region. Nevertheless, the addition or deletion must have occurred between residues 50 and 59 in view of the excellent homology of the sequences of the two subtilisins on either side of this portion of the sequence.

Table IV gives the amino acid composition of subtilisin BPN′ and indicates the replacements in the Carlsberg enzyme. From the presently assigned triplet nucleotide codons for the amino acids (*50*), 61 of the changes can be ascribed to single base changes and 23 to double base changes.

In view of the close similarity of the enzymic properties of the two subtilisins (*39, 40*) it was somewhat surprising to find that they differed in 30% of the structure. The many similarities in their properties suggest that the polypeptide chain conformations are very similar, if not identical, for the two subtilisins and that most substitutions would be con-

50. F. H. C. Crick, *Cold Spring Harbor Symp. Quant. Biol.* **31**, 1 (1966).

TABLE IV
AMINO ACID SUBSTITUTIONS IN SUBTILISINS

Subtilisin BPN′			
Residue	Total	Changes (%)	Subtilisin Carlsberg changed residue (No.)[a]
Lysine	11	36	Asn (2), Thr (1),TYR (1)
Histidine	6	17	Gln (1)
Arginine	2	0	
Aspartic acid	11	45	GLN (1), Gly (1),Glu (1), Ser (2)
Asparagine	17	41	Tyr (1), Asp (1), Ser (3),GLY (1), ALA (1)
Glutamic acid	4	25	SER (1)
Glutamine	11	64	Leu (1), THR (1), SER (2), ASN (1), TYR (1), Glu (1)
Threonine	13	46	Lys (1), Ala (2), Ser (3)
Serine	37	49	Thr (4), Pro (1), Ala (3), Gly (3), VAL (1), Arg (2), Asn (4)
Proline	14	43[b]	ASP (2), Ala (3), deletion (1)
Glycine	33	12	ASN (1), Ala (2),THR (1)
Alanine	37	22	LYS (1), Val (1), Ser (2), Thr (3), GLN (1)
Valine	30	13	Ile (2), TYR (1), Ala (1)
Isoleucine	13	38	Leu (1), Thr (2), Val (2)
Leucine	15	20	Val (1), Met (1), TYR (1)
Methionine	5	20	PHE (1)
Tyrosine	10	20	Phe (1), LEU (1)
Phenylalanine	3	33	ASN (1)
Tryptophan	3	67	Gly (1), Leu (1)
Total	275	30.0[b]	

[a] Residues in capital letters involve at least two base changes in the triplet codon; other changes can be accounted for by single base changes.

[b] Includes deletion.

servative in nature. It is evident from the information in Table IV that at least for the hydrophobic residues most of the substitutions are conservative involving residues of the same type. Furthermore, the complete three-dimensional structure of subtilisin BPN′ shows that all substitutions with but one exception occur in exterior chain segments (*47*); the exception is Ile 31 which is Leu in subtilisin Carlsberg. The side chains of 75 of the 84 substituted residues are pointed outward, and of the 9 inward-pointing side chain substitutions, 8 are conservative. The radical change is Ser 88 which is Val in Carlsberg. A more detailed description of the X-ray work on subtilisin BPN′ appears in this volume (*2*).

Most of the amino acid substitutions are present throughout the linear sequence; however, there are several fairly long segments of the sequence which show no substitution. One of these occurs around the reactive Ser 221 extending from residues 218 through 240, and another is around the reactive His 64 extending from residues 64 through 75. As noted above, there is a conservative substitution of Ser 224 by Thr in subtilisin Amylosacchariticus.

Comparison of the sequences of subtilisins BPN′ and Carlsberg shows the presence of similar sequences in different portions of the molecules. These are obvious in either subtilisin individually but are even more apparent when both protein sequences are considered together (Fig. 3). In addition to those shown in the figure, there are a number of tripeptide

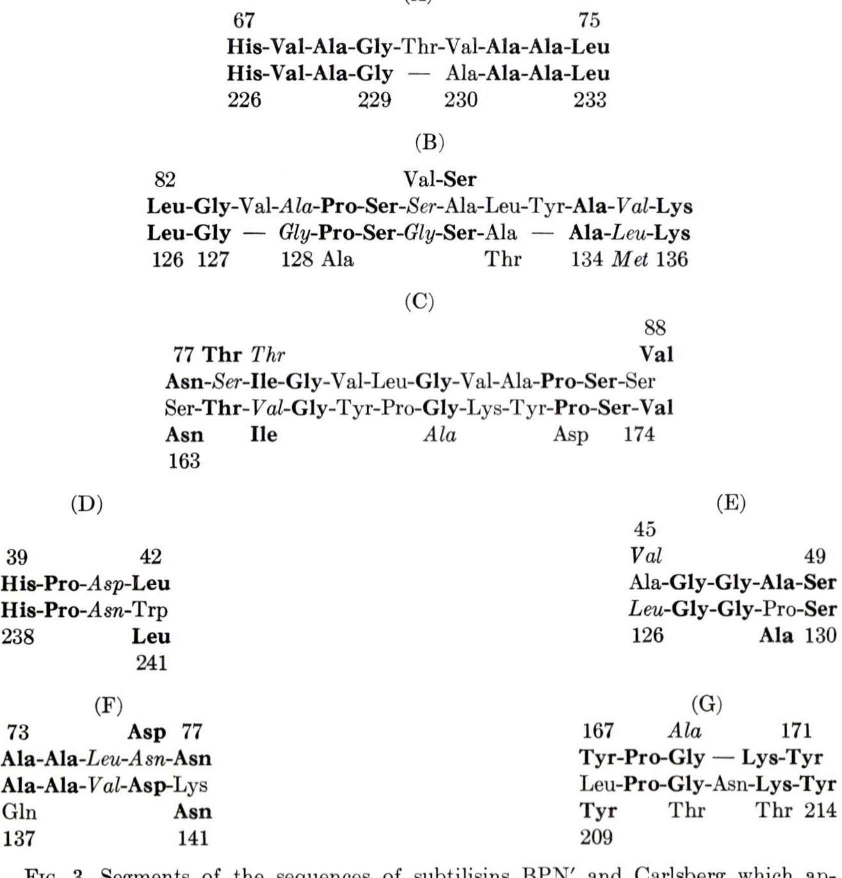

Fig. 3. Segments of the sequences of subtilisins BPN′ and Carlsberg which appear to involve repetition of similar sequences (42). Residues shown in boldface are identical; residues shown in italics would appear to represent conservative replacements. The continuous sequences are BPN′ or both subtilisins; residues above or below these sequences represent differences in subtilisin Carlsberg.

repetitions; e.g., Ala–Ala–Ala (residues 151–153, 230 through 232, and 272 through 274), Val–Lys–Val (residues 26 through 28 and 93 through 95), Gly–Asn–Ser (residues 154 through 156 and 157 through 159 in Carlsberg) and Val–Leu–Gly (residues 81 through 83 and 95 through 97 in BPN′).

The presence of repetitive sequences in other proteins has been interpreted as involving an extension of shorter peptide chains by a process of gene duplication followed by later mutational events (*49*). Such instances appear well documented for the ferrodoxins (*51, 52*) and for the γ-globulins (*53*). Thus it appears that the long polypeptide chain of 275 residues in the subtilisins may have evolved by a duplicated lengthening of the DNA genome.

There are additional features of the sequences in Fig. 3 that merit comment. First, the repetitions are not in exact linear order but for some segments appear to be in inverse order. For example, if we consider the sequences in D, A, and B (Fig. 3), the duplicated sequences are in the order: residues 39 through 42, 67 through 75, and 82 through 94 in the NH_2-terminal part of the molecule, but they are in the inverse order: residues 238 through 241, 226 through 233, and 126 through 136 in the COOH-terminal part. Second, some parts of the sequence are repeated two or more times; e.g., residues 82 through 88 are repeated not only by residues 126 through 131 but also by residues 168 through 174 (Fig. 3, B and C). The finding that the numbers of residues between the repeated segments vary and that the repetitions of some segments occur in inverse order indicates that the subtilisins may have evolved by a complex mechanism with duplications of the sequence occurring several times. Hopefully, more insight into the origin and evolutionary development of these enzymes will be obtained as the sequences of subtilisin Amylosacchariticus and other homologous subtilisins become known.

V. Active Site Studies

A. Serine

The involvement of serine in the active site of subtilisins Carlsberg (*7*) and BPN′ (*12*) was first indicated by the finding that they were in-

51. H. Matsubara, R. M. Sasaki, and R. K. Chain, *Proc. Natl. Acad. Sci. U. S.* **57,** 439 (1967).

52. R. V. Eck and M. O. Dayhoff, *Science* **152,** 363 (1966).

53. R. L. Hill, R. Delaney, R. E. Fellows, Jr., and H. E. Lebovitz, *Proc. Natl. Acad. Sci. U. S.* **56,** 1762 (1966).

activated by DFP. Tsuru and co-workers also showed that DFP is a potent inhibitor of subtilisin Amylosacchariticus (*10*).

Matsubara and Nishimura (*12*) showed that both the esteratic and proteolytic activities of subtilisin BPN′ were inhibited simultaneously by DFP. Furthermore, Matsubara (*54*) crystallized the DIP derivative of subtilisin BPN′ and showed that one mole of the inhibitor was bound to a serine residue per mole of protein. Similar results had been obtained earlier with subtilisin Carlsberg by Ottesen and Schellman (*17*), who also showed that the reaction with DFP was rapid, requiring about 10 min for 50% inactivation at pH 7.6.

Sanger and Shaw (*38*) isolated a peptide containing the reactive serine after partial acid hydrolysis of ^{32}P-DIP-labeled subtilisin Novo. They showed that the sequence around the serine residue Thr–SerP*–Met–Ala differed from the Asp–SerP*–Gly sequence found in several mammalian proteinases sensitive to DFP (*55–58*). This sequence was extended by Noller and Bernhard (*59*) using a specific chromophoric acylating agent, furylacryloylimidazole, to label the reactive serine residue of subtilisins BPN′ and Novo. After enzymic digestion, a furylacrylyl peptide was isolated whose sequence was determined as Asn–Gly–Thr–Ser*–Met. The determination of the complete sequence by Smith *et al.* (*16*) (Fig. 1) demonstrated that the active Ser 221 is located in a long sequence of constant residues from 218 through 240, except for the threonine–serine substitution at residue 224 in subtilisin Amylosacchariticus (Fig. 2) (*24*).

Advantage has been taken of this single highly reactive serine to develop a titration method with the chromophoric acylating agent *N*-*trans*-cinnamoylimidazole to determine the operational normality of subtilisin solutions (*60*).

1. *Serine Sequence in Other Proteinases*

It is interesting that some proteinases from microorganisms, notably those from *Streptomyces griseus* (*61*) and *Sorangium* sp. (*62*) possess

54. H. Matsubara, *J. Biochem.* (*Tokyo*) **46,** 107 (1959).
55. G. H. Dixon, D. L. Kauffman, and H. Neurath, *JACS* **80,** 1260 (1958).
56. N. K. Schaffer, L. Simet, S. Harshman, R. R. Engle, and R. W. Drisko, *JBC* **225,** 197 (1957).
57. B. S. Hartley, M. A. Naughton, and F. Sanger, *BBA* **34,** 243 (1959).
58. J. A. Gladner and K. Laki, *JACS* **80,** 1263 (1958).
59. H. F. Noller and S. A. Bernhard, *Biochemistry* **4,** 1118 (1965).
60. M. L. Bender, M. L. Bequé-Cantón, R. L. Blakeley, L. J. Brubacher, J. Feder, C. R. Gunter, F. J. Kézdy, J. V. Killheffer, Jr., T. H. Marshall, C. G. Miller, R. W. Roeske, and J. K. Stoops, *JACS* **88,** 5890 (1966).
61. S. Wählby and L. Engström, *BBA* **151,** 402 (1968).
62. L. B. Smillie and D. R. Whitaker, *JACS* **89,** 3350 (1967).

the same sequence around the reactive serine as the mammalian proteinases, whereas proteinases from other organisms and plants resemble the subtilisins; e.g., those from *Aspergillus oryzae* (*63*), *Aspergillus flavus* (*64*), and a caseinase from French beans (*65*). Indeed, other sequences have been reported for other DFP-sensitive proteinases, e.g., a proteinase from baker's yeast and phaseolin of French beans possess the sequence Glu–Ser*–Val (*65*) and a proteinase from *Arthrobacter* has the sequence Ser–Ser*–Gly (*66*). Table V shows the presently known sequences around the reactive serine residue of various DFP-sensitive proteinases.

In view of the substantial variations in these sequences, it would appear that the residues in the sequences near the active serine residue may have little influence on the reaction mechanism, the reactivity of the serine hydroxyl, or in determining the specificity of the enzyme; however, these sequences are useful in classifying the various types of enzymes.

2. *Conversion of Active Serine to Cysteine*

The conversion of the reactive serine residue in subtilisin Novo to cysteine has been reported by Polgár and Bender (*70*) and by Neet and

TABLE V
SEQUENCES AROUND REACTIVE SERINE RESIDUES[a]

Source	Sequence[b]
Various mammalian enzymes	Asp-Ser*-Gly
Sorangium sp.	
Streptomyces griseus	
Pseudocholine esterase	Glu-Ser*-Ala[c]
Subtilisins	Thr-Ser*-Met-Ala
Caseinase (beans)	
Aspergillus	
Baker's yeast	Glu-Ser*-Val
Phaseolin (beans)	
Arthrobacter	Ser-Ser*-Gly

[a] Reactive serine residues indicated by an asterisk.

[b] See text for references.

[c] For pseudocholine esterase see Jansz *et al.* (*67*). The same sequence has also been reported for acetylcholine esterase (*68*) and for liver aliesterase (*69*).

63. D. C. Shaw, Ph. D. Dissertation, University of Cambridge (1962).
64. O. Mikeš, J. Turková, N. B. Toan, and F. Šorm. *BBA* **178,** 112 (1969).
65. D. C. Shaw and J. R. E. Wells, *BJ* **104,** 5c (1967).
66. S. Wählby, *BBA* **151,** 409 (1968).
67. H. S. Jansz, D. Brons, and M. G. P. J. Warringa, *BBA* **34,** 573 (1959).

Koshland (*71*). In both cases the proteinase was reacted first with phenylmethyl sulfonyl fluoride to convert the serine to a phenylmethyl sulfonyl derivative (*72*) and completely inactivate the enzyme. This group was then displaced by treatment with thiolacetate at pH 5.2–5.5 and the thiolester thus formed was allowed to decompose spontaneously to form thiolsubtilisin as shown by Eq. (1):

$$\text{Subtilisin-OH} \xrightarrow[-\,\text{HF}]{+\,C_6H_5CH_2SO_2F} \text{subtilisin-O—}SO_2CH_2C_6H_5$$

$$\downarrow -C_6H_5CH_2SO_3^- \quad + CH_3\overset{O}{\overset{\|}{C}}\text{—SH} \tag{1}$$

$$\text{Subtilisin-SH} \xleftarrow[-\,CH_3CO_2^-]{+\,H_2O} \text{subtilisin-S—}\overset{O}{\overset{\|}{C}}\text{—}CH_3$$

Titration with *p*-mercuribenzoate (*73*) or 5,5′-dithiobis-2-nitrobenzoic acid (*74*), as well as amino acid analysis, indicated the formation of 0.7–0.9 mole of cysteine per mole subtilisin. No other changes could be detected by sedimentation analysis, starch gel electrophoresis, fluorescence spectroscopy, ultraviolet absorption, or tryptophan reactivity. However, a small change in the optical rotatory dispersion pattern was observed (*75*).

Study of thiolsubtilisin showed that it had about 33% of the initial activity of native subtilisin toward *p*-nitrophenylacetate (*71*). Thiolsubtilisin also hydrolyzes *N-trans*-cinnamoylimidazole (*76*) and the *p*-nitrophenyl esters of several amino acid derivatives although at drastically reduced rates compared to the native enzyme (*75*). No activity has been detected for the thiol enzyme against other ester, amide, or peptide substrates of subtilisin.

Several findings suggest that the reactions catalyzed by thiolsubtilisin may be enzymic in character: (1). The intermediate in the hydrolysis of *N-trans*-cinnamoylimidazole has the spectral characteristic of a thiol enzyme; (2) the hydrolysis of *p*-nitrophenylacetate follows Michaelis–Menten kinetics; (3) the hydrolysis of the *p*-nitrophenyl esters is inhib-

---

68. D. C. Shaw, unpublished work (1963), cited by F. Sanger, *Proc. Chem. Soc.* p. 76 (1963).
69. H. S. Jansz, C. H. Posthumus, and J. A. Cohen, *BBA* **33**, 396 (1959).
70. L. Polgár and M. L. Bender, *JACS* **88**, 3153 (1966).
71. K. E. Neet and D. E. Koshland, Jr., *Proc. Natl. Acad. Sci. U. S.* **56**, 1606 (1966).
72. D. E. Fahrney and A. M. Gold, *JACS* **85**, 997 (1963).
73. P. D. Boyer, *JACS* **76**, 4331 (1954).
74. G. L. Ellman, *ABB* **82**, 70 (1959).
75. K. E. Neet, A. Nanci, and D. E. Koshland, Jr., *JBC* **243**, 6392 (1968).
76. L. Polgár and M. L. Bender, *Biochemistry* **6**, 610 (1967).

ited by *p*-mercuribenzoate, whereas this hydrolysis catalyzed by native subtilisin is not; and (4) the rate of deacylation of cinnamoyl-thiol-subtilisin and the k_{cat} of the *p*-nitrophenylacetate hydrolysis are dependent on a basic group with pK_a of about 7.2 (*70, 76*).

These arguments are not completely convincing, however, inasmuch as there is complete loss of activity of thiolsubtilisin toward normal ester, amide, and peptide substrates. Furthermore, *p*-nitrophenyl esters react rapidly with cysteine itself and can acylate cysteine residues of proteins nonenzymically. Although the acylation and deacylation of thiolsubtilisin are somewhat more rapid than with cysteine, the effect is less impressive than the total lack of activity toward other ester and amide substrates. Neet *et al.* (*71, 75*) have ascribed this to the changes in the geometry at the active site due to the increased radius and the change in bond angles of the sulfur atom as compared to oxygen.

With the more specific ester substrate *N*-carbobenzyloxyglycinate *p*-nitrophenyl ester, Polgár and Bender (*77*) found thiolsubtilisin to be three orders of magnitude lower in activity than native subtilisin Novo. They showed also that the thiolenzyme has different chromatographic properties on carboxymethyl cellulose than the native enzyme. These results suggest that there may be a subtle change in the conformation of the active site of the thiolenzyme, and these authors argued on kinetic grounds for a hydrogen bond between the sulfhydryl group and the reactive histidine residue in thiolsubtilisin, whereas a similar hydrogen bond between the serine hydroxyl and imidazole does not exist in native subtilisin. The hydrogen bond formation in thiolsubtilisin also could cause the loss of one positive charge accounting for the chromatographic differences between native and thiolsubtilisin. It is of interest that during chromatography of native subtilisin Novo another, as yet unidentified, proteinase impurity was observed. This proteinase, which is presumably present in all preparations of subtilisin Novo, could have caused some of the problems in the sulfhydryl stoichiometry originally observed in the preparation of thiolsubtilisin.

Polgár (*78*) has reported that conversion of the reactive serine to cysteine in subtilisin Carlsberg produces similar changes in the catalytic properties and chromatographic behavior as in subtilisin Novo.

The bifunctional alkylating reagent, 1,3-dibromoacetone, which reacts first with the active cysteine in papain, ficin, and stem bromelain (*79, 80*) and then reacts intramolecularly with an active histidine residue,

77. L. Polgár and M. L. Bender, *Biochemistry* **8**, 136 (1969).
78. L. Polgár, *Acta Biochim. Biophys. Acad. Sci. Hung.* **3**, 397 (1968).
79. S. S. Husain and G. Lowe, *BJ* **108**, 855 and 861 (1968).
80. S. S. Husain and G. Lowe, *BJ* **110**, 53 (1968).

reacts only with the cysteine in thiolsubtilisin (*80*) and does not bridge to the active His 64 (*46*) that, at least in subtilisin BPN′, is close to the active serine (*47*).

B. Histidine

The involvement of histidine in the activity of proteolytic enzymes has been postulated for some years; for the early literature, see, for example, the reviews by Barnard and Stein (*81*) and by Dixon *et al.* (*82*).

The first suggestion of histidine involvement in the active site of subtilisin was provided by photooxidation studies of Oosterbaan and Cohen (*83*). Subsequently, during studies on the sequence around the reactive serine of subtilisin with the chromophoric acylating agent furyl-2-acryloylimidazole, Noller and Bernhard (*59*) suggested that a histidine residue may be acylated prior to acylation of the reactive serine. They inferred the presence of a catalytically active histidine residue in spatial proximity to the serine. The native furylacryloyl enzyme spectrum and the deacylation rates observed for the native acyl enzyme were different from those expected for the *O*-acyl serine derivative, and were more similar to those of an acyl histidine derivative (*84, 85*).

Kinetic evidence for the involvement of histidine was provided by Glazer (*39*) who showed that protonation of a group with a pK' of 6.58 resulted in the inactivation of subtilisin Carlsberg. The heat of ionization of this group, 7.3 kcal/mole at 0°, also fell within the expected range for a histidine residue. Similar results were obtained with subtilisins BPN′ and Novo (Fig. 4) (*86, 87*). Svendsen (*88*) showed that the rate of inactivation of subtilisin Novo by cyanate is influenced by a group with a pK around 7.7. He suggested that a histidine residue might act as the primary receptor from which the carbamyl group would then be transferred to the serine residue at the active site.

Although the above evidence suggested the involvement of a histidine

81. E. A. Barnard and W. D. Stein, *Advan. Enzymol.* **20,** 51 (1958).
82. G. H. Dixon, H. Neurath, and J. Pechère, *Ann. Rev. Biochem.* **27,** 489 (1958).
83. R. A. Oosterbaan and J. A. Cohen, *in* "Structure and Activity of Enzymes," (T. W. Goodwin, J. I. Harris, and B. S. Hartley, eds.), p. 87. New York, 1964.
84. S. A. Bernhard, S. J. Lau, and H. Noller, *Biochemistry* **4,** 1108 (1965).
85. S. Bernhard, Z. Grdinic, H. Noller, and N. Shaltiel, *Proc. Natl. Acad. Sci. U. S.* **52,** 1489 (1964).
86. H. Lineweaver and D. Burk, *JACS* **56,** 658 (1934).
87. B. Myers and A. N. Glazer, unpublished data (1970).
88. I. Svendsen, *Compt. Rend. Trav. Lab. Carlsberg* **36,** 235 (1967).

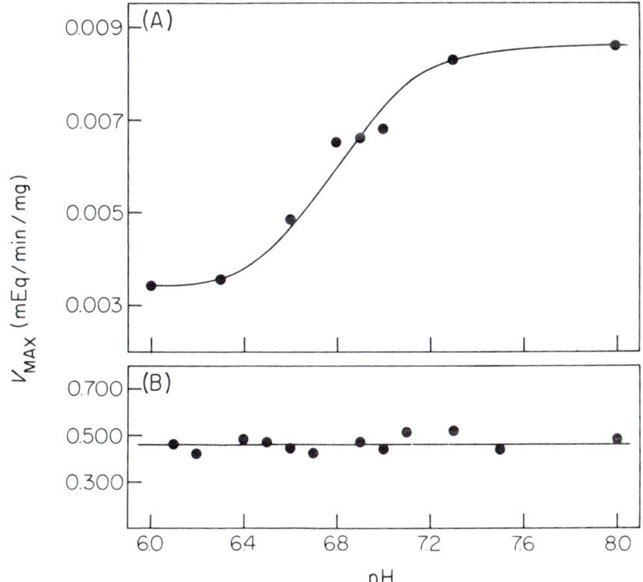

FIG. 4. Variation of V_{max} with pH for (A) the hydrolysis of benzoyl-L-arginine ethyl ester by subtilisin Novo and for (B) the hydrolysis of tosyl-L-arginine methyl ester by subtilisin Amylosacchariticus. Reactions were at 37° with automatic addition of 0.02 *N* base in a pH stat. Initial velocities were determined from the first 5% of the hydrolysis and V_{max} was obtained from the Lineweaver-Burk plot (*86*). A molecular weight of 27,600 was assumed. Data were taken from Myers and Glazer (*87*). Results similar to those with subtilisin Novo were also obtained with subtilisins BPN′ and Carlsberg (*39*).

residue, direct chemical evidence was lacking. Specific reagents used for labeling the histidine residues of trypsin (*89*) and chymotrypsin (*90*) were found to be ineffective in inactivating the subtilisins (*16*). Furthermore, other alkylating agents such as bromopyruvate, bromoacetate, iodoacetate, and iodoacetamide were without effect over a broad range of pH (*24*, *90a*).

Shaw and Ruscica (*91*) synthesized a highly reactive phenylalanine derivative similar to the chymotrypsin inhibitor, tosyl-L-phenylalanine chloromethyl ketone. The new reagent, benzyloxycarbonyl-L-phenylalanine bromomethyl ketone (ZPBK), was shown to react stoichiometrically

89. E. Shaw, M. Mares-Guia, and W. Cohen, *Biochemistry* **4**, 2219 (1965).
90. G. Schoellmann and E. Shaw, *Biochemistry* **2**, 252 (1963).
90a. E. L. Smith, F. S. Markland, and A. N. Glazer, *in* "Structure–Function Relationships of Proteolytic Enzymes" (P. Desnuelle, H. Neurath, and M. Ottesen, eds.), p. 160. New York, 1970.
91. E. Shaw and J. Ruscica, *JBC* **243**, 6312 (1968).

with subtilisin BPN′ with a concomitant loss of activity. The reaction was specific and resulted in the loss of a single histidine residue. Subsequent studies by Markland *et al.* (*46*), using tritiated ZPBK, led to the isolation and identification of a covalently labeled peptide from subtilisin Carlsberg. It was thus unequivocally demonstrated that His 64 is the active histidine. Similarly, a peptide was isolated from subtilisin BPN′ which contained the labeled histidine residue and was also consistent with its being His 64.

Although His 64 is far removed in linear sequence from the active Ser 221, the X-ray analysis of subtilisin BPN′ showed these residues to be in close proximity to one another (*47*). This situation is strikingly similar to that in α-chymotrypsin and trypsin where the reactive histidine residues, 57 and 46, respectively, are also close to the amino-terminal ends and remote from the active serine residues, 195 and 183, respectively (*32*).

Although the reactive serine residue in the subtilisins is in an invariant or nearly invariant sequence, this is not the case for the reactive histidine. Residue 63 on the amino-terminal side of histidine is glycine for the Carlsberg enzyme and serine for BPN′ (see Table VI which also shows for comparison the sequences around the reactive histidine residues in

TABLE VI
SEQUENCE AROUND REACTIVE HISTIDINE RESIDUES[a]

Enzyme	Sequence[b]
Chymotrypsins A and B (bovine)	Val-Thr-Ala-Ala-His*-Cys-Gly-Val-Thr 57
Elastase (porcine)	Met-Thr-Ala-Ala-His*-Cys-Val-Asp-Arg 57
Trypsin (bovine)	Val-Ser-Ala-Ala-His*-Lys-Tyr-Lys-Ser 57
Subtilisin Carlsberg	Asp-Gly-Asn-Gly-His*-Gly-Thr-His-Val 64
Subtilisin BPN′	Asp-Asp-Asn-Ser-His*-Gly-Thr-His-Val 64

[a] The residue numbers are under the sequence; for the mammalian proteinases the numbering system is that of chymotrypsinogen A. The reactive histidine residue is marked by an asterisk.

[b] For review of references for the pancreatic proteinases, see Smillie and Hartley (*92*) and for the subtilisins, see Markland *et al.* (*46*).

92. L. B. Smillie and B. S. Hartley, *BJ* **101**, 232 (1966).

several mammalian proteinases) (*46, 92*). This single amino acid replacement may account in part for the greater activity of subtilisin Carlsberg and some of the small differences in specificity exhibited by these two enzymes.

Subtilisin Amylosacchariticus has a lower specific activity for synthetic ester substrates than the other two subtilisins and is also much less reactive with ZPBK (Fig. 5) (*24*). Since this enzyme also has distinctive kinetic properties, we would expect these differences in reactivity to be reflected in the sequence and conformation of this enzyme.

Morihara and Oka (*92a*) have synthetized the chloromethyl ketone derivatives of two peptide substrates of subtilisin BPN′ (carbobenzoxy-L-Ala-Gly-L-Phe chloromethyl ketone and carbobenzoxy-L-Ala-L-Phe chloromethyl ketone) and have shown that these compounds react much faster with the enzyme than ZPBK. The reaction was stoichiometric and

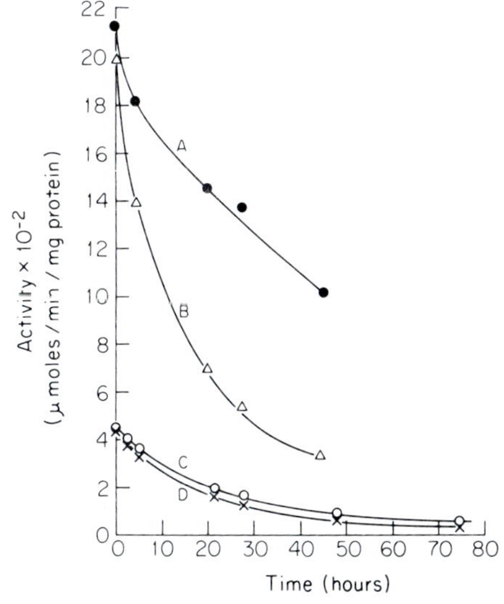

FIG. 5. Comparison of the reactivity of subtilisins Carlsberg and Amylosacchariticus with tritiated ZPBK (*24*). The reaction was carried out at pH 6.8 in a volume of 6.0 ml containing (final concentration): 33% dioxane 0.093 *M* tris·Cl, 0.0093 *M* $CaCl_2$, 12.6 μmoles ZPBK and 0.29 μmole protein. At various times aliquots were removed to assay the esterase activity on acetyl-L-tyrosine ethyl ester by the method of Glazer (*39*). Control samples contained dioxane without ZPBK. (A) Subtilisin Carlsberg control; (B) subtilisin Carlsberg plus ^{3}H-ZPBK; (C) subtilisin Amylosacchariticus control, and (D) subtilisin Amylosacchariticus plus ^{3}H-ZPBK.

92a. K. Morihara and T. Oka, *ABB* **138,** 526 (1970).

inactivation was partly prevented by the presence of hydrocinnamic acid. Amino acid analysis indicated that a histidine residue had been alkylated.

VI. Substrate Specificity and Enzymic Properties

Since the discovery of the subtilisins, it has become evident that these enzymes manifest a broad specificity as proteinases. During purification of the Carlsberg enzyme esterase activity was observed on methyl butyrate, as well as proteinase activity on casein, hemoglobin, ovalbumin, and gelatin. The pH optimum for casein digestion was in the range from pH 10 to 11 (*7*). Graae (*93*) showed that the esterase and proteinase activity was associated with the same enzyme.

Okunuki *et al.* (*94, 95*) reported that subtilisin BPN′ also hydrolyzed a variety of proteins, including casein, hemoglobin, gelatin, and oxidized lysozyme. Cleavage appeared to occur mainly at the NH_2- and COOH-terminal sides of neutral and acidic amino acid residues (*94*). In addition, subtilisin BPN′ was shown to have esterase activity with a pH optimum around 8.5–9.5 (*11*).

A. Protein and Peptide Substrates

Tuppy (*96*), Haugaard and Haugaard (*97*), and Meedom (*98*) studied the hydrolysis of the insulin chains by subtilisin and found extensive cleavage. It was evident that subtilisin hydrolyzed a great variety of peptide bonds, e.g., Meedom found that 20 of the 49 peptide bonds present in porcine insulin were cleaved. Furthermore, hydrolysis occurred not only at internal peptide bonds but also at terminal bonds to yield free amino acids. Early work on the specificity of subtilisins with protein substrates was reviewed by Hagihara (*1*).

Digestion of oxytocin by subtilisin resulted in cleavage at the carboxy-terminal end of glutamine and leucine (*99*) which is consistent

93. J. Graae, *Acta Chem. Scand.* **8,** 356 (1954).

94. K. Okunuki, H. Matsubara, S. Nishimura, and B. Hagihara, *J. Biochem.* (*Tokyo*) **43,** 857 (1956).

95. B. Hagihara, M. Nakai, H. Matsubara, T. Komaki, T. Yonetani, and K. Okunuki, *J. Biochem.* (*Tokyo*) **45,** 305 (1958).

96. H. Tuppy, *Monatsh. Chem.* **84,** 996 (1953)

97 E. S. Haugaard and N. Haugaard, *Compt. Rend. Trav. Lab. Carlsberg, Ser. Chim.* **29,** 350 (1955).

98. B. Meedom, *Compt. Rend. Trav. Lab. Carlsberg, Ser. Chim.* **29,** 403 (1955).

99. H. Tuppy, *BBA* **11,** 499 (1953).

with the specificity for aromatic or apolar residues obtained with small ester and amide substrates (*40, 100*).

Reinvestigation of the subtilisin-catalyzed cleavage of insulin under restricted conditions yielded more definitive information concerning the specificity of subtilisin. Morihara and Tsuzuki (*100*) found that after digestion of the oxidized insulin B chain for 2–3 hr with 0.1% by weight of subtilisin BPN′ at room temperature and pH 9 cleavage occurred only at peptide bonds 4–5 (Gln–His), 9–10 (Ser–His), 15–16 (Leu–Tyr), 16–17 (Tyr–Leu), and 25–26 (Phe–Tyr). This is consistent with the findings of these authors that hydrolysis of small ester and amide substrates such as acetyl-X-ethyl ester and carbobenzoxy (Cbz)–Gly–X–NH_2 occurs at the carboxy-terminal side of residue X (where X represents aromatic or apolar residues such as L-tyrosine, L-phenylalanine, and L-leucine). Earlier, Ottesen and co-workers (*101*) compared the specificities of subtilisins Carlsberg and Novo after hydrolysis of the B chain of oxidized insulin at 30° and pH 8 for 2 hr at very low enzyme-to-substrate ratios. Digestion by subtilisin Carlsberg at an enzyme to substrate ratio of 1:3960 produced 12 peaks after separation of the products of digestion on Dowex 50-X4. The two major peptides, however, obtained in about 50% yields were composed of residues 1–15 and 16–30, indicating that the major point of cleavage was at the Leu–Tyr bond (residue 15–16) of the oxidized insulin B chain. When the enzyme concentration was increased, peptide segments 1–15 and 16–30 disappeared and were replaced by fragments representing hydrolysis at residues 4–5 (Gln–His), 9–10 (Ser–His), and 11–12 (Leu–Val) for fragment 1–15, and at residues 16–17 (Tyr–Leu), 17–18 (Leu–Val), and 26–27 (Tyr–Thr) for fragment 16–30.

Similar results were obtained with subtilisin Novo, but the relative rates of hydrolyses were different. Thus peptide bonds 9–10 (Ser–His) and 26–27 (Tyr–Thr) were hydrolyzed faster by subtilisin Novo than by subtilisin Carlsberg while the opposite was the case with bonds 4–5 (Gln–His) and 11–12 (Leu–Val).

Ottesen and Østergaard (*102*) also observed slight differences between subtilisins Novo and Carlsberg in the conversion of ovalbumin to plakalbumin. At an enzyme-to-protein ratio of one to 10,000 at 30° and pH 8 both subtilisins Novo and BPN′ acted in identical fashion on ovalbumin, as judged by the solubility of the plakalbumin formed, but

100. K. Morihara and H. Tsuzuki, *ABB* **129,** 620 (1969).

101. J. T. Johansen, M. Ottesen, I. Svendsen, and G. Wybrandt, *Compt. Rend. Trav. Lab. Carlsberg* **36,** 365 (1968).

102. M. Ottesen and B. Østergaard, *Compt. Rend. Trav. Lab. Carlsberg* **34,** 187 (1964).

differently from subtilisin Carlsberg. Nonetheless, all three proteinases liberate the same major peptide fragment indicating that the primary site of attack is identical, although the composition of the larger peptide fragments released by the Novo enzyme appeared more complicated than those liberated by subtilisin Carlsberg indicating some differences in specificity.

The well-known action of subtilisin Carlsberg on ribonuclease, under controlled conditions, involves the hydrolysis of a single peptide bond producing the S protein and S peptide (*103, 104*). Exhaustive digestion of ribonuclease by subtilisin Carlsberg, on the other hand, produces cleavage at about 25% of the peptide bonds in the native protein (*105, 106*). Subtilisin BPN′ also cleaves ribonuclease forming the active RNase S derivative (*107*). However, on closer examination it was found that whereas the major (84%) cleavage point with subtilisin Carlsberg was between residues 20 and 21 (Ala–Ser), subtilisin BPN′ produced peptide fragments indicating substantial cleavage at both bonds 20–21 (Ala–Ser) and 21–22 (Ser–Ser) (*108, 109*). In fact, it is possible that the initial site of subtilisin BPN′ cleavage is at bond 21–22 and that only during further digestion is the COOH-terminal serine (residue 21 of the S peptide) partially lost. This would then be a clear-cut difference in specificity inasmuch as BPN′ favors cleavage at bond 21–22 whereas subtilisin Carlsberg cleaves preferentially at bond 20–21.

B. Synthetic Substrates

The subtilisins show low specificity when tested on a wide variety of benzoyl- or acetyl-L-amino acid esters. In comparing subtilisins Novo, BPN′ and Carlsberg, Ottesen and Spector (*9*) and Hunt and Ottesen (*28*) observed that benzoyl-L-tyrosine ethyl ester was a much better substrate than benzoyl-L-leucine ethyl ester for subtilisin BPN′ (and Novo) but both were about the same for subtilisin Carlsberg. For both substrates subtilisin Carlsberg was more active than Novo and BPN′

103. F. M. Richards, *Compt. Rend. Trav. Lab. Carlsberg, Ser. Chim.* **29,** 329 (1955).

104. F. M. Richards and P. J. Vithayathil, *JBC* **234,** 1459 (1959).

105. F. M. Richards, *Compt. Rend. Trav. Lab. Carlsberg, Ser. Chim.* **29,** 322 (1955).

106. M. Ottesen and M. Székely, *Compt. Rend. Trav. Lab. Carlsberg* **32,** 319 (1962).

107. G. Gordillo, P. J. Vithayathil, and F. M. Richards, *Yale J. Biol. Med.* **34,** 582 (1962).

108. M. S. Doscher and C. H. W. Hirs, *Biochemistry* **6,** 304 (1967).

109. E. Gross and B. Witkop, *Biochemistry* **6,** 745 (1967).

whereas the last two were kinetically indistinguishable. By comparison chymotrypsin was 50 times more active on benzoyl-L-tyrosine ethyl ester than subtilisin Carlsberg.

Glazer (*39*), Barel and Glazer (*40*), and Myers and Glazer (*87*) have determined kinetic parameters for the hydrolysis of a number of *N*-acetylamino acid esters by subtilisins Novo, Amylosacchariticus, and Carlsberg. The data are given in Table VII. All of the subtilisins show a markedly higher activity on the esters of aromatic amino acids than on those of aliphatic ones. For subtilisin Carlsberg V_{max} values were higher than for the other enzymes with *N*-acetyl aromatic amino acid esters; subtilisin Amylosacchariticus was the least active. This suggests that the deacylation rate is faster for subtilisin Carlsberg than for the other two subtilisins. This has been verified by the finding that the deacylation rate of *N*-*trans*-cinnamoyl subtilisin Carlsberg is significantly higher than the corresponding rate for subtilisin Novo (*40*). Also the kinetics of the deacylation of indoleacryloyl subtilisins Novo and Carlsberg has been studied as a function of pH (*110*). Essentially the same results were obtained: the deacylation rate for subtilisin Carlsberg was substantially higher than for the Novo enzyme, and both enzymes showed dependency on a group with pK_{app} about 8. The deacylation rate constant for the indoleacryloyl derivatives of α-chymotrypsin, trypsin, and subtilisin Novo were all very similar but much smaller than the corresponding constant for subtilisin Carlsberg. The large differences in the deacylation rate constants were maintained even when using the iodo, nitro, succinyl, and glutaryl subtilisin Carlsberg derivatives and the nitro subtilisin Novo derivative. It would appear, therefore, that the lysyl residues and the accessible tyrosine residues are not involved in the hydrolysis of indoleacryloyl subtilisin Carlsberg. These data also support the findings, which are discussed below, that with low molecular weight substrates modifications of the subtilisins produced no change in enzymic activity (*110*).

From both the kinetic data (Table VII) and the amino acid sequence subtilisin Amylosacchariticus closely resembles the BPN′ and Novo enzymes. Nevertheless, as shown in Fig. 4, V_{max} for Amylosacchariticus is independent of pH over the range 6–8 with both tosyl-L-arginine methyl ester and acetyl-L-tyrosine ethyl ester as substrates (*87*), whereas, over this same pH range, both subtilisins Novo and Carlsberg show a sigmoid dependence of V_{max} on pH (*39*). Apparently, a different rate controlling step dominates the kinetics in the Amylosacchariticus enzyme.

110. J. T. Johansen, R. W. A. Oliver, and I. Svendsen, *Compt. Rend. Trav. Lab. Carlsberg* **37**, 87 (1969).

TABLE VII

Kinetic Constants for Subtilisin-Catalyzed Hydrolysis of *N*-Acetylamino Acid Esters[a]

Substrate	Subtilisin: Amylosacchariticus[b] K_m (M)	Amylosacchariticus[b] V_{max}[e] (sec^{-1})	Novo[c] K_m (M)	Novo[c] V_{max}[e] (sec^{-1})	Carlsberg[c] K_m (M)	Carlsberg[c] V_{max}[e] (sec^{-1})	BPN′[d] K_m (M)	BPN′[d] V_{max}[e] (sec^{-1})
N-Acetyl-L-tyrosine ethyl ester	0.146	548	0.07	731	0.09	1316	0.0222	383.3
N-Acetyl-L-tyrosine methyl ester	0.046	281	0.09	1560	0.07	1930		
N-Acetyl-L-tryptophan methyl ester	0.031	149	0.09	415	0.05	820		
N-Acetyl-L-phenylalanine methyl ester	0.044	93.9	0.06	415	0.03	765		
N-Benzoyl-L-arginine ethyl ester[f]	0.012	19.8	0.007[b]	3.9[b]	0.007	16.1	0.010	3.1
N-Tosyl-L-arginine methyl ester[f]	0.042	17.7	0.07	15.5	0.044	68.6		
N-Acetyl-L-valine methyl ester			0.28	28	0.19	23		0
N-Acetylglycine ethyl ester								0
N-Acetyl-L-alanine methyl ester							0.123	72.0
N-Acetyl-L-norvaline ethyl ester								0
N-Acetyl-L-leucine methyl ester							0.0666	57.5
N-Acetyl-L-phenylalanine ethyl ester							0.0166	30.6
N-Acetyl-L-tryptophan ethyl ester							0.0238	35.8
N-Acetyl-L-lysine methyl ester							0.0909	47.4

[a] The reaction mixture contained, initially, 0.0025 to 0.25 M substrate in 5.0 ml of 0.1 M KCl containing 8% dioxane of volume. The subtilisin concentration was 0.0028 to 0.4 mg/ml. Titrations were performed at pH 8.0, 37°, with 0.02 M base as titrant.

[b] Data from Myers and Glazer (*87*).

[c] Data from Glazer (*39*) and Barel and Glazer (*40*).

[d] Data from Morihara and Tsuzuki (*100*). Conditions differed as follows: measurements at 30° and pH 7.5 using 0.05 N NaOH as titrant. Reaction mixtures contained initially from 0.2 to 35 mM substrate in 5.0 ml of 0.1 M KCl containing 4% ethanol. Subtilisin concentration was adjusted to the range where the rate of reaction was proportional to the enzyme concentration.

[e] Calculated per mole of enzyme, based on a molecular weight of 27,600.

[f] No dioxane in reaction mixture.

The finding (*39*) that certain aromatic compounds, e.g., phenol, indole, hydrocinnamate, and indolepropionate, are competitive inhibitors of the hydrolysis of *N*-acetyl-L-tyrosine ethyl ester but are noncompetitive inhibitors with *N*-α-benzoyl-L-arginine ethyl ester suggests that these two ester substrates are bound in a somewhat different manner at the active sites. Both modes of binding obviously lead to productive complexes, a fact which may in part explain the broad range of specificity of the subtilisins. Thus different substrates have different productive modes of binding and the subtilisins are therefore able to cleave peptide bonds in a wide variety of natural and synthetic substrates.

Although subtilisin Carlsberg has a higher V_{max} for the acetylamino acid esters than either subtilisins Novo or Amylosacchariticus, the K_m values for the substrates (Table VII) and the K_I values for the inhibitors (Table VIII) are similar for all of these enzymes indicating that, despite differences in amino acid sequences they possess similar substrate binding sites. It has been observed (*39*), however, that subtilisin Carlsberg has higher V_{max} values with the *N*-acetylamino acid esters than subtilisin Novo, but that both have similar V_{max} values with the free amino acid esters (*40*). This suggests that the binding site of Carlsberg is less polar than that of the Novo enzyme and thus more sensitive to the polar α-amino group in the free amino acid esters (*40*).

The broad specificity exhibited by the subtilisins is in marked contrast to the high degree of specificity shown by trypsin and chymotrypsin toward synthetic substrates (*111*). Furthermore, the subtilisins have

TABLE VIII
COMPETITIVE INHIBITORS OF HYDROLYSES OF ACETYL-L-TYROSINE ETHYL ESTER BY SUBTILISIN[a]

Inhibitor[b]	Inhibitor conc. (*M*)	pH	K_I Carlsberg (*M*)	K_I Novo (*M*)	K_I BPN′ (*M*)
Indole	0.00625	7.5	0.05	0.05	0.03
Phenol	0.05	7.4	0.10	0.11	0.10
Hydrocinnamate	0.025	8.0	0.14	0.30	0.34
Methyl butyrate	0.088	7.0	0.17	0.17	0.21

[a] The compounds listed in the table were tested as competitive inhibitors of subtilisin-catalyzed hydrolysis of acetyl-L-tyrosine ethyl ester in 0.1 *M* KCl containing 8% dioxane by volume at 37° and at the pH and inhibitor concentration indicated.

[b] Data from Glazer (*39*).

111. N. M. Green and H. Neurath, *in* "The Proteins" (H. Neurath and K. Bailey, eds.), 1st ed., Vol. 2, Part B, p. 1057. Academic Press, New York, 1954.

much higher K_m values with their best ester substrates than do chymotrypsin and trypsin and also act extremely slowly on acetyl-L-amino acid amides (*39, 100*). It would appear that high K_m values for simple ester substrates may be associated with proteinases of low specificity such as papain and ficin (*112*) as well as various bacterial and fungal enzymes (*100*).

Morihara (*113*) and Morihara and Tsuzuki (*100*) have studied the action on synthetic substrates of a variety of alkaline and neutral proteinases from various bacteria and molds. The neutral proteinases hydrolyze mainly at peptide bonds involving the *amino* group of hydrophobic amino acid residues (*113, 114*), whereas the alkaline proteinases appear to hydrolyze mainly at peptide bonds linking the *carboxyl* group of hydrophobic amino acid residues. The data in Table IX indicate that subtilisin BPN′ (as do the other alkaline proteinases studied) hydrolyzes synthetic substrates of the type Cbz–Gly–X–NH_2 where X is Ala, Leu, Phe, Tyr, Ser, and Trp (the last two show less extensive hydrolysis). Benzoyl-Tyr–NH_2 is hardly hydrolyzed at all by BPN′ in distinction to α-chymotrypsin where this amide is hydrolyzed much more rapidly than Cbz–Gly–Tyr–NH_2. To investigate the effect of neighboring groups on the hydrolysis of synthetic substrates a variety of leucine derivatives was studied (Table IX). The hydrolysis of Cbz–Gly–Leu–NH_2 is greatly reduced when the Cbz group is replaced by acetyl and there is almost no hydrolysis when the amino group is unsubstituted. When L-leucine is replaced by D-leucine there is no hydrolysis. With various acylamino acid esters, Morihara and Tsuzuki (*100*) found, as did Glazer (*39*) and Barel and Glazer (*40*), that subtilisin BPN′ has higher activity when aromatic amino acids are present, but there is also activity with esters of the basic amino acids and certain neutral aliphatic amino acids (Table VII).

Additional work by Morihara *et al.* (*114a*) with leucine substrates of the type Cbz-A-$(Gly)_n$-Leu-NH_2 and Cbz-Leu-$(Gly)_n$-B (where A or B are various D or L amino acid residues; n equals 0, 1, and 2; and cleavage occurs at the COOH-side of the leucyl residue) have shown that hydrolysis was influenced by the three amino acid residues on the NH_2-terminal side and the two amino acids on the COOH-terminal side of the sensitive bond. Furthermore, these effects on hydrolysis were mainly related to catalysis rather than substrate binding.

112. E. L. Smith and J. R. Kimmel, "The Enzymes," 2nd ed., Vol. 4, p. 133, 1960.
113. K. Morihara, *BBRC* **26,** 656 (1967).
114. K. Morihara, H. Tsuzuki, and T. Oka, *ABB* **123,** 572 (1968).
114a. K. Morihara, T. Oka, and H. Tsuzuki, *ABB* **138,** 515 (1970).

TABLE IX
HYDROLYSES OF VARIOUS SYNTHETIC SUBSTRATES BY SUBTILISIN BPN′[a]

Substrate	Hydrolysis[b] 20 min (%)	Hydrolysis[b] 16 hr (%)	Substrate	Hydrolysis[b] 20 min (%)	Hydrolysis[b] 16 hr (%)
Cbz-Gly-Gly-NH$_2$		0	Cbz-Gly-Pro-D-Leu-Gly-Pro-OH		0
Cbz-Gly-Ala-NH$_2$	11	>95	Cbz-Gly-Pro-Leu-Gly-OH	20	>95
↑			↑		
Cbz-Gly-Ser-NH$_2$	0	22	Cbz-Gly-Pro-Leu-OH		0
↑			Cbz-Gly-Pro-NH$_2$		0
Cbz-Gly-Val-NH$_2$[c]		<5	Cbz-Gly-Phe-NH$_2$[c]	6	70
Cbz-Gly-Ile-NH$_2$[c]		0	↑		
Cbz-Gly-Leu-NH$_2$	18	>95	Cbz-Gly-Tyr-NH$_2$[c]	32	>95
↑			↑		
Cbz-Gly-D-Leu-NH$_2$		0	Cbz-Gly-Trp-NH$_2$[c]	0	13
Acetyl-Gly-Leu-NH$_2$		11	↑		
↑			Benzoyl-Tyr-NH$_2$		<5
H-Gly-Leu-NH$_2$		<5	Benzoyl-Arg-NH$_2$		<5
Cbz-Leu-NH$_2$		<5			
Cbz-Gly-Pro-Leu-Gly-Pro-OH	37	95			
↑					

[a] All reactions were performed in 0.1 M tris buffer (pH 8) for 20 min and for 16 hr at 40°. The percentage of hydrolysis was determined by ninhydrin analysis. Substrate concentration was 4 mM and the enzyme concentration was 50 μg/ml of reaction mixture. The arrows show the points of cleavage.

[b] Data from Morihara and Tsuzuki (*100*).

[c] The reaction mixture contained 4% methanol because of the low solubility of the substrate.

It was also found that hydrolysis was inhibited when the terminal α-amino and α-carboxyl groups of the peptide substrate were unblocked, if the former group was within four residues on the NH$_2$-terminal side and the latter group within one residue on the COOH-terminal side of the cleavage point. Finally, these authors reported that the side chain specificity at the residue being cleaved is extremely stringent compared with those for each of the five amino acid residues surrounding the sensitive residues, and that bulky residues on either side of the sensitive bond inhibit hydrolysis.

Morihara and Tsuzuki (*100*) have called the alkaline proteinases chymotrypticlike in their specificity. However, in comparing the size and specificity of the active site of α-chymotrypsin and subtilisin BPN′ by using a variety of synthetic peptides, Morihara *et al.* (*115*) found con-

115. K. Morihara, T. Oka, and H. Tsuzuki, *BBRC* **35**, 210 (1969).

siderable differences between these two types of proteinases. By designing synthetic peptides so that hydrolysis occurred at the bond involving the carboxyl group of L-tyrosine the effects of neighboring residues were examined. Thus peptides of the type Cbz–A–Tyr–NH_2, Cbz–A–Gly–Tyr–NH_2, Cbz–Tyr–B–NH_2, and Cbz–Tyr–Gly–B (where A or B are various amino acid residues) were used where the subsite numbering system is

$$S_4—S_3—S_2—S_1—S_1'$$
$$\text{Cbz-A-Gly-Tyr-NH}_2$$

On the assumption that the substrates bind to the enzyme in such a way that the CO–NH linkage being hydrolyzed is always the same and that the amino acid residues occupy adjacent subsites, those toward the amino end occupying subsites S_1, S_2, etc. and those toward the carboxyl end subsites S_1', S_2', etc., it was found that both subtilisin BPN′ and α-chymotrypsin have large active sites which can be divided into at least five subsites, each accommodating one amino acid residue of the substrate. Furthermore, stereospecificity is present in both enzymes at all subsites. Thus for subtilisin BPN′ the rate of hydrolysis of Cbz-L-Ala–Gly-L-Tyr–NH_2 is 15.2 μmoles/min/mg enzyme but with the diastereoisomer Cbz-D-Ala–Gly-L-Tyr–NH_2 the corresponding rate of hydrolysis is only 0.067. Side chain specificity of the subsites indicated that alanine at subsite S_2 greatly promoted the activity of subtilisin BPN′ when compared to glycine, histidine, or tyrosine at this site. Furthermore, elongation of the peptide chain toward the NH_2-terminus in subtilisin BPN′ greatly stimulated the activity as seen by a comparison of the rate of hydrolysis for Cbz–Gly–Tyr–NH_2, 0.098 μmole/min/mg enzyme, versus the corresponding rate for Cbz–Gly–Gly–Tyr–NH_2, 5.00. Changing the glycine in subsite S_3 of the latter peptide to alanine also promotes the rate of hydrolysis from 5.0 to 15.2 μmoles/min/mg enzyme, indicating that subsite S_3 exerts at least some control over the activity of the enzyme. This effect does not appear to be true for α-chymotrypsin. Subsite S_1' does not appear to have much effect on hydrolysis by subtilisin BPN′ whereas it does with α-chymotrypsin. Subsite S_2' may have more of a controlling influence on subtilisin BPN′ although the data are limited.

1. *Unusual Ester Substrates*

Subtilisin also catalyzes the hydrolysis of some triglycerides, including tripropionin and tributyrin (*116*), and the D but not the L isomer of 1-keto-3-carbomethoxy-1,2,3,4-tetrahydroisoquinoline. The enzyme, how-

116. G. Sierra, *Can. J. Microbiol.* **10**, 926 (1964).

ever, retains its normal stereospecificity when the related open chain ester *N*-benzoyl alanine methyl ester is the substrate, hydrolyzing the L and not the D isomers. Since α-chymotrypsin behaves in an identical manner with these substrates, both enzymes have similar specificity with respect to the configuration of their substrates (*117*).

C. Transesterification and Transpeptidation

The subtilisins also catalyze transesterification reactions (*118*) although at a somewhat lower rate than chymotrypsin (*119*) and trypsin (*120*).

No transpeptidation was evident with subtilisins Novo or Carlsberg, but a small amount occurred with subtilisin Amylosacchariticus catalyzed hydrolysis of tetraalanine (*87*). Additionally, like most of the proteolytic enzymes the subtilisins have been shown to catalyze aminolysis reactions (*40*) as shown in Eq. (2):

$$\text{L-Leucine benzyl ester} + \text{glycylglycine } n\text{-amyl ester} \rightleftharpoons \text{L-Leucine glycylglycine } n\text{-amyl ester} + \text{benzyl alcohol} \quad (2)$$

D. Mechanism of Action of the Subtilisins

Much has already been said concerning the mechanism of action of the serine proteinases, in general, and α-chymotrypsin, in particular (*48, 121–124*). Interest was generated in the subtilisins when it was learned that they were inhibited by DFP (*7, 12, 17, 25*). The observation of acyl-enzyme intermediates during the catalysis of hydrolytic reactions (*84*) by the subtilisins, similar to those observed for α-chymotrypsin, has further emphasized the similarities between these and other serine proteinases. Specific acylating agents such as *N*-cinnamoylimidazole (*60*) and *p*-nitrophenylacetate (*76*) have been used to titrate the subtilisins during the attainment of the steady state, and a cinnamoyl enzyme derivative has been isolated (*59, 84*). As already discussed, the involvement of serine and histidine in the activity of the subtilisins is

117. H. Dugas, *Can. J. Biochem.* **47,** 985 (1969).
118. A. N. Glazer, *JBC* **241,** 635 (1966).
119. C. E. McDonald and A. K. Balls, *JBC* **221,** 993 (1956).
120. A. N. Glazer, *JBC* **240,** 1135 (1965).
121. M. L. Bender and F. J. Kézdy, *Ann. Rev. Biochem.* **34,** 49 (1965).
122. J. H. Wang, *Science* **161,** 328 (1968).
123. A. Williams, *Quart. Rev.* (*London*) **23,** 1 (1969).
124. G. P. Hess, "The Enzymes," 3rd ed., Vol. III, Chapter 7, 1971.

now well documented and a role for Asp 32 has been postulated (*47*) similar to the involvement of a specific aspartic acid in the mechanism of α-chymotrypsin and other pancreatic proteinases (*48*).

Keizer and Bernhard (*125*) noted that the protonic equilibrium accompanying acylation of both subtilisin Novo and α-chymotrypsin with N-β-(3-indole) acryloylimidazole is identical, indicating that the involvement of proton dissociable groups in the mechanism of catalysis must be similar for these two enzymes. Thus, hydrolysis presumably occurs via acylation (k_2) and deacylation (k_3) steps according to Eq. (3):

$$\begin{aligned} \mathrm{EH} + \mathrm{RCOX} &\underset{k_{-1}}{\overset{k_1}{\rightleftharpoons}} \mathrm{EH{\cdot}RCOX} \underset{k_{-2}}{\overset{k_2}{\rightleftharpoons}} \mathrm{ECOR} + \mathrm{HX} \\ \mathrm{ECOR} + \mathrm{H_2O} &\underset{k_{-3}}{\overset{k_3}{\rightleftharpoons}} \mathrm{RCO_2H} + \mathrm{EH} \end{aligned} \tag{3}$$

An interesting observation that was made during the studies with thiolsubtilisin (Section V,A) may shed some light on the general mechanism of the serine proteinases. The high reactivity of the serine hydroxyl at the active site of serine proteinases has been generally attributed to hydrogen bonding between this residue and a nearby histidine residue (*122*) which in turn is hydrogen bonded to an aspartic acid residue (*47, 48*). However, Polgár and Bender (*126*) have presented a contrary view. By measurement of the kinetic parameters it was shown that subtilisins Novo and Carlsberg and their thiol derivatives in the acyl-enzyme form possess an ionizable group with a pK_a of about 7.0 in the active site. This group, presumably the active histidine residue, also is present in the unsubstituted enzymes but not in the free thiol derivatives (*76, 78*). The pK has been shifted to 6.15 in free thiolsubtilisin Carlsberg and somewhat lower than pH 5.5 in thiolsubtilisin Novo. To explain this they postulate that a hydrogen bond is present between the thiol group and the nearby histidine residue in the free thiol enzymes, whereas in the serine enzymes, since there was no shift in the pK between the free and acyl enzymes, it appeared that no hydrogen bond is present. Kinetic studies with D_2O supported the above conclusions. It is known that the velocity of reactions with a rate limiting proton transfer decrease to about one-third the normal rate in D_2O. The formation of a hydrogen bond is considered a partial proton transfer; however, if a hydrogen bond were formed between the catalytic groups during the formation of acyl enzyme the rate would not be markedly affected in D_2O. The results indicate that there was a D_2O effect for deacetylation with p-nitrophenyl-

125. J. Keizer and S. A. Bernhard, *Biochemistry* **5**, 4127 (1966).
126. L. Polgár and M. L. Bender, *Proc. Natl. Acad. Sci. U. S.* **64**, 1335 (1969).

acetate both for subtilisin and thiolsubtilisin (Carlsberg) indicating a rate controlling proton transfer in the catalytic step. However, in the formation of the acetyl enzyme there was only a rate determining proton transfer in the serine enzyme. The thiolenzyme showed no effect of D_2O on the rate, supporting the suggestion of a hydrogen bond between the thiol group and the imidazole group. This is contrary to the hypothesis of Wang (*122*) who explained the reactivity of the hydroxyl group of the serine proteinases as resulting from a rigidly and accurately held hydrogen bond of the serine OH to an imidazole group. Furthermore, he postulated that the reason for the inefficiency of the thiol enzyme was that this hydrogen bond was disrupted.

Figure 6 shows the mechanism of formation and hydrolysis of the acyl enzyme as envisioned by Polgár and Bender (*126*) which incorporates features of the mechanism proposed by Wang (*122*) and by Blow *et al.* (*48*). According to this view, all that is needed to initiate the catalytic reaction is a favorable collision between the carbonyl carbon atom of the substrate and the oxygen atom of the serine residue. Bond making, probably the formation of a tetrahedral intermediate, and proton transfer to the imidazole proceed in a concerted manner. This step which involves general base catalysis is followed by a general acid-catalyzed step in which a proton is transferred from the imidazolium to the leaving group in a concerted manner with bond breaking. Between the two general

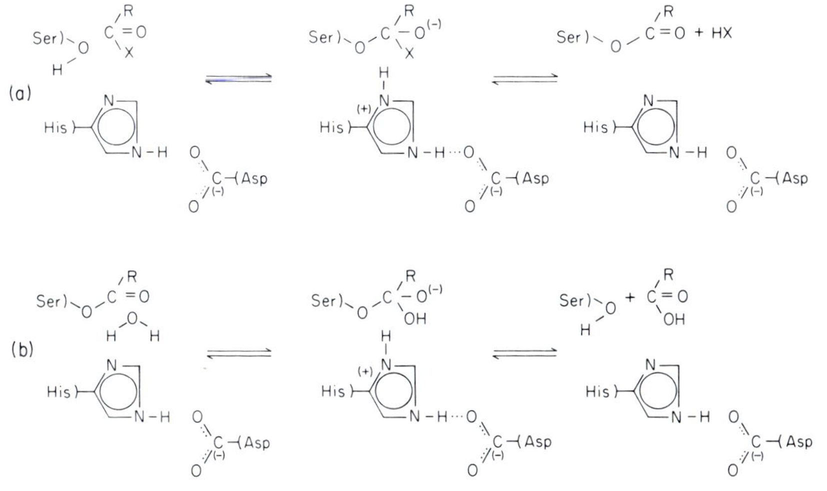

FIG. 6. Possible mechanism of the formation (a), and the hydrolysis (b) of the acyl enzyme in the catalysis by serine proteases; X represents the leaving group. From Polgár and Bender (*126*).

catalyses the imidazolium is stabilized by the aspartate–imidazole hydrogen bond system.

VII. Chemical Modification Studies

Although the subtilisins show variations in primary structure as well as in kinetics and specificity, they are very similar in other ways:

(1) They are resistant to many denaturing agents including 6 *M* urea, 50% ethanol, and anionic detergents at room temperature (*26, 27*).

(2) These enzymes are rapidly and irreversibly denatured below pH 5 (*26*).

(3) Aromatic compounds are competitive inhibitors; this suggests the presence of a hydrophobic binding site (*16, 39*).

These and other findings have led to studies aimed at the specific modification of the subtilisins in attempts to determine which residues are in the substrate binding site, which are involved in the acid lability of the enzyme, and which are otherwise essential for the activity and stability of the enzyme.

A. Lysine Modification

In the pH region where subtilisin is labile, below pH 5, it was supposed that there may be a drastic net increase in positive charge because of the protonation of carboxylate ions or of imidazole groups. Thus, the acid lability might result either from a general charge effect, namely, disruption of the native structure by electrostatic charge repulsion, or from a local charge effect whereby protonation of a particular carboxylate or imidazole group causes a drastic conformational change. By succinylating the lysine residues there would be a drastic net change in the charge of the protein and thus a means would be provided for separating a general charge effect from a local charge effect. Succinic anhydride reacted with the ε-amino groups of 10–11 of the 11 lysine residues in subtilisin Novo essentially without affecting the ionization state of the carboxyl groups or the histidine residues (*26*); however, as will be discussed later, several tyrosine or serine and threonine residues were also succinylated. Nevertheless, with at least 10 lysine residues succinylated, there is a net change in charge of at least minus 20 and the enzymic activity is essentially unchanged with benzoyl-L-tyrosine ethyl ester or benzoyl-L-leucine ethyl ester as substrates over a wide pH range. With the small basic protein clupeine as substrate, a more complex picture was observed and this will be discussed in Section VII,C. The succinylated enzyme appears only slightly less stable than the native enzyme throughout the entire pH range, and it is homogeneous by both moving boundary electrophoresis and ultracentrifugation with no evidence for association or dissociation. Optical rotation measurements, although sparse, suggested no drastic changes in conformation of the succinylated enzyme.

Thus, since succinylation has little effect on catalytic activity or sensitivity to acid pH, it would appear that it is not a general charge effect which is responsible for the acid lability of the subtilisins, but it is presumably a result of specific local effects probably involving either carboxylate or imidazole groups (*126a*). Additionally, since modification of 10–11 out of 11 lysine residues did little to the enzymic activity on ester substrates, it appears that lysine residues are not part of the active center; however, their involvement in binding sites for large molecular weight substrates cannot be excluded. The finding that the lysine residues are readily accessible for reaction with succinic anhydride suggested that they are on the surface of the protein (*26*). This has been confirmed by X-ray analysis which indicates that 10 of the 11 lysine residues are on the surface of subtilisin BPN′ (*47*).

Succinylation of subtilisin Carlsberg modifies 6 of the 9 lysine residues and 14–16 hydroxyl groups of threonine or serine as well (presumably a similar reaction occurs in subtilisin Novo) (*101*).

Carbamylation of subtilisin Novo converted 10–11 of the 11 lysines into homocitrulline residues but almost completely inactivated the enzyme as estimated from its proteolytic activity toward clupeine and its esteratic activity with benzoyl-L-tyrosine ethyl ester (*88*). Since this inactivation was reversed by 1.0 *M* hydroxylamine and resulted in no loss of homocitrulline residues, it appeared that the active serine residue had reacted with potassium cyanate to inhibit the enzyme reversibly. A similar reaction had previously been observed with α-chymotrypsin (*127*). The α-amino group of subtilisin Novo also reacts with cyanate but has no effect on activity since this residue remains carbamylated after reactivation with hydroxylamine.

Although both the ε- and α-amino groups of subtilisin Novo are free to react with cyanate, their carbamylation does not disrupt the native conformation of the molecule as judged by optical rotation and ultra-

126a. M. Ottesen, J. T. Johansen, and I. Svendsen, *in* "Structure–Function Relationships of Proteolytic Enzyme," (P. Desnuelle, H. Neurath, and M. Ottesen, eds.), p. 175. New York, 1970.

127. D. C. Shaw, W. H. Stein, and S. Moore, *JBC* **239,** PC671 (1964).

centrifugation or alter the stability as a function of pH. Similar results were obtained with subtilisin Carlsberg (*101*); carbamylation converted 7–8 lysines into homocitrulline residues and inactivated the enzyme. The inactivation was reversed by hydroxylamine.

No change in specificity on the oxidized B chain of insulin was observed with the succinylated or the carbamylated subtilisin Carlsberg. With subtilisin Novo, however, the initial rate of insulin hydrolysis was decreased several fold for both the succinylated and carbamylated derivative. The specificity was only changed slightly with the succinylated derivative and was identical to that of native subtilisin Novo for the carbamylated derivative.

B. Methionine Modification

Treatment of a 1% solution of subtilisin Carlsberg at pH 8.8 with 0.1 M H_2O_2 resulted in the oxidation of a single methionine residue to methionine sulfoxide (*128*); amino acid analysis indicated no other change in composition. From a combination of tryptic and cyanogen bromide digestions of the oxidized protein the modified residue was identified as Met 222, adjacent to the reactive Ser 221.

Oxidation of the specific methionine residue caused a loss of 91% of the activity with acetyl-L-tyrosine ethyl ester; furthermore, there was a linear relation between the degree of methionine oxidation and loss of activity. In comparing the kinetic parameters, it was found that oxidation increases the K_m slightly for acetyl-L-tyrosine ethyl ester, benzoyl-L-arginine ethyl ester, and cinnamoylimidazole while there is a slight decrease in K_m with Cbz-glycine *p*-nitrophenyl ester. For all of these substrates there is a decrease in k_{cat} which is more pronounced for the substrates with bulky side chains. Although the exact mechanism of the changes in the enzymic activity brought about by oxidation of Met 222 is unknown, there are three possibilities: (1) alteration in conformation of the enzyme, (2) alteration of the electronic environment around the active site caused by changing from the hydrophobic thioether to a hydrophilic sulfoxide, and (3) steric effects. The last possibility appears to be of little importance, at least insofar as formation of the E–S complex is concerned, since there are only slight changes in $K_{m(app)}$ between the native and oxidized enzyme with various substrates. From the X-ray crystallographic data it appears that the side chain of this methionine residue swings away, about 1 Å from the phenylmethane sulfonyl group

128. C. E. Stauffer and D. Etson, *JBC* **244**, 5333 (1969).

when subtilisin BPN′ is sulfonylated (*47*). This indicates that the residue does have freedom of movement and further excludes loss of activity by oxidation on strictly steric grounds.

Earlier studies utilizing photooxidation had suggested that one methionine as well as one histidine and one tryptophan residue were readily oxidized causing rapid loss of enzymic activity (*83*). However, from carboxymethylation of the oxidized derivative it was concluded that the residue photooxidized was not the methionine residue next to the reactive serine.

Ohtsuki *et al.* (*128a*) have shown that *N*-bromosuccinimide inactivated subtilisin BPN′ presumably by oxidation of single tyrosine and methionine residues, although tryptophan was also modified.

C. Tyrosine Modification

The finding by Glazer (*39*) that aromatic compounds act as competitive inhibitors of the hydrolysis of aromatic ester substrates by the subtilisins suggested the presence of a hydrophobic binding site in the subtilisins. Additionally, the finding that the subtilisins exhibit higher activity toward clupeine, but not small positively charged substrates after treatment by reagents that alter the pK values of tyrosine phenolic groups, suggested that certain tyrosine residues are involved in secondary binding sites (*129, 130*).

Svendsen (*130*) reported that both nitration and iodination of subtilisins Carlsberg and Novo increased the apparent activity toward clupeine (for subtilisin Carlsberg a similar effect was also observed with gelatin). Also, succinylation of subtilisin Carlsberg increased the activity with clupeine 6- to 7-fold. The increased activity probably results from improved substrate binding since the K_m for clupeine binding decreases after nitration of either subtilisin Novo or Carlsberg. Both nitration and iodination decrease the pK of the phenolic groups. The decreased pK is probably related to the enhanced activity on the basic protein, clupeine, and suggests that one or more ionized tyrosine residues are at or near the site of formation of the enzyme–substrate complex between clupeine and subtilisin (*126a*).

The activation of subtilisin Carlsberg by succinylation is reversed by treatment with hydroxylamine, and present evidence indicates that 14–

128a. K. Ohtsuki, C. L. Liu, and H. Hatano, *J. Biochem.* (*Tokyo*) **66,** 863 (1969).

129. J. T. Johansen, M. Ottesen, and I. Svendsen, *BBA* **139,** 211 (1967).

130. I. Svendsen, *Compt. Rend. Trav. Lab. Carlsberg* **36,** 347 (1968).

16 threonine and serine residues have been succinylated. It may be significant that in five instances a serine or threonine residue is located next to a tyrosine residue in the sequence of subtilisin Carlsberg (Fig. 1), offering the possibility of introducing a negative charge in the same spatial location either by succinylation or (by lowering the pK of tyrosyl residues) by nitration or iodination (*130*).

There is no promotion of activity, however, with tosyl-L-arginine methylester since measurements of $K_{m(\text{app})}$ and $V_{\max}$ for this substrate indicated only slight changes with derivatives of both subtilisins Novo and Carlsberg. Furthermore, reduction of the nitrotyrosyl residues of subtilisin Carlsberg to aminotyrosyl residues with sodium dithionite caused the enhanced activity on clupeine to revert to that of the unmodified protein. This can be understood since the pK of the phenolic group of aminotyrosine is about 10 (about the same as unmodified tyrosine) whereas that of nitrotyrosine is about 7.

Analysis indicated that 6 of the 13 tyrosyl residues in subtilisin Carlsberg had been modified after 1-hr reaction with a 60-fold molar excess of tetranitromethane at pH 8, with a maximum of 9 residues being modified after 18-hr exposure to a 120-fold excess of the nitrating agent. Even after such long exposure, the enzyme maintained the 6- to 7-fold increased activity and no evidence was obtained for modification of any other amino acids, although tryptophan was not examined (*130*). With subtilisin Novo only 3–4 of the 10 tyrosyl residues were nitrated under the standard conditions. After exhaustive nitration about 8 of the tyrosine residues were nitrated without loss of activity. Iodination produced essentially the same results with about 6 or 7 tyrosine residues in subtilisin Carlsberg iodinated when maximal activation was obtained. Subtilisin Novo, however, behaved somewhat differently on iodination. Four tyrosine residues could be iodinated readily with maximal activation toward clupeine, but when more iodine was added the enzyme was rapidly inactivated. A total of 8 of the 10 tyrosine residues could be iodinated (*130*).

Further characterization of the tyrosine residues in subtilisins was obtained by spectrophotometric titration of the DIP derivatives (*131*). The 10 tyrosine residues in subtilisin BPN′ and the 13 in Carlsberg fall into three groups: 5 or 6 in both enzymes titrate normally with pK_{app} of 9.7–9.9, 3 or 4 in both titrate with pK_{app} of 11.25–11.6, and 2 or 3 in both appear almost completely inaccessible to the solvent with pK_{app} greater than 12.5. In 5 M guanidine hydrochloride all the tyrosine residues titrate normally with a pK_{app} of 9.97 in both proteins. Markland

131. F. S. Markland, *JBC* **244,** 694 (1969).

(*24, 132*) has shown that 5 tyrosine residues in subtilisin Novo react readily with tetranitromethane and an additional 3 or 4 react at least partially. Isolation of the peptides containing nitrotyrosine from combined tryptic and cyanogen bromide digests has allowed precise identification of the nitrated residues (Table X). Tyrosine residues 6, 21, 104, 217, and 262 are fully nitrated, whereas residues 91 and 263 are partially nitrated (see Fig. 1). The tryptic peptide containing tyrosine residues 167 and 171 contains nitrotyrosine; however, it is not clear as yet whether both residues are partially nitrated or only one is nitrated fully. Presumably the 5 readily nitrated residues are the same as those that titrate normally and those that are partially nitrated may be the residues with slightly higher pK_{app}.

Nitration of subtilisin Novo with a fivefold molar excess of tetranitromethane yields only one major nitrated residue which was identified as Tyr 104 (*24*). Nitration of this single residue did not fully promote activity on clupeine indicating that either a combination of several nitrated tyrosine residues is necessary for full increase of activity or that the nitration of a single tyrosine residue controls the activation but that it is not the first to be nitrated. Weber and Kraut (*133*) have reported that the first residue iodinated in subtilisin BPN′ is also Tyr 104. Other

TABLE X
REACTIVITY OF TYROSINE RESIDUES IN SUBTILISIN BPN′ (OR NOVO)

Residue No.[a]	Nitration[b]	Iodination[c]
6	Full	
21	Full	Monoiodo
91	Partial	
104[d]	Full	Diiodo
167	None or ?	One is monoiodo
171	None or ?	
214	None	
217	Full	Diiodo
262	Full	
263	Partial	

[a] Of the 10 tyrosine residues, 5 titrate normally with $pK_{app} = 9.74$, 3 with $pK_{app} = 11.25$, and 2 with $pK_{app} > 12.5$.
[b] Data from Markland (*24, 132*).
[c] Data from Wright *et al.* (*47*).
[d] This is the first residue to be iodinated or nitrated (*24, 133*).

132. F. S. Markland, *Federation Proc.* 28, 877 (1969).
133. B. H. Weber and J. Kraut, *BBRC* 33, 280 (1968).

iodinated residues appear to be 217, 167 or 171, and 21 as determined by X-ray diffraction (*47*) (Table X).

Myers and Glazer (*87*) have observed that *N*-acetylimidazole can acetylate almost all of the tyrosine residue in the phenylmethane sulfonyl derivatives of both subtilisins Novo and Carlsberg. Thus, the tyrosines show no distinction in reactivity with this reagent in contrast to the situation with spectrophotometric titration or nitration. It is conceivable that the tyrosines that titrate in the first two groups (the normally titrating and those with slightly high pK_{app}), and the two classes of tyrosine reactive with tetranitromethane (readily nitrated and only partially nitrated) are all equivalent with respect to *N*-acetylimidazole. In subtilisin Novo this would comprise 8–9 residues leaving at most 1 or 2 buried tyrosine residues. Solvent perturbation studies with 20% ethylene glycol indicated that 8.8 tyrosine and 1.6 tryptophan residues in subtilisin Novo and 8.8 tyrosine and 0.8 tryptophan residues in subtilisin Carlsberg were exposed (*87*). X-ray crystallography indicates that all the tyrosine residues are on the surface of subtilisin BPN′ as well as the 3 tryptophan residues (*47*). The reasons for these differences are not apparent although there may be subtle interactions involving tryptophan and tyrosine which are easier to measure in solution than to visualize in the crystal structure.

VIII. Inhibitors: Dye Binding

Study of the binding of a number of dyes at the active sites of several proteinases and other enzymes has shown that these interactions are highly specific (*90a*). This results in part from the severe limitations imposed on the geometry of the binding site because of the rigid structure of the dyes. Specific binding is found for Biebrich Scarlet ([6-(2-hydroxyl-1-naphthyl)azo]-3,4′-azodibenzene sulfonate) to the active site of α-chymotrypsin (*134*) and for thionine (3,6-diaminophenothiazine) to the active site of trypsin (*135*). Thus, despite considerable similarities in the three-dimensional structures of these enzymes, these protein–dye interactions are sensitive to the differences present in the active site regions. Proflavine (3,6-diaminoacridine) binds to both of these enzymes (*136*, *137*). None of the above dyes interacts with the subtilisins.

134. A. N. Glazer, *JBC* **242,** 4528 (1967).

135. A. N. Glazer, *JBC* **242,** 3326 (1967).

136. S. A. Bernhard and H. Gutfreund, *Proc. Natl. Acad. Sci. U. S.* **53,** 1238 (1965).

137. S. A. Bernhard and B. F. Lee, *Abstr. 6th Intern. Congr. Biochem. New York, 1964* Vol. IV, p. 297.

In contrast, Glazer (*138*) has found that 4-(4′-aminophenylazo)phenylarsonate binds specifically to the active site of the subtilisins and does not bind to the pancreatic proteinases. Spectrophotometric titration indicated that the dye was binding in a time-dependent manner at a single binding site and with a similar spectral shift for the different subtilisins. Assays showed a time-dependent loss in activity which paralleled the spectral shift induced by dye binding. Dye binding was independent of substrate concentration (with benzoyl-L-arginine ethyl ester or acetyl-L-tyrosine ethyl ester). Indole (0.004 *M*), a competitive inhibitor of the subtilisins, did not inhibit the dye–protein interaction suggesting that the *initial* binding site was distinct from the site involved in binding substrates or competitive inhibitors. However, the dye was displaced by phenylmethane sulfonyl fluoride indicating that the binding site was spatially close to the active site of the subtilisins.

On the basis of the stringent stereochemical requirements of the protein–dye interaction, it can be concluded that the conformation of the active site region of subtilisins Novo, Carlsberg and Amylosacchariticus is similar and largely unaffected by the differences in amino acid sequences. However, there are differences in the active site regions since the K_m and V_{max} values for a variety of ester substrates do show both absolute and relative differences (Table VII). Furthermore, the arsonic acid dye failed to interact with trypsin and chymotrypsin, just as thionine, proflavine, and Biebrich Scarlet failed to bind to the subtilisins. Thus, the bacterial and pancreatic proteinases must possess different surface characteristics in the general area of the catalytic sites of these two groups of enzymes.

Certain features of the interaction of the phenylarsonate dyes with the subtilisins appear noteworthy. Generally, the interaction of proteins with small molecules is complete within seconds; the interaction of the subtilisins with the arsonates is very slow, however, requiring from 15 to 60 min for completion depending on the concentrations of the reactants (*138*). This suggested that the interaction involved either a slow conformational change within a weak protein–dye complex that was initially formed very rapidly, or a slow direct interaction of the bound dye with a group involved in catalysis. The protein–dye interaction could be reversed by dilution or by suitable change in pH (*138*).

Studies with simpler phenylarsonate analogs of the dye, such as phenylarsonate and its *p*-nitro, *p*-amino, and *p*-methyl derivatives, showed that these compounds were also time-dependent inhibitors of the subtilisins with an initial site of binding distinct from the substrate binding site. Unlike the large phenylarsonate dye, the smaller phenylarso-

138. A. N. Glazer, *Proc. Natl. Acad. Sci. U. S.* **59**, 996 (1968).

nates were effective inhibitors of α-chymotrypsin and some of them of trypsin also (*139*).

The pH dependence for inhibition of the subtilisin Novo catalyzed hydrolysis of *N*-benzoyl-L-arginine ethyl ester by the phenylarsonate dye followed the theoretical curve for the protonation of a group with a pK_a' of 7.17. With the simpler phenylarsonates as with the dye, the presence of substrate, even at a 100-fold molar excess over the inhibitor, is without effect on the rate and extent of inhibition. This suggests that all of the arsonates bind at a site distinct from the substrate binding site forming an initial complex that is still catalytically active and that this complex then undergoes a slow rearrangement to give an inactive enzyme as shown by Eq. (4):

$$E + I \underset{}{\overset{\text{fast}}{\rightleftharpoons}} (EI)_{\text{active}} \overset{\text{slow}}{\rightleftharpoons} (EI')_{\text{inactive}} \tag{4}$$

From the pH dependence of inhibition of the subtilisins by the phenylarsonates it would appear that the binding of the inhibitor requires the presence of a protonated histidine residue at the active site. A possible scheme for the interaction of the phenylarsonates with the subtilisins is presented in Fig. 7. The initial reaction (Step 1) is envisaged as depending on an electrostatic interaction between the phenylarsonate anion and a protonated histidine residue, as well as a hydrophobic interaction involving the aromatic ring. Subsequently (Step 2), ester formation may take place with the active site serine residue. This mechanism is con-

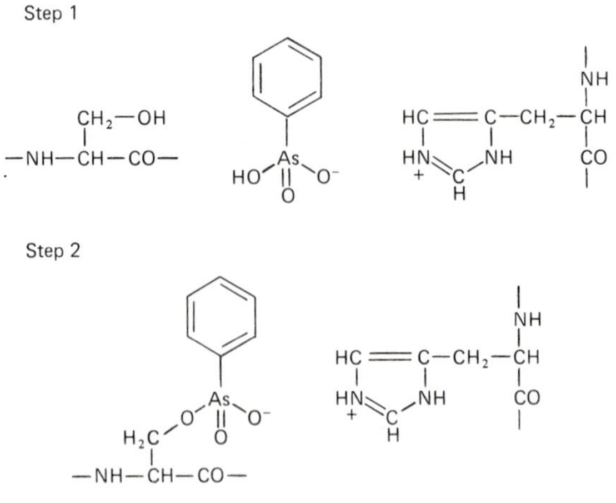

FIG. 7. Possible mechanism for the formation of an arsonic acid ester with the reactive serine in the subtilisins. From Glazer (*139*).

139. A. N. Glazer, *JBC* **243,** 3693 (1968).

sistent with all the available data but awaits definitive proof from X-ray crystallography.

IX. Other *Bacillus* Alkaline Proteinases

As already noted (Section II), Keay and Moser (*13*) were able to group the highly purified subtilisins into two classes on the basis of amino acid composition, serological cross-reactions, and the ratio of esterase to proteinase activity. It remains to be determined whether other alkaline proteinases will fit this classification.

There have been some reports describing the partial characterization of alkaline proteinases from a variety of different strains of *Bacillus*. Among these are *B. cereus*, strain K_p 931 which produces three alkaline proteinases (*140*). *Bacillus licheniformis* culture filtrates also produce three proteinases, one of which is certainly of the subtilisin type since it had high esterase activity with *N*-acetyl-L-tyrosine ethyl ester and much lower activity with *N*-benzoyl-L-arginine ethyl ester, exhibited an alkaline pH optimum, and was inhibited by both DFP and phenylmethane sulfonyl fluoride (*141*).

A transformable strain of *B. subtilis* (strain 168 indole$^-$) produced two different proteinases (designated basic and acidic based on their binding to DEAE-cellulose) at different times in the growth cycle (*142*). Both were inhibited by DFP and phenylmethane sulfonyl fluoride, although the basic one did not possess appreciable activity on benzoyl-L-tyrosine ethyl ester. Amino acid analysis showed significant differences between the two proteinases although the basic proteinase had a composition that was not unlike that of subtilisin Carlsberg. More recently it was shown that the basic proteinase appeared similar kinetically and immunologically to subtilisin Novo as well as having some antigenic regions in common with subtilisin Carlsberg whereas the acidic proteinase appeared altogether different from the other subtilisins (*143*). Under conditions for production of the acidic proteinase a high molecular weight proteinase was also present; this enzyme was devoid of esterase activity on benzoyl-L-tyrosine ethyl ester. Both the acidic and basic proteinases were acid labile and had similar pH optima around pH 7.5–

140. Y. Furukawa, Y. Fujii, and H. Takahashi, *Agr. Biol. Chem.* (*Tokyo*) **32,** 822 (1968).
141. F. F. Hall, H. O. Kunkel, and J. M. Prescott, *ABB* **114,** 145 (1966).
142. H. W. Boyer and B. C. Carlton, *ABB* **128,** 442 (1968).
143. J. H. Hageman and B. C. Carlton, *ABB* **139,** 67 (1970).

8.0. The significance of the production at different times in the growth cycle of the two proteinases is not clear.

Rappaport *et al.* (*144*) isolated a DFP-sensitive proteinase from the same strain of *B. subtilis* as Boyer and Carlton (*142*); however, its amino acid composition was different from either of those described later by Boyer and Carlton. This proteinase in its native form has a sedimentation coefficient of 9.35 S, but at pH 2.3 the aggregated material broke down into 1.8 S inactive subunits consonant with a molecular weight of 28,800 calculated from the amino acid composition. Alanine was obtained as the NH_2-terminal residue although only in 30% yield. Serine and leucine were liberated most rapidly by carboxypeptidase A and reached levels of 0.92 and 0.9 mole per mole of protein. Riggsby and Rappaport (*145*) reported that the active proteinase had a molecular weight of 1.66×10^5 and that either very alkaline or very acidic conditions converted the active species into inactive units of molecular weight 27,600. Some evidence was presented for the occurrence of a 14,000 molecular weight subunit; however, the possibility of autolysis or digestion by a contaminating proteinase cannot be excluded.

Ganno (*146*) has reported a proteinase from another strain of *B. subtilis*. This proteinase is optimally active at pH 7.5–8.5 and is not inhibited by EDTA or common thiol reagents (the effects of DFP were not reported). The molecular weight and amino acid composition are similar to those of the subtilisins but the amino acid analysis indicated only 244 residues. One interesting difference between this enzyme and the subtilisins is the extremely low isoelectric point at pH 3.

Millet (*146a*) characterized three proteolytic enzymes from the Marburg strain of *B. subtilis* during sporulation. One of these enzymes appears similar to the subtilisins in that it is inhibited by DFP, shows optimum pH activity in the neutral to alkaline range, shows esterolytic activity on *p*-tosyl-L-arginine methyl ester and benzoyl-L-tyrosine ethyl ester, and is not inhibited by EDTA.

X. Practical Uses of Subtilisins

In general the applications of the subtilisins have been mainly twofold: for determination of the primary structure of other proteins, and more

144. H. P. Rappaport, W. S. Riggsby, and D. A. Holden, *JBC* **240**, 78 (1965).
145. W. S. Riggsby and H. P. Rappaport, *JBC* **240**, 87 (1965).
146. S. Ganno, *J. Biochem.* (*Tokyo*) **58**, 556 (1965).
146a. J. Millet, *J. Appl. Bacteriol.* **33**, 207 (1970).

recently, as an additive to laundry detergents to improve their cleaning power by hydrolysis of fabric bound proteins.

A. Protein Sequence

Probably the first practical use for subtilisin was in the determination of the amino acid sequence of oxytocin (*99*). Subtilisins have also been used for the hydrolysis of a variety of other proteins and peptides. Among these were the limited digestions, already mentioned, of ovalbumin (*147*) and ribonuclease (*104*). Other uses may be noted briefly: the digestion of native hemoglobin (*148*), the determination of the amino-terminal sequence of α-crystallin (*149*), the identification of citrulline in the medulla protein of the quill of the African porcupine (*150*), the determination of the sequence of the nonapeptide phyllocaerulein from the skin of a South American amphibian (*151*), as an aid in distinguishing the differences in sequence between beef and pig insulins (*98*), and as an aid in determining the positions of the disulfide bridges linking the two chains of insulin (*97*). Foster and co-workers (*152, 153*) have used subtilisin BPN′ to probe the structure of bovine plasma albumin by limited digestion of an albumin–sodium dodecyl sulfate complex. Hill (*154*) has presented an extensive discussion of the hydrolysis of a variety of proteins by the subtilisins, and the articles by Smyth (*155*) described the procedures to be employed in using subtilisin for sequence work.

Because of their broad specificity, the subtilisins have been useful for studying the sequences of small peptides which are not hydrolyzed by more highly specific proteinases.

A water-insoluble active derivative of subtilisin Novo has been prepared by cross-linking with glutaraldehyde. Amino acid analysis indicates that lysine residues are probably involved in the intermolecular cross-linking reaction with glutaraldehyde. This preparation, when mixed with cellulose powder in a chromatographic column, has shown some promise in digesting proteins (*156*).

147. M. Ottesen, *ABB* **65,** 70 (1956).
148. M. Ottesen and W. A. Schroeder, *Acta Chem. Scand.* **15,** 926 (1961).
149. H. J. Hoenders, J. Van Tol, and H. Bloemendal, *BBA* **160,** 283 (1968).
150. P. M. Steinert, H. W. J. Harding, and G. E. Rogers, *BBA* **175,** 1 (1969).
151. A. Anastasia, *Experientia* **25,** 8 (1969).
152. B. J. Adkins and J. F. Foster, *Biochemistry* **4,** 634 (1965); **5,** 2579 (1966).
153. D. M. Pederson and J. F. Foster, *Biochemistry* **8,** 2357 (1969).
154. R. L. Hill, *Advan. Protein Chem.* **20,** 37 (1965).
155. D. G. Smyth, "Methods in Enzymology," Vol. 11, pp. 214 and 421, 1967.
156. K. Ogata, M. Ottesen, and I. Svendsen, *BBA* **159,** 403 (1968).

B. Use in Detergents

The alkaline pH optima, stability to anionic detergents and to relatively high temperatures have made possible the use of the subtilisins as additives to household anionic detergent products. Thus, the housewife has become an amateur enzymologist and has been able to see a demonstrable benefit in the removal of certain formerly stubborn stains and soils. Laboratory tests and practical experience have indicated that the proteolytic enzymes alone do not give a satisfactory cleaning performance; protein spots are not satisfactorily removed by the action of the enzymes alone. Rather, they must be used in combination with detergents and in so doing a marked synergistic cleaning action is obtained, far superior to that resulting from summing the individual cleaning actions of the enzyme and the detergent alone.

Although figures are not readily available it appears that the alkaline proteinases, known collectively as subtilisins, are being produced in far greater quantities for commercial use than all other enzymes together. In view of the increasing practical use of subtilisins and related enzymes, we may expect a great surge in the study of new enzymes of these types in the future.

17

Streptococcal Proteinase

TEH-YUNG LIU • S. D. ELLIOTT

I. Introduction

Proteolytic activity in cultures of hemolytic streptococci was first described by Frobisher (*1*) who noticed blackening of muscle fibers when streptococci were grown in a cooked-meat medium. He named the active principle *histase* but it remained uncharacterized as, indeed, did

1. M. Frobisher, *J. Exptl. Med.* **44,** 777 (1926).

the streptococci concerned in its production, for Lancefield's group classification of streptococci was not introduced until some years later. Whether histase and the protease of group A streptococci are one and the same therefore remains uncertain.

The proteolytic enzyme here to be described first attracted attention by its capacity to destroy the serological reactivity of the type-specific M protein of group A streptococci (*2*). A chance observation led to the finding that with some, though by no means all strains of these microorganisms, M antigen was present in cultures grown at 22° whereas it was absent from the identical strains grown at 37°. Proteolytic activity in the culture supernatants was shown to be responsible for the antigen loss and many strains of group A streptococci, regardless of serological type, were found to produce an extracellular proteinase. This was especially common in "glossy" strains, i.e., strains deficient in type-specific M antigen under all cultural conditions. By contrast, mouse virulent strains rich in M antigen produced little or no proteinase. Passage of glossy, proteolytic strains through mice resulted in the emergence of virulent variants endowed with M antigen but lacking in proteolytic activity.

The protease resembles papain in achieving maximal activity only in the reduced (SH) form (*2*). It first appears in culture fluids as an inactive precursor or zymogen (*3*). This, in the later stage of growth, is transformed into the active enzyme by an autocatalytic reaction initiated by the slight proteolytic activity shown by zymogen reduced through contact with –SH groups in the bacterial cell walls (*4*). Cell wall –SH groups are also responsible for activation of the enzyme in actively growing cultures. Only living organisms are able to achieve this activation; cocci killed by heat or chloroform are unable to do so (*1*).

Transformation of zymogen to enzyme can also be effected by trypsin (*3, 5*). The proteinase so formed is inactive unless reduced by treatment with a thiol compound or cyanide. Unlike active enzyme, the unreduced, trypsin-produced enzyme does not undergo autodigestion and on chromatographic examination it appears to be more homogeneous than the reduced form produced by autocatalytic transformation of zymogen. For this reason tryptic digestion of crystalline zymogen has been used for the preparation of highly purified proteinase (*5–7*).

A. Preparation and Crystallization of Zymogen

The amount of zymogen produced in broth cultures of hemolytic streptococci is proportional to the amount of bacterial growth. A sim-

2. S. D. Elliott, *J. Exptl. Med.* **81,** 573 (1945).

plified form of Dole's dialyzate medium (*8*) supports profuse growth of group A streptococci and its use facilitates purification of zymogen; the latter constitutes the greater part of the nondialyzable material in the culture filtrate. A detailed description of the methods used in preparation of the medium, growth of the streptococci, and assay of the zymogen and proteinase has been given elsewhere (*9*). When a highly productive strain (*10*) is grown under these conditions, yields of zymogen approaching 150 mg/liter may be expected.

Crystallization of the zymogen protein is readily achieved by the following procedure (*10*). The culture filtrate is brought to 0.8 saturation with $(NH_4)_2SO_4$ thereby precipitating from solution all the protein constituents. This precipitate is taken up in chilled, half-saturated $(NH_4)_2SO_4$ in a volume approximately one-hundredth that of the original culture filtrate. Zymogen protein is soluble under these conditions, and this step usually yields an approximate 1% solution. From this the protein crystallizes as fine needles when the temperature is brought to 37° and the pH level to approximately 7.4, isoelectric for zymogen protein (cf. Fig. 1a). Recrystallization is achieved by harvesting the first crop of crystals, dissolving them in chilled half-saturate $(NH_4)_2SO_4$ at approximately pH 6.0, and then bringing the solution to 37° at pH 7.4.

Further purification can be achieved by Sephadex gel filtration followed by chromatography on SE-Sephadex (*5*).

B. Preparation and Crystallization of Proteinase

Incubation of broth cultures of zymogen-producing streptococci under reducing conditions results in the conversion of zymogen to active enzyme. The time required for this change to occur probably depends upon the intensity of the reducing conditions produced. It varies with different brands of peptone used in preparation of the culture medium, and with dialysate medium it may take up to 24 hr after the end of the logarithmic phase of multiplication (*3*).

---

3. S. D. Elliott and V. P. Dole, *J. Exptl. Med.* **85**, 305 (1947).
4. T. Y. Liu and S. D. Elliott, *Nature* 33, 206 (1965).
5. T. Y. Liu and S. D. Elliott, *JBC* **240**, 1138 (1965).
6. T. Y. Liu, N. P. Neumann, S. D. Elliott, S. Moore, and W. H. Stein, *JBC* **238**, 251 (1963).
7. T. Y. Liu, W. H. Stein, S. Moore, and S. D. Elliott, *JBC* **240**, 1143 (1965).
8. V. P. Dole, *Proc. Soc. Exptl. Biol. Med.* **63**, 122 (1946).
9. S. D. Elliott and T. Y. Liu, "Methods in Enzymology," Vol. 19, p. 252, 1970.
10. S. D. Elliott, *J. Exptl. Med.* **92**, 201 (1950).

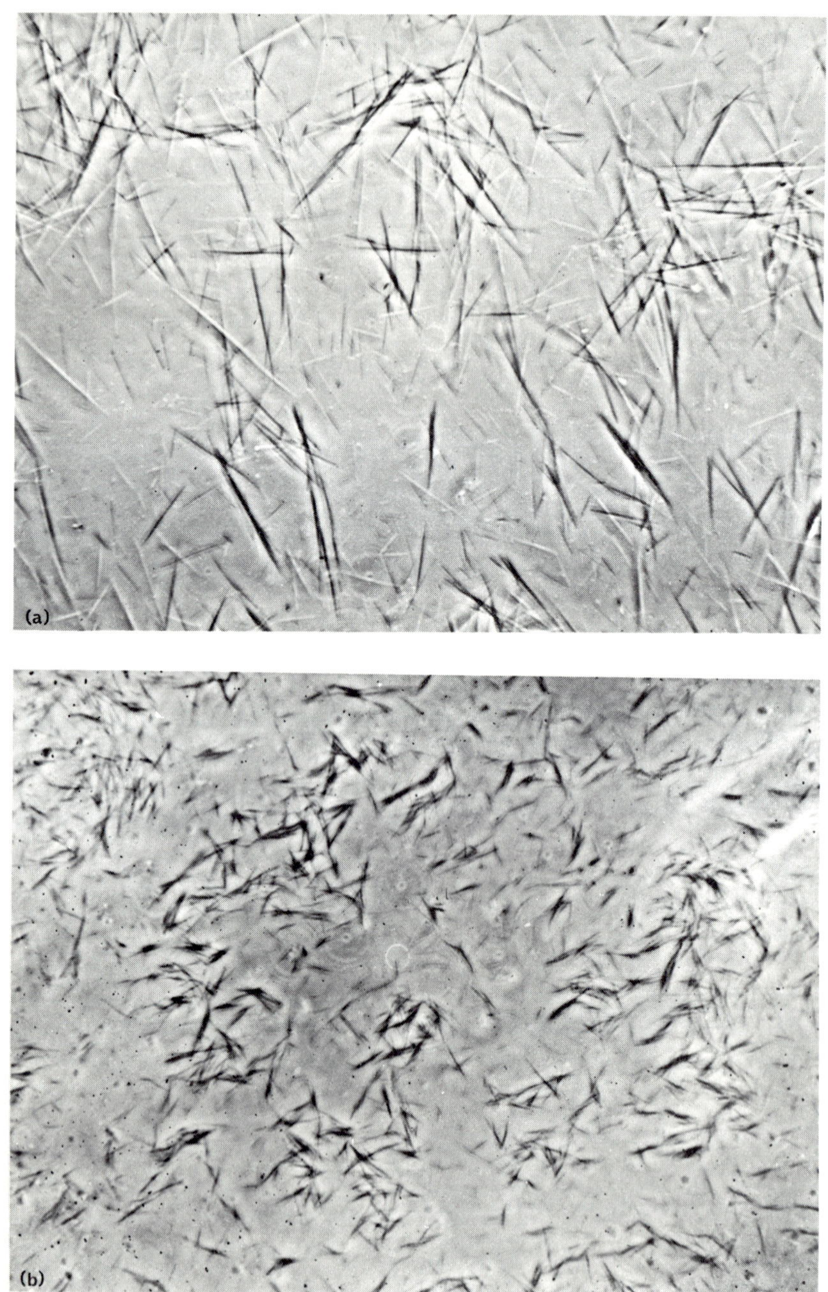

FIG. 1. (a) Crystals of streptococcal proteinase precursor. Phase contrast. ×560. (b) Crystals of streptococcal proteinase. Phase contrast ×560 (*10*).

Crystallization of the proteinase may be accomplished by a procedure similar in general to that already described for purification of the zymogen (*10*). Incubation of the streptococcal culture in dialyzate broth is continued until zymogen has been converted to active enzyme; the latter is then precipitated by the addition of $(NH_4)_2SO_4$ to 0.8 saturation. By redissolving the precipitate in an appropriate volume of water, an approximate one-hundredfold concentration of the enzyme is achieved and the concentrated preparation is then dialyzed against 0.4 saturated $(NH_4)_2SO_4$ in the cold at approximately pH 8. Crystallization of the enzyme protein in the form of short needles usually commences when the $(NH_4)_2SO_4$ concentrate reaches approximately 0.15 saturation and more than 70% of the active material is recovered in crystalline form by the time 0.4 saturation is reached (cf. Fig. 1b). In a typical experiment (*10*) the ratio of proteolytic activity to total protein in the enzyme preparation was doubled after the first crystallization; deoxyribonuclease, a contaminating streptococcal product, was eliminated after five recrystallizations. The purified proteinase is now generally prepared by controlled proteolysis of the zymogen by trypsin, as described in Section IV.

II. Physical Properties

A. Gel Filtration and Chromatography

The twice recrystallized zymogen is essentially homogeneous on gel filtration and in several chromatographic systems. The zymogen emerges as a single peak upon gel filtration on Sephadex G-100 (*11*) or when chromatographed on CM-cellulose or SE-Sephadex (*6*). The once crystallized zymogen contains from 5 to 20% fraction which may be an aggregate form of the zymogen (*11*).

In contrast to crystalline zymogen, the twice crystallized proteinase isolated from the culture medium shows three major components when subjected to chromatography on SE-Sephadex. These, although they differ in specific activity, are indistinguishable from each other on the basis of amino acid composition, other than by the small differences in half-cystine content (*6*). Proteinase produced by the action of trypsin on the zymogen under controlled condition is homogeneous when chromatographed on the SE-Sephadex (*5*) or on Sephadex G-100 (*11*).

11. T. Y. Liu, A. A. Kortt, F. T. Dunne, and S. D. Elliott, in preparation.

B. Electrophoresis

Migration in a Longsworth scanning apparatus (*12*) established the isoelectric points of the zymogen at pH 7.35 and that of the enzyme at pH 8.42 in buffers of 0.1 ionic strength. The electrophoretic properties of the zymogen, the modified zymogen, and the enzyme can also be conveniently compared by zone electrophoresis on strips of cellulose acetate (*5*).

C. Molecular Weight

The determinations of the molecular weights by high-speed equilibrium ultracentrifugation have yielded a value of 44,000 $\pm$ 500 for the zymogen and 32,000 $\pm$ 800 for the enzyme produced by the action of trypsin on the zymogen (*6*). These results are in excellent agreement with the minimum molecular weights calculated from the amino acid composition which gave 44,603 for the molecular weight of the zymogen and 33,589 for the molecular weight of the enzyme (Table I) (*6, 11, 13, 14, 15*).

D. Stability

A crystalline suspension of the zymogen in half-saturated ammonium sulfate stored at —20° retained potential enzymic activity and chromatographic behavior for at least 4 years. If, however, the zymogen or the enzyme was kept as a crystalline suspension in half-saturated ammonium sulfate at room temperature there was a gradual modification to yield inactive protein.

A crystalline zymogen preparation assayed against an ester substrate (*16*) in the absence of an activator such as dithioerythritol or mercaptoethanol exhibits between 0.01 and 0.05% of the potential esterase activity. This activity can be completely abolished by exposure to tetrathionate or iodoacetamide. Zymogen solutions so treated are stable for several days at 4° or for 2 years at —20°. Activated proteinase free from sulfhydryl reagents can be stored under nitrogen in the presence of 10^{-5} *M* EDTA without loss of more than 3–5% of activity per week.

12. T. Shedlovsky and S. D. Elliott, *J. Exptl. Med.* **94,** 363 (1951).
13. S. Moore and W. H. Stein, "Methods in Enzymology," Vol. 6, p. 819, 1963.
14. S. Moore, *JBC* **238,** 235 (1963).
15. T. W. Goodwin and R. A. Morton, *BJ* **40,** 628 (1946).
16. T. Y. Liu, N. Nomura, E. K. Jonsson, and B. G. Wallace, *JBC* **244,** 5745 (1969).

Proteinase produced by the action of trypsin on the zymogen is stable at $-20°$ for at least 2 years without loss of activity. Both the zymogen and the enzyme are unstable at pH values below 4.5 and above 9.5 (*3*). They are denatured by treatment with 0.1 *N* HCl or 0.01 *N* NaOH for 2 hr at room temperature.

At a temperature of $60° \pm 1°$, the proteinase is irreversibly inactivated with precipitation (*11*).

After exposure of the zymogen or the enzyme to 8 *M* urea or 4 *M* guanidium, full enzymic activity can be restored by dialysis (*11*). In 8 *M* urea, the zymogen is activable by autocatalysis (*6*). The enzyme exhibits about 20% of its proteolytic activity in 8 *M* urea when provided with a reducing reagent such as mercaptoethanol (*17*). Exposure of the zymogen or the enzyme to 0.5% sodium dodecyl sulfate results in denaturation (*11*). Although the zymogen is denatured by exposure to 100° for 10 min at pH 6.7 or to 0.1 *N* HCl for 2 hr at 25°, 90% of the enzymic activity can be restored provided that the denatured zymogen was solubilized with urea or guanidium chloride and given time to refold (*18*). However, under the identical conditions, the heat or acid denatured enzyme is irreversibly inactivated.

III. Immunological Properties

The zymogen and enzyme proteins differ in immunological specificity (*10*). Each induces in rabbits the formation of antibodies which precipitate with the homologous antigen. Zymogen gives delayed precipitation with antiserum to the proteinase but the proteinase is readily precipitated by zymogen antiserum, though less strongly than is the homologous antigen. The serological reactions of zymogen and enzyme have been of practical value in assaying these proteins in culture filtrates.

Absorption of zymogen antiserum with proteinase removes all cross-reacting antibody and leaves a zymogen-specific antiserum. In contrast, absorption of proteinase antiserum with zymogen removes all the antibody. These findings suggest that zymogen may have two immunologically distinct determinant groups—one specific, the other shared by proteinase. The latter group is masked possibly by the configuration of the zymogen molecule which hinders its combination with enzyme antibodies but does not prevent it from eliciting their production in rabbits.

Direct evidence for the dual specificity of zymogen protein has been

17. T. Y. Liu, *JBC* **242,** 4029 (1967).
18. M. C. Lin and M. Bustin, *JBC* **245,** 3384 (1970).

TABLE I
AMINO ACID COMPOSITIONS OF ZYMOGEN, MODIFIED ZYMOGEN, AND TRYPSIN ENZYME OF STREPTOCOCCAL PROTEINASE[a,b]

	Zymogen				Modified zymogen				Trypsin enzyme			
	Expressed to nearest	Hydrolysis time (hr)			Expressed to nearest	Hydrolysis time (hr)			Expressed to nearest	Hydrolysis time (hr)		
Amino acid[c]	integer	22	48	72	integer	22	48	72	integer	22	48	72
Aspartic acid	53	53.5	53.1	53.2	52	50.9	51.7	52.2	40	40.1	40.4	39.7
Glutamic acid	41	40.7	40.8	40.3	38	37.8	38.1	38.3	29	29.0	28.9	28.8
Glycine	47	46.7	47.0	46.9	44	44.0	44.2	44.3	38	38.2	37.9	38.3
Alanine	32	31.7	31.2	32.0	28	27.8	27.8	28.3	22	22.3	22.0	22.3
Valine	28	24.7	26.7	27.5	27	23.9	25.7	26.5	23	20.1	22.1	22.7
Leucine	21	21.3	20.9	21.1	20	19.8	19.8	19.9	17	17.0	17.4	17.3
Isoleucine	23	21.5	22.8	22.8	21	20.0	20.9	21.1	14	13.0	13.8	13.5
Serine	37	32.2	27.9	22.6	36	32.4	28.6	26.2	26	22.0	18.4	16.8
Threonine	17	16.4	15.2	13.5	16	14.5	14.3	13.8	11	10.9	9.9	9.7
Half-cystine[d]	1	1.0	—	—	1	1.0	—	—	1	1.0	—	—
Methionine[d]	7	7.1	6.5	6.4	6	5.9	5.8	5.9	5	4.8	4.7	5.1

Proline	15	14.7	15.0	14.4	15	13.5	14.6	15.1	15	15.5	—	15.3
Phenylalanine	17	16.9	17.0	16.6	16	15.2	15.4	15.6	12	12.4	12.0	12.3
Tyrosine	22	21.6	21.2	21.2	21	20.7	20.0	20.2	18	17.7	17.7	17.6
Tryptophan[e]	5	4.9	—	—	5	—	—	—	5	5.1	—	—
Histidine	8	7.9	8.2	7.9	8	7.7	7.7	7.9	8	8.2	8.4	8.5
Lysine	29	29.2	29.0	29.5	29	29.0	29.0	29.0	17	17.0	17.2	17.3
Arginine	11	10.9	10.9	10.6	10	10.0	10.1	9.9	9	8.9	8.4	8.2
Amide[f]	52	—	—	—	50	—	—	—	48	—	—	—

[a] Determined by ion exchange chromatography. The results are expressed as the calculated number of residues per molecule (based on half-cystine equals 1.0) for a molecular weight of 33,589, which is compatible with the ultracentrifugally determined value of 32,000 ± 800 for the enzyme formed by the action of trypsin on the zymogen and a molecular weight of 44,604, which is compatible with the ultracentrifugally determined value of 43,500 ± 800 for the zymogen.

[b] From Liu *et al.* (*6, 11*).

[c] Averages of at least three determinations of 22, 48, and 72 hr hydrolyzates. The values for serine, threonine, and tyrosine were not corrected for decomposition during acid hydrolysis (*13*).

[d] Measured as cysteic acid or methionine sulfone after performic acid oxidation (*14*).

[e] Tryptophan was determined spectrophotometrically (*15*).

[f] The 22, 48, and 72 hr values for ammonia have been extrapolated to zero time to obtain the amide content.

provided by experiments in which the conversion of zymogen to enzyme by trypsin was carried out under controlled conditions. In the early stages of the reaction an intermediate protein, "modified zymogen," is formed (*5, 12*). Modified zymogen reacts serologically equally well with zymogen and with enzyme antisera, but it appears as a single substance on immunoelectrophoresis (cf. Fig. 2). During this step of the reaction, the tryptic digestion probably produces an unfolding of the zymogen molecule with coincident cleavage of a small peptide of about 20 amino acid residues (see below). Further proteolysis by the trypsin converts modified zymogen to enzyme with the loss of about 100 residues. Separation of peptide fragments of molecular weight below 14,000 gave a component that reacted with zymogen antiserum but failed to react with

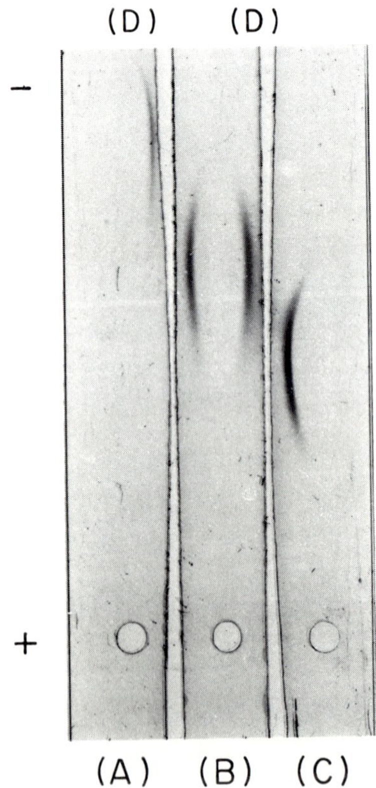

Fig. 2. Immunoelectrophoresis of streptococcal proteinase: (A) enzyme, (B) zymogen, (C) modified zymogen, and (D) zymogen antiserum. Electrophoresis was performed for 3 hr at 25° on a microscope slide (1 by 3 inches) in 0.05 M Veronal buffer at pH 8.6 with a voltage gradient of 0.5 V/cm. Rabbit antiserum against the zymogen was used in the slot to detect the protein (*5*).

enzyme antiserum. This fragment was thought to carry zymogen specificity. The conversions are indicated in Fig. 3.

IV. Chemical Properties

A. Zymogen-to-Enzyme Transformation

Partial proteolytic cleavage of the zymogen to the enzyme can be effected by the action of trypsin, subtilisin, or the streptococcal proteinase itself. In all instances, the conversion of the zymogen to the enzyme is accompanied by the removal of about 100 amino acid residues. The enzyme preparations obtained by different methods have similar amino acid compositions (Table II). Moreover, analyses of the NH_2-terminal residues gave glutamic acid as the major residue for all preparations. Since trypsin, subtilisin, and the streptococcal proteinase cleave the zymogen to yield nearly the same product, there must be certain bonds that are particularly available to attack by all three enzymes. Even though the resulting enzyme contains only a single chain which is not stabilized by any disulfide cross-links, the molecule possesses a structure which is resistant to further autodigestion or to further digestion by trypsin.

1. *Proteolysis by Trypsin*

During proteolysis by trypsin, as zymogen begins to disappear and enzyme to appear, a product with intermediate properties may be demonstrated (Fig. 2). This product is more acidic than the zymogen by the electrophoretic test, whereas the final product, the enzyme, is more basic. The intermediate can be separated from the enzyme and the zymogen by chromatography on carboxymethyl cellulose. Amino acid analyses (Table I) shows that it has only about 20 amino acid residues less than the zymogen. About 90% maximal enzymic activity is generated upon reduc-

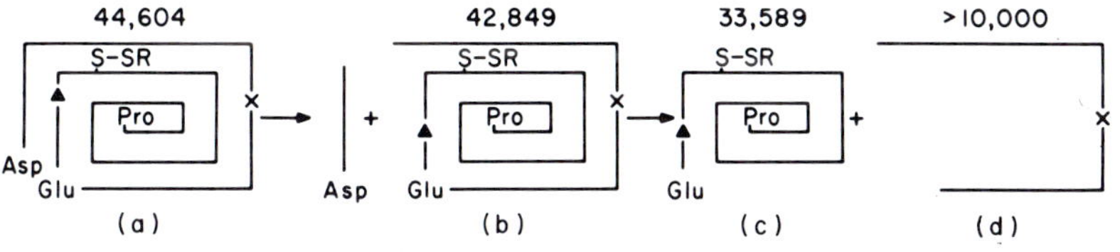

Fig. 3. Zymogen-to-enzyme transformation catalyzed by trypsin (*5*): (a) zymogen, (b) modified zymogen, (c) enzyme, and (d) fragment. Antigenic site: (▲) enzyme and (X) zymogen.

TABLE II
AMINO ACID COMPOSITIONS OF DIFFERENT PREPARATIONS OF STREPTOCOCCAL PROTEINASE[a,b]

	Enzymes formed by				
Amino acid[c]	Trypsin	Activated proteinase	Subtilisin	Autodigestion	Isolation from culture
Aspartic acid	40.1	39.3	39.2	40.5	38.8
Glutamic acid	29.0	28.5	28.6	28.1	28.9
Glycine	38.2	38.5	30.9	35.4	37.2
Alanine	22.3	22.2	20.7	21.1	22.2
Valine	20.1	20.1	19.3	20.1	20.4
Leucine	17.0	17.0	17.0	17.0	17.0
Isoleucine	13.0	12.5	12.7	12.3	12.4
Serine	22.0	23.0	22.4	23.4	23.4
Threonine	10.9	10.6	9.8	10.5	10.8
Half-cystine[d]	1.0	1.0	0.9	0.6–1.0	0.7
Methionine[d]	4.8	4.8	4.9	4.6	5.0
Proline	15.5	13.7	14.0	14.1	14.2
Phenylalanine	12.4	12.2	11.0	11.7	12.0
Tyrosine	17.7	18.2	16.9	17.4	17.9
Tryptophan[e]	5.1	—	—	—	—
Histidine	8.2	8.0	8.1	8.0	7.8
Lysine	17.0	16.2	17.0	16.5	16.9
Arginine	8.9	8.0	8.0	8.1	8.6

[a] Results are given for the enzymes formed from zymogen by the action of trypsin, activated proteinase, subtilisin, and by autodigestion. The results are expressed as the calculated number of residues per molecule (based on leucine equals 17).

[b] From Liu and Elliott (*5*).

[c] Averages of at least three determinations of 22 hr hydrolyzate. The values for serine, threonine, and tyrosine were not corrected for decomposition during acid hydrolysis.

[d] Measured as cysteic acid or methionine-sulfone after performic acid oxidation (*14*).

[e] Tryptophan was determined spectrophotometrically (*15*).

tion of this modified zymogen with thiol compounds. Thus, the removal of amino acid residues from the zymogen has made the active site of the enzyme available to the substrate and has, at the same time, exposed the antigenic site characteristic of the enzyme without losing the one characteristic of the zymogen (Fig. 3).

When an aggregated form of the zymogen is digested by trypsin, transformation to enzyme is usually incomplete even after prolonged incubation (up to 72 hr), and the products formed are heterogeneous (*11*).

2. *Proteolysis by Subtilisin*

Transformation of zymogen by subtilisin under conditions as used for transformation by trypsin is relatively rapid, being complete in 1 hr, as judged by serological tests, electrophoresis, and assay for proteolytic activity (*5*). Longer treatment results in a gradual decrease in the proteolytic activity of the reaction mixture. The product, which had amino acid composition as given in Table II, was heterogeneous, however, as judged by the NH_2-terminal analysis; it gave 0.70 residue of glutamic acid, 0.22 of alanine, and 0.15 of valine.

3. *Proteolysis by Preformed Streptococcal Proteinase*

Conversion of zymogen to proteinase can also be effected by activated streptococcal proteinase obtained by dithiothreitol treatment of trypsin-produced enzyme and freed from SH reagents by gel filtration or dialysis under nitrogen (*5*). The product has the same chromatographic behavior and enzymic activity per milligram as that obtained when trypsin is used. End group analysis gives the following NH_2-terminal residues: glutamic acid, 1.00; glycine, 0.12; methionine, 0.06; and isoleucine, 0.06.

4. *Autocatalytic Transformation*

The speed of the zymogen-to-enzyme conversion in the presence of 0.05 *M* mercaptoethanol is dependent upon the zymogen concentrations (*5*). At a concentration of 5–10 μg of protein per milliliter, there was practically no detectable autodigestion within a few hours but at 5–6 mg/ml, the transformation was complete in 120 min at 37°. The specific activity reached did not decrease upon longer incubation. The product, when chromatographed on SE-Sephadex, showed heterogeneity similar to that observed with the crystalline enzyme isolated from the streptococcal cultures (*6*). Analysis for NH_2-terminal residues gave the following results: glutamic acid, 0.94; cysteic acid, 0.08; and valine, 0.06.

5. *Activation by Bacterial Cell Walls*

The conversion of zymogen to enzyme occurs under reducing conditions provided by streptococcal cell walls free of detectable amounts of preformed proteinase (*4*). The process might be initiated by the presence of performed enzyme in the zymogen preparation or streptococcal culture in amounts too small for detection but capable of triggering the autocatalytic reaction; a hypothesis was proposed that zymogen itself when reduced may have slight but significant proteolytic activity (*4*). Subsequently, it has been shown that the reduced zymogen is indeed active (*19*). In streptococcal cultures, the transformation of zymogen to enzyme may proceed as shown in Table III. Under these conditions such a process would result in a mixture of the reduced and unreduced forms of

19. M. Bustin, M. C. Lin, W. H. Stein, and S. Moore, *JBC* **245,** 846 (1970).

TABLE III
CONVERSION OF ZYMOGEN TO PROTEINASE[a,b]

(1)	Zymogen (—SSR)[c] (inactive)	$\xrightarrow[\text{(reduction)}]{\text{cell wall (—SH)}}$	Zymogen (—SH) (slight activity)
(2)	Zymogen (—SSR)	$\xrightarrow[\text{(proteolysis)}]{\text{zymogen (—SH)}}$	Proteinase (—SSR) (inactive)
(3)	Proteinase (—SSR)	$\xrightarrow[\text{(reduction)}]{\text{cell wall (—SH)}}$	Proteinase (—SH) (active)
(4)	Zymogen (—SSR)	$\xrightarrow[\text{(proteolysis)}]{\text{proteinase (—SH)}}$	Proteinase (—SSR) (inactive)

[a] Series of linked reactions probably occurring in streptococcal cultures.
[b] From Liu and Elliott (*4*).
[c] Unreduced form of protein (*20*).

the enzyme in a ratio dependent on the opportunities for contact with the reducing groups on the bacteria in suspension.

B. THE NATURE OF THE SULFHYDRYL GROUP

Streptococcal proteinase and its zymogen have the unique property of containing only a single half-cystinyl residue per molecule (*6*). The –SH group of the potential cysteinyl residue is liberated only when the enzyme is activated by reduction. These findings raise the question as to the nature of the linkage in which the half-cystinyl residue participates in the unreduced protein.

The presence of only one half-cystinyl residue is well established (*6*). Performic acid oxidation followed by acid hydrolysis of the zymogen or of the unreduced enzyme yields only one residue of cysteic acid per molecule and no other strongly acidic, ninhydrin-positive amino acid. After reduction by thiols or sodium borohydride, the presence of a single, highly reactive sulhydryl group is readily demonstrated; brief treatment of the reduced protein with iodoacetate, for example, leads to the formation of one residue of *S*-carboxymethylcysteine per molecule. The reductive step causes no detectable change in molecular weight from dimer to monomer (*6*).

In the zymogen or in the unreduced enzyme, even in the presence of 8 *M* urea or 4 *M* guanidium chloride, the potential –SH group is not available for reaction with iodoacetamide, iodoacetic acid, or *N*-ethylmaleimide (*6*). EDTA at a concentration of 0.001 *M* and other chelators such as *o*-phenanthroline and 8-hydroxyquinoline are similarly ineffective in the release of the potential –SH group from the zymogen and the unreduced enzymes (*6*). Spectroscopic analyses, performed by Dr. B. L. Vallee, have demonstrated the absence of more than a few tenths of an atom of heavy metal per molecule of the enzyme. These results exclude the possibility that the potential –SH group in the proteinase is linked to a heavy metal (*6*).

Several other possibilities were considered in attempting to define the unusual nature of the potential sulfhydryl group present in streptococcal proteinase. First, the potential –SH group on the proteinase could exist as a thioester; second, it might exist as part of a mixed disulfide of the type protein S–X in which X is a ninhydrin negative compound such as a volatile mercaptide, –SR, or –SOH, or $-SO_2H$; third, it could also exist simply as a sulfenic or sulfinic acid; or, finally, it could be a sterically inaccessible –SH group. All such compounds will be oxidized to a sulfonic acid by performic acid and all of them could give rise to a free –SH group after treatment with a large excess of a thiol such as mercaptoethanol. In general, however, sterically inaccessible –SH groups are freed by 8 *M* urea or even more effectively by guanidium chloride, while thioesters are converted to the free thiol by hydroxylamine, ammonia, or other bases. That disulfides are unaffected by such treatment provides a basis for distinguishing among the possibilities, but no methods have been established for detecting sulfenic or sulfinic acid groups in proteins.

The nature of the masking of the potential –SH groups was established by the experiment of Ferdinand *et al.* (*20*). When zymogen was reduced by sodium borohydride or cysteine under a stream of nitrogen and the nitrogen then passed through a solution of mercuric chloride, nearly a full residue of a volatile mercaptan was trapped in aqueous mercuric chloride. This result gives strong evidence that the zymogen and the unreduced enzyme contain a mixed disulfide, protein–S–SR where RSH is volatile.

The identity of the volatile mercaptan and the pathway by which it becomes combined with the protein in the mixed disulfide remain to be determined.

C. The Reactivity of the Sulfhydryl Group

Of the functional groups present in proteins, the sulfhydryl group is usually the one most reactive toward alkylating agents; and, in the sulfhydryl enzymes, some particular sulfhydryl group is often more reactive than the others. The presence of a highly reactive sulfhydryl group in the reduced form of streptococcal proteinase was first demonstrated by

20. W. Ferdinand, W. H. Stein, and S. Moore, *JBC* **240, 1150** (1965).

Liu *et al.* (*7*), who showed that the sulfhydryl group on the native enzyme could compete successfully with a 100- to 1000-fold excess of β-mercaptoethanol or cysteine for a 10-fold excess of iodoacetate. Later, a more detailed study was conducted by Gerwin (*21*) to investigate the nucleophilic reactivity of the sulfhydryl group of the proteinase. Using chloracetic acid and chloracetamide, the rates of alkylation of the sulfhydryl group of streptococcal proteinase and a reduced glutathione were investigated over a pH range of 4–10. The pK of the sulfhydryl group of the enzyme was found to be higher than 8.0. The protein sulfhydryl group is 50–100 times as reactive as the sulfhydryl in glutathione. The curve of k_2 with respect to pH is sigmoid when the protein is alkylated by chloracetamide and bell-shaped when chloracetic acid is the alkylating agent. Both reactions lead solely to the formation of the corresponding S-substituted cysteine derivative. The rate curves with respect to pH for the alkylation of reduced glutathione by the two reagents do not show disparities. These observations suggest that the ionic environment of this sulfhydryl group in the proteinase is very complex.

D. Amino Acid Compositions

The amino acid compositions of the zymogen and the modified zymogen are listed in Table I. Zymogen contains a total of 460 residues of which 94 residues are either aspartic or glutamic acid or their amides, 48 residues are basic amino acids, and the rest of the 318 residues are neutral amino acids. Of the acidic amino acids, 52 residues are in the amide form leaving a net 42 residues as anionic residues. The greater number of cationic groups (48 vs. 42) is in accord with the slightly alkaline isoelectric point (pH 7.35) of the zymogen molecule. The change from zymogen to the modified zymogen is accompanied by the removal of only about 20 amino acid residues. The number of ionic groups remains essentially identical. Yet, on electrophoresis, modified zymogen is much more acidic than the zymogen. This probably implies that a significant conformational change has taken place upon the removal of about 20 amino acids from the zymogen. The finding that the removal of these amino acid residues from the zymogen has made the active site of the enzyme available to the substrate and has, at the same time, exposed the antigenic site characteristic of the enzyme (without losing that characteristic of the zymogen) is in agreement with this supposition. The amino acid compositions of the enzyme obtained by partial proteolysis of the zymogen using five different methods is given in Table II. The enzyme

21. B. I. Gerwin, *JBC* **242,** 451 (1967).

produced by the action of trypsin on zymogen contains a total of 351 residues of which 34 residues are basic amino acids and 21 are anionic residues. The greater number of cationic groups is in accord with the alkaline isoelectric point (pH 8.42) of the enzyme.

E. Amino-Terminal and Carboxyl-Terminal Residues

Amino-terminal residue determination (*5, 6*) by the cyanate procedure of Stark and Smyth (*22*) showed that the crystalline zymogen had 0.94–1.04 moles of NH_2-terminal aspartic acid per mole. Only traces (less than 0.05 mole) of other terminal residues were found except for glutamic acid, for which a value of 0.19 mole was obtained even after applying the blank correction [cf. Stark and Smyth (*22*)]. The yield of glutamic acid was not reduced by prior treatment of the zymogen with 0.1 *N* HCl to eliminate the possibility of proteolysis during carbamylation. The significance of the small quantity of NH_2-terminal glutamic acid remains for study.

The enzyme produced by the action of trypsin on the zymogen yielded 0.9–1.0 mole of glutamic acid as the sole amino-terminal residue. Occasional preparations had less than 0.05 equivalent of amino-terminal valine and glycine.

The carboxyl-terminal amino acid residue of the zymogen and the enzyme were determined (*11*) by the hydrazinolysis procedure of Fraenkel-Conrat and Tsung (*23*). Both the zymogen and the enzyme gave 0.4–0.6 mole of proline as the sole carboxyl-terminal residue. Carboxypeptidases A and B failed to liberate any detectable free amino acids from the zymogen, the enzymes or their denatured product under a variety of conditions. The failure of the carboxypeptidases to liberate carboxyl-terminal residues from these proteins is in accord with the finding by chemical methods that proline is the carboxyl-terminal residue for both zymogen and the enzyme.

These results indicate that streptococcal proteinase consists of a single polypeptide chain and that in the transformation of zymogen to enzyme by trypsin proteolysis has occurred only at the amino-terminal end of the peptide chain of zymogen to reduce the molecular weight from 44,603 to 33,589. Evidence for proteolysis at the amino end is provided by the disappearance of the NH_2-terminal aspartic acid characteristic of the zymogen and the appearance of NH_2-terminal glutamic acid. Both the

22. G. R. Stark and D. G. Smyth, *JBC* **238**, 214 (1963).
23. H. Fraenkel-Conrat and M. T. Tsung, "Methods in Enzymology," Vol. 11, p. 151, 1967.

zymogen and the enzyme have proline as their carboxyl-terminal residue; therefore, it is likely that proteolysis has not taken place from the carboxyl end of the zymogen.

F. Amino Acid Residues at the Active Center

The sequence of amino acid residues around the single cysteinyl residue of the proteinase prepared from zymogen by the action of trypsin has been elucidated by Liu *et al.* (*7*). Enzyme labeled with iodoacetic-1-^{14}C acid when hydrolyzed with trypsin yielded a single ^{14}C-containing peptide of 27 amino acid residues with a sequence as shown in Fig. 4.

A procedure for selective alkylation of a histidine residue in streptococcal proteinase without concomitant reaction with sulfhydryl group has been described (*17*). The reduced active form of the enzyme was first treated with sodium tetrathionate to protect the sulfhydryl group and to introduce a negative charge near the active center. The resulting inactive derivative was not detectably alkylated by iodoacetic acid, iodoacetamide, or *N*-chloroacetyltryptophan. After treatment with these reagents, activity could be regenerated by reduction with a thiol. If, however, the *S*-sulfenylsulfonate derivative of the enzyme was exposed to a positively charged alkylating agent, N^{α}-bromoacetylarginine methyl ester, one molecule of the reagent was incorporated per molecule of protein, as indicated by an increase in the amount of arginine by one residue, a decrease in the amount of histidine by one residue, and the formation of one residue of 3-carboxymethylhistidine. The modified protein had no enzymic activity when the sulfhydryl group was regenerated by reduction. These experiments thus furnish direct chemical evidence for the involvement of a histidine residue, as well as of a sulfhydryl group, in the active site of streptococcal proteinase. The position in the peptide chain of the histidine which has been alkylated has not been established.

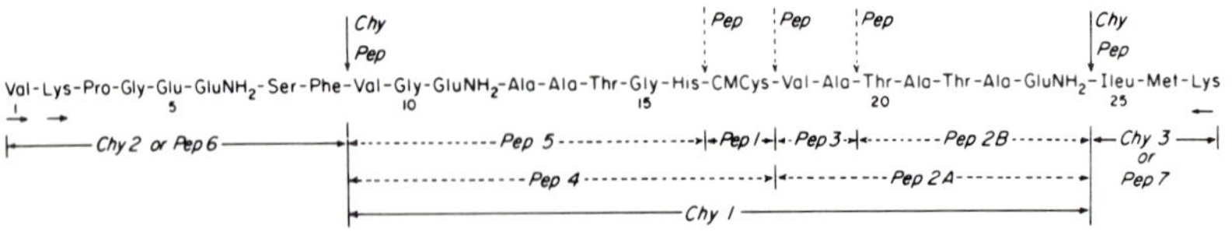

FIG. 4. Sequence of amino acid residues around the sulfhydryl group at the active site of streptococcal proteinase. The points of hydrolysis by chymotrypsin (*Chy*) and pepsin (*Pep*) are indicated by solid and dashed vertical arrows, respectively. Short horizontal arrows that point to the right underneath amino acid residues denote sequences elucidated by Edman degradation. Arrows that point in the opposite direction denote sequences elucidated by the use of carboxypeptidases A and B (*7*).

The enzyme has 8 histidine residues, one of which is adjacent to the sulfhydryl, occurring in a –His–Cys– sequence (see Fig. 4). The histidine which has been alkylated is probably not this particular one next to the cysteinyl residue in the peptide chain.

V. Enzymic Properties

A. Methods of Assay

Principal methods of measurement of streptococcal proteinase are (1) measurement of its hydrolytic activity; (2) estimation of its serological reactivity; and (3) for highly purified material, measurement of its absorbancy at 280 nm.

The milk-clotting action of streptococcal proteinase affords a quick means of assay reproducible over a narrow range of enzyme concentration (*3*). Alternatively, assay with protein substrates such as casein can be performed by the method of Kunitz modified by Liu *et al.* [see Elliott and Liu (*9*)]. The enzyme can be assayed with a peptide substrate such as N^{α}-benzyloxycarbonyllysylphenylalanine (*16*) or N^{α}-benzyloxycarbonylphenylalanyltyrosine (*24*); the amino acids released from the peptide are quantitatively measured either by the amino acid analyzer or by the ninhydrin procedure. Assay with an ester substrate (*16*) such as N^{α}-benzyloxycarbonyllysine·HCl *p*-nitrophenyl ester may be followed spectrophotometrically at 340 mμ.

Both the enzyme and the inactive zymogen from which it is derived are antigenic in rabbits. Zymogen and proteinase have distinct immunological specificities, and precipitation reactions with the appropriate antisera afford a rapid means of assessing the concentration of either protein (*9*). Serological tests such as the test for milk-clotting activity are of particular value in the rapid assay of crude culture fluids.

For highly purified preparations such as result from chromatographic treatment of crystalline material, the enzyme concentration can be estimated from the amount of protein in solution as determined by its absorbance at 280 nm (*6*). An absorbance of 1.64 (1 cm 280 nm) corresponds to 1 mg proteinase per milliliter. The corresponding specific absorbancy of the zymogen is 1.37.

B. Activation

Streptococcal proteinase is a sulfhydryl enzyme like papain and ficin. After conversion from zymogen to enzyme, reduction is essential for catalytic activity as a proteinase, peptidase, or esterase. Reduction may be brought about by thiol compounds such as thioethanol, thioglycolic acids, and dithiothreitol or by cyanide (*2, 6*) or bisulfite (*20*). EDTA, in a concentration of 0.001 *M*, stabilizes but does not enhance the enzymic activity of the activated proteinase (*6*). Once the enzyme is activated by reducing agent, the presence of the activator is no longer necessary for the catalytic activity of the enzyme. Thus, the activators cannot be functioning as coenzymes.

The reduced enzyme is reversibly inactivated by sodium tetrathionate or dinitrofluorobenzene to form the *S*-sulfenylsulfonate enzyme and the *S*-dinitrophenyl enzyme (*11, 17*). These two inactive forms of the enzyme will regenerate full enzymic activity upon exposure to a sulfhydryl reducing agent.

The reduced enzyme is irreversibly inactivated by iodoacetic acid (*2*), iodoacetamide, *N*-ethylmaleimide (*6*), and methyliodide (*11*) by modification of the single sulfhydryl group of the enzyme.

C. Specificity

1. *Action on Peptide and Amide Bond*

The ability of streptococcal proteinase to function as a proteolytic enzyme under reducing conditions was first demonstrated by its capacity to destroy the serological reactivity of the type-specific M protein of group A streptococci (*2*). Human and rabbit fibrin were found to be attacked and streptococcal fibrinolysin was also found to be inactivated by the enzyme. Other susceptible protein substrates include casein, milk, and gelatin (*2*).

When casein is hydrolyzed by streptococcal proteinase, turbidity develops as the proteolysis proceeds. This phenomenon is not found when either trypsin or papain is used in the hydrolysis of casein, and it may reflect differences in the modes of action of these proteolytic enzymes toward protein substrates such as casein. Streptococcal proteinase has been shown to favor the cleavage of hydrophobic residues in the peptide chain (*24*). Therefore, the appearance of turbidity in the casein assay probably indicates the release of insoluble hydrophobic peptides into the assay media.

A closer examination of the specificity of streptococcal proteinase was made by Mycek *et al.* in 1952 (*25*) when crystalline preparations of the

24. B. I. Gerwin, W. H. Stein, and S. Moore, *JBC* **241,** 3331 (1966).
25. M. J. Mycek, S. D. Elliott, and J. S. Fruton, *JBC* **197,** 637 (1952).

proteins became available. They observed that the enzyme hydrolyzed typical trypsin substrates such as benzoylarginine amide and benzoyllysine amide. It also hydrolyzed benzoylhistidine amide, benzyloxycarbonyllyisoglutamine, and benzyloxycarbonylisoasparagine. Later, Liu *et al.* (*6*) noted that benzyloxycarbonyl glycylphenylalanine, a typical substrate for carboxypeptidase A was also hydrolyzed by the enzyme. In every instance, however, the rate of hydrolysis was slow; for example, streptococcal proteinase is only about 1% as effective as trypsin in hydrolyzing benzoylarginine amide. On the other hand, protein substrates such as casein are hydrolyzed twice as rapidly by the proteinase as by trypsin.

Gerwin *et al.* (*24*) used the reduced, carboxymethylated phenylalanyl chain of insulin as substrate to investigate the specificity of streptococcal proteinase. By examining the peptides formed during proteolysis they showed that one of the bonds most rapidly hydrolyzed was the Phe–Tyr linkage in the –Phe–Phe–Tyr– sequence, residues 24 through 26 as shown in Fig. 5 (*24*). With this as a lead, a series of synthetic peptides was tested. The results showed that the chief requirement for hydrolysis is the presence of a bulky side chain on the amino acid adjacent (on the amino-terminal side) to the residue contributing the carbonyl to the susceptible peptide bond. The nature of the NH_2-blocking group has a decisive effect upon the ratio of hydrolysis of the dipeptide substrates. When such a group is lacking altogether, or when it is small, hydrolysis is impaired. This can be seen from the very slight cleavage of L-phenylalanyl-L-phenylalanine (5% in 24 hr) and acetyl-L-phenylalanyl-L-phenylalanine (less than 1% in 24 hr). The substitution of benzyloxycarbonyl for acetyl leads to a dramatic increase in the susceptibility to enzymic hydrolysis. Benzyloxycarbonylalanylphenylalanine is an excellent substrate which is hydrolyzed to the extent of 30% in 1 hr.

Liu *et al.* (*16*) continued specificity studies of the proteinase with a series of carefully designed synthetic peptide substrates. The results of

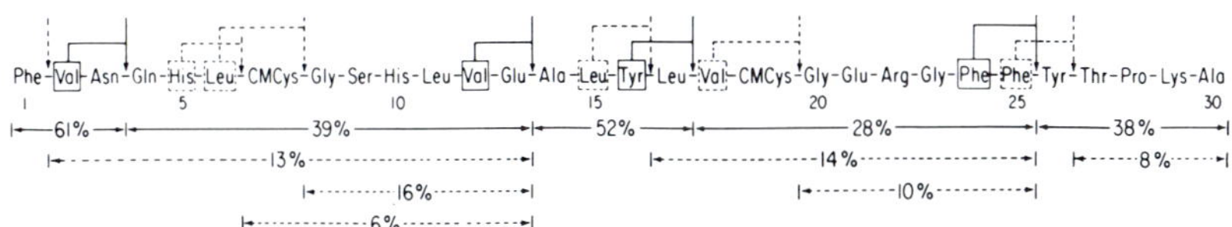

FIG. 5. Hydrolysis of the reduced and carboxymethylated phenylalanyl chain of insulin by streptococcal proteinase. This figure indicates the points of cleavage as determined by peptide analysis. Major cleavages are indicated by solid arrows and minor ones by dashed arrows. The amino acid residues enclosed in boxes occupy positions which are analogous to those of the NH_2-blocking groups insusceptible to acyl dipeptides (*24*).

this study are given in Table IV. The kinetic data indicate that for the series of peptides substrates Z–X–Y listed, where Y = phenylalanine, the relative specificity as expressed by k_{cat}/K_m of streptococcal proteinase for the X group is N^{ϵ}-Boc-Lys >Nle >Glu >Lys. However, these variations are small, especially for the series where X = Nle, Glu, and Lys, considering the drastic change in the ionic environment of X that accompanies the change from a neutral group to a cationic or an anionic group. Evidently, with this series of substrates, where the length of the side chain of the amino acid is similar, the enzyme only poorly recognizes the difference between a polar group and a hydrophobic group. The data indicate that variation in the relative specificity of

TABLE IV
HYDROLYSIS OF SYNTHETIC PEPTIDE SUBSTRATES BY STREPTOCOCCAL PROTEINASE[a,b]

Substrate[c]	K_m (mM)	k_{cat} (sec^{-1})	k_{cat}/K_m (mM^{-1} sec^{-1})	Relative reactivity
Z-Lys-Phe	5.05 ± 0.65	2.25 ± 0.10	0.446	1.00
N^{α}-Z-N^{ϵ}-Boc-Lys-Phe	2.64 ± 0.62	8.60 ± 0.60	3.26	7.31
Z-Glu-Phe	3.88 ± 0.49	2.30 ± 0.10	0.593	1.33
Z-Nle-Phe	2.28 ± 0.13	2.48 ± 0.04	1.09	2.44

[a] Experimental conditions were pH 7.6; 0.2 M N-ethylmorpholine acetate buffer; 37°; 1×10^{-5} M EDTA, 5×10^{-6} M dithioerythritol; enzyme concentration, 8.75×10^{-7} M; S_0 = 0.40 to 3.70×10^{-3} M. Values for K_m and V_{max} were calulated with CDC 6600 computer according to the method of Hanson *et al.* (*26*) as modified by K. Thompson.

[b] From Liu *et al.* (*16*).

[c] Abbreviations: Z, benzyloxycarbonyl; Boc, tertialbutyloxycarbonyl; and Nle, norleucine.

these substrates is due to K_m and not k_{cat}. The most susceptible peptide, Z–Nle–Phe, is the one that binds strongest, since it has the smallest value of K_m; and the least susceptible peptide, Z–Lys–Phe, is the one that binds least, with the highest K_m value. A substantial increase in susceptiblity was observed when the ϵ-amino group of lysine in X was substituted with a t-butyloxycarbonyl group, a bulky group, as in N^{α}-Z-N^{ϵ}Boc-Lys–Phe. The data in Table IV indicate that the increase in susceptibility results from an increase in k_{cat} and not a decrease in K_m. The compound has about the same K_m value as Z-Nle–Phe, but a k_{cat} value 3.5 times as large, which suggests that for the two substrates with hydrophobic neutral side chains the observed difference in the relative rate of hydrolysis probably results from the enhanced rate of deacylation of the bulky hydrophobic residue. An acyl-enzyme intermediate of the type N^{α}-Z-N^{ϵ}-Boc-Lys–$\overset{\text{O}}{\overset{\|}{\text{C}}}$–S enzyme might undergo hydrolysis faster than Z-Lys–$\overset{\text{O}}{\overset{\|}{\text{C}}}$–S enzyme.

2. *Esterase Activity*

Ester substrates such as benzoylarginine ethyl ester, tosylarginine methyl ester, and benzoyltyrosine ethyl ester are not hydrolyzed by streptococcal proteinase at a measurable rate.

A spectrophotometric method was described by Liu *et al.* (*16*) to demonstrate that streptococcal proteinase is an efficient esterase in the cleavage of phenyl and *p*-nitrophenyl esters of the type Z–X–Y, where Z = benzyloxycarbonyl, X = lysine, norleucine, or glutamic acid, and Y = phenyl or *p*-nitrophenyl. The results are given in Table V. The

TABLE V
RELATIVE RATES OF HYDROLYSES OF SYNTHETIC ESTER SUBSTRATES BY STREPTOCOCCAL PROTEINASE[a,b]

Substrate[c]	pH V	Initial rate (mole/sec/mole enzyme)
Z-Lys-OP	5.5	2.63
	7.6	2.38
Z-Nle-OP	5.5	1.84
	7.6	1.84
Z-Glu-OP	5.5	0.965
	7.6	0.995
Tos-Lys-OP	5.5	<0.1
	7.6	<0.1
Tos-Nle-OP	5.5	<0.1
	7.6	<0.1
Z-Lys-ONp	5.5	5.66
Z-Nle-ONp	5.5	2.10
Z-Glu-ONp	5.5	1.40
N^{ϵ}-Z-Lys-ONp	5.5	0.186
N^{α}-Z-N^{ϵ}-Tos-Lys-ONp	5.5	0.346
Z-Gly-ONp	5.5	0.226

[a] Experimental conditions were: 25°; initial substrate concentration, 1.65×10^{-4} *M*; enzyme concentration, 3.75×10^{-7} *M*, 20% acetonitrile-water (v/v), ionic strength, 0.200, 0.20 *M* sodium acetate buffer for pH 5.50, 0.20 *M* *N*-ethylmorpholine acetate buffer for pH 7.6, 1×10^{-5} *M* EDTA, 5×10^{-6} *M* dithioerythritol.

[b] From Liu *et al.* (*16*).

[c] Abbreviations: Z, benzyloxycarbonyl; Tos, tosyl; ONp, *p*-nitrophenoxy; and OP, phenoxy.

phenyl ester substrates gave essentially identical rates at pH 5.5 and 7.6. These results are in agreement with the pH-activity study with Z–Lys–ONp. Streptococcal proteinase exhibits a preference for X, the acyl component of the sensitive bond, in the decreasing order Lys >NLe >Glu. This sequence is not the same as for peptide bond cleavage, where the most susceptible peptide was the one in which X = Nle; and the least susceptible one was where X = Lys.

When X is glycine, as in the case of *p*-nitrophenyl Z-glycinate, the susceptibility to enzymic hydrolysis decreases dramatically. This is in agreement with the finding of Mycek *et al.* (*25*), who found glycine-containing dipeptides are resistant to streptococcal proteinase. Removal of Z from X and substitution of ϵ-NH_2 of Lys with Z results in an ester substrate N^{ϵ}-Z–Lys–ONp, which was found to be a very poor substrate. These results are in general agreement with those of Gerwin *et al.* (*24*), who found that for peptidase activity there exists on the enzyme a hydrophobic binding site which interacts with A in a given substrate A–X–Y. The requirement for esterase activity with respect to A is therefore similar to that for peptidase activity.

3. *Transferase Activity*

Mycek *et al.* (*25*) have studied streptococcal proteinase-catalyzed hydroxamic acid formation from benzoylarginine amide. Only small amounts of hydroxamic acid were demonstrated at pH 7.5, (see Table VI) (*16, 26*). The low yield of hydroxamic acid was attributed to its rapid hydrolysis at this pH. A similar result was noted with trypsin and chymotrypsin when the reaction was carried out at pH 7.5, which is near their pH optima for their proteolytic activity. Previous studies

TABLE VI

CATALYSIS OF HYDROXAMIC ACID FORMATION BY STREPTOCOCCAL PROTEINASE[a,b]

Time (min)	Ammonia liberation (μmoles/ml)	Hydroxamic acid present (μmoles/ml)
30	8.4	0.8
60	12.6	0.96
120	18.7	1.2
180	20.3	1.1

[a] Substrate, benzoyl-L-arginine amide (0.05 *M*); hydroxylamine, 0.05 *M*; enzyme concentration, 0.074 mg of protein N per milliliter of test solution; pH 7.5 [tris(hydroxymethyl)aminomethane, 0.02 *M*]; temperature, 37.5°.

[b] From Mycek *et al.* (*25*).

26. K. R. Hanson, R. Ling, and E. Havier, *BBRC* **29,** 194 (1967).

by Fruton and his associates (*27*) have shown that cysteine-activated papain catalyzed the reaction between benzolarginine amide and hydroxylamine to form a hydroxamic acid in high yield. The reaction was performed at pH 5.0 which is near the pH optimum for papain catalyzed hydrolysis of amide and ester substrates.

The synthesis of anilides by streptococcal proteinase from benzyloxycarbonylalanine and aniline at pH 5.5 was demonstrated by Liu *et al.* (*11*). The yield of the anilide was 16% in 48 hr at 37°. When aniline was replaced with phenylalanyl ethyl ester, under identical experimental conditions, benzyloxycarbonylalanylphenylalanyl ethyl ester was obtained in 26% yield after 72 hr (*11*). These experiments demonstrate that streptococcal proteinase is an efficient transferase like papain and ficin.

D. Kinetics of Hydrolysis

1. *Effect of pH*

Hydrolysis of Z–Lys–ONp between pH 4 and 6.9 gave kinetic constants k_{cat}, K_m and k_{cat}/K_m as shown in Table VII. The k_{cat} values are

TABLE VII
Kinetic Constants of the Streptococcal Proteinase Catalyzed Hydrolyses of Z-Lys-*p*-Nitrophenyl Ester[a,b]

pH	K_m (mM)	k_{cat} (sec^{-1})	k_{cat}/K_m (sec^{-1} mM^{-1})
4.27	0.222 ± 0.028	29.7 ± 0.21	134
4.49	0.494 ± 0.073	89.8 ± 0.86	182
4.68	0.188 ± 0.043	55.0 ± 0.68	292
4.96	0.259 ± 0.045	83.0 ± 0.95	359
5.09	0.192 ± 0.086	84.9 ± 2.10	441
5.29	0.247 ± 0.083	115.0 ± 2.2	465
5.48	0.314 ± 0.031	148.0 ± 0.89	471
5.69	0.248 ± 0.030	133.0 ± 0.92	531
5.95	0.265 ± 0.078	132.0 ± 0.97	498
6.15	0.253 ± 0.085	127.0 ± 1.25	502
6.35	0.262 ± 0.075	142.0 ± 0.89	531
6.85	0.277 ± 0.038	135.0 ± 1.35	487

[a] Experimental conditions were 1.6% acetonitrile-water (v/v), ionic strength, 0.2, sodium acetate buffer; 1×10^{-5} M EDTA, 5×10^{-6} M dithioerythritol; 25°; enzyme concentration, 1.25×10^{-7} M; $S_0 = 2.14$ to 13.31×10^{-5} M. Values for K_m and V_{max} were calculated with the CDC 6600 computer. The program was written by Hanson *et al.* (*26*) and modified by K. Thompson for use with the CDC 6600.

[b] From Liu *et al.* (*16*).

27. J. Durell and J. S. Fruton, *JBC* **207,** 487 (1954).

fairly constant above pH 5.0 and the changes in the K_m values for this substrate are not appreciable over the pH range studied. The pH dependence of k_{cat}/K_m for Z–Lys–ONp is shown in Fig. 6. The ratio k_{cat}/K_m depends on an ionizable group with p$K = 4.8$. This pK value represents the ionization of the enzyme and thus should be identical for all substrates provided there is no pH-dependent interaction of the charged substrate with noncatalytic groups on the enzyme which would effect this value (*28*). Such an effect, if present at all, is small above pH 4.5 since both cationic substrates such as Z–Lys–ONp and neutral substrates such as Z–Ala–ONp (*29*) exhibit similar pH dependences of k_{cat}/K_m. From the values of K_m and V_{max} for the streptococcal proteinase-catalyzed hydrolysis of Z–Phe–Tyr over the pH range 5–10, Gerwin *et al.* (*24*) have inferred a pK_a of 6.4 for a catalytically important group in the free enzyme (Fig. 7). This group was thought possibly to be an unprotonated histidine residue at the active site of the enzyme. Direct chemical evidence for the involvement of a histidine residue in the active site of the enzyme was demonstrated by Liu (*17*).

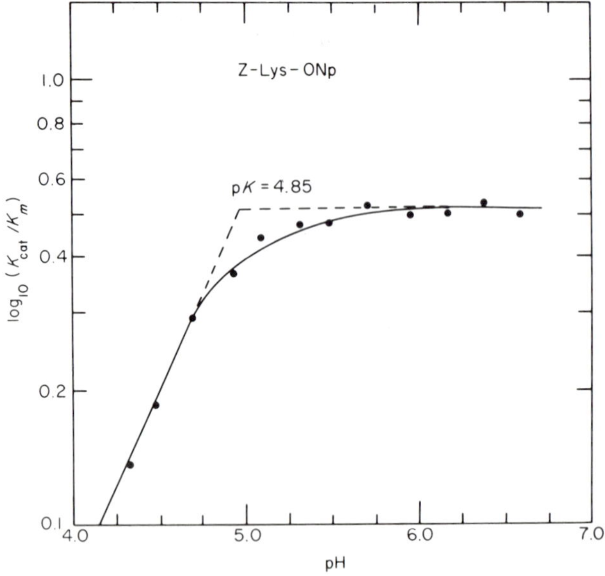

FIG. 6. pH-rate profiles for the streptococcal proteinase-catalyzed hydrolysis of benzyloxycarbonyl lysine·HCl *p*-nitrophenyl ester at 25°. Reaction conditions are those described in Table VII (*16*).

28. A. Kortt and T. Y. Liu, *Fed. Proc.* **29,** 3695 (1970).
29. L. Peller and R. A. Albertz, *JACS* **81,** 5907 (1959).

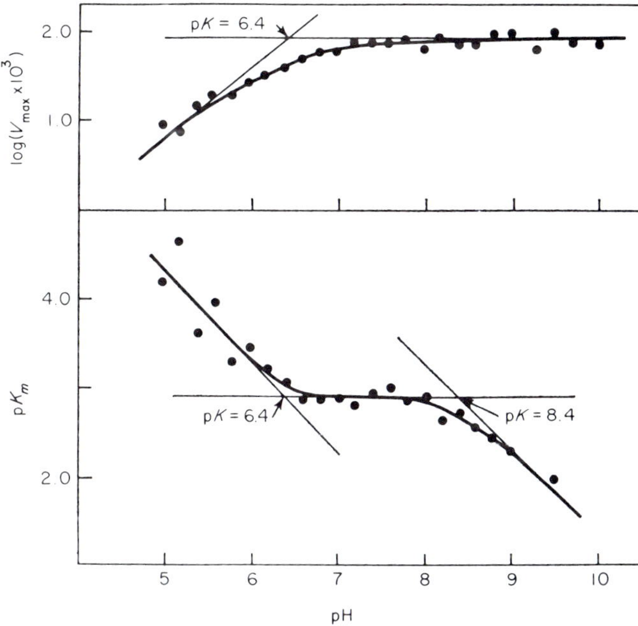

FIG. 7. pH-activity curve for streptococcal proteinase. Ordinate values are derived from values of V_{max}, expressed as millimoles of tyrosine per liter per minute, and from values of K_m, expressed on a molar basis. The substrate was benzyloxycarbonyl-L-phenylalanyl-L-tyrosine; enzyme concentration, 1 μM. Hydrolyses were performed at 37° in a wide range buffer which was 0.05 M each in boric acid, monosodium phosphate, and sodium acetate, and 1 mM in EDTA (disodium salt). At each pH, the ionic strength of the buffer was adjusted with NaCl to a final value of 0.188. Aliquots (200 μl) were removed at appropriate intervals for ninhydrin analysis (*24*).

The apparent discrepancy of 1.6 pK units (6.4 vs. 4.8) between the results of Gerwin *et al.* (*24*) and the studies of Liu *et al.* (*17*) on the pK of a catalytic entity in the streptococcal proteinase-catalyzed hydrolysis of peptide and ester substrates emphasizes the necessity of having independent chemical evidence for assigning a particular functional group at the active site of an enzyme. It should be noted that the conditions used by Gerwin *et al.* differ from Liu *et al.* in that the former used a peptide substrate at 37° whereas the latter used an ester substrate at 25°. On the basis of the kinetic data alone, one would hesitate to assign to an imidazole group a p$K = 4.8$. However, on the basis of the chemical evidence cited, an imidazole group may be present in the active site of streptococcal proteinase for the hydrolysis of benzyloxycarbonyllysine *p*-nitrophenyl ester. It is not possible at this stage to rule out the alternative that this group might actually be a carboxyl group. On the other hand, peptide and ester substrates of the type mentioned may not combine with the enzyme in the same manner, as shown by the different order of susceptibility of various peptide substrates and ester substrates to streptococcal proteinase.

The existence in the streptococcal proteinase of a group with a pK of around 8.4 was suggested by the studies of Gerwin *et al.* (*24*) as shown in Fig. 7 and that of Kortt and Liu (*28*). Using benzyloxcarbonyl-phenylalanyltyrosine as a substrate, Gerwin *et al.* noted in a graph of pK_m with respect to pH an inflection at pH 8.4. Kortt and Liu studied the effect of pH on the relative value for K_{cat} for benzoyloxycarbonyl-lysine phenyl ester and noted the pK value on the alkaline limb of the curve to be near pH 8.5. A plot of pK_m with respect to pH also indicated an inflection at pH 8.5. This inflection, which indicates the existence of a group with pK of around 8.5 probably reflects the ionization of the single –SH group in streptococcal proteinase. There is ample evidence that this –SH group is essential for activity (*6, 7*).

2. *Effect of Dielectric Constant*

Lowering of the dielectric constant has a profound effect on the enzymic cleavage of ester substrates by streptococcal proteinase (*16*). Results of studies of the effect of acetonitrile on the rate of hydrolysis of benzyloxycarbonylalanyl *p*-nitrophenyl ester by Kortt and Liu (*28*) are shown in Table VIII. The values of K_m appear to be unaffected by changes in dielectric constant. On the other hand, k_{cat} values decreases with decrease in the dielectric constant. Earlier Liu *et al.* (*16*) showed that the rate of hydrolysis of benzyloxycarbonyllysine *p*-nitro-

TABLE VIII
EFFECT OF ACETONITRILE ON PROTEINASE-CATALYZED HYDROLYSIS OF BENZYLOXYCARBONYLALANYL *p*-NITROPHENOLATE[a,b]

Acetonitrile (v/v) in the assay mixture (%)	k_{cat}(sec^{-1})	$K_m \times 10^{-4}$ M
3.13	75.6 ± 0.835	1.32 ± 0.223
8.00	40.7 ± 1.12	0.986 ± 0.052
12.8	23.4 ± 1.18	0.989 ± 0.092
17.6	12.9 ± 0.843	0.807 ± 0.108
21.4	8.02 ± 0.24	0.978 ± 0.055

[a] Experimental conditions were 25°; ionic strength 0.200; pH 5.50, enzyme concentration, 1.13×10^{-8} M, substrate concentration, S_0 = 0.353 to 1.41×10^{-4} M.
[b] From Kortt and Liu (*28*).

phenyl ester, a positively charged substrate, was retarded almost sixfold when 20% acetonitrile was present in the assay mixture. Further studies by Kortt and Liu indicate that for this cationic substrate both the values of K_m and k_{cat} are affected by changes in dielectric constant. The reaction between streptococcal proteinase and its substrates is probably more complex than the simple formula (*1*) (see Section VI) would indicate.

3. *Structural Inhibitors*

It has been observed that tosyllysylphenylalanine (Tos-Lys-Phe) and tosylnorleucylphenylalanine (Tos-Nle–Phe) are resistant to proteolytic cleavage by streptococcal proteinase (*16*). Moreover, Tos-Nle–Phe was found to be a competitive inhibitor of benzyloxycarbonylnorleucylphenylalanine (Z–Nle–Phe). It has a K_i value (4.2 mM) quite similar to the K_m value for Z–Nle–Phe. The fact that a relatively small change in the nature of A (Tos in place of Z) in a dipeptide A–X–Y can produce such a large effect on the cleavage of the X–Y bond supports the finding of Gerwin *et al.* (*24*) who noted the critical specificity of streptococcal proteinase with respect to A.

In this instance it is clear from the inhibition studies that the binding step is not the cause of the decreased rate of hydrolysis of Tos peptide by streptococcal proteinase since the K_i values were quite similar to the K_m values of the corresponding Z peptides.

On the other hand, the acylation by the Tos peptide proceeded too slowly to be detected. Two lines of evidence show that this step is the major cause of the lack of susceptibility of the Tos peptide to the proteinase. If deacylation had not occurred and acylation were appreciable, the acyl-enzyme intermediate would build up; in fact, no acyl-enzyme intermediate could be detected. Also, a rate determining deacylating step should be accelerated by the addition of nucleophiles such as hydroxylamine, but no such acceleration was found.

Gerwin *et al.* (*24*) showed that benzyloxycarbonyl glycylphenylalanine behaved kinetically as a competitive inhibitor with benzyloxycarbonylphenylalanyltyrosine (Z–Phe–Tyr) with a K_i value of 3.3 mM, almost equal to the K_m for Z–Phe–Tyr ($K_m = 1.6$ mM). Benzyloxycarbonylphenylalanylglycine, which does not bind strongly enough to the enzyme, is neither a good substrate nor a competitive inhibitor of the enzyme against Z–Phe–Tyr as a substrate. Z–Lys–OP behaved as a competitive inhibitor for hydrolysis of Z–Lys–OP or Z–Lys–ONp, an inhibition not of the competitive type was observed. A decrease of k_{cat} in the presence of inhibitor indicated that the inhibition may be noncompetitive or of a mixed type. These results suggest that the ester and anilide substrates both use the same catalytic cite on streptococcal proteinase for their hydrolysis but do so perhaps by different mechanisms.

4. *Esterase vs. Peptidase Activity*

Streptococcal proteinase belongs to a group of enzymes that exhibit a relatively high ratio of esterase to peptidase activity. Table IX lists ratios of esterase–peptidase activity for some representative proteoyltic enzymes (*16, 24 30–46*). Comparison of the rate of hydrolysis of an ester substrate with the corresponding amide or anilide indicates that proteolytic enzymes generally fall into two major categories. In one, they hydrolyze the ester substrates at a rate at least 1×10^3 times faster than the corresponding amide, and in the other they hydrolyze amide and ester substrates at comparable rates. Most of the "serine" enzymes such at trypsin, chymotrypsin, and subtilisin have been shown to belong to the first group, whereas the sulfhydryl enzymes such as papain and ficin belong to the second group. Streptococcal proteinase is unique in this respect since it is a sulfhydryl enzyme but belongs in the first group of enzymes. Comparison of the rate of hydrolysis of Bz–Arg–OEt and Bz–Arg–A catalyzed by papain and trypsin reveals that these enzymes hydrolyze the ester substrate Bz–Arg–OEt at about the same rate ($k_{cat} = 15.7$ sec^{-1} vs. 8.4 sec^{-1} for papain and trypsin), but the rate of hydrolysis of the amide substrate Bz–Arg–A differs considerably ($k_{cat} = 8.48$ sec^{-1} vs. 0.043 sec^{-1} for papain and trypsin). The observed difference between these two types of enzyme with respect to ester and amide bond cleavage results, therefore, from the lower efficiency of the trypsin-type enzymes for the hydrolysis of amide bonds

30. J. R. Whitaker and M. L. Bender, *JACS* **87,** 2728 (1965).
31. E. L. Smith and M. J. Parker, *JBC* **233,** 1387 (1958).
32. A. Stockell and E. L. Smith, *JBC* **227,** 1 (1957).
33. B. R. Hammond and H. Gutfreund, *BJ* **72,** 349 (1959).
34. T. Inagami and T. Murachi, *Biochemistry* **2,** 1439 (1963).
35. K. Inouye and J. S. Fruton, *JACS* **89,** 187 (1967).
36. T. G. Rajogopalan, W. H. Stein, and S. Moore, *JBC* **241,** 4295 (1966).
37. G. R. Delpierre and J. S. Fruton, *Proc. Natl. Acad. Sci. U. S.* **54,** 1161 (1965).
38. T. Inagami, *JBC* **235,** 1019 (1960).
39. S. A. Bernhard, *BJ* **59,** 506 (1955).
40. M. L. Bender and F. J. Kézdy, *Ann. Rev. Biochem.* **34,** 49 (1965).
41. E. Shaw, "Methods in Enzymology," Vol. 11, p. 667, 1967.
42. R. J. Foster and C. Niemann *JACS* **77,** 1886 (1955).
43. L. W. Cunningham and C. S. Brown, *JBC* **221,** 287 (1956).
44. A. N. Glazer, *JBC* **242,** 433 (1967).
45. E. Shaw and J. Ruscica, *JBC* **243,** 6312 (1968).
46. W. O. McClure and H. Neurath, *Biochemistry* **5,** 1425 (1966).

rather than from the higher efficiency of the papain-type enzymes for the hydrolysis of ester substrates.

All the serine enzymes mentioned have been shown also to have at least one histidine residue at the active site (*40, 41*), whereas for the sulfhydryl enzymes papain and ficin, in addition to the –SH group, a carboxyl group has been shown kinetically to participate in the mechanism (*30–33*). Streptococcal proteinase is a "histidine enzyme" like trypsin, chymotrypsin, and subtilisin, and at the same time it is an "–SH" enzyme like papain and ficin.

It is particularly noteworthy that pepsin hydrolyzes ester and amide substrates at a comparable rate. The presence in pepsin of at least one carboxyl group which is essential for its catalytic action is well documented (*36, 37*). It thus appears that for the proteolytic enzymes that utilize imidazole as a general base and the protonated form of the same imidazole as a general acid for their catalytic function, a high ratio of esterase to peptidase activity is observed. For the proteolytic enzymes that show comparable rates of hydrolysis of a specific ester and amide substrate, a carboxyl group has been proposed as the catalytic entity.

VI. Mechanism of Action

The high efficiency of the proteinase in catalyzing transferase activity gives indirect support for an acyl enzyme as an intermediate. With many ester substrates decomposition of the acyl enzymes may be slower than its formation. The acyl enzyme could, therefore, exist long enough for some attacking group such as aniline or an amino acid to compete successfully with water. In these circumstances, transpeptidation reactions would be expected to be catalyzed more effectively by an enzyme with a thiolester intermediate than by one such as chymotrypsin with an oxygen ester intermediate, for the former are more susceptible than the latter to attack by nitrogen nucleophile. Both classes of ester have similar susceptibilities to attack by oxygen nucleophile (*47*).

The overall reaction for the streptococcal proteinase-catalyzed hydrolysis of specific substrates may be represented by the equation:

$$E + S \underset{k_{-1}}{\overset{k_1}{\rightleftharpoons}} ES \overset{k_2}{\rightarrow} ES' + P_1 \overset{k_3}{\rightarrow} E + P_2 \tag{1}$$

where k_2 and k_3 are the catalytic rate constants for the acylation and deacylation steps, respectively, and P_1 and P_2 are products.

47. T. C. Bruice, *in* "Organic Sulfur Compounds" (N. Kharasch, ed.), Vol. I, p. 421. Pergamon Press, Oxford, 1961.

TABLE IX

COMPARISON OF SPECIFIC ESTER AND AMIDE SUBSTRATES HYDROLYZED BY SOME PROTEOLYTIC ENZYMES

Enzyme	Substrate[a]	k_{cat} (sec^{-1})	$\frac{[k_{cat}]_{ester}}{[k_{cat}]_{peptide}}$	k_{cat}/K_m (sec^{-1} mM^{-1})	$\frac{[k_{cat}/K_m]_{ester}}{[k_{cat}/K_m]_{peptide}}$	Proposed central catalytic entity
Streptococcal proteinase	Z-Lys-OP[b]	45.8	283	95.2	4.23×10^3	His[b]
	Z-Lys-AN[b]	0.162		0.0225		
Papain	BAEE[c]	15.7	1.85	10.8	4.12	—COOH[c,d]
	BAA[c]	8.48		2.62		
Ficin	BAEE[c]	2.5	1.15	0.1	2.18	—COOH[e]
	BAA[c]	2.2		0.046		
Bromelain	BAEE[f]	0.50	140	2.94×10^{-3}	1.02	
	BAA[f]	0.0035		2.92×10^{-3}		
Pepsin	Z-His-Phe(NO_2)-Pla-OMe[g]	77×10^{-2}	2.66	19.2	3.0	—COOH[h,i]
	Z-His-Phe-(NO_2)-OMe[g]	29×10^{-2}		6.3		
Trypsin	BAEE[j]	8.4	195	8.4×10^5	6.05×10^4	His[l,m]
	BAA[k]	0.043		13.9		
Trypsin	Z-Lys-OP[n]	82	280	1.13×10^3	9.42×10^3	His[l,m]
	Z-Lys-AN[n]	0.293		0.12		
Chymotrypsin	Ac-Phe-OMe[o]	52.5	1.62×10^3	42	2.75×10^5	His[l,m]
	Ac-Phe-A[o]	0.046		1.53×10^{-3}		

Subtilisin	AcTyrOEt[p]	$>1 \times 10^5$[p]	His[q]
	AcTyrA[p]		
Carboxypeptidase A	CiPla[r]	9.45×10^4[r]	
	Ci-Phe[r]		

[a] Abbreviations: Z, benzyloxycarbonyl; OP, phenyl ester; AN, anilide; OEt, ethyl ester; and A, amide.

[b] pH 5.5, 25°, Liu *et al.* (*16*), Gerwin *et al.* (*24*).

[c] pH 5.2, 25°, Whitaker and Bender (*30*).

[d] Smith and Parker (*31*), Stockell and Smith (*32*).

[e] pH 6.06, 25°, Hammond and Gutfreund (*33*), k_3 was used instead of k_{cat}.

[f] pH 6.0, 25°, Inagami and Murachi (*34*), k_3 was used instead of k_{cat}.

[g] Pla, β-phenyl-L-lactyl; OMe, methyl ester, pH 4.0, 37°; Inouye and Fruton (*35*).

[h] Rajagopalan *et al.* (*36*).

[i] Delpierre and Fruton (*37*).

[j] pH 8.0, 25°, Inagami (*38*).

[k] pH 7.6, 25°, Bernhard (*39*).

[l] Bender and Kézdy (*40*).

[m] Shaw (*41*).

[n] pH 7.6, 25°, 0.2 *M* *N*-ethylmorpholine acetate buffer in 0.2 *M* NaCl, 1×10^{-3} *M* $CaCl_2$. For Z-Lys-OP, E_0 was 7.39×10^{-3} *M*, S_0 was 0.5 to 5.5×10^{-4} *M*. For Z-Lys-An, E_0 was 1.67×10^{-6} *M*, S_0 was 1.8 to 2.9×10^{-4} *M*.

[o] pH 7.9, 25°, Foster and Niemann (*42*); pH 8.0, 25°, Cunningham and Brown (*43*) and Bender and Kézdy (*40*).

[p] pH 7.6, 40°, Glazer (*44*). Expressed as the relative rate of hydrolysis.

[q] Shaw and Ruscica (*45*).

[r] Ci-Pla, cinnamoylphenyllactate; Ci-Phe, cinnamoylphenylalanine; pH 7.5, 25°, McClure and Neurath (*46*). Values are relative, taking the initial rate of carboxypeptidase A with cinnamoylphenylalanine as unity.

The sulfhydryl group may be hydrogen-bonded to the basic imidazole group at the active site. The mechanism proposed for the streptococcal proteinase as shown in Fig. 8 is adapted from a reaction scheme postulated for chymotrypsin by Wang (*48*). In the figure, the general substrate is represented by RCOXR′. For the hydrolysis of phenyl and *p*-nitrophenyl esters, XR′ = OC_6H_5 or $OC_6H_4NO_2$, respectively, and for anilide and *p*-nitroanilide, XR′ = NHC_6H_5 or $NHC_6H_4NO_2$, respectively. This mechanism has been used to explain streptococcal proteinase-catalyzed hydrolysis of a number of esters and anilide and *p*-nitroanilide substrates. Since for chymotrypsin the acylation step is rate limiting for amide and anilide substrates (*49, 50*), it may be studied in detail by varying the leaving group $-P_1$ in Eq. (1). Thus by measuring the rate constants k_2, in Eq. (1), for a series of substituted anilides of *N*-acetyl-L-tyrosine, Inagami *et al.* (*51*) concluded that a proton transfer was involved in the acylation step. In view of this experimental result, as well as the observation that k_2 is relatively constant above pH 7, Parker and Wang (*52*) proposed a simple theory of chymotrypsin-catalyzed hydrolysis of peptide, amide, and anilide substrates. In this mechanism, a pre-transition state protonation of the substrate followed by the nucleophilic

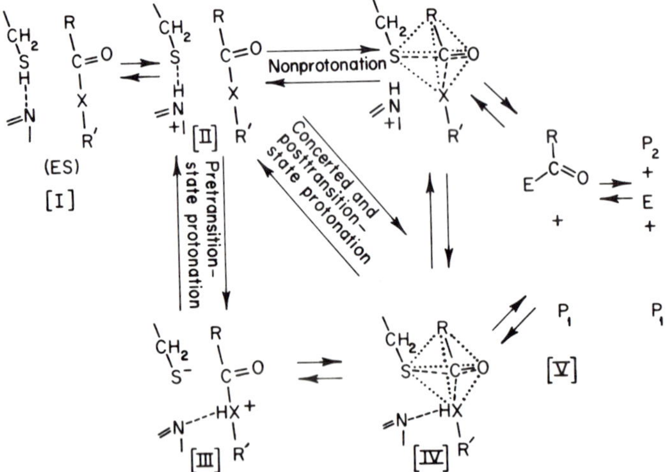

FIG. 8. General reaction scheme for catalysis by streptococcal proteinase.

48. J. H. Wang, *Science* **161**, 328 (1968).
49. B. Zerner and M. L. Bender, *JACS* **85**, 356 (1963).
50. T. Inagami and J. M. Sturtevant, *BBRC* **14**, 69 (1964).
51. T. Inagami, S. S. York, and A. Patchornik, *JACS* **87**, 126 (1965).
52. L. Parker and H. J. Wang, *JBC* **243**, 3729 (1968).

attack of the protonated substrate by the alkoxide group of serine is postulated.

For a mechanism as given in Fig. 8 the first-order rate constants k_c and k_a may be defined by the following equivalent reaction scheme:

$$[\mathrm{I}] \rightleftharpoons [\mathrm{II}] \rightleftharpoons [\mathrm{III}], \quad [\mathrm{II}] \rightarrow [\mathrm{V}], \quad [\mathrm{III}] \rightarrow [\mathrm{V}]$$

Concerted and posttransition state protonation ([I] ⇌ [II] → [V]); Pretransition state protonation ([III] → [V])

The following rate equation is obtained according to the reasoning given by Wang (*48*):

$$\begin{aligned} \text{Rate} &= k_c[\mathrm{II}] + k_a[\mathrm{III}] \\ &= \left[k_c \frac{K_{\mathrm{IMH}^+}}{K_{\mathrm{subH}^+}} + k_a \frac{K_{\mathrm{CysH}'}}{K_{\mathrm{subH}}} \right] [\mathrm{I}] \end{aligned} \tag{2}$$

where K_{CysH}', $K_{\mathrm{IMH}}{}^+$, and $K_{\mathrm{subH}}{}^+$ are acid dissociation constants of cysteine and histidine in streptococcal proteinase and of the protonated substrate, respectively.

Parker and Wang (*52*) showed that the overall acylation rate constant for chymotrypsin is given by

$$k_2 = \frac{K_{\mathrm{Ser}}}{K_{\mathrm{subH}^+}} \cdot k_a \tag{3}$$

where K_{Ser} and $K_{\mathrm{subH}}{}^+$ represent the acid dissociation constants of Ser 195 and the protonated substrate, resceptively; k_a is the first-order rate constant characteristic of the attack on the protonated substrate by the adjacent alkoxide group. For the streptococcal proteinase-catalyzed hydrolysis

$$k_2' = \frac{K_{\mathrm{CysH}'}}{K_{\mathrm{subH}^+}} \times k_a \tag{4}$$

since not only does the mercaptide ion have a nucleophilic action 10^2–10^3 times as fast as that of alkoxide ion, but at pH 5.5–7.6 the concentration of RS^- in streptococcal proteinase is higher than that of $\mathrm{R{-}O}^-$ in chymotrypsin by a factor of $(10^{-9})/(10^{-13})$ or $\sim 10^4$. Therefore,

$$k_2' \approx k_2 \times 10^4 \tag{5}$$

This implies that for "sulfhydryl" enzymes such as streptococcal proteinase the contribution of $K_{\mathrm{CysH}}'/K_{\mathrm{subH}}{}^+$ to the overall rate of acylation according to Eq. (2) is significantly larger than that of $K_{\mathrm{Ser}}/K_{\mathrm{subH}}{}^+$ to the rate of the "serine" enzymes such as chymotrypsin or trypsin. Thus, for streptococcal proteinase-catalyzed hydrolysis of peptide, amide, or anilide, the pretransition state protonated path $[\mathrm{I}] \rightleftharpoons [\mathrm{III}] \rightarrow [\mathrm{V}]$ will be more important than the concerted and post-transition state protonation [I] ⇄ [II] → [V]. The data shown in Table X for the relative rates of hydrolysis of anilide and *p*-nitroanilide substrates can best be explained by this mechanism. The unusual feature of these data shown in Table X is the considerable decrease in the rate of hydrolysis of the *p*-nitroanilide (NpA) substrate compared with the anilide substrate. For, if acylation is the rate limiting step for these substrates, and the rate of acylation is enhanced by an electron withdrawing group, a nitroanilide substrate should be more susceptible to enzymic hydrolysis than the anilide substrate. The finding that Z–Lys–NpA is hydrolyzed 36-fold more slowly than Z–Lys–AN by streptococcal proteinase is, however, in accord with the supposition that for streptococcal proteinase catalyzed hydrolysis of peptide, amide, or anilide the pretransition state protonated path [I] ⇄ [III] → [V] is more important than the concerted and post-transition state path [I] ⇄ [II] → [V]. Since the nitrogen in the susceptible C–N bond of Z–Lys–NpA is a much weaker base than the nitrogen in the susceptible C–N bond of Z–Lys–AN, protonation of nitroanilide substrate is more difficult than that of anilide substrate, hence, the decrease in the rate of hydrolysis on the nitroanilide substrate by the enzyme.

Bundy and Moore (*53*) have studied the chymotrypsin-catalyzed

TABLE X
COMPARISON OF ESTERASE AND PEPTIDASE ACTIVITIES OF STREPTOCOCCAL PROTEINASE[a]

Substrate[b]	pH	K_m (mM)	k_{cat} (sec^{-1})	k_{cat}/K_m (mM^{-1} sec^{-1})	Relative reactivity
Z-Lys-OP[c]	5.5	0.826	80.8	97.6	1
	7.6	0.483	45.8	95.2	0.98
Z-Lys-ONp[d]	5.5	0.314	148	472	4.83
Z-Lys-AN[e]	5.5	7.19	1.62×10^{-1}	2.25×10^{-2}	2.31×10^{-2}
	7.6	4.53	1.03×10^{-1}	2.26×10^{-2}	2.32×10^{-2}
Z-Lys-NpA[f]	5.5	4.28	2.67×10^{-3}	6.23×10^{-4}	6.37×10^{-6}
	7.6	3.35	1.91×10^{-3}	5.88×10^{-4}	6.01×10^{-6}

[a] From Liu *et al.* (*16*).

[b] Abbreviations: Z, benzyloxycarbonyl; OP, phenoxy; ONp, *p*-nitrophenoxy; AN, anilide; and NpA, *p*-nitroanilide.

[c] Enzyme concentration, 3.75×10^{-7} M; $S_0 = 1.07$ to 8.37×10^{-4} M.

[d] Enzyme concentration, 1.25×10^{-7} M; $S_0 = 2.14 \times 10^{-5}$ to 1.33×10^{-4} M.

[e] Enzyme concentration, 1.25×10^{-6} M; $S_0 = 2.0 \times 10^{-4}$ to 2×10^{-3} M.

[f] Enzyme concentration, 8.9×10^{-6} M; $S_0 = 2.0 \times 10^{-4}$ to 2×10^{-3} M.

53. H. F. Bundy and C. L. Moore, *Biochemistry* **5**, 808 (1966).

hydrolysis of the anilide and nitroanilides of *N*-benzoyltyrosine and reported that the rate of acylation is enhanced by an electron-withdrawing group. This suggests that the two substrates are hydrolyzed by different mechanisms. For the hydrolysis of benzoyltyrosine *p*-nitroanilide catalyzed by chymotrypsin, the concerted posttransition state path [I] $\rightleftarrows$ [II] $\rightarrow$ [V] dominates. In this mechanism, the effect of the electron-withdrawing *p*-NO_2 group is to weaken the susceptible C–N bond and enhance the rate of acylation. For the hydrolysis of the anilide substrate catalyzed by chymotrypsin, the pretransition state protonate path [I] $\rightleftarrows$ [III] $\rightarrow$ [V] is the major mechanism. Thus, the hydrolysis of the *p*-nitroanilide by chymotrypsin proceeds by a mechanism similar to that of the esters. The observed difference between streptococcal proteinase, a sulfhydryl enzyme and chymotrypsin, a hydroxyl enzyme in the relative rate of hydrolysis of anilide, and *p*-nitroanilide substrates is therefore in agreement with the mechanism predicted from Eqs. (2)–(5).

The kinetic constants listed in Table X for ester substrates of the type Z–X–Y show that the change of Y from phenyl to *p*-nitrophenyl produces relatively little change in K_m, but k_{cat}/K_m increases fivefold. Since K_m is related to the enzyme–substrate complex dissociation constant, it can be concluded that the binding of the substrate Z–X–Y to the enzyme is not affected appreciably by the presence or absence of the nitro group on Y. The fact that a *p*-nitrophenyl ester is hydrolyzed nearly five times as fast as the phenyl ester can be attributed to the powerful electron-withdrawing property of the nitro group, which enhances the rate of acylation of the enzyme by these substrates. The results are consistent with the supposition that for the hydrolysis of the *p*-nitrophenyl ester the rate limiting step is presumably the deacylation step (*29, 30*), and that for the acylation step in the streptococcal proteinase catalyzed hydrolysis of both phenyl and *p*-nitrophenyl esters the direct transformation of [II] to [IV] through the concerted and posttransition state protonation mechanism illustrated in Fig. 8 becomes more important.

In the light of the mechanism shown in Fig. 8, further examination of the data shown in Table IX is instructive. For the pair of ester and amide substrates hydrolyzed by each enzyme k_3, the rate constant for deacylation, will be the same since both the amide and the ester substrate will form the same acyl-enzyme intermediate. The overall rate constant k_{cat} is dependent on both k_2 and k_3 as shown in Eq. (6):

$$k_{cat} = \frac{V_{max}}{(E)} = \frac{k_2 + k_3}{k_2 k_3} \tag{6}$$

and since,

$$[k_3]_{ester} = [k_3]_{amide} \tag{7}$$

therefore,

$$\frac{[k_{cat}]_{ester}}{[k_{cat}]_{amide}} \approx \frac{[k_2]_{ester}}{[k_2]_{amide}} \tag{8}$$

It is evident from Eq. (8) that for the pair of ester and amide substrates hydrolyzed by each enzyme the difference in the overall rate constant k_{cat} is dependent only on the rate constant k_2, the acylation step. The unusual facts are that for some enzymes such as papain and ficin the k_2 value for an amide substrate is quite similar to that for the corresponding ester substrate. Carboxylic acid esters are known to be much more susceptible than the amide to hydrolysis catalyzed by nucleophiles (such as hydroxide ion) by a factor of the order of one thousand. Consequently, it is surprising that there is such a small difference in k_2 for the papain- and ficin-catalyzed hydrolysis of Bz–Arg–OEt and Bz–Arg–A. Whitaker and Bender (*30*) postulated either that there is a substantial electrophilic component involved in the acylation or else the sulfhydryl group present in the active site of papain and ficin reacts equally effectively with carboxylic acid ester and amide.

The data given in Table IX show, that pepsin, a nonsulfhydryl enzyme, has a similar low $[k_2]_{ester}/[k_2]_{peptide}$ ratio as do papain and ficin, while other sulfhydryl enzymes such as streptococcal proteinase and bromelin have a high ratio of $[k_2]_{ester}/[k_2]_{amide}$. Proteolytic enzymes such as trypsin, chymotrypsin, and subtilisin also show a high ratio of $[k_2]_{ester}/[k_2]_{amide}$. For trypsin and chymotrypsin all the evidence available to date indicates a histidine residue to be the catalytic entity of these enzymes (*40, 41, 45*). For pepsin, at least one carboxyl residue has been implicated to function as the catalytic entity (*36, 37*). If amide and peptides bonds are cleaved preferentially by the pretransition state protonation path [I] $\rightleftarrows$ [III] $\rightarrow$ [V] as shown in Fig. 8, these data would suggest that the histidine–cysteine system of streptococcal proteinase, and the histidine–serine system of trypsin and chymotrypsin are less efficient in carrying out the protonation of the nitrogen of these substrates during the acylation step [I] $\rightleftarrows$ [II] $\rightleftarrows$ [III] than the carboxyl–X (where X = other functional group whose nature is as yet unknown) system of pepsin or the carboxyl–cysteine system of papain and ficin. The fact that ester substrate such as Bz–Arg–OEt are hydrolyzed by trypsin, papain, ficin, and bromelin at a comparable rate is in accord with the supposition that for the ester substrates, because or their much lower basicity and higher reactivity of the susceptible C–O bond, the acylation step proceeds by the concerted and post-transition state protonation path (I) $\rightleftarrows$ (II) $\rightleftarrows$ (V).

Alternatively, the data shown in Table IX could be taken to indicate, as Whitaker and Bender have suggested, that an electrophilic component in papain, ficin, and pepsin equalizes the two rates of nucleophilic attack on the ester and the corresponding amide or peptide substrate and that this effect is lacking in streptococcal proteinase, bromelin, trypsin, chymotrypsin, and subtilisin. The electrophile suggested to be at the active site of papain, ficin, and pepsin could come from a positive group such as a lysine or an arginine residue, or from an imidazole group of a histidine residue. It could also be a negative group of an aspartic or a glutamic acid residue. It is not inconceivable that this electrophile is assisting in the protonation of the amide or peptide substrate in some way at the pretransition state to facilitate the acylation of the substrate by the enzymes. For papain, X-ray crystallographic studies (*54*) [as well as chemical (*55*) and kinetic (*30–32*) studies] have revealed the presence of a histidine as well as an aspartyl residue to be within 5 Å from the active site sulfhydryl group of cysteine. For pepsin, a number of investigations (*40, 55, 56*) have implicated the presence of several carboxyl groups to be at the active site of this enzyme.

54. S. S. Husain and G. Lowe, *Chem. Commun.* p. 310 (1968).
55. T. R. Hollands and J. S. Fruton, *Biochemistry* **1,** 2045 (1968).
56. G. E. Clement and S. L. Snyder, *JACS* **88,** 5338 (1966).

18

The Collagenases

SAM SEIFTER • ELVIN HARPER

I. Introduction

To qualify as a collagenase an enzyme must be capable of causing hydrolytic cleavage of molecules of collagen in their native conformation. As will be discussed later, a further restriction in the definition is that cleavage must occur in the domains of the substrate pervaded by the collagen helix. This qualification is made necessary by the fact that native molecules of collagen have regions of their polypeptide chains that remain outside the collagen fold, and these are amenable to scission by general proteases such as trypsin or chymotrypsin. Accordingly, a discussion of collagenases must be prefaced with some consideration of the nature of the substrate.

A. Nature of the Collagen Substrate

Collagens from a variety of sources, vertebrate and invertebrate, have been studied as substrates for collagenases. All are made up of polypeptide chains that as a consequence of special amino acid composition and sequence assume a characteristic helical structure for which a prototypic model is that of poly-L-proline (form II) (*1, 2*). The responsible composition is the occurrence, in those regions bearing the characteristic helix, of a glycine residue as every third residue, and a content of proline plus hydroxyproline residues that approximates 20% or more of the total residues. Relative to the residues of glycine, those of the imino acids are spaced so that triad sequences of –Gly–Pro–X– and –Gly–Pro–Hyp– occur with sufficiently great frequency to cause the characteristic helix to pervade the major part of the polypeptide chain, even to be "driven" into regions where these sequences do not dominate (X merely represents any of many kinds of amino acids but is Ala about one-fourth of the repeats). Along each polypeptide chain these triads repeat in groups, one group of these sequences then being followed by a region in which Gly is generally still in the first position of a triad but the second and third positions are held by residues other than those of imino acids, often residues of polar amino acids.

In each collagen molecule, often called the tropocollagen macromolecule, there are three parallel polypeptide chains, each wound in a left-handed poly-L-proline type of helix. The individual chains are called $\alpha 1$, $\alpha 2$, and again $\alpha 1$ (*3*). In some tissues, collagen occurs in which the second $\alpha 1$ chain appears to be replaced by a chain with somewhat different composition, called $\alpha 3$. Although $\alpha 1$ has distinct compositional differences from the other two chains, all three would appear to have a large degree of homology, and certainly bear the types of repeating sequences described above. The terminal regions of each chain, in particular the amino-terminal regions as described, constitute a segment of

1. G. N. Ramachandran, *in* "Treatise on Collagen" (G. N. Ramachandran, ed.), Vol. 1, p. 103. Academic Press, New York, 1967.
2. P. H. von Hippel, *in* "Treatise on Collagen" (G. N. Ramachandran, ed.), Vol. 1, p. 253. Academic Press, New York, 1967.
3. K. A. Piez, *in* "Treatise on Collagen" (G. N. Ramachandran, ed.), Vol. 1, p. 207. Academic Press, New York, 1967.

about 15 residues that do not have the characteristic repeating sequences and indeed have a composition resembling more that of a globular protein. Precisely these domains remain out of the collagen helix and in fact are susceptible to the actions of general proteases. Each of the α chains is stabilized in its conformation by the pyrrolidine side chains of the imino acid residues and the unique distribution of glycine.

The three chains are then wound together to form a triple helical structure, much as the strands of a rope. The superstructure in turn has a characteristic supertwist of opposite disposition (right-handed). This gives the whole structure a coiled coil conformation. Stabilization of this structure is in large measure the result of interchain hydrogen bonding. There is some controversy concerning the number of hydrogen bonds per triad that permit hydrogen bonding, the number in any case being one or two (*1*). The association of the three chains in this manner apparently also gives further stabilization to the helices of the individual chains, even in those regions not dominated by –Gly–Pro–X– and –Gly–Pro–Hyp– sequences. The whole structure, then, constitutes a fairly tight, stable arrangement that at the same time can accommodate bulky side chains of residues of amino acids other than glycine.

The above sketch describes most of the collagens that have been studied, particularly those found in the connective tissues of vertebrates. There are special collagenous structures, both among vertebrates and invertebrates, in which the collagen may have differences in composition and superstructure from those just described. Thus, basement membranes in general contain a protein with many of the compositional features of collagen but with notable differences. Yet these are capable of digestion by the clostridial collagenases. The cuticular collagens of the earthworm and of *Ascaris* have unique structural features, but both are digestible by collagenases.

A characteristic of collagens is the relative ease with which they are denatured by mild heating and by certain salts. The processes involved are discussed with great clarity by von Hippel (*2*). For certain acid-extracted collagens such as that of the carp swim bladder, ichthyocol, the collagen→gelatin transition occurs at about 30°, the exact value depending on the nature and concentration of the salt present in the solution. Denaturation under mild conditions results in the formation of "parent gelatin," i.e., a gelatin in which there has been no scission of peptide bonds. The process consists of the unwinding of the triple-stranded helix and the disorganization of the poly-L-proline type of structure in each chain. The individual kinds of α chains can be isolated from parent gelatin, and it is found that both $\alpha 1$ and $\alpha 2$ chains have molecular weights of about 95,000. Parent gelatin also contains dimers of $\alpha 1$-$\alpha 1$ (β_{11}) and $\alpha 1$-$\alpha 2$ (β_{12}) in which linkage is through nonpeptidic covalent bonds. Trimers, known as γ components, are also found in certain parent gelatins; these consist of α chains joined by intramolecular cross-links. A detailed discussion of this subject is available in a review by Piez (*3*).

The cross-linkages just alluded to may be of several kinds, but only one has been identified with some certainty. After biosynthesis, the collagen molecule appears to undergo a process of "maturation" and chemical modification. In one of these transformations, certain residues of lysine, within five or so residue positions from the amino-terminal ends of α chains, undergo oxidative deamination to residues of α-aminoadipic acid δ-semialdehyde. Residues of the aldehyde on adjoining chains then undergo aldol condensation to form a covalent cross-link. In the special case of *Ascaris* cuticular collagen, cross-linking is by disulfide bridges. At this point in the development of the subject, no collagenase is known to catalyze the scission of any cross-link in collagen.

Certain physical characteristics of solutions of collagen are of great importance in the investigation of collagenases. The characteristic helical forms of the collagen molecule endow it with special optical rotatory properties. Thus the acid-extracted collagens of calfskin and of carp swim bladder have high negative specific rotations, at the D line, in the vicinity of —400°. Denaturation with the formation of parent gelatin results in a decrease of the negative specific rotation to about —100°. Another useful property of solutions of collagens is their high intrinsic viscosity. Acid-extracted collagens may have values for intrinsic viscosity of approximately 16 dl/g, whereas the parent gelatins have values of about 0.5 dl/g. One can readily infer that the actions of collagenases will cause changes in molecular weight, viscosity, and perhaps specific optical rotation. For the latter to occur, enzymic cleavage must impose an instability on the fragments of collagen formed thereby causing their unfolding.

B. Definition of Collagenases

In terms of the above features of structure of collagen, a collagenase is defined as an enzyme capable of causing hydrolytic scission of peptide bonds located in the characteristic poly-L-proline type of helical regions when the substrate is in the undenatured state. This definition permits inclusion of enzymes acting in these regions and perhaps making only one cut per α chain, as in the case of some tissue collagenases, and other enzymes acting in the same regions but making multiple scissions, as in the case of clostridial collagenases. The definition does not require that the enzyme cleave bonds in the repeating "nonpolar," "crystalline" regions of the α chains responsible for the typical collagen structure. In fact it allows inclusion of an enzyme, if such is discovered, that acts on the more "polar" regions located in the helical domains. So far too little is known of the specificities of the many collagenases described to state whether these indeed act only in nonpolar regions. On the contrary side of this requirement, a peptidase that cleaves bonds in a synthetic oligopeptide containing sequences identical with those in a nonpolar region of collagen cannot be classified as a collagenase unless it also acts on undenatured collagen (*4*, *5*). Furthermore, a general protease such as trypsin that readily digests parent gelatin extensively cannot be considered a collagenase since it does not attack the same bonds held in the context of the typical helical regions of the undenatured substrate. Yet many general proteases, as already mentioned, can conduct limited proteolysis of undenatured collagen, but only in the terminal regions of α chains outside the helical domains.

In summary, then, enzymes do not qualify as collagenases if they digest native collagen in terminal regions of the chains outside the collagen fold, if they digest denatured collagen, or if they cleave sequences in oligopeptides resembling sequences found in the helical regions of α chains. They do qualify, however, if in addition to one or all of the above they cleave sequences in the typical helical regions when the collagen is undenatured. Great pains have been taken in elaboration of this definition, not for the sake of semantic purity but to make meaningful an ever-expanding class of very unusual proteases. This group of proteases seemingly has evolved to deal with a highly specialized protein structure in its native (undenatured) state. An examination of the inferred functions of the known collagenases accordingly should be instructive.

C. The Known Collagenases and Their Inferred Functions

At present, approximately 20 enzymes have been described that classify as collagenases. These, together with some of their properties, are listed in Table I (see p. 655). In most cases little more is known about each of these enzymes other than its occurrence, extent of action on a collagen substrate, and mode of activation. Only a few of these have been

4. E. Harper and J. Gross, *BBA* **198,** 286 (1970).
5. H.-G. Heidrich, D. Prokopová, and K. Hannig, *Z. Physiol. Chem.* **350,** 1430 (1969).

prepared in purified form, and of these only one or two in sufficient amount to allow fuller characterization.

Inspection of Table I reveals that broadly there are two groups of collagenases. One group is elaborated by microorganisms that in themselves do not contain collagen. A second group of collagenases, the so-called tissue collagenases, are produced by multicellular organisms that in fact have collagen as a major extracellular component of their tissues. The eventuality that a third group of collagenases will be recognized as serving strictly nutritional (digestive) functions in multicellular organisms is presaged by the discovery of an extractable collagenase of the hepatopancreas of the crab (*6*). There are no rigorous demonstrations of mammalian pancreatic collagenases, although one should anticipate that such enzymes, homologous with the known pancreatic proteases, should have evolved. These might be expected to occur in carnivores, and to be DFP-sensitive.

One can infer two sorts of functions for the collagenases of microorganisms, those associated with mechanisms of invasion of a host and those that are nutritional. An example of the first would seem to be the collagenases of *Clostridia*. These enzymes are capable of weakening and destroying the connective tissue barriers of a host. Not infrequently the organisms fabricate elastases and polysaccharidases to facilitate this task and even produce phospholipases for the additional function of disorganizing cell membranes. Most of the collagenases of this sort are truly extracellular, i.e., they are secreted and do not appear in the growth media merely as the result of cell death and lysis. At least one microbial collagenase, however, is found in association with a particle of the elaborating cell and is not secreted. Only recently described (*7, 8*), the collagenase of *Bacteroides melaninogenicus* appears to be responsible for the destruction of fibers encountered in human chronic periodontal disease. The organism is found in the gingival crevices where most probably a secreted collagenase would be washed away in a flow of saliva. The advantage for such an organism to have evolved with a cell-bound collagenase would appear real.

The second function of collagenases of microorganisms could be digestive in the nutritional sense. The collagen of the host would be cleaved to smaller peptides by the collagenase, and associated peptidases could act to provide amino acids for nutritional purposes. The peptidases in fact do occur, lending validity to this proposed function.

That some multicellular organisms elaborate collagenases with the

6. A. Z. Eisen and J. J. Jeffrey, *BBA* **191,** 517 (1969).
7. R. G. Gibbons and J. B. MacDonald, *J. Bacteriol.* **81,** 614 (1961).
8. E. Hausmann and E. Kaufman, *BBA* **194,** 612 (1969).

TABLE I
KNOWN COLLAGENASES AND SOME OF THEIR PROPERTIES[a, d]

Source	Enzyme[b]	Substrates and assay	Products	Activators	Inhibited by[c]	Miscellaneous[c]	Ref.
1. *Clostridium histolyticum* (extracellular)	Collagenase A (clostridiopeptidase A—EC 3.4.4.19)	Collagen (decrease of viscosity of solution). Gelatin (ninhydrin reactivity). Many synthetic peptides (see text and Table IV)	With collagen or gelatin, small peptides + core. Average molecular number of 500 and $\overline{MW}$ of 800. Peptides of N-terminal Gly residues	Ca^{2+}. Also has intrinsic metal, Zn probably. Co^{2+}, Mn^{2+} and Fe^{2+} may activate	EDTA, cysteine; see Table VI	MW 105,000. Can yield inactive subunits and perhaps active ones. Not inhibited by DFP	*6, 36, 39, 40a,b, 42, 43, 46, 93*
2. *Clostridium histolyticum* (extracellular)	Collagenase B (clostridiopeptidase B)	Same as for collagenase A	Same as for collagenase A	Ca^{2+}, Zn probably intrinsic	EDTA; cysteine	MW 57,400	*36, 42*
3. *Clostridium histolyticum* (extracellular)	Collagenase A-α	Digests insoluble collagen weakly				Probably same as collagenase A	*40c*
4. *Clostridium histolyticum* (extracellular)	Collagenases B-α and B-β	Neither digests insoluble collagen. Added to A-α, digests strongly				Probably related to B and C enzymes of Grant and Alburn (*36*)	*40c*
5. *Clostridium histolyticum* (extracellular)	"Pseudocollagenase"	Digests gelatin but not collagen. Digests Z-Gly-Pro-Gly-Gly-Pro-Ala-OH. Test by ninhydrin reactivity					*40d*
6. *Mycobacterium tuberculosis* (extracellular)	Two separable collagenases	Collagen (by viscosity). Z-Gly-Pro-Leu-Gly-Pro-OH (by ninhydrin)	Soluble peptides	Ca^{2+} used in assay		MW estimated at 77,000 (A). pH optimum, 6–8	*40e*
7. *Pseudomonas aeruginosa* (extracellular)	Two separable collagenases	Azocoll (treated collagen that yields soluble peptides with enzyme). Z-Gly-Pro-Gly-	Soluble peptides			Suggestion that there may be subunits	*40f*

TABLE I (*continued*)

Source	Enzyme[b]	Substrates and assay	Products	Activators	Inhibited by[c]	Miscellaneous[c]	Ref.
		Gly-Pro-Ala-OH (ninhydrin). Not tested against untreated collagen					
8. *Bacteroides melaninogenicus* (particulate)	Collagenase	Collagen (by viscosity)	Disc electrophoresis shows few bands. More limited action than clostridiopeptidase A	Ca^{2+} not used in assay. Cysteine; dithiothreitol	EDTA and H_2O_2		*7, 8*
9. *Streptomyces madurae* (extracellular)	Collagenase	Collagen (viscosity). Azocoll; peptides similar to those used for clostridiopeptidase A (ninhydrin)	Products similar to those obtained with clostridiopeptidase A	Ca^{2+}	EDTA (reversed by Ca^{2+}). Cysteine; urea. Not by PMB or iodoacetate	MW 35,000. $s_{20,w}$ = 3.285 S $D_{20,w}$ = 8.1 × 10^{-7} cm^2 sec^{-1}. (¯) = 0.731 cm^3 g^{-1}	*40g*
10. *Trichophyton schoenleinii* (extracellular)	Collagenase	Collagen (viscosity). Azocoll (soluble peptides)	Products similar to those with clostridiopeptidase A. Specificity similar	Ca^{2+} used in assay but does not reverse inhibition by EDTA	EDTA; cysteine; urea. Not by PMB or iodoacetate	MW 20,100. ($\bar{v}$) = 0.689 cm^3 g^{-1}. pH optimum = 6.5	*40h*
11. *Aspergillus oryzae* (extracellular)	Collagenase	Collagen (decrease in optical rotation). Gelatin; synthetic peptides; denatured hemoglobin (ninhydrin). *N*-acetyltyrosine ethyl ester			DFP. Not by cysteine, KCN, PMB, NEM	For hemoglobin, pH optimum is 6.0–6.5. For others, 9–10. MW 20,000. Amino acid analysis given. May contain a chymotrypsin-like enzyme	*40i*
12. Tadpole, *Rana* (*catesbiana*) (cell culture)	Collagenase	Collagen (viscosity; release of labeled peptides from ^{14}C-labeled collagen). Gelatin	From collagen, two triple-stranded fragments, $\frac{1}{4}$ and $\frac{3}{4}$ of molecular weight. See text and Fig. 1. From gelatin, smaller peptides	Ca^{2+}	EDTA (reversed by Ca^{2+}); cysteine; serum	Optimum pH 8–9	*4, 40j, 87–91*

13. Human skin (cell culture)	Collagenase	Collagen (viscosity). Gel inhibition	Same as for tadpole enzyme	Ca^{2+}	EDTA; cysteine; human serum	Optimum pH, 7–8	*23*
14. Rat skin and fibroblasts (*in situ* and culture extracts)	Collagenase	Skin collagen *in situ* (disappearance of collagen used as assay). Insoluble collagen digestion *in vitro*		*In vivo*, cortisone		*In vitro* assay done at pH 5.5	*40k*
15. Guinea pig or rabbit carrageenin granulomas (tissue culture)	Collagenase	Collagen *in situ* and *in vitro*. Collagen gel assay	By electrophoresis on acrylamide gel, two bands observed			Active at physiological pH	*40l*
16. Bones from man, goats, rats (tissue culture)	Collagenases	Collagen (viscosity). Collagen gel inhibition assay		Ca^{2+}	EDTA; cysteine	pH optimum, 7–9	*40m*
17. Rat uterus, postpartum resorbing (tissue culture)	Collagenase	Collagen (viscosity). Gelatin	With collagen, gives fragments like those given by tadpole enzyme. But above 37° gives dialyzable peptides	Ca^{2+}	EDTA. Not inhibited by cysteine	Optimum pH, 7–7.4. EDTA inhibition not restored by Ca^{2+}. Perhaps contains a second metal	*24*
18. Human rheumatoid synovium (tissue culture)	Collagenase	Collagen (viscosity). ^{14}C-peptide release from gels	Same as for tadpole enzyme	Probably Ca^{2+}	EDTA; human serum	Optimum pH, 7.6	*96–98*
19. Human rheumatoid synovial fluid (direct)	Collagenase A	Collagen (viscosity). ^{14}C-peptides from collagen gel	Same as for tadpole enzyme		Very weakly inhibited (if at all) by human serum	MW estimated 40,000–50,000	*69*
20. Human rheumatoid synovial fluid (direct)	Collagenase B	Same as for synovial collagenase A	Same as for tadpole enzyme		Strongly inhibited by human serum	MW estimated at 20,000–25,000. Seems to resemble the collagenase of synovium	*69*

TABLE I (*continued*)

Source	Enzyme[b]	Substrates and assay	Products	Activators	Inhibited by[c]	Miscellaneous[c]	Ref.
21. Human granulocytic leukocytes (culture and direct extraction)	Collagenase	Collagen (viscosity). ^{14}C-Peptides from collagen gel	Same as for tadpole enzyme	Ca^{2+}	EDTA; cysteine and reduced glutathione. Not inhibited by serum	Does not act on fibers	*94*
22. Miscellaneous tissues in culture (diseased human skin; edges of healing wounds; inflamed human gingiva; regenerating newt limbs)	Collagenases	Collagen (viscosity). Collagen gel inhibition	Many yield products like those of tadpole enzyme				*40n–r*
23. Mouse fibroblasts; HeLa cells; culture and direct extraction	Collagenases or peptidases	Assayed by using peptides known to be split by clostridiopeptidase A					*11, 40s*
24. Hepatopancreas of the crab, *Uca pugilator* (direct extraction)	Collagenase	Collagen (viscosity)	Same as for tadpole collagenase		DFP. Phenylmethylsulfonyl fluoride	Not inhibited by EDTA or cysteine. Seems to be serine hydrolase	*6*
25. Larvae, warble fly, *Hypoderma bovis*	Collagenase	Collagen *in situ*					*9*
26. Larvae, blowfly, *Lucilia sericata*	Collagenase	Collagen *in situ*					*10*

[a] The enzymic activities are arranged to go from microorganisms to multicellular organisms. Within the multicellular organisms, the enzymes are arranged for the most part as groups that can only be obtained after surviving cell culture and those that can be obtained by direct extraction of a tissue or fluid. The hepatopancreatic digestive collagenase is set off toward the end of the table. In terms of functions, then, the general order is a group of enzymes used for purposes of invasivity and/or nutrition by microorganisms, another group used by multicellular organisms for tissue remodeling, repair, or removal, and the one hepatopancreatic digestive enzyme. The collagenases of the larval flies have not been studied sufficiently, but their functions would be largely nutritional.

[b] Some of the collagenases listed have not been tested with native collagen as a substrate, and there must be some doubt yet as whether they should be classified here. Some may be peptidases, and some may be general proteases (No. 5 is frankly called a *pseudocollagenase* because it does not digest native collagen; No. 11 digests hemoglobin as well as collagen; No. 23 may be peptidase in nature).

[c] Abbreviations: EDTA, ethylenediaminetetraacetate; DFP, diisopropylphosphorofluoridate; PMB, *p*-mercuribenzoate; and NEM, *N*-ethylmaleimide.

[d] References *40a–w* are found on p. 666.

functions just described is suggested by occurrence of such enzymes in the larvae of the warble fly, *Hypoderma bovis* (*9*), and blowfly, *Lucilia sericata* (*10*), respectively.

The action of most tissue collagenases, however, appears to be more limited and directed to different ends. The collagenases such as that of the resorbing tail of the tadpole or the involuting postpartum uterus are exquisitely controlled. They function in a program of remodeling of specialized tissues at a particular time of development or physiological expression. Other tissue collagenases are associated with a mechanism of repair of tissues, removing injured collagen and perhaps promoting regeneration or wound healing in an orderly manner. Whether there are tissue collagenases involved with invasiveness of tumor cells, as has been implied [see, for instance, Strauch (*11*)], is yet to be firmly established. One may recall that hyaluronidases also have been ascribed this "function," and of interest is the fact that frequently cells elaborating collagenases also produce hyaluronidases. That tumor cells have peptidases capable of digesting peptides derived from a preliminary destruction of collagen would appear to be unquestionable.

All tissue collagenases, whether part of a system of general turnover of collagen or part of a mechanism of remodeling or repair of tissues, may provide secondarily a means of conservation of amino acids and in this sense serve a nutritional function as well.

The overall action of collagenases is to initiate disruption or destruction of the native collagen molecule. The disrupted molecule may suffer a decrease in stability of its helical structures and thereby become more susceptible to the action of general proteases or peptidases that may be present. Clostridial collagenases, and perhaps other collagenases of microorganisms, not only initiate the disruption of collagen but also carry out its extensive degradation as well. They are sufficient for the function of invasivity. For the nutritional function, their action would have to be supplemented by the actions of other proteases and peptidases.

D. Further Generalizations about the Properties of Collagenases

Although most of the collagenases reported have not been studied in a highly purified state, certain generalities emerge. Bacterial collagenases appear to promote more extensive cleavage of collagen than do tissue collagenases. For example, clostridiopeptidase A may catalyze approxi-

9. E. Lienert and W. Thorsell, *Wien. Tieraerztl. Monatschr.* **43,** 746 (1956).
10. D. F. Waterhouse and H. Irzykiewicz, *J. Insect Physiol.* **1,** 18 (1957).
11. L. Strauch, *Mitt. Max-Planck Ges.* **1,** 40 (1968); *Umschau* **69,** 310 (1969).

mately 200 cleavages per α chain, but the collagenase of the tadpole tail or that of the resorbing uterus may cause only one scission per α chain. However, it should be noted that some tissue collagenases (or impure preparations of these) can digest collagen to dialyzable peptides if incubation is conducted at 37° or above the temperature of the collagen → gelatin transition.

Almost all of the collagenases shown in Table I are inhibited by ethylenediaminetetraacetate (EDTA), and of these all except that of *Trichophyton schoenleinii* appear to require Ca^{2+} for activity. Several collagenases are inhibited by cysteine, and of these a group appear to contain zinc or some other metal in their structures. However, there are both bacterial and tissue collagenases that appear not to be inhibited by cysteine; in fact, that of *Bacteroides melaninogenicus* is stimulated by compounds containing the –SH group. Of interest is the knowledge that clostridial collagenases like many other extracellular bacterial enzymes contain no cysteine or cystine in their composition, and it has been speculated that this condition is important in a general mechanism by which enzymes are secreted. It is thus of interest whether the collagenase of *B. melaninogenicus,* which is not secreted, contains these amino acids.

Several of the listed collagenases have been tested with diisopropylphosphorofluoridate (DFP) and found not to be inhibited. Accordingly, these would be excluded from the class of "serine proteases." However, the collagenase of *Aspergillus oryzae,* an enzyme not fully characterized, is reported to be inhibited by this reagent.

Of special interest are reports that several microorganisms and tissues appear to elaborate two forms of collagenase. This is true of *Clostridium histolyticum,* and human rheumatoid synovial fluid has two collagenases, perhaps arising from synovial lining cells. The relationship between the two collagenases of one cell or tissue has not been clearly established although the suggestion has been made that they might bear a parent–subunit relationship. The specificities of the two kinds of collagenase in a particular case have not been sufficiently delineated.

E. Uses of Collagenases

Laboratory applications of collagenases have been limited largely to clostridiopeptidase A, the only one available in at least partially purified form. In certain limited experiments, tissue collagenases such as that of the tadpole have been used with dramatic effect to prepare and map large fragments of the collagen molecule retaining the triple helical structure. Commercial collagenases are frequently accompanied by variable amounts of other proteases, by peptidases, and perhaps by mucopolysaccharidases and glycosidases. Interpretations of results can only safely be made when the enzyme preparation used has been well defined in its catalytic capacities.

Clostridial collagenases have been employed to study the "crystalline" regions of the collagen molecule and also to provide fragments for study from the "amorphous" regions. Clostridiopeptidase A has been used effectively to study the unfolding and refolding of the collagen helix (*12*). The enzyme in the past has been used for studying repeated sequences in collagen (*13–15*), but now the studies have advanced to sequences in individual α chains (*16*). The enzyme also has helped locate specific structural features of the collagen molecule such as the aldol cross-linkages (*17*, *18*), the carbohydrate moieties attached to residues of hydroxylysine (*19*, *20*), and the hydroxylamine-sensitive linkages (*21*). The tissue collagenases have served to provide larger fragments of α chains that permit a study of the self-assembly process in collagen (*22–25*).

Collagenases are useful as a specific means of identifying collagen. Clostridiopeptidase A can digest native collagen *in situ* and thereby, by a subtractive procedure, help locate the collagenous component in a tissue. The use of collagenase has helped classify the basement membrane proteins and the reticulins as particular arrays of collagen (*26*, *27*). In experiments dealing with biosynthesis of collagen, use of collagenase has assured that the formed protein indeed is collagen (*28*, *29*). Collagenase may be used to dissolve immune aggregates containing col-

12. P. H. von Hippel and K. Wong, *Biochemistry* **2**, 1399 (1963).
13. P. M. Gallop and S. Seifter, *in* "Collagen" (N. Ramanathan, ed.), p. 249. Wiley (Interscience), New York, 1962.
14. R. E. Schrohenloher, J. D. Ogle, and M. A. Logan, *JBC* **234**, 58 (1959).
15. W. Grassmann, H. Hörmann, and A. Nordwig, *in* "Collagen" (N. Ramanathan, ed.), p. 263. Wiley (Interscience), New York, 1962.
16. P. Bornstein, *Biochemistry* **6**, 3082 (1967).
17. M. Rojkind, O. O. Blumenfeld, and P. M. Gallop, *JBC* **241**, 1530 (1966).
18. M. Rojkind, L. Rhi, and M. Aguirre, *JBC* **243**, 2266 (1967).
19. W. T. Butler and L. W. Cunningham, *JBC* **241**, 3882 (1966).
20. L. W. Cunningham, J. D. Ford, and J. P. Segrest, *JBC* **242**, 2570 (1967).
21. O. O. Blumenfeld and P. M. Gallop, *Biochemistry* **1**, 947 (1962).
22. J. Gross and Y. Nagai, *Proc. Natl. Acad. Sci. U. S.* **54**, 1197 (1965).
23. A. Z. Eisen, J. J. Jeffrey, and J. Gross, *BBA* **151**, 637 (1968).
24. J. J. Jeffrey and J. Gross, *Biochemistry* **9**, 268 (1970).
25. T. Sakai and J. Gross, *Biochemistry* **6**, 518 (1967).
26. A. H. T. Robb-Smith, *Recent Advan. Gelatin Glue Res., Proc. Conf., Univ. Cambridge, 1957* p. 38. Pergamon Press, Oxford, 1958.
27. R. G. Spiro, *JBC* **242**, 1915 and 1923 (1967).
28. M. Urivetsky, V. Kranz, and E. Meilman, *ABB* **100**, 478 (1963).
29. A. A. Gottleib, A. Kaplan, and S. Udenfriend, *JBC* **241**, 1551 (1966).

lagen and therefore assumes some importance in studies related to the immunogenicity of collagen. Or, in immunological studies, it may be used to remove collagen from a tissue before reaction with an anticollagen antibody (*30, 30a*).

Collagenases are being used with considerable effect as a means of loosening cells held in a net of connective tissue, thereby allowing subsequent isolation of specific cell types. Thus, in studies relating to the biosynthesis of insulin, collagenase has been applied successfully for the preparation of islet cells of the pancreas (*31*).

The separation of connective tissue elements is also aided by treatment of a tissue with collagenase. Thus, elastin can be prepared subtractively by the digestion of collagen in the same tissue. Certain carbohydrate components may also be freed by preliminary digestion of a tissue with a collagenase.

Finally, collagenases may be employed in the study of certain disease processes and of wound-healing mechanisms.

In this chapter we shall consider in some detail (1) the collagenases of *Clostridium histolyticum,* (2) the collagenase of the tadpole of *Rana catesbiana,* and (3) the collagenases of human rheumatoid synovial fluid. Other collagenases are described in Table I and will receive only occasional additional comment.

II. The Collagenases of *Clostridium histolyticum*

The occurrence of an extracellular collagenase of *Clostridium histolyticum* was first demonstrated rigorously by Maschmann (*32*). This investigator clearly differentiated between "gelatinase" activities of general proteases acting on denatured collagen and highly specific collagenases acting on an insoluble, undenatured collagen substrate. Following these observations, Bidwell *et al.* (*33*), MacLennan *et al.* (*34*), and DeBellis *et al.* (*35*) made significant contributions in the

30. S. Rothbard and R. F. Watson, *J. Exptl. Med.* **129,** 1145 (1969).

30a. S. Fuchs and W. F. Harrington, *BBA* **221,** 119 (1970).

31. S. Moskalewski, *Gen. Comp. Endocrinol.* **5,** 342 (1965); also, D. F. Steiner, J. L. Clark, C. Nolan, A. H. Rubenstein, E. Margoliash, B. Aten, and P. E. Oyer, *Recent Progr. Hormone Res.* **25,** 207 (1969).

32. E. Maschmann, *Biochem. Z.* **295,** 1 (1937); **300,** 89 (1938).

33. E. Bidwell and W. E. van Heyningen, *Biochem. J.* **42,** 140 (1948).

34. J. D. MacLennan, I. Mandl, and E. L. Howes, *J. Clin. Invest.* **32,** 1317 (1953).

35. R. DeBellis, I. Mandl, J. D. MacLennan, and E. L. Howes, *Nature* **174,** 1191 (1954); see also I. Mandl, *Advan. Enzymol.* **23,** 163 (1961).

preparation of what is now known as clostridiopeptidase A (EC 3.4.4.19). Grant and Alburn (*36*) showed the multiplicity of clostridial enzymes and in particular reported the isolation of a B enzyme. In this discussion the two enzymes will be referred to as collagenase A and collagenase B.

A. Culture of Organism and Purification of Enzymes

Several different strains of *Clostridium histolyticum* have been used for production of collagenases. Several investigators have begun their studies with crude or partially purified enzyme preparations obtained from commercial sources. The authors have prepared collagenases from strain 47Q5 grown in the medium of Warren and Gray (*37*) as modified by Takahashi and Seifter (*38*). Use of this medium has resulted in highly reproducible yields and specific activities of collagenases and in a low production of general proteases. A difficult problem in purification of clostridial collagenase is separation from other proteases in the growth medium.

The collagenases constitute about 10% of the total protein precipitated from the cell-free growth medium by saturation with ammonium sulfate. Most present methods employ column techniques with Sephadex and DEAE-cellulose. For details of purification, the reader may consult references given in Table II and a review by Seifter and Harper (*39*). Since the work of Grant and Alburn (*36*), investigators have recognized that in fact two collagenases, called A and B, can be separated by fractionation using DEAE-cellulose.

Both collagenases A and B have been obtained in a high state of purity (*40*), but the possibility that trace contaminants are present cannot be ruled out. The gel electrophoretic patterns for both show single major bands with perhaps a faint suggestion of impurity or dissociated protein. By criteria of immunodiffusion and immunoelectrophoresis, employing antibodies made to purified preparations of collagenases A and B and to the proteins of the growth medium, significant homogeneity is demonstrated with perhaps a trace of collagenase B present in preparations of collagenase A. Sedimentation equilibrium studies by the Yphantis techniques also show a large degree of homogeneity for each enzyme.

36. N. H. Grant and H. E. Alburn, *ABB* **82,** 245 (1959).
37. G. H. Warren and J. Gray, *Nature* **192,** 755 (1961).
38. S. Takahashi and S. Seifter, *J. Bacteriol.* (submitted for publication); see also Seifter and Harper (*39*).
39. S. Seifter and E. Harper, "Methods in Enzymology," **19,** 613 (1970).
40. E. Harper and S. Seifter, *JBC* (1970) (in press).

TABLE II
A SUMMARY OF COLLAGENASES PREPARED FROM FILTRATES OF *Clostridium histolyticum*

Procedure of purification[a]	Enzymes isolated	Molecular weight	Reference
Adsorption to alumina $C\gamma$; elution with 0.2 *M* phosphate, pH 7.5	One	109,000[b]	Seifter *et al.* (*41*)
Elution from DEAE-cellulose with (a) 0.06 *M* phosphate, pH 7.4; (b) 0.6 *M* acetate, pH 5.6; and (c) phosphate acetate, pH 6.0	A B C	Not determined Not determined Not determined	Grant and Alburn (*36*)
Elution from DEAE-Sephadex A-50 with (a) 0.01 *M* tris-HCl and 0.01 *M* Ca acetate, pH 7.2; and (b) 0.06 *M* Na acetate and 0.1 *M* Ca acetate, pH 5.6	I II	112,000[c] 112,000[c]	Mandl *et al.* (*40a*)
Starch electrophoresis followed by elution from Sephadex G-200 with (a) 0.05 *M* tris, pH 7.5 and 0.004 *M* Ca chloride; (b) 0.005 *M* tris-0.1 *M* Na chloride, pH 7.5, and 0.004 *M* Ca chloride	I II	95,000[d] 79,000[d]	Yoshida and Noda (*43*)
Carrier-free electrophoresis on acetylated-methylated polyamide followed by elution from Sephadex G-100 with 0.05 *M* tris, pH 8.0, and 0.01 *M* Ca acetate	One	109,000[b]	Strauch and Grassmann (*40t*)
Elution from Sephadex G-50 followed by elution from DEAE-cellulose with (a) 0.06 *M* phosphate, pH 7.4; (b) 0.6 *M* acetate, pH 5.6	A B	105,000[d] 57,400[d]	Harper *et al.* (*42*)

[a] All purifications were begun with a fraction of the culture filtrate made by precipitation with ammonium sulfate.
[b] Determined by sedimentation diffusion.
[c] Determined by location in a sucrose gradient.
[d] Determined by sedimentation equilibrium.

Frequently, preparations of clostridial collagenases are obtained that seem to show homogeneity by certain physical methods but nevertheless exhibit extraneous catalytic activities. The latter may have the nature of amidases, general proteases, or carbohydrases, and all purified preparations should be tested for their presence. Preparations free of one or all these activities can be obtained by the methods referred to in this discussion (*39*).

The problem of purity nevertheless remains a significant one for future studies of structure of the two enzymes and their mechanisms of action. Neither enzyme, despite repeated attempts, has been obtained in crystalline form. Until the use of the newly described medium containing bisulfite (*38*), enzyme preparations with reproducible high specific activity were not obtained. Difficulties in these respects may stem from the following circumstances. First, the collagenases are in a medium containing several potent general proteases, allowing the possibility that limited proteolysis occurs. Consequently, purified preparations of collagenases obtained from the growth medium may be variable with respect both to activity and composition. Second, the culture medium is undefined, not only with regard to organic constituents but also to the nature of metals present. Since, as will be discussed below, the collagenases appear to require zinc or a transition element, and these seem to be interchangeable, the possibility exists that the harvested enzymes are not unique molecular species in terms of metal component. Third, the culture medium contains sodium thioglycolate (to assure anaerobiosis), and in fact the organism generates some hydrogen sulfide in the course of its growth. Yet the collagenases produced are known to be inhibited by SH-containing compounds and, in the case of cysteine, are able to form stable, inactive derivatives. These derivatives can elute from a column of DEAE-cellulose, for instance, at the same positions where the corresponding active collagenases appear. The possibility exists, therefore, that variability in specific activity could result from the presence of inactive enzyme protein. The first and third difficulties are largely ameliorated by incorporation of bisulfite in the culture medium as the agent for anaerobiosis. The second difficulty will be solved by application of media defined in metal composition.

B. Composition

The amino acid compositions of collagenases A and B are shown in Table III. Generally, the two enzymes have compositional similarity, but the few differences may prove to be significant. On the other

TABLE III
AMINO ACID COMPOSITIONS OF COLLAGENASES A AND B[a]

Amino acid	Collagenase A	Collagenase B
Aspartic acid	159	170
Threonine	54	60
Serine	27	32
Glutamic acid	85	93
Proline	56	40
Glycine	96	96
Alanine	73	68
Valine	66	60
Isoleucine	56	61
Leucine	73	74
Tyrosine	50	48
Phenylalanine	50	49
Lysine	103	103
Histidine	16	18
Arginine	37	29
Tryptophan	7	6
Amide	67	Not done

[a] Residues per 1000 total residues; uncorrected for losses during hydrolysis.

40a. I. Mandl, S. Keller, and J. Manahan, *Biochemistry* **3,** 1737 (1964).

40b. J. Manahan and I. Mandl, *ABB* **128,** 6, 19 (1968).

40c. T. Kono, *Biochemistry* **7,** 1106 (1968).

40d. W. M. Mitchell, *BBA* **159,** 554 (1968).

40e. S. Takahashi, *J. Biochem.* (*Tokyo*) **61,** 258 (1967).

40f. G. Schoellmann and E. Fisher, Jr., *BBA* **122,** 557 (1966).

40g. J. W. Rippon, *BBA* **159,** 147 (1968).

40h. J. W. Rippon, *J. Bacteriol.* **95,** 43 (1968).

40i. A. Nordwig and W. F. Jahn, *European J. Biochem.* **3,** 519 (1968).

40j. A. H. Kang, Y. Nagai, K. A. Piez, and J. Gross, *Biochemistry* **5,** 509 (1966).

40k. J. C. Houck, V. K. Sharma, Y. M. Patel, and J. A. Gladner, *Biochem. Pharmacol.* **17,** 2081 (1968).

40l. R. Pérez-Tamayo, *Lab. Invest.* **22,** 137 (1970).

40m. H. M. Fullmer and G. Lazarus, *Israel J. Med Sci.* **3,** 758 (1967).

40n. A. Z. Eisen, *J. Clin. Invest.* **46,** 1052 (1967).

40o. H. C. Grillo and J. Gross, *Dev. Biol.* **15,** 300 (1967).

40p. D. G. Walker, C. M. Lapiere, and J. Gross, *BBRC* **15,** 397 (1964).

40q. H. M. Fullmer and W. Gibson, *Nature* **209,** 728 (1966).

40r. H. C. Grillo, C. M. Lapiere,, M. H. Dresden, and J. Gross, *Develop. Biol.* **17,** 571 (1968).

40s. L. Strauch and H. Vencelj, *Z. Physiol Chem.* **348,** 465 (1967).

40t. L. Strauch and W. Grassmann, *Z. Physiol. Chem.* **344,** 140 (1966).

40u. Y. Nagai, S. Sakakibara, H. Noda, and S. Akabori, *BBA* **37,** 567 (1960).

40v. E. Harper, *in* "Collgenases—First Interdisciplinary Symposium" (I. Mandl, ed.) 1970 (in press); also E. Harper, A. Berger, and E. Katchalski, (in preparation).

40w. I. Mandl, H. Zipper, and L. T. Ferguson, *ABB* **74,** 465 (1958).

hand, the differences may result from the circumstances discussed in the previous section dealing with purity of preparations. Contents of collagenases A and B are almost identical in glycine, tyrosine, phenylalanine, lysine, and histidine. Among the preparations analyzed, differences are most notable in the case of serine and of proline. Whether the differences are the result of experimental procedures has not yet been established. They cannot be readily rationalized on a possible parent–subunit relationship between A and B that will be considered in another section of this chapter.

Of special significance for many studies with the clostridial collagenases is the absence of sulfur-containing amino acids, particularly of cysteine and cystine. This was first noted by Grant and Alburn (*36*) when they did total sulfur analysis of their preparations. Because this matter is fundamental to studies on inactivation by SH-containing agents, Harper and Seifter (*40*) carefully investigated the possible presence of even one cysteine residue in collagenase A. They were able to demonstrate that oxidation of the enzyme with performic acid did not cause appearance of any cysteic acid nor did treatment with iodoacetate reveal formation of any carboxymethylcysteine.

C. Molecular Size

Seifter *et al.* (*41*), using a preparation now known to correspond to collagenase A, found a value of 109,000 for the molecular weight by sedimentation-diffusion analysis. Other workers since that time (see Table II), using different preparations and various methods of determining molecular weight, have published values of 112,000 (by sucrose gradient), 95,000 (by sedimentation equilibrium), and 109,000 (by sedimentation diffusion). In all of the above the state of purity of the enzyme was not fully defined, and in some cases the collagenases were prepared from different strains of *Clostridium histolyticum*. In more recent studies, Harper and Seifter (*40*) and Harper *et al.* (*42*) separated collagenases A and B with the purity noted in a previous section of this chapter, and obtained a molecular weight of 105,000 for the A enzyme and 57,400 for the B enzyme. In Table II, one may note that Yoshida and Noda (*43*) obtained two collagenases with molecular weights of 95,000 and 79,000, respectively. Although the identification of the first enzyme with col-

41. S. Seifter, P. M. Gallop, L. Klein, and E. Meilman, *JBC* **234,** 285 (1959).
42. E. Harper, S. Seifter, and V. Hospelhorn, *BBRC* **18,** 627 (1965).
43. E. Yoshida and H. Noda, *BBA* **105,** 562 (1965).

lagenase A is almost certain, the identification of the second with collagenase B is not.

D. Possible Subunits

Levdikova *et al.* (*44, 45*), starting with a preparation that corresponds to collagenase A, showed that the parent molecule of 100,000 molecular weight could be dissociated into four inactive subunits of 25,000 molecular weight each. To accomplish this they employed EDTA or *o*-phenanthroline, chelating agents seemingly directed against the calcium ions known to activate the enzyme and to the putative zinc component of the enzyme, respectively. Dissociation into the smaller units was obtained at both low and high pH values, and even the use of alkali alone at pH 12 was sufficient. Harper *et al.* (*42*) and Harper and Seifter (*40*) then presented the idea that collagenase A could be the parent molecule of an active subunit, collagenase B. The progression would be, collagenase A (active, molecular weight of 105,000) to collagenase B (active, molecular weight of 57,400) to the inactive subunits (molecular weight, 25,000). From the work of Harper and his colleagues, evidence in favor of a parent–subunit relationship between collagenases A and B is (1) collagenase B is approximately half of the molecular weight of collagenase A; (2) the amino acid compositions of the two enzymes have a large measure of identity; (3) antisera prepared against the individual purified enzymes show significant serological cross-reactivity by immunodiffusion and immunoelectrophoretic techniques; and (4) under certain circumstances to be described, ^{35}S-cysteine combines stoichiometrically with collagenase A in a molar ratio of 2:1, whereas it combines with collagenase B in a molar ratio of approximately 1:1.

Against the idea of identical subunits is the fact that there do appear to be some significant differences in amino acid compositions of collagenases A and B. Admittedly, though, the preparations analyzed might not have been sufficiently pure and, as already noted, both enzymes could have undergone a degree of proteolytic degradation in the culture medium.

The possibility has been considered that collagenase A is in equilibrium with collagenase B and that recycling the one through the DEAE-cellulose column used in purification would in fact reveal a dissociation

44. G. A. Levdikova, N. Orekhovich, N. I. Solov'eva, and V. O. Shpikiter, *Dokl. Akad. Nauk. SSR* **153,** 725 (1963); *English Transl.* **153,** 1429 (1964).

45. G. A. Levdikova, N. Orekhovich, N. I. Solov'eva, and V. O. Shpikiter, *Proc. 6th Intern. Congr. Biochemistry, New York, 1964* p. 94, Vol. IV, 1964.

of one to the other. This does not seem to be borne out, although occasionally preparations of A, after recycling, do seem to give rise to B. However, it was not established that B was not present in small amount in the preparation of A used.

Also of interest is the fact that among the other collagenases shown in Table I there are examples of a single cell or tissue producing two enzymes with collagenolytic activity. Thus the collagenases of human synovial fluid occur in A and B species, both grossly showing similar specificity. Interestingly, the molecular weights of these show an apparent 1:2 ratio.

E. Physical Properties

The molecular weights of collagenases A and B have been already considered. In the preparation characterized by Seifter *et al.* (*41*), most likely collagenase A, a sedimentation constant of $s_{20,w} = 5.4\,S$ and a diffusion constant of $D_{20,w} = 4.3 \times 10^{-7}$ cm^2/sec were obtained. Using another preparation, Levdikova *et al.* (*44*) reported a diffusion constant of 5×10^{-7} cm^2/sec. An isoelectric point of pH 8.6 has been reported by Nordwig (*46*).

Both collagenases are soluble in water and in dilute salt solutions. They are stable to lyophilization and can be preserved in the dry form for several months without serious deterioration in activity. The purer the preparation, however, the less stable are both dissolved and dry forms.

F. Chemical and Biosynthetic Modifications

The major modification so far achieved is the formation of inactive derivatives of collagenases A and B with cysteine. This matter will be discussed under catalytic properties of the enzymes. Also, the destruction of histidine residues of collagenase A by photooxidation in the presence of methylene blue or Rose Bengal will be discussed below.

Modifications of the enzyme that prove extremely useful can be achieved by changing the conditions of culture. Thus inclusion of cobalt chloride in the medium causes an increase of specific activity of the enzyme of 180% of that obtained when the usual medium, undefined with respect to composition of metals, is used. At present, studies are under way in the authors' laboratories to determine the degree of incorporation

46. A. Nordwig, *Leder* **13,** 10 (1962).

of cobalt. An enzyme labeled with ^{14}C can be obtained by using a ^{14}C-algal hydrolyzate in the culture medium. An enzyme of this sort is most useful for studying interactions with a collagen substrate.

G. Catalytic Properties of Collagenases A and B

Most studies in this regard have been concerned with collagenase A. In this discussion the separate properties of the two enzymes will be clearly delineated.

1. *Reactions Catalyzed*

Both collagenases A and B can catalyze the hydrolysis of "insoluble" tissue collagen, acid-extracted collagen, neutral salt-extracted collagen, or the gelatin derived from any of these. As considered under the definition of collagenases, both enzymes, but A in particular, have been studied for action against special collagens such as the cuticular collagens of the earthworm and *Ascaris* and against reticulin and proteins of basement membrane; indeed, all were substrates.

2. *Specificity*

a. The Specificity of Collagenase A. This enzyme has been studied extensively with preparations of varying degrees of purity, and the subject is ready for rigorous experimental reconsideration. Even the best preparations used in those investigations were not of the uniformly high specific activity now possible to achieve, and the possibility of admixture with small amounts of collagenase B was not excluded. Yet enough experiments have been performed on newer types of preparations to indicate that the general specificity assigned to clostridiopeptidase A (collagenase A) is correct. As will be discussed, the bond specificity of collagenase B has never really been properly investigated.

Using as typical substrates either tendon collagen in a suspended form or an acid-extracted collagen of skin or the gelatin derived from one of them, digestion to completion by collagenase A causes a 6- to 7-fold increase in ninhydrin reactivity. Depending on the kind of collagen used, collagenase A causes from about 150–200 scissions per α chain. Seifter *et al.* (*41*), using various methods of end group determination, showed that the digestion of ichthyocol or calfskin acid–extracted collagen, or the gelatin of either, with collagenase A yielded peptides of number-average molecular weight of about 500. In comparison, trypsin acting to completion on the gelatins produced peptides of number-average molecular weight of about 1700. Von Hippel and Wong (*12*) determined the weight-average molecular weights of peptides obtained after complete digestion of ichthyocol gelatin and found a value of about 810 when collagenase was used and about 2400 when trypsin was employed. Notable is the fact that a complete digest of collagen or gelatin with collagenase produces a "core" that is retained when the mixture is dialyzed. The core may account for 10% of the collagen molecule and is characterized by a fairly large number of peptides that appear to have originated both from terminal regions of the collagen molecule and from the so-called polar (amorphous) regions. The production and nature of the core is considered in papers by von Hippel *et al.* (*47*) and Franzblau *et al.* (*48*).

Michaels *et al.* (*49*) and Kasakova *et al.* (*50*) established that the action of collagenase on collagen resulted in formation almost solely of peptides with a glycyl residue in the amino-terminal position. This indicated that the minimum requirement for specificity for the enzyme is a peptide bond involving the amino group of glycine. There was no defined specificity for the residue in the carboxyl-donating position of a susceptible peptide bond. Gallop *et al.* (*51*), on the basis of analysis of peptides found after digestion of purified collagen with purified collagenase, assigned a specificity for the enzyme of –Y–Gly–Pro–X– since most of the tripeptides found consisted of Gly–Pro–X (Y could be any of approximately 17 kinds of amino acids, and X most frequently was Ala or Hyp). At about the same time, Schrohenloher *et al.* (*14*) found a similar distribution of tripeptides in a digest of collagen with a less well characterized preparation of collagenase. Thus the requirements for specificity were extended to include a condition that the susceptible glycine residue had to be bonded through its carboxyl function to the imino nitrogen of a proline residue. In this connection, von Hippel *et al.* (*47*) determined the average pK value of the liberated α-amino groups and found it to be characteristic of Gly–Pro– peptides. Peptides with Pro in this position generally have higher pK' values than peptides with amino acid residues in this position.

Seifter *et al.* (*41*) then reported cooperative experiments with von

47. P. H. von Hippel, P. M. Gallop, S. Seifter, and R. S. Cunningham, *JACS* **82,** 2774 (1960).

48. C. Franzblau, S. Seifter, and P. M. Gallop, *Biopolymers* **2,** 185 (1964).

49. S. Michaels, P. M. Gallop, S. Seifter, and E. Meilman, *BBA* **29,** 450 (1958).

50. O. V. Kasakova, V. N. Orekhovich, and V. O. Shpikiter, *Proc. Acad. Sci. USSR* (*English Transl.*) **122,** 657 (1958).

51. P. M. Gallop, S. Seifter, S. Michaels, L. Klein, and E. Meilman, *Connective Tissue, Thrombosis, Atherosclerosis, Proc. Conf., Princeton, N. J., 1958* p. 46. Academic Press, New York, 1959.

Hippel and Harrington indicating that random copolymers of glycine and proline were weakly acted on by collagenase. In critical experiments, Nagai and Noda (*52*) synthesized oligopeptides embodying sequences predicted to be specific for collagenase and found that these indeed were substrates for the enzyme. Further important pioneering studies in this area with synthetic peptides were performed by Heyns and Legler (*53*), Seifter *et al.* (*54*). Grassmann and Nordwig (*55*), and Poroshin *et al.* (*56*). Table IV lists some of the oligopeptides and polymeric peptides that have been tested as substrates. Grassmann and his colleagues (*57*) were then able to define the requirements for specificity of the enzyme as a sequence –P–Y–Gly–P–, where P may be Pro, Hyp, or a residue of sarcosine; Y is probably any other amino acid residue. Experiments with collagen have shown that indeed Y could be another residue of Hyp. Scission, as already indicated, occurs between Y and Gly. Only a few peptides have been tried in which Hyp follows Gly, and the action against them appears to be slower. In this regard one should mention that collagenase can digest the cuticular collagen of the earthworm, a protein in which about 90% of the pyrrolidine residues are of hydroxyproline (*58*).

The matter of the specificity of collagenase A would seem to have been laid to rest with the study of synthetic peptides. Certainly there would seem to be sufficient repetition in collagen of sequences with the defined specificity to account for the approximately 200 scissions per α chain (*12, 48*). Yet earlier Seifter *et al.* (*59*) had suggested that the strict specificity deduced from the studies with peptides might not fully define the action of collagenase on collagen. From the time of the earliest studies on collagenolytic digests of the protein substrate, investigators have recognized that certain peptides can be isolated that must represent escapes from this specificity. For instance, significant quantities of Gly–Ala–Pro

52. Y. Nagai and H. Noda, *BBA* **34,** 298 (1959).

53. K. Heyns and G. Legler, *Z. Physiol. Chem.* **315,** 288 (1959).

54. S. Seifter, P. M. Gallop, and E. Meilman, *Recent Advan. Gelatin Glue Res., Proc. Conf. Univ. Cambridge, 1957* p. 164. Pergamon Press, Oxford, 1958.

55. W. Grassmann and A. Nordwig, *Z. Physiol. Chem.* **322,** 267 (1960).

56. K. T. Poroshin, T. D. Kuzarenko, V. A. Shibnev, and V. G. Debabov, *Ber. Acad. Wiss. USSR* p. 550 (1960); *Biokhimiya* **26,** 244 (1961).

57. W. Grassmann, H. Hörmann, A. Nordwig, and E. Wünsch, *Z. Physiol. Chem.* **316,** 287 (1959).

58. D. Fujimoto and E. Adams, BBA **107,** 232 (1965); **154,** 183 (1968); see also, for discussion of specificity, E. Adams, S. Antoine, and A. Goldstein, *ibid.* **185,** 251 (1969).

59. S. Seifter, P. M. Gallop, and C. Franzblau, *Trans. N. Y. Acad. Sci.* [*2*] **23,** 540 (1961).

TABLE IV
THE ACTION OF CLOSTRIDIOPEPTIDASE A ON SOME SYNTHETIC OLIGOPEPTIDES AND AMINO ACID POLYMERS

Compound[a]	Action	Ref.
Gly-Pro-Leu-Gly-Pro-NH_2	Hydrolyzed; Gly-Pro-NH_2 appears	*52*
Z-Gly-Pro-Leu-Gly-Pro-OH	Hydrolyzed	*40u*
Z-Gly-Pro-Gly-Gly-Pro-Ala-OH	Hydrolyzed; Gly-Pro-Ala-OH appears	*55*
Z-Pro-Ala-Gly-Pro-NH_2	Hydrolyzed	*53*
PZ-Pro-Leu-Gly-Pro-D-Arg-OH	Hydrolyzed; Gly-Pro-D-Arg-OH appears	*80*
Z-Hyp-Gly-Gly-Pro-OCH_3	Hydrolyzed; Gly-Pro-OCH_3 appears	*57*
(Pro-Ala-Gly)$_n$-OH	Hydrolyzed	*59*
(Pro-Gly-Gly)$_n$-OH	Hydrolyzed	*59*
(Pro-Gly-Hyp-*O*-acetyl)$_n$-OH	Hydrolyzed	*59*
(Pro-Gly-Pro)$_n$-OH (about 6000 MW)	Hydrolyzed; yields Gly-Pro-Pro-OH and Gly-Pro-OH	*40v*
(Gly-Pro-Ala)$_n$-OH	Hydrolyzed; yields only Gly-Pro-Ala-OH	*40v, 58*
(Gly-Ala-Pro)$_n$-OH	Hydrolyzed	*58*
(Pro-Leu-Gly)$_n$-OH	Hydrolyzed; yields Gly-Pro-Leu-OH	*52*
Gly-Pro-Leu-Leu-Pro-OH	Not hydrolyzed	*40u*
Pro-Leu-Gly-OH	Not hydrolyzed	*52*
Gly-Pro-Leu-OH	Not hydrolyzed	*52*
Leu-Phe-Pro-Leu-Phe-Pro-OH	Not hydrolyzed	*40w*
(Pro)$_n$-OH (Polyproline)	Not hydrolyzed	*49*
(Pro-Gly-Ala)$_n$-OH	Not hydrolyzed	*40v*
Glycylproline diketopiperazine	Not hydrolyzed	*70*

[a] Abbreviations: Z, benzyloxycarbonyl; PZ, phenylazobenzyloxycarbonyl. A listing of additional oligopeptides may be found in Nordwig (*46*). Additional considerations of specificity are discussed by Adams *et al.* (*58*).

TABLE V
BINDING OF ^{14}C-LABELED COLLAGENASES A OR B TO A SUSPENSION OF COLLAGEN (ICHTHYOCOL) AT 0°: EFFECT OF EDTA

Reaction mixture	cpm bound to collagen	
	Collagenase A	Collagenase B
Enzyme + collagen (no EDTA)	2230	3950
Enzyme + collagen + 0.001 *M* EDTA	190	372

were found by Ogle *et al.* (*60*). Papers by Hannig and Nordwig (*61*) and Greenberg *et al.* (*62*) should be consulted for further consideration of such peptides. Now studies with separated α chains would seem to raise the specter of true specificity again.

From the work of Kang *et al.* (*63*) and of Bornstein (*64*) the amino acid sequence in a relatively long segment of the amino-terminal region is known for the $\alpha 1$ chain of acid-extracted rat skin collagen. These workers also have provided evidence for sites of cleavage in this sequence catalyzed by several proteases including a commercial preparation of clostridiopeptidase A. The sequence and cleavage points by collagenase are shown here because they are highly illustrative of the problem:

```
-Gly-Pro-Ser|Gly-Pro-Arg|Gly-Leu-Hyp|Gly-Pro-Hyp|Gly-Ala-Hyp|Gly-Pro-Gln
 16  17  18 | 19  20  21| 22  23  24| 25  26  27| 28  29  30| 31  32  33

-Gly-Phe-Gln|Gly-Pro-Hyp-Gly-Glu-Hyp-Gly-Glu-Hyp-Gly-Ala-Ser|Gly-Pro-Met-
 34  35  36 | 37  38  39  40  41  42  43  44  45  46  47  48 | 49  50  51
```

Collagenase makes seven scissions in this sequence. If the deduced specificity rule were being obeyed, only one should be expected (between residues 18 and 19). This in fact occurs but so do six others. Of the seven cleavages, six occur in accordance with the minimum specificity deduced from the studies using digests of collagen (*15, 49, 50, 51*). Thus five peptides in a digest of the above fragment contain Y–Gly–Pro. The bonds cleaved between residues 21 and 22 and again between 27 and 28 do not furnish peptides with the Y–Gly–Pro– sequence.

One could then justifiably conclude that perhaps an insufficient variety of synthetic oligopeptides has been studied to allow a definition of absolute specificity for collagenase A or, alternatively, that the true specificity should be deduced only from the nature of products of digestion of collagen. However, the matter is complicated by the inadequate characterization of collagenase preparations used. Thus, more recently, Kang *et al.* obtained a peptide from chick skin collagen (designated $\alpha 1$ CB2) that has a sequence identical with the one above beginning at residue 16, except that the residue of Ser at position 18 is replaced by a residue of Ala. Then, with Harper (*64a*), the scission of this by a purified Worthington clostridiopeptidase A was studied. Instead of one proteolytic split expected from the defined specificity, seven were obtained

60. J. D. Ogle, R. B. Arlinghaus, and M. A. Logan, *ABB* **94,** 85 (1961).
61. K. Hannig and A. Nordwig, *in* "Treatise on Collagen" (G. N. Ramachandran, ed.), Vol. 1, p. 73 Academic Press, New York, 1967.
62. J. Greenberg, L. Fishman, and M. Levy, *Biochemistry* **3,** 1826 (1964).
63. A. H. Kang, P. Bornstein, and K. A. Piez, *Biochemistry* **6,** 788 (1967).
64. P. Bornstein, *Biochemistry* **6,** 3082 (1967).
64a. E. Harper and A. Kang, *BBRC* **41,** 482 (1970).

as in the sequence above. Harper then subjected the enzyme preparation to further purification and obtained two collagenases. One indeed gave only the anticipated cleavage with production of Gly–Pro–Ala, and the second gave this as well as the six others obtained with the unfractionated enzyme. As stated before the time is propitious for reconsideration of the specificity of clostridiopeptidase A using highly purified preparations.

Although collagenase A does act on oligopeptides containing the defined sequence, frequently these peptides can be cleaved by peptidases to produce fragments identical with those produced by collagenase. The fact that such peptidases are often elaborated with collagenases invites the extreme caution proposed here in interpretation of results.

Finally, one should mention that the general specificity of collagenase for the regions of collagen rich in imino acid residues can be visualized by electron microscopy. Studies of this kind have been made by Nishigai *et al.* (*65*), Nordwig *et al.* (*66*), and Kühn and Eggl (*67*).

b. Specificity of Collagenase B. Few studies have been made to define the specificity of this enzyme, and frequently the assumption is made that it is identical with the specificity of collagenase A. Harper *et al.* (*42*) studied the relative activities of the two enzymes on a synthetic oligopeptide and found that they caused the same cleavage but at different rates. Collagenase B acts on acid-extracted collagen and on gelatin to produce grossly the same number of cleavages at the end of the reaction as does collagenase A. However, end group analysis has not been performed, nor peptides separated and identified, so that one cannot say that a glycine residue is always amino terminal.

An interesting and perhaps important difference was noted by Harper *et al.* (*42*) when the action of collagenases A and B on a suspension of insoluble beef tendon collagen was compared. Equal numbers of units of the two collagenases, determined by standard viscometric assay, were incubated with identical quantities of tendon collagen in suspension. When the collagenase A had just caused complete dissolution of the collagen, the collagen in the tube with collagenase B was still suspended but had undergone marked physical change and assumed an hyaline appearance. Yet when this material was converted to gelatin by autoclaving, it yielded an identical number of leucine equivalents by ninhydrin reaction as did a control sample of the collagen. This indicated that at

65. M. Nishigai, Y. Nagai, and H. Noda, *J. Biochem.* (*Tokyo*) **48,** 152 (1960).
66. A. Nordwig, H. Hörmann, K. Kühn, and W. Grassmann, *Z. Physiol. Chem.* **325,** 242 (1961).
67. K. Kühn and M. Eggl, *Biochem. Z.* **346,** 197 (1966).

this point, despite the physical change noted, few or no proteolytic scissions had been catalyzed by collagenase B. If now the gelatin were digested with collagenase B, the characteristic 6-fold increase in ninhydrin reactivity was obtained as when collagenase A was used. Either these phenomena were simply a consequence of different rates of reactivity of the two enzymes that become much more apparent when an insoluble substrate is used or collagenase B has two separately discernible reactions. If the former is true, then the use of collagenase B permits the isolation of a first step in collagenolysis that has not been described accurately by the kind of end group analysis that has been applied, that is, a step in which the physical character of the collagen is changed without discernible proteolysis.

This phenomenon might be ignored or relegated to the curious if there were not other indications that collagenases promote a physical change before proteolysis is wide-scale. Thus even with collagenase A, Gallop *et al.* (*68*) observed that it promoted a profound change in the aspect of suspended fibers made from acid-extracted collagen. Under certain circumstances, solutions of calfskin collagen suffer a small increase in negative specific optical rotation at the D line when mixed with collagenase, even before changes in viscosity are measured; subsequently, of course, the negative rotation becomes less negative. Concentrated solutions of gelatin (approximately 8%) for a short period of time after mixing with collagenase often assume a cloudy aspect that clears with subsequent proteolysis. Finally, in another case Harris *et al.* (*69*) noted that the human leukocytic collagenase, having several properties in common with the human synovial fluid collagenase A, differs from the latter in that it is relatively ineffective in degradation of reconstituted collagen fibrils.

From the preceding discussion, it is apparent that collagenases cannot be talked of only in terms of bond specificity. Their special characteristic, if not their uniqueness, would seem to be that they can bind to the undenatured substrate and induce profound reorganization of the molecule, or parts of it, as a prelude to proteolysis. Since many enzymes can digest collagen once it is denatured, the key to the action of collagenases may be a capacity to cause these local physical changes. Different collagenases may possess this property to different degrees exhibited in part by variable rates with which they attack substrates most difficult to disorganize, those in suspension as fibers and those truly insoluble by virtue of cross-linking.

68. P. M. Gallop, S. Seifter, and E. Meilman, *JBC* **227,** 891 (1957).

69. E. D. Harris, Jr., D. R. DiBona, and S. M. Krane, *J. Clin Invest.* **48,** 2104 (1969).

Another implication of the seeming difference between collagenases A and B is that perhaps the B enzyme functions like certain of the tissue collagenases. This is to say that its actions *in situ* on tissue collagen could be accomplished primarily with only a few proteolytic scissions, thereby initiating a disorganization of the structure and allowing other enzymes to finish the process.

3. *Cofactors*

Gallop *et al.* (*68*) and Seifter *et al.* (*54*) first showed that calcium ions are required both for the binding of collagenase A to collagen and for full catalytic activity. They also demonstrated that Mg^{2+} could not substitute for Ca^{2+} in these functions. Harper (*70*) showed that this requirement held for collagenase B acting with a collagen substrate. Thus, both enzymes are inactivated by addition of EDTA and reactivated by addition of Ca^{2+}.

A suspension of collagen fibers at neutral pH will absorb either collagenase A or collagenase B at 0° (*68, 70*). The enzyme–substrate complex can then be isolated by centrifugation. If the temperature is then raised, the low temperature complex is converted to the kinetic complex and proteolysis occurs. If the fixation of enzyme is attempted in the presence of EDTA, no binding occurs at low temperature; thus, isolation of the treated collagen and warming it in the presence of Ca^{2+} does not lead to proteolysis. This binding of the collagenases is illustrated in Table V (see p. 673).

Whether the requirement for Ca^{2+} pertains also to the hydrolysis by collagenase of synthetic oligopeptides and polymers has not been rigorously demonstrated. The requirement with respect to certain peptides in fact has been questioned by Yagisawa *et al.* (*71*).

There are several studies that indicate the function of Ca^{2+} is to hold the enzyme in conformation proper and necessary for binding to substrate and for catalytic activity. For example, the presence or absence of Ca^{2+} appears to determine the nature of the interaction of collagenase A with cysteine. As already mentioned, cysteine inhibits collagenase and, in fact, an inactive complex of the two may be isolated. Furthermore, the binding of cysteine is considered to occur at the "active site" of the enzyme. If the binding is permitted to occur in the absence of added Ca^{2+} or in the presence of low concentrations of this ion, the binding is firm

70. E. Harper, Doctoral Dissertation, Albert Einstein College of Medicine, Bronx, New York, 1965; see also, E. Harper and S. Seifter, *Federation Proc.* **24,** 359 (1965).

71. S. Yagisawa, F. Morita, Y. Nagai, H. Noda, and Y. Ogura, *J. Biochem.* (*Tokyo*), **58,** 407 (1965).

and cannot be reversed by dialysis against water or passage through suitable columns. On the contrary, if the interaction proceeds in the presence of concentrations of $CaCl_2$ of the order of 0.05 *M*, the complex that forms, although inactive, undergoes significant reactivation merely by dialysis against water. As stated earlier, in the presence of Ca^{2+}, the enzyme assumes a conformation in which apparently its active center is disposed properly for binding substrate and promoting catalysis. The experiments just discussed imply that this same conformation allows only weak binding of the inhibitor, cysteine. In the absence of Ca^{2+} the enzyme is inactive, and apparently the conformation is such that cysteine is allowed more or firmer binding sites. The relationship of these observations to the possible presence of a zinc atom in the enzyme will be considered in another section.

Another set of pertinent experiments by Takahashi and Seifter (*72*) showed that the inactivation of collagenase A by photooxidation in the presence of methylene blue is significantly protected against by the presence of Ca^{2+}; Mg^{2+} did not protect.

Finally, in this regard, Takahashi *et al.* (*73*) have shown that the activity of collagenase A is enhanced after equilibrium dialysis against solutions of salts of certain transition elements, notably, cobalt, manganese, and iron. The enhancement of activity is considerably greater when the dialysis is conducted with these metal ions in the presence of Ca^{2+}.

4. *Evidence for Occurrence of an Intrinsic Metal Component*

From experiments with inhibitors, Seifter *et al.* (*41*) inferred that in addition to requiring activation by Ca^{2+} collagenase A contains an intrinsic metal component. From the kinds of inhibitors that were most effective, they concluded that the most likely metal was zinc or iron. Table VI (p. 680) gives a list of the substances tested for inhibition and shows effective concentrations. In studying inhibition, both the concentration of enzyme and the kind of assay used are important. For the cases illustrated, usually the agent was preincubated with the enzyme (approximately four viscometric units), and 1 ml of the mixture transferred to a viscometer holding 4 ml of collagen substrate dissolved in buffered 0.5 *M* $CaCl_2$. Under these conditions the possibility exists that an agent in fact might be strongly inhibitory but that its inhibitory action was

72. S. Takahashi and S. Seifter, *BBA* **214,** 556 (1970).
73. S. Takahashi, E. Harper, and S. Seifter, *Intern. Congr. Rheum.* No. 382a, 1969.

relieved on dilution with substrate. Thus the inhibition recorded in the table would be that unrelieved by dilution. In a number of instances, however, the inhibition has been checked in an assay with suspended collagen with identical results. The concentration of enzyme is important not only because in some instances one is dealing with stoichiometric attachment of the agent but also because of the possible reversibility of this attachment.

Many of the inhibitors used are agents that could react with carbonyl functions as well as with metal atoms. Because it is known now that several enzymes in fact contain pyruvyl residues in their primary structures, the authors have tested collagenase A, in a preliminary way, for occurrence of such groups. The results have been negative. On the other hand, inhibition by some of the agents shown could not be explained by reaction with carbonyl groups.

The most effective inhibitor studied when the enzyme is used in low concentration is 2,3-dimercaptopropanol (BAL). Against both collagenases A and B it has been found effective in concentrations as low as $10^{-5}\,M$, but with higher concentrations of enzyme (200 viscometric units), it appears to be effective only in concentrations of $10^{-3}\,M$. Whether this difference has any significance concerning the mode of inhibition or merely represents the enhanced oxidation of the easily oxidized agent when more protein is present is not sure.

The inhibitor studied in greatest detail is cysteine. Since the enzyme contains neither cysteyl nor cystyl residues in its composition, inhibition by cysteine (or other SH-containing compounds) cannot be through reduction of disulfide bridges or disulfide exchange. The fact that mercaptoethanol is an order of magnitude less effective an inhibitor than cysteine also speaks against interpretations of that kind. That the effect of cysteine is on a metal atom in the enzyme is supported by parallel inhibition by such chelating agents as *o*-phenanthroline, α,α-bipyridyl, and 8-hydroxyquinoline sulfonate.

Harper and Seifter (*70*) reacted collagenase A and collagenase B respectively with ^{35}S-cysteine in the presence of low concentrations of Ca^{2+} and recycled the enzymes through columns used in the last steps of purification. Inactive enzyme appeared in the eluate from DEAE-cellulose in the same position as active enzyme. Radioactivity exactly coincided with the enzyme peaks with approximately two cysteine molecules bound to one molecule of collagenase A and somewhat less than one cysteine molecule bound to one molecule of collagenase B. Takahashi and Seifter (unpublished) reacted collagenase A and cysteine in the presence of relatively high concentration of Ca^{2+}, and observed that dialysis and recycling through the column reduced the amount of bound

TABLE VI
THE EFFECTS OF VARIOUS COMPOUNDS ON THE ACTIVITY OF CLOSTRIDIOPEPTIDASE A[a]

Compound	Final molarity in assay system	Inhibition (%)	Remarks
EDTA	1×10^{-3}	100	Reversed by addition of Ca^{2+}; not by Mg^{2+}
o-Phenanthroline	5×10^{-3}	100	Not reversed by Ca^{2+}
8-Hydroxyquinoline-5-sulfonate	5×10^{-3}	100	Not reversed by Ca^{2+}
α, α-Bipyridyl	2×10^{-3}	94	Not reversed by Ca^{2+}
8-Hydroxyquinoline	1×10^{-3}	70	Not reversed by Ca^{2+}
Potassium cyanide	1×10^{-2}	40	
Dithiobiuret	5×10^{-3}	20	
Salicylaldehyde	5×10^{-3}	20	
Thioacetamide	5×10^{-4}	0	Agent has limited solubility
Sodium azide	1×10^{-2}	0	
Hydroxylamine	1×10^{-2}	0	
Ethylenediamine	1×10^{-2}	0	
Hydrogen peroxide	1×10^{-2}	0	
DFP	3×10^{-2}	0	Also does not inhibit action on gelatin. Enzyme not a "serine protease"
2,3-Dimercaptopropanol (BAL) (4 units enzyme)	5×10^{-5}	85	Not reversed by Ca^{2+}. Removes Zn from enzyme labeled with ^{65}Zn. Also inhibits collagenase B
2,3-Dimercaptopropanol (200 units enzyme)	1×10^{-3}	100	
L- (or D,L-)Cysteine, pH 7	1×10^{-2}	100	Also inhibits collagenase B. Not reversed by Ca^{2+}, but firmness of binding and degree of reversibility depends on whether Ca^{2+} is present during preincubation with agent. No change of enzyme in ultracentrifuge
Cysteine, pH 7	5×10^{-4}	33	
Cysteine, pH 9	5×10^{-4}	53	
Cysteine, pH 10	5×10^{-4}	85	
Glutathione (reduced)	1×10^{-2}	90	
Mercaptoethanol	1×10^{-2}	50	No change of enzyme in ultracentrifuge
Mercaptoethanol	1×10^{-3}	5	
Sodium thioglycolate	1×10^{-2}	50	Agent often used in culture medium
Dithiothreitol	1×10^{-3}	50	
Sodium bisulfite	1×10^{-2}	0	Added as reducing agent in new culture medium
Penicillamine (8 units of enzyme)	1×10^{-3}	40	
L- (or D,L-)Histidine	1×10^{-2}	50	Also inhibits collagenase B. Reversible on dialysis

TABLE VI (*continued*)

Compound	Final molarity in assay system	Inhibition (%)	Remarks
Imidazole	1×10^{-2}	20	Same degree of inhibition for collagenase B. Fully reversed by dialysis against water
Diketopiperazine of glycylproline		0	Also not a substrate

[a] In most cases the enzyme was preincubated with the agent for 30 or 60 min at room temperature and then added to a viscometer. Unless otherwise indicated the pH was at 7 or 7.4. The usual collagen (ichthyocol) substrate, dissolved in 0.5 *M* $CaCl_2$, was added to the viscometer. By this process the enzyme and agent underwent a 1:5 dilution. The molarity shown is the concentration of agent in the viscometer, that is, during the assay. In some cases the calcium or the substrate may have caused some reversal of inhibition during the assay, thus, the inhibition noted here would be that which was not reversed. This is especially true in the case of histidine, in which this agent has been shown to be competitive with the substrate. However, for many of these agents tests have also been done with collagen "suspension assays," providing independent verification of the action of the agent. This is true, in particular, of EDTA, *o*-phenanthroline, cysteine, mercaptoethanol, 2,3-dimercaptopropanol, penicillamine, histidine, and DFP. Also to be noted is the fact that neither cysteine nor mercaptoethanol causes any dissociation of the enzyme discernible by ultracentrifugal analysis. *o*-Phenanthroline at pH 11 appears to cause clostridiopeptidase A to dissociate to inactive subunits of 25,000 molecular weight. At neutral pH it does not cause either the A or B enzymes to dissociate.

cysteine closer to one molecule per molecule of enzyme with partial return of enzymic activity.

Cysteine complexes of both collagenases were found incapable of engaging in the low temperature binding to a suspended collagen substrate. Thus, cysteine covers a site on the enzyme required for substrate binding. Cysteine complexes of collagenases A and B, respectively, were titrated with *p*-mercuribenzoate by the method of Boyer (*74*). Results showed that the –SH group of the bound cysteine was not available for this titration. Similar results were obtained with 5,5′-dithio-bis(2-nitrobenzoic acid). Thus, not only was a site on the enzyme necessary for substrate binding covered, but coincidentally the –SH group of the cysteine was firmly bound. In addition, Seifter *et al.* (*75*) demonstrated that when a metal component was lost from collagenase A, the enzyme no longer had capacity to bind cysteine. They prepared and isolated

74. P. D. Boyer, *JACS* **76**, 4331 (1954).
75. S. Seifter, S. Takahashi, and E. Harper, *BBA* **214**, 559 (1970).

^{65}Zn-containing collagenase A and rendered it inactive by treatment with 2,3-dimercaptopropanol. The inactive enzyme was then passed through a column of Sephadex G-100, causing removal of virtually all the ^{65}Zn as a thiol complex and separation of a protein free of radio label. The inactive protein was then treated with ^{35}S–cysteine and the mixture again passed through Sephadex. This procedure resulted in the recovery of nonlabeled, inactive, enzyme protein separate from ^{35}S–cysteine, indicating that in the absence of zinc the protein could not bind cysteine.

The last experiment described above demonstrates a difference in mode of reaction of the two SH-containing compounds, cysteine and 2,3-dimercaptopropanol. The former combines with the putative metal atom and is fixed to the protein through this interaction. The latter agent apparently combines with the metal atom to abstract it from the protein. This behavior is consistent with the known chelation properties of the two agents.

The requirement of relatively high concentrations of cysteine for inhibition of the enzyme, despite the apparently stoichiometric and irreversible binding of the agent, is largely a reflection of the pH value chosen for the reaction. With an increase in pH from pH 7 to 10 less cysteine was required to inhibit the enzyme. The results suggest that binding of cysteine involves the amino group as well as the –SH group since the effect of increasing pH corresponded to a titration of the amino group. The higher concentrations of cysteine required at pH 7 merely reflect the lower concentration of chelating species present at this pH compared to that present at higher pH value.

The highest values of stability constants of complexes of zinc with various amino acids are with cysteine, intermediate values with histidine, and significantly lower values with most of the other amino acids. Accordingly, Harper (*76*) studied inactivation of collagenases by several imidazole derivatives. Histidine was found to inhibit both collagenases A and B reversibly when used at neutral pH. Other imidazole compounds were less inhibitory than histidine, again showing the multifunctional nature of the presumed chelation to the metal atom of the enzyme. The availability of a reversible inhibitor for collagenase offers many experimental advantages. Inhibition by histidine was competitive with the collagen substrate; thus, increasing the concentration of collagen used in the viscometric assay proportionately decreased the inhibition by a constant concentration of the amino acid. Again this suggests that the metal atom is at the site of substrate binding. Results with histidine and

76. E. Harper, *Federation Proc.* **25**, 790 (1966).

imidazole are shown in Table VI.

5. *Activation of Collagenase by Transition Elements*

Yagisawa *et al.* (*71*) studied the effects of ions of several metals on the activity of clostridiopeptidase on a variety of synthetic substrates. They found inhibition by zinc in relatively high concentration but also observed activation by Co^{2+}. Takahashi and Seifter (*73*) studied the effects of metals in detail using, however, a collagen substrate. The following observations were made. First, if cobalt ions are incorporated into the growth medium of *Clostridium histolyticum,* collagenase A can be isolated with a specific activity 1.8 times that found when the unfortified culture medium is used. Second, if collagenase A, prepared from the usual unfortified medium, is subjected to equilibrium dialysis in the presence of both Ca^{2+} and Co^{2+}, the enzymic activity is increased 1.8 times. Significant but less striking results are obtained with manganous or ferrous ions in place of cobaltous ions. Thus, as in the case of certain other zinc-containing enzymes, collagenase activity is improved by presumed replacement of the metal atom. (That the metal atom is indeed replaced is now being studied with the ^{65}Zn-containing enzyme.) In this case, however, Ca^{2+} appears to exert a strong directing effect on the replacement. As noted before, the variable specific activity of different collagenase preparations may be owing in part to the fact that the culture media used are not defined in metal composition. The possibility exists that various collagenase preparations are mixtures of molecular species differing in the kinds of metal present. This is not to say that the enzyme is synthesized with different metals, but variable exchange may occur as the synthesized enzyme is exposed to other metals in the medium.

6. *Attempts to Isolate Zinc-Containing Peptides from the Active Center*

Harper and Seifter (*77*) and Takahashi and Seifter (*78*) have used a double-label technique in an attempt to isolate portions of the collagenase molecule at the active center. Collagenase was prepared using a growth medium containing ^{65}Zn. The enzyme then was reacted with ^{35}S–cysteine, and the doubly labeled enzyme reisolated. The enzyme was digested with trypsin or sequentially with trypsin and chymotrypsin. Although difficulties were encountered in the completeness of digestion,

77. E. Harper and S. Seifter, *Abstr. 152nd ACS Meeting, New York* p. C-39 (1966).
78. S. Takahashi and S. Seifter, in preparation.

the investigators isolated several basic peptides containing both ^{65}Zn and ^{35}S–cysteine, and among other amino acids, histidine, glutamic acid, and lysine.

7. *Assays*

Screening assays utilizing one kind or another of denatured collagen or synthetic peptide should always ultimately be confirmed by an assay utilizing undenatured collagen. In one kind of definitive assay, the enzyme is mixed with a suspension of powdered "insoluble" collagen (such as that of bovine Achilles tendon) and proteolysis monitored by appearance of ninhydrin-reactive or peptide material in the soluble phase. Gallop *et al.* (*68*) developed a similar suspension assay with the difference that the substrate was a suspension of reconstituted fibers prepared from acid-extracted collagen, in particular, of ichthyocol. They defined a unit of collagenase as that amount causing the solution of 1 mg of suspended collagen under the conditions of the assay (37°, incubation for 20 min at pH 7.4). Then Gallop *et al.* (*68*) and Seifter *et al.* (*59*) showed that the activity also could be monitored using the same acid-extracted collagen in solution with $0.5\,M$ $CaCl_2$ buffered at neutral pH. Parameters measured, however, were either the change in optical rotatory activity or change in viscosity; the basis for these changes has been considered already in the introduction to this chapter. Gallop *et al.* (*68*) correlated the action of collagenase A in a single-timed assay using a suspension of fibers of ichthyocol with a kinetic viscometric assay using the same substrate in solution in $CaCl_2$. This allowed them to calculate a constant to interconvert results obtained in the two assays, so that in fact the unit of the viscometric assay is identical to that of the suspension assay. Because the assay depends on use of ichthyocol the conversion constant applies only to it; if another collagen such as that of rat skin is used another constant must be derived and applied. Variations of the viscometric assay and assays taking advantage of the change in optical rotation have since been employed with considerable success in identification and quantification of many collagenases of both microorganisms and tissues.

Screening assays are varied. In one case, a 2% solution of gelatin at neutral pH is allowed to be acted on by enzyme. At various times, aliquots are removed and placed in an ice bath. Failure of the solution to gel upon cooling is indicative of proteolysis. Extent of proteolysis can then be determined by reaction with ninhydrin, recalling that completion occurs when there is an approximate 7-fold increase of leucine equivalents from the initial time value.

A second and important screening procedure is with the use of synthetic oligopeptides. Several of these are available commercially. The peptides have been developed specifically for assay of collagenases on the basis of the assigned specificity of clostridial collagenase A. Useful substrates are those of Wünsch (*79*), Wünsch and Heidrich (*80*), and Schöllmann (*81*). The first of these is Z–Gly–Pro–Gly–Gly–Pro–Ala–OH and can be cleaved by either collagenase A or B at the –Gly–Gly– bond; activity can be measured by development of ninhydrin reactivity resulting from appearance of the amino-terminal glycine residue. A second peptide is PZ–Pro–Leu–Gly–Pro-D-Arg–OH, scission occurring at the –Leu–Gly– bond. The original peptide is yellow, and scission releases a yellow fragment that after acidification can be extracted into ethyl acetate and determined spectrophotometrically. A previously expressed caution in the use of these peptides should be reiterated. For instance, Heidrich *et al.* (*5*) showed that an extract of rat liver mitochondria could cleave the yellow substrate but was unable to digest native collagen. Harper and Gross (*4*), using culture media of tadpole tissues, could separate a peptidase capable of cleaving the same peptide but unable to digest native collagen. On the other hand, the purified *bona fide* tadpole collagenase could not cleave the synthetic substrate. (Z represents the benzyloxycarbonyl group, and PZ the *p*-phenylazobenzyloxycarbonyl group.)

8. *Thermodynamics and Kinetics*

Yagisawa *et al.* (*71*) have performed interesting kinetic studies with a purified clostridiopeptidase on a variety of synthethic oligopeptides. This paper should be consulted especially for the effects of specific substituents on rate of enzymic activity. The discussion below, however, will be limited to a consideration of the action of collagenase on collagen and gelatin.

Gallop *et al.* (*82*) introduced the use of neutral solutions of 0.5–1.0 *M* $CaCl_2$ for dissolving acid-extracted collagen, thereby permitting study of the action of enzymes in solution as opposed to suspension. They demonstrated the time course of the reaction of collagenase and of trypsin on collagen by changes in specific viscosity. Seifter *et al.* (*54*) also monitored the time course of the collagenase by changes in specific optical rotation. Then Seifter *et al.* (*41*) noted an unusual advantage of study offered by this system, namely, that it enabled one to follow the separate time courses of collagenolytic action on both the native and denatured

79. E. Wünsch, *Z. Physiol. Chem.* **332**, 295 (1963).
80. E. Wünsch and H. G. Heidrich, *Z. Physiol. Chem.* **333**, 149 (1963).
81. G. Schöllmann, *Z. Physiol. Chem.* **348**, 1629 (1967).
82. P. M. Gallop, S. Seifter, and E. Meilman, *J. Biochem. Biophys. Cytol.* **3**, 545 (1957).

forms of collagen present together in solution. Thus, the action on the native form could be followed by change in viscosity, and the total action on both forms could be followed by change in ninhydrin reactivity. From such a study with calfskin acid-extracted collagen and the gelatin prepared from it, they concluded that at 20° the binding of collagenase A to either substrate was probably the same but that the rate of proteolysis was approximately 10 times greater for the gelatin as opposed to the native collagen. They thought that the different rates in competing reactions, therefore, were not a consequence of preferential binding to gelatin but that potentially susceptible peptide bonds were more vulnerable in gelatin because of the absence of a stabilizing lattice structure.

Harrington *et al.* (*83*), studying the kinetics of proteolysis of several enzymes on native and denatured forms of fibrous proteins, concluded that in each case, including the collagen–collagenase system, the overall kinetics could be reduced to the sum of two concurrent apparent first-order reactions proceeding at markedly different rates. Denaturation of collagen appeared to cause the transfer of all susceptible bonds to a single reaction class exhibiting the thermodynamic parameters usually associated with cleavage of synthetic substrates and denatured substrates. In that communication the authors considered that this could imply a differentiation, by collagenases, between "amorphous" and "crystalline" regions in collagen.

Revision of the latter suggestion came from subsequent detailed studies in which collagenase was used to probe the configurational and conformational aspects of the collagen-fold transition. These studies have served to provide excellent kinetic information concerning the action of collagenase A. First, Harrington and von Hippel (*84*) found that above the transition temperature (collagen → gelatin) collagenase acts on collagen in a manner consistent with first-order kinetics. On the other hand, when a gelatin solution was recooled below the transition temperature and measurement taken within 5 min, the substrate was now cleaved with complex kinetics indicative of two events corresponding to fast and slow reactions. These events occurred before any observable change in optical rotation and viscosity of the gelatin solution, and they could not be explained by reformation of the collagen fold or aggregation of the molecules. Von Hippel and Harrington inferred that below the transition temperature the pyrrolidine residues became locked into a more stable configuration. They then proposed that the sequence of events in the recooling of gelatin was: First, the pyrrolidine residues begin to lock into the more stable trans-configuration; second, when this event has

83. W. F. Harrington, P. H. von Hippel, and E. Mihalyi, *BBA* **32**, 303 (1959).
84. W. F. Harrington and P. H. von Hippel, *BBA* **36**, 427 (1959).

proceeded to a sufficient degree, helices begin to reform, as evidenced by mutarotation; and, third, hydrogen bonding and aggregation occur with formation of gelatin gels. These investigators found in actuality that about 20% of the bonds in recooled gelatin (just below the transition temperature) were split in a fast reaction and about 80% in a slow reaction, and inferred that collagenase was sensitive to the initial configurational change occurring in the recooling process.

The second study seeking to explain the special sensitivity of collagenase to the intramolecular events occurring in the substrate in the neighborhood of the transition temperature was that of von Hippel *et al.* (*47*). They found that the ratio of susceptible bonds cleaved in the two reaction classes (fast, 20%; slow, 80%) was the same for gelatin cooled below the transition temperature as for collagen that had not been heated. Yet the apparent enthalpy and entropy of activation of proteolysis for the two systems differ markedly. Whereas, after initial "gelatinization" values of ΔH of about +22 and +29 kcal/mole and ΔS of approximately +30 and +51 eu were obtained for the fast and slow reaction classes in cold gelatin, respectively, the corresponding quantities obtained with unheated collagen (ichthycol) were considerably larger. These were, ΔH about +41 and +46 for the fast and slow reactions, respectively, and ΔS of +90 and +110, respectively. If indeed the enzyme distributes the susceptible bonds into the two observed reaction classes on the basis of local configuration, then the uniformly higher ΔH and ΔS values found for each class with unheated collagen could be a measure of the "tighter" folding of the polypeptide chains in the collagen macromolecule superimposed on the effect of local configurational differences. This conclusion is consistent with conclusions reached by Seifter *et al.* (*41*) in the competing substrate study already discussed.

Von Hippel *et al.* (*47*) suggested that collagenase operated on a collagen substrate in an approximately random attack. Von Hippel and Wong (*12*) then obtained experimental evidence giving reasonable support to this mechanism. They determined molecular number, molecular weight, and M_Z values for the fragments of a gelatin substrate present at different times in the course of enzymic proteolysis. If the process were random, the ratio of these values should be approximately 1:2:3. The data obtained were in fair accord with this ratio and therefore with the proposed mechanism. Furthermore, as the molecule weight decreased during the course of digestion, no trend in the ratio was discernible, indicating that an almost constant relative distribution of molecular sizes was maintained throughout most of the degradation process. (Of interest is the fact that trypsin, acting particularly on the polar regions of gelatin, gave similar kinds of data showing random attack.) Thus each enzymic scission would appear to occur as an independent event, and all chains appear to be degraded simultaneously rather than sequentially.

9. *Substrate Interactions*

As noted previously, an essentially nonproductive binding of collagenase to a suspended collagen substrate can occur at 0°. Certain special properties of both enzyme and substrate facilitate measurement of the extent of this interaction. First, the enzyme contains considerable numbers of residues of tryptophan and tyrosine, whereas collagen has no tryptophan and very few residues of tyrosine. Thus binding can be followed spectrophotometrically either in the soluble phase or in the solid phase after its gelatinization. Second, collagen contains hydroxyproline whereas the enzyme does not. Thus nonproductive binding can be distinguished from productive binding by determination of hydroxyproline in the soluble phase. Third, ^{14}C-labeled enzyme, produced biosynthetically, can be used for quantification of both soluble and solid phases in an absorption experiment. In addition to measurements of these kinds, one can simply measure residual enzymic activity in the soluble phase.

Takahashi and Seifter (*85*), employing several of these devices, found that at 0°, in experiments in which enzyme concentration was kept constant while substrate concentration was varied, 50% of the enzyme was bound when the molar ratio of collagen to enzyme was about 50:1. Considerable barriers likely exist to the penetration of enzyme to molecules located innermost in the fibers. Accordingly, advantage was taken of the reversible inhibition of collagenase by histidine to make a similar determination with collagen in solution. The ratio of collagen to enzyme when 50% of the enzyme was bound was about 25:1, a definitive improvement in the binding ratio.

10. *Mechanism of Catalysis*

Knowledge in this area is still primitive. Changes of configuration and conformation that have been studied in the enzyme–substrate interaction are virtually limited to defining changes in the substrate—a rather unusual state of affairs.

An early event in bond cleavage would seem to be movement of the collagenase polypeptide chain into a conformation necessary for binding and catalysis. This requires the specific participation of Ca^{2+}. The active conformation apparently positions an intrinsic metal atom, presumably zinc, so that one ligand is provided the substrate.

When the enzyme is inactivated by photooxidation in the presence

85. S. Takahashi and S. Seifter, unpublished results (1970).

of methylene blue, 4 of 16 residues of histidine are destroyed (*72*). No residues of tyrosine are affected at the point at which enzymic activity is just abolished. Either histidine residues are involved in chelation to the metal or are involved independently in the catalytic activity. Added zinc ions have a moderately protective effect against photoinactivation, but whether zinc is released from the enzyme during photoinactivation has not been determined. A strong protective effect is exerted by Ca^{2+}, but again it has not been determined whether fewer histidine residues are destroyed in the presence of this ion.

Models of collagen structure may give information concerning the binding and action of the enzyme. Dickerson (*86*) gives a picture of a space-filling model of the collagen molecule, and diagrams by Ramachandran (*1*) project this structure. Both in the Rich and Crick and in the Madras structures proposed for collagen, in a single polypeptide chain the glycine residues repeating in every third residue are projecting into the triple-stranded structure. The P residues are pointing outward; X residues fall toward the center. Thus a pocket is seemingly made with the X and glycine residues on the bottom and the projecting P residues on the sides. The zinc atom of the collagenase might fit into the pocket and make a ligand with the Gly peptide bond.

These concepts of the action of collagenase A do not explain why the action of the tissue collagenases is so limited, often to only one proteolytic scission per α chain. Perhaps each of this group of enzymes has evolved with absolute bond specificity for a peptide sequence that occurs only once in a chain or alternatively, it need not have this absolute specificity except in terms of exposure of this sequence in the collagen structure. The susceptible peptide bond might repeat, but a uniqueness of environment allows only one of the repeats to be available to the enzyme. As will be discussed below, tadpole collagenase acts more extensively on gelatin than on native collagen, illustrating that specificity includes uniqueness of environment.

III. The Collagenase of the Tadpole of *Rana catesbiana*

A. Introduction

Starting about 10 years ago, Gross and his colleagues (*87–90*) began a series of investigations that considerably increased our understanding

86. R. E. Dickerson, *in* "The Proteins" (H. Neurath, ed.), 2nd ed., Vol. 2, p. 603. Academic Press, New York, 1964.

of the biological degradation of collagen. First, they introduced methods by which collagenolytic enzymes in tissues could be detected by their lytic actions on native collagen substrates in tissue culture at neutral pH. Second, using these methods, they demonstrated the existence of a collagenase in the tails, gills, and back skin of tadpoles undergoing metamorphosis. Third, they demonstrated that the appearance of collagenolytic activity was under very fine physiological control, including regulation by endocrine secretions. This kind of control had been suggested previously for several tissues including mammalian skin. Fourth, they introduced the possibility of studying the assembly and organization of the collagen molecule by application of a group of enzymes with defined but extremely limited action on the native collagen molecule.

Following these discoveries, most of the tissue collagenases shown in Table I were isolated using the methods described or related procedures. The surviving cell tissue culture method was an important point of departure because most frequently simple extractive procedures cannot remove a collagenase from the collagen in the tissue; the binding is especially tight as in the case of the clostridial collagenases when they are mixed with tissue collagen. Also, in many cases the enzymes are elaborated only on "physiological demand" and disappear under further biological control. Thus without the kind of culture method used, collagenases may be too transient for detection.

B. Preparation of Enzyme

Full details of the preparation may be found in original articles by Nagai *et al.* (*90*) and Harper and Gross (*4*) and in a review by Seifter and Harper (*39*). Tissues (e.g., tail and back skin) are allowed to metabolize under conditions of surviving cell tissue culture. A medium is employed that contains balanced salts, glucose, and antibiotics to limit bacterial contamination. The enzyme can be detected in the medium perhaps in two and usually in three days of incubation. Apparently the enzyme digests endogenous collagen to which it is bound and is leached into the medium. This process recalls the method of purification by adsorption to collagen used for clostridial collagenase by Gallop *et al.* (*68*).

---

87. J. Gross and C. M. Lapiere, *Proc. Natl. Acad. Sci. U. S.* **48,** 1014 (1962).
88. C. M. Lapiere and J. Gross, *in* "Mechanisms of Hard Tissue Destruction," Publ. No. 75, p. 663. Am. Assoc. Advance. Sci. Washington, D. C., 1963.
89. J. Gross and Y. Nagai, *Proc. Natl. Acad. Sci. U. S.* **54,** 1197 (1965).
90. Y. Nagai, C. M. Lapiere, and J. Gross, *Biochemistry* **5,** 3123 (1966).

In an incubation, then, when enzymic activity appears to be maximum, the medium is separated, dialyzed, and lyophilized. Purification may involve steps of fractionation with ammonium sulfate, gel filtration with either Sephadex G-200 or 8% Agarose, and gel slab electrophoresis. The enzyme also has been purified using columns of DEAE-cellulose but appears to be less stable under these conditions. In general the enzyme does not withstand repeated manipulations such as lyophilization or, in solution, freezing and thawing.

Frequently, purified preparations of the tadpole enzyme have been accompanied by general protease and peptidase activities, but Harper and Gross (*4*) now report a collagenase free of caseinolytic activity and of peptidases able to cleave certain peptides used in the assay of clostridial collagenases. Of interest is that at one point in the purification, after the Agarose step, two peaks of collagenolytic activity were observed (*4*).

The enzyme has not yet been characterized sufficiently by physicochemical methods to warrant discussion.

C. Catalytic Properties

1. *Reactions Catalyzed and Specificity*

The enzyme degrades collagens from several species. In one case the products found had carboxyl-terminal glycine residues and amino-terminal leucine or isoleucine residues Acting on denatured collagen, the tadpole enzyme had more extensive effect and yielded peptides again with carboxyl-terminal glycine residues but with amino-terminal residues of leucine, isoleucine, valine, alanine, or phenylalanine (*91*). The enzyme does not cleave the synthetic peptides, PZ–Pro–Leu–Gly–Pro-D-Arg–OH or Z–Gly–Pro–Gly–Gly–Pro–Ala–OH. The differences from clostridial collagenases are striking: The bacterial collagenase A makes many scissions, almost all resulting in peptides with amino-terminal glycine and has general nonspecificity in the carboxyl-terminal position; also, it cleaves the synthetic peptides pictured.

2. *Activation and Inhibition*

Like most of the described tissue collagenases (and clostridial), the tadpole enzyme has activity over a broad range of pH values but

91. Y. Nagai, C. M. Lapiere, and J. Gross, *Abstr. 6th Intern. Congr. Biochem., New York,* 1964 Vol. II, p. 170. 1964.

optimum activity between pH of 8–9. The enzyme is activated by Ca^{2+}. From studies by Nagai *et al.* (*90*), the enzyme is known to be inactivated by heating in solution to 60° and by lowering the pH to 3.5. It is inhibited by 0.002 *M* EDTA, but the inhibition is reversible by dialysis against solutions of $CaCl_2$. The enzyme is irreversibly inhibited by cysteine in a concentration of 0.005 *M*. Diisopropylphosphorofluoridate in concentration as high as 10^{-2} *M* is without effect. In all of these properties the tadpole enzyme is similar to the clostridial collagenases.

3. *Assays*

At least three types of assay systems have been described and would appear to be useful for most tissue collagenases. These are described in detail by Nagai *et al.* (*90*) and reviewed elsewhere (*39*). In one type of assay, called the change of opacity assay, the collagen is acted on by the collagenase, and the extent of activity is determined by the subsequent failure of the collagen to form fibrils under defined conditions. This method is similar to that devised by Houck and Patel (*92*). A second assay, called the viscosity assay is a variation of the viscosity method introduced by Gallop *et al.* (*68*). Modifications include utilization of acid-extracted collagen of guinea pig skin and a tris buffer, pH 7.4, containing 0.4 *M* NaCl. The reaction is conducted at 27° (below the transition temperature), and the decrease in specific viscosity with time is recorded. The assay does not provide a measure of units of enzymic activity; if this is required, considerations described by Seifter and Gallop (*93*) may be employed.

The third assay described by Gross and his colleagues is rapid and extremely useful but requires the use of collagen labeled biosynthetically with ^{14}C-glycine. Aliquots of this kind of collagen, prepared from skins of appropriately treated guinea pigs, are converted to gels at 37°. The gels are disrupted and reincubated with enzyme under test for various times. Reaction is stopped by addition of EDTA, the mixture is centrifuged to separate undissolved collagen, and the supernatant fluid is assayed for radioactivity. Another kind of radioactivity assay can be carried out at 33° with the collagen in solution instead of in the gel form. In this case, at the end of enzymic incubation, the protein must be precipitated with phosphotungstic acid. In both assays, enzymic activity is expressed in terms of counts per minute released minus the counts per minute obtained with suitable controls.

92. J. C. Houck and Y. M. Patel, *Ann. N. Y. Acad. Sci.* **93,** 333 (1962).
93. S. Seifter and P. M. Gallop, "Methods in Enzymology," Vol. 5, p. 659, 1962.

4. *Action as Viewed by Electron Microscopy*

The purified tadpole collagenase cleaves the collagen molecule across the three chains at one point in each chain (*63*). Thus, native collagen of rat skin is cleaved into two pieces, both retaining the triple helical collagen structure. A larger piece, called TC^A (denoting a portion of the tropocollagen molecule containing the end designated as A by electron microscopists), represents 75% of the molecule; while the smaller piece, TC^B (arising from the B end of the tropocollagen molecule as designated by microscopists), represents 25% of the original molecule. The two pieces of the cleaved collagen molecule can be separated, or alternatively, the mixture can be denatured by mild heat and fractionated to yield corresponding pieces of cleaved α chains, named according to the chain from which they derive and the end of the molecule (A or B). Thus, the smaller pieces, designated $\alpha 1^B$ and $\alpha 2^B$, are each of 24,000 molecular weight, and the larger pieces, $\alpha 1^A$ and $\alpha 2^A$, are each 71,000 molecular weight. The A ends contain the original amino-termini of the α chains. Figure 1, kindly furnished by Dr. J. Gross, demonstrates exquisitely the action of tadpole collagenase on collagen as viewed by electron microscopy. The technique of using enzymes to cleave collagen, then reconstituting the pieces as in this instance, and finally aligning electron photomicrographs as shown here, is proving of aid in locating the site of enzymic action. By now it has been used for virtually all of the known tissue collagenases; thus, pictures similar to that in Fig. 1 are available, for instance, for an enzyme of human granulocytes (*94*), for one from human rheumatoid synovium (*69*), and for one from resorbing postpartum uterus of the rat (*95*). Figure 2 shows a short fragment of TC^B as viewed by an electron microscope.

IV. The Collagenases of Human Synovial Fluid

A brief discussion of this subject is given because it may offer an example of the possible importance of collagenases in processes of human disease.

94. G. S. Lazarus, J. R. Daniels, R. S. Brown, H. A. Bladen, and H. M. Fullmer, *J. Clin. Invest.* **47**, 2622 (1968).
95. J. J. Jeffrey and J. Gross, *Biochemistry* **9**, 268 (1970).

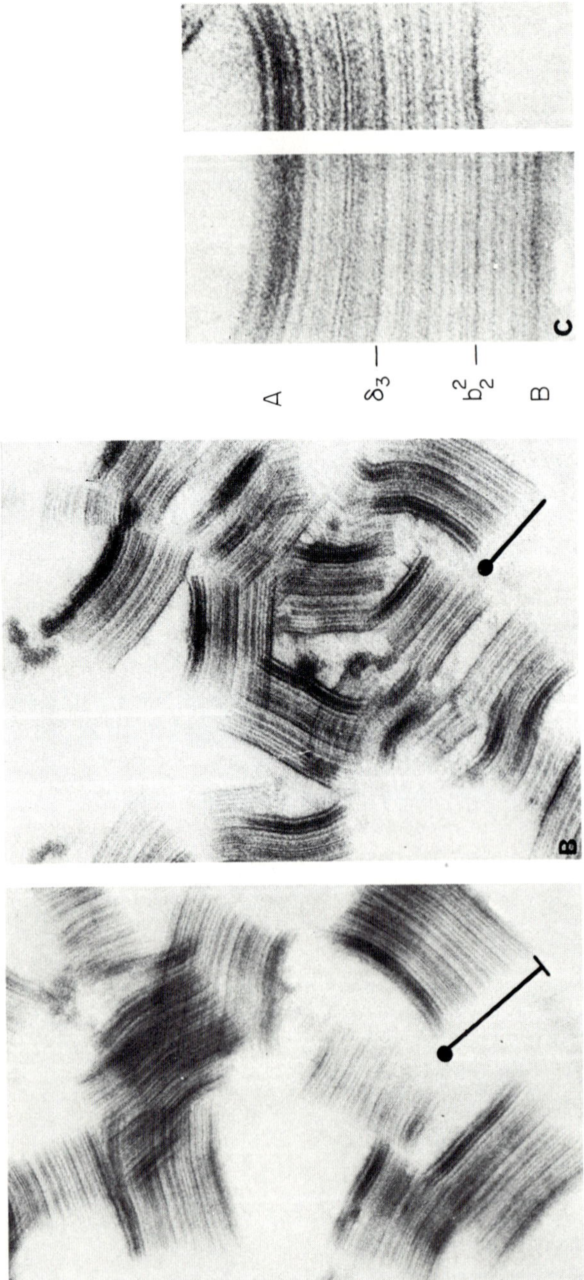

FIG. 1. (A) Segment long spacing of normal calfskin collagen; (B) SLS of three-quarter lengths TC^A fragments from reaction mixture. (C) SLS of normal collagen compared with that of TC^A. From Gross and Nagai (*89*).

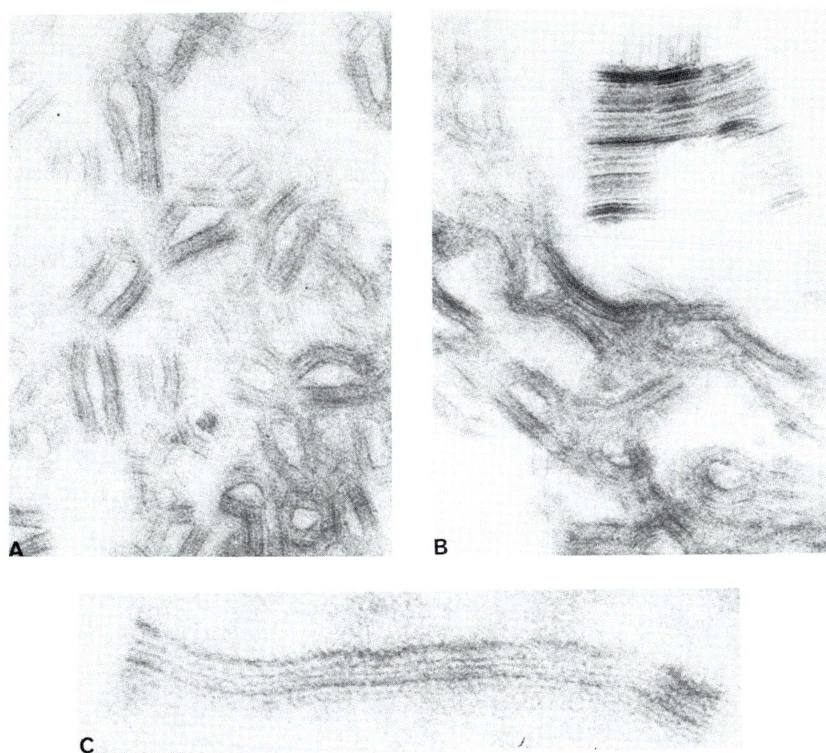

FIG. 2. (A) One-quarter length SLS of the short fragment TC^B; (B) a similar field showing segments of TC^B, and in addition a group of segments of the large fragments TC^A aggregated in a dimeric fashion; (C) an enlarged picture of a single broad SLS of the short fragments of TC^B. From Gross and Nagai (*89*).

Evanson *et al.* (*96, 97*) prepared cultures of rheumatoid synovium and demonstrated the presence of a collagenase similar in properties to the collagenases of other tissues such as normal human skin and tadpole tail. Like other tissue collagenases, that of the synovium is inhibited by certain serum proteins. Additional characteristics of synovial collagenolytic activity have been discussed by Harris *et al.* (*69*) and Lazarus *et al.* (*98*).

96. J. M. Evanson, J. J. Jeffrey, and S. M. Krane, *Science* **158,** 499 (1967).
97. J. M. Evanson, J. J. Jeffrey, and S. M. Krane, *J. Clin. Invest.* **47,** 2639 (1968).
98. G. S. Lazarus, J. L. Decker, C. H. Oliver, J. R. Daniels, C. V. Multz, and H. M. Fullmer, *New Engl. J. Med.* **279,** 914 (1968).

Extracts made directly from tissues known to elaborate collagenases show no collagenolytic activity except in the case of crab hepatopancreas or human granulocytes. Homogenates of human rheumatoid synovium do not. Yet the collagenases are found in culture media containing surviving cells. Accordingly, Harris *et al.* (*69*) examined synovial fluids for collagenolytic activity since these represent a medium bathing cells manufacturing collagenase. The proteins of synovial fluid derive mostly by exudation from blood serum, but some are added by the synovial lining cells. These investigators were able to isolate, in fact, two collagenases from some rheumatoid synovial fluids. The enzymes were purified partially by gel filtration with Sephadex G-200. The two enzymes have been labeled A and B, respectively. The B enzyme seemed to be similar to, if not identical with, that isolated from cultures of synovium. It had an estimated molecular weight (by gel filtration) of 20,000–25,000. The A enzyme had an estimated molecular weight of 40,000–50,000. Both are capable of degrading collagen in solution and in the form of fibrils. From electron micrographs and disc-gel electropherograms, both enzymes seem to act like other tissue collagenases including that of the tadpole (see Fig. 1). That is, they cleave the collagen molecule in three chains in regions of each chain about three-quarters of the length removed from the amino-terminal ends. Thus they produce pieces of collagen, still retaining helical structure, that represent three-quarters and one-quarter of the molecular weight, respectively. Whether one or multiple cleavage occurs in the sensitive region of each chain is not known, nor is the nature of peptide bonds cleaved.

The B enzyme is strongly inhibited by serum, and the inhibitor appears to be protein in nature. This behavior is similar to that of the collagenase obtained by culture of synovium but different from that of the A enzyme of synovial fluid. The latter is only weakly inhibited, if at all, by serum used in concentrations that produce inhibition of the B enzyme.

The relationship between the A and B enzymes is not known, nor is their possible interplay in the destruction of collagen. Harris *et al.* (*69*) have speculated that the B enzyme might be of importance at the sites where the rheumatoid panus is in direct contact with cartilage and other collagenous structures. This area is less accessible to serum and presumably the inhibitor in serum. The A enzyme, unaffected by the inhibitor, supposedly could act on collagen in any part of the joint. The several investigators are now studying whether the amount of collagenase in a given synovial fluid correlates with the rate of articular destruction or clinical activity of the disease.

Acknowledgments

The authors acknowledge a long-standing debt of appreciation to Professors Paul M. Gallop and Jerome Gross, pioneers in the subject of this chapter. One of us (S.S.) gives special thanks to Dr. Marcos Rojkind and Dr. Shizuko Takahashi for many stimulating discussions concerning material presented here. This chapter was written while the authors were under auspices of grants from the U. S. Public Health Service, AM 3172, AM 11913, and AM 3564.

19

Clostripain

WILLIAM M. MITCHELL • WILLIAM F. HARRINGTON

I. Introduction

A. Definition

Clostripain [clostridiopeptidase B (EC 3.4.4.20)] is a sulfhydryl proteolytic enzyme derived from the culture filtrate of *Clostridium histolyticum*. This enzyme possesses amidase-esterase as well as proteolytic activity over a narrow specificity range with the major activity directed toward the carboxyl peptide linkage of arginine. Lysyl bonds are also attacked but at rates generally much slower than hydrolysis at arginine bonds.

B. Historical Perspective—The Multiplicity of Proteases in *C. histolyticum* Culture Filtrates

The first description of the organism now known as *Clostridium histolyticum* was apparently made by Weinberg and Séguin (*1*) in 1916 from wound cultures of eight cases of gas gangrene. The prevalence of the latter and its dire consequences to the wounded during World War I provided the first major impetus for research on the *Clostridia* (*1a*). Although the organism (*C. histolyticum*) is only an occasional pathogen (*2*), the lesions produced by culture innoculation into guinea pig muscle are striking and led to the identification in 1917 of the organism's proteolytic activity in cell-free culture filtrates (*3*). In 1931, Weinberg and Randin (*4*) described the digestion of horse Achilles' tendon at 37° by *C. histolyticum* and a year later (*5*) identified the causative agent as an exotoxin which they named *ferment fibrinolytique*. It is now known that the proteolytic activity observed represented a variety of proteolytic enzymes. That the digestive properties of *C. histolyticum* culture filtrate resulted from a multiplicity of enzymes first became apparent in 1938 when Kocholaty (*6*) described a secreted protease which was cysteine activated and Maschmann (*7*) reported the presence of a noncysteine-activated protease. Later studies of Van Heyningen (*8*) also provided evidence for a multiplicity of proteases.

The cysteine-activated enzyme was first isolated in relatively pure form by Kocholaty and Krejci (*9*), and their findings were extended and

1. Weinberg, M. and Séguin, P., *Compt. Rend.* **163,** 449 (1916).

1a. *"La peau est ouverte; une bouillie hémorragique s'échappe de la lésion et l'on aperçoit le squelette du membre complétement dénudé. Souvent l'articulation du genou est attaquée; les ligaments, la capsule, le revêtement cartilagineux de l'articulation sont détruits, et le tibia se détache spontanément du fémur (autoamputation inflammatoire). La lésion n'est pas putride, il n'y a pas formation de gaz. Le cobaye peut survivre 12 à 24 heures à cette horrible mutilation."* See Weinberg and Séguin (*1*).

2. MacLennan, J. D., *Bacteriol. Rev.* **26,** 177 (1962).
3. Weinberg, M. and Séguin, P., *Compt. Rend. Soc. Biol.* **80,** 157 (1917).
4. Weinberg, M. and Randin, A., *Compt. Rend. Soc. Biol.* **107,** 27 (1931).
5. Weinberg, M. and Randin, A., *Compt. Rend. Soc. Biol.* **110,** 352 (1932).
6. Kocholaty, W., Weil, L., and Smith, L., *Biochem. J.* **32,** 1685 (1938).
7. Maschmann, E., *Biochem. Z.* **295,** 391 (1938).
8. Van Heyningen, W. E., *Biochem. J.* **34,** 1540 (1940).
9. Kocholaty, W. and Krejci, L. E., *Arch. Biochem.* **18,** 1 (1948).

modified by the later work of Ogle and Tytell (*10*) who reported for the first time the remarkable specificity of clostripain (*11*). Further evidence for multiple exotoxins in the culture filtrates of *C. histolyticum* was advanced by Bard and McClung (*12*) and Oakley and Warrack (*13*) using a differential immunological technique (*14*). Three antigenic components were identified: the α-antigen as a lethal, necrotizing toxin, β-antigen as a collagenase, and γ-antigen as a cysteine-activated protease capable of attacking azocoll (a dye-labeled gelatin) but not native collagen. MacLennan *et al.* (*15*) also provided evidence for a δ-antigen, a protease not activated by cysteine but whose activity is enhanced by calcium and is capable of benzoylarginine amide hydrolysis. Bowen (*16*) described a hemolytic factor which was nonidentical with the lethal toxin and was labile to oxidation with partial recovery on subsequent reduction. This substance was found to be susceptible to proteolytic attack by trypsin, papain, or pepsin, but it was not itself assayed for proteolytic activity. Isolation and/or partial characterization of the δ-antigen (i.e., cysteine-inhibited protease) has been reported by Mandl *et al.* (*17, 18*) and Nordwig and Strauch (*19*). The latter workers also described an amidase-esterase which, they stated, is nonproteolytic but possesses many of the properties of the cysteine-activated (γ-antigen) protease. Several peptidases have also been reported by Mandl *et al.* (*20*) with activity against various di- and tripeptides.

In addition to these general proteases and peptidases, the culture filtrate of *C. histolyticum* contains a variety of specific collagenases. The

10. Ogle, J. D. and Tytell, A. A., *Arch. Biochem. Biophys.* **42,** 327 (1953).

11. Of a number of simple substrates examined, the cysteine-activated protease readily hydrolyzed only the esters and amides of α-benzoyl-L-arginine and arginine methyl ester.

12. Bard, R. C. and McClung, L. S., *J. Bacteriol.* **56,** 665 (1948).

13. Oakley, C. L. and Warrack, G. H., *J. Gen. Microbiol.* **4,** 365 (1950).

14. The immunological assignments discussed above are somewhat confusing in that they tend to imply isolation of the antigen. However, this is not the case since assignments were made on differential response from crude or partially purified preparations to multiple immune sera.

15. MacLennan, J. D., Mandl, I., and Howes, E. L., *J. Gen. Microbiol.* **18,** 1 (1958).

16. Bowen, H. E., *Yale J. Biol. Med.* **25,** 131 (1952).

17. Mandl, I., MacLennan, J. D., and Howes, E. L., *J. Clin. Invest.* **32,** 1323 (1953).

18. DeBellis, R., Mandl, I., MacLennan, J. D., and Howes, E. L., *Nature* **174,** 1191 (1954).

19. Nordwig, A. and Strauch, L., *Z. Physiol. Chem.* **330,** 145 (1963).

20. Mandl, I., Ferguson, L. T., and Zaffuto, S. F., *Arch. Biochem. Biophys.* **69,** 565 (1957).

utilization of these enzymes as molecular probes of collagen structure and collagen biosynthesis demands an enzyme free of contaminating proteases, a goal that has proven difficult to achieve. In 1957, Gallop *et al.* (*21*) reported a partial purification of collagenase from *C. histolyticum* and introduced the viscometric method of assay. The same authors (*22*) later described an improved method of preparation using C γ gel. No attempt was made to assay cysteine-activated proteolysis or cysteine-activated hydrolysis of synthetic substrates. In the following year Mandl and her colleagues (*23*) reported purification of collagenase by hanging curtain electrophoresis. This preparation was considered to be free of general protease activity but contained cysteine-activated activity against benzoylarginine amide. Keller and Mandl (*24*) have also described the purification of collagenase utilizing Sephadex G-200 chromatography in addition to a procedure using DEAE-Sephadex A-50. The loss of all noncollagenolytic proteolytic activity was claimed, but a significant activity against benzoylarginine naphthylamide as well as benzoylarginine amide was observed in material isolated by both techniques. More recently, Mandl *et al.* (*25*) have reported an altered gradient method on DEAE yielding multiple fractions with differing activity against ichthyocol and a synthetic collagenase hexapeptide.

The demonstration of a multiplicity of collagenolytic enzymes in *C. histolyticum* culture filtrates has been repeatedly confirmed in the past five years. Harper *et al.* (*26*) utilized column chromatography on Sephadex G-200, and more recently Sephadex G-50, followed by chromatography on DEAE-cellulose (*27*). Evidence of a multiplicity of collagenases was presented with the suggestion that one species is a dimer of the other from ultracentrifugation studies. However, no data were reported on the presence or absence of cysteine-activated protease or esterase activity. Two collagenases of differing activity against native collagen and synthetic peptides have also been described by Yoshida and Noda (*28*) using DEAE-cellulose chromatography. Evidence for the

21. Gallop, P. M., Seifter, S., and Meilman, E., *J. Biol. Chem.* **227,** 891 (1957).
22. Seifter, S., Gallop, P. M., Klein, L., and Meilman, E., *J. Biol. Chem.* **234,** 285 (1959).
23. Mandl, I., Zipper, H., and Ferguson, L. T., *Arch. Biochem. Biophys.* **74,** 465 (1958).
24. Keller, S. and Mandl, I., *Arch. Biochem. Biophys.* **1,** 81 (1963).
25. Mandl, I., Keller, S., and Manahan, J., *Biochemistry* **3,** 1737 (1969).
26. Harper, E., Seifter, S., and Hosplehorn, V. D., *Biochem. Biophys. Res. Commun.* **18,** 627 (1965).
27. Harper, E., Doctoral Dissertation, Albert Einstein College of Medicine, New York, 1965.

presence of multiple collagenases has also been advanced by Schaub and Strauch (*29*). They found a protease in their amidase-esterase fraction (the latter inhibited by hydrogen peroxide) which degrades calfskin collagen at 30°C at high substrate:enzyme ratios but which is unable to degrade the synthetic substrate Cbz–Gly–Pro–Gly–Gly–Pro–Ala. The recent work of Kono (*30*) has clearly established not only the multiplicity *C. histolyticum* collagenases but also their synergistic effect. An additional enzyme (pseudo-collagenase) related to the collagenase family on the basis of the primary sequence requirements for hydrolysis as well as size and electrophoretic mobility has been isolated and characterized by Mitchell (*31*). This enzyme, present in culture filtrates in small yield, rapidly attacks synthetic collagenase substrates and gelatin, but it has no hydrolytic capacity against native collagen. The existence of this specific gelatinase, distinct from collagenase, was suggested in 1958 by Mandl and Zaffuto (*32*) who used a differential immunological method. Ogle and Logan (*33*), in a short communication two years earlier, had described the gelatinase action of an ethylenediaminetetraacetate (EDTA)-sensitive enzyme from an alcohol precipitate which yielded only N-terminal glycine on analysis of the liberated peptides, although the unique specificity for gelatin with the exclusion of the parent collagen was not shown.

Thus the culture filtrate of *C. histolyticum* serves as a rich source of proteolytic enzymes. Although most investigators have focused attention on the group of enzymes within the collagenase family during the past 20 years, the recent demonstration of the narrow substrate specificity of clostripain has generated interest in this protease as well. As a result of the various levels of purity obtained in preparations and methodologies used in assessment of proteolytic activity, some confusion exists in the literature regarding the identification of the enzyme, and it has been described at various times as clostripain (*34, 35*), as cysteine-activated proteinase (*10*), γ-protease (*12, 13*), amidase-esterase (*19*), and clostridiopeptidase B (*35*).

---

28. Yoshida, E. and Noda, H., *Biochim. Biophys. Acta* **105**, 562 (1965).
29. Schaub, M. C. and Strauch, L., *Biochem. Biophys. Res. Commun.* **21**, 34 (1965).
30. Kono, T., *Biochemistry* **7**, 1106 (1968).
31. Mitchell, W. M., *Biochim. Biophys. Acta* **159**, 554 (1968).
32. Mandl, I. and Zaffuto, S. F., *J. Gen. Microbiol.* **18**, 13 (1958).
33. Ogle, J. D. and Logan, M. A., *Federation Proc.* **15**, 323 (1956).
34. Labouesse, B. and Gros, P., *Bull. Soc. Chim. Biol.* **42**, 543 (1960).
35. Mitchell, W. M. and Harrington, W. F., *J. Biol. Chem.* **243**, 4683 (1968).

II. Purification and Assay

Methods for the purification of clostripain have been developed utilizing *C. histolyticum* culture filtrates directly (*34, 36*), but commercially available collagenase preparations afford a more convenient starting material (*35, 36*). A sensitive and accurate assay of clostripain activity is obtained by measurement of the initial hydrolytic rate against benzoylarginine ethyl ester (BAEE) (*35, 36*). Since the esterolysis of BAEE results in an increased ultraviolet absorption by the product, benzoylarginine, the initial rate can be monitored conveniently through the change in optical density (at 253 nm) with time (*37*). By using the partially purified commercial collagenase preparations, approximately a twelvefold increase in specific activity by chromatography on hydroxypapatite and Sephadex is obtained. The product shows apparent homogeneity as judged by sedimentation velocity and sedimentation-equilibrium ultracentrifugation, immunodiffusion, immunoelectrophoresis, and polyacrylamide gel electrophoresis (*35*).

III. Structural Properties

A. Physical Constants

The physical properties which characterize clostripain are listed in Table I. The molecular weight of 50,000 g/mole obtained by Mitchell and Harrington (*35*) is large for a protease but half the value reported by Labouesse and Gros (*34*) from sedimentation-diffusion data. This discrepancy may be the result of dimerization in the low salt medium employed in the studies of Labouesse and Gros, an interpretation consistent with the measured sedimentation coefficients $s^0_{20,w} = 4.43$ (high salt) (*35*) and $s^0_{20,w} = 6.70\,S$ (low salt) (*34*). Hydrodynamic evidence and electron microscopy reveal the protein to be spherical with a radius of approximately 33–35 Å (*38*). Clostripain possesses an acid isoelectric point (*39*) from isoelectric focusing techniques (Fig. 1) and a low α-

36. Mitchell, W. M. and Harrington, W. F., *in* "Methods in Enzymology" (G. Perlmann and L. Lorand, eds.), Vol. XIX, p. 635. Academic Press, New York, 1970.
37. Schwert, G. W. and Takenaka, Y., *Biochim. Biophys. Acta* **16,** 570 (1955).
38. Mitchell, W. M. and Schmidt, J. J., *Biochim. Biophys. Acta* **175,** 207 (1969).
39. Mitchell, W. M., *Biochim. Biophys. Acta* **178,** 194 (1969).

TABLE I
PHYSICAL PROPERTIES OF CLOSTRIPAIN

Molecular weight, apparent[a]	50,000
Sedimentation coefficient ($s^0_{20,w}$)[a]	4.43 S
Partial specific volume, $\bar{V}$, calculated	0.72 ml/g
Intrinsic viscosity [η]	0.053 dl/g
Particle shape	Spherical
Particle radius, hydrodynamic and electron microscopy	33–35 Å
Isoelectric point (8°C)	4.8–4.9
Optical rotatory dispersion (4.5°C)	
Cotton minima	233 n*M*
b_0(Moffitt)	−197°
λ_c(one term Drude)	229°
α-Helicity, % estimated	<30
Helix-coil transition	~50°C

[a] See text for reported values significantly different from those reported here.

helical content as judged by optical rotatory dispersion in the 220–300-nm range (*35*). A small but definite decline in the mean specific residue rotation $[m']_\lambda$ is observed when the temperature is raised from 4.5° to 39°, but over the temperature interval, 39–51°, the molar rotation changes rapidly and a major fraction of the helical content is lost. The loss in structure coincides with the complete loss of enzymic activity which occurs near 50°.

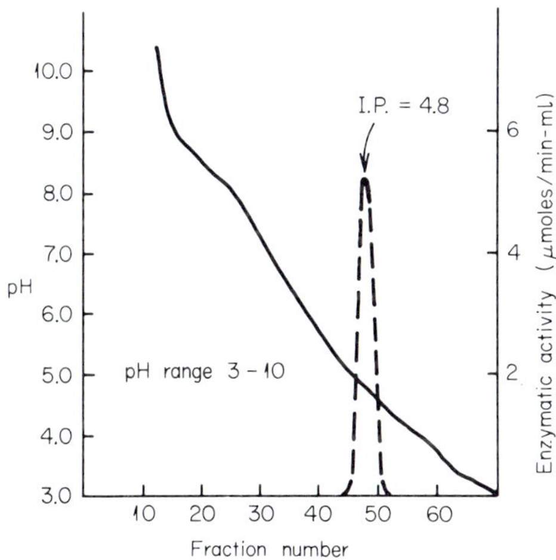

FIG. 1. Isoelectric focusing of clostripain at 8° over a pH gradient of 3–10. The abscissa represents the fraction number (3.1 ml/fraction), the left ordinate the pH gradient (——), and the right ordinate the activity against BAEE.

B. Amino Acid Composition

The amino acid composition of clostripain (*35*), summarized in Table II, shows several differences from the collagenolytic enzymes making up the bulk of proteolytic activity of *C. histolyticum* culture filtrates. Serine, histidine, tryptophan, half-cystine, and methionine are markedly higher than the reported values for collagenase A or collagenase B, whereas proline, alanine, valine, isoleucine, and tyrosine are present in distinctively lower amounts.

IV. Enzymic Properties

A. General Features

1. *pH Optimum and Ion Effects*

The pH optimum for hydrolysis of L-arginine methyl ester in phosphate buffer is about 7.2 whereas maximum activity against α-benzoyl-

TABLE II
Comparison of Amino Acid Contents of Clostripain, Collagenase A, and Collagenase B

Amino acid	Clostripain[a]	Collagenase A[b] (residues/1000 residues[c])	Collagenase B[b]
Aspartic acid	148	159	170
Threonine	49	54	60
Serine	74	27	32
Glutamic acid	99	85	93
Proline	30	56	40
Glycine	82	96	96
Alanine	51	73	68
Valine	47	66	60
Isoleucine	43	56	61
Leucine	77	73	74
Tyrosine	35	50	48
Phenylalanine	39	50	49
Lysine	91	103	103
Histidine	24	15	18
Arginine	24	37	29
Tryptophan	19	5	6
Half-cystine	42	0	0
Methionine	27	0	0

[a] As reported by Mitchell and Harrington (*35*).
[b] As reported by Harper (*27*) and Harper *et al.* (*26*).
[c] Uncorrected for hydrolytic losses.

arginine ethyl ester occurs in the range pH 7.4–7.8 in phosphate or tris HCl buffer (*35*) (Fig. 2). Calcium is unique among metal ions in enhancing the esterase activity (50% increase in initial rate of cleavage at 10^{-2} *M*) and also act to depress thermal inactivation. The work of Labouesse and Gros (*34*) indicates that binding of calcium ion is an essential prerequisite of activity whereas significant inhibition of esterase activity is observed in the presence of other cations such as Co^{2+}, Cu^{2+}, Cd^{2+}, Na^{+} and K^{+} (Fig. 3) (*35*). Conversely, EDTA completely inhibits enzymic activity possibly secondary to Ca^{2+} chelation. Citrate, borate, Veronal, and tris anions partially inhibit esterase activity.

2. *Sulfhydryl Requirement*

The enzymic activity of clostripain is rapidly lost on incubation in the presence of oxidizing agents such as hydrogen peroxide, but activity may be regenerated completely by reduction with dithiothreitol (DTT) (*35*). Labouesse and Gros (*34*) have investigated the effect of cysteine concentration on the esterase activity of the enzyme and report that one molecule of cysteine is required per mole of enzyme for activation; their studies also demonstrate inactivation of the enzyme by *p*-chloromercuri-

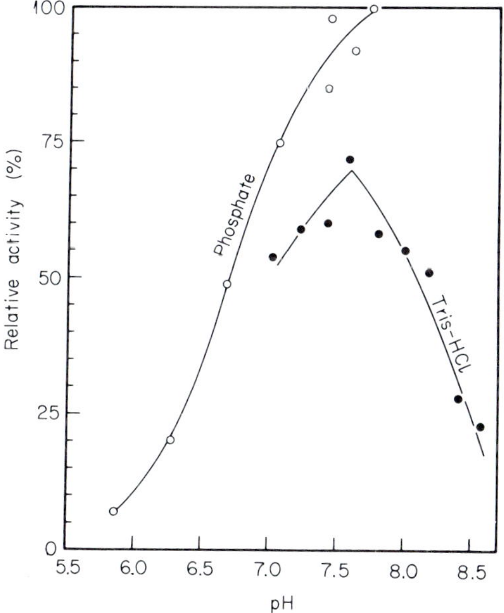

FIG. 2. Clostripain pH optimum profile against BAEE hydrolysis. Substrate concentration is 0.25 m*M* with 0.5 m*M* dithiothreitol (DTT) and $\Gamma/2 = 0.045$.

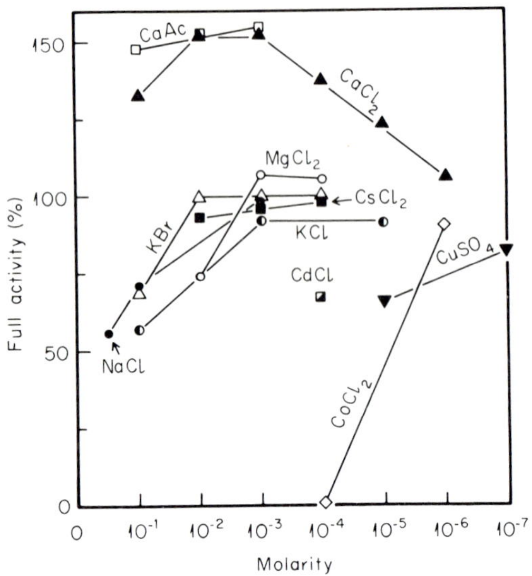

FIG. 3. Effect of various mono and divalent cations on the clostripain-catalyzed hydrolysis of BAEE (0.25 mM). All assays in 0.5 mM DTT and $\Gamma/2 = 0.03$ tris buffer.

benzoate (PCMB) with complete loss of activity occurring at a 1:1 molar ratio enzyme: inhibitor.

B. SPECIFICITY

1. *Synthetic Esters and Amides*

The relative rate of hydrolysis of a number of simple substrates is presented in Table III. Among the synthetic esters and amides which have been examined, arginine-containing substrates are much more rapidly hydrolyzed than identically substituted lysine compounds. The cleavage rate of L-lysine methyl ester is anomalously low and the amides of L-phenylalanine, L-leucine, glycine, and L-proline as well as the peptides L-leucylglycine, glycylglycine and L-leucylglycine are not attacked. No demonstrable activity can be detected against the standard chymotryptic substrates N-acetyl-L-tyrosine ethyl ester and N-benzoyl-L-tyrosine ethyl ester. Thus the enzyme appears to have a specificity similar to that of trypsin in that cleavage of the simple substrates occurs at the carboxyl-terminal bonds of arginine and lysine in peptide linkage, but a striking preference is shown for cleavage at arginine residues.

TABLE III
COMPARISON OF ESTERASE AND AMIDASE ACTIVITIES FOUND IN *C. histolyticum* CULTURE FILTRATES

	I[a,b]	II[a]	III[b]	IV	V	VI
A. Arginine						
N-Benzoyl-L-arginine ethyl ester		+		+	+	+
N-Benzoyl-D,L-arginine naphthylamide	+			+		
N-Benzoyl-L-arginine amide	+	+	+	+	+	+
N-Benzoyl-L-arginine methyl ester	+					
p-Toluenesulfonyl-L-arginine methyl ester	+			+	+	+
L-Arginine methyl ester		+	+		+	
D-Arginine methyl ester	−					
N-Benzoyl-L-arginine benzyl ester		+				
N-Benzoyl-L-arginine isopropyl ester		+				
p-Toluenesulfonyl homarginine methyl ester						−
p-Toluenesulfonyl norarginine methyl ester						−
B. Lysine						
p-Toluenesulfonyl-L-lysine methyl ester				−	±[d]	
L-Lysine ethyl ester	−					
L-Lysine methyl ester		(±)[c]			−	
N-Benzoyl-L-lysine methyl ester						±[d]
C. Others						
N-Acetyl-L-tyrosine ethyl ester				−	−	
N-Benzoyl-L-tyrosine ethyl ester				−		
D,L-Phenylalanine methyl ester		−				
N-Benzoyl-D,L-phenylalanine ethyl ester		−				
N-Benzoyl-D,L-phenylalaninamide		−				
L-Leucine methyl ester		−				
L-Leucinamide		−				
Benzoylglycinamide		−				
Glycyglycine ethyl ester		−				
L-Proline benzyl ester		−				
L-Prolinamide		−				
Benzoylglycine phenylalaninamide	−					
Benzoylglycine leucinamide	−					

[a] I, Nordwig and Strauch (*19*); II, Ogle and Tytell (*10*); III, DeBellis *et al.* (*18*); IV, Mitchell and Harrington (*35*); V, Gros and Labouesse (*52*); VI, Inagami and Cole (*40*).

[b] Assays on relatively crude preparations.

[c] Conversion was found with this substrate. However, all other hydrolyses were reported virtually complete so that initial rates and relative magnitudes of activity cannot be compared.

[d] Conversion at ~8% the rate of the arginine homolog.

Moreover, hydrolysis is drastically sensitive to the length of the guanido chain. Tosylhomoarginine methyl ester (with one side chain methylene unit longer than arginine) and tosylnorarginine methyl ester (with one side chain methylene unit shorter than arginine) are inactive as substrates when compared to tosylarginine methyl ester.

More detailed kinetic data have recently been completed by Inagami and Cole (*40*). Table IV summarizes their data on the binding and hydrolytic constants for various arginine esters and amides. Although V_{max} is similar in magnitude between the ester and amide of the two series studied, the K_m indicates a significantly tighter binding affinity for the ester. It is of interest that the active site titrant for trypsin (*41*), *p*-nitrophenyl-*p*-guanidobenzoate, serves as a substrate for clostripain, further indicating significant differences between the active sites of trypsin and clostripain.

2. *Proteins and Polypeptides*

The preference of clostripain for an arginine moiety in synthetic esters and amides is also present in naturally occurring and synthetic polypeptides and proteins where peptide hydrolysis occurs most rapidly at the carboxyl side of arginine. Although lysine serves as a substrate, hydrolytic rates are generally one to two orders of magnitude less. Moreover, this unusually specific protease is distinguished by its ability to hydrolyze the arginyl-prolyl peptide bond (*42*), a bond usually resistant to proteolysis. Figure 4 illustrates the marked preference for arginine residues in naturally occurring proteins. In these experiments the indicated protein was exposed for various lengths of time to clostripain (100/1, w/w). Following proteolysis peptide fragments were further digested

TABLE IV

Clostripain Kinetic Constants[a]

Substrate(λ)	K_m(mM)	V_{max}(μmoles/sec/mg)
$BAEE_{253}$	0.22	18.2
BAA_{253}	1.3	17.3
$TAME_{247}$	0.022	1.5
TAA_{247}	0.25	2.0
p-$NPGB_{400}$	2.5	6.2
$BLME_{253}$	2.7	4.4

[a] Conditions: tris·KCl·DTT::40:100:3 mM, pH 7.8, 25°.

40. Inagami, T. and Cole, P. W., personal communication, Vanderbilt University (1970).
41. Chase, T., Jr. and Shaw, E., *Biochem. Biophys. Res. Commun.* **29**, 508 (1967).
42. Mitchell, W. M., *Science* **162**, 374 (1968).

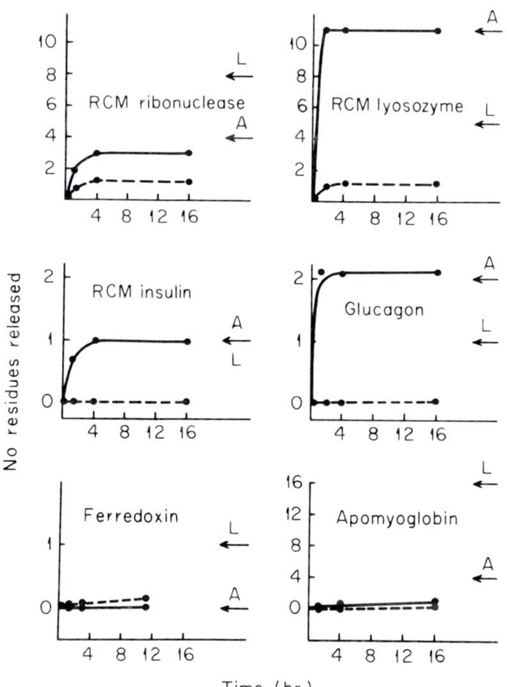

FIG. 4. Number of COOH-terminal residues exposed per molecule during timed hydrolysis of various polypeptides in the presence of clostripain. COOH-terminal arginine and lysine exposed during clostripain digestion are released as free amino acids by carboxypeptidase B. The arrows indicate the arginine and lysine residues susceptible to tryptic hydrolysis. Illustration from Mitchell and Harrington (*35*).

with carboxypeptidase B (DFP and soybean trypsin inhibitor treated) to liberate carboxyl-terminal arginine and lysine residues with subsequent analysis by the Spackman *et al.* (*43*) short column method. The solid line plots as a function of time the liberated arginine while the dotted line indicates lysine liberation. In each case the rapid appearance of arginine with respect to lysine confirms the general specificity of clostripain. The solid arrows represent the tryptic release of arginine (A) and lysine (L), respectively. One residue of lysine was released in clostripain-treated RCM (reduced carboxymethylated) lysozyme, but this substrate contains a lysylarginine sequence which would be released by carboxypeptidase B following clostripain hydrolysis at arginine. The only apparent exception is RCM ribonuclease where one residue of lysine was found per three residues of arginine. Figure 4 further illustrates that not

43. Spackman, D. H., Stein, W. H., and Moore, S., *Anal. Chem.* **30,** 1190 (1958).

all potential substrates are susceptible to clostripain hydrolysis. Apomyoglobin contains four arginine residues per molecule with less than one arginine bond split in 16 hr. A similar phenomenon has been observed with arginine-rich histone (*44*), in which less than 10% of the total arginine sites were split. Thus not only is the enzyme highly selective in its side chain acceptance (i.e., arginine), but also it is probably sensitive to the local environment of the potential substrate site as well.

C. Inhibitors

1. *Competitive Inhibitors*

The specificity for the arginine moiety in potential substrates is also reflected in competitive inhibitor studies. Table V illustrates K_I values

TABLE V
Comparative Inhibition Constants of Various Competitive Inhibitors on the Clostripain-Catalyzed Hydrolysis of BAEE

Inhibitor	K_I(mM)
I. *Arginine homologs, derivatives*[a,b]	
L-Arginine	1.0
D-Arginine	2.4
L-Lysine	44.0
N-α-Methyl-L-arginine	>10
N-α-Acetyl-L-arginine	>10
L-Homoarginine	0.23
N-α-Tosyl-L-homoarginine methyl ester	0.02
II. *Alkylguanidines*[b] $\left(RHNC\begin{matrix}\diagup NH \\ \diagdown NH_2\end{matrix}\right)$	
Methylguanidine	0.09
Ethylguanidine	0.08
Propylguanidine	0.03
Butylguanidine	0.003
III. *Alkylamines*[b] ($R—NH_2$)	
Methylamine	28.0
Ethylamine	27.0
Propylamine	8.5
Butylamine	27.0
IV. *Acylamidine*[a,b]	
Benzamidine	0.4

[a] Unpublished data from Rainey, Porter, and Mitchell (*45, 46*).
[b] Unpublished data from Inagami and Cole (*40*).

44. Mitchell, W. M., Doctoral Dissertation, Johns Hopkins University, 1966.

for arginine and various arginine homologs versus lysine on the hydrolysis of BAEE by clostripain (*40, 45, 46*). Both L-arginine and L-lysine inhibit the enzyme, although L-arginine is more effective by some one to two orders of magnitude. D-Arginine is slightly less effective as an inhibitor than L-arginine. However, methylation or acetylation of the α-NH_2 of L-arginine results in a molecule devoid of significant inhibitory activity. The importance of the length of the guanido side chain on hydrolysis previously mentioned is also reflected in inhibitory capabilities. L-Homoarginine with one additional carbon in the side chain is a very effective inhibitor of BAEE hydrolysis. The most effective competitive inhibitors, however, are the alkylguanidines which also serve as competitive inhibitors for trypsin. The most effective competitive inhibitor found to date is butylguanidine with approximately 2½ orders of magnitude better binding affinity than trypsin. The alkylamines are significantly poorer inhibitors than the alkylguanidines (Table V) which again complements the general specificity data. Table VI compares the K_I values for the alkylguanidines and alkylamines for trypsin and clostripain. The K_I determined for propyl and butyl alkylguanidines and amines on trypsin BAEE hydrolysis are roughly equivalent, while there is a marked disparity in the ability of the alkylguanidines versus the alkylamines to inhibit clostripain, a feature consistent with the specificity of both enzymes.

Studies with synthetic homopolymers of arginine and lysine add further evidence regarding the uniqueness of the active site of clostripain. Polyarginine is completely resistant to hydrolysis by clostripain whereas

TABLE VI
THE INHIBITION CONSTANTS OF ALKYLGUANIDINES AND AMINES IN THE TRYPSIN- AND CLOSTRIPAIN-CATALYZED HYDROLYSIS OF BAEE

Enzyme	Methyl	Ethyl	1-Propyl	*n*-Butyl
	$K_I(mM)$ of Alkylguanidines			
Trypsin[a]	7.0	1.4	0.53	1.3
Clostripain[b]	0.09	0.08	0.03	0.003
	$K_I(mM)$ of Alkylamines			
Trypsin[a]	260.0	62.0	8.7	1.7
Clostripain[b]	28.0	27.0	8.5	27.0

[a] Data from Inagami and York (*55*).
Conditions: Alkylguanidines K_I at pH 8.0 and 25°C; alkylamides K_I at pH 6.6 and 25°C.
[b] Data from Inagami and Cole (*40*).
Conditions: pH 7.8 and 25°C.

45. Rainey, J. and Mitchell, W. M., unpublished results (1970).
46. Porter, W. and Mitchell, W. M., unpublished results (1970).

this homopolymer is digested to di- and tripeptides by trypsin (*35*). Although neither polyarginine nor polylysine is hydrolyzed by clostripain, both are effective inhibitors of the esterase activity against BAEE. Polyarginine functions as an inhibitor some two orders of magnitude more efficiently than polylysine (Table VII), a phenomenon consistent with a preferential binding of the arginine moiety to the active site of the enzyme.

2. *Affinity Labeling*

The active site of clostripain apparently owes its hydrolytic properties to a reactive sulfhydryl, which is amenable to specific labeling. As previously mentioned, clostripain is inactivated on a mole per mole basis by PCMB, although definitive evidence for active site attack was not established. Recently, tosyllysinechloromethyl ketone (TLCK) has proven to be a specific active site label (*46*). Figure 5 illustrates the loss of clostripain activity against BAEE, a phenomenon which is easily prevented when the enzyme is incubated in the presence of a competitive inhibitor such as benzamidine or tosylhomoarginine methyl ester (THAME). Moreover, tosylphenylalaninechloromethyl ketone (TPCK) is without effect except at high concentrations. Although the site of attack has not been elucidated to date, a sulfhydryl reactive group appears to be involved in view of clostripain's free sulfhydryl requirement, the known sites of attack by these reagents (i.e., histidine and cysteine) in other systems, a pronounced lability of labeled site in clostripain to acid hydrolysis, and an observed $pK_{es} \simeq 8.25$ in log V_{max} vs. pH plots (*40*).

The rapid attack by TLCK on the active site of clostripain would appear to contradict the general phenomenon of restricted arginine specificity. However, in retrospect, the rapid and specific affinity of TLCK for the active site of clostripain is reasonable. Although the rate of TLCK inactivation is faster in clostripain than trypsin, the expected greater reactivity of a sulfhydryl in the active site of clostripain over

TABLE VII
EFFECT OF THREE POTENTIAL SUBSTRATES ON THE HYDROLYSIS OF BAEE BY CLOSTRIPAIN[a]

Inhibitor	Inhibitor conc. (μg/ml)	Relative activity (%)
Polyarginine	0.36	53
Polylysine	83	44
RCM RNase[b]	100	51

[a] Enzymic activity assay at 25°C as described in methods.
[b] The RNase concentration is equivalent to approximately 3 μg/ml arginine residues.

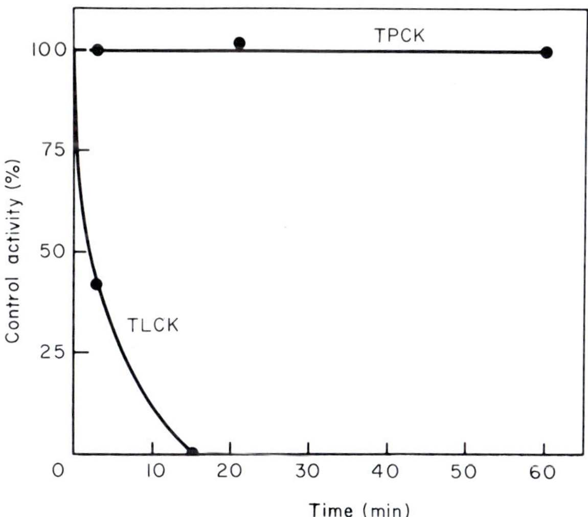

FIG. 5. The inhibition of esterase activity in clostripain by TLCK and the lack of inhibition of TPCK in Na phosphate buffer, pH 6.5, TLCK at 5-fold molar excess and TPCK at 27-fold molar excess over enzyme concentration.

trypsin's histidyl active nucleophile is compatible with the known major specificity for guanidine groups by clostripain. Unfortunately, no arginine homolog (i.e., tosylargininechloromethyl ketone) is presently available.

D. As a Contaminant in Purified Collagenase Preparations

The polyproline II type of helical configuration (*47*) renders collagen relatively immune to hydrolysis by most proteolytic enzymes. This feature has been utilized by numerous investigators in various collagen structure studies using specific collagenases (especially from *C. histolyticum*) which readily attack the native collagen molecule. The large bulk of these studies achieve maximal collagen digestion by long incubation time and/or large enzyme–substrate ratios. Under these conditions contaminating proteases which can potentially digest degraded collagen peptides or noncollagenous proteins assume significant roles. A ubiquitous contaminant of various highly purified *C. histolyticum* collagenase preparations is clostripain (*48*). Table VIII documents this level of clostripain contamination from several diverse sources, both

47. Harrington, W. F. and von Hippel, P., *Advan. Protein Chem.* **16**, 1 (1961).
48. Mitchell, W. M., *Johns Hopkins Med. J.* **127**, 192 (1970).

TABLE VIII
CONTAMINATION OF *C. histolyticum* COLLAGENASE PREPARATIONS BY CLOSTRIPAIN

Collagenase source	Clostripain activity[a] DTT(5×10^{-3} *M*)	Clostripain activity[a] H_2O_2(10^{-2} *M*)	Contamination (%)	Collagenase activity[b] No addition	Collagenase activity[b] H_2O_2(10^{-2} *M*)
I. Commercial					
Worthington purified CLSPA preparation	3	0	5	186	174
II. Noncommercial					
A α [Ref. (*30*)]	0.56	0	<1	63	81
B α [Ref. (*30*)]	0.57	0	<1	226	185
B β [Ref. (*30*)]	0.28	0	<0.5	261	234
Fr. I [Ref. (*31*)]	0.23	0	<0.5	217	236
Pseudocollagenase [Ref. (*31*)]	3	0	5	—	—

[a] Activity against benzoylarginine ethyl ester (5×10^{-4} *M*) in 0.1 *M* Na phosphate buffer, pH 7.8 as μmole min^{-1} mg^{-1} at 25° [reference (*35*)].
[b] Viscometric assay—units/mg [reference (*21*)].

commercial and noncommercial in origin. This activity can be completely removed by small concentrations of H_2O_2 which have no effect on collagenase activity. Thus it would appear prudent that in all studies using *C. histolyticum* collagenase as a structural and/or analytical tool, clostripain be inactivated by H_2O_2 or its absence from the collagenase preparation be documented.

E. AS A TOOL IN SEQUENCE ANALYSIS AND STRUCTURE–FUNCTION STUDIES

The unique specificity of clostripain and the judicious use of controlled enzymic hydrolysis should facilitate sequence analysis by allowing the isolation of larger initial peptide fragments without chemical modification of the substrate. The first application of clostripain to date in such a sequence problem has been its utilization on lysozyme Ch (chalaropsis), a fungal β-1, 4-*N*, *O*-diacetylmuramidase showing striking structural differences to egg white lysozyme (*49*). Figure 6 illustrates the sequence of a cyanogen bromide peptide (CNBr III) from lysozyme Ch.

49. Mitchell, W. M. and Hash, J. H., *J. Biol. Chem.* **244,** 17 (1969).

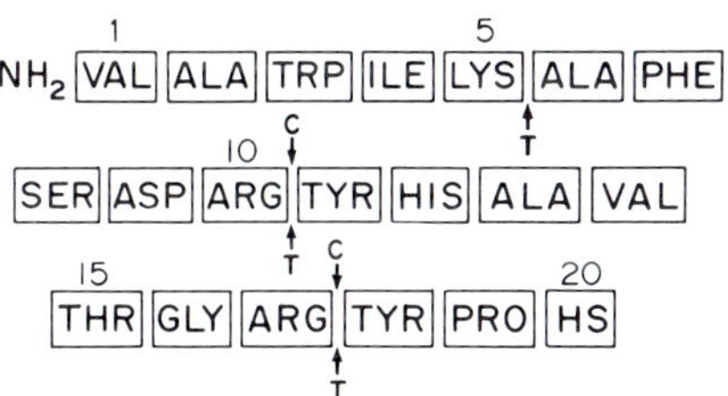

FIG. 6. The sequence of CNBr III fragment of lysozyme Ch. Arrow over T indicates trypsin-catalyzed hydrolysis and arrow under C indicates clostripain-catalyzed hydrolysis; HS stands for homoserine.

Digestion with clostripain resulted in hydrolysis only at the indicated arginine residues allowing a nonambiguous overlap of tryptic peptides (*50*). In a similar manner the limited hydrolytic potential of clostripain should prove of unique value in various structure–function applications. Clostripain has been used for this purpose recently in delineating the essential region of glucagon (a 29 amino acid polypeptide) for its hormonal activity of stimulation of lipolysis (*51*). Glucagon possesses a diarginine sequence at positions 17 and 18 in the chain. Hydrolysis with clostripain results in the rapid production of two pieces approximately two-thirds and one-third of the original molecule (*35, 52*) with the complete loss of lipolytic activity in isolated fat cells (*51*).

Unfortunately, no commercial preparation of purified clostripain has appeared to date. A more easily available enzyme should result in its utilization on a wide scale.

V. The Origin of Active Site Specificity—A Working Hypothesis

The affinity of various guanidine compounds for the active site of clostripain is striking (Sections IV,B and C). Moreover, since the length of the guanidine side chain is of critical importance in positioning the carboxyl carbon of potential substrates (Table III) for attack by the enzyme's active sulfhydryl nucleophile, and since the molecular dimensions between the center of the single positive charge in various lysine vs. arginine homologs is identical (Fig. 7), then the principal origin of clostripain's specificity must arise from atomic differences in the ionic

50. Shih, J. and Hash, J. H., *JBC* **246,** 994 (1971).
51. Felts, P. W., Ferguson, M. E. C., Hagey, K. A., Stitt, E. S., and Mitchell, W. M., *Diabetologia* **6,** 44 (1970).
52. Gros, P. and Labouesse, B., *Bull. Soc. Chim. Biol.* **42,** 559 (1960).

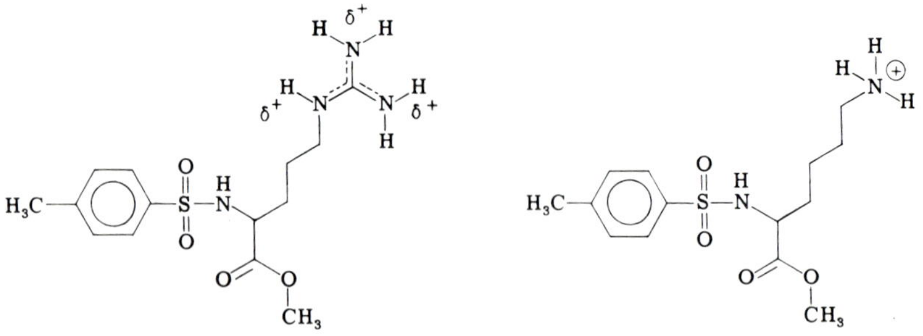

FIG. 7. Comparative charge distribution in (a) TAME (tosylarginine methyl ester) and (b) TLME (tosyllysine methyl ester).

bonding of the active site with the single positive charge of lysine or arginine. The singular difference in this charge is that the lysyl charge is localized while the arginyl charge is delocalized owing to the tautomeric state of the guanidine group (*53*). Furthermore, the partial double bond character of each bond results in greatly limited degrees of rotational freedom for this group.

One might expect that clostripain utilizes the delocalization of charge in attaining its unusual specificity. Recently, Eyl and Inagami (*54*) have succeeded in specifically labeling the source of substrate ionic bonding in the active site of trypsin. This was accomplished by masking trypsin's carboxyls with glycinamide in the presence of a competitive inhibitor, benzamidine, removing the latter, and labeling all remaining carboxyls with ^{14}C glycinamide. Aspartic acid 177 was specifically modified and is undoubtedly the origin of ionic bonding between substrate and active site (Fig. 8). A single carboxyl is further inferred by examination of the heats of binding of various alkylamines and alkylguanidines to the active site of trypsin. Since the guanidine group is in a tautomeric delocalized state, a single carboxyl binding site in the enzyme would be expected to have a significantly smaller heat of binding. The experimental value for the heat of binding in trypsin for alkylamines is —12 kcal/mole while the alkylguanidine value is —4.5 kcal/mole (*55*). Thus the binding energy in clostripain for the guanidine group could be significantly enhanced by postulating two cooperative carboxyls

53. Pauling, L., "The Nature of the Chemical Bond" 3rd ed., Cornell Univ. Press, Ithaca, New York, 1960.
54. Eyl, A. and Inagami, T., *Biochem. Biophys. Res. Commun.* **38,** 149 (1970).
55. Inagami, T. and York, S. S., *Biochemistry* **7,** 4045 (1968).

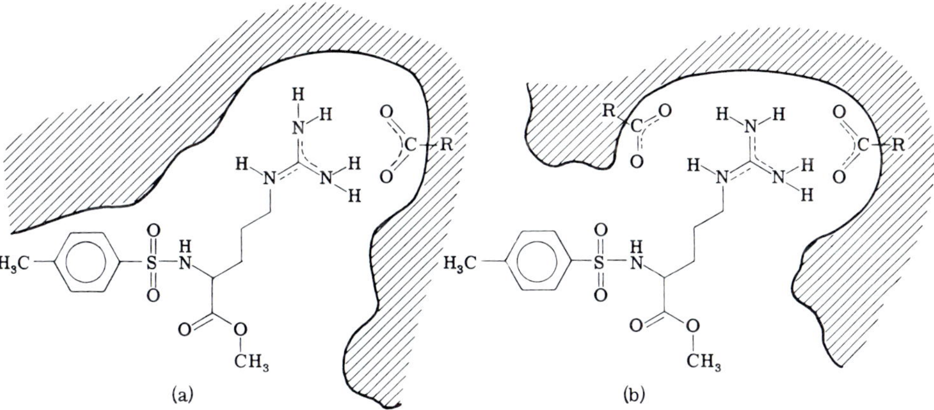

FIG. 8. Diagrammatic representation of active site of (a) trypsin with Asp 177 as ionic bonding residue for substrate, and diagrammatic representation of active site of (b) clostripain with postulated two carboxyl binding mechanism.

in the active site (*56*) as depicted in Fig. 8. Furthermore, a two carboxyl site might be expected to decrease the hydrolytic capacity for lysyl substrates by competition for the single localized charge with a resultant imperfect fit for the most efficient catalysis. Although this model is lacking in direct evidence at this time, it does offer a reasonable mechanism for clostripain's remarkable specificity and offers a direct experimental approach to its acceptance or rejection.

ACKNOWLEDGMENTS

The authors are indebted to Dr. Tadashi Inagami and Dr. John Hash for their generosity in allowing utilization of significant data from their respective laboratories prior to publication. The following grants have actively supported the authors during manuscript preparation: AM-10833, 5-F3-AM-34798, and AM-04349.

56. It is evident from the K_I values for the various alkylguanidines (Table V) that hydrophobic bonding is operative but to a minor extent when compared to the ionic interactions.

20

Other Bacterial, Mold, and Yeast Proteases

HIROSHI MATSUBARA • JOSEPH FEDER

I. Introduction

The past decade has witnessed considerable interest in microbial proteases both for basic understanding of enzyme mechanisms and industrial application. These enzymes are predominantly extracellular, isolated in active form from the culture filtrates of the appropriate organism.

Most of the various proteolytic enzymes of animal origin are represented among these enzymes including amino- and carboxypeptidases and a variety of endopeptidases. Certain of these proteases including the subtilisins, streptococcal protease, clostripain, and collagenases have been studied quite extensively and are reviewed in other chapters of this volume. The opportunity for comparative studies between these enzymes and similar ones from higher organisms in terms of structure–function relationships, evolutionary and genetic relationships, and the possibility of genetic manipulation makes this an area of considerable interest.

An attempt has been made to classify the proteases reviewed in this chapter by their pH dependence, effect of inhibitors, metal involvement, specificity, and kinds of catalytic properties (peptidase, amidase, or esterase). Except for a few carboxypeptidases (the aminopeptidases are reviewed by Delange and Smith, Chapter 3, this volume) the majority of the enzymes are endopeptidases which have been divided into four groups in this chapter: (1) acid proteases, (2) diisopropylphosphofluoridate (DFP)-sensitive alkaline proteases, (3) metal chelator-sensitive neutral proteases, and (4) thiol proteases.

The acid proteases which include the *Aspergillus saitoi* acid protease, *Rhizopus chinensis* acid protease, and *Paecilomyces varioti* acid protease are characterized by low pH activity and stability profiles, insensitivity to metal chelating agents, phosphorylating compounds such as DFP, and thiol poisons. These enzymes show limited esterolytic activity and are reminiscent of the animal pepsins and rennins. Some are inhibited by the specific pepsin inactivators. The DFP-sensitive alkaline proteases are best represented by the subtilisins which appear to be serine proteinases similar to trypsin and chymotrypsin. Included in this group will be the *A. sojae* alkaline proteases, *Sorangium* α-lytic protease, and the trypsinlike enzyme from *Streptomyces griseus*. These enzymes generally are active at neutral and alkaline pH, have broad substrate specificities including considerable esterolytic activity toward a wide variety of ester substrates, and are sensitive to the reagents such as DFP which implicate a serine residue in their mechanism of action. The neutral proteases are sensitive to metal chelating agents such as EDTA and *o*-phenanthroline and have been shown in some cases to require a metal atom such as zinc for activity. They exhibit pH optima near neutrality and have rather narrowly defined substrate specificity toward peptide bonds but no esterolytic or amidase activity toward most ester and amide substrates. These include the *Bacillus subtilis* neutral protease, thermolysin, and the neutral proteases from *Streptomyces naraensis* and *Aspergillus oryzae*. Only a few enzymes have been found that are sensitive to sulfhydryl reagents, and these will be briefly discussed. (Streptococcal proteinase and clostripain are discussed by Teh-Yung Liu and Elliott, Chapter 17, and Mitchell and Harrington, Chapter 19, this volume.) In some instances, particularly where the enzyme has not been extensively characterized, an enzyme will be classified in a particular group while apparently displaying only some of the characteristic features of that group but none of any other group. The attempt here has been to avoid basing the classification of proteases upon particular protein or peptide substrates unless significant reaction type and mechanism are implied with the particular substrate. Proteases lysing bacterial cell walls are described in a separate section. Brief descriptions on yeast enzymes are also included in this chapter although they are not the primary concern in this review.

II. Acid Proteases

As early as the 1930's proteases possessing activity at acid pH were discovered in the culture media of various species of molds (*1*). The distinct characteristic of these acid proteases has been their pH optima in the region of pH 2.0–5.0 and their insensitivity to such reagents as DFP, metal chelating agents, and thiol poisons. The enzymes are neither activated by reducing agents nor metal ions. This type of enzyme might therefore correspond to the animal pepsins (*2*) (see also Fruton, Chapter 4, this volume). The major source for the isolation of these acid proteases has been the culture media of molds. In contrast to these extracellular mold enzymes a few intracellular acid proteases have been isolated from yeast.

A closely related group of acid proteases possess particular specificity characteristics which make them suitable for the clotting of milk in a manner similar to the animal rennins. These enzymes might be considered as a distinct group of microbial rennins. The distinguishing characteristic of these enzymes is their limited cleavage of only particular bonds to permit the milk clotting without continued extensive proteolysis of casein fragments (*3–6*). These two groups of acid proteases, pepsinlike and renninlike, will be discussed separately.

1. B. Hagihara, "The Enzymes," 2nd ed., Vol. 4, p. 193, 1960.
2. F. A. Bovey and S. S. Yanari, "The Enzymes," 2nd ed., Vol. 4, p. 63, 1960.
3. P. Jollès, C. Alais, and J. Jollès, *ABB* **98,** 56 (1962).
4. J. Jollès, C. Alais, and P. Jollès, *BBA* **168,** 591 (1968).
5. A. Delfour, J. Jollès, C. Alais, and P. Jollès, *BBRC* **19,** 452 (1965).
6. B. Foltmann, *Compt. Rend. Trav. Lab. Carlsberg* **35,** 143 (1966).

A. Pepsinlike Acid Proteases

1. *Distribution and Isolation*

The pepsinlike acid proteases are widely distributed among the molds and include the *Aspergillus, Penicillium, Paecilomyces, Rhizopus,* and *Trametes.* The enzymes were isolated from the culture medium and some of them are available in commercial scale. The purification procedures of the enzymes are mainly based on various combinations of salt fractionation, organic solvent fractionation, ion exchange column chromatography, gel filtration, and electrophoresis. Purification steps are carried out where most acid proteases are stable in the pH range from 2 to 6. A brief discussion of some of the individual mold and yeast enzymes follows below. Additional enzymes not included in this discussion are listed in the comprehensive table at the end of this chapter (Table XV).

a. Aspergillus. A number of enzymes have been isolated from *Aspergilli* including *A. saitoi* (*1, 7–10*), *A. niger* (*1, 11*), *A. awamori* (*1, 12*), *A. oryzae* (*1, 13–16*), and *A. fumigatus* (*17, 18*). The enzyme from *A. saitoi* (aspergillopeptidase A) was purified from the crude starting preparation (*7–9*) by chromatography on Duolite CS-101, DEAE-cellulose, and SE-Sephadex (*10*). The enzyme was electrophoretically homogeneous and sedimented in the ultracentrifuge as a single peak, yielding a molecular weight between 30,000 and 40,000 (*10*). An $A_{1\,\text{cm}}^{1\%}$ at 280 nm of 13.15 was reported for this enzyme (*19*). At least two distinct acid proteases have been isolated from *A. niger* by SE-cellulose and gel filtration on Sephadex (*1, 11*). The two enzymes have optimum activity at pH 2.6 and 2.0. An acid protease was isolated by Bergkvist from *A. oryzae* by

7. E. Ichishima and F. Yoshida, *Agr. Biol. Chem.* (*Tokyo*) **26,** 547 and 554 (1962).
8. E. Ichishima. Y. Gomi, T. Watarai, and F. Yoshida, *Agr. Biol. Chem.* (*Tokyo*) **27,** 302 (1963).
9. E. Ichishima, M. Funabashi, and F. Yoshida, *Agr. Biol. Chem.* (*Tokyo*) **27,** 310 (1963).
10. E. Ichishima and F. Yoshida, *BBA* **99,** 360 (1965).
11. Y. Koaze, H. Goi, K. Ezawa, U. Yamada, and T. Hara, *Agr. Biol. Chem.* (*Tokyo*) **28,** 216 (1964).
12. V. M. Stepanov, L. S. Lobareva, N. F. Frolova, and L. I. Oreshchenko, *Izv. Akad. Nauk SSSR, Ser. Khim.* **12,** 2840 (1968); *CA* **70,** 64589 (1969).
13. R. Bergkvist, *Acta Chem. Scand.* **17,** 1521, 1541, 2230, and 2239 (1963).
14. Y. Nunokawa, *Agr. Biol. Chem.* (*Tokyo*) **29,** 687 (1965).
15. K. Nakanishi, *J. Biochem.* (*Tokyo*) **46,** 1263, 1411, and 1553 (1959).
16. K. Nakanishi, *J. Biochem.* (*Tokyo*) **47,** 16 (1960).
17. A. G. Jönsson and S. M. Martin, *Agr. Biol. Chem.* (*Tokyo*) **28,** 734 (1964).
18. S. M. Martin and A. G. Jönsson, *Can. J. Biochem.* **43,** 1745 (1965).
19. E. Ichishima and F. Yoshida, *Nature* **207,** 525 (1965); *BBA* **110,** 155 (1965).

tannin precipitation followed by chromatography over CM- and DEAE-cellulose (*13*). The enzyme which had an optimum activity at pH 4.3–4.5 and 45° was stable over the range of pH 3–6, insensitive to EDTA, cysteine, KCN, and soybean trypsin inhibitor (*13*). The enzyme was, however, inhibited by 10^{-2} *M* of Zn^{2+}, Cu^{2+}, Cd^{2+}, and Fe^{2+}. Nunokawa (*14*) reported the presence of at least two acid proteases from a crude preparation of *A. oryzae,* Takadiastase, using SE- and DEAE-Sephadex. The activation of trypsinogen to trypsin by the *A. oryzae* acid protease has been studied by Nakanishi (*15, 16*). This reaction, which is common to a number of these acid proteases, will be discussed in detail in a separate section (Section II,A,4). The production of acid proteases from *A. fumigatus* has been reported (*17, 18*).

b. Penicillium. The study of the activation of trypsinogen to trypsin by the pepsinlike enzyme from *Penicillium janthinellum* resulted in the purification of the enzyme (peptidase A) by ammonium sulfate fractionation, AE-cellulose column chromatography, and phospho-cellulose treatment (*20*). This enzyme was inhibited by the pepsin inhibitor, diazoacetyl-D,L-norleucine methyl ester (*21*) in the presence of cupric ions but not by *p*-bromophenylacyl bromide. The significance of these studies will be discussed in detail in the section on inhibitors (Section II,A,4). Another acid protease was isolated from *Penicillium notatum* culture media by tannin precipitation followed by a combination of ammonium sulfate fractionation, ion exchange chromatography, gel filtration, and electrophoresis (*22, 23*). The enzyme had an isoelectric point of 4.92 (*23*).

c. Rhizopus. Fukumoto *et al.* (*24*) isolated an acid protease from *Rhizopus chinensis* by the use of ammonium sulfate and barium acetate fractionations, gel filtration on Sephadex, CM-cellulose chromatography, and crystallization from dilute aqueous acetone. The enzyme was stable from pH 2.8 to 6.5 and exhibited optimum activity at pH 2.9–3.3. Metal chelating agents and thiol reagents did not affect the enzymic activity. An $A_{1\,\text{cm}}^{1\%}$ at 280 nm of 12.6 was reported (*24*). The culture filtrates of *Rhizopus oligosporus* was found to contain two proteolytic enzymic activities with optima at pH 3.0 and 5.5 (*25*).

20. T. Hofman and R. Shaw, *BBA* **92,** 543 (1964).
21. T. G. Rajagopalan, W. H. Stein, and S. Moore, *JBC* **241,** 4295 (1966).
22. W. E. Marshall, R. Manion, and J. Porath, *BBA* **151,** 414 (1968).
23. B. Makonnen and J. Porath, *European J. Biochem.* **6,** 425 (1968).
24. J. Fukumoto, D. Tsuru, and T. Yamamoto, *Agr. Biol. Chem.* (*Tokyo*) **31,** 710 (1967).
25. H. L. Wang and C. W. Hesseltine, *Can. J. Microbiol.* **11,** 727 (1965).

d. Paecilomyces. The acid protease from *Paecilomyces varioti* Bainier TPR-220 was isolated by Sawada in crystalline form by a series of fractionation procedures involving the treatments with ethanol, corn starch, calcium acetate, ammonium sulfate, and aqueous acetone crystallization (*26*). The enzyme was stable from pH 3.0 to 6.0 below 40° and had optimum activity at pH 3.0 and 60°. Several divalent cations afforded stability toward heat treatment (*27*).

e. Mucor. Mucor pusillus, a thermophilic mold, produces an acid protease when grown on chemically defined medium (*28*). The enzyme has been purified from wheat bran medium by ammonium sulfate and ethanol fractionation, gel filtration, and ion exchange chromatography (*29*). The enzyme was homogeneous by electrophoresis. The enzyme was not inhibited by DFP, reducing agents, thiol poisons, EDTA, or *o*-phenanthroline, but a number of metal ions were inhibitory. Optimum activity toward hemoglobin was observed at pH 3.8 and the temperature optimum was at 55°. An acid protease has been reported from *Mucor hiemalis* NRRL 3103 (*30*). The enzyme has a pH optimum from 3.0 to 3.5. The protease, which appears to be loosely bound to the mycelium, can be readily released by 0.5 *M* NaCl (*30*).

f. Other Fungi. Two acid proteases were isolated in homogeneous forms from the plant pathogenic fungi, *Trametes sanguinea* (*31, 32*) and *Acrocylindrium* sp. (*33*). The *Trametes* protease was purified by acetone and ammonium sulfate fractionations, chromatography on Duolite CS-101 and A-7, and crystallization from dilute acetone solution (*31*). The enzyme was homogeneous by ultracentrifugation, electrophoresis, and chromatography (*32*). The *Acrocylindrium* acid protease seemed to have similar physical and enzymic properties to those of the *Trametes* enzyme. A $s_{20,w}$ value of 3.2 was reported (*33*).

The culture media of *Alternaria tenuissima* E-34 were found to produce at least three proteases including an acid protease with pH optimum of 2.8 (*34*).

g. Yeast. The presence of various intracellular yeast proteases has

26. J. Sawada, *Agr. Biol. Chem.* (*Tokyo*) **27,** 677 (1963).
27. J. Sawada, *Agr. Biol. Chem.* (*Tokyo*) **28,** 348 (1964).
28. G. A. Somkuti and F. J. Babel, *J. Bacteriol.* **95,** 1415 (1968).
29. G. A. Somkuti and F. J. Babel, *J. Bacteriol.* **95,** 1407 (1968).
30. H. L. Wang, *J. Bacteriol.* **93,** 1794 (1967).
31. K. Tomoda and H. Shimazono, *Agr. Biol. Chem.* (*Tokyo*) **28,** 770 (1964).
32. K. Tomoda, *Agr. Biol. Chem.* (*Tokyo*) **28,** 774 (1964).
33. F. Uchino, Y. Kurono, and S. Doi, *Agr. Biol. Chem.* (*Tokyo*) **31,** 428 (1967).
34. A. G. Jönsson and S. M. Martin, *Agr. Biol. Chem.* (*Tokyo*) **29,** 787 (1965).

been known for some time. Dernby (*35*) first reported proteolytic activity in yeast autolyzates in 1917, and a number of studies have appeared in the literature (*36–52*) since then. The yeast enzymes include an acid protease, DFP- and *p*-mercuribenzoate-sensitive endopeptidases with alkaline and more neutral pH optima, and EDTA-sensitive carboxypeptidaselike enzyme. The various enzymes will each be discussed along with the other enzymes of the appropriate groups.

The presence and isolation of an acid protease (proteinase A) from *Saccharomyces cerevisiae* autolysates have been reported by several authors (*41, 43, 45, 47–49*). Unlike the DFP-sensitive alkaline proteases, proteinases B and C from this organism, which appear to exist as a zymogen or inhibitor bound form in the cell, the acid protease is found in an active form (*48*). The various yeast proteases have been separated from one another by chromatography on DEAE-Sephadex A-50. The homogeneity of the purified acid protease was demonstrated by ultracentrifugation and electrophoresis (*48*). The enzyme has a pH optimum toward hemoglobin of pH 3.0 and no synthetic substrates have been reported for this enzyme (*48*). No activity has been demonstrated toward Ac–Tyr–OEt and Bz–Arg–OEt (Ac, acetyl; Bz, benzoyl; OEt, ethylester) nor is the enzyme inhibited by DFP, *p*-mercuribenzoate, EDTA, or serum inhibitors. A molecular weight of about 60,000 was obtained by the Archibald method (*47*) and the enzyme is quite acidic having an isoelectric point at pH 3.8. The isolated enzyme contained about 10% hexose by dry weight. An $A_{1\,\text{cm}}^{1\%}$ at 280 nm was reported to

35. K. G. Dernby, *Biochem. Z.* **81**, 107 (1917).
36. R. Willstätter and W. Grassmann, *Z. Physiol. Chem.* **153**, 250 (1926).
37. W. Grassmann and W. Haag, *Z. Physiol. Chem.* **167**, 188 (1927).
38. W. Grassmann and H. Dyckerhoff, *Z. Physiol. Chem.* **179**, 41 (1928).
39. M. Hecht and H. Civin, *JBC* **116**, 477 (1936).
40. M. J. Johnson, *JBC* **137**, 575 (1940).
41. J. F. Lenney, *JBC* **221**, 919 (1956).
42. F. Felix and J. Labouesse-Mercouroff, *BBA* **21**, 303 (1956).
43. J. F. Lenney and J. M. Dalbec, *ABB* **120**, 42 (1967).
44. F. Felix and N. Brovillet, *BBA* **122**, 127 (1966).
45. T. Hata, R. Hayashi, and E. Doi, *Agr. Biol. Chem.* (*Tokyo*) **31**, 150 (1967).
46. E. Doi, R. Hayashi, and T. Hata, *Agr. Biol. Chem.* (*Tokyo*) **31**, 160 (1967).
47. T. Hata, R. Hayashi, and E. Doi, *Agr. Biol. Chem.* (*Tokyo*) **31**, 357 (1967).
48. R. Hayashi, Y. Oka, E. Doi, and T. Hata, *Agr. Biol. Chem.* (*Tokyo*) **32**, 359 (1968).
49. R. Hayashi, Y. Oka, E. Doi, and T. Hata, *Agr. Biol. Chem.* (*Tokyo*) **32**, 367 (1968).
50. J. F. Lenney and J. M. Dalbec, *ABB* **129**, 407 (1968).
51. J. F. Lenney and J. M. Dalbec, *Federation Proc.* **27**, 788 (1968).
52. E. Juni and G. A. Heym, *ABB* **127**, 89 (1968).

be 11.9 for this enzyme. The physical properties of this yeast enzyme are listed in Table II.

An extracellular acid protease from *Candida albicans* has been isolated by Fasold and Staib (*53*) by gel filtration procedure on Sephadex G-75 followed by DEAE-Sephadex A-25 chromatography. The molecular weight was found to be 40,000–42,000 by gel filtration and ultracentrifugation ($s_{20,w} = 3.5$, 3.2 S). The pH optimum was 3.2 with bovine serum albumin. The enzyme lost activity on standing. No inhibition was observed with EDTA ($10^{-2}\,M$), *p*-mercuribenzoate ($5 \times 10^{-2}\,M$), and the pepsin inhibitors, *p*-bromophenylacyl bromide ($5 \times 10^{-3}\,M$) and *N*-diazoacetylnorleucine methyl ester ($5 \times 10^{-3}\,M$). The enzyme cleaved the oxidized insulin B chain at the following positions: Val 2–Asn 3, Leu 11–Val 12, Val 12–Glu 13, Ala 14–Leu 15, Val 18–$CySO_3H$ 19, and Phe 25–Tyr 26.

2. *Chemical Properties*

a. Amino Acid Composition and Terminal Residues. The amino acid composition of several mold acid proteases is given in Table I (*19, 20, 54, 55*) together with that of pepsin (*56*). Several common features are seen in the compositions as follows:

(1) The content of basic amino acids is fairly low compared to that of acidic amino acids after correction for the amide ammonia content, indicating the strongly acidic nature of the protein. This agrees with the low isoelectric point described below. Pepsin, however, is still more strongly acidic as a result of the very low content of basic amino acids.

(2) The content of the sulfur-containing amino acids, methionine and half-cystine, is considerably lower in the mold acid proteases than in pepsin.

(3) The total numbers of amino acid residues range from 290 to 340 for all the acid proteases including pepsin.

One or more amino acid residues are missing in some of the acid proteases. The *A. saitoi* acid protease has no methionine but does have two half-cystine residues which have been reported to be in the form of a disulfide linkage (*19*). The enzyme from *P. janthinellum* lacks arginine, methionine, and half-cystine residues (*20*). These two acid proteases, in

53. R. H. Fasold and F. Staib, *BBA* **167**, 399 (1968).
54. J. Sawada, *Agr. Biol. Chem.* (*Tokyo*) **30**, 393 (1966).
55. D. Tsuru, A. Hattori, H. Tsuji, and J. Fukumoto, *J. Biochem.* (*Tokyo*) **67**, 415 (1970).
56. T. G. Rajagopalan, S. Moore, and W. H. Stein, *JBC* **241**, 4940 (1966).

TABLE I
AMINO ACID COMPOSITION OF VARIOUS ACID PROTEASES

Amino acid	*A. saitoi* [Ref. (*19*)]	*P. janthinellum* [Ref. (*20*)]	*P. varioti* [Ref. (*54*)]	*R. chinensis* [Ref. (*55*)]	Pepsin [Ref. (*56*)]	*M. pusillus* (Mucor rennin) [Ref. (*92*)]	Rennin [Ref. (*6*)]
Lysine	11	5	28	12	1	11–12	8
Histidine	3	3	7	0	1	1–2	4
Arginine	1	0	3	9	2	4	5
Tryptophan	1[a]	4–5	3	5	6	2–3	4
Aspartic acid	34–35	36	45	43	40	44	30
Threonine	25	28	27	32	25	21	19
Serine	42–43	42	39	26	43	22	27
Glutamic acid	22	28	23	21	26	20	29
Proline	10	12	20	14	16	14	12
Glycine	31–32	39	33	39	34	34	25
Alanine	20	23	26	23	16	16–17	13
Half-cystine	2[a]	0	1	4	6	2	6
Valine	22	22	18	19	20	24	21
Methionine	0[a]	0	1	3	4	3	7
Isoleucine	11–12	12–13	15	21	23	12	15
Leucine	19–20	20	21	23	28	15	19
Tyrosine	17–18	14	16	14	16	13	15
Phenylalanine	13	19	14	16	14	19	14
Ammonia	31–32	43	—	32	27	—	31
Total	284–290	307–309	340	324	321	277–281	273

[a] Bioassay.

spite of their fairly large taxonomical difference, have similar compositions except for lysine, tryptophan, arginine, and half-cystine residues, and except for the basic residues are similar to pepsin. The *Paecilomyces varioti* acid protease has the highest number of basic residues (*54*), but the low isoelectric point suggests a low number of amides in the molecule. The sole cysteine residue was titratable by *p*-mercuribenzoate (*54*). Histidine was completely absent in the *Rhizopus chinensis* acid protease (*55*).

No studies have been reported on the primary structures of any of these acid proteases except for the determination of the N- and C-terminal residues of some of the enzymes. The *A. saitoi* acid protease yielded stoichiometric amounts of serine and alanine at the N- and C-termini of the molecule, respectively, indicating a molecule of one polypeptide chain (*57*). *Paecilomyces* acid protease had either leucine or isoleucine at the N-terminus and possibly alanine at the C-terminus (*54*). The *R. chinensis* enzyme had an N-terminal alanine, and only a single polypeptide chain was suggested (*55*).

b. Nonprotein Components. Very few studies have been reported on the phosphate content of the acid proteases. Pepsin contains one phosphate group per mole (*2*). The *A. oryzae* enzyme was shown to have a different phosphate content than pepsin but contained neither hexose nor hexosamine (*19*). The intracellular acid protease from *Saccharomyces cerevisiae* (proteinase A) was reported to contain 9.7% of hexose (*47*).

c. Chemical Modification. Studies on the chemical modification of various amino acid residues and the effect on activity have been reported only for the *A. saitoi* acid protease (*58*). *N*-Bromosuccinimide and photooxidation destroyed tryptophan, tyrosine, and histidine with considerable loss of activity. Upon acetylation with a 100 *M* excess of *N*-acetyl imidazole at pH 7.0, 4 out of the 18 tyrosine residues were modified with a 90% loss of activity. Reaction of 2-hydroxy-5-nitrobenzyl bromide with the sole tryptophan in the molecule was obtained only in the presence of 7 *M* urea, indicating that the residue was buried. This is in agreement with other physical and ultraviolet spectral studies (*59–61*) (see Section II,A,3). A role for tyrosine participation in catalysis was suggested (*58*).

57. E. Ichishima and F. Yoshida, *J. Biochem.* (*Tokyo*) **59,** 183 (1966).
58. E. Ichishima and F. Yoshida, *Nature* **215,** 412 (1967).
59. E. Ichishima and F. Yoshida, *BBA* **128,** 130 (1966).
60. E. Ichishima and F. Yoshida, *BBA* **147,** 341 (1967).
61. E. Ichishima and F. Yoshida, *Agr. Biol. Chem.* (*Tokyo*) **31,** 507 (1967).

TABLE II
MOLECULAR CHARACTERISTICS OF VARIOUS ACID PROTEASES

Organism	MW	$s_{20,w} \times 10^{13}$	$D_{20,w} \times 10^7$	$\bar{V}$	f/f_0	$[\alpha]_D$	$-a_0$	$-b_0$	λ_c (nm)	$-[m']$ (nm)	pI
A. saitoi (*10, 19, 59–61*)	34,200[a]	3.33	—	0.722	—	−35	211	0	207	2400	3.65
	34,900[b]	—	—	—	—	—	—	—	—	(220)	—
	35,200[d]	—	—	—	—	—	—	—	—	—	—
A. oryzae A (*14*)	—	2.55	—	—	—	—	—	—	—	—	Above 3
A. oryzae B (*14*)	—	2.87	—	—	—	—	—	—	—	—	3
P. janthinellum (*20*)	32,100[a]	3.1	8.4	0.718	—	—	—	—	—	—	Below 3.8
	32,000[c]	—	—	—	—	—	—	—	—	—	—
P. varioti (*54*)	37,300[a]	2.75 (15°)	—	—	—	—	—	—	—	—	3.8
	35,400[d]	—	—	—	—	—	—	—	—	—	—
T. sanguinea (*32*)	33,500[a]	2.95 (17°)	9	—	—	—	—	—	—	—	3.5
	34,000[c]	—	—	—	—	—	—	—	—	—	—
R. chinensis (*55, 62*)	35,000[a]	2.83	—	—	—	−33	200	0	208	2100	5.2
	—	—	—	—	—	—	—	—	—	(227)	—
S. cerevisiae (*47*)	60,000[a]	3.3	—	—	—	—	—	—	—	—	3.8
Pepsin (*2, 63–66*)	32,700[a]	2.88	—	0.726	1.08	−76	—	0	212	2800	Below 1
	27,000[c]	3.0–3.3	8.7–11.2	0.75	1.13	−69	—	0	216	(226)	—
	35,000[a]	—	—	—	—	—	—	—	—	—	—
M. pusillus (Mucor rennin) (*92, 94, 95*)	30,600[a]	2.39	7.9	0.742	1.33	—	—	—	—	—	3.7
	32,500[e]	—	—	—	—	—	—	—	—	—	—
	29,000[a]	—	—	—	—	—	—	—	—	—	—
	30,000[d]	—	—	—	—	—	—	—	—	—	—
E. parasitica renninlike (*89, 90, 100*)	39,000[d]	—	—	—	—	—	—	—	—	—	5.5
	34,000[e]	—	—	—	—	—	—	—	—	—	Below 4.6
	37,500[e]	—	—	—	—	—	—	—	—	—	—
Rennin (*2, 102–104*)	40,000[c]	3.2	9.0	0.749	0.98	—	—	—	—	—	4.6
	31,000[e]	4.0	9.5	0.73	—	—	—	—	—	—	—
	30,700[d]	—	8.5	—	—	—	—	—	—	—	—

[a] Ultracentrifugation.
[b] Sedimentation velocity.
[c] Sedimentation diffusion.
[d] Amino acid composition.
[e] Gel filtration.

3. *Physical Properties*

a. Isoelectric Point. The isoelectric points of various acid proteases was found generally in the pH range from 3 to 3.8 (*10, 14, 20, 32, 54*) as determined by free boundary electrophoresis, paper electrophoresis, or the ampholyte method (Table II). These values, although in general agreement with those predicted from the amino acid compositions, are much higher than that of pepsin which is pH 1.0 or below (*2*). The isoelectric points of *R. chinensis* (*62*) and *P. notatum* (*23*) were exceptionally high, pH 5.2 and 4.9, respectively.

b. Molecular Weight. All of the acid proteases investigated to date have very similar molecular weights, around 35,000, which is also close to that of pepsin. The amino acid compositions and the number of terminal residues in several cases indicate that the molecular weights are for a monomeric unit of the acid protease molecule. Table II gives a number of physical constants for several acid proteases including sedimentation constant $s_{20,w}$, diffusion constant $D_{20,w}$, partial specific volume $\bar{V}$, molecular weight MW, and other physical parameters (*14, 19, 20, 32, 54, 55, 59–61, 63–66*). The *A. saitoi* protease had an intrinsic viscosity $[\eta]$ of 0.032 dl/g (*19*).

c. Stability. The acid proteases are generally stable at acid pH in the range from pH 2 to 6. Some examples are *A. saitoi*, pH 2.0–5.0 (*67*); *P. janthinellum*, pH 3.0–5.5 (*20*); *P. varioti*, pH 3.0–6.0 (*26*); *R. chinensis*, pH 2.8–6.5 (*24*); *A. oryzae*, pH 2.5–6.0 (*14*); *T. sanguinea*, pH 2.0–6.0 (*32*); *Acrocylindrium* sp., pH 2.5–5.0 (*33*); and *M. pusillus*, pH 3.0–6.0 (*29*). At higher pH the enzymes are rapidly inactivated. This stability of the mold acid proteases is very similar to pepsin (*2*) which displays much greater stability at pH even below 2. The acid stability may be attributed in part to the low content of basic groups which could contribute strong electrostatic repulsive forces at low pH and in part to the participation of hydrogen bonding between unprotonated carboxyl groups (*15*).

Although the heat stability of the acid proteases was studied under

62. D. Tsuru, *Kagaku To Kogyo* (*Tokyo*) **43,** 17 (1969).
63. R. C. Williams, Jr. and T. G. Rajagopalan, *JBC* **241,** 4951 (1966).
64. G. E. Perlmann, *Proc. Natl. Acad. Sci. U. S.* **45,** 915 (1959).
65. P. Urnes and P. Doty, *Advan. Protein Chem.* **16,** 401 (1961).
66. B. Jirgensons, *JBC* **238,** 2716 (1963).
67. F. Yoshida and N. Nagasawa, *Bull. Agr. Chem. Soc. Japan* **20,** 257 and 262 (1956).

various conditions, generally about 50% inactivation of the enzyme was observed after 10–15 min at 60°–65° within the stable pH region (*24, 27, 29, 32, 33*). The *A. niger* acid protease was reported to be fairly resistant to heat treatment (*11*). The *A. saitoi* acid protease underwent only 5% autodigestion after 16 hr at pH 5.5 and 22°–25° (*60*). The *Paecilomyces* acid protease (*27*) was protected against thermal inactivation by several divalent cations, Ca^{2+}, Co^{2+}, Mg^{2+}, Sr^{2+}, and Zn^{2+}, unlike the *A. saitoi* enzyme which was not protected (*68*).

d. Conformation. Some studies have been reported on the conformation of the acid proteases from *A. saitoi* (*19, 58, 61*) and *R. chinensis* (*55*). The physical parameters obtained from optical rotatory dispersion (ORD) studies are given in Table II. The optical rotation $[\alpha]_D = -35°$ (*59–61*) and $[\alpha]_D = -33°$ (*55*) suggested a globular nature of the acid protease, while the low intrinsic viscosity $[\eta] = 0.032$ dl/g (*19*) suggested complete folding of the peptide chain. The ORD constants $\lambda_c = 207$ nm (*59–61*) and $\lambda_c = 208$ nm (*55*), the Moffitt and Yang parameters $b_0 = 0$ (*55, 59–61*), and the observation that $-a_0 = 211$ (*59–61*) and $-a_0 = 200$ (*55*) increase in 4 *M* urea and become greater at pH 5.5 without change of the b_0 value indicated that the enzyme had no α-helical content (*55, 59–61*). The enzymic activity decreased parallel with the change of $-a_0$, 46% inactivation after 30 min in 5 *M* urea. The λ_c value changed to 224 nm in 5 *M* urea. The change in $-a_0$ and activity accompanied the parallel change of difference spectrum at 294 nm, indicating that the sole tryptophan residue in the *A. saitoi* enzyme was buried within the molecule (*59–61*). This was supported by the chemical modification studies (Section II,A,2). The infrared spectral studies of the *A. saitoi* acid protease suggested that the deutrium-exchanged enzyme was present in the form of antiparallel β structure. A strong amide I absorption appeared at 1632 cm^{-1} and the amide II absorption was at 1540 cm^{-1} (*59–61*). This enzyme was completely, irreversibly inactivated after standing at pH 8.0 and 0° for 20 hr with a change in the $-a_0$ value from 211 to 446. A similar irreversible inactivation occurred when the enzyme was treated with 7 *M* urea at pH 5.5 and 20° for 1 hr (*59–61*). No change in optical properties was observed in the pH region from 2.7 to 5.7 (*59–61*). This is in agreement with the observed stability of the enzyme (*68*), but at higher pH and temperature, 55°, the conformation of the protein changed to a random coil structure (*59–61*). In the presence of sodium lauryl sulfate, an anionic detergent, the *A. saitoi* enzyme changed from a non α-helical to a partial α-helical conformation (*59–61*).

68. F. Yoshida, *Bull. Agr. Chem. Soc. Japan* **20,** 252 (1956).

4. *Enzymic Properties*

a. Optimum pH. The pH optima of all the mold acid proteases lie within the pH range 2–5 with lower values for the protein substrates as compared to synthetic substrates. The *A. saitoi* acid protease exhibited maximum activity toward protein substrates at pH 2.5–3.0 (*68*) and toward peptide substrates at pH 4.0–4.5 (*67*). Similar differences were reported for protein and peptide substrates, respectively, for a number of other acid proteases including *P. varioti* protease, pH 3.0 (*27*) and pH 3.5–5.5 (*69*), and *R. chinensis* protease, pH 2.9–3.3 (*24*) and pH 5.0 (*70*). *Penicillium janthinellum* protease had an optimum activity toward protein substrates at pH 3.0–4.0 (*20*), *T. sanguinea* protease at pH 2.3–2.5 (*32*), *Acrocylimdrium* sp. protease at pH 2.0 (*33*), *Mucor pusillus* protease at pH 5.6 toward casein and pH 3.8–4.0 toward hemoglobin (*29*), *A. niger* var. *macrosporus* at pH 2.0 and 2.6 (*11*), and *A. oryzae* protease at pH 3.0 (*14*). Pepsin has an optimum activity in the range of pH 2–4 (*2*).

b. Substrate Specificity. The mold acid proteases primarily catalyze the hydrolysis of peptide bonds in proteins and some small synthetic peptides. Unlike the DFP-sensitive proteases they exhibit only limited esterolytic activity, a property shared with pepsin (*2, 71, 72*). Table III shows a number of synthetic substrates which have been studied with the mold acid proteases. Most of these reports give only qualitative comparisons of the various substrates without evaluations of kinetic parameters such as the K_m or catalytic rate constants. Those compounds which have proved not to be substrates for any of the enzymes have been omitted from the table. The most common simple dipeptide substrate for the acid proteases was Z–Glu–Tyr. The *Paecilomyces* acid protease had a value for K_m of 4.08×10^{-3} *M* for this compound at pH 3.7 (*69*). A value for K_m of 5×10^{-4} *M* was obtained for the *P. janthinellum* peptidase B with this substrate (*73*), while peptidase A from this organism was inactive toward all the synthetic peptide substrates (*20*). Peptidase A, however, did hydrolyze bovine serum albumin to a considerable extent as was the case with *P. notatum* (*22*). Amidase activity has been demonstrated toward Bz–Arg–NH_2 for a few of the acid proteases including those from *P. varioti* (*69*), *A. saitoi* (*67, 70*), and *T. sanguinea*

69. J. Sawada, *Agr. Biol. Chem.* (*Tokyo*) **28,** 869 (1964).
70. D. Tsuru, A. Hattori, H. Tsuji, T. Yamamoto, and J. Fukumoto, *Agr. Biol. Chem.* (*Tokyo*) **33,** 1419 (1969).
71. K. Inouye and J. S. Fruton, *JACS* **89,** 187 (1967).
72. K. Inouye and J. S. Fruton, *Biochemistry* **6,** 1765 (1967).
73. R. Shaw, *BBA* **92,** 558 (1964).

(*70*) but not for those enzymes from *R. chinensis* (*24, 70*), *A. oryzae* (*16*), *P. janthinellum* (*20, 73*), and pepsin (*70*).

An apparent K_m of $6.5 \times 10^{-3} M$ was reported for the *P. varioti* acid protease-catalyzed hydrolysis of Bz–Arg–NH_2 with an activation energy of 6.86 kcal/mole (*69*). The pH optimum for this substrate was 5.5 at 45° and that for Z–Glu–Tyr was 3.5 (*69*). The *A. saitoi* acid protease had a K_m of $2.5 \times 10^{-2} M$ for this compound with a pH optimum of 4.5 for both Z–Glu–Tyr and Bz–Arg–NH_2 (*68, 74*).

The *R. chinensis* acid protease was shown to have esterolytic activity toward Tyr–OEt and tosyl–Arg–OEt (TAEE) (*24*), a property not shared by most of the other mold acid proteases.

The substrate specificities of the *A. saitoi* (*74*) and *R. chinensis* (*70*) acid proteases were studied using oxidized insulin B chain, and Fig. 1 compares the cleavage points by these enzymes with those for pepsin and rennin. Although the two mold proteases appear to have slightly broader specificity than pepsin or rennin, there are some common bonds split by all of these enzymes. The *A. saitoi* enzyme cleaved bonds adjacent to alanine, leucine, phenylalanine, and tyrosine (*67, 74*), while the *Rhisopus* enzyme hydrolyzed peptide bonds in which the amino acid residues contributing the amino groups of the bonds cleaved had bulky hydrophobic side chains (*70*).

Fukumoto *et al.* (*24, 70*) have suggested that the *R. chinensis* acid protease catalyzes a transpeptidation reaction when Lys–Tyr–Glu was used as a substrate. The implication of such a reaction and its acid pH dependence on the mechanism of action of this enzyme would be most significant, particularly if one uses pepsin for comparison.

Some comparative studies of casein hydrolysis and milk clotting have been reported for some of the microbial acid proteases with reference to pepsin and rennin as shown in Table IV (*24*). Although quite similar proteolytic activities were observed for the mold acid proteases their milk-clotting activities were quite different. The *A. saitoi* acid protease was reported to have 4–5-fold greater proteolytic activity than pepsin (*10*) and the *P. varioti* enzyme twice as much activity (*26*).

c. Activation of Trypsinogen and Chymotrypsinogen. The activation of trypsinogen at low pH by a protease from *Penicillium* sp. was reported many years ago by Kunitz (*2, 75*). Because of the stability of trypsin and chymotrypsin at low pH toward autodigestion this system has attracted several workers to investigate the activation mechanism by acid proteases. Nakanishi (*15*) isolated an acid protease, trypsinogen-

74. E. Ichishima and F. Yoshida, *Agr. Biol. Chem.* (*Tokyo*) **25,** 102 (1961).
75. M. Kunitz, *J. Gen. Physiol.* **21,** 601 (1938).

TABLE III
HYDROLYSIS OF SYNTHETIC SUBSTRATES BY VARIOUS ACID PROTEASES

Substrates	*A. saitoi* [Refs. (*67, 70*)]	*A. oryzae* [Ref. (*16*)]	*P. varioti* [Refs. (*69, 70*)]	*R. chinensis* [Refs. (*24, 70*)]	*T. sanguinea* [Ref. (*70*)]	*P. janthinellum* [Ref. (*20*)]	Pepsin [Refs. (*2, 70*)]	Rennin [Ref. (*70*)]
Z-Glu-Tyr	+++	+++	+++	+++		0	+++	+++
Z-Glu-Phe				+++		0	+++	+++
Z-Phe-Tyr				+++			+++	+++
Z-Leu-Arg		+++				0		
Z-Gly-Phe				+		0	+	+
Bz-Arg-NH_2	+++	0	+++	0	+	0	0	0
Bz-Arg-Leu		+						
Gly-Leu-Tyr				+			+	+
Gly-Glu-Tyr				+			+	+
Leu-Gly-Phe				+			+	+++
Glu-Gly-Phe				+			0	+++
Ala-Gly-Gly	+	+	0	0			0	0
Leu-Gly-Gly	+		0	0		0		

Lys-Tyr-Glu				+++			0	+++
Gly-Pro-Ala				0			0	+++
His-Phe-Arg-Trp				+++				
Leu-Tyr			+	+	+	0	+	+
Leu-Phe				+++			+	+
Leu-Gly	+		+	0	+	0	0	0
Leu-Arg		+++						
Glu-Tyr				+	0		+	+
Gly-Tyr			+	0	+	0	0	0
Gly-Val	0		+	0	+	0		
Gly-Asp	+		0	0	0		0	0
Gly-Leu	+		+	0	0	0	0	0
Gly-Gly	0		+	0	+		0	0
Ala-Gly	0		+	0	+	0	0	0
Phe-Arg		+++						
Poly-Glu				+			+	+
Poly-Lys				+			+	+
Tyr-OEt				+++				
Tosyl-Arg-OEt				+++		0		
Orn-Leu-D-Phe-Pro-Val				+++				0

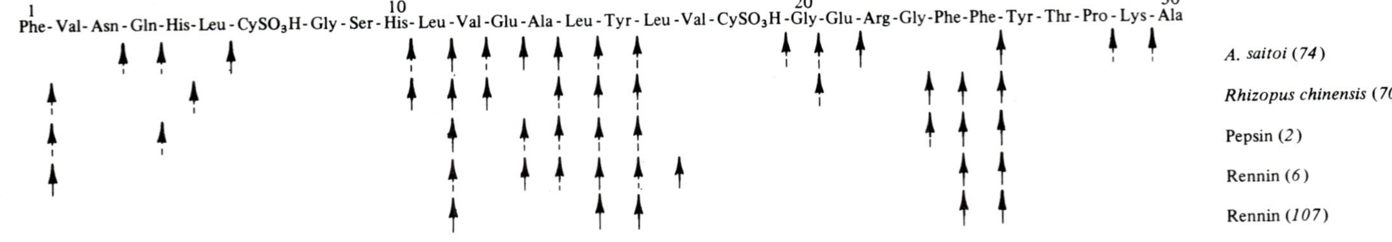

FIG. 1. Hydrolysis of the oxidized insulin B chain by various acid proteases. Solid and broken arrows represent the primary and secondary hydrolysis, respectively.

kinase, from *A. oryzae* and showed a maximum activation of trypsinogen with this enzyme at pH 3.0 and 0° within 4 hr. The active enzyme had similar properties to trypsin activated by other processes (*76*). The acid protease released the amino-terminal hexapeptide, Val–Asp_4–Lys, revealing the new amino-terminal isoleucine in trypsin. This peptide was further cleaved by the acid protease to yield valine, Asp–Asp–Asp–Asp, and lysine (*15*). A similar phenomenon was observed with the *A. saitoi* acid protease (*77*). The activation which was carried out at pH 4.5 and 0° was rapid and followed first-order kinetics. It was suggested, however, that the *A. saitoi* enzyme cleaved a bond other than Lys–Ile in trypsinogen without affecting the trypsin activity. Chymotrypsinogen A was also activated by this enzyme (*77*). The *P. janthinellum* acid protease, peptidase A, also activated trypsinogen in a manner similar to the *Aspergillus* enzymes (*20*). The activation followed Michaelis–Menten kinetics yielding a K_m value of $7.6 \times 10^{-6}\,M$ at 0° and pH 3.4 (*20*). It has also been reported (*78*) that this activation was accompanied by the splitting of a few peptide bonds other than Lys–Ile but this seems kinetically unimportant. It is interesting to note that both the *A. oryzae* and *Penicillium* acid proteases have no activity toward catalyzing the hydrolysis of Bz–Arg–NH_2, a reaction which is exhibited by the *A. saitoi* enzyme, yet all activate trypsinogen by cleavage of Lys–Ile bond.

The activation of trypsinogen by the acid proteases at low pH was much faster than by trypsin and did not require calcium. This agrees with the suggestions of several workers that the role of calicum for activation by trypsin was the neutralization of negative charges of the aspartyl residues to increase the affinity of trypsin for the amino-terminal region of trypsinogen (*79–81*). Similarly, the suggestion by Nakanishi (*16*) that low pH increased the affinity of the *A. oryzae* trypsinogen-kinase for trypsinogen and enhanced activation because of the protonation of the carboxyl groups fits the thesis proposed above.

d. Inhibitors. No inhibition has been reported for the acid proteases by DFP, thiol poisons such as *p*-mercuribenzoate and iodoacetic acid, and metal chelating agents such as EDTA and *o*-phenanthroline

76. P. Desnuelle, "The Enzymes," 2nd ed., Vol. 4, p. 119, 1960.
77. C. Gabeloteau and P. Desnuelle, *BBA* **42,** 230 (1960).
78. T. Hofman, *Bull. Soc. Chim. Biol.* **42,** 1279 (1960).
79. T. M. Radhakrishnan, K. A. Walsh, and H. Neurath, *JACS* **89,** 3059 (1967); *Biochemistry* **8,** 4020 (1969).
80. M. Delaage, P. Desnuelle, M. Lazdunski, E. Bricas, and J. Savrda, *BBRC* **29,** 235 (1967).
81. J. P. Abita, M. Delaage, M. Lazdunski, and J. Savrda, *European J. Biochem.* **8,** 314 (1969).

(*14, 15, 20, 24, 27, 29, 68, 78, 82*) with the exception of the *P. varioti* acid protease (*27*) which was inhibited by *p*-mercuribenzoate, iodoacetic acid, $AgNO_3$, and oxidizing agents which apparently reacted with the sole cysteine residue in the molecule (*54*). The *P. janthinellum* trypsinogen activating peptidase A which had no activity toward a number of synthetic substrates was shown to be inhibited by the acid protease substrate, Z–Glu–Tyr, and Z–Glu, Z–Val–Tyr, ε-amino caproic acid, and lysine (*20, 73*). Peptidase B, which might be more like an animal dipeptidase, was inhibited by a series of fatty acid salts, with the inhibition proportional to the molecular weight of the fatty acid (*73*). Šodek and Hofman have suggested that another common feature of the acid proteases besides a low pH optimum may be the presence of a carboxyl group at the active site (*83*). The *P. janthinellum* peptidase A was shown to be completely inactivated by the pepsin inhibitor diazoacetyl-D,L-norleucine methyl ester in the presence of cupric ions at pH 5.0 (*21, 83*). It was found that one equivalent of norleucine was incorporated per mole of enzyme. Treatment with hydroxylamine removed the norleucine group, suggesting that the reaction involved the formation of an ester bond between the inhibitor and a carboxyl group at the active site (group A) as proposed for pepsin (*83*). Another inactivating reagent for pepsin, *p*-bromophenacyl bromide (*84, 85*) had no effect on peptidase A (*83*). This reagent was shown by Erlanger *et al.* (*86*) to react with a β-carboxyl group of an aspartic acid residue (group B) which does not play a primary role in catalysis. Another pepsin inhibitor, *N*-diazoacetyl-*N'*-2,4-dinitrophenylethylenediamine, was shown to inhibit the acid protease from *A. awamori* in the presence of cupric ions (*12*).

These studies strongly suggest similarities in the catalytic groups at the active sites of both animal pepsins and mold acid proteases and can serve as the basis for the classification of these enzymes under the general group of pepsinlike enzymes.

B. Renninlike Acid Proteases

The renninlike acid proteases share a number of common properties with the pepsinlike enzymes such as acid pH optima, stability, and in-

82. T. Hofman, *BJ* **74,** 41p (1960).
83. J. Šodek and T. Hofman, *JBC* **243,** 450 (1968).
84. B. F. Erlanger, S. M. Vratsonas, N. Wasserman, and A. G. Cooper, *JBC* **240,** PC3447 (1965).
85. E. Gross and J. L. Morell, *JBC* **241,** 3638 (1966).
86. B. F. Erlanger, S. M. Vratsonas, N. Wasserman, and A. G. Cooper, *BBRC* **28,** 2 (1967).

sensitivity to inhibitors such as DFP, metal chelating agents, and thiol poisons. They are classified as a separate group on the basis of their specificity toward casein which permits hydrolysis leading to the clotting of milk.

1. *Distribution and Isolation*

The acid proteases exhibiting renninlike activity have been isolated from a variety of microorganisms (*87, 88*) including *Endothia parasitica* (*89–91*), *Mucor pusillus* (*88, 92–98*), *Bacillus cereus* (*99*), *Aspergillus candidus* (*100*), and *Byssochlamys fulva* (*101*).

a. Mucor pusillus. The renninlike protease from *Mucor pusillus* var. *Lindt* was isolated from wheat bran medium by a variety of salt and solvent fractionations (*96*) followed by column chromatography on Amberlite IRC-50 and DEAE-Sephadex A-25 to eliminate traces of cellulase activity (*93, 94*). The homogeneity of the enzyme was demonstrated by ultracentrifugation and electrophoresis (*94*). The isoelectric point of the enzyme was at pH 3.5–3.8. Unlike a number of acid pro-

87. I. J. Babbar, R. A. Srinivasan, S. C. Chakroverty, and A. T. Dudani, *Indian J. Dairy Sci.* **18,** 89 (1965); I. Emuauilor, *Intern. Diary Congr., Proc. 14th, Rome, 1956* pp. B401 and 506 (1956); K. Morihara, *Agr. Biol. Chem.* (*Tokyo*) **39,** 514 (1965); N. S. Paleva and N. V. Popova, *Fermentnaya Spirt. Prom.* **31,** 6 (1965); Z. Puhan, *Intern. Dairy Congr., 17th, 1966* p. D199; G. A. Somkuti and F. J. Babel, *J. Dairy Sci.* **49,** 700 (1966); R. A. Srinivasan, *Appl. Microbiol.* **12,** 475 (1964).

88. K. Arima, S. Iwasaki, and J. Tamura, *Agr. Biol. Chem.* (*Tokyo*) **31,** 540 (1967).

89. K. Hagemeyer, I. Fawwal, and J. R. Whitaker, *J. Dairy Sci.* **51,** 1916 (1968).

90. J. L. Sardinas, *Appl. Microbiol.* **16,** 248 (1968).

91. J. L. Sardinas, U. S. Patent 3,275,453 (1966).

92. J. Yu, G. Tamura, and K. Arima, *BBA* **171,** 138 (1969).

93. K. Arima, J. Yu, S. Iwasaki, and G. Tamura, *Appl. Microbiol.* **16,** 1727 (1968).

94. S. Iwasaki, T. Yasui, G. Tamura, and K. Arima, *Agr. Biol. Chem.* (*Tokyo*) **31,** 1421 (1967).

95. S. Iwasaki, T. Yasui, G. Tamura, and K. Arima, *Agr. Biol. Chem.* (*Tokyo*) **31,** 1427 (1967).

96. S. Iwasaki, G. Tamura, and K. Arima, *Agr. Biol. Chem.* (*Tokyo*) **31,** 546 (1967).

97. G. H. Richardson, J. H. Nelson, and R. E. Lubnow, *J. Dairy Sci.* **50,** 1066 (1967).

98. J. Yu, G. Tamura, and K. Arima, *Agr. Biol. Chem.* (*Tokyo*) **32,** 1048 and 1051 (1968).

99. N. Melachouris and S. L. Tuckey, *J. Dairy Sci.* **51,** 650 (1968).

100. I. Y. Veselov, D. Y. Tipograf, and T. A. Petina, *Prikl. Biokhim. i Mikrobiol.* **1,** 52 (1965).

101. S. G. Knight, *Can. J. Microbiol.* **12,** 420 (1966).

teases the *Mucor* enzyme did not exhibit trypsinogen activating properties which is shared by animal rennin and pepsin (*94*).

b. Endothia parasitica. The protease from *Endothia parasitica* was purified from broth filtrate by several methods including simple ammonium sulfate and acetone fractionations (*90, 91*) and by the use of a combination of these fractionations with gel filtration on Sephadex G-100 and DEAE-cellulose chromatography (*89*). Two different isoelectric points, pH 5.5 (*90*) and below 4.6 (*89*), have been reported for this enzyme.

c. Others. The enzyme isolated from culture medium of *B. cereus* (*99*) has not been purified, but the crude enzyme behaves similarly of calf rennin with respect to heat inactivation and specific proteolytic activity on κ-casein.

2. *Physical and Chemical Properties*

The *M. pusillus* renninlike enzyme (Mucor rennin) was shown to be stable over a larger pH range (pH 3–9), when kept at 5°, with respect to both proteolytic and milk-clotting activity (*94*). The enzyme was more thermostable at pH 6.0 than at pH 3.0, retaining about 90% of its activity after 10 min at 55° and the higher pH compared to less than 60% at the lower pH (*94*). The enzyme from *E. parasitica* exhibited a maximum stability at pH 4.5. This enzyme was also relatively thermolabile, complete inactivation occurring at 60° in 5 min (*90*). At 20° an initial loss of 40% of the activity within 10 hr was observed followed by stability for days (*90*). The *B. cereus* enzyme was also rapidly inactivated above 60°, a property shared by all of the enzymes with the animal rennin (*99*).

The amino acid composition of the *Mucor* enzyme was similar to that of calf rennin (*92*) as shown in Table I. The protease from *E. parasitica,* in contrast, contained no tryptophan (*90*). The molecular weights and various other physical parameters of the fungal enzymes are listed in Table II (*89, 90, 92, 94, 95, 100*). Although the molecular weight of the *M. pusillus* enzyme was close to that of rennin, the molecular shape was probably different as reflected in the frictional ratios (*92, 98, 102–104*). Antiserum prepared against the renninlike protease from *M. pusillus* did not cross-react with rennin, pepsin, and the acid proteases from *A. saitoi, A. oryzae,* and *T. sanguinea.* Likewise, no cross-reaction was observed

102. B. Foltmann, *Acta Chem. Scand.* **17,** 872 (1963).
103. B. Foltmann, *Acta Chem. Scand.* **13,** 1927 (1959).
104. H. Schwander, P. Zahler, and H. Nitschman, *Helv. Chim. Acta* **35,** 553 (1952).

with culture filtrates of other strains of *Mucor* or *Rhizopus* including *Mucor hiemalis* (*95*).

3. *Enzymic Properties*

The pH optima for the *Mucor pusillus* proteolytic activity and milk-clotting activity were different. Optima of pH 4.5, 4.0, and 3.5 were reported for the hydrolysis of κ-casein, hemoglobin, and Hammersten casein, respectively (*92*). This is similar to what has been reported for animal rennin (*103–106*). The optimum pH for the milk-clotting activity, however, was pH 5.5. No clotting was observed at pH 7.0 within 30 min (*92*). A comparison of the effect of pH on the milk-clotting activity of animal rennin and the *B. cereus* renninlike enzyme revealed that although in both cases decreased clotting times were observed as the pH was lowered from 6.9 to 5.9, the *B. cereus* enzyme was much less sensitive to the pH change in this region (*99*). Considerable difference was observed in the effect of temperature between the two enzymes. While the calf rennin yielded maximum coagulation rates at 40°–45°, the microbial enzyme had a maximum activity at 75°–80° (*99*).

The *M. pusillus* enzyme was shown not to be inhibited by *p*-mercuribenzoate, EDTA, or a number of metal ions except for Hg^{2+} and Fe^{3+} (*94*). Likewise no inhibition of the *Endothia parasitica* protease was reported with tetrathionate, $HgCl_2$, *N*-ethylmaleimide, *p*-mercuribenzoate, cysteine, chloromethyl ketones of *N*-tosyl-L-phenylalanine (TPCK) and *N*-tosyl-L-lysine (TLCK), iodoacetamide, and EDTA (*89*). This enzyme was shown to be mildly proteolytic similar to rennin (*90*). Table IV compares the relative ratios of milk-clotting activity over proteolytic activity for a number of proteases. As can be seen in the table, calf rennin has the highest ratios followed by the *M. pusillus* enzyme. This value, rather than the clotting activities, seems to be the important criterion in establishing renninlike activity. The *Rhizopus chinensis* acid protease and pepsin both possess considerably greater clotting activity than rennin, but their proteolytic activities are likewise high. Similar properties are exhibited by papain and ficin (*24*, *93*).

Although degradation products of different sizes are produced by calf rennin from casein, the *B. cereus* enzyme produced smaller fragments (*99*). Nonspecific proteolysis by the microbial rennin increased rapidly and continued to produce lower molecule weight fragments, while the calf rennin initially liberated nitrogen products rapidly [as a result of liberation of glycomacropeptide (*3–6*)] but then leveled off. The data suggested

105. N. J. Berridge, *BJ* **39**, 179 (1945).
106. E. Schram, S. Moore, and E. J. Bigwood, *BJ* **57**, 33 (1954).

TABLE IV
PROTEOLYTIC AND MILK-CLOTTING ACTIVITY OF SEVERAL PROTEASES[a]

Enzyme	Proteolytic activity at optimal pH (A) (unit/mg protein)	Milk-clotting activity at optimal pH (B) (unit/mg protein)	(B)/(A)
R. chinensis acid protease	2,200	6,400	2.91
A. saitoi acid protease	2,000	700	0.35
T. sanguinea acid protease	2,200	2,400	1.09
Pepsin	3,000	8,800	2.94
M. pusillus (Mucor rennin)[b]	275	2,800	10.2
Rennin	100	1,600	16.0
B. subtilis var. *amylosacchariticus*			
Neutral protease II	12,500	14,400	1.15
Alkaline protease	2,200	50	—

[a] Taken from Fukumoto *et al.* (*24*).
[b] Recalculated values from Arima *et al.* (*93*).

that the *B. cereus* enzyme displayed specific proteolytic cleavage of the κ-casein, very similar to calf rennin, but the action on the other casein fractions, particularly β-casein, was nonspecific (*99*). The production of nonprotein nitrogen from κ-casein by *M. pusillus* protease was slower than that produced by calf rennin, suggesting that the milk-clotting mechanisms were slightly different (*96*). However, the electrophoretic analysis of casein products from the two enzymes were similar (*96*). Hydrolysis of insulin B chain by rennin (*6*, *107*) is included in Fig. 1.

III. DFP-Sensitive Alkaline Proteases

The most widely distributed group of proteolytic enzymes of both microbial and animal origin are the serine proteases (*108*), generally identified by their sensitivity to organophosphorus reagents such as DFP and isopropylmethylphosphofluoridate (Sarin). The most familiar of the bacterial DFP-sensitive enzymes, the extracellular subtilisins of *B. subtilis* (*1*), are discussed by Kraut, Chapter 15, and Markland and Smith, Chapter 16, this volume. These microbial enzymes, sometimes referred to as alkaline proteases, share with the animal enzymes of this group, in addition to a serine residue at the active site, esterolytic activity,

107. J. C. Fish, *Nature* **180,** 345 (1957).
108. B. S. Hartley, *Ann. Rev. Biochem.* **29,** 45 (1960).

alkaline pH optima, and in some particular cases similar specificities. They are generally neither inhibited by metal chelating reagents and thiol poisons nor activated by metal ions or reducing agents. These enzymic properties set them apart from the other mold and bacterial endopeptidases, both acid and neutral. Broad substrate specificities are generally exhibited by these enzymes with the exception of the cell wall lytic proteases of *Sorangium* and the trypsinlike enzymes of *Streptomyces,* which will be discussed as unique subgroups.

A. Distribution and Isolation

The DFP-sensitive endopeptidases, widely distributed among fungi and bacteria, are almost entirely extracellular and are obtained from the culture media of these organisms. The exceptions are a few intracellular yeast enzymes. Isolation and characterization of these enzymes have been reported from *Bacillus* (Markland and Smith, Chapter 16, this volume), *Arthrobacter* (*109–114*), *Streptomyces griseus* (*115, 116*), *Streptomyces fradiae* (*117, 118*), *Streptomyces rectus* (*119–121*), *Aspergillus oryzae* (*13, 122–124*), *Aspergillus flavus* (*125*), *Aspergillus fumigatus* (*17, 18*), *Aspergillus sydowi* (*126*), *Aspergillus sojae* (*127*), *Penicillium*

109. B. V. Hofsten, H. V. Kley, and D. Eaker, *BBA* **110,** 585 (1965).

110. B. V. Hofsten and C. Tjeder, *BBA* **110,** 576 (1965).

111. S. Wählby, L. Engström, U. Bjare, and B. V. Hofsten, *Acta Chem. Scand.* **20,** 1993 (1966).

112. H. Eklund, M. Zeppezauer, and C.-I. Brändén, *JMB* **34,** 193 (1968).

113. S. Wählby, *BBA* **151,** 409 (1968).

114. B. V. Hofsten and B. Reinhammar, *BBA* **110,** 599 (1965).

115. S. Wählby, Ö. Zellerqvist, and L. Engström, *Acta Chem. Scand.* **19,** 1247 (1965).

116. A. Hiramatsu, *J. Biochem.* (*Tokyo*) **61,** 168 (1967).

117. K. Morihara, T. Oka, and H. Tsuzuki, *BBA* **139,** 382 (1967).

118. K. Morihara and H. Tsuzuki, *ABB* **126,** 971 (1968).

119. K. Mizusawa, E. Ichishima, and F. Yoshida, *Agr. Biol. Chem.* (*Tokyo*) **28,** 884 (1964).

120. K. Mizusawa, E. Ichishima, and F. Yoshida, *Agr. Biol. Chem.* (*Tokyo*) **30,** 35 (1966).

121. K. Mizusawa, E. Ichishima, and F. Yoshida, *Appl. Microbiol.* **17,** 366 (1969).

122. K. Morihara and H. Tsuzaki, *ABB* **129,** 620 (1969).

123. A. R. Subramanian and G. Kalnitsky, *Biochemistry* **3,** 1861 (1964).

124. A. Norwig and W. F. Jahn, *Z. Physiol. Chem.* **345,** 284 (1966).

125. J. Turková, O. Mikeš, K. Gančev, and M. Boublík, *BBA* **178,** 100 (1969).

126. G. Danno and S. Yoshimura, *Agr. Biol. Chem.* (*Tokyo*) **31,** 1151 and 1159 (1967).

127. K. Hayashi, D. Fukushima, and K. Mogi, *Agr. Biol. Chem.* (*Tokyo*) **31,** 642, 1171, and 1237 (1967).

cyaneofulvum (*128–131*), *Bacteroides amylophilus* (*132*), *Sorangium* sp. (*133–136*), *Alternaria tenuissima* (*23, 137, 138*), *Cliocladium roseum* (*139*), and *Saccharomyces cerevisiae* (*45, 47–50, 140, 141*).

In many cases the same organism produced a number of peptidases and proteases of various kinds. The reader is referred to the appropriate enzyme section for a discussion of the other enzymes.

1. *Streptomyces*

Streptomyces griseus K-1 produces a mixture of proteases (Pronase) including DFP- and EDTA-sensitive proteases and some exopeptidases (*142–146*). Two or three DFP-sensitive enzymes are present in the culture filtrates of this organism (*115, 147–150*) as well as in preparations from *S. griseus* ATCC3463 (*116, 151*) and *S. fradiae* (*117, 118*). Two similar enzymes have been reported from a thermostable strain, *S. rectus* (*119–121*). An intracellular protease produced by *Streptomyces moderatus* sp. N was purified and shown to be inhibited by DFP (*152*).

128. K. Singh and S. M. Martin, *Can. J. Biochem. Physiol.* **38,** 969 (1960).
129. S. M. Martin, K. Singh, H. Ankel, and A. H. Kahn, *Can. J. Biochem. Physiol.* **40,** 237 (1962).
130. H. Ankel and S. M. Martin, *BJ* **91,** 431 (1964).
131. P. Hill and S. M. Martin, *Can. J. Microbiol.* **12,** 243 (1966).
132. T. H. Blackburn, *J. Gen. Microbiol.* **53,** 27 and 37 (1968).
133. D. C. Gillespie and F. D. Cook, *Can. J. Microbiol.* **11,** 109 (1965).
134. D. R. Whitaker, F. D. Cook, and D. C. Gillespie, *Can. J. Biochem.* **43,** 1927 (1965).
135. D. R. Whitaker, *Can. J. Biochem.* **43,** 1935 (1965).
136. D. R. Whitaker, *Can. J. Biochem.* **45,** 911 (1967).
137. A. G. Jönsson, *Appl. Microbiol.* **15,** 319 (1967).
138. A. G. Jönsson, *ABB* **129,** 62 (1969).
139. T. Kishida and S. Yoshimura, *Agr. Biol. Chem.* (*Tokyo*) **30,** 1183 (1966).
140. T. Hata, R. Hayashi, and E. Doi, *Agr. Biol. Chem.* (*Tokyo*) **31,** 160 (1967).
141. R. Hayashi, Y. Oaka, and T. Hata, *Agr. Biol. Chem.* (*Tokyo*) **33,** 196 (1969).
142. A. Hiramatsu and T. Ouchi, *J. Biochem.* (*Tokyo*) **54,** 462 (1963).
143. M. Nomoto, Y. Narahashi, T. Ouchi, and A. Hiramatsu, *Abstr. 6th Intern. Congr. Biochem., New York, 1964* Vol. IV, p. 325 (1964).
144. Y. Narahashi and M. Yanagita, *J. Biochem.* (*Tokyo*) **62,** 633 (1967).
145. T. Murachi, *BBA* **71,** 239 (1963).
146. S. Wählby and L. Engström, *BBA* **151,** 402 (1968).
147. Y. Narahashi, K. Shibuya, and M. Yanagita, *J. Biochem.* (*Toyko*) **64,** 427 (1968).
148. M. Trop and Y. Birk, *BJ* **109,** 475 (1968).
149. S. Wählby, *BBA* **151,** 394 (1968).
150. S. Wählby, *BBA* **185,** 178 (1969).
151. T. Ouchi, *Agr. Biol. Chem.* (*Tokyo*) **26,** 723 and 734 (1962).
152. F. Reusser, *ABB* **112,** 156 (1965).

Thermostable proteases (temperature optima at 60° and 70°) have been isolated from culture filtrates of thermophilic *Actinomycetes, Thermomonospora fusca* (A29), and *Thermoactinomyces vulgaris* (A60), which also possess bacterial lytic activity (*153*). Although no information is available on the effect of DFP on these enzymes they do have quite alkaline pH optima, pH 8.0 and 9.0, which suggest that they might belong to this group of enzymes (*153*).

Generally, the enzymes were purified from culture filtrates of the organisms by a combination of salt and solvent fractionations, gel filtration, and ion exchange chromatography. A number of the enzymes have been obtained in crystalline form (*13, 109, 117, 122, 126, 138, 139*). A summary of the methods employed for the purification of a number of DFP-sensitive alkaline proteases is given in Table V. This includes the enzymes from all of the various microbial sources discussed in this section. Table V also includes the information of homogeneity of various enzymes in this group.

2. *Aspergillus*

The various *Aspergillus* species including *A. oryzae* (aspergillopeptidases B and C) (*13, 122–124*), *A. sydowi* (*126*), *A. flavus* (aspergillopeptidase F_1) (*125*), *A. sojae* (*127*), *A. melleus* (*154*), and *A. fumigatus* (*17, 18*) produce a variety of proteolytic enzymes including alkaline proteases (DFP-sensitive), neutral proteases (EDTA-sensitive) (Section IV), and acid proteases (Section II). The isolation procedures and evidence for homogeneity for these enzymes are shown in Table V.

3. *Sorangium*

Two proteolytic enzymes (α- and β-lytic proteases) which also possess bacterial cell wall lytic activity have been isolated from culture filtrates of a species of *Sorangium* (*Myxobacter* 495) (*133–136*). These enzymes have been isolated in pure form, the details of which are given in Table V. The unique cell wall lytic properties of the α-lytic enzyme, which is DFP-sensitive, will be discussed together with the β-lytic enzyme, which is not DFP-sensitive, and others in a separate section (Section V).

4. *Others*

An extracellular, serine-type protease has been isolated in crystalline form from culture filtrates of *Arthrobacter* B_{22} (*109–113*) (Table V).

153. A. J. Desai and S. A. Dhala, *J. Bacteriol.* **100,** 149 (1969).
154. M. Ito and M. Sugiura, *Yakugaku Zasshi* **88,** 1576, 1583, and 1591 (1968).

TABLE V
SUMMARY OF PURIFICATION AND HOMOGENEITY OF DFP-SENSITIVE PROTEASES

Organism	Type	Purification method[a]	Homogeneity[b]
S. griseus K-1 (*115, 142, 144, 149*)		3, 5	
S. griseus (*147*)		1, 2, 3, 4	
S. griseus (*115, 149, 158*)	Trypsinlike	2	1, 4
S. griseus (*148*)	Trypsinlike	2	2, 3
S. griseus (*115, 149, 150*)	PNPA-hydrolase I	1, 2	1, 2, 3
S. griseus (*115, 149, 150*)	PNPA-hydrolase II	1, 2	1, 2, 3
S. griseus ATCC 3463 (*116, 151*)		2, 3, 4, 5	
S. fradiae ATCC 3535 (*117*)		2, 3, 4, 6	2
S. fradiae ATCC 3535 (*118*)	Trypsinlike	1, 2, 3, 4	2
S. rectus var. *proteolytics* (*119–121*)	Thermostable	2, 3, 4	1
S. moderatus sp. N (*152*)	Intracellular	1, 2, 3	
T. fusca A29 (*153*)	Thermostable	1, 4	
T. vulgaris A60 (*153*)	Thermostable	1, 4	
A. oryzae (*123*)	Aspergillopeptidase B	1, 2, 3	1, 2
A. oryzae (*124*)	Aspergillopeptidase C	4, 5	1, 2
A. oryzae (*13*)		2, 4, 5, 6, 7	1, 2
A. oryzae OUT5038 (*122*)		1, 2, 6	
A. sydowi (*126*)		2, 3, 6, 7	1, 2
A. flavus (*125*)	Aspergillopeptidase F_1	1, 2	1, 2
A. sojae (*127*)		1, 2, 3, 4	1, 2, 3
A. melleus (*154*)		1, 3, 4, 6, 7	1, 2
A. fumigatus (*17, 18*)		2, 3, 7	2
Sorangium sp. (*Myxobacter* 495) (*133–136*)	α-Lytic	2, 3	1, 2
Arthrobacter B_{22} (*109, 110*)		1, 2, 3, 4, 6	1, 2
P. cyaneo-fulvum (*128–131*)		1, 2, 3, 4	2
A. tenuissima (*23, 137, 138*)		1, 3, 4, 6, 7	1, 2
C. roseum (*139*)		2, 4, 6, 7	2
P. omnivorum (*155–157*)		2, 4	2
S. cerevisiae (*45, 47–50, 140, 141*)	Intracellular	1, 2, 3, 4, 7	1, 2

[a] 1, Gel filtration; 2, ion exchange; 3, salt fractionation; 4, organic solvent fractionation; 5, electrophoresis; 6, crystallization; and 7, other methods.
[b] 1, Ultracentrifugation; 2, electrophoresis; 3, chromatography; and 4, other methods.

The enzyme was shown to be quite stable over the pH range from 3 to 11 and in the presence of up to 8 *M* urea (*114*). An alkaline protease with broad specificity has also been isolated from *Penicillium cyaneo-fulvum* culture filtrates (*128–130*) (Table V). Also, a similar intracellular enzyme has been reported from this organism (*131*). Two intracellular enzymes isolated from *Saccharomyces cerevisiae* (proteinases B and

C) exhibit alkaline pH optima, esterolytic activity, and are inhibited by both *p*-mercuribenzoate and DFP (*45, 47–49, 140, 141*). The enzymes exist in an inactive form as either a zymogen or enzyme–inhibitor complex (*48–50*). Extracellular DFP-sensitive alkaline proteases have also been isolated from *Alternaria tenuissima* (*23, 137, 138*) and *Cliocladium roseum* (*139*). *Phymatotrichum omnivorum*, a plant pathogenic fungus, produces an extracellular protease which is both DFP- and EDTA-sensitive (*155, 157*).

B. Chemical Properties

1. *Amino Acid Composition, Terminal Residues, and Nonprotein Components*

The amino acid compositions of various DFP-sensitive proteases are shown in Table VI (*109, 125, 127, 138, 158–165*). The *A. oryzae* enzymes, aspergillopeptidases B (*160*) and C (*161*), and aspergillopeptidase F_1 from *A. flavus* (*125*) have amino acid compositions similar to each other and differ from the others with respect to distinct features. Unlike the other enzymes these have no half-cystine residues, a feature associated with the subtilisins (*162*) (see also Kraut, Chapter 15, this volume). The lysine, arginine, histidine, and threonine contents which are almost identical for these enzymes are also distinctly different from the others. Aspergillopeptidases B (*166*) and F_1 (*125*) have identical N- and C-

155. A. A. Burgum, J. M. Prescott, and R. J. Hervey, *Proc. Soc. Exptl. Biol. Med.* **115**, 39 (1964).

156. J. M. Prescott and A. A. Burgum, *Federation Proc.* **23**, 241 (1964).

157. A. A. Burgum and J. M. Prescott, *ABB* **111**, 391 (1965).

158. L. Jurášek, D. Fackre, and L. B. Smillie, *BBRC* **37**, 99 (1969).

159. L. Jurášek and D. R. Whitaker, *Can. J. Biochem.* **45**, 917 (1967).

160. A. R. Subramanian and G. Kalnitsky, *Biochemistry* **3**, 1868 (1964).

161. A. Norwig and W. F. Jahn, *European J. Biochem.* **3**, 519 (1968).

162. E. L. Smith, R. J. DeLange, W. H. Evans, M. Landon, and F. S. Markland, *JBC* **243**, 2184 (1968).

163. D. M. Shotton and B. S. Hartley, cited by M. O. Dayhoff, ed., "Atlas of Protein Sequence and Structure," Vol. 4, p. D-120. Published by The National Biomedical Research Foundation, 1969.

164. K. Walsh and H. Neurath, *Proc. Natl. Acad. Sci. U. S.* **52**, 884 (1964); O. Mikeš, V. Holeyšovský, V. Tomášek, and F. Šorm, *BBRC* **24**, 346 (1966); B. S. Hartley, cited by M. O. Dayhoff, ed., "Atlas of Protein Sequence and Structure," Vol. 4, p. D-115. Published by The National Biomedical Research Foundation, 1969.

165. B. S. Hartley and D. L. Kaufman, *BJ* **101**, 229 (1966).

166. A. R. Subramanian, S. Spadari, and G. Kalnitsky, *Federation Proc.* **24**, 593 (1965).

TABLE VI

AMINO ACID COMPOSITION OF VARIOUS DFP-SENSITIVE PROTEASES

Amino acid	*Sorangium* α-lytic [Refs. (*158*, *159*)]	Elastase [Ref. (*163*)]	*S. griseus* trypsinlike [Ref. (*158*)]	*A. tenuissima* [Ref. (*138*)]	*Arthrobacter* B_{22} [Ref. (*109*)]	Trypsin [Ref. (*164*)]	*A. oryzae*[a] B [Ref. (*160*)]	*A. oryzae*[a] C [Ref. (*161*)]	*A. flavus*[b] [Ref. (*125*)]	*A. sojae* [Ref. (*127*)]	Subtilisin BPN′ [Ref. (*162*)]	Subtilisin Carlsberg [Ref. (*162*)]	α-Chymotrypsin [Ref. (*165*)]
Lys	2	3	6.5	1	9	14	11–12	12	11	14	11	9	14
His	1	6	1.0	1	2	3	4	4	3–4	5	6	5	2
Arg	12	12	8.4	10	1	2	2	3	2	3	2	4	3
Try	3–4	7	—	4	5	4	2	2	2	2	3	1	8
Asp	15	24	18.1	18	24	22	21	21–22	21	31	28	27	22
Thr	17–18	19	16.4	33	19	10	11	13	11	18	13	19	22
Ser	19–20	22	14.2	35	27	34	19	23	20	28	37	32	27
Glu	13	19	17.4	12	12	14	12	13	12–13	19	15	12	15
Pro	4–5	7	8.0	4	12	8	4	5–6	4–5	6	14	9	9
Gly	31–32	25	28.4	55	29	25	19	21	20	27	33	35	23
Ala	24	17	26.0	22	23	14	23	23	23	32	37	41	22
Half-Cys	6	8	6.1	6	4	12	0	0	0	2	0	0	10
Val	19	27	17.8	19	12	17	15	16	15	18	30	32	23
Met	2	2	2.7	0	4	2	0	1	1	2	5	5	2
Ile	8	10	8.0	7	5	15	9–10	10–11	9–10	14	13	10	10
Leu	10	18	11.0	7	8	14	9	10	9	14	15	16	19
Tyr	4	11	8.2	13	15	10	5	4–5	5	8	10	13	4
Phe	6	3	5.7	6	10	3	5	6	5	7	3	4	6
Ammonia	—	32	—	23	29	29	15	—	17	20	28	25	23
Total	198–199	240	203.9	253	221	223	171–173	187–191	173–177	250	275	274	241

[a] B, Aspergillopeptidase B; C, aspergillopeptidase C.

[b] Aspergillopeptidase F_1.

terminal residues, glycine and alanine, respectively. The *A. sojae* protease, with the exception of one to two half-cystine residues, exhibits a very similar relative ratio of amino acids to those of the other aspergillopeptidases, although the apparent composition (for 250 residues) is different (*127*).

The existence of a single disulfide bond in the *A. sojae* enzyme was suggested because of the nontitratable nature of the cysteine residues by heavy metals in urea (*127*). The six half-cystine residues in the *Sorangium* α-lytic enzyme are all nontitratable by heavy metals and probably form three disulfide bridges (*159*). The *Arthrobacter* enzyme also exhibits nontitratable half-cystine residues suggesting two disulfide bridges (*114*). An N-terminal valine has been reported for this enzyme.

The presence of nonprotein components in some of these enzymes has been reported. Small amounts of carbohydrate were found in aspergillopeptidases B, C, and F_1 (*125, 166*). There have been some question as to whether or not aspergillopeptidase B contains phosphate (*160, 166*). The *A. sojae* protease was shown to be devoid of any sugar or phosphate compounds (*127*), and the absence of hexosamine was reported for the *Alternaria* enzyme (*138*). No metal was found in aspergillopeptidase B (*166*) and the *Sorangium* α-lytic protease (*159*).

2. *Active Serine Peptides*

As discussed above, the most distinguishing characteristic of this group of enzymes is the inhibition by organophosphorus compounds such as DFP which implicates the role of an active serine in the mechanism of action of these enzymes. This incorporation of organophosphorus compound into the enzyme molecules is stoichiometric (*115, 125, 149, 166–168*), and the reaction products, such as diisopropylphosphoryl (DIP) enzymes are fairly stable, similar to the animal serine proteinases (*108, 169*), permitting in some cases the isolation of the peptide containing the DIP-serine residue (*113, 146, 167–171*). Table VII compares the sequences of the serine active site peptides of variation DFP-sensitive proteases from mold and bacterial origin with those of animal origin. Both the *Sorangium* α-lytic protease (*167, 168*) and the *S. griseus* protease (*146*) have a sequence of Asp–Ser*–Gly–Gly which is common with that for the animal proteases, trypsin, chymotrypsin, and elastase (*169*). This is particularly interesting because of the very

167. D. R. Whitaker and C. Roy, *Can. J. Biochem.* **45,** 911 (1967).
168. D. R. Whitaker, L. Jurášek, and C. Roy, *BBRC* **24,** 173 (1966).
169. M. L. Bender and F. J. Kézdy, *Ann. Rev. Biochem.* **34,** 49 (1965).
170. O. Mikeš, J. Turková, N. B. Toan, and F. Šorm, *BBA* **178,** 112 (1969).
171. D. C. Shaw, cited by F. Sanger, *Proc. Chem. Soc.* p. 76 (1963).

TABLE VII
SERINE PEPTIDES OF VARIOUS DFP-SENSITIVE PROTEASES

Enzyme	Amino acid sequence[a]
Trypsin (*169*)	
Chymotrypsin (*169*)	-Gly-Asp-Ser*-Gly-Gly-Pro-
Elastase (*169*)	
Sorangium α-lytic (*167, 168*)	-Asp-Ser*-Gly-Gly-
S. griseus trypsinlike (*146, 158*)	-Gly-Asp-Ser*-Gly-Gly-Pro-
Subtilisins (*169*)	-Gly-Thr-Ser*-Met-Ala-
A. flavus (*170*)	-Gly-Thr-Ser*-Met-Ala-
A. oryzae (*171*)	-Thr-Ser*-Met-Ala-
Arthrobacter B_{22} (*113*)	Ser-Ser*-Gly-

[a] Ser* indicates active serine residue.

similar specificity and enzymic properties of the *Streptomyces* alkaline proteases to trypsin (*115, 118, 147–149, 158*). A second group of common sequences, Thr–Ser*–Met–Ala typical of subtilisins (*169*), is shared by the *A. flavus* protease (aspergillopeptidase F_1) (*170*) and the *A. oryzae* enzyme (*171*). The *Arthrobacter* protease has a sequence Ser–Ser*–Gly (*113*) which is in part similar to each of the types above.

3. *Cystine Peptides of Sorangium and Streptomyces Proteases*

The amino acid sequences of one of the three cystine bridges of the *Sorangium* α-lytic protease (*172*) and of all three cystine bridges in the *S. griseus* trypsinlike protease (*158*) were established by applying the diagonal method (*173*). The specificity of the *Sorangium* enzyme has been shown to be similar to that of elastase while the *S. griseus* protease exhibits specificity similar to trypsin as described in Section III,D,2. Complete amino acid sequences of these enzymes for comparative purposes will be significant for the elucidation of catalytic mechanisms as well as evolutionary relationships. The pancreatic serine proteinases such as chymotrypsin, trypsin, and elastase have large portions of common sequences in their molecules. Therefore, a comparison of the sequences around the DFP-sensitive serine residue, the active histidine residue, and around various cystine bridges might yield interesting inferences.

Figure 2 shows the sequences for the *S. griseus* and *Sorangium* enzymes which correspond to the region in trypsin, chymotrypsin, and elastase around the two histidine residues at 40 and 57 and the cystine bridge

172. L. B. Smillie and D. R. Whitaker, *JACS* **89,** 3350 (1967).
173. J. R. Brown and B. S. Hartley, *BJ* **101,** 214 (1966).

	39 … 46		52 … 63
Trypsin (*164*)	-Tyr-His-Phe-Cys-Gly-Gly-Ser-Leu-	/	-Val-Val-Ser-Ala-Ala-His-Cys-Tyr-Lys-Ser-Gly-Ile-
Chymotrypsin (*165*)	-Phe-His-Phe-Cys-Gly-Gly-Ser-Leu-	/	-Val-Val-Thr-Ala-Ala-His-Cys-Gly-Val-Thr-Thr-Ser-
S. griseus trypsinlike (*158*)	-Ser-Met-Gly-Cys-Gly-Gly-Ala-Leu-	/	-Thr-Ala-Ala-His-Cys-Val (Asn_2,Gly_2)
Elastase (*163*)	-Ala-His-Thr-Cys-Gly-Gly-Thr-Leu-	/	-Val-Met-Thr-Ala-Ala-His-Cys-Val-Asp-Arg-Glu-Leu-
Sorangium α-lytic (*172*)	Cys-Ser-Val-Gly-Phe-	/	-Phe-Val-Thr-Ala-Gly-His-Cys-Gly-Thr-Val-Asn-Ala-

FIG. 2. Comparison of the amino acid sequences containing histidine residues in various DFP-sensitive alkaline proteases. The residue numbers are based on those in trypsin.

formed by cystine residues 42 and 58. Remarkable similarity is observed for these proteins in this region, particularly between the *S. griseus* enzyme and the pancreatic proteases. The role of histidine 57 in the animal enzymes as indicated by alkylation with TLCK and TPCK *174–176*) and the similarity in structure pointed out above between the microbial and animal enzymes clearly indicate the significance of such comparisons for function inferences. Only one histidine residue is present in both the *S. griseus* and *Sorangium* proteases. Histidine 40 of the animal sequence is replaced by a methionine residue in the *S. griseus* protease. This fact questions the functional role for His 40 in catalysis, as suggested by kinetic studies (*177, 178*).

The active histidine residue (*179*) and serine residue (*180*) (see Markland and Smith, Chapter 16, this volume) in subtilisins are located differently from those mentioned above, and from these studies it was suggested that the bacterial proteases have evolved along at least two independent pathways (*172*).

Significant structural similarities have been shown for other cystine bridge areas between *S. griseus* and trypsin with 33 identical residues out of a total of 55 residues compared (*158*). The importance of the aspartic acid residue corresponding to residue 189 of trypsin was pointed out for the binding site of the cationic substrates (*158*).

Comparative studies of the sequences of the microbial proteases as they become available and those of animal origin are important tools for obtaining information not only concerning structure–function relationship but also evolutionary implications.

C. Physical Properties

1. *Molecular Weight, Physical Constants, and Isoelectric Point*

The molecular weights of the various proteases in this group are within the range of 20,000–25,000, similar to trypsin and chymotrypsin as shown in Table VIII (*109, 117, 120, 125, 127, 135, 138, 147, 149, 150, 158–161, 181*). A few enzymes, however, were reported with higher molecular

174. E. B. Ong, E. Shaw, and G. Schoellmann, *JBC* **240,** 694 (1965).
175. K. J. Stevenson and L. B. Smillie, *JMB* **12,** 937 (1965).
176. V. Tomášek, E. S. Severin, and F. Šorm, *BBRC* **20,** 545 (1965).
177. R. J. Kézdy, G. D. Clement, and M. L. Bender, *JACS* **86,** 390 (1964).
178. H. Kaplan and D. R. Whitaker, *Can. J. Biochem.* **47,** 305 (1969); *JACS* **89,** 3352 (1967).
179. F. S. Markland, E. Shaw, and E. L. Smith, *Proc. Natl. Acad. Sci. U. S.* **61,** 1440 (1968).
180. E. L. Smith, F. S. Markland, C. B. Kasper, R. J. DeLange, M. Landon, and W. H. Evans, *JBC* **241,** 5974 (1966).

weights. *Phymatotrichum omnivorum* (*157*), *P. cyaneo-fulvum* (*129*), and *A. oryzae* OUT5038 (*122*) proteases were reported to have molecular weights of 33,000, 44,000 and 52,000, respectively.

The isoelectric points for these proteases are in the range of pH 9–10.5. An exception is the slightly acidic value of pH 5.1 for *A. sojae* protease as based on the amino acid composition (*127*).

It appears from these properties and amino acid compositions that the *Sorangium* α-lytic protease, *S. griseus* protease, the *Alternaria* protease, and probably the *Arthrobacter* protease more closely resemble the animal-type enzymes, while the *Aspergillus* proteases more closely resemble the subtilisins.

The *A. sojae* protease exhibited a small intrinsic viscosity $[\eta] = 0.027$ dl/g, indicating a compact nature of the molecule. The frictional ratio f/f_0 was reported to be 1.12 and the axial ratio a/b, 3.0 (*127*). The partial specific volumes of the proteases in this group are similar to each other, ranging from 0.70 to 0.72.

2. *Stability*

Most of the enzymes of this group of alkaline proteases are stable in the pH range from 5 or 6 to 9 or 10 at lower temperatures (*13, 114, 117, 119, 126, 129, 147, 151*) with a few reported to be stable even at pH as low as 2.5–3.0 (*114, 129, 147, 151*). Nearly half of the proteases are unstable at temperatures of 50°–60° with losses of 50% of the activity within 10–15 min or less (*13, 17, 126, 139, 144, 147, 151, 157, 166*). The *Arthrobacter* protease (*114*) and several *Streptomyces* proteases (*117, 119–121*) are moderately heat stable. Considerable thermal stability has been reported for the *S. fradiae* protease IIb (*117*); after incubation at 80° for 10 and 60 min, 100 and 72% activity, respectively, still remained.

In some cases Ca^{2+} afforded some protection against thermal inactivation (*17, 119–121, 126*). For the *A. sydowi* protease (*126*), the Ca^{2+} effect was reported to be pH dependent. Similar stabilizations by divalent cations was observed with *Arthrobacter* (*114*) and *S. griseus* (*151*) proteases. The alkaline proteases were denatured by 5–8 *M* urea (*114, 144, 159, 166*).

3. *Conformation Studies*

A 10–15% of α-helical conformation was reported for the *A. sojae* alkaline protease molecule (*182*) based upon the optical rotation $[\alpha]_D =$

181. L. Jurášek and D. R. Whitaker, *Can. J. Biochem.* **43,** 1955 (1965).
182. K. Hayashi, D. Fukushima, and K. Mogi, *Agr. Biol. Chem.* (*Tokyo*) **32,** 988 (1968).

TABLE VIII
MOLECULAR CHARACTERISTICS OF VARIOUS DFP-SENSITIVE PROTEASES

Enzyme	MW	$s_{20,w} \times 10^{13}$	$D_{20,w} \times 10^{7}$	$\bar{V}$	pI	$A_{1\,cm}^{1\%}$ (280 nm)
Sorangium α-lytic (*135, 159, 181*)	19,000[a,b]	2.2		0.72	Above 10.5	8.9
	20,100[c]					
A. tenuissima (*138*)	23,000[a]	2.35		0.706	Above 10.5	13.3[g]
	24,780[c]					
S. griseus trypsinlike (*149, 158*)	19,000[a]	2.3		0.72		
	20,850[c]					
S. griseus K-1 III (*147*)						11.5
IV (*147*)						10.2
S. griseus K-1 PNPA						
Hydrolase I (*150*)	16,000[a]			0.72		
Hydrolase II (*150*)	18,000[a]			0.71		
S. fradiae II (*117*)	17,700[f]	2.24	11.5	0.72		
S. fradiae III (*117*)	14,000[b]					Around 9
S. fradiae trypsinlike IV (*117*)	16,500[b]	2.54				Around 9
S. rectus A (*120*)		2.95				
S. rectus B (*120*)		2.96				

Arthrobacter B_{22} (*109*)	22,000[a]	2.7			Around 9	18.6
	23,040[c]					
A. oryzae[h] B (*160*)	18,000[a]			0.682		9.0
	17,700[c]			0.72		
A. oryzae[h] C (*161*)	20,500[a]			0.70		
	19,650[c]					
A. flavus[i] (*125*)	19,800[b]	3.05				9.04
	18,000[c]					
	19,000[d]					
A. sojae (*127*)	25,000[b]	2.82	9.8	0.726	5.1	8.98
	25,750[c]					
	25,500[e]					

[a] Ultracentrifugation.
[b] Gel filtration.
[c] Amino acid composition.
[d] Light scattering.
[e] Sedimentation viscosity.
[f] Sedimentation diffusion.
[g] Calculated from the extinction coefficient.
[h] B, aspergillopeptidase B; C, aspergillopeptidase C.
[i] Aspergillopeptidase F_1.

18.3°, the reduced mean residue rotation $[m']_{233} = -2400$, the optical dispersion constant $\lambda_C = 256$ nm, and the Moffitt-Yang parameters $a_0 = -78$ and $b_0 = -74$. In 6 *M* guanidine hydrochloride at pH 2 these values were $[\alpha]_D = 70.6°$, $\lambda_c = 216$ nm, $a_0 = -381$, and $b_0 = 0$, indicating a denaturation resulting not only from disruption of α-helical structure but also a disruption of other compactly folded structures suggested by its low intrinsic viscosity (*127*). A very low, if any, antiparallel β structure was suggested in this molecule because of the large difference in spectra between this protease and other proteins with typical antiparallel β structure. Infrared spectra of the enzyme suggested the presence of a predominant amount of disordered structure with minor α-helical and β structures (*182*). Ultraviolet difference spectra suggested that some tyrosine residues and two tryptophan residues were buried within the molecule. The reaction of 2 hydroxy-5-nitrobenzyl bromide with the two tryptophans only when the molecule was denatured supported this suggestion. Participation of these residues in maintaining the fairly compact structure of *A. sojae* alkaline protease was suggested (*182*).

The specific optical rotation of the *Sorangium* α-lytic protease, $[\alpha]_{335} = -207°$, exhibited no pH dependence over the range of pH 5.0–10.5, a phenomenon which is similar to that observed with chymotrypsinogen but not chymotrypsin (*183*). Similar features were observed between the ORD spectrum of the α-lytic enzyme and that for chymotrypsin. However, appreciably different Cotton effects were observed around 260 to 300 nm from chymotrypsin (*183*). A value of $[m']_{230} = -1650$ suggested the presence of very little α-helical conformation (*183*), similar to that reported for the *A. sojae* enzyme (*182*). The Moffitt-Yang parameter of $b_0 = 0$ also suggested the absence of any α-helical structure in this protein.

A preliminary crystallographic study of two different crystalline preparations of the *Arthrobacter* protease has been reported (*112*). The two crystals, square bipyramidal and needle, yielded unit cell values of $a = 75.1$ and 73.3 Å, $b = 85.7$ and 122.6 Å, $c = 115.6$ and 106.4 Å, and $V = 7.44 \times 10^5$ and 9.59×10^5 Å^3, respectively.

D. Enzymic Properties

1. *Optimum pH and Hydrolysis of Proteins*

The DFP-sensitive proteases are in general optimally active at akaline pH. They exhibit fairly broad pH profiles for the hydrolysis of proteins,

183. G. M. Patterson and D. R. Whitaker, *Can. J. Biochem.* **47**, 317 (1969).

peptides, and synthetic substrates covering the pH range from neutrality to pH 11 (*13, 111, 114, 117, 119–124, 126, 132, 135, 138, 139, 144, 147, 151, 161, 166*). Some of the enzymes, however, were shown to have fairly sharp pH profiles such as the *Streptomyces* alkaline protease, pH 9–10 (*117, 147, 151*), pH 10.6–10.8 (*119–121*), pH 10–11 (*122*), and pH 11–11.5 (*117*), *A. oryzae* alkaline proteases, pH 10.3–10.4 (*123, 166*) and pH 9.5–10.5 (*122*), and *Cliocladium* protease, pH 10–11 (*139*). The *P. cyaneo-fulvum* protease exhibited an optimum activity at pH 4.4 for polyglutamic acid and at pH 10.7 for polylysine (*138*). The *A. fumigatus* enzyme also showed two pH optima, pH 6.6 and 10 for casein and pH 4 and 9 for hemoglobin (*17*). The possibility of multiple enzymes for the two optima for each substrate is not necessarily excluded. The *Phymatotricum omnivorum* protease had optima at around pH 5 for hemoglobin and pH 7–8 for Ac–Tyr–OEt (*157*). Buffer effects on the pH optimum of the *A. oryzae* enzyme have been reported (*124, 161*). Generally, the catalytic activity of the microbial DFP-sensitive alkaline proteases toward the hydrolysis of protein substrates was two- to sixfold greater than that of trypsin and chymotrypsin (*111, 114, 123, 126, 166*). A wide variety of proteolytic activities have been reported for these enzymes including milk-clotting (*17*), fibrinolytic (*13, 17*), collagenase (*124, 161*), gelatinase (*147*), and keratinase activities (*117*). The keratin-degrading enzyme from *S. fradiae* will be discussed separately at the end of this section as one of the DFP-sensitive alkaline proteases with particular substrate characteristics.

Some of the enzymes, however, were reported to possess limited activity toward some proteins (*138, 157*).

2. *Substrate Specificity*

a. Hydrolysis of Synthetic Substrates and Insulin B Chain. A broad range of substrate specificities is exhibited by the DFP-sensitive alkaline proteases from the trypsinlike enzymes produced by *S. fradiae* (*117, 118*) and *S. griseus* (*115, 147–149, 151, 158*), and the elastaselike *Sorangium* α-lytic enzyme (*178, 184*) to the *P. cyaneo-fulvum* (*129*) and *A. oryzae* (*122, 185*) enzymes which cleave bonds characteristics of the combined specificities of trypsin, chymotrypsin, and elastase.

An enzyme isolated from *S. griseus* was shown to catalyze the hydrolysis of trypsin substrates such as Bz–Arg–OEt and –OMe (*115, 147–149, 151, 158*), tosyl–Arg–OMe (*147*), and tosyl–Lys–OMe (*147*), and also the chymotrypsin substrate Ac–Tyr–OEt (*148, 151*). The *S. fradiae*

184. D. R. Whitaker, C. Roy, C. S. Tsai, and L. Jurášek, *Can. J. Biochem.* **43,** 1961 (1965).

185. F. Sanger, E. D. P. Thompson, and K. Kitai, *BJ* **59,** 509 (1955).

enzyme also catalyzed the hydrolysis of the trypsin substrates Bz–Arg–OEt, Bz–Arg–NH_2, and Ac–Lys–OMe (*117, 118*) but not Ac–Tyr–OEt (*118*). Specificity identical to trypsin was exhibited by these two *Streptomyces* proteases toward oxidized insulin B chain, cleaving only at the carboxyl sites of lysine and arginine (*118, 158*) (Fig. 3). The hydrolysis of oxidized lysozyme by the *S. fradiae* enzyme also was shown to be similar to that of trypsin by peptide mapping techniques (*118*). Considerable kinetic differences have been observed between the *S. fradiae* enzyme and trypsin (*118*). However, both enzymes react similarly with TLCK and TPCK. These observations together with the structural studies described previously (Sections III,B,2 and 3) suggest that the active sites and specificity sites of both of the *Streptomyces* enzymes and trypsin must be quite similar.

A close relationship was observed between the *Sorangium* α-lytic protease and pancreatic elastase. The *Sorangium* protease did not hydrolyze trypsin and chymotrypsin substrates such as Bz–Arg–OEt, Bz–Arg–NH_2, and Ac–Tyr–OEt, carboxypeptidase substrates such as Z–Glu–Tyr and Z–Gly–Phe, and various di- and tripeptides, but it did catalyze the hydrolysis of certain *N*-substituted neutral, aliphatic L-amino acid esters including Bz–Ala–OMe, Ac–Val–OMe, *p*-nitrophenyl acetate, and *p*-nitrophenyltrimethyl acetate (*178, 184*). These studies, together with the results observed for the *Sorangium* protease cleavage of oxidized insulin B chain (*184*) (Fig. 3), suggested that the specificity of the *Sorangium* α-lytic enzyme was rather similar to that of the pancreatic elastase. Kinetic properties of the two enzymes, however, were considerably different (*178*).

The requirement for the L configuration of the amino acid was demonstrated (*122, 157, 161, 178*). A chymotrypsinlike activity was observed for the *Phymatotrichum omnivorum* protease (*156, 157*). It is interesting to note that among the DFP-sensitive microbial proteases are enzymes which upon the basis of specificity might be classed as trypsinlike, chymotrypsinlike, and elastaselike in addition to those with very broad overall specificities. One of the *A. oryzae* proteases, aspergillopeptidase C, exhibited a uniquely different specificity on insulin B chain from the rest (*161*). Studies with glucagon have suggested that aspergillopeptidase C hydrolyzed peptide bonds adjacent to acidic amino acids and serine residues (*161*).

A broad specificity was observed for the *S. griseus* ATCC3463 protease with baker's yeast cytochrome c (*116*). The position of the susceptible bond within the peptide chain was shown to be important for the specificity of the *Sorangium* α-lytic enzyme (*186*).

186. C. S. Tsai, D. R. Whitaker, and L. Jurášek, *Can. J. Biochem.* **43**, 1971 (1965).

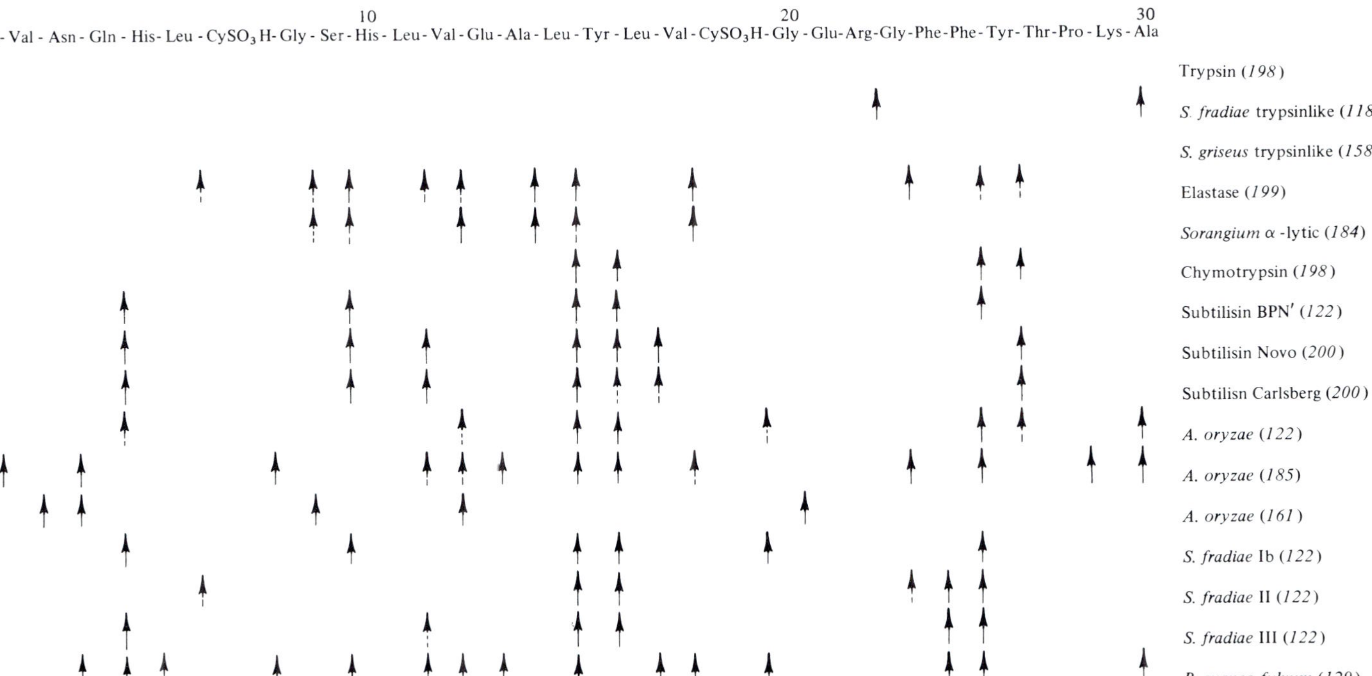

FIG. 3. Hydrolysis of the oxidized insulin B chain by various DFP-sensitive alkaline proteases. Solid and broken arrows represent the primary and secondary hydrolysis, respectively.

b. Activation of Chymotrypsinogen A. The activation of chymotrypsinogen A was observed with several bacterial and mold proteases including those from *B. subtilis* (*187*), *Aspergillus* (*15, 16, 77*), and *Penicillium* (*20, 75, 78*) (Section II,A,4,C). *Streptomyces griseus* K-1 preparation, Pronase, was used for chymotrypsinogen A activation (*188*) and yielded δ-chymotrypsin as the major product.

3. *Inhibitors*

Diisoprophylphosphofluoridate is the most common inhibitor for the proteases in this group (*17, 111, 114, 115, 117, 119–122, 125, 126, 129, 132, 138, 142, 147–149, 151, 156, 157, 161, 166–168*). A stoichiometric amount of the DIP group is incorporated into the enzyme (*115, 125, 149, 166*) at the serine residue of the active site (Section III,B,2). Other serine specific reagents such as sarin (*167, 168*), phenylmethane sulfonyl fluoride (PMSF) (*157, 161*), and diphenylcarbamyl chloride (*161*) were also inhibitory. In general, TLCK and TPCK, the site specific reagents for chymotrypsin and trypsin (*189*), did not inhibit the microbial alkaline proteases (*117, 122*). An exception was the trypsinlike enzyme from *S. fradiae* (*117, 118*) which was inactivated by TLCK. Bromoacetophenone was not inhibitory for aspergillopeptidase C (*125*). Among the naturally occurring inhibitors, the soybean trypsin inhibitor was inhibitory to the *Streptomyces* trypsinlike protease (*117, 148*) but not to others such as that of *Arthrobacter* B_{22} (*111, 114*), *S. fradiae* proteases Ia, Ib, II, and crystals (*117*), and *A. oryzae* proteases including aspergillopeptidase B (*13, 166*). The potato inhibitor was shown to strongly inhibit any of the proteases tested including the *Streptomyces* enzymes (*117, 119–121, 142, 147, 151*) and the *A. sydowi* protease (*126*). Metal chelating agents such as EDTA and *o*-phenanthroline were not inhibitory to any of the proteases in this group with the exception of the enzyme from *P. omnivorum* (*156, 157*) and *S. fradiae* proteases III and IV (*117*). Only with this latter enzyme was some metal activation observed particularly with Co^{2+}, Ni^{2+}, and Zn^{2+} (*157*). Thiol poisons such as iodoacetate and *p*-mercuribenzoate were not inhibitory to any of these enzymes except for those of the thermostable *S. rectus* (*119–121*) which was sensitive only to *p*-mercuribenzoate and not iodoacetate or *N*-ethylmaleimide and could not be reactivated by reducing agents.

187. A. Abrams and C. F. Jacobsen, *Compt. Rend. Trav. Lab. Carlsberg* **27,** 447 (1951).

188. W. M. Awad and R. E. Wilcox, *BBA* **73,** 285 (1963).

189. E. Shaw, "Methods in Enzymology," Vol. 11, p. 677, 1967.

Cyanide and cysteine generally were without effect on these enzymes (*111*, *114*, *119–121*, *151*) except for the *P. omnivorum* protease (*157*).

4. *Kinetics and Mechanism*

The acylation of the active site serine residue to form an acyl-enzyme intermediate in the enzymic catalysis has been reported by Bernhard *et al.* (*190*) and Bender *et al.* (*191*) for subtilisin Novo. This mechanism is shared with a number of other animal and plant proteinases (*191*). It is probably safe to anticipate the demonstration of a similar mechanism for most of the other microbial enzymes of this group. The role of an unprotonated histidine residue in the mechanism has also been suggested for the elastaselike *Sorangium* α-lytic enzyme (*178*) from the pH dependence of the enzyme-catalyzed hydrolysis of Ac–Val–OMe. The k_{cat} was shown to be dependent on an ionization with a pK_a of 6.7 in H_2O and 7.3 in D_2O while an isotope effect of two was observed for k_{H_2O}/k_{D_2O} (*178*). This, too, is a property shared with chymotrypsin and other animal serine proteinases. This is particularly interesting in view of the similar sequence around the sole histidine residue of the *Sorangium* enzyme to those in animal enzymes (Sections III,B,2 and 3). The k_{cat} for the α-lytic enzyme-catalyzed hydrolysis of *p*-nitrophenyl trimethyl acetate was fivefold greater than that of elastase (*178*). The α-lytic enzyme was unaffected by acetylation of the α- and ε-amino groups as reflected by its activity, and unlike chymotrypsin, the substrate binding was not dependent on the ionization of an N-terminal α-amino group (*178*). A constant K_m was obtained over the pH range 5–10 for the enzyme-catalyzed hydrolysis of Ac–Val–OMe (*178*).

Although the trypsinlike protease from *S. fradiae* exhibited a specificity exactly similar to that of trypsin, the enzymes were considerably different by kinetic parameters (*118*). A K_m of 2.8 m*M* and a k_{cat} of 160 sec^{-1} were obtained for the *S. fradiae* enzyme-catalyzed hydrolysis of Bz–Arg OEt while trypsin yielded values of K_m and k_{cat} of 0.004 m*M* and 14.5 sec^{-1}, respectively (*118*). Similar differences were also observed using the substrates Bz–Arg–NH_2 and Ac–Lys–OMe (*118*).

E. KERATINASE

A few enzymes have been described which catalyze the hydrolysis of native keratinous material (*192*). Both an EDTA-sensitive enzyme

190. S. A. Bernhard, S. J. Lau, and H. Noller, *Biochemistry* **4**, 1108 (1965).

191. M. L. Bender, M. L. Bejué-Canton, R. L. Blakeley, L. J. Brubacher, J. Feder, C. R. Gunter, F. J. Kézdy, J. V. Killheffer, Jr., T. H. Marshall, G. G. Miller, R. W. Roeske, and J. K. Stoops, *JACS* **88**, 5890 (1966).

from *Trichophyton granulosum* (*193, 194*) and a DFP-sensitive alkaline protease from *S. fradiae* (*195–197*) have been reported to possess keratinase activity. The *S. fradiae* enzyme, which appears to belong to the serine proteinase group, will be discussed here as a type with unique substrate properties. Native keratin is resistant to proteolytic enzyme unless its S–S linkages or its polymeric structure is chemically or mechanically modified. A protease secreted by *S. fradiae* which has been grown in keratin-salt medium was able to degrade native keratin (*195, 196*). The enzyme was conjugated to an acidic polymer which did not possess any proteolytic activity. The keratinase conjugate, which was purified by ammonium sulfate fractionation and treatment with a diatomaceous material Microcel E, was able to solubilize more than one-third of native keratin at pH 9.0 and likewise digested casein and hemoglobin. It had 20-fold greater activity on wool than trypsin which could solubilize only 5% of the total weight. The digested wool was found to consist of various sized peptides. The adsorption of the soluble enzyme onto wool was suggested to be a factor determining the kinetics of digestion by the keratinase conjugate (*196*). Although the role of the acidic polymer is not known, a carbohydrate nature was suggested (*197*). The keratinase, which was readily separated from the acidic polymer by treatment with DEAE-cellulose (*197*), was crystallized and the purity demonstrated by electrophoresis and ultracentrifugation. Its solubility was minimal at pH 8.0 in contrast to pH 3.6 for the conjugate (*196*). The isoelectric point of the keratinase was pH 8.9, and a molecular weight of 27,000 was calculated from its sedimentation constant, $s_{20,w} = 2.4\,S$. The keratinase was about fivefold more active toward hemoglobin than the conjugate or trypsin, and polylysine proved to be the best synthetic substrate yielding oligomeric products. The enzyme was not inhibited by soybean trypsin inhibitor (*197*). It is interesting to study the specificities of these enzymes with insulin B chain and compare them with those of other DFP-sensitive proteases shown in Fig. 3 (*118, 122, 129, 158, 161, 184, 185, 198–200*).

Another keratinase from *S. fradiae* has been reported which was produced by growth of the organism both in keratin-salt medium or regular

192. J. B. Sumner and G. F. Somers, "Chemistry and Methods of Enzymes," 3rd rev. ed. Academic Press, New York, 1953.

193. R. J. Yu, S. R. Harmon, and F. Blank, *J. Bacteriol.* **96**, 1435 (1968).

194. W. C. Day, P. Toncic, S. L. Stratman, U. Leeman, and S. R. Harmon, *BBA* **167**, 597 (1968).

195. J. J. Noval and W. J. Nickerson, *J. Bacteriol.* **77**, 251 (1959).

196. W. J. Nickerson, J. J. Noval, and R. S. Robison, *BBA* **77**, 73 (1963).

197. W. J. Nickerson and S. C. Durand, *BBA* **77**, 87 (1963).

rich medium (*117*). The enzyme, which also was inhibited by DFP, had a small molecular weight, 17,700. It was suggested that a combination of an S–S reduction system and keratinase activity accounted for the keratin digestion of the bacterial culture broth (*117*).

IV. Metal Chelator-Sensitive Neutral Proteases

The neutral proteases are primarily a group of metallo-endopeptidases which share a common specificity and have pH optima near neutrality. All of these enzymes are sensitive to metal chelating reagents and insensitive to DFP and thiol reagents. Generally, the presence of a hydrophobic side chain on the amino acid donating the amino group of the peptide bond cleaved is the specificity requirement of the enzyme. This has led to the suggestion that these enzymes be termed *amino-endopeptidases* (*201*). To date, no amidase activity has been exhibited by these enzymes nor any esterolytic activity toward a host of *p*-nitrophenyl esters or other proteolytic ester substrates. The *B. subtilis* neutral proteases and the *B. thermoproteolyticus* thermolysin have been shown to contain a gram-atom of zinc per 38,000–40,000 molecular weight of enzyme which is essential for activity. Calcium has been shown to be important for the maintenance of the enzyme stability. Among the enzymes to be discussed within this group by virtue of extensive characterization or because of preliminary indications of similarity to these characteristics will be the *Bacillus subtilis* neutral proteases (*1, 202–219*), the

198. F. Sanger and H. Tuppy, *BJ* **49,** 481 (1951).
199. M. A. Naughton and F. Sanger, *BJ* **70,** 4p (1958).
200. J. T. Johansen, M. Ottesen, I. Svendsen, and G. Wybrandt, *Compt. Rend. Trav. Lab. Carlsberg* **36,** 365 (1968).
201. J. Millet and R. Acher, *European J. Biochem.* **9,** 456 (1969).
202. D. Tsuru, J. D. McConn, and K. T. Yasunobu, *BBRC* **15,** 367 (1964).
203. J. Fukumoto and H. Negoro, *Proc. Japan Acad.* **27,** 441 (1951).
204. J. Fukumoto, T. Yamamoto, and K. Ichikawa, *J. Agr. Chem. Soc. Japan* **31,** 331 (1957); **32,** 233 (1958).
205. J. D. McConn, D. Tsuru, and K. T. Yasunobu, *JBC* **239,** 3706 (1964).
206. D. Tsuru, J. D. McConn, and K. T. Yasunobu, *JBC* **240,** 2415 (1965).
207. J. D. McConn, D. Tsuru, and K. T. Yasunobu, *ABB* **120,** 479 (1967).
208. J. Feder and C. Lewis, *BBRC* **28,** 318 (1967).
209. J. Feder, *Biochemistry* **6,** 2088 (1967).
210. K. Morihara, *BBRC* **26,** 656 (1967).
211. K. Morihara, H. Tsuzuki, and T. Oka, *ABB* **123,** 572 (1968).
212. J. Feder, *BBRC* **32,** 326 (1968).
213. K. Morihara and T. Oka, *BBRC* **30,** 625 (1968).

Bacillus thermoproteolyticus thermolysin (*210, 220–239*), the *Bacillus megaterium* megateriopeptidase (*201, 240–246*), the *Streptomyces naraensis* neutral protease (*247*), the *S. griseus* neutral proteases (*142, 144, 147, 248, 249*), the *Pseudomonas aeruginosa* enzymes (*231, 250–261*), and the *Aspergillus oryzae* protease II (*13, 16, 262, 263*).

---

214. K. Morihara, T. Oka, and H. Tsuzuki, *ABB* **132,** 489 (1969).
215. L. Keay, *BBRC* **36,** 257 (1969).
216. A. M. Benson and K. T. Yasunobu, *ABB* **126,** 653 (1968).
217. D. Tsuru, T. Yamamoto, and J. Fukumoto, *Agr. Biol. Chem.* (*Tokyo*) **30,** 651 (1966).
218. D. Tsuru, H. Kira, T. Yamamoto, and J. Fukumoto, *Agr. Biol. Chem.* (*Tokyo*) **30,** 856 and 1164 (1966).
219. D. Tsuru, H. Kira, T. Yamamoto, and J. Fukumoto, *Agr. Biol. Chem.* (*Tokyo*) **31,** 718 (1967).
220. S. Endo, *J. Ferment. Technol.* **40,** 346 (1962); *CA* **62,** 5504g (1965).
221. T. Ohta and Y. Ogura, *J. Biochem.* (*Tokyo*) **58,** 607 (1965).
222. H. Matsubara, A. Singer, R. M. Sasaki, and T. H. Jukes, *BBRC* **21,** 242 (1965).
223. H. Matsubara, *BBRC* **24,** 427 (1966).
224. H. Matsubara, R. M. Sasaki, A. Singer, and T. H. Jukes, *ABB* **115,** 324 (1966).
225. K. Morihara and H. Tsuzuki, *BBA* **118,** 215 (1966).
226. H. Matsubara, R. M. Sasaki, and R. K. Chain, *Proc. Natl. Acad. Sci. U. S.* **57,** 439 (1967).
227. T. Ohta, Y. Ogura, and A. Wada, *JBC* **241,** 5919 (1966).
228. T. Ohta, *JBC* **242,** 509 (1967).
229. T. Ando and K. Suzuki, *BBA* **140,** 375 (1967).
230. K. Morihara and M. Ebata, *J. Biochem.* (*Tokyo*) **61,** 149 (1967).
231. K. Morihara and H. Tsuzuki, *ABB* **120,** 68 (1967).
232. H. Matsubara, *in* "Molecular Mechanisms of Temperature Adaptation," Publ. No. 84, p. 283. Am. Assoc. Advance Sci., Washington, D. C., 1967.
233. R. P. Ambler and R. J. Meadway, *BJ* **108,** 893 (1968).
234. H. Matsubara, A. Singer, and R. M. Sasaki, *BBRC* **34,** 719 (1969).
235. S. A. Latt, B. Holmquist, and B. L. Vallee, *BBRC* **37,** 333 (1969).
236. H. Drucker and J. T. Yang, *Federation Proc.* **28,** 849 (1969).
237. R. A. Bradshaw, *Biochemistry* **8,** 3871 (1969).
238. H. Matsubara, "Methods in Enzymology," Vol. 19, p. 642, 1970.
239. J. Feder and J. M. Schuck, *Abstr. 158th Meeting Am. Chem. Soc., New York* MICR p. 13 (1969); BIOL. p. 320 (1969).
240. J. Chaloupka and P. Kreckova, *BBRC* **8,** 120 (1962).
241. J. Chaloupka, P. Kreckova, and L. Rihova, *BBRC* **12,** 380 (1963).
242. J. Millet and J. P. Aubert, *Ann. Inst. Pasteur* **98,** 282 (1960).
243. J. Millet and R. Acher, *BBA* **151,** 302 (1968).
244. L. Gorini and L. Audrain, *BBA* **6,** 472 (1951).
245. J. Millet, *Bull. Soc. Chim. Biol.* **51,** 61 (1969).
246. J. Millet and J. P. Aubert, *Compt. Rend.* **259,** 2555 (1964).
247. A. Hiramatsu, *J. Biochem.* (*Tokyo*) **62,** 353 and 364 (1967).
248. M. Nomoto and Y. Narahashi, *J. Biochem.* (*Tokyo*) **46,** 653 and 1481 (1959).

A. Distribution and Isolation

1. *Bacillus*

The *B. subtilis* var. *amylosacchariticus* and *B. subtilis* NRRLB3411 (also referred to as strain AM) each produce large quantities of extracellular neutral protease in addition to an alkaline protease and an α-amylase. Both enzymes have been isolated, and physical–chemical and enzymic studies indicate a very close similarity between the two enzymes (*217, 218*). McConn *et al.* (*202, 205*) isolated the *B. subtilis* neutral protease from crude preparations of the enzyme obtained from fermentation liquors by a combination of DEAE-cellulose treatment, ammonium sulfate fractionation, and CM-cellulose chromatography. The homogeneity of the enzyme was demonstrated by rechromatography on CM-cellulose, ultracentrifugation, and electrophoresis (*205*). The *B. subtilis* var. *amylosacchariticus* enzyme was purified in a similar manner by Tsuru *et al.* (*217*) except that Duolite A-2, an anion exchange resin, was used to remove the pigments and α-amylase, and separation of the neutral and alkaline proteases was achieved over DEAE-Sephadex A-50. The enzyme, which was crystallized from 50% acetone-calcium acetate at pH 8.0, was shown to be homogeneous by rechromatography on DEAE-Sephadex, SE-Sephadex, CM-cellulose, and in the ultracentrifuge (*217*). The use of hydroxylapatite for the purification of both enzymes has been reported (*215*). A thermostable neutral protease, thermolysin

---

249. M. Nomoto, Y. Narahashi, and M. Murakami, *J. Biochem.* (*Tokyo*) **48,** 453 and 906 (1960).

250. K. Morihara, *Agr. Biol. Chem.* (*Tokyo*) **26,** 842 (1962).

251. K. Morihara, *Bull. Agr. Chem. Soc. Japan* **24,** 464 and 467 (1960).

252. K. Morihara, *BBA* **73,** 113 (1963).

253. H. Inoue, T. Nakagawa, and K. Morihara, *BBA* **73,** 125 (1963).

254. K. Morihara, *J. Bacteriol.* **88,** 745 (1964).

255. K. Morihara and H. Tsuzuki, *BBA* **92,** 351 (1964).

256. K. Morihara, H. Tsuzuki, T. Oka, H. Inoue, and M. Ebata, *JBC* **240,** 3295 (1965).

257. K. Morihara and H. Tsuzuki, *ABB* **114,** 158 (1966).

258. G. G. Johnson, J. M. Morris, and R. S. Berk, *Can. J. Microbiol.* **13,** 711 (1967).

259. M. F. Li and C. Jordan, *Can. J. Microbiol.* **14,** 875 (1968).

260. R. H. Suss, J. W. Fenton, II, F. F. Muraschi, and K. D. Miller, *BBA* **191,** 179 (1969).

261. J. D. Mull and W. S. Callahan, *J. Bacteriol.* **85,** 1178 (1963).

262. T. Yasui, *Nippon Nogeikagaku Kaishi* **38,** 361 (1964).

263. V. M. Stepanov, E. A. Timokhina, and A. M. Zyakun, *BBRC* **37,** 470 (1969).

(*222*), was isolated from the culture medium of the thermophilic bacteria, *B. thermoproteolytics* Rokko by Endo (*220, 238*). The enzyme was shown to be homogeneous as judged by electrophoresis, sedimentation, and Sephadex gel filtration (*220*). Except for its thermostability (*220, 228, 232*), thermolysin is quite similar to the *B. subtilis* neutral proteases with respect to substrate specificity (Section IV,D,3) and the presence of zinc in the molecule which is required for activity (*215, 235, 239*). An extracellular neutral protease has also been reported from *B. megaterium,* megateriopeptidase (*240–246*). The enzyme was purified from the culture filtrate by ethanol and ammonium sulfate fractionations and gel filtration over Sephadex G-200 (*243, 245*). Homogeneity was shown by constant specific activity upon repeated gel filtration over Sephadex G-200 and G-75, and electrophoresis on cellulose polyacetate over the pH range of 5.8–10.5 (*243, 245*). Culture filtrates from *B. cereus* have been shown to contain a variety of proteases (*264–267*). Salter (*264*) reported the partial purification of neutral protease from *B. cereus* NCTC 945 by ammonium sulfate precipitation and DEAE-cellulose chromatography. The enzyme had a pH optimum near neutrality and was completely inhibited by $10^{-4}\,M$ EDTA and partially inhibited by NaCN and cysteine. Furukawa *et al.* (*265*) separated three proteolytic fractions from *B. cereus* KP 931 fermentation beers, all of which had pH optimum between pH 10 and 11 against casein. However, fraction III was inhibited by EDTA. This enzyme was also markedly stabilized by Ca^{2+}, a property shared with the neutral proteases. This might prove to be a metalloprotease with a high alkaline pH dependence.

2. *Streptomyces*

The *Streptomyces* produce a number of both DFP-sensitive alkaline proteases (Section III,A,1) and EDTA-sensitive neutral proteases in addition to carboxypeptidase-like and aminopeptidase-like enzymes (*144, 147, 247*). Hiramatsu (*247*) isolated an EDTA-sensitive protease from *S. naraensis* culture broth by acetone fractionation, chromatography on Amberlite CG-50 to remove dipeptidases, and chromatography on DEAE-cellulose. The purity of the final enzyme was shown to be about 90% by zone electrophoresis on starch and free-boundary electrophoresis (*247*). Column chromatography of the crude enzyme mixture from *S. griseus* K-1 (Pronase) with CM-cellulose, DEAE-Sephadex, and

264. D. N. Salter, *BJ* **72,** 23p (1959).
265. Y. Furukawa, Y. Fujii, and H. Takahashi, *Agr. Biol. Chem.* (*Tokyo*) **32,** 822 and 907 (1968).
266. R. Neumark and N. Citri, *BBA* **59,** 749 (1962).
267. S. Levisohn and A. I. Aronson, *J. Bacteriol.* **93,** 1023 (1967).

Amberlite CG-50 also yielded an EDTA-sensitive protease (*144 147*). The *S. griseus* and *S. naraensis* proteases were quite similar with regard to pH and temperature optima, pH and thermal stability, and inhibition by EDTA, but they exhibited physical differences with respect to adsorption on DEAE-cellulose and electrophoretic mobility (*247*).

3. *Aspergillus*

A number of different proteolytic enzymes are produced by *A. oryzae* including EDTA-sensitive neutral proteases (*13, 16, 262, 263, 268*). Bergkvist isolated a neutral protease (protease II) from a tannin precipitate of the culture broth by use of DEAE-cellulose and CM-cellulose (*13*). The enzyme which was readily inhibited by EDTA and cysteine also exhibited fibrinolytic and fibrinogenolytic activity (*13*). The activation of trypsinogen by a neutral protease from *A. oryzae* has been reported by Nakanishi (*16*). Additional EDTA-sensitive neutral proteases have been reported in the culture filtrates of *A. ochraceus* (*269–272*), *A. flavus* (*273*), and *A. parasiticus* (*274*).

4. *Pseudomonas*

A neutral proteaselike enzyme exhibiting elastolytic activity has been described by Morihara and co-workers (*231, 250, 251, 254, 256, 257*). The enzyme was isolated from the culture broth of *P. aeruginosa* IFO 3455 by ammonium sulfate fractionation, acetone precipitation, DEAE-cellulose chromatography, and crystallization with ammonium sulfate (*254, 256*). Although referred to as an elastase, based on its hydrolytic activity toward elastin, the enzyme shares an almost identical substrate specificity with the *B. subtilis* and *B. thermoproteolyticus* proteases toward insulin B chain (*231, 257*) (see Section IV,D,3). It is likewise inhibited by EDTA and *o*-phenanthroline but not by DFP (*256*). The *B. thermoproteolyticus* enzyme, thermolysin, also exhibits considerable elastolytic activity (*231*). Another EDTA-sensitive extracellular enzyme from *P. aeruginosa* IFO 3080 which does not hydrolyze elastin has also

268. Y. Miura, *J. Agr. Chem. Soc. Japan* **32,** 486 (1958).
269. T. Kishida and S. Yoshimura, *J. Biochem.* (*Tokyo*) **55,** 95 (1964).
270. S. Miyake, S. Yoshimura, Y. Matsuoka, and K. Kimura, *Hyogo Noka Daigaku Kenkyu Hokoku* **3,** 142 (1958).
271. S. Miyake, S. Yoshimura, Y. Nashitani, and H. Ueda, *Hyogo Noka Daigaku Kenkyu Hokoku* **3,** 147 (1958).
272. S. Miyake, S. Yoshimura, and T. Kishida, *Hyogo Noka Daigaku Kenkyu Hokoku* **4,** 23 (1959).
273. M. V. Lisenkov and O. S. Tsiperovich, *Ukr. Biokhim. Zh.* **41,** 157 (1969).
274. S. C. Dhar and S. M. Bose, *Enzymologia* **28,** 89 (1964).

been described by Morihara and co-workers (*252, 253, 255, 275*). A role for calcium has been suggested for this enzyme which has a more alkaline pH optimum (*252, 255*). Other strains of *P. aeruginosa* have also been reported to produce elastaselike enzymes (*258–261*).

5. *Others*

The isolation of an extracellular protease from the dermatophyte *Trychophyton granulosum* which exhibits keratinolytic activity has been reported (*193, 194*). The enzyme which was purified by ammonium sulfate fractionation and chromatography on CM-cellulose and Sephadex G-100 was inactivated by EDTA and activated by various metal ions and had an alkaline pH optimum. *Clostridium histolyticum* produces a proteinase which can be separated from the collagenase and thiol activated protease by gel filtration and ECTEOLA cellulose. The enzyme exhibits a substrate specificity quite similar to that of the neutral proteases from *B. subtilis, B. thermoproteolyticus,* and *P. aeruginosa* (*276, 277*). The production of an EDTA-sensitive protease with an optimum activity in the neutral pH region has been reported for some unclassified psychrophilic bacteria (*278–280*). The production of neutral proteaselike enzymes has also been reported for *Proteus mirabilis* (*281*), *Micrococcus caseolyticus* (*282*), *Microsporum* sp. (*283*), *Aeromonas hydrophilia* (*284*), and *Halobacterium salinarium* (*285*).

B. Chemical Properties

1. *Amino Acid Composition and Terminal Residues*

The amino acid compositions of various EDTA-sensitive neutral proteases are given in Table IX (*194, 206, 215, 227, 256, 275*). The *Bacillus*

275. K. Morihara, N. Yoshida, and K. Kuriyama, *BBA* **92,** 361 (1964).
276. A. B. McQuade and W. G. Crewther, *BBA* **167,** 619 (1968).
277. O. V. Kazakova, *Biokhimiya* **122,** 657 (1958).
278. I. J. McDonald, C. Quadling, and A. K. Chombers, *Can. J. Microbiol.* **9,** 303 (1963).
279. I. J. McDonald and A. K. Chombers, *Can. J. Microbiol.* **9,** 871 (1963).
280. Y. Nunokawa and I. J. McDonald, *Can. J. Microbiol.* **14,** 215 and 225 (1968).
281. S. E. Hampson, G. L. Mills, and T. Spencer, *BBA* **73,** 476 (1963).
282. M. Desmazeaud and J. Hermier, *Ann. Biol. Animale, Biochim., Biophys.* **8,** 565 (1968).
283. F. F. Roberts and R. N. Daetsch, *Antonie van Leeuwenhoek, J. Microbiol. Serol.* **33,** 145 (1967).
284. R. Gross and N. W. Coles, *Australian J. Sci.* **31,** 330 (1969).
285. P. Norberg and B. V. Hofsten, *J. Gen. Microbiol.* **55,** 251 (1969).

TABLE IX
AMINO ACID COMPOSITION OF VARIOUS METAL CHELATOR-SENSITIVE PROTEASES

Amino acid	*B. thermoproteolyticus* [Ref. (*227*)]	*B. subtilis* [Ref. (*206*)]	*B. subtilis* NRRLB3411 [Ref. (*215*)]	*B. subtilis* var. *amylosaccharíticus* [Ref. (*215*)]	*P. aeruginosa* (elastolytic) [Ref. (*256*)]	*P. aeruginosa* (nonelastolytic) [Ref. (*275*)]	*T. granulosum* [Ref. (*194*)]
Lys	12	20	17	17	11.9	15.5	9
His	9	7	5	5	7.4	5.8	6
Arg	10	10	8	9	16.3	7.0	14
Try	5	4	3	3	4.9	6.1	25
Asp	43	61	48	49	48.7	65.7	46
Thr	23	37	29	31	19.6	24.0	22
Ser	23	42	31	32	25.5	42.4	19
Glu	20	33	27	26	19.9	35.4	16
Pro	8	14	9	12	11.8	10.6	17
Gly	36	37	30	30	38.7	65.0	34
Ala	28	35	28	28	31.9	57.8	27
Half-Cys	0	0	0	0	4.6	0.0	—
Val	24	24	19	17	19.7	24.5	9
Met	2	5	4	2	8.6	0.0	0
Ile	18	18	13	13	9.6	16.8	8
Leu	17	26	21	23	14.9	37.4	15
Tyr	29	29	22	24	24.0	20.6	14
Phe	10	14	11	10	18.9	19.5	13
Ammonia	38	35	—	—	—	—	—
Total	315	416	325	331	336.9	454.1	(284)

enzymes from *B. thermoproteolyticus* (*227*), *B. subtilis* NRRLB3411 (*206, 215*), and *B. subtilis* var. *amylosacchariticus* (*215*) have remarkably similar amino acid compositions when compared on the basis of an equal number of total residues. No half-cystine residues are found, and all the evidence indicates a single polypeptide chain structure. The *P. aeruginosa* (*256*) enzyme though similar with respect to the overall amino acid composition has about five half-cystine and nine methionine residues. The *T. granulosum* (*194*) keratinaselike enzyme is devoid of both half-cystine and methionine residues, but has a very high content of tryptophan. The *B. subtilis* NRRLB3411 neutral protease yielded molar amounts of glycine and leucine for the N- and C-terminal residues, respectively (*206*). Alanine was found to be the penultimate amino acid at the C-terminus (*206*). Alanine was reported to be the N-terminal amino acid of the *B. subtilis* var. *amylosacchariticus* enzyme (*218*). Thermolysin was shown to have the N-terminal sequence to be Ile–Thr–Gly–Thr (*238*).

2. *Metal Content*

The metal requirement for the activity of the neutral proteases is primarily zinc, which has been directly demonstrated by analysis for some of the enzymes and indirectly indicated from chelator inhibition for others. It is the major criterion for grouping these proteases together. McConn *et al.* (*202, 205*) demonstrated a direct proportionality between the zinc content of the *B. subtilis* neutral protease and the activity. Removal of the zinc by EDTA resulted in an inactive apoenzyme which was reactivated upon the addition of Zn^{2+} (*202, 205*). The zinc content of the purified enzyme was about 2.0 μg/mg of enzyme which corresponds to about one atom of zinc per molecule of enzyme (*202, 205*). Replacement of the zinc with Cu^{2+}, Pb^{2+}, Hg^{2+}, and Cd^{2+} by a dialysis exchange technique and the effect on the enzymic activity has been reported (*207*). The *B. subtilis* var. *amylosacchariticus* enzyme was also shown to contain 1.9 μg of zinc per milligram of protein by Tsuru *et al.* (*218*). The zinc-free inactive enzyme was reactivated by addition of Zn^{2+} (*218*). The crystalline enzyme was shown to also contain 4.3 μg of calcium per milligram of protein (*218*). This might prove quite significant in view of the very effective stabilization of the enzyme by Ca^{2+} (*217*). The *B. thermoproteolyticus* thermolysin has been shown to contain about 2.0 μg/mg of zinc which corresponds to about one gram-atom of zinc per 37,500 g of protein (*215, 235, 239*). Latt *et al.* (*235*) demonstrated reactivation of the inactive apoenzyme upon addition of Zn^{2+}. They also reported a calcium content of about 3.8 μg/mg for the crystalline enzyme (*235*). Although no direct metal analyses have been reported for the *S. naraensis* (*247*) and *A. oryzae* (*13*) neutral proteases, reactivation of the chelator-inhibited enzymes by metals have been demonstrated.

C. Physical Properties

1. *Molecular Weight, Physical Constants, and Isoelectric Point*

The molecular weights and some physical constants for a few neutral proteases are given in Table X (*206, 215, 218, 227, 252, 253, 256*). The molecular weights range between about 35,000 and 40,000. The *B. subtilis* neutral protease exhibited a time dependence for the apparent weight-average molecular weight. A negative slope of the plot of apparent molecular weight vs. time with increasing slopes at higher protein concentration suggested the occurrence of autolysis with this enzyme (*206*). An increase rather than a decrease in the apparent molecular weight with time was observed with the *B. subtilis* var. *amylosacchariticus* enzyme (*218*). The intrinsic viscosity $[\eta]$ of the *B. thermoproteolyticus* thermolysin was reported as 3.3 ml/g, similar to that of ordinary globular proteins (*227*).

The crystalline enzyme with elastolytic activity from *P. aeruginosa* was quite insoluble in water and showed decreasing solubility in buffers of decreasing pH (*256*). The frictional ratio of the nonelastolytic enzyme from *P. aeruginosa* (*253*) was 1.17 from which axial ratios a/b of 0.52 and 2.0 were calculated for oblate ellipsoid and prolate ellipsoid shapes, respectively. The isoelectric point (pI) of the *B. subtilis* enzymes were quite basic (pI = 8.5–8.9) compared to that for the *P. aeruginosa* enzymes which were acidic (pI = 5.9 and less than 4.08 for the elastolytic and nonelastolytic enzymes, respectively). A molecular weight of 34,300 was estimated for the keratinaselike enzyme from *Trichophyton granulosum* (*194*). A value of 22.15 was found for $A_{1\,\text{cm}}^{1\%}$ at 280 nm which correlated with the high tryptophan content (*194*).

2. *Stability*

A wide range of stabilities is exhibited by the neutral proteases from the thermostable *B. thermoproteolyticus* enzyme to the rather unstable *B. subtilis* enzyme. The *B. subtilis* neutral protease is stable between pH 6.5 and 9.7 with up to 80% of the activity being retained after 24 hr at pH 7.0 and 25° (*205*). The enzyme is quite heat labile with loss of more than 90% of its activity after 15 min at 60° (*205*). Calcium ions stabilized the enzyme considerably (*205*). Also effective were Sr^{2+}, Mg^{2+},

TABLE X
MOLECULAR CHARACTERISTICS OF VARIOUS METAL CHELATOR-SENSITIVE PROTEASES

Enzyme	MW	$s_{20,w} \times 10^{13}$	$D_{20,w} \times 10^7$	$\bar{V}$	pI	$A^{1\%}_{1\,cm}$ at 280 nm	f/f_0
B. subtilis NRRLB3411 (*206, 215, 218*)	44,700[a] 35,100[b]	3.24		0.746	8.95	13.6	1.3
B. subtilis var. *amylosacchariticus* (*215, 218*)	33,800[a] 40,500[b]	3.02	8.2	0.735	8.5	13.8	1.21
B. thermoproteolyticus (*227*)	37,500[a]	3.54–3.63		0.73		17.65	
P. aeruginosa (elastolytic) (*256*)	39,500[c]	3.38	7.58	0.72	5.9	14.52	
P. aeruginosa (*252, 253*)	48,400[c]	3.99	7.4	0.73	<4.08	16.0	1.17

[a] Ultracentrifugation.
[b] Amino acid composition.
[c] Sedimentation diffusion.

Mn^{2+}, and Ba^{2+}. A calcium concentration in excess of $2 \times 10^{-3} M$ was necessary to minimize autolysis (*205*). The *B. subtilis* var. *amylosaccharitcus* enzyme was also stabilized by Ca^{2+} and to a lesser extent by Co^{2+} and Sr^{2+} (*217*). In the presence of $3.3 \times 10^{-3} M$ Ca^{2+} the enzyme was stable up to 52° for 15 min and retained 50% of activity at 60° after 15 min. Complete inactivation was obtained under the latter conditions in the absence of Ca^{2+} (*217*). The zinc-free inactive enzyme was very unstable but stabilized between pH 6.5 and 8.5 in the presence of Ca^{2+} (*218*). Similarly, the thermal sensitivity of the zinc-free enzyme was considerable, retaining only 70% of the activity at 30° after 15 min incubation (*218*). The *B. megaterium* neutral protease was stable up to about 45° in tris calcium buffer at near neutral pH but at higher temperatures lost activity (*245*). About 80% of the activity was lost upon incubation at 60° for 15 min (*245*).

In marked contrast, the *B. thermoproteolyticus* thermolysin retains about 50% of its activity after 1 hr at 80° (*220, 227, 228, 232*). It has been suggested that the thermostability results from extensive hydrophobic regions in the enzyme which contribute to a tight stable structure (*228*). Calcium seems to also play an important role in this stability since its removal causes loss of activity and conformation (*228*). The *P. aeruginosa* elastaselike neutral protease was stable in the pH range from 6 to 10 and was resistant to heat treatment, retaining 85% of both proteolytic and elastolytic activity after 10 min incubation at 70° in tris buffer at pH 8.0 (*256*). The nonelastolytic protease from *P. aeruginosa* was stable from pH 5.0 to 9.0 but very unstable below pH 3.5 and above pH 11 (*252*). The enzyme was stable up to 50° and then lost its activity at 60° (pH 7.0) after 10 min. Calcium also contributed to the increased thermal stability of this enzyme (*252*). The *A. oryzae* protease II was stable over the range of pH 4–10 but became totally inactivated after 2 min at 60° and lost about 50% of its activity at 50° after 20 min incubation (*13*). Both the *S. griseus* (*147*) and *S. naraensis* (*247*) neutral proteases were stabilized by Ca^{2+}.

3. *Conformational Studies*

Ohta and co-workers (*227*) reported the Moffitt constants $b_0 = -168.5$ and $a_0 = -3.3$ to -96.3 (neutral pH) ($\lambda_0 = 212$ nm), for the thermostable thermolysin. The ultraviolet rotatory dispersion curve of the enzyme exhibited a trough at 232 nm, and an average specific rotation $[m'] = -2870$ was obtained at this wavelength in 0.05 M tris buffer, pH 7. A very low helical content therefore was suggested for the enzyme (*227*). The presence of abnormal tyrosine and tryptophan residues that were buried within the nonpolar environment of the protein was shown by spectrophotometric, fluorescence spectral, and ORD experiments (*227*). When the thermolysin was modified by heating at 80° for 1 hr the V_{max} was unchanged but the Michaelis constant was increased (*228*). This modified enzyme differed from the native molecule in the absence of the abnormally ionizing tyrosine residues and the Cotton effect near 275 nm, but it showed no difference in the absorption and fluorescence spectrum or in intrinsic viscosity (*228*). It was suggested that the molecule had a rigid structure resulting from hydrogen bonding of the phenol group of tyrosine and hydrophobic interactions which could contribute significantly to the thermal stability (*228*). It is particularly significant that this enzyme has no disulfide bonds and appears to be a single polypeptide chain. Drucker and Yang (*236*) estimated about 20% helical content for thermolysin based on the rotation at 233 nm and the b_0 of the Moffitt equation which decreased in the absence of Ca^{2+}. Morihara and co-workers (*275*) reported a helical content of about 5% for the *P. aeruginosa* IFO 3080 enzyme as calculated from $b_0 = -30$ ($\lambda_0 =$ 212 nm).

D. Enzymic Properties

1. *Optimum pH and Hydrolysis of Proteins*

The EDTA-sensitive neutral proteases generally exhibit maximum proteolytic activity at a more neutral pH region than the DFP-sensitive enzymes. The *B. subtilis* neutral protease (*205*) and the enzyme from *B. subtilis* var. *amylosacchariticus* (*217*) have pH optima near neutrality with casein substrate and temperature optima of 50° and 52°, respectively. In the presence of 5.5×10^{-4} *M* calcium acetate the temperature optima shifted to 58° for both enzymes with a 20–40% increase in reaction rates (*205, 217*). Morihara *et al.* (*214*) reported a bell-shaped pH dependence for the *B. subtilis* var. *amylosacchariticus* neutral protease–catalyzed hydrolysis of Z–Gly–Leu-NH_2, Z–Ala–Leu–Ala, and Z–Gly–Pro–Leu–Gly–Pro. This is similar to that for casein with a pH optimum near neutrality. The *B. megaterium* neutral protease, megateriopeptidase, also has a sharp pH optimum toward casein at pH 7.2 and 55° (*245*). Thermolysin, however, has a broad pH optimum between about 7 and 9 with casein substrate (*211, 220*).

The elastaselike enzyme from *P. aeruginosa* has an optimum for both caseinolytic and elastolytic activity at pH 8.0 (*256*). The nonelastolytic enzyme from *P. aeruginosa* has a pH optimum at 8–9 toward casein, and 7–9 for egg albumin and denatured hemoglobin with a temperature optimum of 60° (*252*). A pH optimum of 7.5 and a temperature optimum of 40° was reported for the *S. naraensis* neutral protease with both hemoglobin and casein (*247*). The *S. griseus* K-1 neutral proteases also exhibited pH optima at 7.5–8.0 with casein substrate (*147*). A temperature optimum of 50° and pH optima of 6.8, 6.8, and 6.3 were reported for the *A. oryzae* protease II with casein, hemoglobin, and gelatin, respectively (*13*).

A rather alkaline pH optimum of 9.5–9.8 was reported for the caseinolytic activity of the keratinaselike enzyme from *T. granulosum* (*194*). An optimum activity was obtained at an ionic strength of 0.1 and a temperature of 45° (*194*). Some of the neutral proteases have been shown to hydrolyze elastin (*231, 250, 251, 254, 256–261*), keratin (*193, 194*), and fibrin and fibrinogen (*13*).

2. *Inhibitors*

The sensitivity to metal chelating reagents, indicative of a metal's role in catalysis, has been one of the major criteria for grouping the neutral proteases together as one class of enzymes. EDTA was the most effective inhibitor of the neutral proteases followed by *o*-phenanthroline (*1, 13, 147, 194, 202–205, 210, 217, 218, 220, 225, 235, 242, 245, 247, 252, 256, 264, 265, 278, 281, 284*). None of these enzymes was inhibited by DFP, sulfhydryl reagents, soybean trypsin inhibitor, or potato protease inhibitor. Yasunobu and co-workers (*202, 205*) reported the inhibition of the *B. subtilis* enzyme by EDTA, *o*-phenanthroline, dithizone, cyanide, and sodium diethyldithiocarbamate with 50% inhibition at concentrations of $2.2 \times 10^{-6} M$, $5.0 \times 10^{-5} M$, $4.5 \times 10^{-4} M$, $1.8 \times 10^{-3} M$, and $2.5 \times 10^{-3} M$, respectively. The inactive enzyme was reactivated to 38–82% by the addition of Zn^{2+}, Co^{2+}, and Mn^{2+} (*202, 205*). The extent of reactivation declined with the preincubation time of the EDTA with the enzyme (*205*). The *B. subtilis* var. *amylosacchariticus* neutral protease when treated with $2.5 \times 10^{-4} M$ EDTA at pH 5.8 was also inactivated. The enzyme could be reactivated by the addition of Zn^{2+} or Co^{2+}, partially reactivated by Fe^{2+}, Mn^{2+}, and Ni^{2+}, but not reactivated by Ca^{2+}, Mg^{2+}, or Cu^{2+} (*218*). The need for short time treatment with EDTA and the subsequent addition of Ca^{2+} strongly suggests that in addition to the inactivation owing to removal of Zn^{2+} from the enzyme, a process which can be reversed upon readdition of the metal ion, the removal of Ca^{2+} by EDTA results in an irreversible inactivation. The latter process, perhaps results from autolysis of the unstable Ca^{2+}-free protein (*205*). Initial inhibition without removal of the metal by the chelator but as a result of the formation of an inactive enzyme–chelator complex has been suggested for the *P. aeruginosa* nonelastolytic enzyme (*255*) and might prove to be a more common phenomenon. Both the elastolytic and caseinolytic activities of the elastaselike enzyme from *P. aeruginosa* were almost completely inhibited by millimolar concentrations of EDTA and *o*-phenanthroline but unaffected by soybean trypsin inhibitor, DFP, or *p*-mercuribenzoate. (*256*).

Latt *et al.* (*235*) reported values of pK_1 for 2,2′-bipyridine, mercaptoethylamine, mercaptoacetic acid, and cyanide of 3.4, 3.1, 2.4, and 2.2, respectively, for the inhibition of thermolysin. The addition of Zn^{2+} to the enzyme completely inhibited by *o*-phenanthroline or 2,2′-bipyridine resulted in complete reactivation (*235*). The *B. megaterium* enzyme was also inhibited by EDTA, 2,2′-bipyridine, and *o*-phenanthroline (*245*). Fifty per cent inactivation of the *S. naraensis* neutral protease was obtained with concentrations of $1 \times 10^{-7}\,M$, $1 \times 10^{-5}\,M$, $1 \times 10^{-4}\,M$, and $1 \times 10^{-2}\,M$ of EDTA, *o*-phenanthroline, 8-hydroxyquinoline, and citrate, respectively (*247*). Reactivation of the EDTA-inactivated enzyme was completely effected by Zn^{2+} and slightly restored by Co^{2+} and Mn^{2+} (*247*). No irreversible effect was observed with preincubation times with EDTA of up to 24 hr (*247*).

The *A. oryzae* protease II was inhibited by ascorbic acid in addition to EDTA, cysteine, and laurylamine (*13*). Two types of inhibitors, one reversible and the other irreversible, were found in the α_1- and α_2-globulin fractions of human serum (*13*). A number of animal sera including rat, sheep, horse, ox, and pig also inhibited the *A. oryzae* neutral protease (*13*).

Morihara and co-workers (*214*) reported competitive inhibition of the *B. subtilis* var. *amylosacchariticus* neutral protease by Z–Ala–D-Leu–Ala, Z–D-Ala–Leu–Ala, Z–Gly–Leu, and Z–Ala–Leu with K_1 values of 140, 115, 70, and 38 mM, respectively.

3. *Substrate Specificity and Kinetics*

The substrate specificities of the neutral proteases were studied with both simple di- and tripeptides (*201, 205, 209–214, 218, 221, 223, 234, 243, 276*) and protein substrates (*208, 216, 219, 222, 224–226, 229–233, 243, 247, 269*). Both types of studies indicated a substrate specificity which required a hydrophobic amino acid such as leucine or phenylalanine as the amino acid whose amino group was involved in the bond to be cleaved.

Table XI summarizes the results reported by Feder (*209*) for the *B. subtilis* NRRLB3411 neutral protease catalyzed hydrolysis of some dipeptide substrates. No hydrolysis was observed for Z–Gly–Gly–NH_2,

TABLE XI
HYDROLYSIS OF DIPEPTIDES BY THE *B. subtilis* NRRLB3411 NEUTRAL PROTEASE[a,b]

Substrate	$k/E \times 10^4$ sec^{-1} (mg/ml)$^{-1}$ [c]	$\frac{k(\text{substrate})}{k(\text{Z-Gly-Leu-NH}_2)} \times 100$	$K_m \times 10^2\ M$
Z-Gly-Gly-NH_2	No reaction		
Z-Gly-Ala-NH_2	0.43	0.8	1.04
Z-Ala-Gly-NH_2	No reaction		
Z-Gly-Val-NH_2	6.95	12.7	2.16
Z-Gly-Nle-NH_2	11.7	21.4	3.60
Z-Gly-Phe-NH_2	3.47	6.4	0.32
Z-Gly-Tyr-NH_2	Negligible		
Z-Gly-Leu-NH_2	54.57	100.0	2.94
Bz-Gly-Leu-NH_2	33.41	61.0	2.13
Z-Leu-Gly-NH_2	No reaction		
Z-Ala-Leu-NH_2	147.7	271.0	1.01
Z-Ser-Leu-NH_2	249.2	458.0	0.47
Z-Thr-Leu-NH_2	406.5	747.0	1.22
Z-Met-Leu-NH_2	27.18	48.0	—
Z-His-Leu-NH_2	341.5	626.0	0.38
Z-Tyr-Leu-NH_2	381.47	699.0	—
Gly-Leu-NH_2	Negligible		
Z-Gly-Leu	≦1.38	2.5	—
Z-Gly-Leu-OMe	Negligible		

[a] Summarized from Feder (*209*).
[b] Reactions monitored by pH stat at pH 8.0 in 0.1 *M* KCl, 25° ($S \sim 10^{-3}\ M$).
[c] k is pseudo-first order rate constant ($k = k_{cat}\ (E)/K_m$); (E) given in (mg/ml).

but Z–Gly–Ala–NH_2, Z–Gly–Val–NH_2, and Z–Gly–Leu–NH_2 were hydrolyzed with progressively greater rates respectively (*209*). Reversing the position of the amino acids such as Z–Ala–Gly–NH_2, Z–Leu–Gly–NH_2, and Z–Phe–Gly–NH_2 resulted in loss of the enzyme-catalyzed hydrolysis. This was indicative of the absolute specificity requirement for the amino donating amino acid of the peptide bond. Secondary effects resulting from the side chain of the amino acid contributing the carbonyl group of the peptide bond were also observed. For example, Z–Thr–Leu–NH_2 and Z–Tyr–Leu–NH_2 were catalyzed with rate constant sevenfold greater than Z–Gly–Leu–NH_2. The presence of a free amino or carboxyl group or the ester essentially eliminated hydrolysis, indicating the endopeptidase nature of the enzyme (*209*). The need of an α-amino acid peptide chain and an unhindered amide bond adjacent to the bond cleaved was indicated from the lack of enzyme-catalyzed hydrolysis of Z-β-Ala–Phe–NH_2, Z-β-Ala-β-Ala–NH_2, Z–Ala-β-Ala–NH_2, Z-β-Ala–

Phe–NH_2, Z–Ser–Ala–NH_2, and Z–Ser–Leu–NH_2 (*209*). Similar specificities have been reported by Matsubara (*223*) for thermolysin and by Morihara and co-workers (*210, 211, 230*) for a number of neutral proteases including the enzymes for *B. subtilis, B. thermoproteolyticus, P. aeruginosa, S. griseus,* and *A. oryzae.* Table XII compares the relative rates of hydrolysis of a few dipeptides as catalyzed by various neutral proteases (*209–211*). Similar specificities with dipeptide substrates have been reported by Millet and Acher (*201, 243, 245*) for the *B. megaterium* megateriopeptidase. The noncollagenolytic neutral proteaselike enzyme from *C. histolyticum* also exhibited similar specificity toward dipeptides (*276*).

The effect of amino acids adjacent to the amino acids forming the sensitive peptide bond on the neutral protease-catalyzed hydrolysis of a peptide bond have recently been reported (*213, 214, 235*). Matsubara and co-workers (*234*) demonstrated that thermolysin was unable to hydrolyze peptide bonds at the amino site of hydrophobic amino acids whose carbony group was attached to a proline residue. This has been observed with both the *B. subtilis* neutral protease (*216*) and thermolysin (*222, 233, 286–289*) hydrolysis of protein substrates for sequence studies.

TABLE XII
COMPARISON OF CHELATOR-SENSITIVE PROTEASE-CATALYZED HYDROLYSIS OF DIPEPTIDES

	Relative rates (substrate/Z-Gly-Leu-NH_2) × 100				
Substrate	*B. subtilis*	*B. thermoproteolyticus*	*P. aeruginosa*	*S. griseus*	*A. oryzae*
Z-Gly-Leu-NH_2	100	100	100	100	100
Z-Ala-Leu-NH_2	197[a] 271[b]	207[a]	1010[a]	380[a]	900[a]
Z-Tyr-Leu-NH_2	1070[a] 699[b]	900[a]	167[a]	960[a]	600[a]
Z-Gly-Phe-NH_2	6.4[b]	100[c]	164[c]	168[c]	289[c]

[a] From Morihara *et al.* (*211*).
[b] From Feder (*209*).
[c] From Morihara (*210*).

286. H. Matsubara and R. M. Sasaki, *JBC* **243,** 1732 (1968).
287. K. Sugeno and H. Matsubara, *JBC* **244,** 2979 (1969).
288. J.-Y. Lin, C. M. Tsung, and H. Fraenkel-Conrat, *JMB* **24,** 1 (1967); M. Wallis and M. A. Naughton, cited by M. O. Dayhoff and R. V. Eck, *in* "Atlas of Protein Sequence and Structure," (M. O. Dayhoff, ed.), p. 283 (1967–1968).
289. T. C. Vanaman, S. J. Wakil, and R. L. Hill, *JBC* **243,** 6409 (1968).

Recently, Morihara *et al.* (*213, 214*) have reported that the effects of the amino acids adjacent to those constituting the peptide bond cleaved on the *B. subtilis* neutral protease, thermolysin, and the *P. aeruginosa* enzyme-catalyzed hydrolysis of simple substrates suggest that these enzymes have a compound active site accommodating up to six amino acids. Table XIII as reported by Morihara *et al.* (*214*) presents the kinetic effects on K_m, k_{cat} and k_{cat}/K_m by various peptide substrates which can fit the various subsites proposed. The *B. subtilis* neutral protease, thermolysin, and the *P. aeruginosa* enzyme-catalyzed hydrolysis of Z–Gly–Leu–Ala was 27-, 42-, and 72-fold greater, respectively, than that of Z–Gly–Leu–NH_2 (*213*). Little change was observed in the K_m, however, between these substrates, but the k_{cat} increased considerably (*214*). The subsites have specificity with respect to the nature of the side chain and the stereochemistry of the residue. This can be seen by comparing the relative rates for the *B. subtilis* var. *amylsacchariticus* catalyzed hydrolysis of Z–Gly–Leu–NH_2, Z–Gly–Leu–Gly, Z–Gly–Leu–Ala, and Z–Gly–Leu-D-Ala which are 1, 0.76, 28.7, and 0.007, respectively (*214*). The major differences here too were reflected in the k_{cat} while the K_m varied but very little (*214*).

The enzyme-catalyzed hydrolyses of protein substrates yielded a specificity similar to what was projected from the dipeptide studies. Figure 4 compares the cleavage points on insulin B chain obtained with a number of neutral proteases (*201, 208, 211, 219, 224, 225, 257, 269, 281*).

Most of the enzymes including the neutral proteases from *B. thermoproteolyticus* (*224, 225*), *B, subtilis* (*208, 211, 219*), *P. aeruginosa* (*257*), *S. griseus* (*211*), *A. oryzae* (*211*), and *B. megaterium* (*201*) yielded similar points of cleavage on the insulin B chain. The primary bonds attacked were between His 5–Leu 6, His 10–Leu 11, Ala 14–Leu 15, Tyr 16–Leu 17, Gly 23–Phe 24, and Phe 24–Phe 25 while some cleavages at Leu 17–Val 18 and Phe 25–Tyr 26 occurred with some of the enzymes. The enzymes from *Proteus mirabilis* (*281*) and *A. ochraceus* (*269*) behaved somewhat differently.

In addition to the insulin B chain, Matsubara and co-workers reported specificity studies of the *B. thermoproteolyticus* thermolysin with beef heart cytochrome c (*222*) and tobacco mosaic virus protein (*224*) which yielded similar results. The *S. naraensis* neutral protease cleaved baker's yeast cytochrome c primarily on the amino end of the bulky hydrophobic residues of phenylalanine, isoleucine, leucine, methionine, valine, and tyrosine with most of the cleavage at leucine (*247*).

Because of this unique specificity neutral proteases have been used in amino acid sequence determinations. Particularly thermolysin has widely been used in the sequencing of various proteins and peptides such

TABLE XIII
HYDROLYSIS OF Z-GLY-Leu-$(GLY)_n$-B AND Z-A-$(GLY)_n$-LEU-ALA BY *B. subtilis* NEUTRAL PROTEASE[a,b]

Category	Peptides									K_m (mM)	k_{cat} (sec^{-1})	$\frac{k_{cat}}{K_m}$	Relative activity
	P_4	P_3	P_2	P_1	‡	P'_1	P'_2	P'_3	P'_4				
I			Z-	Gly-		Leu-	NH_2			50	51.2	1.02	1
			Z-	Gly-		Leu-	Gly			50	39.0	0.78	0.76
			Z-	Gly-		Leu-	Ala			42	1231.2	29.3	28.7
			Z-	Gly-		Leu-	D-Ala			40	0.3	0.0075	0.007
			Z-	Gly-		Leu-	Leu			6.7	180.6	27.0	26.5
			Z-	Gly-		Leu-	Phe			8.3	266.6	32.1	31.5
II			Z-	Gly-		Leu-	Gly-	Gly		55.5	94	1.7	1.7
			Z-	Gly-		Leu-	Gly-	Ala		50	520.8	10.4	10.2
			Z-	Gly-		Leu-	Gly-	D-Ala		57.1	22.3	0.39	0.38
			Z	Gly-		Leu-	Gly-	Phe		14.3	78.3	5.5	5.4
III			Z-	Gly-		Leu-	Gly-	Gly-	Ala	62.5	125.1	2.0	2.0
			Z-	Gly-		Leu-	Gly-	Gly-	D-Ala	63.0	118.0	1.9	1.9

IV			Z-	Ala-	Leu-	Ala	16.3	1408.4	86.4	84.7
			Z-	D-Ala-	Leu-	Ala	Negligibly small reaction			
			Z-	Phe-	Leu-	Ala[c]	1.3	1251.0	962.3	943.4
			H-	Phe-	Leu-	Ala	Negligibly small reaction			
V		Z-	Gly-	Gly-	Leu-	Ala	23.3	225.2	9.7	9.5
		Z-	Ala-	Gly-	Leu-	Ala	17.0	2210.0	130.0	127.4
		Z-	D-Ala-	Gly-	Leu-	Ala	30.3	250.4	8.2	8.0
		Z-	Phe-	Gly-	Leu-	Ala[c]	2.5	2460.0	984.0	964.7
		H-	Phe-	Gly-	Leu-	Ala[c]	7.9	206.0[d]	26.1	25.6
VI	Z-	Ala-	Gly-	Gly-	Leu-	Ala	26.6	363.2	13.8	13.5
	Z-	D-Ala-	Gly-	Gly-	Leu-	Ala	27.7	148.2	5.4	5.2
	Z-	Phe-	Gly-	Gly-	Leu-	Ala	11.8	181.6	15.4	15.1
	H-	Phe-	Gly-	Gly-	Leu-	Ala[c]	12.8	175.9[d]	14.5	14.2

[a] Categories I, II, III, IV, V, and VI show the peptide series Z-Gly-Leu-B, Z-Gly-Leu-Gly-B, Z-Gly-Leu-Gly-Gly-B, Z-A-Leu-Ala, Z-A-Gly-Leu-Ala, and Z-A-Gly-Gly-Leu-Ala (A or B stands for various amino acid residues), respectively. The arrow shows the bond split. Except where specified, the initial substrate concentrations used were in the ranges of 1–25 m*M*.

[b] Taken from Morihara *et al.* (*214*).

[c] Initial substrate concentrations used were below 7 m*M*.

[d] The calculation was made on assumption that the ninhydrin color yield of Phe-Gly-Leu-Ala (or Phe-Gly-Gly-Leu-Ala) and Phe-Gly (or Phe-Gly-Gly) were the same; thus, the increase of the color value in the reaction mixture was exclusively ascribed to the Leu-Ala released. However, a rough comparison of the color yields of these peptides, using a densitometer, revealed that Phe-Gly-Leu-Ala: Phe-Gly: Leu-Ala is about 0.85:1.35:1.0 and Phe-Gly-Gly-Leu-Ala: Phe-Gly-Gly: Leu-Ala is about 0.9:1.5:1.0. Therefore, the real value must be smaller than the k_{cat} values listed (perhaps by a factor of about 0.65).

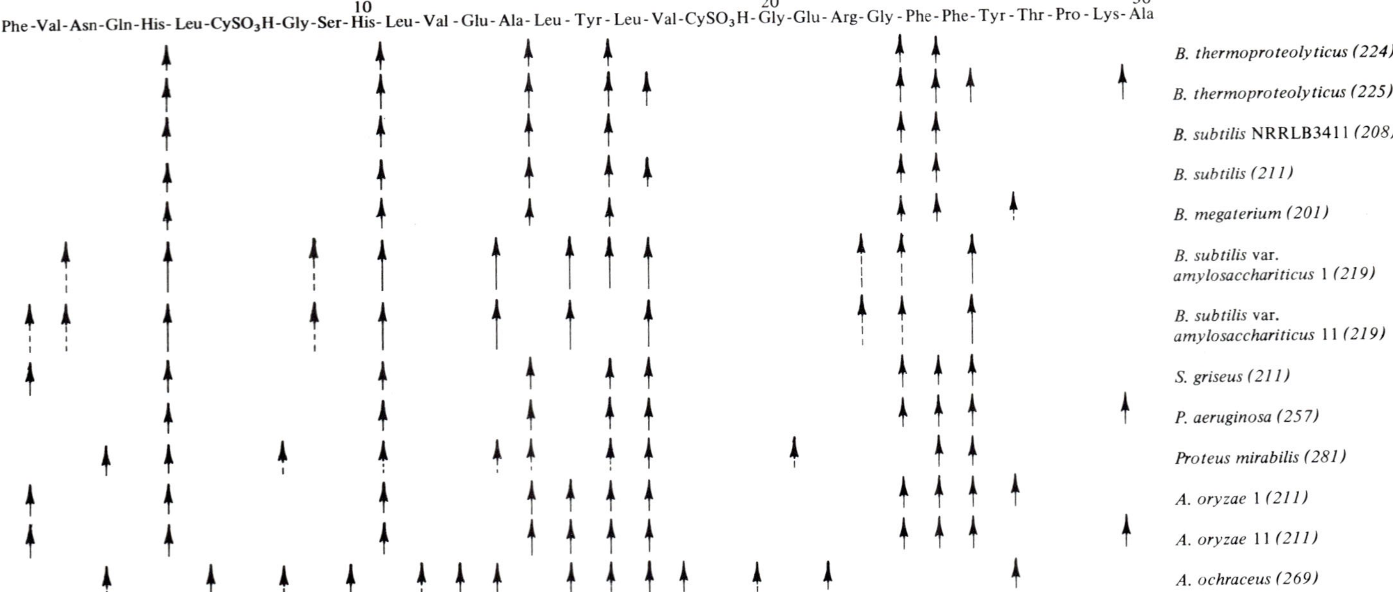

FIG. 4. Hydrolysis of the oxidized insulin B chain by various metal chelator-sensitive neutral proteases. Solid and broken arrows represent the primary and secondary hydrolysis, respectively.

as ferredoxins (*226, 286, 287, 290–294*), azurin (*233*), acyl carrier protein (*289*), histones (*295, 296*), clupeine (*229*), cytochromes (*297–299*), chymotrypsin (*300*), and carboxypeptidase (*237*), and also in studying the susceptibility of staphylococcal nuclease under the various conditions (*301*). The *B. subtilis* neutral protease was also used in this field but to a minor extent (*216, 294*). These studies have confirmed with only minor exceptions (*238*) the earlier proposal for the specificity requirement of thermolysin (*222, 224*).

None of the neutral proteases exhibited esterase activity. The *B. subtilis* neutral protease did not catalyze the hydrolysis of *p*-nitrophenyl acetate, Z–Gly–ONp, Bz–Arg–OEt, Ac–Tyr–OEt, Ac–Leu–NH_2 Bz–Leu–NH_2, *N-trans*-cinnamoylimidazole, hippuryl-D,L-phenyllactic acid, and a series of aliphatic esters (*205, 209*). The *B. thermoproteolyticus* (*225, 232*) and *B. megaterium* (*201*) enzymes were also inactive against these substrates. In light of the strict specificity requirements reported above for dipeptides, the lack of esterase activity with esters studied is understandable. Examination of esters analogous to the dipeptide substrates such as the amide of benzyloxycarbonylglycylleucic acid might yield esterase activity (*209*). The need for simple methods to monitor the neutral protease-catalyzed hydrolysis of dipeptides has led to the synthesis of furylacryloyl (FA) blocked substrates which permit spectrophotometric monitoring of dipeptide hydrolysis (*212, 239*). The use of 3-(2-furylacryloyl) (FA)–Gly–Leu–NH_2 for the study of the *B. subtilis* neutral protease and thermolysin kinetics has been described (*212, 215, 239*). A comparative kinetic study of the *B. subtilis* neutral protease and thermolysin-catalyzed hydrolysis of FA–Gly–Leu–NH_2, FA–Ala–Phe–NH_2, and FA–Phe–Phe–NH_2 has been reported (*239*). At pH 7.2 values of $5.06 \times 10^3\ M^{-1}\ \text{sec}^{-1}$ and $9.46 \times 10^3\ M^{-1}\ \text{sec}^{-1}$ have been obtained, respec-

290. H. Matsubara and R. M. Sasaki, *Federation Proc.* **26,** 723 (1967).

291. K. K. Rao and H. Matsubara, *BBRC* **38,** 500 (1970).

292. H. Matsubara, R. M. Sasaki, D. K. Tsuchiya, and M. C. W. Evans, *JBC* **245,** 2121 (1970).

293. N. J. Tsunoda, K. T. Yasunobu, and H. R. Whiteley, *JBC* **243,** 6262 (1968).

294. A. M. Benson and K. T. Yasunobu, *JBC* **244,** 955 (1969).

295. R. J. DeLange, D. M. Fambrough, E. L. Smith, and J. Bonner, *JBC* **244,** 319 (1969).

296. G. Moskowitz, Y. Ogawa, W. C. Starbuck, and H. Busch, *BBRC* **35,** 741 (1969).

297. A. Tsugita, M. Kobayashi, T. Kajihara, and B. Hagihara, *J. Biochem.* (*Tokyo*) **64,** 727 (1968).

298. K. Dus, L. Sletten, and M. D. Kamen, *JBC* **243,** 5507 (1968).

299. K. Wada and K. Okunuki, *J. Biochem.* (*Tokyo*) **66,** 249 (1969).

300. D. M. Blow, J. J. Birktoft, and B. S. Hartley, *Nature* **221,** 337 (1969).

301. H. Taniuchi, L. Moravek, and C. B. Anfinsen, *JBC* **244,** 4600 (1969).

tively, for the neutral protease- and thermolysin-catalyzed hydrolysis of FA–Gly–Leu–NH_2 with identical bell-shaped pH profiles having maxima at pH 7.1 (*239*).

The neutral proteases represent an interesting new group of enzymes possessing unique specificity characteristics together with a mechanism of action involving a metal ion yet to be investigated.

V. Other Proteases of Microbial Origin

In addition to the three broad classes of endopeptidases discussed above, a number of different kinds of proteolytic enzymes are produced by bacteria and molds. Both aminopeptidases (see DeLange and Smith, Chapter 3, this volume) and carboxypeptidases have been reported. A few thiol enzymes have been reported and of these the streptococcal proteinase and clostripain are discussed in Chapters 17 and 19 of this volume, respectively. Another unique group of proteases, which in some cases shares properties with one of the other groups, is the bacterial cell wall lysing proteases. These enzymes by virtue of the specific bonds cleaved have uniquely different characteristics from the other enzymes discussed.

A. Bacterial Cell Wall Lytic Proteases

A number of enzymes including some proteolytic enzymes capable of lysing cell walls have been reported. A few of the more characterized bacterial cell wall lysing proteases will be described here. The reader is referred to some excellent reviews which have appeared in recent years (*302–304*) on the structure and enzymology of bacterial cell walls.

1. *Myxobacter AL-1 Protease*

During the study of cell wall structures, Ensign and Wolfe (*305*) isolated the soil bacterium *Myxobacter* AL-1 which lysed the cells of *A. crystallopoietes*. The lytic enzyme was isolated from the culture medium by $ZnCl_2$ treatment, tris EDTA and citrate washings, and chromatography over DEAE-cellulose and Sephadex. Homogeneity of the enzyme

302. J. L. Strominger and J. M. Ghuysen, *Science* **156**, 213 (1967).

303. J. M. Ghuysen, *Bacteriol. Rev.* **32**, 425 (1968).

304. J. M. Ghuysen, J. L. Strominger, and D. J. Tipper, *Comp. Biochem.* **26A**, 53 (1968).

305. J. C. Ensign and R. S. Wolfe, *J. Bacteriol.* **90**, 395 (1965); **91**, 524 (1966).

was demonstrated by sucrose-density gradient centrifugation, ultracentrifugation, and electrophoresis on cellulose acetate and polyacrylamide gel. Both the cell wall lytic and proteolytic activities were maximal at pH 9.0. The enzyme was stable in the range of pH 6.5–9.5 and retained 50% of its activity after 5 hr at 50°. No inhibition was observed with *p*-mercuribenzoate, DFP, fluorodinitrobenzene, and trypsin inhibitor; but EDTA, citrate, and phosphate were strongly inhibitory at concentrations above 10^{-2} *M*, although no evidence is currently available for the presence of a metal in the molecule (*306*). Although an early study reported a molecular weight of 9800 based on gel filtration method (*305*), Jackson and Wolfe (*306*) found a value of 14,000 using various physical and chemical procedures for an enzyme isolated by a modified procedure. Various molecular characteristics of the *Myxobacter* AL-1 protease are listed in Table XIV. A very compact molecule was suggested by a frictional coefficient of 1.0. The enzyme, which was stable in 8.0 *M* urea (*306*), contained a disulfide bond, no phosphate, but did have 0.88 mole of hexose per mole of enzyme (*306*). The amino acid composition was Lys_2, His_6, Arg_3, Trp_3, Asp_{16}, Thr_{11}, Ser_{17}, Glu_7, Pro_5, Gly_{27}, Ala_{10}, half-Cys_2, Val_3, Met_2, Ile_3, Leu_7, Tyr_8, Phe_4, and $(NH_3)_{15}$ for a total of 136 residues (*306*). Since iodoacetate reacted with the enzyme only after reduction, the two half-cystine residues were assumed to form a disulfide bond. The molecule contained a very high content of aspartic acid, serine and glycine (*306*). An isoelectric point of 10.0 was reported (*305*) which does not agree with the analytical results. No carbohydrase activity was exhibited by the AL-1 protease (*305*). However, in addition to the cell lytic activity it digested 37, 33, and

TABLE XIV
MOLECULAR CHARACTERISTICS OF *Myxobacter* AL-1 PROTEASE[a]

MW	$s_{20,w} \times 10^{13}$	$D_{20,w} \times 10^7$	$\bar{V}$	f/f_0	pI	$A_{1\,cm}^{1\%}$ at 280 nm
14,300[b]						
14,000[c]	2.40	14	0.698	1.0	10[f]	15.8
13,977[d]	1.04[f]					
8,700[e,f]						

[a] From Jackson and Wolfe (*306*).
[b] Sedimentation equilibrium.
[c] Sedimentation-diffusion.
[d] Amino acid composition.
[e] Gel filtration.
[f] From Ensign and Wolfe (*305*).

306. R. J. Jackson and R. S. Wolfe, *JBC* **243**, 879 (1968).

15% of albumin, casein, and gelatin, respectively. Only three bonds in insulin B chain were cleaved, Ala 14–Leu 15, Val 18–$CySO_3H19$, and Gly 23–Phe 24 (*306*). Although a number of di- and tripeptides were not hydrolyzed by the enzyme, peptides containing over four glycine residues were easily cleaved (*306*) as shown in Fig. 5. However, polyglycine was not hydrolyzed while penta- and hexaglycine were cleaved (*306*). It is interesting to note that the AL-1 protease cleaved Z–Gly–Pro–Leu–Gly–Pro at the Pro–Leu bond but did not attack Z–Ala–Leu–NH_2 or Z–Gly–Phe–NH_2 (*306*). The specificity of this enzyme appears, therefore, to be quite different from that of the metalloaminoendopeptidases (Section IV) and displays no clear side chain specificity (*306*). The enzyme, however, did cleave cell wall peptides at specific positions (*306–310*) as shown in Fig. 5.

2. *Sorangium α- and β-Lytic Proteases*

Sorangium sp. produced two proteases, α- and β-lytic proteases (*133, 135*). The α-lytic protease which differed from the β-lytic enzyme in amino acid composition and specificity has been discussed previously in Section III with the DFP-sensitive alkaline proteases. The β-lytic

```
Synthetic peptides                  Peptideglycans of cell walls
----------------------------        ------------------------------------
Gly–Gly–Gly–Gly                     D–lactyl–Ala–D–Gly (Amidase action)
         ↑                                      ⇡
Gly–Gly–Gly–Gly–Gly                 D–Ala–Gly–Gly–Gly–Gly–Gly
         ↑   ↑                           ⇡              ↑   ↑
Gly–Gly–Gly–Gly–Ala                    Gly–Gly–Gly–Gly
             ↑                                ⇡
Ala–Gly–Gly–Gly–Gly                 D–Ala–Gly–Ser–Gly–Gly–Gly
             ↑                           ↑   ↑          ↑
Gly–Gly–Ala–Gly–Gly                 D–Ala–Gly–Gly–Ser–Gly–Gly
         ↑                               ↑       ⇡       ↑
Cbz–Gly–Gly–Gly–Gly                 D–Ala–Ser–Gly–Ser–Gly–Gly
             ↑                           ↑       ⇡       ↑
Gbz–Gly–Gly–Gly–Gly–Gly             D–Ala–Ala
                     ↑                   ↑
Cbz–Gly–Pro–Leu–Gly–Pro             D–Ala–D–IsoAsn
             ↑                           ↑
```

Fig. 5. Hydrolysis of synthetic peptides and peptideglycans of bacterial cell walls by *Myxobacter* AL-1 protease. Solid and broken arrows represent the primary and secondary hydrolysis, respectively. The figure is summarized from Tipper *et al.* (*306–310*).

307. D. J. Tipper, J. L. Strominger, and J. C. Ensign, *Biochemistry* **6**, 906 (1967).
308. D. J. Tipper, *Biochemistry* **8**, 2192 (1969).
309. W. Katz and J. L. Strominger, *Biochemistry* **6**, 930 (1967).
310. K. D. Hungerer, J. Fleck, and D. J. Tipper, *Biochemistry* **8**, 3567 (1969).

enzyme was purified and crystallized, and its homogeneity was demonstrated (*181*). A molecular weight of about 19,000 has been reported for the β-lytic protease with $s_{20,w} = 2.2$ S, $\bar{V} = 0.72$, and $A^{1\%}_{1cm}$ at 280 nm = 20.5 (*135, 159, 181*). The amino acid composition of the β-lytic enzyme was Lys_3, His_8, Arg_5, Trp_5, Asp_{21}, Thr_{13}, Ser_{22}, Glu_{10}, Pro_8, Gly_{24}, Ala_{13}, half-Cys_4, Val_5, Met_4, Ile_4, Leu_9, Tyr_{13}, and Phe_6 for a total of 177 residues (*159*). One atom of zinc per molecule has been reported, but no essential role for the metal was observed (*167*). The enzyme was slightly more acidic than lysozyme (*135*), indicating that it was a basic protein with more than 50% of the amino acids amidated. The four half-cystine residues were presented as two disulfide bonds (*159*). Both the α- and β-lytic proteases hydrolyzed cell walls of *Arthrobacter* and *Micrococcus* at D-lactyl-Ala and N^{ϵ}-(D-Ala)–Lys bonds in the peptideglycan structures (*186*). A higher activity was exhibited by the β-lytic protease than the α-lytic enzyme with these substrates. The enzymes were active on other protein substrates (*135*), but they had no exopeptidase or amidase activity (*184*). The β-lytic protease cleaved the insulin B chain primarily at Gly 23–Phe 24 and to a lesser extent at Val 18–$CySO_3H$19 (*184*), similar to the AL-1 protease (*306*), but different from the α-lytic enzyme (Section III,D,2). The specificity of the enzyme in protein hydrolysis might be determined by the distance of the susceptible bonds within the chain.

3. *Other Endopeptidases*

A number of other cell wall lysing proteases have been studied with respect to their specificities on cell wall materials and utilized as excellent tools to study the structure of cell walls.

Streptomyces species is a good source of enzymes which lyse various bacterial cell walls. Among them are five different endopeptidases, SA endopeptidase (*133*), ML endopeptidase (*311*), MR endopeptidase (*312*), KM endopeptidase (*313*), and L_3 enzyme (*314, 315*). Recently, the former four enzymes were purified from one culture using a combination of ion exchange chromatography, gel filtration on Sephadex, and electrophoresis (*313*). The SA endopeptidase has been shown to hydrolyze such bonds as D-Ala–Ala, D-Ala–Gly, and D-Ala-D-isoAsn while the ML endopeptidase cleaves N^{ϵ}-(D-Ala)–Lys and the KM endopeptidase cleaves D-

311. J. M. Ghuysen, D. J. Tipper, C. H. Birge, and J. L. Strominger, *Biochemistry* **4**, 2245 (1965).
312. J. F. Petit, E. Muñoz, and J. M. Ghuysen, *Biochemistry* **5**, 2764 (1966).
313. J. M. Ghuysen, L. Dierickx, J. Coyette, M. Leyh-Bouille, M. Guinard, and J. N. Campbell, *Biochemistry* **8**, 213 (1969).
314. Y. Mori, K. Kato, T. Matsuhara, and K. Kotani, *Biken's J.* **3**, 139 (1960).
315. K. Kato, J. L. Strominger, and S. Kotani, *Biochemistry* **7**, 2762 (1968).

Ala-(D)-*meso*-diaminopimelic acid, N^{α}-(D-Ala)-D-Lys, and N^{α}-(D-Ala)-D-Orn. The MR endopeptidase hydrolyzes bonds such as Ala–Thr, Ala–Ala, Gly–Gly, and various proteins, but the bonds cleaved in the various proteins have not been determined. The purity of these enzymes has not been conclusively established. The L_3 enzyme hydrolyzed the D-Ala-*m*-diaminopimelic acid bond which is also cleaved by an *E. coli* endopeptidase (*316*). An enzyme, L_{11}, has been isolated from *Flavobacterium* and found to have a similar specificity to that of the AL-1 protease (*317–320*). However, *N*-acetylmuramyl–Ala bond is less susceptible to cleavage by the L_{11} enzyme than by the AL-1 enzyme (*320*). A similar specificity was exhibited by the *Pseudomonas* enzyme (*308, 321*). Lysostophin isolated from *Micrococcus* sp. cleaved the Gly–Gly bond and was found to have a molecular weight of about 30,000 (*322, 323*).

Aeromonas hydrophilia produced an enzyme (*324*) which lysed cell walls of *S. aureus* at the pentaglycine cross-links with the liberation of oligoglycine peptides. Although it had a specificity similar to that of the AL-1 protease, it did not digest gelatin or albumin. Dipeptides and tripeptides containing leucine were hydrolyzed by this enzyme. A very small molecular weight, $s_{20,w} = 1.1$ S, was reported for the enzyme which was a basic protein (*324*). An endopeptidase isolated from sporulating *B. thuringiensis* was found to cleave cell wall peptides at Ala-D-Glu (*325*).

B. Fungal Carboxypeptidases

A few fungal enzymes with carboxypeptidase-like activity have been reported (*42, 44, 73, 156, 157, 326, 327*). The isolation of an extracellular

316. W. Weidel and H. Pelzer, *Advan. Enzymol.* **26,** 193 (1964).
317. K. Kato, S. Kotani, T. Matsuhara, J. Kogami, S. Hasimoto, M. Chimori, and I. Kazekawa, *Biken's J.* **8,** 155 (1962).
318. J. M. Ghuysen, J. F. Petit, E. Muñoz, and K. Kato, *Federation Proc.* **25,** 410 (1966).
319. K. Kato and J. L. Strominger, *Biochemistry* **7,** 2754 (1968).
320. K. Kato, T. Hirata, T. Murayama, K. Suginaka, and S. Kotani, *Biken's J.* **11,** 1 (1968).
321. J. W. Zyskind, P. A. Patec, and M. Lache, *Science* **147,** 1458 (1965).
322. C. A. Schindler and V. T. Schuhardt, *Proc. Natl. Acad. Sci. U. S.* **51,** 414 (1964).
323. H. P. Browder, W. A. Zygmunt, J. R. Young, and P. A. Tavormina, *BBRC* **19,** 383 (1965).
324. N. W. Coles, C. M. Gilbo, and A. J. Broad, *BJ* **111,** 7 (1969).
325. S. L. Kingan and J. C. Ensign, *J. Bacteriol.* **96,** 629 (1968).
326. J. M. Prescott and J. D. Boston, *ABB* **121,** 555 (1967).
327. J. D. Boston and J. M. Prescott, *ABB* **128,** 88 (1968).

carboxypeptidase from culture filtrates of *Phymatotrichum omnivorum* has been described by Prescott and co-workers (*156*, *157*, *326*, *327*). The enzyme was purified by a combination of acetone precipitation, DEAE-cellulose chromatography, and gel filtration on Sephadex G-75 and G-200 (*326*). A molecular weight of 31,400 was obtained by sedimentation velocity with values of $s_{20,w} = 2.35 \times 10^{-13}$ sec, $D_{20,w} = 6.49 \times 10^{-7}$ cm^2 sec^{-1}, and $\bar{V} = 0.715$ cm^3 g^{-1} (*327*). The enzyme had an isoelectric point near pH 2.2. The amino acid composition yielded Lys_5, His_4, Arg_7, Trp_{17}, Asp_{26}, Thr_{23}, Ser_{22}, Glu_{20}, Pro_{15}, Gly_{25}, Ala_{23}, half-Cys_4, Val_{13}, Met_{15}, Ile_{12}, Leu_{16}, Tyr_{17}, Phe_{12}, $(NH_3)_{11}$ for a total of 276 residues (*327*). Marked difference in amino acid composition between the fungal and the animal enzymes was noted particularly with respect to methionine, lysine, histidine, arginine, and tryptophan. The enzyme exhibited both peptidase and esterase activity toward substrates with free carboxyl groups such as Z–Gly–Phe, Z–Gly–Leu, Z–Gly–Tyr, Z–Gly–Trp, and hippuryl phenyllactic acid (*326*). The fungal carboxypeptidase was maximally active over the range of pH 6–8. The enzyme probably requires a metal ion as evidenced from the inactivation by EDTA or *o*-phenanthroline at 10^{-3} *M* and the restoration of activity upon the addition of Co^{2+}, Zn^{2+}, or Ni^{2+} (*326*).

Felix and Labouesse-Mercouroff (*42*) described a carboxypeptidase which was later isolated from autolyzed brewer's yeast by alcohol precipitation, chromatography on DEAE-cellulose and DEAE-Sephadex A-50, and gel filtration on Sephadex G-200 (*44*). The enzyme (peptidase α) had a pH optimum of 6.0–6.2 for the hydrolysis of Z–Gly–Leu (*44*). Chelating agents such as EDTA and *o*-phenanthroline inhibited the enzyme activity which was restored by the addition of Zn^{2+}, Co^{2+}, and certain other bivalent cations (*44*).

A D-alanine carboxypeptidase from *E. coli* has been reported by Izaki *et al.* (*328*). Carboxypeptidase-like activity has been reported for the extracellular peptidase (peptidase B) of *P. janthinellum* which has a pH optimum at 4.7 (*73*).

C. Thiol Proteases

The two well-characterized thiol proteases of microbial origin, streptococcal proteinase and clostripain are discussed in Chapters 18 and 20 of this volume, respectively. Only a few additional thiol enzymes have been

328. K. Izaki, M. Matsuhashi, and J. L. Strominger, *Proc. Natl. Acad. Sci. U. S.* **55**, 656 (1966).

TABLE XV
DISTRIBUTION OF VARIOUS MICROBIAL PROTEASES[a]

Organism	Reference			
	Acid protease	DFP-sensitive alkaline protease	Metal chelator-sensitive neutral protease	Others
Acrocylimdrium sp.	*33*			
Aeromonas hydrophila			*284*	*324*[b]
Alternaria tenuissima	*34*	*23, 137, 138*		
Arthrobacter B_{22}		*109, 110–114*		
Aspergillus awamori	*1, 12*			
Aspergillus candidus	*100*[c]			
Aspergillus flavus		*125, 170, 171*	*273*	
Aspergillus fumigatus	*17, 18*	*17, 18*		
Aspergillus melleus		*154*		
Aspergillus niger	*1, 11, 14*			
Aspergillus ochraceus			*269–272*	
Aspergillus oryzae	*13–16, 19*	*13, 122–124, 160, 161, 166, 170, 171, 185*	*13, 16, 262, 263, 268*	
Aspergillus parasiticus			*274*	
Aspergillus saitoi	*1,7–10, 19, 57–61, 67, 68, 70, 74, 77*			
Aspergillus sojae		*127, 182*		
Aspergillus sydowi		*126*		
Bacillus cereus	*99*[c]		*264, 265*	
Bacillus megaterium			*201, 240–246, 268*	
Bacillus subtilis			*1, 202–219*	
Bacillus thermoproteolytics			*220–239*	
Bacillus thuringiensis				*325*[b]
Brewer's yeast				*42*,[d] *44*[d]

Byssochlamys fulva	101[c]			
Candida albicans	53			
Cliocladium roseum		139		
Clostridium histolyticum			276, 277	276[e]
Endothia parasitica	89–91[c]			
Escherichia coli				328[d]
Flavobacterium				317–320[b]
Halobacterium salinarium			285	
Micrococcus caseolyticus			282	
Micrococcus sp.				322,[b] 323[b]
Microsporum sp.			283	
Mucor hiemalis	30			
Mucor pusillus	28, 29, 88,[c] 92–98[c]			
Myxobacter AL-1				304–310[b]
Paecilomyces varioti	26, 27, 54, 69, 70			
Penicillium cyaneo-fulvum		128–131, 138		
Penicillium janthinellum	20, 21, 73, 78, 83			73[d]
Penicillium notatum	22, 23			
Phymatotrichum omnivorum		155–157		156,[d] 157,[d] 326,[d] 327,[d]
Proteus mirabilis			281	
Pseudomonas aeruginosa			231, 250–261	
Rhizopus chinensis	24, 55, 62, 70, 74			
Rhizopus oligosporus	25			
Saccharomyces cerevisiae	41, 43, 45, 47–49	45, 47–50, 140, 141		
Sorangium sp.		133–136, 158, 159, 167, 168, 172, 178, 183, 184, 186		133,[b] 135,[b] 159,[b] 167,[b] 181,[b] 186[b]
Streptococcus lactis				329,[e] 330[e]
Streptomyces fradiae		117,[f] 118, 195–197[f]		
Streptomyces griseus		1, 115, 116, 142–151, 158, 191	142, 144, 147, 248, 249	
Streptomyces moderatus		152		

TABLE XV (*Continued*)

Organism	Reference			
	Acid protease	DFP-sensitive alkaline protease	Metal chelator-sensitive neutral protease	Others
Streptomyces naraensis			*247*	
Streptomyces rectus		*119–121*		
Streptomyces sp.				*133*,[b] *311–315*[b]
Thermoactinomyces vulgaris		*153*		
Thermomonospora fusca		*153*		
Trametes sanguinea	*31, 32, 70*			
Trichophyton granulosum		*193*,[f] *194*[f]	*193, 194*	

[a] Organisms are listed in alphabetical order.
[b] Cell wall lytic protease.
[c] Renninlike acid protease.
[d] Carboxypeptidase-like enzyme.
[e] Thiol protease.
[f] Keratinase.

reported. A protease isolated from *Streptococcus lactis* (*329, 330*) exhibited an optimum pH of 8.5 after activation by thioglycolic acid, sodium sulfate, or cysteine. It was inhibited by *p*-mercuribenzoate and NaF but not by DFP and might be similar to the streptococcal proteinase. Another enzyme isolated from *C. histolyticum* during the isolation of a neutral protease was reported to have a molecular weight of about 50,000 (*276*). It is probably the same enzyme or very similar to clostripain (*331*). The *Paecilomyces* acid protease (*26*) which was reported to be inhibited by thiol poisons and to have a sulfhydryl group was discussed together with the acid proteases in Section II.

The microorganisms producing various proteases discussed in this chapter are given in Table XV for the convenience of the reader.

ACKNOWLEDGMENT

The authors express their thanks to Dr. K. Morihara for his kind cooperation.

329. R. Sasaki and T. Nakae, *Japan. J. Zootech. Sci.* **30,** 7 (1959).
330. W. T. Williamson, S. B. Tove, and M. L. Speck, *J. Bacteriol.* **87,** 49 (1964).
331. W. M. Mitchell and W. F. Harrington, *JBC* **243,** 4683 (1968).

Author Index

Numbers in parentheses are reference numbers and indicate that an author's work is referred to although his name is not cited in the text.

C

E

F

G

L

N

O

P

S

T

Subject Index

A

B

C

D

F

G

J

K

L

M

N